International Conference on Innovation in Polymer Science and Technology (IPST 2016)

IOP Conference Series: Materials Science and Engineering
Volume 223

Medan, Indonesia
7 - 10 November 2016

Editors:

Rike Yudianti
Jun-ichi Azuma

ISBN: 978-1-5108-4645-6
ISSN: 1757-8981

Printed from e-media with permission by:

Curran Associates, Inc.
57 Morehouse Lane
Red Hook, NY 12571

Some format issues inherent in the e-media version may also appear in this print version.

Printed by Curran Associates, Inc. (2017)

For permission requests, please contact the Institute of Physics
at the address below.

Institute of Physics
Dirac House, Temple Back
Bristol BS1 6BE UK

Phone: 44 1 17 929 7481
Fax: 44 1 17 920 0979

techtracking@iop.org

Additional copies of this publication are available from:

Curran Associates, Inc.
57 Morehouse Lane
Red Hook, NY 12571 USA
Phone: 845-758-0400
Fax: 845-758-2633
Email: curran@proceedings.com
Web: www.proceedings.com

Table of contents

Volume 223

Innovation in Polymer Science and Technology 2016 (IPST 2016)
7–10 November 2016, Medan, Indonesia

Accepted papers received: 10 July 2017
Published online: 5 August 2017

Preface

011001
OPEN ACCESS
Innovation in Polymer Science and Technology 2016 (IPST 2016)

011002
OPEN ACCESS
List of Committee

011003
OPEN ACCESS
Photographs

011004
OPEN ACCESS
Peer review statement

Papers

012001
OPEN ACCESS
Performance of maleated castor oil based plasticizer on rubber: rheology and curing characteristic studies

I N Indrajati and I R Dewi.....1

012002
OPEN ACCESS
Swelling behaviour in n-pentane and mechanical properties of epoxidized natural rubber with different epoxide content

N A Kinasih, M I Fathurrohman and D A Winarto.....11

012003
OPEN ACCESS
Mechanical properties and mophology natural rubber blend with bentonit and carbon black

E M Ginting, N Bukit, Muliani and E Frida.....22

012004
OPEN ACCESS
Practical application of thermoreversibly Cross-linked rubber products

L M Polgar, F Picchioni, E de Ruiter and M van Duin.....29

012005
OPEN ACCESS
A comparative study of ground tire rubber devulcanization using twin screw extruder and internal mixer

O Ujianto, D B Putri, Jayatin and D AWinarto.....39

012006
OPEN ACCESS
The effects of the addition of alkanolamide on carbon blacks filled natural rubber compounds

I Surya, M Ginting and H Ismail.....48

012007
OPEN ACCESS
Effect of filler water absorption on water swelling properties of natural rubber

J Trakuldee and K Boonkerd.....56

012008
OPEN ACCESS
Effect of carbon black composition with sludge palm oil on the curing characteristic and mechanical properties of natural rubber/styrene butadiene rubber compound

R Mohamed, N Mohd Nurazzi and M Huzaifah.....63

012009
OPEN ACCESS
The effect of carbon black loading and structure on tensile property of natural rubber composite

S Savetlana, Zulhendri, I Sukmana and F A Saputra.....71

012010
OPEN ACCESS
Effect of urea deproteinization on catalytic hydrogenation of natural rubber latex

A Cifriadi, M Chalid and S Puspitasari.....80

012011
OPEN ACCESS
Effect of type and content of tackifier on adhesion of natural rubber and reclaimed natural rubber based sealant

P Raethong and K Boonkerd.....90

012012
OPEN ACCESS
The effect of alkanolamide addition on cure and tensile properties of unfilled natural rubber compounds

I Surya, SF Siregar and H Ismail.....95

012013
OPEN ACCESS
The influence of natural rubber – butadiene rubber and carbon black type on the mechanical properties of tread compound

L A Wisojodharmo, R Fidyaningsih, D A Fitriani, D K Arti, Indriasari and H Susanto.....101

012014
OPEN ACCESS
Effect of maleated natural rubber on tensile strength and compatibility of natural rubber/coconut coir composite

O Ujianto, R Noviyanti, R Wijaya and B Ramadhoni.....109

012015
OPEN ACCESS
Utilization of amylopectin-grafted-poly(hexyl methacrylate) as bio-compatible agent for polypropylene/starch polymers blend

I S Purwaningsih and A S Handayani.....120

012016
OPEN ACCESS
Effects of organoclay to miscibility, mechanical and thermal properties of poly(lactic acid) and propylene-ethylene copolymer blends

S Wacharawichanant, C Ounyai and P Rassamee.....128

012017
OPEN ACCESS
Effect of vulcanization temperature on curing characteristic, physical and mechanical properties of natural rubber/palygorskite composites

K C Lee, N A Md Yusoff, N Othman and N A Mohamad Aini.....138

012018
OPEN ACCESS
Physico-mechanical properties of silanized-montmorillonite reinforced chitosan-co-poly(maleic anhydride) composites

O A Saputra, A Fajrin, M Nauqinida, V Suryanti and E Pramono.....149

012019
OPEN ACCESS
Novel route of synthesis for cellulose fiber-based hybrid polyurethane

F H Ikhwan, S Ilmiati, H Kurnia Adi, R Arumsari and M Chalid.....159

012020
OPEN ACCESS
The effect of modified ijuk fibers to crystallinity of polypropylene composite

I Prabowo, J Nur Pratama and M Chalid.....170

012021
OPEN ACCESS
Improvement of acoustical characteristics : wideband bamboo based polymer composite

M Farid, A Purniawan, A Rasyida, M Ramadhani and S Komariyah.....179

012022
OPEN ACCESS
Effect of lignin on morphology, biodegradability, mechanical and thermal properties of low linear density polyethylene/lignin biocomposites

M Ghozali, E Triwulandari, A Haryono and E Yuanita.....186

012023
OPEN ACCESS
Structure and properties of organically modified poly(butylene adipate-co-terephthalate) based nanocomposites

A Rasyida, K Fukushima and M-C Yang.....199

012024
OPEN ACCESS
Izod impact performance of hybrid carbon nanotube-alumina filled epoxy nanocomposites

M R Zakaria, H Md. Akil, M H Abdul Kudus and N S A Ahmad Bakhtiar.....211

012025
OPEN ACCESS
Mechanical, thermal and morphological properties of poly(lactic acid)/ethylene-butyl acrylate copolymer nanocomposites

S Wacharawichanant, P Hoysang and S Ratchawong.....218

012026
OPEN ACCESS
Effects of number of ply, compression temperature, pressure and time on mechanical properties of prepreg kenaf-polypropilene composites

H S S Tomo, O Ujianto, R Rizal and Y Pratama.....228

012027
OPEN ACCESS
Synthesis and morphology of polyvinyl alcohol /zinc sulfide nanocomposite

M Sirait and Motlan.....236

012028
OPEN ACCESS
Improvement of polypropylene (PP)-modified bitumen through lignin addition

E Yuanita, B E Hendrasetyawan, D F Firdaus and M Chalid.....242

012029
OPEN ACCESS
Thermal conductivity and tensile properties of tin oxide filled UPR/EPS composites with and without organic nanocrystal

R Mohamed, S A Syed Mustafa, Mohd N Norizan and L Suraya Amerudin.....250

012030
OPEN ACCESS
Effect of expanded organoclay by stearic acid to curing, mechanical and swelling properties of natural rubber nanocomposites

A Ramadhan, M Irfan Fathurrohman, A F Falaah, N Setyawan and B Soegijono.....258

012031
OPEN ACCESS
Effects of natural zeolite and ferric oxide to electromagnetic and reflection loss properties of polyurethane nanocomposite

G Gultom, B Wirjosentono, M Ginting and K Sebayang.....268

012032
OPEN ACCESS
Improvement of thermal and mechanical properties of composite based on polylactic acid and microfibrillated cellulose through chemical modification

L Suryanegara, R A Nugraha and S S Achmadi.....275

012033
OPEN ACCESS
Added value of lignin as lignin-based hybrid polyurethane for a compatibilizing agent

S Ilmiati, J Haris Mustafa, A Yaumal, F Hanum and M Chalid.....285

012034
OPEN ACCESS
The effect of ratio chitosan-bentonite and processing time on the characterization of chitosan-bentonite composite

E Savitri and A Budhyantoro.....293

012035
OPEN ACCESS
Unsaturated polyester/expanded polystyrene composite : thermal characteristics and flame retardancy effects

R Mohamed, S A Syed Mustafa, Mohd N Norizan and L S Amerudin.....301

012036
OPEN ACCESS
Utilization of bagasse and coconut fibers waste as fillers of sandwich composite for bridge railway sleepers

K A Soehardjo and A Basuki.....309

012037
OPEN ACCESS
Preparation of catalyst coated membrane by modified decal transfer method for proton exchange membrane fuel cell

Indriyati, Y Irmawati and B Prihandoko.....320

012038
OPEN ACCESS
Effects of pore forming agents on chitosan-graft-poly(N-vinylpyrrolidone) hydrogel properties for use as a matrix for floating drug delivery

E Budianto, M F Al-Shidqi and A H Cahyana.....327

012039
OPEN ACCESS
Synthesis and characterization of PVA blended $LiClO_4$ as electrolyte material for battery Li-ion

I Gunawan, Deswita, B. Sugeng and Sudaryanto.....337

012040
OPEN ACCESS
Provision of micro-nano bacterial cellulose as bio plastic filler by sonication method

Maryam, D. Rahmad, Yunizurwan, A. Kasim, Novelina and Emriadi.....344

012041
OPEN ACCESS
Calcium modified edible Canna (*Canna edulis* L) starch for controlled released matrix

A P Putri, M Ridwan, T A Darmawan, F Darusman and A Gadri.....352

012042
OPEN ACCESS
Preparation of micro-fibrillated cellulose based on sugar palm ijuk (*Arenga pinnata*) fibres through partial acid hydrolysis

A Saputro, I Verawati, G Ramahdita and M Chalid.....360

012043
OPEN ACCESS
Formulation and synthesis of hydrogels having lower critical solution temperature near body temperature

A Z Abidin, H P R Graha and D A Trirahayu.....369

012044
OPEN ACCESS
Study of the influence of ZnO addition on the properties of chitosan-banana starch bioplastics

L Sapei, K S Padmawijaya, O Sijayanti and P J Wardhana.....380

012045
OPEN ACCESS
Water absorbency of chitosan grafted acrylic acid hydrogels

N Astrini, L Anah and A Haryono.....387

012046
OPEN ACCESS
Effect of glycerol and zinc oxide addition on antibacterial activity of biodegradable bioplastics from chitosan-kepok banana peel starch

Y E Agustin and K S Padmawijaya.....395

012047
OPEN ACCESS
Sengon wood (*Paraserianthes falcataria* (*L.*) *Nielsen*) carbon as supporting material for electrochemical double layer capasitor

Wulandhari, N Syarif, I. Waruwu and M. Ridho Prayogo.....402

012048
OPEN ACCESS
Effect of epoxidised soybean oil loading as plasticiser on physical, mechanical and thermal properties of polyvinylchloride

M Rahmah, N Mohd Nurazzi, A R Farah Nordyana and S M Syed Anas.....409

012049
OPEN ACCESS
Physicochemical properties of sugar palm starch film: Effect of concentration and plasticizer type

D J Prasetyo, W Apriyana, T H Jatmiko, Hernawan, S N Hayati, V T Rosyida, Y Pranoto and C D Poeloengasih.....419

012050
OPEN ACCESS
Preparation of immobilized glucose oxidase wafer enzyme on calcium-bentonite modified by surfactant

R K Widi, D C Trisulo, A Budhyantoro and R Chrisnasari.....428

012051
OPEN ACCESS
Synthesis and micowave characterization of conductive polyaniline prepared by continous polymerization process

A Manaf, A. Bimantoro, M AE Hafizah and Andreas.....435

012052
OPEN ACCESS
Mechanical strength and ionic conductivity of polymer electrolyte membranes prepared from cellulose acetate-lithium perchlorate

T Sudiarti, D Wahyuningrum, B Bundjali and I Made Arcana.....442

012053
OPEN ACCESS
Urea-formaldehyde resins: production, application, and testing

A Nuryawan, I Risnasari, T Sucipto, A Heri Iswanto and R Rosmala Dewi.....450

012054
OPEN ACCESS
Incorporation of titanate nanosheets to enhance mechanical properties of water-soluble polyamic acid

C Harito, Dmitry V Bavykin and Frank C Walsh.....460

012055
OPEN ACCESS
Transesterification of sago starch and waste palm cooking oil in densified CO_2

H Muljana, A K Sugih, N Christina, K Fangdinata, J Renaldo, Rudy, H J Heeres and F Picchioni.....471

012056
OPEN ACCESS
Physicochemical properties and characteristics of microcrystalline cellulose derived from the cellulose of oil palm empty fruit bunch

H Nasution, Yurnaliza, Veronicha, Irmadani and S Sitompul.....481

012057
OPEN ACCESS
Preparation of micro-fibrillated cellulose from sorghum fibre through alkalization and acetylation treatments

Ismojo, P H Simanulang, A Zulfia and M Chalid.....488

012058
OPEN ACCESS
Performance comparison of plastic shopping bags in modern and traditional retail

F A Radini, R Wulandari, S J A Nasiri and D A Winarto.....495

012059
OPEN ACCESS
Swelling and tensile properties of starch glycerol system with various crosslinking agents

R Mohamed, N Mohd, N Nurazzi, M I Siti Aisyah and F Mohd Fauzi.....505

012060
OPEN ACCESS
Analysis physical properties of composites polymer from cocofiber and polypropylene plastic waste with maleic anhydrate as crosslinking agent

E Pelita, T R Hidayani and A Akbar.....512

012061
OPEN ACCESS
The microwave assisted-synthesis of carboxymethyl cellulose from nata de-coco bacterial cellulose

L O A N Ramadhan, M Nur Rahmat, P E Susilowati, L O Ahmad and U Edy Rusbandi.....521

012062
OPEN ACCESS
Effect of bench time polymerization on depth of cure of dental composite resin

K Harahap, A Yudhit and F Sari.....528

012063
OPEN ACCESS
Effect of chitosan addition to characteristic and antimicrobial activity of zinc doped hydroxyapatite

A Rasyida, S T Wicaksono, N N Pradita, H Ardhyananta and A Purnomo.....534

012064
OPEN ACCESS
Effect of oil palm empty fruit bunches fibers reinforced polymer recycled

B Hermawan, S Nikmatin, Sudaryanto, H Alatas and S G Sukaryo.....541

012065
OPEN ACCESS
Effects of additional nanosilica of compressive strength on mortar

N Retno Setiati.....550

012066
OPEN ACCESS
Water absorption and its effect on the tensile properties of tapioca starch/polyvinyl alcohol bioplastics

H Judawisastra, R D R Sitohang, L Marta and Mardiyati.....562

IOP Conf. Series: Materials Science and Engineering **223** (2017) 011001 doi:10.1088/1757-899X/223/1/011001

PREFACE

International Conference on Innovation in Polymer Science and Technology 2016 (IPST2016)

It is with great pleasure that we introduce this special issue of the IOP Conference Series: Materials Science and Engineering, which publishes papers selected from the presentations made at the International Conference and Exhibition on Innovation in Polymer Science and Technology 2016 (IPST2016) in Medan, Indonesia. This event which was attended by about 260 partisipants from all continents, was organized by the Indonesian Polymer Association (HPI), which was supported by Universitas Sumatra Utara (USU) and Universitas Negeri Medan (UNIMED) on November 7 – 10, 2016.

This special edition encompasses a range of complementary research topics from excellent scientists concentrated in the southern hemisphere. The increasing global interest in sustainability is highlighted in this issue, with novel methods, new materials, applications and products. This international conference covered some topics, viz. biomass based and biodegradable polymers, advanced rubber process and technology, nanostructure, polymer composites, polymer in renewable energy, synthesis and functional polymer. We applaud the efforts of the conference participants and encourage their current and future personal and collaborative efforts in this important field of scientific endeavor.

We would like to express our sincere gratitude and our highest appreciation to K.U. (Katja) Loos (University of Groningen, The Netherlands), Mochamad Chalid (Indonesia), Ishak bin Ahmad (Malaysia), Sunit Hendrana (Indonesia), Rameshwar Adhikari (Nepal), Lorenzo Massimo Polgar (The Netherlands), Wu Guozhong (China), Hermawan Judawisastra (Indonesia), Md Akil Hazizan (Malaysia), Ji Heung Kim (Korea), Md. Abu Bin Hasan Susan (Bangladesh), I Made Arcana (Indonesia), Nadras Othman (Malaysia), Wilairat Cheewasedtham (Thailand), Agus Haryono (Indonesia), Sudaryanto (Indonesia), Sudirman (Indonesia), for reviewing and giving recommendation on manuscripts from all participants of IPST 2016.

With our warm regards,
Rike Yudianti (Indonesian Institute of Science)
Jun-ichi Azuma (Osaka University, Japan)

Editor-in-Chief
Innovation in Polymer Science & Technology 2016

COMMITTEE

IOP Conf. Series: Materials Science and Engineering **223** (2017) 011002 doi:10.1088/1757-899X/223/1/011002

ORGANIZING COMMITTEE

Chairman : Mochamad Chalid
Co-chair : Eddyanto
Secretary : Yenny Meliana
Members :

Evi Triwulandari
Sri Fahmiati
Melati Septiyanti
Elsy Rahimi Chaldun
Evana Juanita
Riastuti Fidyaningsih
Tri Partuti
Yuyun Irmawati
Sri Budi Harmami
Lisman Suryanegara
Yuyun Nurhamiyah
Wiwik Pujiastuti
Guntarti Supeni
Dita Adi Saputra
Anung Syampurwadi
Ika Juliana
Nuri Astrini
Arie Listyarini
Heru Susanto
Indriyati
M. Ghozali
Tita Puspitasari
Rike Yudianti
Myrtha Karina
Witta Kartika Restu
Oka Pradipta Arjasa
Ghizka

IOP Conf. Series: Materials Science and Engineering **223** (2017) 011003 doi:10.1088/1757-899X/223/1/011003

INNOVATION IN POLYMER SCIENCE AND TECHNOLOGY
(IPST) 2016
JW MARRIOTT HOTEL AND RESORT MEDAN
NORTH SUMATERA - INDONESIA
NOVEMBER 7-10, 2016

PERHIMPUNAN POLIMER INDONESIA (HPI)

IPST 2016

Using
LIPI
Method
WASEDA
Kinetic and Thermodynamic Study of Fluorenol
Introduction
Background
Purpose
Hydrogen-Releasing Rate of Fluorenol Polymer
Conclusion
PARTICIPANT
PP4-09

Peer review statement

All papers published in this volume of *IOP Conference Series: Materials Science and Engineering* have been peer reviewed through processes administered by the proceedings Editors. Reviews were conducted by expert referees to the professional and scientific standards expected of a proceedings journal published by IOP Publishing.

IOP Conf. Series: Materials Science and Engineering **223** (2017) 012001 doi:10.1088/1757-899X/223/1/012001

Performance of maleated castor oil based plasticizer on rubber: rheology and curing characteristic studies

I N Indrajati, I R Dewi
Center for Leather, Rubber, and Plastics, Ministry of Industry, Republic of Indonesia
Sokonandi St. Numb. 9 Yogyakarta 55166, Indonesia

E-mail: i-novia@kemenperin.go.id

Abstract. The objective of this study was to evaluate the performance of maleated castor oil (MACO) as plasticizer on natural rubber (NR), ethylene propylene diene monomer (EPDM), and nitrile butadiene rubber (NBR). The parameter studied were involving rheological, curing and swelling properties. The MACOs were prepared by an esterification reaction between castor oil (CO) and maleic anhydride (MAH) with the help of xylene as water entrainer to improve water removal. Resulting oils then applied as a plasticizer in each of those rubbers within a fixed loading of 5 phr. Comparison has been made to evaluate the performance of MACO and conventional plasticizer (paraffinic oil for NR and EPDM, DOP for NBR) on each rubber. Rheology, curing characteristic and swelling of each rubber were studied. The results showed that rubber (NR/EPDM/NBR) plasticized with MACO had given similar flow characteristic to conventional plasticizers. MACO exhibited slow curing, confirmed by higher t_{90}, but the scorch safety was of the same magnitude. MAH loading tended to decrease the flow properties and curing rate, while scorch time (ts_2) was independent.

1. Introduction

The plasticizer, a low molecular weight substance, often added into rubber both to modify processing steps and final product properties. Petroleum-based plasticizers are widely used in rubber industries even though they give negative effect to the environment because of their toxic-polycyclic aromatic compound content. Some of those plasticizers also exhibit the tendency of leach out of material over time since they are not bonded covalently within polymer matrices [1]. The depletion of fossil feedstock and global warming issue are other problem that leads to exploring alternative materials. Vegetable-based oils are the potential candidates to substitute petroleum-based oil in this regard. This is mainly due to their nature of sources are renewable. Castor oil is potential candidates within this regard. It is a triacylglycerol containing one fatty acid, i.e. ricinoleic acid or 12-hydroxyoleic acid, that comprises over 89% of the fatty acid [2][3]. Castor oil is promising because it has unique molecular structure. Three functional groups are presented in its molecule, i.e. carboxylic groups, carbon double bonds (C=C) and hydroxyl groups [4]. These functional groups make the oil suitable for many chemical reactions and modifications.

Virgin castor oil needs to increase its compatibility with polymer matrices by modification [5]. Maleinization treatment through esterification with maleic anhydride is one of the modification route [2,6,7]. Maleated castor oil (MACO) is produced by esterification reaction of castor oil with maleic anhydride (MAH). Since esterification is an equilibrium reaction, thus high reaction yield can be achieved by shifting the chemical equilibrium toward the ester production. According to Le Chatelier principle, the yield and selectivity of the reaction product can be increased by the byproduct removal from the reaction system [8]. Previous studies have been successfully done but without water removal process[2,6,9]. Mazo *et al.*, (2012) [10] compared conventional to microwave heating in MACO production. This study resulted in faster esterification reaction and showed the presence of the non-thermal effect of microwave heating. Unfortunately, this technology is rather difficult and expensive. Byproduct removal from reaction system was unclear. These are constituting a strong driving force to search for alternatives. Kulkarani and Sawant (2003) [11] carried out esterification of fatty acid using entrainer to enhance water removal. Direct esterification without solvent requires high temperature and resulted in the poor color product.

Recently, MACO is applied either as polymer matrices [6,12] or as polymer bio modifier [9,13], but application in therubber fields is limited. Dewi *et al.,* (2016) [14] synthesized MACO and applied it on NBR/PVC blends. In the present study, MACO was synthesized from castor oil and maleicanhydride in various concentration, with the help of xylene as water entrainer. Xylene was used both to facilitate esterification reaction at low temperature and to enhance water removal. The resulting oils (MACO) were then applied as a plasticizer in natural (NR) and synthetic rubber (EPDM and NBR). Their effect on rubber rheology, curing characteristic and swelling properties in various solvents were evaluated.

2. Materials and Method

2.1. Material

Castor oil (Thai castor oil grade) and xylene were purchased from local market (PT. Brataco), while maleic anhydride (MAH) was obtained from PT. Justus Kimia Raya Indonesia. Natural Rubber (NR) grade Ribbed Smoke Sheet (RSS 1) supplied by PT. Perkebunan Nusantara IX (Indonesia). Nitrile Butadiene Rubber (NBR) as grade Krynac 4975F and Ethylene Propylene Diene Monomer (EPDM) as grade Keltan 4551A. Each type of rubber was employed in 100 phr. Other materials used for compound preparation consist of carbon black grade N-330 (OCI, Korea) 30 phr and grade N-774 (OCI, Korea) 50 phr, zinc oxide(ZnO) 5 phr, aflux 42M (Rhein Chemie) 1 phr, polymerized 2,2,4-trimethyl-1,2-dihydroquinoline (TMQ) (Kemai) 1 phr, paraffin wax Antilux 65 A 0.5 phr, MBTS (Shandong Sianxian) 1.2 phr, TMTD 0.3 phr, and sulfur SP-325 (Miwon) 1.2 phr. All of rubber compound materials were obtained from local market in industrial grade. Toluene and hexane (technical grade) were used as solvent.

2.2. Synthesis of Maleated Castor Oil

Our previous work [14] had been successfully synthesized MACO from castor oil and maleic anhydride with the aid of xylene as water entrainer. In this work, the mole ratio of CO and MAH was varied within a fixed ratio of CO/xylene 1:1 by weight. MC1-MC5 were denoted for MACO with various MAH loading as listed on Table 1. MACO was produced by an esterification reaction between CO and MAH. The reactions were carried out in a three-necked flask equipped with a dean stark apparatus and cold-water condenser. A mixture of castor oil and xylene were added into the three-necked flask and heated into 146°C. Then, add a portion of MAH incrementally into the mixture and the reaction was preceded for 5 h with continuous stirring 300 rpm. After the reaction completed, the solvent was removed by vacuum distillation. The effectiveness of esterification reaction was monitored by measuring acid value. Its determination was carried out by titration method using ethanolic KOH. The acid value was defined as expressed by equation 1. The functional groups before and after esterification were investigated using FTIR.

$$acid\,number = \frac{56.1\,V\,N}{w_{sample}} \quad \text{(1)}$$

Table 1. Mole ratio of CO/MAH

Sample ID	Mole Ratio CO/MAH
MC1	1:1.0
MC2	1:1.5
MC3	1:2.0
MC4	1:2.5
MC5	1:3.0

Table 2. Notation of rubber compounds in the experiments

Types of Plasticizer	Rubber		
	NR	EPDM	NBR
Conventional	R0	E0	N0
MC1	R1	E1	N1
MC2	R2	E2	N2
MC3	R3	E3	N3
MC4	R4	E4	N4
MC5	R5	E5	N5

2.3. Rubber Compound Preparation

Compounding recipes of mixes are following to previous section (material). Variations were made on the types processing oil of each rubber. For comparison, petroleum-based plasticizers were used, i.e. paraffinic oil for NR and EPDM, and DOP for NBR. All types of plasticizers were used in fixed amount of 5 phr. Compounding was carried out using laboratory two-roll mill. Finished compounds were then immersed in water in order to interrupt the premature crosslinking, which might be formed. Compounds were then allowed to mature for 24 h and stored in a dry place ata temperature of 25°C.

Notation of rubber compounds as seen in Table 2. Notation R0-R5 were denote for NR, while E0-E5 and N0-N5 were represented EPDM and NBR, respectively.

2.4. *Curing Characteristic and Rheological Properties Measurement*

Rheology and curing characteristics were studied using Moving Die Rheometer (MDR) Gotech 3000A at a frequency 100 cpm (cycle per minutes) and an angle 3 deg. For NR compounds, curing tests were performed at 150°C for NR and 170°C for EPDM and NBR. MDR experiment provided curing data such as scorch time (ts_2) and optimum cure time (t_{90}). Rheology was studied by evaluating the viscous to the elastic ratio (V/E).V/E ratio was represented by viscous (S") and elastic (S') torques obtained from MDR [15].

2.5. *Swelling Experiment*

The swelling experiment was performed on 2x2 cm cut sample prepared by compression molding. The samples were immersed in two separated solvents, i.e. toluene and hexane, at 25°C for 24 h. The properties of solvent used in this experiment were listed on Table 3.

Table 3. Properties of solvent and rubber

Solvent	Molecular Weight	Molar Volume	Density (ρ), g/cm^3	Solubility Parameter (δ), MPa$^{1/2}$
Toluene	92.14	106.29	0.867	17.8
Hexane	86.18	131.60	0.655	13.2
NR			0.920	16.6
EPDM			0.860	15.0
NBR			0.970	19.4

After the period was completed; the samples were taken out and blotted with filter paper to remove liquid on the surface. Swelling ratio (Qt) is defined by Equation 2,

$$Qt\% = \frac{wo\text{-}wt}{wo} \times 100 \quad \ldots\ldots\ldots\ldots\ldots\ldots\ldots\ldots \quad (2)$$

where w_0 and w_t are the mass of the test piece before swelling and after time "t" immersion, respectively. The mass of the sample weighed using electronic balance having precision 0.0001 *g*.

3. Result and Discussion

3.1. *Properties of MACO*

The effectiveness of esterification reaction was monitored by measuring acid value. Figure 1 depicts the changes in acid value during esterification reaction. The initial increase of acid number found in the early stage of reaction indicates the formation of free acid groups. Further progress reaction, thesegroups are decreased because they took part in the formation of dimer, trimer or even oligomer [12]. Water formation might be indicated by cloud formation in the dean stark and is collected as separated phase [14]. With this regard, the resultant MACO was in the form of dimer or oligomer. Further analysis is needed to determine the molecular weight of the oil. An increasing mole of MAH causes the acid value to increase. The higher MAH loading, the higher amounts of the free acid group formed. Thus, higher acid value at one reaction time.

The MACO formation is also confirmed by the FTIR spectra as depicted in Figure 2. A broad peak is found at 3433 cm^{-1} which belong to hydroxyl (-OH) groups presents on the ricinoleic acid molecule. Another characteristic peak of CO is found at 1662 cm^{-1} which can be assigned to C-double bond stretching (–C=C–) frequency. When CO reacts with MAH, the –OH peak is evidently decreased, while the double bond peak is enhanced. The peak at 1662 is found to be stronger along with increasing MAH mole in the reaction system, while –OH peak is more reduced even nearly disappear on CO:MAH=1:2 mole. These clearly reflect the more ester groups attached to the CO molecule chains. This finding is agreed with previous studies [2,6,13]. The cyclic anhydride peak at 1779 and 1849 cm^{-1}[12] is not observed in all spectra, indicating almost all of MAH were reacted with CO.

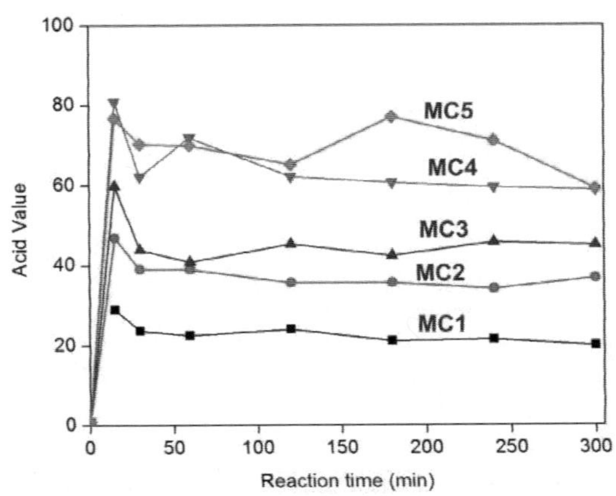

Figure 1. Changes of acid value during esterification reaction

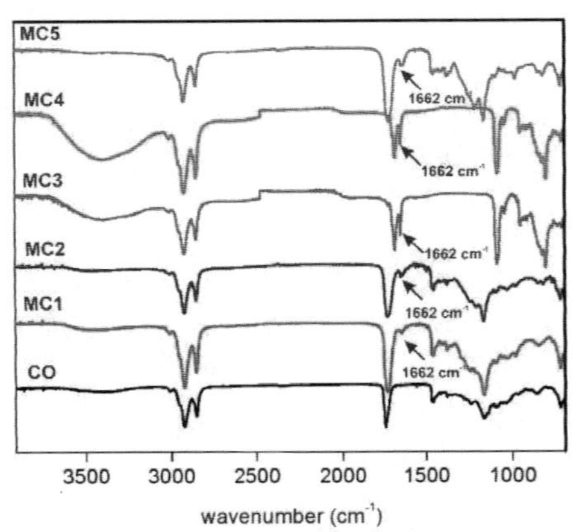

Figure 2. Infrared spectra of pure CO, MC1, MC2 and MC3, MC4 and MC5

3.2. Curing Characteristic and Rheology

Figure 3a-c depicts the curing curve (plot torque versus time) of sample rubber compounds, while Table 4 summarizes the extracted data from curing curve on Figure 3. Curing curve is a powerful tool to study the crosslinking process and rheology of rubber compounds. This tool provides valuable information related to the processing characteristic that should be considered for subsequent processes. It is generally known that the curing curves can be divided into three stages (Figure 3a), i.e. induction

(I), curing (II) and over curing (III). The induction period is dominating by the viscous (plastic) behavior, thus give a brief description related to the processing safety. Both plasticizer, control, and MACOs, give similar induction period for each type of rubber. It can be said that MACO do not significantly affect to this period. EPDM samples exhibit longer induction period than two other rubbers.

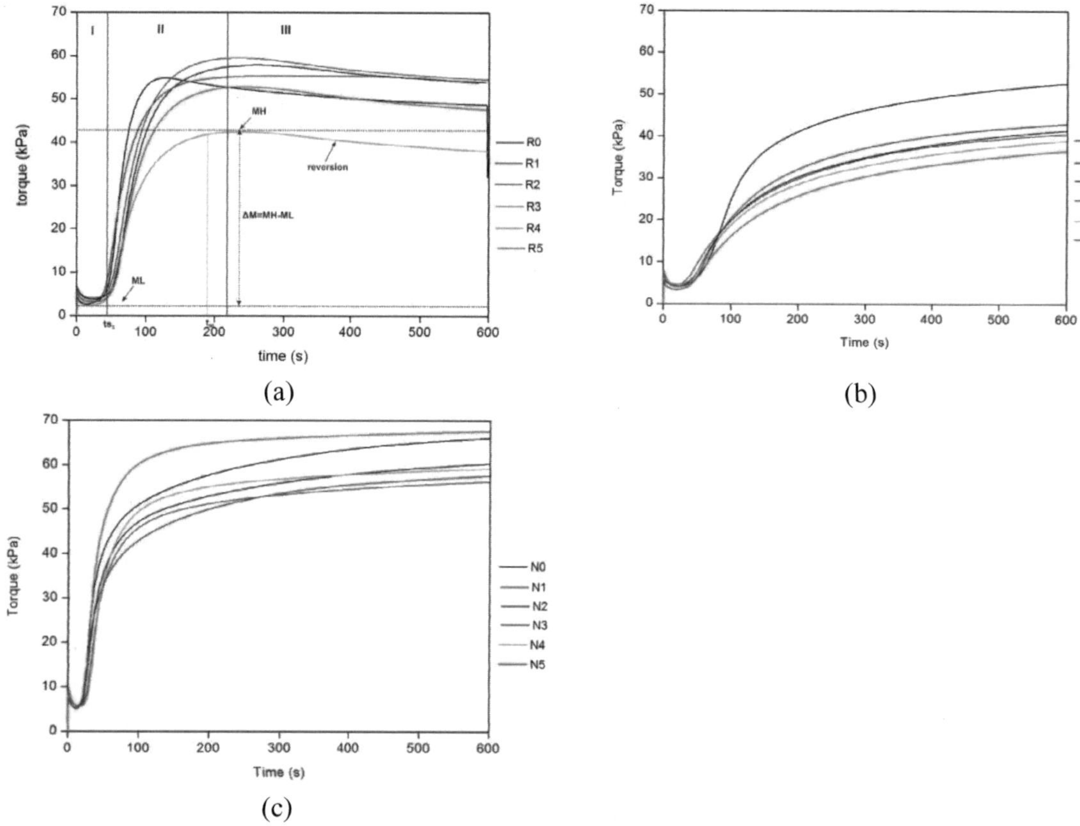

Figure 3. Curing curve (plot of rheometer torque vs time) a) NR, b) EPDM, and c) NBR

During induction period, there is no chemical crosslinking formed. Interaction of rubber molecule chains with the filler particle surface or by chain entanglements produces physical crosslinking. This crosslink is reflected by the minimum elastic torque (ML) which is also related to the compound viscosity [16]. One can understand rubber flow behavior during processing by assessing this parameter [17]. It can be seen that MACO provides slightly higher viscosity than conventional plasticizer in NR compounds. This probably due to the interaction of functional groups on MACO with the rubber molecule chains. Bocqué *et al.,* (2016) [5] explained that ester groups existed in vegetable-based oil interact via van der Waals bonding with the polymer chains and lead to a compatibilization with the polymer. But this effect is less pronounced in EPDM or NBR. MACO gives different response in NBR where the compounds viscosity is comparable. Compounds with higher viscosity should be treated carefully to avoid the defect during processing.

Table 4. Rheometer torques and cure time

Sample	MH kgf.cm	ML kgf.cm	MH-ML kgf.cm	S"MH kgf.cm	S"ML kgf.cm	ts_2 sec	t_{90} sec	CRmax kgf.cm/s
R0*	54.83	2,66	52.17	1.74	3.28	40	87	82.46
R1	55.34	3.33	52.01	3.59	3.89	39	117	60.47
R2	57.61	3.76	53.85	3.15	4.45	52	140	54.88
R3	57.13	3.96	53.17	3.24	4.56	48	135	56.89
R4	42.48	2.41	40.07	3.93	3.25	47	136	39.76
R5	52.78	3.13	49.65	3.42	3.82	50	137	51.15
E0*	56.47	4.06	52.41	4.13	4.74	51	505	27.50
E1	45.16	3.37	41.79	6.37	4.54	47	449	19.41
E2	44.50	4.23	40.27	7.30	5.21	46	485	17.18
E3	43.14	4.46	38.68	7.01	5.23	39	493	15.29
E4	42.29	4.59	37.70	7.39	5.40	47	565	15.41
E5	40.33	4.16	36.17	7.08	5.15	51	611	12.84
N0*	68.09	5.24	62.85	15.28	6.24	21	319	90.75
N1	59.76	5.17	54.59	15.91	5.94	20	331	62.91
N2	62.84	5.72	57.12	16.73	6.52	22	358	70.64
N3	58.21	5.61	52.60	15.84	6.37	25	283	65.84
N4	60.63	5.55	55.08	15.69	6.35	27	202	72.59
N5	68.58	5.64	62.94	16.43	6.34	23	125	75.21

*control: for NR and EPDM used paraffinic oil, while NBR used DOP

Chemical crosslinks are formed during curing period. The highest crosslink can be produced during vulcanization_is described by the maximum elastic torque (MH). The differences between MH and ML thus qualitatively represents the apparent crosslink density of the rubber compound [18]. It is evidence that MHs are varied for each rubber used in this experiment. MACO provides lower MH than conventional plasticizer in EPDM, while gives comparable value in NR and NBR. Accordingly, the apparent crosslink densities are not influenced by the changes of MAH loading for NR and NBR, but it is greatly affected in EPDM. In over cure period, NR compounds exhibit typical reversion phenomenon. The introduction of MACO into NR mixes do not remove the reversion, but it slower the reversion occurrence (Figure 3a). Reversion is not occurred in EPDM and NBR.

Scorch time (ts_2) is another important parameter to characterized the induction period because it gives a sign where vulcanization reaction begins. Scorch time has a strong correlation with the processing safety. The higher scorch time, the safer a process. Scorch time is also independent of the MAH concentration in MACO for non-polar rubber (NR and EPDM), but it tends to increase in polar rubber (NBR). The extent of curing often evaluates by assessing the optimum cure time (t_{90}), in which most of the desired properties of the final product are achieved. Acidic nature of resultant MACO affects the curing time retarded reaction. Slow reaction of MACO plasticized-NR is confirmed by lower cure rate maximum (CR_{max}) achieved during vulcanization. EPDM show typical slow curing because of its low unsaturation on the backbone.

Flow properties of rubber compounds correlate to their viscosity. In rubber, viscosity rather complex since rubber is a viscoelastic material. Viscous (V) and elastic (E) elements are both existed at one time._During induction period, rubber compounds melt and flow once they are exposed to heat and pressure. Here, flow characteristic plays an important role and in turn determines the properties of the final product. Figure 4a-c shows the plot of S"/S' versus time of the rubber samples. As expected, the domination of viscous element is observed in all rubber samples as observed from S">S'. Rubber compound will have a better flow to fill the mold cavities before vulcanization started. When vulcanization begins, this domination is changed. Elastic element starts to rise as a result from chemical crosslinking. Compounds nearly stop to flow because the movements of rubber molecule chains are restricted by the crosslinks (S"<S'). The extent of each element domination finds to be different. The concentration of MAH in MACO influences either the S"/S' ratio or the extent of domination of each element. It is probably affected by the higher compound viscosity and scorch time.

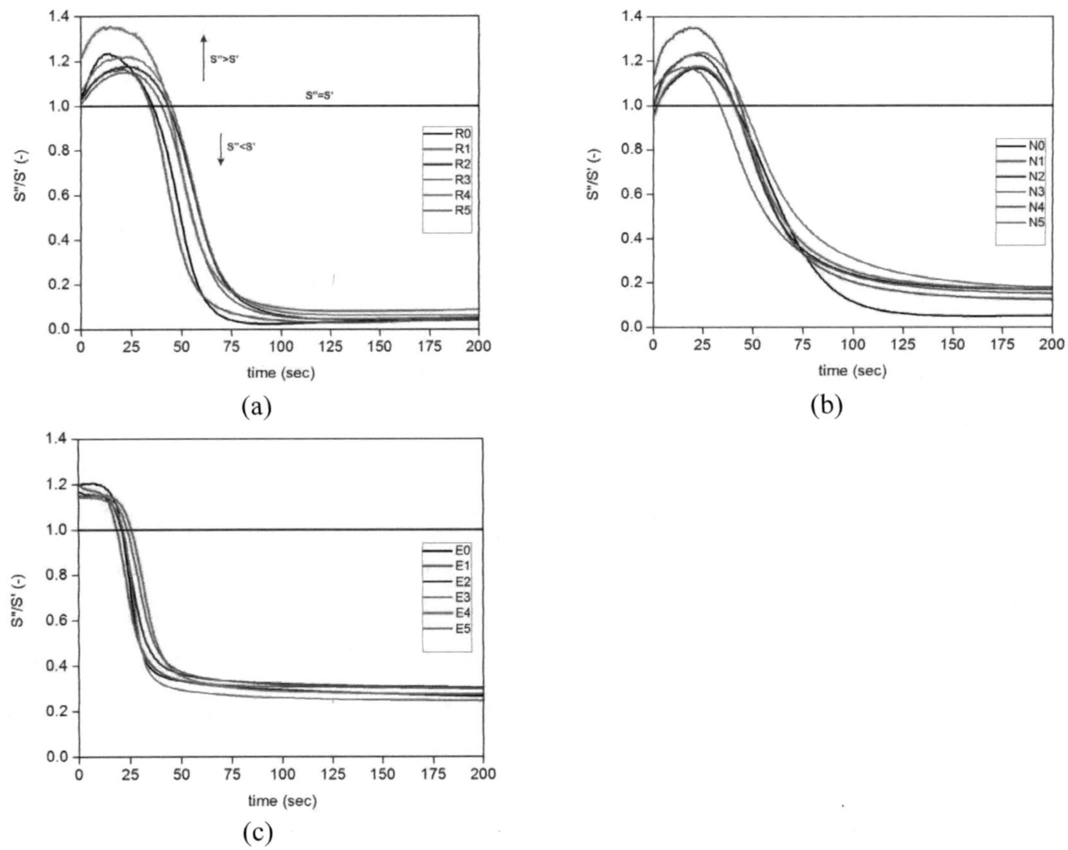

Figure 4. Plot S"/S' versus reaction time a) NR,b) EPDM, and c) NBR

3.3. Swelling Characteristic

Swelling by immersion in the liquid is a powerful tool to evaluate the effect of liquid in rubber samples. When rubber contact with solvent, the molecules of solvent will penetrate into rubber matrix and causes rubber to swell. Two types of solvents were used in this study, i.e. toluene and hexane. The choice of these solvent was based on the nature of them. Toluene is aromatic, while hexane is aliphatic solvent. These two solvents have different ability in dissolving rubber [19]. Figure 5a-b shows the swelling results of all rubber samples in toluene and hexane after 24 h immersion. It is evident that all rubber samples exhibit positive mass changes after swelling for both solvents. It indicates no plasticizer leaching occurs.

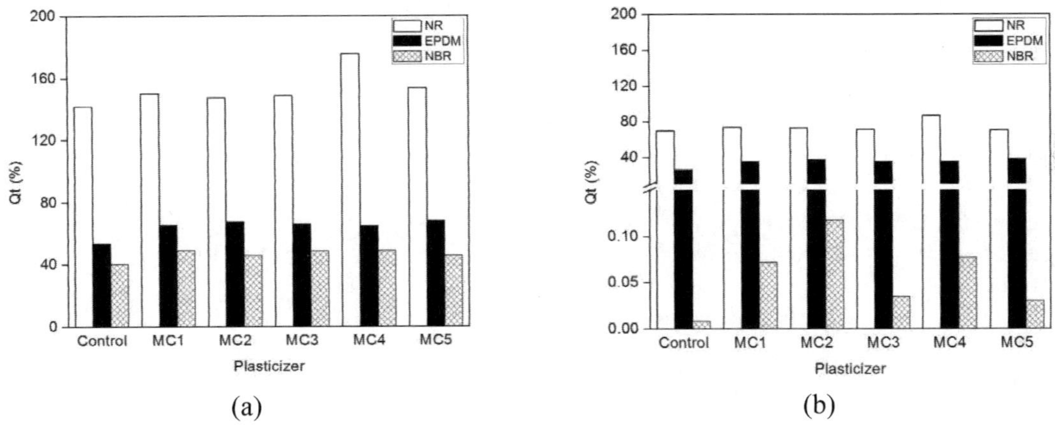

Figure 5. Swelling percentage of rubber samples in a) toluene and b) hexane

All rubber samples show great affinity with toluene than that of hexane. This result may be attributed by higher molar volume of hexane compared to that of toluene (Table 3). Polar group of acrylonitrile (C≡N) group presents in NBR molecule chains_interact strongly with the aromatic solvent [20] and results in high solvent sorption (high swelling percentage). On the other hand, NBR only provides weak interaction with the nonpolar hexane. Thus, low swelling percentage is observed.

4. Conclusion

MACO has been successfully synthesized and applied in NR, EPDM, and NBR. The resultant oils are predicted to be dimer or oligomer form, but acidic in nature. Application of MACO provides the different responseson rubber rheology, curing characteristic and swelling properties_to each rubber, but those are comparable with the conventional plasticizer (paraffinic or DOP). On the rheology and curing properties, all rubber (NR/EPDM/NBR) plasticized with MACO had given similar flow characteristic to conventional plasticizers._MACO provides slightly higher viscosity than conventional

IOP Conf. Series: Materials Science and Engineering **223** (2017) 012001 doi:10.1088/1757-899X/223/1/012001

plasticizer in NR compounds, but less pronounced in EPDM or NBR. The apparent crosslink densities are not influenced by the changes of MAH loading for NR and NBR, but it is greatly affected in EPDM._Scorch safety seems to be independent upon plasticizer types and MAH loading in MACO and the acidic nature of resultant MACO affects the curing time retarded reaction. While on teh swelling properties, MACO gives comparable swelling percentage to the conventional plasticizer._No plasticizer is found to leach out during swelling experiments.

Acknowledgement

Authors would like to thank to the head of Center for Leather. Rubber and Plastics for the financial support and Prof. Dr. Rochmadi, SU for intensive discussion related to MACO synthesis.

References

[1] Erythropel H C *et al* 2015 *Chemosphere* **134** 106–12.
[2] Saied M A *et al* 2008 *Eur. J. Lipid Sci. Technol.* **110** (10) 926–934.
[3] Ogunniyi D S 2006 *Bioresour. Technol.* **97** (9) 1086–91.
[4] da Costa H M *et al* 2003 *Polym. Test.*, vol. 22, no. 7, pp. 769–777, Oct. 2003.
[5] Bocqué M *et al* 2016 *J. Polym. Sci. Part A Polym. Chem.* **54** (1) 11–33.
[6] Wang H J *et al* 2008 *Biomacromolecules* **9** (2) 615–23.
[7] Mazo P C *et al* 2011 *Lat. Am. Appl. Res.* **41** (1) 11–15.
[8] Sert E and Atalay F S 2014 *Chem. Eng. Process. Process Intensif* **81** 41–47.
[9] Ray D *et al* 2012 *Ind. Eng. Chem. Res.*, **51** (6) 2603–2608
[10] Mazo P *et al* 2012 *Chem. Eng. J.* **185–186** 347–351.
[11] Kulkarni M G and Sawant S B 2003 *J. Am. Oil Chem. Soc.* **80** (10) 1033–1038.
[12] Mistri E *et al* 2011 *Ind. Crops Prod.* **34** (1) 900–906.
[13] Ghorui S *et al* 2011 *Ind. Crops Prod.* **34** (1) 893–899.
[14] Dewi I R *et al* 2016 *Maj. Kulit, Karet, dan Plast.* **32** (1) 51.
[15] Nabil H *et al* 2013 *Polym. Test.* **32** (2) 385–393.
[16] Ahmadi M and Shojaei A 2013 *Thermochim. Acta* **566** 238–248.
[17] Konar B B and Saha M 2012 *J. Macromol. Sci. Part A* **49**(3) 214–226.
[18] Oh J *et al* 2014 *J. Appl. Polym. Sci.* **131** (2) .
[19] Ahmed K *et al* 2012 *Adv. Mater. Phys. Chem.* **2** (2) 90–97.
[20] Maria H J *et al* 2013 *J. Mater. Sci.* 5373–5386.

IOP Conf. Series: Materials Science and Engineering **223** (2017) 012002 doi:10.1088/1757-899X/223/1/012002

Swelling behaviour in n-pentane and mechanical properties of epoxidized natural rubber with different epoxide content

N A Kinasih [1], M I Fathurrohman [1], D A Winarto[2]
[1]Indonesian Rubber Research Institute, Bogor, Indonesia
[2] Center for Polymer Technology (STP-BPPT), South of Tangerang, Indonesia

E-mail: norma.kinasih88@gmail.com

Abstract. Epoxidized natural rubber (ENR) with different level of epoxidation (i.e. 10, 20, 30, 40 and 50 mol% indicated as ENR ENR10, ENR20, ENR30, ENR40 and ENR50, respectively) were prepared. They were then vulcanized by using efficient system vulcanization. The effect of epoxide content on curing characteristic, swelling and mechanical properties in N-pentane was investigated. The Attenuated Resonance Fourier Transform Infrared (ATR-FTIR) and H-Nuclear Magnetic Resonance (H-NMR) were used to determine the epoxidation level. Glass transition (Tg) of ENR samples was determined by using Direct Scanning Calorimetry (DSC). The result revealed that the resistance of ENR in N-pentane increased with increasing epoxidation level, which indicated by decreasing equilibrium mol uptake and diffusion coefficient. The compression set of ENR and aging resistance increased with increasing epoxide content, except ENR50 was due to ENR 50 have two Tg value. However, the value of hardness and tensile strength were not effected by epoxidation level.

1. Introduction

Epoxidized natural rubber (ENR) is chemical modified rubber from natural rubber (NR), whereby some of the unsaturation double bond are converted into epoxides groups, which are randomly distributed along the polymer chain. The most interesting effects of epoxidation on NR lie in the physical properties of vulcanizates. Increasing of epoxide level in ENR vulcanizate has contributed to shorter schorch time [1], better rubber filler interaction and mechanical properties [2], higher polarity (oil resistance) and glass-transition temperature (Tg) [3], lower resilience (more damping), air permeability (comparable to butyl rubber and medium-acrylonitrile-content NBR), higher hysteresis, and better wet traction [4]. However, ENR has lower thermal stability, it was degraded along with the increment of heating rate [5].

The application of ENR in rubber compound and rubber blend has increase the oil and non-polar solvent resistant properties. The oil and non-polar solvent resistant properties was reflected on swelling behavior test result. Tanrattanakul *et al.,* (2003) [3] studied the swelling behavior ENR vulcanizate in toluene, ether, engine oil, gear oil, ASTM oil and motor oil. Furthermore, Ismail *et al.*, (2001) [6] investigated the swelling behavior of silica and carbon black filled ENR compound in toluene. However, there is very little information about swelling behavior of ENR vulcanizate with different epoxy content in n-pentane. N-pentane is the appropriate solution to replace LPG, according

IOP Conf. Series: Materials Science and Engineering **223** (2017) 012002 doi:10.1088/1757-899X/223/1/012002

to the requirements of Indonesian rubber seal product standard (SNI 7655:2010). The investigation about the swelling of ENR in n-pentane is very important to inform the possibility utilization of ENR in LPG rubber seal.

In this study, effect of epoxy content (10, 20, 30, 40 and 50%) in silica filled ENR was investigated. The epoxide content, Tg and viscosity properties of ENR was investigated. Moreover, the curing characteristic, mechanical properties and swelling behavior of silica filled ENR were studied.

2. Experimental Research

2.1 Chemicals

Fresh natural rubber latex was collected from local plantation, Ciomas-Indonesia. Anionic surfactant (Emulgen) was distributed by KAO Indonesia Chemicals. Silica of Zeosil175Gr was supplied from Solvay. Vulcurent as anti-reversion and rhenofit 1987 were supplied from Rhein Chemi. Protease enzyme (Papain), sulfur, silane-Si69, TiO_2, pigmen, antilux, DTDM, TMTD, CBS, zinc oxide, stearic acid, and DOP were supplied from local suppliers. n-pentanewas obtained from Merk.

2.2 ENR Preparation

ENR was prepared from deproteinized natural rubber latex (DPNR) phase by enzymatic process. Followed by adding 0.3 mol/mol isoprene unit of formic acid to suspend the enzymatic reaction of DPNR. Afterward, ENR was produced by forming the *insitu* performic acid in DPNR stage to produce ENR with epoxide level 10% (ENR10), 20% (ENR20), 30% (ENR30), 40% (ENR40) and 50% (ENR50). The epoxidation reaction condition was tabulated in Table 1.

Table 1. Epoxidation reaction condition of ENR

Sample	Temperature (°C)	Time (h)	H_2O_2 concentration (mol/mol isoprene unit)
ENR10	60	6	0.4
ENR20	60	6	0.5
ENR30	60	6	0.75
ENR40	65	6	0.75
ENR50	65	5.5	1

2.3 ENR Characterization

The ENR characterization was done to determine epoxide level, Tg and viscosity of ENR. The epoxy content of ENR was determined by ATR and H-NMR instrument. IR spectra were obtained on Thermo Fisher Scientific, Madison,Wisc within the wave number range from 4000 to 400 cm^{-1}. Epoxy content of IR spectra was analyzed by OMNIC software version 8.0 and epoxide level was determined according to Chakraborty *et al.,* (2010) [7]. H-NMR measurement was carried out by a JEOL JNM ECA 500, 500 MHz nuclear magnetic resonance spectrometer using $CDCl_3$ as solvent. The determination of epoxide contents of nmr spectra was determined according to Phinyochep and Boonjarairaak (2006) [8].

Transition temperature (Tg) was obtained by DSC 8000 Perkin Elemer instrument. The confirmation of epoxide content estimation was done by calculating % epoxide content based on Tg value, as shown below [9]. Given that Tg of NR is -62.26, according to experimental Tg measurement of DPNR. Moreover, the Tg value increases by 0.92°C/mol % [9].

$$Epoxidecontent(mol\,\%) = \frac{T_g of ENR + 62.26}{0.92} \quad (1)$$

Moreover, viscosity of ENR was recorded by mooney viscometer. The rubber were heated at 100°C for 1 min preheating and 5 min heating.

2.4 Compounding and Curing

The formulation of ENR compounds are given in Table 2. Each compounds of ENR were mixed using a two-roll mill (Berstorf), which was maintained at 65 ± 5°C. The mastication of ENR was done on 5 min at 55-60°C.

$$\mathrm{CRI} = \frac{100}{t_{90} - t_{s2}} \quad (2)$$

Table 2. ENR compound formulation

Material	Part per Hundred Rubber (phr)				
	ENR10	ENR20	ENR30	ENR40	ENR50
ENR	100	100	100	100	100
Silica	50	50	50	50	50
Silane	2	2	2	2	2
Rhenofit	2.5	2.5	2.5	2.5	2.5
Pigmen	1	1	1	1	1
Antilux	2	2	2	2	2
MMB	1	1	1	1	1
DOP	5	5	5	5	5
Vulcuren	1	1	1	1	1
ZnO	5	5	5	5	5
Stearic acid	1	1	1	1	1
CBS	1	1	1	1	1
TMTD	2	2	2	2	2
DTDM	1	1	1	1	1
Sulfur	0.3	0.3	0.3	0.3	0.3

Then, curing characteristic of ENR compound was evaluated by moving die rheometer (MDR 2000). The curing characteristic evaluation was included t_{90} (curing time), t_{s2} (scorch time), maximum torque (S'_{max}), minimum torque (S'_{min}), delta torque and curing time index (CRI). The CRI was calculated by equation 2. Afterward, the ENR compound was cured in compression molded under a pressure approximately 100 kg/cm^2 with an optimum cure time of t_{90}. The ENR vulcanizate was mold as square (15x15 cm) with 2 mm thickness.

2.5 Mechanical Properties

Hardness (Shore A) was determined using durometer type A according to ISO 48. The test specimens with 6 mm thick were made up of three layers of rubber from the compression-molded sheets. Tensile strength and elongation at break test was held at room temperature on LyoidTensometer, with crosshead rate 500 mm/min. The test samples were cut from compression-mold sheets using dumbbell shaped type 2 according to ISO 37. Compression set testing was conducted at room temperature (RT) during three days respectively. That test was determined according to ISO 815-2 using the standard test specimen of cylindrical shape of 12 mm in diameter and 6.3 mm thick by compression method.

2.6 Thermal Aging

The ENR vulcanizate conditioned at 70°C for 168 h in an oven according to ISO 188 and the mechanical properties changes before and after the thermal aging were then assed.

2.7 N-pentane Resistance Properties

N-pentane resistance was obtained by swelling test of NR vulcanizate in n-pentane, conducted at room temperature. The measurement according to the immersion – weight gain method. The samples were removed from the solvent at the specified time, quickly blotted with a filter paper and weighed using electronic balance. The immersing and weighing was continued till equilibrium swelling was attained. The molar percentage uptake (Q_t) and sorption coefficient (S) of n-pentane per gram of NBR vulcanized was determined using equation 3 and 4, respectively [10].

$$Q_t = \frac{(M_t - M_0)}{\text{Molecular weigth of n-pentane}/M_0} \times 100 \qquad (3)$$

$$S = \frac{M_\infty}{M_0} \qquad (4)$$

Where M_t is mass uptake at time t, M_o is the initial mass of ENR vulcanized and M_∞ is mass of n-pentane absorbed at equilibrium.

3. Result and Discussion

3.1 ENR Characterization

The chemical structure of ENR was analyzed by ATR and H-NMR. The IR spectra of DPNR and ENR is shown in Figure 1. Figure 1 shown that the characteristic signal of epoxy ring was shown after epoxidation process, due to IR spectra of DPNR and ENR. The IR spectrum of ENR indicates the appearance of characteristic signals of oxirane group at 1,249 and 870 cm^{-1}, signals of isoprene NR at 1600 and 835 cm^{-1}, signals of opening epoxy ring at 3400 cm^{-1} (asymmetric stretching of O-H), 1727 cm^{-1} (asymmetric stretching of C=O) and 1064 cm^{-1} (asymmetric stretching of hydrofuran) [11]. However, the present of O-H in DPNR was indicated H_2O content in DPNR, due to the fresh latex as material based. Furthermore, the epoxy content of ENR was increase with increase absorption of 1,249 cm^{-1} and 870 cm^{-1} wave number and decrease absorption of 1600 cm^{-1} and 835 cm^{-1} wave number. The opening epoxy ring was shown slightly decrease with increasing of epoxide level.

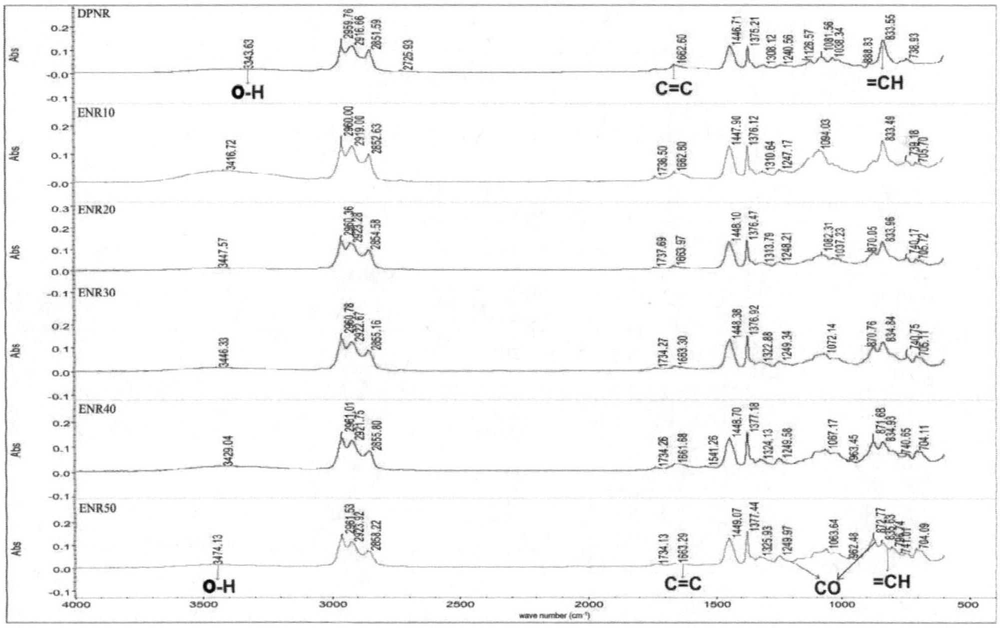

Figure 1. IR spectra of DPNR and ENR

The H-NMR spectrum of ENR in Figure 2 shows the presence of signals at 0.84 ppm and 5.1 ppm assigned to methyl and isoprene units, respectively [8]. After epoxidation of DPNR, other two signals appeared at 2.7 and 1.2 ppm, which were assigned to epoxy group. The increasing epoxide level was linearly with wide integration at 2.7 ppm. The estimated epoxy group content of the ENR was tabulated in Table 3. The estimate epoxide content was also determined by Tg value of DSC measurement. Based of Table 3, it can be shown that the estimation epoxy group content was relatively same, as an expectation.

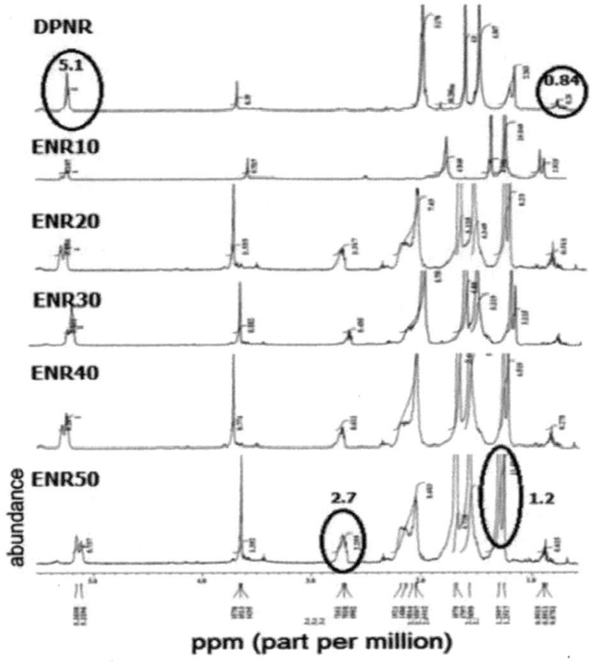

Figure 2. NMR spectra of DPNR and ENR

Table 3. ENR characteristic

Samples	Tg (°C)	epoxy content (% mol)		
		FTIR	HNMR	DSC
DPNR	-62.26	0	0	0
ENR 10	-46.12	12.52	2.14	17.54
ENR 20	-42.65	20.36	29.19	21.32
ENR 30	-31.88	34.53	31.00	33.02
ENR 40	-25.79	42.56	39.04	39.64
ENR 50	-41.77 &-19.42	52.38	53.91	22.27 & 46.57

The DSC thermograph was shown in Figure 3. As previously reported [4] the Tg of ENR varies linearly with molar epoxy content. Tg value was increased averagely 0.68°C/mol %. Epoxidation process was made natural rubber need more energy for mixing, which due to mooney viscosity value (Table 4). The width range of Tg shown the homogeneity of the samples, which is normally it is somewhat 10°C wide [3,9]. In the present study, ENR samples showed homogeneity in the epoxidation, indicated by a narrow width (<10°C) of the Tg as shown in Figure 3. However, the thermograph of ENR50 was shown two broad Tg peaks which has >10°C. Therefore, in term of ENR50 the result shows the inhomoginity of epoxidation process.

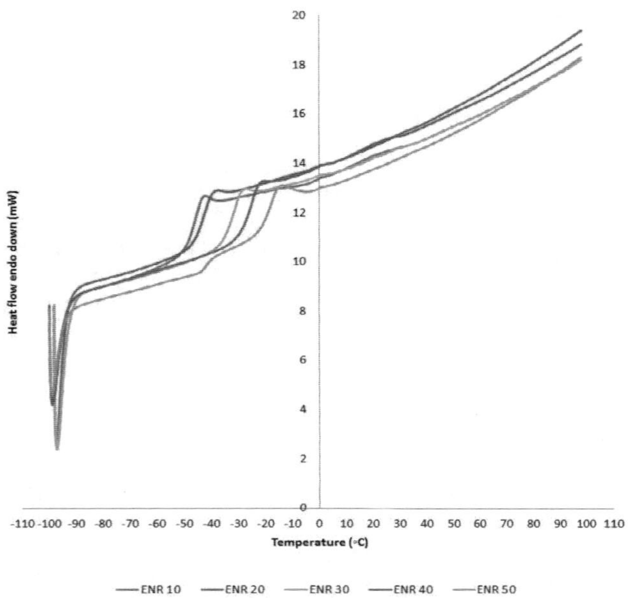

Figure 3. DSC thermograph of ENR

The mooney viscosity of ENR are presented in Table 4. The mooney viscosity of ENR was increased with increasing epoxy content, which fit with Rohadi *et al.* (2014) [2]. This trend was probably due to ENR's themselves possess a relatively high polarity in their structures and possible to form self association via hydrogen bonding or polar interactions between either the same or neighbouring molecules, and even self cross-link due to the opening of the oxirane rings [12]. Moreover, the degradation of ENR mooney viscosity was shown in ENR50. This phenomena probably due to inhomogeneity of epoxidation process, which was shown in DSC thermograph (Figure 3).

3.2 Curing Characteristic

The curing characteristic of ENR compound with different epoxy content are given in Table 5. The minimum torque was indicate the viscosity of the compound [13]. The result of minimum torque was increase as epoxide level, which was linier with mooney viscosity result (Table 3). Moreover, maximum and delta torque was increase as increase of epoxide level of the compound. Delta torque is related to crosslink density and stiffness of the rubber vulcanized [14]. Based on Table 2, the crosslink density is expected to increase with increase of epoxide level of the compound. The increasing level of epoxide content was faster the cure time, which due to high cure rate index.

IOP Conf. Series: Materials Science and Engineering **223** (2017) 012002 doi:10.1088/1757-899X/223/1/012002

Table 4. Mooney Viscosity of ENR

Samples	Mooney viscosity (ML 1+4 at 100°C)
ENR 10	124
ENR 20	103
ENR 30	123
ENR 40	138
ENR 50	128

Table 5. Curing characteristic of ENR vulcanizate

Curing Characteristic	ENR10	ENR20	ENR30	ENR40	ENR50
S_{max}(kg-cm)	11.31	11.3	10.39	11.13	9.92
S_{min}(kg-cm)	0.2	0.26	0.44	0.51	0.26
ΔS (kg-cm)	11.11	11.04	9.95	10.62	9.66
T_{90} (min)	4.31	3.29	4.28	4.02	5.05
T_{s2} (min)	1.15	1.18	1.31	1.4	1.55
CRI	31.65	47.39	33.67	38.17	28.57

3.3 Mechanical Properties

Table 6 represent the mechanical properties of ENR vulcanizate. The mechanical properties of the rubber was related to the amount of crosslink density [14]. The hardness and tensile strength of the ENR vulcanizate was relatively same, due to the crosslink amount, which was shown on maximum torque (Table 4). The elongation at break and compression set properties indicated the elasticity of the vulcanizate, which reflect on higher elongation at break and lower compression set. ENR50 was reached highest tensile strength. This result was observed that ENR's with a greater degree of polarity has contributed to lead better mechanical properties. Surprisingly that ENR50 have the lowest compression set, normally the compression set increased with increasing the epoxide level [17]. This phenomenon occurred due to inhomoginity of epoxidation process produced ENR50 that contain two epoxide level 22.27% mol and 46.57 % mol (Table 3).

Table 6. The mechanical properties of ENR vulcanizate

Mechanical properties	ENR10	ENR20	ENR30	ENR40	ENR50
Hardness (Shore A)	55	59	55	57	55
Tensile strength (MPa)	13.9	12.9	15.2	12.5	16.1
Elongation at break (%)	360	360	420	360	420
Compression set (%)	8.97	9.76	9.62	10.36	7.79

3.4 Thermal Resistance Properties

The effect of increasing epoxy content on ENR thermal resistance properties are presented in Figure 4. The thermal resistance properties were observed based on the value change of hardness, tensile strength and elongation at break on before and after thermal degradation. The lower value change indicated increasing vulcanizate resistant to thermal degradation. It clearly observed that increasing of epoxide level was decrease the value change of thermal degradation to the maximum level, which was ENR40. Furthermore, it slightly increased on ENR50 due to the inhomogeneity epoxidation process of ENR 50. This result indicate that ENR40 vulcanizate has optimum resistant to thermal degradation.

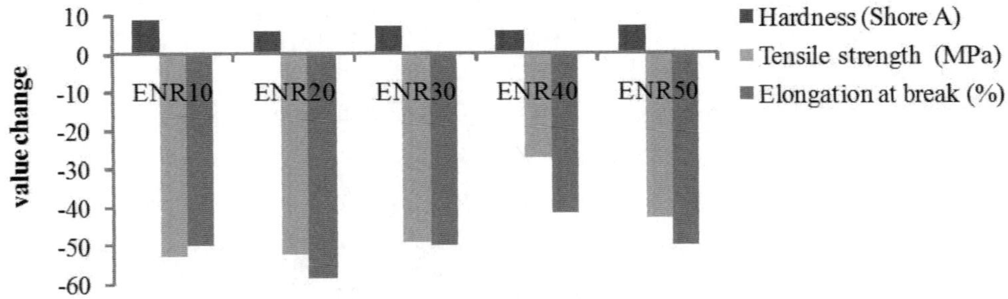

Figure 4. The thermal resistance properties of ENR vulcanizate

3.5 N-pentane Resistance Properties

The n-pentane resistance properties are clearly manifested in Figure 5 and Table 7. Swelling in n-pentane was governed by the competition between the driving force toward "dissolution of the rubber in the low-molecular-weight solvent", *i.e.* swelling, and the elastic force which increases upon solvent uptake [15]. The figures indicate that the molar percentage uptake increases initially and pass through a maximum and then stable. This trend means that a number of n-pentane diffuse into rubber vulcanizates and the distance between the rubber molecules reach equilibrium.

ENR10 has the maximum uptake whereas ENR50 the lowest. Usually, decreased solvent uptake ability was related to increase crosslink. However, the crosslink formed of ENR50, which was shown on maximum torque (Table 2), was lowest. This phenomenon was probably due to high epoxide level can increase the polarity of natural rubber and the solubility parameter of ENR increases with increasing epoxide level [3]. The closer solubility parameter of rubber to solvent, had low driving force to inhibit solvent uptake. The solubility parameter of ENR 0, 26, and 48% was 16.7, 17.4, and 18.2 $MPa^{1/2}$ respectively [16]. Whereas, the solubility parameter of n-pentane was 14.40 $MPa^{1/2}$ [19]. Thus, the epoxide level increases with decreasing the solvent uptake of ENR vulcanizate.

The sorption and equilibrium mol uptake of ENR was linier to molar percentage uptake (Q_t) of ENR. Table 6 presented that the sorption and equilibrium mol uptake of ENR was decreased with increasing epoxide content of the vulcanizate. Increasing of epoxide level in ENR vulcanizate has contributed to higher polarity [3]. ENR's with a greater degree of polarity has contributed to lower sorption and equilibrium mol uptake.

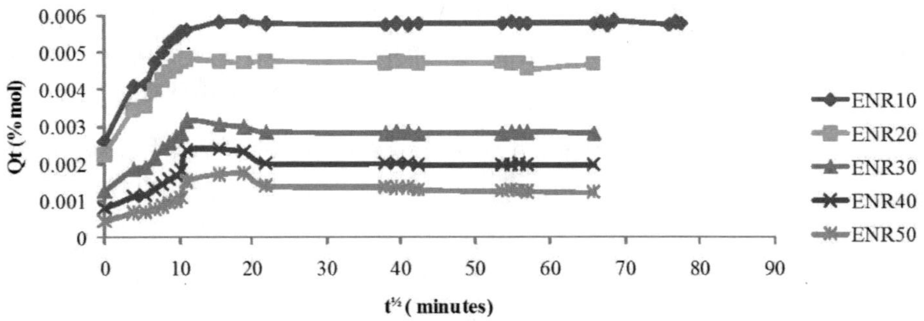

Figure 5. Molar percentage uptake (Q_t) of ENR with different epoxy level.

Table 7. Sorption and equilibrium mol uptake of ENR vulcanized with different epoxy content

Sample	Sorption (S) (%)	Equilbriummol uptake (Q∞) (%)
ENR10	1.4235	0.0058
ENR20	1.3418	0.0048
ENR30	1.2160	0.0029
ENR40	1.1677	0.0020
ENR50	1.1256	0.0014

4. Conclusion

Effect of epoxide level on mechanical properties and n-pentane resistance of ENR vulcanizate has been studied The chemical structure of ENR trough IR spectrum showed that increasing absorption of 1,249 cm^{-1} and 870 cm^{-1} wave number as increase epoxide content of ENR. Moreover, NMR spectrum was signal at 2.7 and 1.2 ppm, which were represented epoxy group, after epoxidation process. Tg and Mooney viscosity of ENR increase linearly with molar epoxy content. The increasing level of epoxide content was faster the curing time and lead better mechanical properties and thermal resistance of vulcanizate. The increasing thermal resistant was shown in ENR50. In addition, the increasing level of epoxy content was increased the n-pentane resistance of the vulcanizate, which was determined lowest molar uptake, sorption and equilibrium mole uptake on ENR50.

Acknowledgment

It is a pleasure to acknowledge the SINAS financial support by State Minister for Research and Technology fund for Indonesian Rubber Research Institute (RT-2015-0207)

References

[1] Poh B T and Tan K T 1991 *J. Appl. Poly. Sci.* **42** 1407-1416.
[2] Rohadi A A *et al* 2014 *Appl. Mech. Mater* **554** 71-75.
[3] Tanrattanakul V *et al* 2003 *J. Appl. Poly. Sci.* **90** 261-269.
[4] Gelling I R and Porter M 1988 Natural rubber science and technology Roberts, A.D. (Ed.) Oxford University Press. Oxford 359-456.
[5] Canzhong H Wang *et al* 2013 *J.Poly. Eng.* **33** 331-335.
[6] Ismail H *et al* 2001 *Int. J. Polymeric Mater.* **49** 191-204.
[7] Chakraborty S *et al* 2010 Quantitative application of FTIR in rubber. Rubber Word, **241** 33-38. Phinyochep P and Boonjarairaak K 2006 *Proc. Int. Rub. Conf.* IRC, Lyon, France, May 16-18, 2006.
[8] Loadman M J R, Tidd B K 1990 Natural rubber science and technology. Robberts. A. D. (Ed). Oxford University Press. Oxford. UK. Chapter 22.
[9] Obasi H C *et al* 2009 *Int. J. Poly. Sci.* **2009** 1-6.
[10] Alwaana I M *et al* 2013 *J. Poly Mater.* 30 1 117-130.
[11] Ng S C and Gan L H 1981 *Eur. Poly. J.* **17** 1073-1077.
[12] Marković G *et al* 2013 *Composites Part B: Engineering* **55** 368-373.
[13] Movahed S O *et al* 2015 *J. Appl. Poly. Sci.* **132** 41512-415121.
[14] Van Duin M and Dikland H 2007 *Polym. Degrad. Stabil.* **92** 2287-2293.
[15] Brandrup J *et al* 1998 Polymer Handbook. Wiley. New York.
[16] Gelling I R 1991 *J. Nat. Rub. Res.* **6** 184-205.

IOP Conf. Series: Materials Science and Engineering **223** (2017) 012003 doi:10.1088/1757-899X/223/1/012003

Mechanical properties and mophology natural rubber blend with bentonit and carbon black

E M Ginting[1], N Bukit[1], Muliani[1] and E Frida[2]
[1]Department of Physics ,University State of Medan ,Indonesia
[2]Quality University, Indonesia

E-mail : evamarlina67@yahoo.com

Abstract. Purpose of this study was to determine the mechanical properties and morphology of composite natural rubber and natural bentonite with addition of Na-bentonite filler and carbon black to natural rubber. The method is carried material mixed with filler composition variations (0,10,20,30) phr using open mill for 6 min. Results of the open mill is vulcanized at a temperature of 170°C. Further testing mechanical properties and morphology. Results showed that the addition of Na-bentonite filler and carbon black influence on the mechanical properties of tensile strength, elongation at break, modulus of elasticity, hardness, and strong tear. Morphological results showed cavities in the rubber compound and the occurrence agglomeration.

1. Introduction

Natural rubber has properties of plasticity or viscosity, elasticity, tensile strength, and. But that is not polar nature and content of the high unsaturated bond in the molecule, natural rubber is not resistant to oxidation, ozonation, heat and expand in the oil. To improve the quality of the rubber, engineering conducted on the rubber usually by adding filler with the aim to improve the performance of the rubber. One of the most common fillers used is carbon black. Carbon black filler material is active or reinforcing fillers are able to add hardness and resilience tear, abrasion resistance, and high tensile strength on the goods produced [1-2]. Reinforcing fillers generally have a small particle size, surface active chemical, which has a porous surface and a non-uniform shape can improve adhesion [3]. Particles of inorganic filler material for polymers have been used widely because generally cheaper financing. Fillers are frequently used, fiber glas, mica, talc, SiO_2 and $CaCO_3$, usually forming micro-composites with improved properties [4]. Bentonite is a type of clay that contains mineral montmorillonite more than 85% with the chemical formula is $[(OH_4)Si_8\ Al_4O_{20}\ n\ H_2O]$. Bentonite is divided into easy bentonite expands and can not expand. Bentonite is sodium bentonite expands is (Na-bentonite).

Clay is one of the excipients non charcoal is often used as an ingredient fillers in the rubber industry. Clay is cost minerals and has become part important in the rubber industry where its use as a filler economical to modify the creation and the performance of natural rubber and synthetic rubber. There are many types of clay, but montmorillonite has a long record as the most important inorganic

IOP Conf. Series: Materials Science and Engineering **223** (2017) 012003 doi:10.1088/1757-899X/223/1/012003

materials was added as a filler material into natural rubber [5-6]. Improved properties depend on many factors including the aspect ratio of the filler material, the degree of dispersion and orientation within the matrix, and adhesion at the interface matriks- excipients .

In this study aims to determine the mechanical properties and morphology ,of composites with Na -bentonit natural rubber and carbon black as filler.

2. Experimental

2.1. Materials and Instrument

Natural rubber SIR 20, bentonite, carbon black, ZnO of PT Brataco, sulfur, stearic acid from PT brataco, wax from China, butyl hydroxy toluene (BHT), Diethyl Zinc Ditio carbamate (ZDEC), distilled, saturated NaCl. The tools used in the characterization and manufacturing test samples is open mill, Scanning Electron Microscopy (SEM) Zeiss models, tensiometer MosantoT-10 Durometer type-A, X-Ray Fluorescence (XRF), XRD (6100 Shimadzu)

Table 1. The Composition of The Mixture of Natural Rubber Compound

Materials	Formula Compound (phr)				
	1	**2**	**3**	**4**	**5**
Natural Rubber	100	100	100	100	100
Wax	1.5	1.5	1.5	1.5	1.5
Na -bentonite	0	30	20	10	0
Carbon black	0	0	10	20	30
Stearic acid	2	2	2	2	2
BHT	2	2	2	2	2
ZnO	5	5	5	5	5
ZDEC	1.5	1,5	1.5	1.5	1.5
Sulfur	3	3	3	3	3

2.2. Preparation of Bentonite.

Bentonite natural crushed into powder using a mortar. 150 g of bentonite which had been sieved to 200 mesh was added to the glass beaker and add 500 mL saturated NaCl. The mixture is stirred with a magnetic stirrer at a temperature of 70 °C for 2 h. Then, into the mixture was added 500 mL of distilled water. The mixture is stirred for 10 min. Then allowed to stand until precipitation occurs, then the mixture is filtered and the filtrate is separated using a vacuum filter. Furthermore sediment obtained saturated NaCl was added for the second time as much as 500 mL and the mixture was stirred again for 2 h at 70 °C. Sediment obtained is separated from the solution by means filtered. Sediment obtained washed with hot distilled water. Sediment obtained dried in an oven at 120 °C for 6 h.

Preparation of rubber compound. The mixture of materials as shown in Table 1, was added to the open-mill, and then milled until completely solid rubber, rubber While the passage of the grinding process, the materials inserted one by one gradually for 6 min. Results of the open mill is vulcanized at 170°C

3. Results and Discussion

3.1 Mechanics Analysis

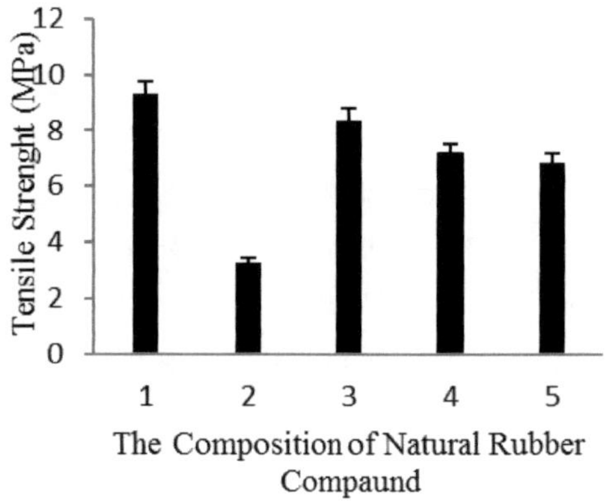

Figure 1. Tensile Strength of the Composition of Natural Rubber Compound

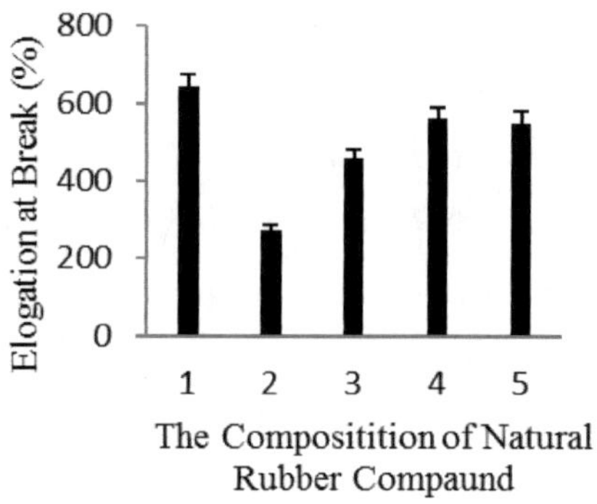

Figure 2. Elongation at Break of the Composition of Natural Rubber Compound

From Figure 1 tensile strength values obtained rubber compound with fillers decreased compared with the tensile strength of rubber compounds without fillers.This is due to the rubber compound is not affected fillers, so it does not slow down the vulcanized rubber.And the value of tensile strength is the smallest in the sample 2, namely rubber compound with 30 phr filler na-bentonite. This is because Na-bentonite siliceous which have high acidity that can inhibit vulcanized rubber resulting in lower density crosslinked with vulcanized natural rubber as filler silica. Continued decline in the density of cross causing chains of the rubber molecules are easy to move and not rigid. This is according to [7] ,also contains carbon that is high enough so as to reduce the value of tensile strength.

Elongation at break is one of the physical properties of the finished goods of rubber, to determine the nature of the elasticity of the product that will show to what extent the product may be stretched to the right in place. In Figure 2 the greatest value of the elongation at break is on the sample 1 is a rubber compound without filler materials with an elongation at break of 643.33%, this is due to the sample 1 does not use filler material so that the resulting rubber compound is not rigid .serta easy to stretch and have The value of the elongation at break.

The value of the elongation at break is the smallest in the sample 2, namely rubber compound with 30 phr filler na-bentonite with an elongation at break of 273.33%, this was due Na-bentonite siliceous which is rigid So that the resulting compound is difficult to stretch the resulting value of the

IOP Conf. Series: Materials Science and Engineering **223** (2017) 012003 doi:10.1088/1757-899X/223/1/012003

extension breaking low. also contains carbon that is high enough and can reduce the value of the elongation at break.

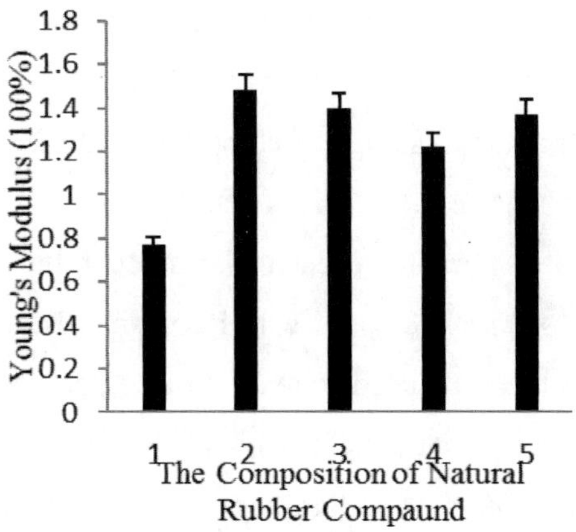

Figure 3. Young's Modulus (100%) of the Composition of Natural Rubber Compound

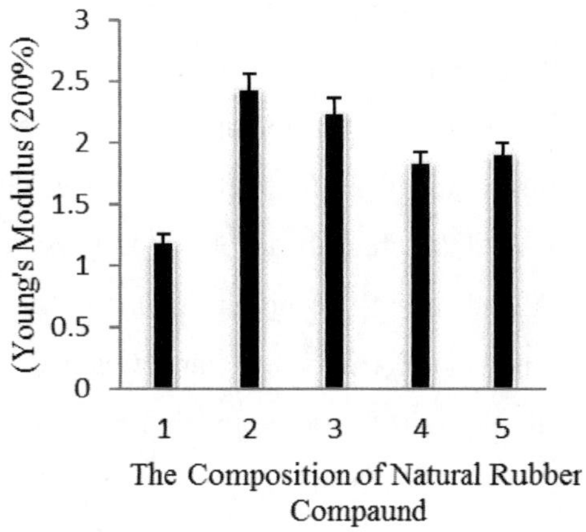

Figure 4 .Young's Modulus (200%) of the Composition of Natural Rubber Compound

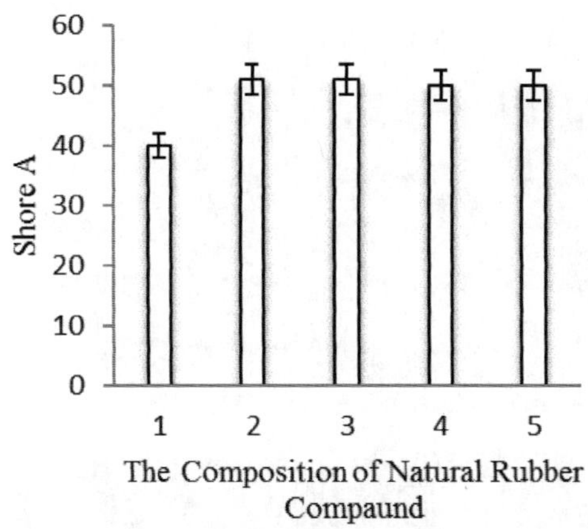

Figure 5. Shore A of the Composition of Natural Rubber Compound

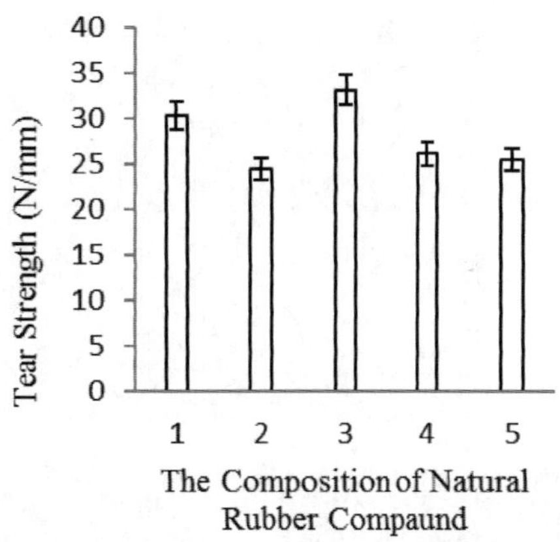

Figure 6. Tear Srength A of the Composition of Natural Rubber Compound

Figure 3 modulus of elasticity (100%) at the highest rubber compound with 30 phr filler Na-bentonite, .Value smallest elastic modulus rubber compound it is because the rubber compound is not affected by the filler so that it has a high crosslinking density, causing the value of the extension rupture increases, so that the resulting elastic modulus decreased. Similarly, the modulus of elasticity of 200% is shown in the Figure 4. In cabon black filler modulus of elasticity lower than the compound with fillers Na-bentonit. is consistent with research [8] that the addition of carbon black structure lowers the modulus value, in addition to the need for the addition of some excipients with an appropriate comparison.

Figure 5 shows that the greatest hardness in the rubber compound with 20,30 phr Na-bentonite and 10 phr carbon black, due to the high hardness rubber compound with silica fillers. This is possible because the silica as a filler material that amplifies that which would add to the violence on finished goods of rubber and can provide improvement of physical properties of rubber, increase in hardness due to the interaction of silica in the mixture tends to increase the stiffness of the mix.

Tear resistance of Figure 6 the highest value obtained in sample 3 compound with 20 phr filler Na-bentonite and 10 phr carbon black. It is possible tear resistance influenced by the inclusion of carbon black. increasing the amount of intake of carbon black then tear resistance can be decreased rapidly. The results were obtained tear resistance values were decreased with increasing the amount of intake of carbon black. From the results of this research can be used to make finished goods molding rubber soles when seen from his physical nature which include tensile strength, elongation at break, resilience tear, meet the standards of SNI(0778: 2009). But for the hardness properties not meet the standards.

3.2 Morphology Analysis

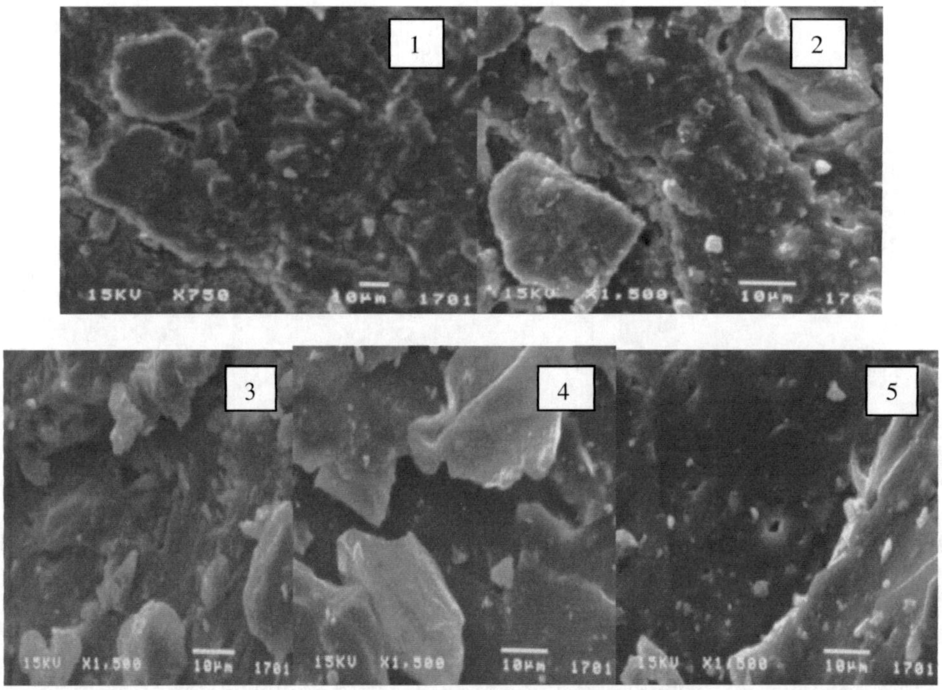

Figure 7. Morphology of the Composition of Natural Rubber Compound

The presence of cavities on the surface due to interaction between the rubber compound phase is still not good around the filler particles. This is due bentonite for their hydrophilic OH groups .It is in accordance with the results of the study [1] states that the weak interaction between these phases, is caused by the tendency of molecules of silica in rubber compounds that have hydroxyl groups form hydrogen bonds between the molecules of silica or other chemical material which is polar. For rubber compound with 20 phr filler Na-bentonite and 10 phr of carbon black in Figure 7(3) is obtained form the surface of the rubber compound that lack of homogeneity

This is due to the uneven distribution of particles is possible because mixing chemicals rubber compound rubber uneven. The presence of cavities on the surface due to interaction between the rubber compound phase is still not good on the filler. While carbon black is hydrophobic because of their carboxyl group of around porous.compound rubber composition Na-bentonite more than carbon black affecting its mechanical properties, in this case the tensile strength, hardness, resilience tear higher compared with the composition of the rubber compounds Na -bentonit fewer than carbon black.

In the rubber compound with the composition of the Na-bentonite fewer than carbon black affecting its mechanical properties, in this case the tensile strength, hardness, tear powerful lower than the Na-bentonite compound composition is more than carbon black.

For rubber compound with 30 phr filler carbon black in Figure 7(5) is obtained form the surface of the rubber compound that lack of homogeneity. This is due to the uneven distribution of particles is possible because mixing chemicals uneven rubber compound at the time open mill.

3.3 XRD Analysis

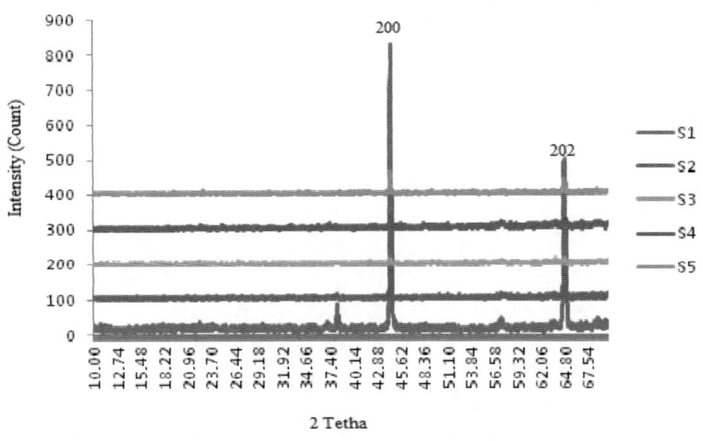

Figure 8. Difraction of the Composition of Natural Rubber Compound

Figure 8 shows the diffraction pattern of XRD for-bentonite mixture of natural rubber and carbon black. The addition of natural bentonite into the polymer matrix of natural rubber is carried out on a different variation is (0, 10, 20, 30) phr. The figures show that there is a basal spacing of bentonite at 44.43 theta angle that is equal to 2.03 Å with d_{hkl} 200 on natural rubber bentonite (10, 20, 30) phr, whereas the second peak at d_{hkl} 202 with angle tetha 64.76, d spacing 1.43 Å. In general, the addition of a mixture of bentonite and carbon black is almost the same diffraction pattern.

4. Conclusion

The results showed that the addition of Na-bentonite filler and carbon black influence on the mechanical properties of tensile strength, elongation at break, modulus of elasticity, hardness, and strong tear. Morphological results showed cavities in the rubber compound and the occurrence agglomeration , thereby reducing the tensile strength and elongation at break

Acknowledgments

We thank the Physics Laboratory Unimed and Laboratory of Polymer Physics LIPI Bandung and rubber research center Sungai Putih Lubuk pakam on the facilities provided research.

References

[1] Bahruddin Z I dan Zul Amraini S 2010 *J. Teknik Kimia Indonesia* **9**(2) 62.
[2] Rattanasom N *et al* 2007 *J.Polym. Testing* **26**(3) 369.
[3] Hanafi I *et al* 2005 *Solid State Sci.Technol.* **13** (1) 184.
[4] Ray S S and Okamoto M 2003 *Progress in Polym. Sci.* **28** 1539.
[5] Dong W *et al* 2006 *J. European Polym.* **42** 2515.
[6] Frounchi M *et al* 2006 *J. Membrane Sci.* **282** 142.
[7] Bukit N and Frida E 2013 *Makara J. Technology* **17**(3) 113.
[8] Rahmaniar 2006 *J. Dinimika Penelitian BIPA* **17** 29.

Practical application of thermoreversibly Cross-linked rubber products

L M Polgar [1,2], F Picchioni[1,2], E de Ruiter[1] and M van Duin[1,3]
[1]University of Groningen, Departement of Chemical Engineering, Nijenborgh 4, 9747 AG Groningen, The Netherlands
[2]Dutch Polymer Institute (DPI), P.O. Box 902, 5600 AX Eindhoven, The Netherlands
[3]ARLANXEO Netherlands, Keltan R&D, P.O. Box 1130, 6160 BC Geleen, The Netherlands

E-mail: l.m.polgar@rug.nl

Abstract. Currently, rubber products cannot simply be reprocessed after their product life, due to the irreversible cross-linking methods traditionally applied. The purpose of this work is to investigate how thermoreversible cross-linking of rubbers via Diels Alder chemistry can be used for the development of recyclable rubber products. Unfortunately, the applicability of the thermoreversible EPM-g-furan/BM system appears to be limited to room temperature applications, because of the rapid deterioration of the compression set at elevated temperatures compared to irreversibly cross-linked EPM. However, the use of EPM rubber modified with thiophene or cyclopentadiene moieties may extend the temperature application range and results in rubber products with acceptable properties. Finally, rubber products generally comprise fillers such as silica, carbon black or fibers. In this context, the reinforcing effect of short cut aramid fibers on the material properties of the newly developed thermoreversibly cross-linked EPM rubbers was also studied. The material properties of the resulting products were found to be comparable to those of a fiber reinforced, peroxide cured reference sample.

1. Introduction

Rubber products have become an integral and essential part of our daily lives over the last 150 years since the invention of sulphur vulcanization by Charles Goodyear. Cross-linking is crucial as it provides good elasticity and high strength to rubber products. Unfortunately, cross-linking also prevents melt processability, making it difficult to recycle cross-linked rubber products, at least in a cradle-to-cradle context. Devulcanization allows for the cleavage of cross-links, but also causes bond scission in the polymer main chain, which is detrimental for the performance of the rubber recyclate compared to the virgin rubber material [1,2]. An alternative to devulcanization of rubbers is found in thermoreversible cross-linking with Diels-Alder (DA) chemistry being particularly useful [3]. Thermoreversibly cross-linking is applied to obtain materials that combine the elastic properties of a permanently cross-linked rubber at service conditions with the melt (re)processability of a non-cross-linked rubber or thermoplastic. In our previous studies a commercial ethylene-propylene rubber (EPM) grafted with maleic anhydride has first been modified with furfurylamine to introduce furan groups along the rubber backbone. Next, these pendant furan groups were cross-linked with a bismaleimide (BM) via a DA coupling reaction (Figure 1).

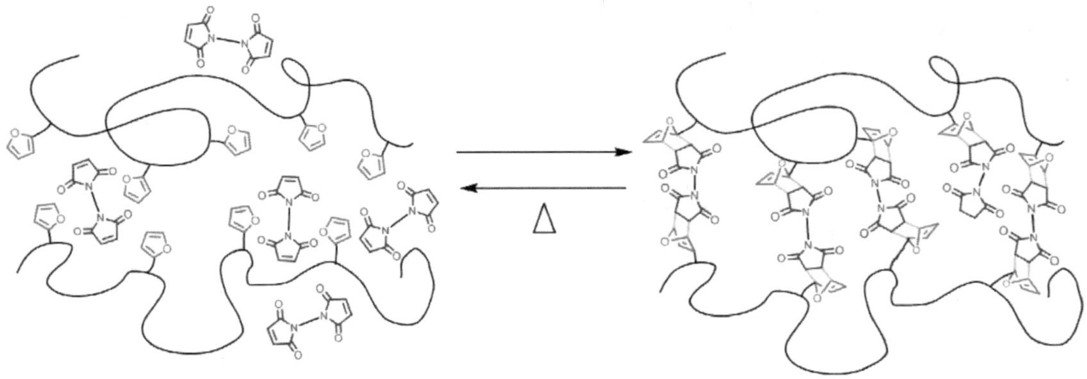

Figure 1. Diels-Alder Cross-Linking and Retro-Diels-Alder De-Cross-Linking of Furan-Functionalized Elastomers with Bismaleimide Cross-Linkers.

The thermoreversibly cross-linked products can be compression moulded into samples that display material properties similar to peroxide cured and sulphur vulcanized EPDM reference samples. Furthermore, upon cutting and re-moulding samples with similar properties as the original samples were obtained (Figure 2). This demonstrates the thermoreversibility of the DA reaction and the recyclability of DA cross-linked rubber.

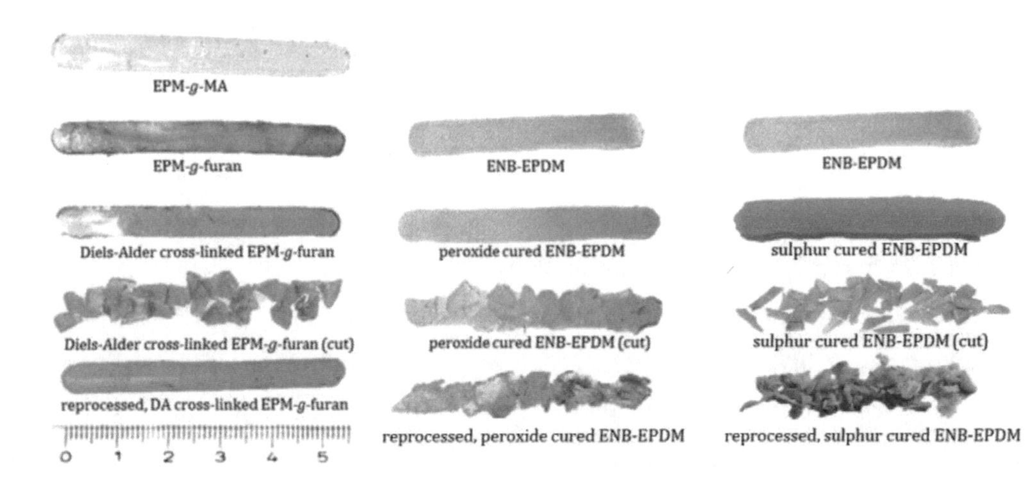

Figure 2. Reprocessing of Bismaleimide Cross-Linked, Peroxide Cured and Sulphur Vulcanized EP(D)M Rubbers [3]

The goal of this study is to investigate the conditions at which these newly developed thermoreversibly DA cross-linked rubber products can be applied with a focus on the effects of compression and relaxation times and temperature on elasticity. Some alternative DA diene-dienophile couples are studied and suggested for further research in order to overcome any shortcomings of the original EPM-g-furan/BM combination. Some rubber products are compounded with strong fibers, such as aramid fibers, for reinforcement [4,5]. It will be explored whether benefits of applying thermoreversibly cross-linking in the presence of aramid fibers can be found, for example in the recovery of the fibers by de-cross-linking the rubber matrix, the ability to homogeneously redistribute the fibers by reprocessing the product and/or the possibility of inducing (thermoreversible) interactions between the fibers and the rubber matrix.

2. Experimental

2.1. Materials

Non-cross-linked, peroxide cured and sulphur vulcanized EPDM reference samples with a medium ethylidene (ENB-EPDM, *Keltan 8550C,* 5 3 wt% ethylene, 5.5 wt% ENB) and non-cross-linked, maleated EPM (EPM-g-MA, *Keltan DE5005*, 49 wt% ethylene, 2.1 wt% MA, M_n = 50 kg/mol, PDI = 2.0) were kindly provided by ARLANXEO Netherlands B.V. Prior to the reaction, EPM-g-MA was dried in a vacuum oven for one hour at 175 °C to convert present diacids into anhydride. Short cut aramid fibers (3 mm) were kindly provided by Teijin Aramid. Furfurylamine (FFA, Sigma-Aldrich, ≥99%) and 2-thiophenemethylamine (2-TMA, Sigma-Aldrich, 97%) were freshly distillated before use. 3-Cloropropylamine (APCl, > 99%), lithium cyclopentadiene (LiCp, > 99%), 1,1-(Methylenedi-4,1-phenylene)bismaleimide (BM, 95%), Bis(t-butylperoxy-i-propyl) benzene (*Perkadox14-40*, AkzoNobel), Octadecyl-1-(3,5-di-tert-butyl-4-hydroxyphenyl) propionate (99%), tetrahydrofuran (THF, >99.9%), decalin (98%) and acetone (>99.5%) were all bought from Sigma-Aldrich.

2.2. Methods

2.2.1. Furan-functionalization and BM Cross-linking of EPM-g-MA

The EPM-g-MA precursor was converted into EPM-*g*-furan according to a reported procedure [6]. 40.0 g of EPM-*g*-furan (8.6 mmol furan content) and 40 mg anti-oxidant were dissolved in 500 mL THF. 0.5 eq. (based on furan content of EPM-g-furan as determined from characterization by infrared spectroscopy and elemental analysis) of the BM was dissolved in THF and added to the 10 wt% rubber solution under stirring. The majority of the solvent was evaporated in the fume hood by blowing over air. The residual solvent was removed in a vacuum oven at 50 °C and the resulting product was compression molded at 150 °C and 100 bar for 1 h and thermally annealed in an oven at 50 °C for 7 days. Compression molding was performed on a Taunus Ton Technik V8UP150A press, equipped with a temperature controller.

2.2.2. Cyclopentadiene Modification and Cross-linking

The method for the functionalization of EPM-g-MA with furfurylamine was slightly modified to facilitate grafting of thiophene and cyclopentadiene moieties to the EPM backbone. Cyclopentadiene was grafted onto the EPM by first grafting 3-Cloropropylamine and using the halide as a leaving group for lithium cyclopentadiene according to a reported procedure [7,8]. 45.0 g EPM-g-MA rubber (9.6 mmol MA) was dissolved in 500 mL THF at room temperature. 4.5 g of APCl (28.9 mmol; 3.0 eq. based on MA content in EPM-g-MA) was then added to the 10 wt% rubber solution. The reaction mixture was stirred for 5 h in a closed system at room temperature and then precipitated by pouring it slowly into a tenfold amount of acetone (5 L) under mechanical stirring, yielding the polymer product as white threads. The product (EPM-g-Cl) was dried to constant weight in an oven at 35 °C. Subsequently, the product was briefly compression molded at 175 °C and 100 bar to ensure the conversion of all intermediate maleimide acid products to imide products. 40.0 g of the resulting EPM-g-Cl (8.6 mmol pendant Cl groups as determined from characterization by infrared spectroscopy and elemental analysis) and 40 mg anti-oxidant were dissolved in 500 mL THF under N_2. An equimolar amount of LiCp was added under stirring and left to react at 30 °C for 18 h. Afterwards, the reaction mixture was precipitated in a 20:1 molar excess of demi-water under mechanical stirring, yielding the polymer product as white threads. The product (EPM-g-Cp) was dried in a vacuum oven at 150°C overnight to remove any traces of water and unreacted LiCp. The resulting product was compression molded at 150 °C and 100 bar for 1 h and thermally annealed in an oven at 90 °C for 3 days.

2.2.3. Thiophene-functionalization and BM Cross-linking of EPM-g-MA

45.0 g EPM-*g*-MA rubber (9.6 mmol MA) was dissolved in 500 mL THF at room temperature. 3.2 g of freshly distillated 2-TMA (28.9 mmol; 3.0 eq. based on MA content in EPM-g-MA) was then added to the 10 wt% rubber solution. The reaction mixture was stirred for 5 h in a closed system at room temperature and then precipitated by pouring it slowly into a tenfold amount of acetone (5 L) under mechanical stirring, yielding the polymer product as white threads. The product (EPM-g-thiophene) was dried to constant weight in an oven at 35 °C. Subsequently, the product was briefly compression molded at 175 °C and 100 bar to ensure the conversion of all intermediate maleimide acid products to imide products. 40.0 g of the resulting EPM-g-thiophene (8.6 mmol thiophene content as determined from characterization by infrared spectroscopy and elemental analysis) and 40 mg anti-oxidant were dissolved in 500 mL THF. 0.5 eq. (based on thiophene content of EPM-g-thiophene) of the BM was dissolved in THF and added to the 10 wt% rubber solution under stirring. The majority of the solvent was evaporated in the fume hood by blowing over air. The residual solvent was removed in a vacuum oven at 50 °C and the resulting product was compression molded at 150 °C and 100 bar for 1 h and thermally annealed in an oven at 90 °C for 3 days.

2.2.4. Preparation of fiber-filled, BM Cross-linked EPM-g-furan Rubber Compounds

18 g of EPM-g-furan was fed to an internal mixer (70% fill factor) and homogenized at 50 rpm and 130 °C for 4 min. Then, 1, 3 or 5 phr of short cut aramid fibers was added. When the torque was stable (after typically 2-3 min), the BM cross-linker was added (0.2 to 1.5 molar eq. based on furan content). Mixing was continued for 2 min before the compound was removed from the mixer. Sample bars of the obtained products were obtained by pre-heating the materials in a mold at 140 °C for 5 min and compression molding them at 140 °C and 100 bars for 15 min. The resulting sample bars were thermally annealed in an oven at 50°C for a minimum of 72 h to ensure complete cross-linking.

2.2.5. Preparation of fiber-filled, Peroxide Cured ENB-EPDM Rubber Compounds

18 g of ENB-EPDM was fed to an internal mixer (70% fill factor) and homogenized at 50 rpm and 70 °C for 4 min. Then, 1, 3 or 5 phr of short cut aramid fibers was added. When the torque was stable (after typically 2-3 min), 0.5, 1 or 2 phr of pure peroxide was added slowly to the mixture. The rubber compound was mixed for 4 more min until it was removed from the mixer. This rubber compound was then vulcanized by pre-heating in a mold at 160 °C for 5 min and compression molding at 160 °C and 50 bars for 35 min.

2.3 Characterization

The conversion of EPM-g-MA to EPM-g-furan, EPM-g-thiophene and EPM-g-Cl was followed by Fourier Transform Infrared spectroscopy (FT-IR) and elemental analysis (EA). FT-IR spectra were recorded on a Perkin-Elmer Spectrum 2000. Rubber films with a thickness of 0.1 mm were compression molded at 150 °C and 100 bar for 30 min, thermally annealed to ensure maximum DA cross-linking and measured in a KBr tablet holder. Measurements were performed over a spectral range from 4000 to 600 cm^{-1} at a resolution of 4 cm^{-1}, co-averaging 32 scans. Deconvolution was used to quantify the areas under the individual FT-IR peaks (R^2> 0.95). The differences in relative peak areas were used to calculate the reaction conversions. The methyl rocking vibration peak at 723 $cm^{-1}$was used as an internal reference, as it originates from the EPM backbone and is not affected by chemical modification. The decrease of the absorbance of the C=O symmetrical stretch vibration of the anhydride groups at 1856 cm^{-1} was used to calculate the conversion of the reaction from EPM-g-MA to EPM-g-furan, according to a reported procedure [6]. EA for the elements N, C and H was performed on a Euro EA elemental analyzer. The nitrogen content was related to the anhydride conversion according to a reported procedure [3].

Tensile tests were performed on an Instron 5565 with a clamp length of 15 mm, according to the ASTM D4-112 standard. A strain rate of 500 ± 50 mm/min was applied. Test samples of 45 mm long, 5 mm wide and 1 mm thick were prepared by compression molding. For each measurement 10 samples were tested and the two outliers with the highest and the lowest values were excluded. Data presented are averages of the other 8 tests. The median stress-strain curves in the figures were selected to represent the entire series of a sample. The compression set test is used to measure the elasticity as it is affected by the time and temperature of compression and/or relaxation. Compression set tests were generally performed according to the ASTM D931 standard, using a home-made device and cylindrical samples with a thickness of 6.0 ± 0.1 mm and a diameter of 13.0 ± 0.1 mm, which were prepared by compression molding. The samples were compressed to 75% of their original thickness for 70 h at room temperature and relaxed for 30 min at 50 °C. Another series of compression set experiments was performed in which the samples were compressed to 75% of their original thickness for 70 h at 25, 60, 100 and 140 °C and relaxed for 30 min at room temperature.

3. Results

3.1. Application Window

The furan-functionalization of EPM-g-MA and subsequent BM cross-linking was successfully performed according to the reported procedure [6]. The cross-link density of this thermoreversibly cross-linked rubber was found to be $2.07 \cdot 10^{-4} \pm 0.21 \cdot 10^{-4}$ mol/cm^3, which is similar to that of the irreversibly peroxide cured and sulphur vulcanized EPDM reference samples ($1.98 \cdot 10^{-4} \pm 0.18 \cdot 10^{-4}$ and $1.85 \cdot 10^{-4} \pm 0.16 \cdot 10^{-4}$ mol/cm^3, respectively).Compression set experiments were performed on these samples under different conditions to probe the effects of the dynamics of the thermoreversibly cross-linked rubber with the thermoset rubber references (Figure 3). The compression set was studied as a function of relaxation time at room temperature for the 3 mentioned rubber products after 70 h of compression at room temperature (Figure 1A) and as a function of the compression time at room temperature followed by 30 min relaxation at room temperature (Figure 1B).

The compression set of the BM cross-linked EPM-g-furan, the sulfur vulcanized and peroxide cured EPDM reference samples decreases with the time after the imposed compression. It appears that the decay in compression set is linear up to 50 min after the measurement at which the initial compression set value seems to have halved for all three samples. Similarly, longer compression times result in higher compression set values for all three samples. Meanwhile, the compression set differs significantly between the different samples despite their similar cross-link densities. The results suggest that the dynamic character of the thermoreversible cross-links does not have a significant effect on the elastic properties of the reprocessable rubbers at room temperature and that the DA cross-links act as covalent cross-links at room temperature.

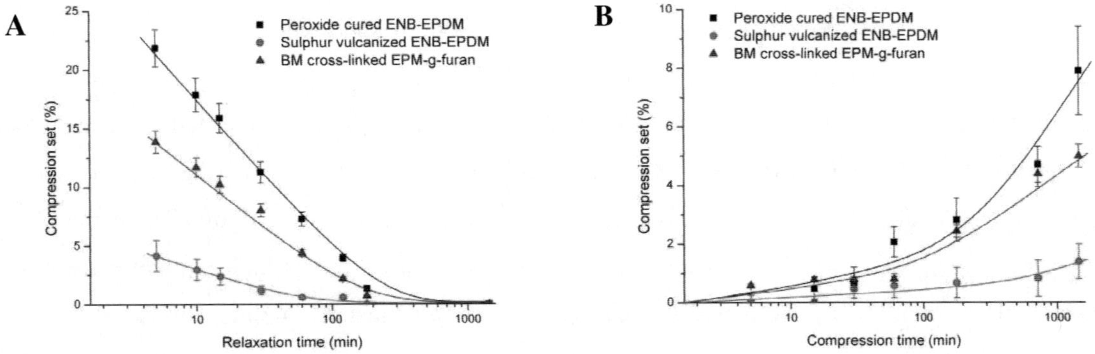

Figure 3. Compression set of BM cross-linked EPM-g-furan samples and of peroxide cured and sulphur vulcanized EPDM reference samples with similar cross-link densities applying A: relaxation for different times at room temperature after 70 h of compression at room temperature and B: 30 min relaxation at room temperature after compression for different times at room temperature. Lines are guides to the eye. Error bars indicate ± 1% standard deviation.

The compression set of BM cross-linked EPM-g-furan and the peroxide cured and sulphur vulcanized EPDM references have also been studied for different compression temperatures (Figure 4).

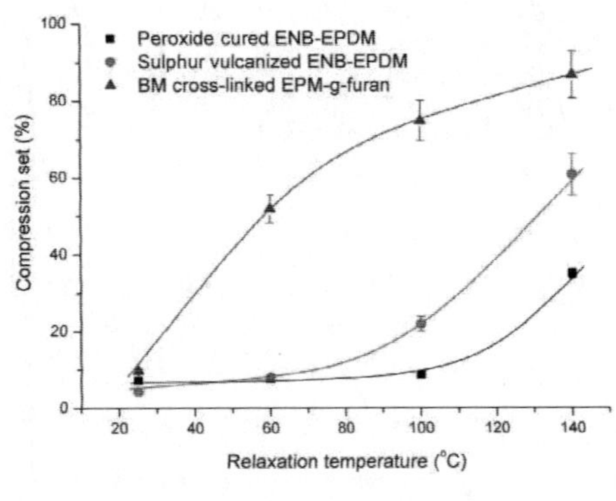

Figure 4. Compression set at different temperatures of BM cross-linked EPM-g-furan samples and of peroxide cured and sulphur vulcanized EPDM reference samples with similar cross-link densities. The lines are guides to the eye. The error bars indicate ± 1 % standard deviation, taken here as relative error on the measured data.

Although the room temperature compression set is similar for the 3 rubber samples, it appears that a slight increase in compression temperature has a strongly detrimental effect on the compression set of the thermoreversibly cross-linked EPM-g-furan sample compared to the irreversibly cross-linked reference samples. This is probably due to a shift of the DA reaction equilibrium towards the retro-DA reaction at elevated temperatures. The dynamic equilibrium allows for a rearrangement of the cross-links and, thus accommodating the sample to the imposed compression. This means that the use of the thermoreversibly cross-linked EPM-g-furan is strictly limited to (below) room temperature applications, such as wire and cable insulation or sealant applications. Away to overcome this can be found in using a different diene-dienophile couple for the DA cross-linking reaction.

3.2 Alternatives Yielding Improved High Temperature Performance

Other diene-dienophile couples can be used to thermoreversibly cross-link modified elastomers with bismaleimide via DA chemistry [9]. An example is the modification of EPM-g-MA with 2-thiophenemethylamine and the thermoreversible BM cross-linking of the resulting EPM-g-thiophene. The product was found to be cross-linked based on its insolubility (cross-linked polymers are non-soluble with respect to their non-cross-linked precursors) and the high Young's modulus and low elongation at break with respect to its non-cross-linked precursor (Figure 5).

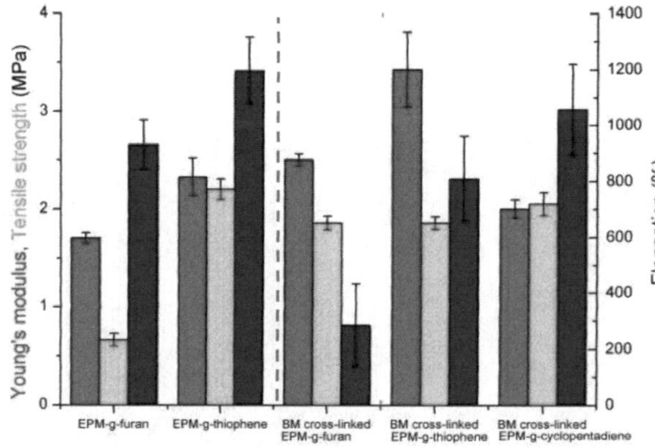

Figure 5. Tensile Test Results of Non-Cross-Linked EPM-G-Furan and EPM-G-Thiophene, Their BM Cross-Linked Products and A BM Cross-Linked EPM-G-Cyclopentadiene rubber. Error Bars Indicate ± 1 Standard Deviation.

Cyclopentadiene also seems to be a good alternative for furan as the temperature at which the retro-DA reaction is known to starts to occurring is higher (150-215°C for cyclopentadiene [10,11] versus 110-170 °C for furan [12-14]). Some preliminary experiments were performed in which EPM-g-MA was functionalized with 3-chloropropylamine and the resulting pendant chloride group was reacted with lithium cyclopentadiene [7]. Although this approach was successful for the highly functionalized polyketones used in a previously reported study [7], the application to EPM-g-MA elastomers with a low degree of functionalization was less successful. The high reactivity of the cyclopentadiene groups is responsible for a number of side-reactions [15], resulting in a low conversion. More importantly, the formation of cross-links between the pendant cyclopentadiene groups upon reaction with bismaleimide was found to proceed much slower compared to the furan/bismaleimide couple. This is probably because the DA reaction kinetics for cyclopentadiene requires higher temperatures, but may also be related to the lower concentration of functional groups, but is also likely to be a result of the low mobility of the polymer backbones that make it hard for the functional groups to recombine. The relatively small, bifunctional BM cross-linking agents were found to have a relatively high mobility in the rubber matrix, allowing for a faster recombination of diene and dienophile. As a result, the cyclopentadiene cross-linked EPM has a lower Young's modulus and higher elongation at break with than the BM cross-linked EPM-g-furan (Figure 5). It is clear that further optimization of this system is actually required before being able to draw any conclusion on its applicability.

3.3. Aramid Fiber Reinforced Rubber Composites

Finally, rubber products are usually reinforced with fillers such as aramid fibers. For the BM cross-linked EPM-g-furan, the Hardness and Young's modulus increase and the elongation at break of the rubber compounds decreases upon compounding with larger amounts of short cut aramid fiber (Figure 6). No significant changes were observed in the tensile strength and the compression set. With the exception of the compression set, the effect of fiber loading of the thermoreversibly cross-linked rubbers on the material properties was the same as for fiber-loaded, peroxide cured ENB-EPMD rubbers with the same cross-link density ($2.45 \cdot 10^{-4} \pm 0.28 \cdot 10^{-4}$ and $2.63 \cdot 10^{-4} \pm 0.27 \cdot 10^{-4}$, respectively). The inferior compression set of the thermoreversibly cross-linked rubbers with respect to that of the peroxide cured reference sample might be a result of the rearrangement of cross-links under compression. Nevertheless, fiber filled rubber products may prove to be a suitable application for thermoreversibly cross-linked rubber as the addition of aramid fibers significantly increases the hardness and Young's modulus.

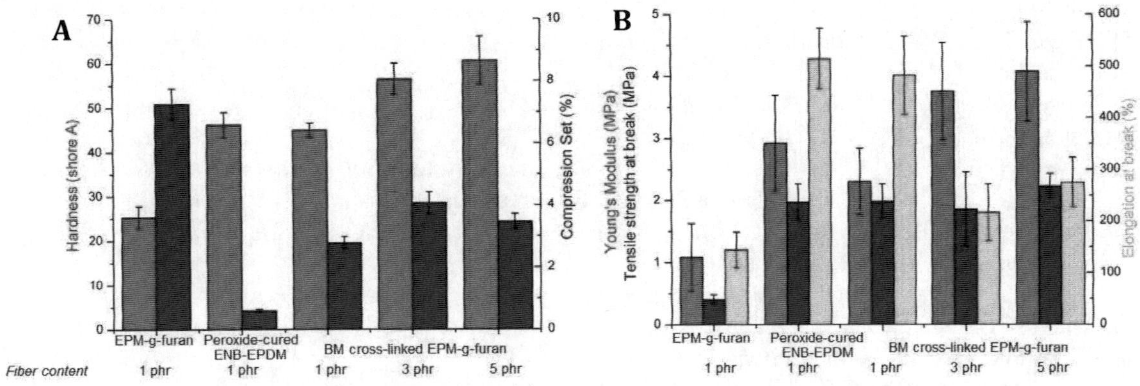

Figure 6. A: Hardness and Compression Set at Room Temperature and B: Young's Modulus, Tensile Strength and Elongation at Break of Non-Cross-Linked EPM-G-Furan and Peroxide Cured ENB EPDM with 1 phr of Short-Cut Aramid Fibers and BM Cross-Linked EPM-G-Furan with Different Fiber Loading. Error Bars Indicate ± 1 Standard Deviation.

4. Conclusions

The applicability of a newly developed, reprocessable EPM rubber modified with furan groups and cross-linked with bismaleimide was studied by determining the material properties under different conditions. The dynamic character of the thermoreversible Diels-Alder cross-linking chemistry does not have a significant effect on the elastic properties of the reprocessable rubber at room temperature. However, the application of this thermoreversibly cross-linked elastomer remains limited to room temperature applications, because of the rapid deterioration of the elasticity upon an increase in temperature compared to irreversibly cross-linked rubbers. Alternative diene-dienophile couples, such as bismaleimide cross-linking of EPM modified with thiophene or direct cross-linking with cyclopentadiene are proposed to overcome these issues regarding the temperature stability. Using such components should increase the threshold temperature for the retro-DA de-cross-linking reaction. Finally, the BM cross-linked EPM-g-furan rubber was used for the preparation of aramid fiber reinforced rubbers. The material properties of the resulting products are comparable to those of a fiber reinforced, peroxide cured reference sample and improve with the fiber loading.

Acknowledgements

This research forms part of the research program of the Dutch Polymer Institute, project #749.

References

[1] Sutanto P *et al* 2006 *Int. Polym. Proc.* **21** 211.
[2] Sutanto P *et al* 2006 *J. Appl. Polym. Sci.* **102** 5028.
[3] Polgar LM *et al* 2015 *Macromol.* **48** 7096.
[4] Hintze C *et al* 2014 *Polym. Eng. Sci.* **54** 2958.
[5] Hintze C *et al* 2013 *Rubber Chem. Technol.* **86** 579.
[6] Polgar LM *et al* 2016 *J. Vis. Exp.* **e54496**.
[7] Toncelli C *et al* 2016 *J. Appl. Polym. Sci.* **133** 42924.
[8] Patet RE *et al* 2015 *Acs Catalysis* **5** 2367.
[9] Toncelli C *et al* 2011 Thermoreversibility in Polymeric Systems*Self-Healing at the Nanoscale* ed Amendol V and Meneghetti M (Boca Raton: CRC Press) 197-247.
[10] Peterson AM, Palmese GR 2009 *Click Chemistry for Biotechnology and Materials Science* (Hoboken: John Wiley & Sons, Ltd).
[11] Stille J, Plummer L 1961 *J. Org.Chem.* **26** 4026.
[12] Chen XX *et al* 2002 *Science* **295** 1698.
[13] Teramoto N *et al* 2006 *Carbohydr. Polym.* **64**78.
[14] Gandini A *et al* 2008 *Eur. Polym. J.* **44** 4029.
[15] Kobuke Y *et al* 1970 *J. Am. Chem. Soc.* **92**6548.

IOP Conf. Series: Materials Science and Engineering **223** (2017) 012005 doi:10.1088/1757-899X/223/1/012005

A comparative study of ground tire rubber devulcanization using twin screw extruder and internal mixer

O Ujianto[1], D B Putri[2], Jayatin[1], D AWinarto[1]

[1]Center for Polymer Technology, Agency for Assessment and Application of Technology, Banten, Indonesia
[2]Chemical Engineering, Polytechnic STMI, Jakarta, Indonesia

E-mail: onny.ujianto@bppt.go.id

Abstract. Devulcanization of ground tire rubber (GTR) was done using twin screw extruder (TSE) and internal mixer (IM). Processing parameters were varied to analyze its effect on gel content. Fourier Transform Infrared Spectroscopy (FTIR) analysis was performed as qualitative technique to confirm structural change. The devulcanized rubbers with the least gel content percentage produced in both TSE and IM were then used as filler in natural rubber (NR)/coconut coir (CC) composite preparation. Effects of gel content percentage on NR/CC composite tensile strength and elongation at break were analyzed. The results show that the gel content decreased by 41% for sample processed in TSE and 50% in IM compared to control sample. Overall, the devulcanization is influenced by high energy generated by thermal or thermo-mechanical process. FTIR spectra show chemically structural changes of GTR as C=C, CH_2, $CH_3$with higher intensity for IM sample than its counterpart indicated devulcanization. The replacement of GTR to DGTR on NR/CC/GTR composites provided less network structures and resulted better tensile strength and elongation at break.

1. Introduction

Waste tire disposal has been quite a crucial issue worldwide. It was proven that approximately 17 million of waste tire generated each year [1]. Waste tires could be considered as a wealthy material, due to its contain such as natural rubber, and synthetic rubber (generally SBR and BR) [2-4]. Recovery of waste tire can be classified in two ways : material recovery [1, 5] and energy recovery [1] used as source of energy in cement kilns and paper mills. Energy recovery is easier and cheaper than its companion, but it poses dangerous threat to environment and human health for gases produced throughout the process. Hence, material recovery is an alternative to reduce waste tire.

Material recovery could be done with various methods including shredding, grinding that leading to transformation of tire into a powder form [2,6,7], devulcanizing [8-11,4,12] and blending with another polymer [11,4,2]. The key parameter of material recovery was the breakage of three-dimensional cross-linked structure leading to compatibilities improvement with other polymer or filler.

Various methods have been proposed for reclaiming or devulcanizing waste tire, including thermomechanical using twin , and other methods using microwave [15-18], and supercritical CO_2 [10, 11,19]. From the available techniques, twin screw extruder is more attractive due to its ease and continuity. However, the effect of processing time in extruder cannot be analyzed due to its dependency on screw rotation. On the other hand, the processing time is an independent factor in an internal mixer, so effect of time and the interaction to another variable can be explored.

In order to measure degree of devulcanization, techniques as gel/sol content [2,20-23] and crosslink density [20, 22] were applied. However, gel content is simpler and good measurement and calculation to indicate devulcanization efficiency [21]. Meanwhile, the structural change may be analyzed using FTIR [22], DSC and TGA [21].

In this research, comparative study on the effect of twin screw extruder and internal mixer for ground tire rubber devulcanization was done. Effects of processing temperature_on gel content were explored. The structural analysis was observed using FTIR for samples with the lowest gel content. The use of devulcanized rubber as filler on NR composites was compared and analyzed from tensile properties data to that of GTR.

2. Experimental

2.1. Materials

GTR was donated by PT. Sinar Jaya Rubber (Bandung, Indonesia). It is ground rubber from vehicle car tires. The maximum particle size of GTR was 30 meshes and processed as received. No other chemical or additives were added as promotor agents during devulcanization.

Natural rubber (NR), *Standard Indonesian Rubber* (SIR) 20 purchased from Aneka Bumi Pratama Company (Palembang, Indonesia) was used as composite matrix. Coconut coir from Indonesian local producer was used as reinforcement. Maleated Natural Rubber (MNR) produced in our laboratory, used as compatibilizer. The additives used for composite preparation were Butylatedhidroxy toluene (BHT) as antioxidant, zinc oxide (ZnO) as activator, N-cyclohexyl-2-benzothiazole sulphenamide (CBS) as accelerator, and sulphur (S) as vulcanization agent.

2.2. Devulcanization of GTR

Devulcanization was carried out using two equipment, intermeshing co-rotating Collins twin-screw extruder ZK 25E x 36D (TSE) and Haake (IM) with Banbury rotors. In IM 80% filling capacity was applied to produce optimum shear. Process parameters were varied according to two-level Full Factorial Design with additional mid-point using Minitab 16, summarized in Table 1. Parameters in TSE and IM have a different variable since time could not be controlled in TSE, and limitation on IM maximum rotor rotation.

Table 1. Parameter applied in TSE and IM

	TSE		IM		
Run	T (°C)	SS (rpm)	T(°C)	SS (rpm)	t (min)
1	100	100	100	100	10
2	100	400	100	100	30
3	300	100	100	200	10
4	300	400	100	200	30
5	200	250	300	100	10
6	200	250	300	100	30
7	-	-	300	200	10
8	-	-	300	200	30
9	-	-	200	150	20
10	-	-	200	150	20

2.3. *Quantitative and Qualitative Analysis of Devulcanized GTR (DGTR)*

Devulcanization performance was analyzed quantitatively from crosslink degree by Gel Content determination (GC). It was chosen as a good indicator of devulcanization efficiency [21]. It was performed by soxhlet extraction with toluene as solvent. 0.050 ± 0.005 gr of samples were extracted in 100 mL of toluene at 110 °C for 24 h. The extracted samples were dried in oven at 80 °C for 5 h. Percentage of gel content was calculated as follows:

$$\%\, Gel\ Content = 100\% - \%\, Extracted \tag{1}$$

$$\%\, Extract = \frac{w_2 - w_3}{w_2 - w_1}\ x\ 100\% \tag{2}$$

Where w_1 is weight of wire net, w_2 is weight of sample in wire net before extraction, w_3 is weight of sample in wire net after extraction.

Qualitative analysis on structural changes of DGTR was analyzed using Bruker Tensor FTIR. Approximately 2 g of samples were individually put in tubes to undergo pyrolysis method. Each prepared tubes were filled with nitrogen gas (N_2) and placed above the burner until it reaches pyrolysis term and get the oily polymer of samples. The oil produced were places in natrium chloride glass to analyze the chemical structure in samples.

2.4. *Composite Preparations*

The coconut coir (CC) was soaked in NaOH solution at 5% concentration for 1 h. It was then washed by distillated water and dried at 100 °C for 1 h. The treated coconut coir was then cut and sieved maximum1 cm long. The composite formulations are summarized in Table 2. The processes were divided into two steps. All formulations in step 1 were done in TSE at screw speed 40 rpm, and temperature 40 °C for zone 1, and 160 – 190 °C for zone 2 – 8. The extruded samples were continued to step 2 process using Huicai roll mill at 40 °C and 60 rpm for the first 10 min. It was then followed by the addition of ZnO for 10 min and CBS for the next 10 min. Vulcanization of the blend was ended up by mixing the blends and S for 30 min.

Table 2. Parameter applied in TSE and IM

Sample	Formulation (phr)							
	Step 1 (TSE)					Step 2 (Roll Mill)		
	GTR	DGTR	CC	MNR	BHT	ZnO	CBS	S
NR Control	0	0	0	0	2	5	2	5
NR/CC composites	0	0	30	6.5	2	5	2	5
NR/CC/GTR composites	10	0	30	6.5	2	5	2	5
NR/CC/DGTR-TSE composites	0	10	30	6.5	2	5	2	5
NR/CC/DGTR-IM composites	0	10	30	6.5	2	5	2	5

Vulcanized compounds were processed in Collins Hot Press P300 P. It was preheated at 150 °C with no pressure for 5 min and pressed at the same temperature, 15 bar for 10 min. The pressed compounds were cooled at 25 °C for 25 min and conditioned for 24 h.

2.5. *Composite Characterization*

Tensile test were carried out according to ASTM 412 type C using universal testing machine Shimadzu AGS10 kNG at 500 mm/min of crosshead rate. Tensile strength and elongation at break of NR, NR/CC composite without GTR, NR/coconut coir composite with GTR as well as DGTR were compared and analyzed. The DGTR added in composites were those produced in TSE or IM with the lowest gel content.

3. Results and Discussions

3.1. Effect Process on Gel Content Reduction

The effect of process conditions on gel content was analyzed because it is potentially break the three dimensional network structures. The lower gel content, the less percentage of crosslink in the rubber compound, the more effective devulcanization process. It may be influenced by thermal, shear or combination between these two mechanisms. Therefore, the analysis on the effect of process conditions on gel content is crucial.

Table 3 shows gel content percentage of samples processed either in TSE or IM. In general, gel content percentages decreased by 41% (from 85% to 44%) for TSE, and 50% (from 85% to 35%) for IM. The decrease on gel content percentage in this research was higher than reported results using two roll mill with additional reclaiming agent [24] and microwave techniques [16, 21]. The better results achieved in this research might be due to generated shear from screw in TSE or rotor in IM. This shear would function as three dimensional scissors which break crosslink structure in normal, radial and tangential direction [25, 26].

Table 3. Gel Content Percentage Samples Processed in TSE or IM

	TSE			IM			
Runs	T (°C)	SS (rpm)	GC (%)	T (°C)	SS (rpm)	t (min)	GC (%)
Untreated	-	-	85	-	-	-	85
1	100	100	78	100	100	10	85
2	100	400	70	100	100	30	84
3	300	100	53	100	200	10	80
4	300	400	44	100	200	30	74
5	200	250	65	300	100	10	46
6	200	250	67	300	100	30	85
7	-	-		300	200	10	51
8	-	-		300	200	30	35
9	-	-		200	150	20	72
10	-	-		200	150	20	72

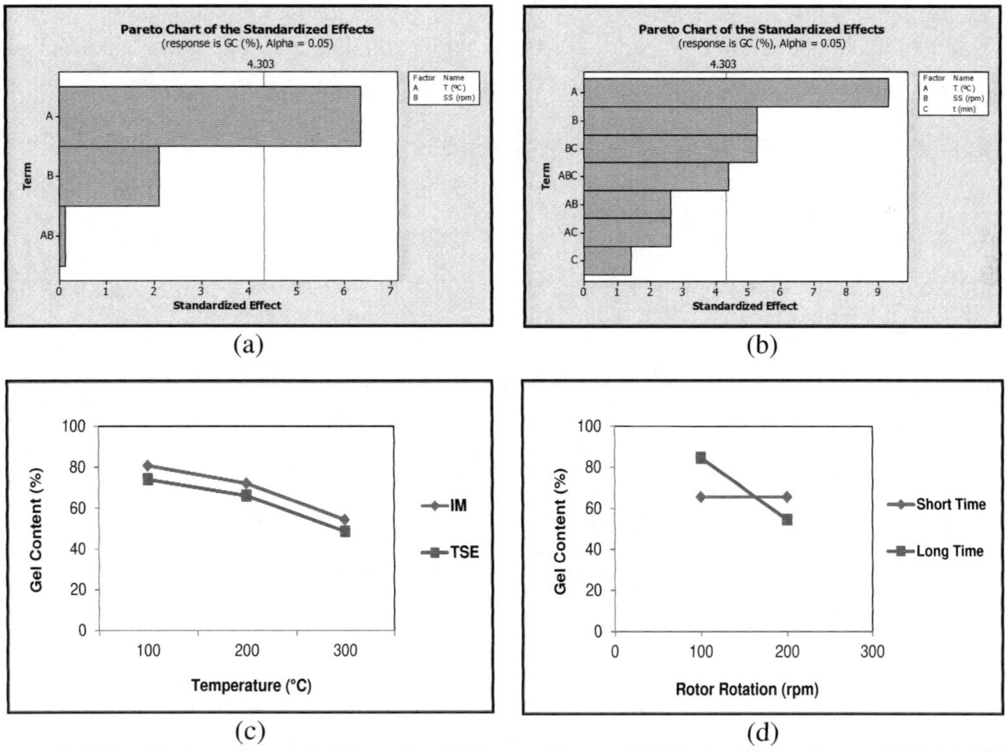

Figure 1. Effect Parameters on Gel Content: (a) Pareto Charts of TSE, (b) Pareto Chart of IM, (c) Effect Temperatures in TSE and IM, (d) Effect Factor Interactions in IM

The least gel content was produced at high settings for either in TSE (high temperature, high screw rotation) or IM (high temperature, high rotor rotation, and long processing time) attributed to high energy needed for devulcanization.

Effects of processing parameters were analyzed from Pareto Chart (Figure 1). The charts were generated using Minitab software from collected data. The Pareto charts (Figure 1 (a) and (b)) are useful to analyze the significant parameters applied in process equipment. The significant factors are limited by a vertical red line. All factors on the right side of this line have significant effect on the reduction of gel content percentage. In contrast, variables on the left side may not significantly influence the gel content. Figures 1_(a) and (b) show that temperature is the most significant parameter in either TSE or IM to reduce gel content. The effect on gel content is negative (reducing gel content) (Figure 1 (c). This finding aligned to previous research using another technique as microwave reported the reduction of gel content due to higher temperature [21].This also suggests that temperature is the main factor to reduce gel content in devulcanization process.

In contrast, Figure 1 (a) and (b) shows differenteffect of screw rotation in TSE and in IM. The effect was not significant in TSE, but significant in IM. Insignificant effect of screw rotation in TSE may be due to uncontrollable processing time. High screw rotation may cut cross-linked chain_[27], however it is also generate short processing time.These would not provide optimum conditions for temperature, as combination with generated shear, to break the network structures maximally. On the other hand, low screw rotation is able to provide longer processing time and facilitate thermal chain scissoring. As a result, the difference effect of screw rotation in TSE was not significant. In contrast, in IM, the variation on rotor rotation shows significant effect on gel content due to interaction to controllable processing time.

The effect of factor interactions on gel content are shown in Figure 1 (d). It is shown that effect rotor rotation is not significant at short processing time. The difference is clearly shown at longer processing time. At this condition, higher gel content on lower rotor rotation suggests that shear generated during process may not enough to perfectly break –S– bond, and produce secondary crosslinking. However, higher rotor rotation produced less gel content, suggested the effect of shear in devulcanization process.

3.2. FTIR Analysis

FTIR spectra for untreated GTR as well as the lowest gel content processed in TSE (44%) and IM (35%) are shown in Figure 2. In general, there is no peak difference on FTIR spectra for all samples. However, % transmittances (2925, 2857, 1644, 1452 and 1376 cm^{-1}) increased as lower gel content, indicates additional functional group at those peaks. More functional group of C=C (1644 cm^{-1}), CH_2

(2857 and 1452 cm^{-1}) and CH_3(2925 and 1376 cm^{-1}) indicated that the devulcanization occurred, while addition C=O (1707 cm^{-1}) structure suggested that there was some degradation during process.

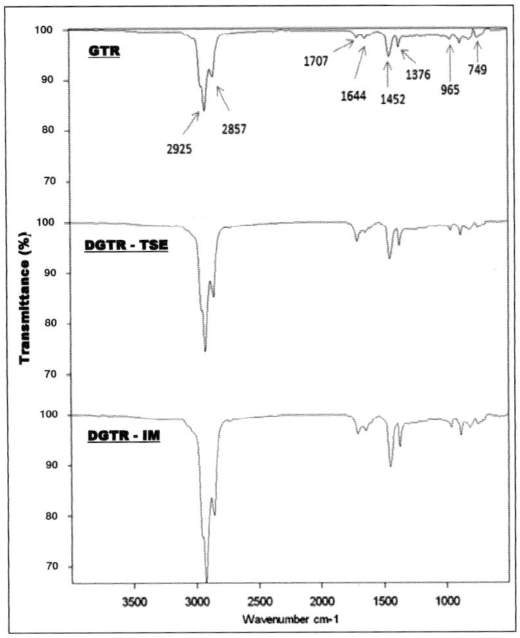

Figure 2. FTIR Spectra of GTR Processed on TSE and IM

The change on % transmittance may be a good indication of devulcanization, however, there is difficulty to determine the actual structures. So, another technique as Nuclear Magnetic Resonance (NMR) may be needed to be done in future research.

3.3. Analysis on Tensile Properties of NR/CC Composites

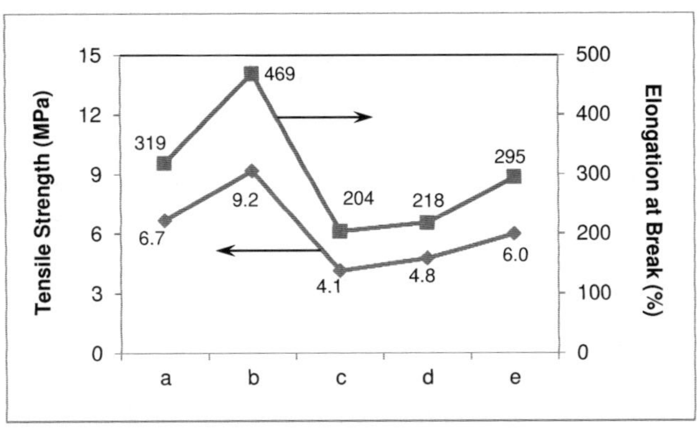

Figure 3. Effect of DGTR on Tensile Properties of NR/CC Composites: (a) NR Control, (b) NR/CC Composites, (c) NR/CC/GTR Composites, (d) NR/CC/GTR-TSE Composites, (e) NR/CC/GTR-IM Composites

Figure 3 shows effect of DGTR on tensile strength and elongation at break of NR/CC composites. DGTR added in composites were those with the least gel content processed in TSE (44%) or IM

(35%). These percentages were chosen to give maximum magnitude of effect DGTR on the NR property recovery. The tensile strength of control NR is 6.7 MPa which is close to previous report using SIR 20 as binder [28]. The strength is also close to another study using different NR [29], but lower than others [30, 31]that may be due to differences on NR type and formulations.

In general, tensile strength and elongation at break of composites with GTR (85% gel content) or DGTR are lower than that of without DGTR attributed to availability of cross-linked structures in GTR or DGTR, indicated by gel content. The higher gel content may produce less interaction between GTR/DGTR and NR chain. This is confirmed by higher mechanical properties of IM sample than TSE's and GTR's

.

4. Conclusions

A comparative study on the effect of processing equipment on GTR devulcanization using TSE and IM was done. Overall, gel content decreased after processed in TSE or IM. In TSE, devulcanization is only significantly influenced by temperature. In contrast, in IM, temperature, rotor rotation, and rotor rotation-processing time interaction, as well as temperature-rotor rotation-processing time interaction significantly reduce gel content. The least gel content was produced at high settings for either in TSE (high temperature, high screw rotation) or IM (high temperature, high rotor rotation, and long processing time) attributed to high energy needed for devulcanization. Structural changes on DGTR processed in TSE or IM suggested the effect of devulcanization. The use of DGTR providing less crosslink network and improve the mechanical properties of NR/CC/GTR.

Acknowledgment

Authors would like to thank to Center for Polymer Technology, Agency for Assessment and Application of Technology, Indonesia for financial support and Mr. Yogi A.S for conducting FTIR experiment.

References

[1] Chyan J *et al* 2013 *Int. Biodeterior. Biodegrad.* **85** 638-645.
[2] Fan P and Lu C 2011 Journal of Polymer and the Environment **19**(4) 943-949.
[3] Holst O *et al* 1998 *Biodegradation* **9**(3) 301-310.
[4] Shi J et al 2014 Polym. Degrad.Stabil. **99** 166-175.
[5] Li W *et al* 2014 *The International Journal of Life Cycle Assessment.* **19**(11) 1833-1842.
[6] Rungrodnimitchai S and Kotatha D, 2015 *Chem. Eng. J.* **282** 161-169.
[7] Zhu J *et al* 2011 *J. Polym. Res.* **18**(4) 533-539.
[8] Balasubramanian M 2009 *J. Polym. Res.* **16**(2) 133-141.
[9] Formela K *et al* 2015 *Iran Polym. J.* **24**(4) 289-297.
[10] Liu Z et al 2015 Polym. Degrad.Stabil. **119** 198-207.
[11] Mangili I et al 2014 Polym. Degrad.Stabil. **102** 15-24.

[12] Wang S *et al* 2015 *Constr.Build. Mater.* **93** 678-684.
[13] Scaffaro R *et al* 2005 *Polym. Degrad.Stabil.* **90**(2) 281-287.
[14] Kebritchi A *et al* 2013 *Fuel* **112** 319-325.
[15] Karabork F and Akdemir A 2015 *J. Appl. Polym. Sci.* **132**(33).
[16] Molanorouzi M and Mohaved S O 2016 *Polym. Degrad.Stabil.* **128** 115-125.
[17] Mangili I et al 2014 The J. Supercrit. Fluids **92** 249-256.
[18] Mangili I *et al* 2015 *Chemometr. Intell. Lab.* **144** 1-10.
[19] Scuracchio C H *et al* 2007 *J. Therm. Anal. Calorim.* **87**(3) 893-897.
[20] Tao G *et al* 2013 *J. App. Polym. Sci.* **129**(5) 2598-2605.
[21] Yazdani H et al 2013 J. Vinyl Addit. Techn. **19**(1) 65-72.
[22] De D *et al* 2006 *Eur. Polym. J.* **42**(4) 917-927.
[23] Zhang X *et al* 2008 *J. Polym. Res.* **16**(4) 411-419.
[24] Zhang X *et al* 2013 *J. Appl. Polym. Sci.* **127**(5) 4006-4014.
[25] Si H, Chen T, and Zhang Y 2013 *J. Appl. Polym. Sci.* **128**(4) 2307-2318.
[26] Frida E 2013 Makara Journal of Technology. **16** 171-179.
[27] Xu C, Chen Y, and Zeng X 2012 *J. Appl. Polym. Sci.* **125**(3) 2449-2459.
[28] Anggaravidya M *et al* 2015 *Macromol. Symp.* **353**(1) 77-87.
[29] Fathurrohman M I *et al* 2015 *Macromol. Symp.* **353**(1) 62-69.

IOP Conf. Series: Materials Science and Engineering **223** (2017) 012006 doi:10.1088/1757-899X/223/1/012006

The effects of the addition of alkanolamide on carbon blacks filled natural rubber compounds

I Surya[1], M Ginting[2], H Ismail[3]
[1]Department of Chemical Engineering, University of Sumatera Utara, Medan, Indonesia
[2]Department of Science of Chemistry, University of Sumatera Utara, Medan, Indonesia
[3]School of Materials and Mineral Resources Engineering, Universiti Sains Malaysia, Penang, Malaysia

E-mail: indradanas@yahoo.com

Abstract. The effects of the addition of alkanolamide (ALK) on cure characteristics and filler dispersion of the two types of carbon black (CB) filled natural rubber (NR) compounds were investigated. The N-550 and N-220 types of CB were utilised, and each of them was incorporated separately into NR compound with a fixed loading, 30.0 phr. The ALK was synthesized from Refined Bleached Deodorized Palm Stearin and diethanolamine; and incorporated into the CB filled NR compounds as an additive ingredient. The ALK loadings were 3.0, 5.0, 7.0 and 9.0 phr. It was found that ALK could function as a secondary accelerator and also an internal plasticiser. As a secondary accelerator; ALK decreased the scorch and cure times of the N-550 and N-220 CBs filled NR compounds. The higher the ALK loading, the shorter the scorch and cure times. As an internal plasticiser; ALK decreased the minimum torque and improved the dispersions of both types of CB in NR phases. At a similar ALK loading; N-220 exhibited higher cure rate, minimum and maximum torques and torque difference values. However, degree of filler dispersion of N-220 was poorer.

1. Introduction

In rubber technology, a rubber compound contains a group of base rubber and rubber additives. Each of them has a specific function either in processing, curing or end use of the rubber products [1]. Rubber or blend of rubbers will provide rubbery behavior to the compound; rubber additives such as plasticisers are used to reduce viscosity, enhance deformability and alter properties; vulcanising agents can crosslink the rubber chains; and accelerators improve the action of vulcanising agents to speed up the cure rate.

Rubber additives are classified into curative and non curative additives. Curative additives play an important role in determining the cure characteristics of rubber compounds. They affect the rate and nature of the vulcanisation process. The popular curative additives are sulphur or other vulcanising agents, zinc oxide, stearic acid, and accelerators.

Fillers are considered as non curative additives. The main purpose of utilising them is to enhance the mechanical properties of rubber vulcanisates such as tensile strength, resistances to abrasion, tearing and flexing or cheapen the cost of products. Carbon black (CB) is the most popular filler in manufacturing of various rubber products, and normally CB is used to reinforce the rubber compounds [2-3]. However at a relatively higher loading, the filler particles tend to form agglomeration and will poor the filler dispersion. Practically, to overcome the filler dispersion problem, special additives such as processing aid, dispersant aid, oil etc. are used.

Alkanolamide (ALK) as a relatively new rubber additive is suggested to solve the filler dispersion problem. It was synthesized from Refined Bleached Deodorized Palm Stearin (RBDPS) and ethanolamine. RBDPS is one type of vegetable oils which comes from palm tree. Physically, the RBDPS-based ALK is an oily substance. Oil is one type of plasticising agents [4] and hence, ALK potentially can be utilised to lubricate, plasticise or soften the CB filled NR compounds and improve the CB dispersion in the rubber phases.

Some previous works on ALK have been reported. The ALK enhanced the mechanical properties of the silica-filled NR vulcanisates [5]. The enhancement of the properties was attributed to the improvement of silica dispersion and the excelling crosslink density that stemmed from the incorporation of ALK. The results also indicated that ALK could function as an accelerator and internal plasticiser. Studied on the comparison of ALK and aminopropyltriethoxy Silane (APTES)-silane coupling agent on the properties of silica-filled NR compounds [6], it was reported that due to its unique function as an accelerator and internal plasticiser, ALK produced a higher reinforcing efficiency than APTES at a similar loading. Studied on the effect of ALK loading on properties of CB-filled NR, epoxidised natural rubber (ENR) and styrene butadiene rubber (SBR) compounds revealed that ALK gave cure enhancement, better filler dispersion and greater rubber–filler interaction to three different types of rubbers [7].

In this study, the effects of ALK loading on the cure characteristics of CBs filled NR compounds and degree of CB dispersion in the filled NR compounds were investigated. Two types of CBs, N-550 and N-220, were used as the fillers. They are different in particle sizes. N-220 has smaller particle size than N-550 [4].

2. Materials and Methods

2.1 Materials

NR grade SMR-L was obtained from Guthrie (M) Sdn. Bhd., Seremban, Malaysia. N-550 and N-220 grades of CB were supplied by Cabot Corporation. Other compounding ingredients such as suphur, zinc oxide, stearic acid, N-isopropyl-N'-phenyl-p-phenylenediamine (IPPD) and Mercapto Benzothiazolyl disulfide (MBTS) were supplied by Bayer (M) Ltd., Sdn. Bhd., Petaling Jaya, Selangor, Malaysia. All materials were utilised as supplied. The ALK was synthesized in laboratory using RBDPS and diethanolamine. The chemical reaction of the preparation of ALK is presented in Figure 1.

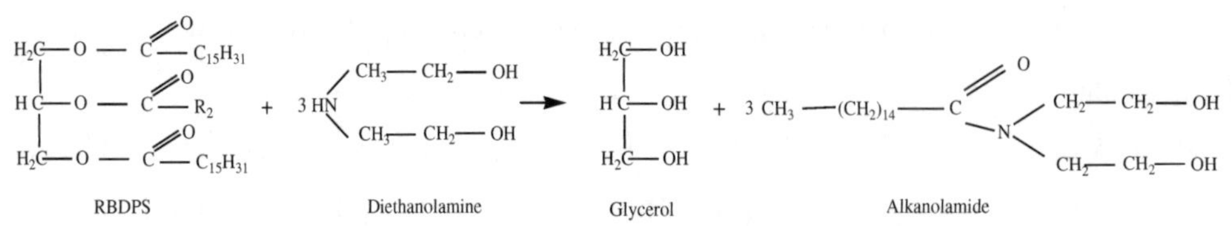

Figure 1. Chemical reaction of the preparation of Alkanolamide.

2.2 Laboratory Preparation of Alkanolamide

The reaction was carried out at atmospheric pressure in a 1000 mL reaction vessel fitted with a stirrer. Typically 1.0 mol of RBDPS, and 3.0 mol each of diethanolamine, sodium methoxide and methanol (as catalysts) were placed in the reaction flask. The mixture was stirred and heated; the reaction temperature was kept constant at 70°C for about 5 h. The resultant mixture was extracted with diethyl ether and washed with saturated sodium chloride solution. The crude ALK was purified with anhydrous sodium sulphate and then concentrated by a rotary evaporator. The cream-coloured and oily ALK was characterized by Fourier transform infra-red (FTIR) spectroscopy. The infrared spectrum of ALK is presented in Figure 2.

2.3 Compounding

A semi-efficient vulcanisation system was used for the rubber compounding. The compounding procedure was done in accordance with the American Society for Testing and Material (ASTM) – Designation D 3184 – 80. The compounding was performed on a two-roll mill (Model XK-160). Table 1 displays the compound formulation of CBs filled NR compounds with various ALK loading.

Table 1. The composition of the CBs filled NR compounds.

Ingredients	Content (phr)*
SMR-L	100.0
Zinc oxide	5.0
Stearic acid	2.0
IPPD	2.0
MBTS	1.5
Sulphur	1.5
CB (N-220/N-550)	30.0
ALK	0.0; 3.0; 5.0; 7.0; 9.0

* *parts per hundred part of rubber*

2.4 Cure Characteristics

The cure characteristics of the CBs filled NR compounds with and without ALK were obtained using a Monsanto Moving Die Rheometer (MDR 2000) which was employed to determine the scorch

IOP Conf. Series: Materials Science and Engineering **223** (2017) 012006 doi:10.1088/1757-899X/223/1/012006

time (ts_2), cure time (t_{90}), minimum torque (M_L), maximum torque (M_H) and torque difference (M_H—M_L) according to ISO 3417. Samples of the respective compounds were tested at 150°C.

3. Results and Discussion

3.1 Characterization of ALK

Figure 2 shows the typical infrared spectrum of the ALK which was obtained from RBDPS and ethanolamine. The spectrum clearly shows the presence of wavenumbers of the functional groups of ALK such as O-H, C=O and C-N groups.

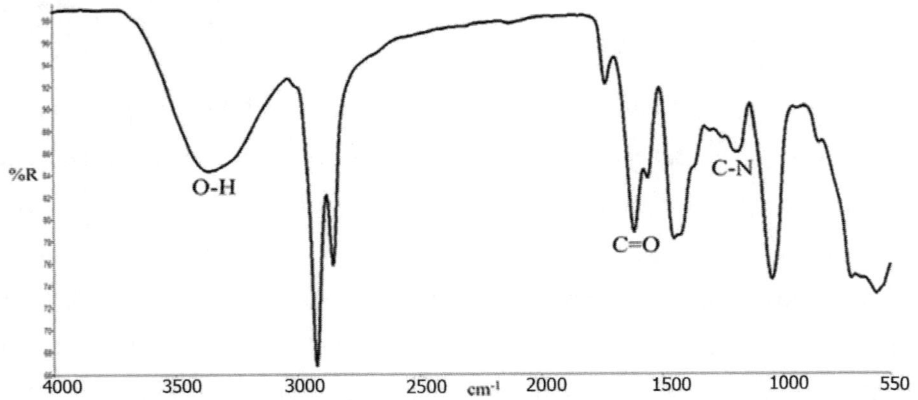

Figure 2. The infrared spectrum of ALK from RBDPS and ethanolamine.

3.2 Cure Characteristics

The effects of ALK loading on the cure characteristics of the two different types of CBs (N-550 and N-220) filled NR compounds are presented in Table 2. The cure characteristics were measured at 150°C including scorch and cure times, minimum and maximum torques and torque differences.

Scorch time is the required time to reach the two torque units rise above the minimum torque. Scorch time is considered as the required time at which vulcanisation begins, and cure time is the required time for a rubber compound to reach 90% of the state of cure (8-9).

As presented in Table 2, it was observed that the scorch times (ts_2) of CBs filled NR compounds with ALK were lower than those of without ALK (control compounds). The higher the ALK loading, the lower were the scorch times. The curing systems of those two rubber compounds used MBTS as the accelerator, and it is functionally classified as a primary accelerator which usually provides scorch delay to a rubber compound (4, 10). The lower the mass ratio of MBTS, the lower was the scorch safety. The addition of ALK reduced mass ratios of the base rubber (NR) and also rubber additives. This explained why the scorch times tended to slightly decrease when the ALK was added.

Table 2. Cure characteristics of CBs filled NR compounds at various ALK loadings

CBs filled NR Compounds	ALK Loadings (phr)				
	0.0 (Control)	3.0	5.0	7.0	9.0
N-550 filled					
ts_2, min	3.36	1.79	1.35	1.20	1.14
t_{90}, min	7.21	4.10	3.50	3.26	3.17
M_H, dN.m	6.66	8.40	8.66	8.56	8.18
M_L, dN.m	0.33	0.30	0.16	0.14	0.12
M_H—M_L, dN.m	6.33	8.10	8.50	8.42	8.06
N-220 filled					
ts_2, min	2.13	1.42	1.26	1.12	1.08
t_{90}, min	5.56	3.45	3.16	2.88	2.80
M_H, dN.m	8.42	11.95	9.80	9.52	9.42
M_L, dN.m	0.39	0.37	0.33	0.32	0.29
M_H—M_L, dN.m	8.03	11.58	9.47	9.20	9.13

From Table 2, it was also observed that the addition of each 3.0 phr of ALK into the control compounds decreased the cure time (t_{90}). Lower cure time means higher cure rate and hence, a cure enhancement phenomenon was observed. It was attributed to the function of ALK as a curative agent (secondary accelerator) in the curing or vulcanisation process. In rubber compounding, amine is usually utilised not only as an accelerator but also an accelerator activator [7,11]. In the case of the cure enhancement phenomenon, the amine part of ALK together with zinc oxide and fatty acid, activated the MBTS-accelerator more pronouncedly and hence, improved the rate of sulphur reaction of CBs filled NR compounds. It was also observed that the higher the ALK loading, the more pronounced the cure enhancement. It was simply attributed to a higher amount or mass ratio of amine in the CBs filled NR compounds.

At a similar ALK loading, the scorch and cure times of N-220 were shorter than those of N-550. It was simply attributed to the difference in sizes of their particles. According to Fetterman [12-13], CB will react with sulphur during vulcanisation and form sulphur bonds that link the rubber chains and also rubber – filler bonds. The nature of rubber – filler bond is determined by some factors and one of them is the surface activity of filler. Greater surface area tends to provide greater surface activity [14]. N-220 has a smaller particle-size or greater surface area than N-550 [4] and hence, shorter scorch and cure times of N-220 were due to its greater surface activity which reduced time to complete the rubber – filler bonds formation.

Table 2 shows that the additions of 3.0 phr of ALK decreased the minimum torques (M_L) of both control compounds. The M_L value is used to measure of the viscosity of a rubber compound ([5]. The lower the M_L value, the lower is the viscosity. The addition of ALK reduced the viscosity of the CBs filled NR compounds. This was attributed to the additional function of ALK as an internal plasticiser which plasticised, softened and reduced the viscosity of the filled NR compounds, respectively. The higher the ALK loading, the lower was the viscosity.

Table 2 also shows that the addition of each 3.0 phr of ALK increased the torque difference (M_H—M_L) values of both control compounds. N-550 and N-220 have different trends of torque differences with further increases the ALK loading. The torque difference of N-550 was further increased with the addition of ALK up to 5.0 phr, and decreased beyond the loading. Whilst, the torque difference of N-220 was increased by the addition of 3.0 phr, and decreased beyond the loading. The torque difference indicates the degree of crosslink density of a rubber compound [7, 16]. The greater the value, the higher is the crosslink density. The increasing of torque difference or crosslink density up to the optimum loadings of ALK (5.0 phr for N-550; 3.0 phr for N-220) was attributed to the chemical and physical properties of ALK. The nitrogen atom or the amine constituent of ALK activated chemically not only the rubber but also the elemental suphur during suphuration or curing reaction. Together with the others curatives, amine could form intermediate complexes which attached the available elemental suphur to rubber chains more efficient [4], causing higher degree of cure. The nitrogen atom also can act as a hydrogen acceptor and amine accelerators can activate the elemental suphur and/or the rubber for the curing reaction [4].

Physically, ALK is an oily material and has the function as an internal plasticiser agent. As discussed earlier, ALK reduced the viscosity of the CBs filled NR compounds. It improved the CB dispersion and rubber - filler interaction/bond, respectively. The rubber – filler bond is defined as additional physical crosslinks [17], and together with sulphide crosslinks contribute to total crosslink density of a rubber vulcanisate [10, 18].

The reductions of the torque differences, beyond the optimum loadings, were most probably attributed to the dilution effect of the excessive amounts of ALK which formed some oily layers. These layers might absorb a part of CB filler and other curatives and coated them together. By this mechanism, the total crosslink density was reduced.

3.3 Filler Dispersion

Based on torque properties data in Table 2, the degree of CBs dispersion in NR phases, with and without ALK, can be quantitatively calculated by Equation (1) [7, 19-20].

$$L = \eta_r — m_r \qquad (1)$$

where: η_r= [M_{Lf}/ M_{Lg}], and m_r= [M_{Hf}/ M_{Hg}]; where M_{Lf} and M_{Hf} are the minimum and maximum torques of the filled compounds, and M_{Lg} and M_{Hg} are the minimum and the maximum torques of the unfilled/gum rubber compound. A lower value of L, at a particular CB loading, meant a better degree of CB dispersion. The minimum and maximum torques of gum NR compounds were 0.05 and 4.85, respectively.

Figure 3 shows the values of L for CBs dispersion in the NR phases at various ALK loading. It can be seen that the L values of CB-filled compounds with ALK were lower than those of CB-filled compounds without ALK. The higher the ALK loading, the lower was the L value. This was attributed to the plasticisation effect of ALK which reduced minimum torque or viscosity the rubber compounds and hence, improved the CB dispersion.

At a similar ALK loading, The L values of N-220 were higher than those of N-550. A higher value of L, meant a lower degree of filler dispersilibity. Again, it was simply attributed to the particle-sizes of those CBs. According to Bhakuni [21], decreasing particle-size decreases the dispersilibity of filler in a rubber compound. The N-220 has smaller particle-size than N-550 and hence, its smaller particle-size decreased its degree of dispersilibity.

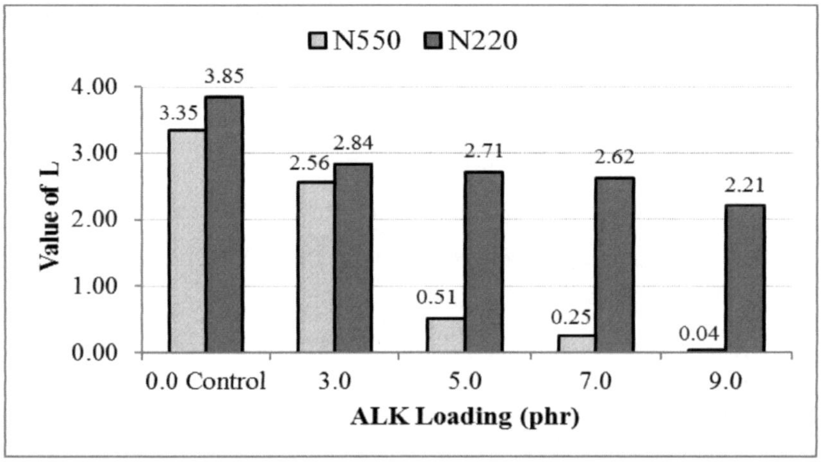

Figure 3. The effect of ALK loading on value of L of the carbon blacks filled NR compounds.

4. Conclusions

From this study, the following conclusions were drawn:

1. Alkanolamide can be function as a curative additive in carbon blacks filled natural rubber compounds. It acted as a secondary accelerator and internal plasticiser in the carbon blacks filled natural rubber compounds. Alkanolamide reduced not only the scorch and cure times but also minimum torque of the carbon blacks filled natural rubber. The higher the loading, the more pronounced the acceleration of curing and plasticising effects.
2. Alkanolamide improved the torque difference of the carbon blacks filled natural rubber up to the optimum loadings. A 5.0 phr was the optimum loading of Alkanolamide for N-550, and 3.0 phr was for N-220 filled natural rubber compounds.

3. Alkanolamide also improved the filler dispersions of N-550 and N-220 in natural rubber phases. Increasing the Alkanolamide loading further increased the dispersion degrees of the fillers. At a similar Alkanolamide loading, the filler dispersion of N-220 war poorer than that of N-550.

Acknowledgement

The authors would like to thank Universiti Sains Malaysia for providing the research facilities for carrying out the experiment and for making this research work possible. Indra Surya and Mimpin Ginting are grateful to University of Sumatera Utara for the Penelitian BPPTN research grant under contract No. 6049/UN5.1.R/PPM/2016.

References

[1] Dick JS 2001 *Rubber technology : compounding and testing for performance*. Hamser, Munich.
[2] Bueche F 1965 *Reinforcement of elastomers.* NewYork: Interscience.
[3] Blow CM 1971 *Rubber technology and manufacture*. Newnes-Butterworths Ltd London,(SW/120) 527.
[4] Rodgers B 2004 *Rubber Compounding: Chemistry and Applications:* CRC 2004.
[5] Surya I *et al* 2013 *Polym. Testing* **32**(8) 1313-21.
[6] Surya I *et al* 2014 *Polym. Testing* **40** 24-32.
[7] Surya I *et al* 2015 *Polym. Testing* **42** 208-14.
[8] Ngamsurat S *et al* 2011 *Energy Procedia* **9** 452-8.
[9] Surya I and Ismail H 2016 *Plast Rubber Compos* **45(7)** 287-93.
[10] Surya I and Ismail H 2016 *Polym. Testing* **50** 276-82.
[11] Long H 1985 *Basic Compounding and Processing of Rubber, Rubber Division.* American Chemical Society Inc. USA
[12] Fetterman MQ 1984 Elastomerics **116**(9) 18-31.
[13] Fetterman M 1986 *Rubber World* **194**(1) 38-40.
[14] Dannenberg E 1975 *Rubber Chem. Technol.* **48**(3) 410-44.
[15] Ismail H *et al* 2002 *Polym. Testing* **21**(5) 565-9.
[16] Teh P *et al* 2004 *Eur. Polym. J.* **40**(11) 2513-21.
[17] Nunes R *et al* 2000 *Polym. Testing* **19**(1) 93-103.
[18] Polmanteer K and Lentz C 1975 *Rubber Chem. Technol.* **48**(5) 795-809.
[19] Lee B 1979 *Rubber Chem. Technol.* **52**(5) 1019-29.
[20] Pal P and De S 1982 *Rubber Chem. Technol.* 55(5) 1370-88.
[21] Bhakuni R *et al* `1989 *Tires"in``Encyclopedia of Polymer Science and Engineering*,"JI Kroschwitz, Editor. John Wiley & Sons, New York.

IOP Conf. Series: Materials Science and Engineering **223** (2017) 012007 doi:10.1088/1757-899X/223/1/012007

Effect of filler water absorption on water swelling properties of natural rubber

J Trakuldee[1], K Boonkerd[2]
[1]Multidisciplinary Program in Petro Chemistry & Polymer Science, Faculty of Science, Chulalongkorn University, Thailand
[2]Department of Materials Science, Faculty of Science, Chulalongkorn University, Thailand

E-mail: Janyarak_trakuldee@hotmail.com

Abstract. The efficient water swelling rubber can be obtained by using high hydrophilic rubber such as chloroprene rubber. However, chloroprene rubber is synthetic rubber developed from the petroleum. Recently, many researches try to replace the usage of synthetic rubber with natural rubber. This is not only due to the concerning of environment but the cost reduction as well. However, natural rubber is hydrophobic, thus not absorbing water. To develop the water swelling rubber from natural rubber, the addition of water absorption filler is needed. The study was aimed to formulate water swelling rubber from natural rubber filled with sodium polyacrylate (SA)/sodium bentonite clay (SBC) hybrid filler used to water absorbent. The filler loading was kept constantly at 150 phr. The effect of SA/SBC ratio varied from 1:0, 1:1, 1:2 and 1:3 on the water absorption of the hybrid filled natural rubber was determined. The obtained result showed that the water adsorption proportionally increased with increasing SA loading but decreased with increasing SBC loading. The effect of glycidyl methacrylate (GM) and poly ethylene glycol (PEG) on the water absorption was studied later. The result from a scanning electron microscope depicted that the presence of GM can depress the falling out of SA from the rubber matrix while the presence of PEG increased water absorption.

1. Introduction

Water swelling rubber products were used widely as sealing of construction joint, preventing water leakage from pipe or blocking connection in civil construction such as subway and subsea tunnel. In general, a water swelling rubber is installed between concrete joints. When concrete joints are ruptured or leaked, the water swelling rubber will contact with water, and then swells and automatically fills up the void between the two portions of concrete join. Commercially, the effective water swelling rubber are prepared from common hydrophilic rubbers such as chloroprene rubber, chlorohydrins rubber, polyurethane and polyvinyl chloride filled with various kinds of superabsorbent polymer particles [1-4]. However, water swelling rubbers made from synthetic rubber has higher cost than the one made from natural rubber.

IOP Conf. Series: Materials Science and Engineering **223** (2017) 012007 doi:10.1088/1757-899X/223/1/012007

Currently, there are many researchers pay attention to develop the water swelling rubber from natural rubber. However, the structure of natural rubber is consisting of carbon and hydrogen atom only, making it has hydrophobic property [5]. Either sodium polyacrylate or sodium bentonite clay generally is used as water absorbent [6, 7]. However, to develop the effective water swelling rubber from natural rubber, the high amount of water adsorbent filler is needed. However, when rubber loaded with high amount of filler, it fails under external force easily. Many researches showed that the addition of glycidyl methacrylate and poly ethylene glycol enhanced not only water absorption but also mechanical strength of the water swelling rubber [8].

In this study, it was aimed to develop water swelling rubber with high water absorption efficiency and reasonable mechanical strength from natural rubber. Both SA and SBC are used as water absorbent hybrid filler. The effect of GM and PEG on the water adsorption was investigated.

2. Experiment

2.1. Materials.

Natural rubber grade STR5L was purchased Sakdarungreaungkij Co., Ltd. Sodium polyacrylate (SA) with absorbability 300 g/g was bought from STP Chem Co., Ltd., Thailand. Sodium bentonite clay (SBC) with grade SAC and particle size 200 mesh was obtained from Thai Nippon Chemical Co., Ltd., Thailand. Coupling agents including glycidly methacrylate (GM) and poly ethylene glycol (PEG) with molecular weight 4000 g/gmol were purchased from Sigma-Aldrich Co., Ltd., Singapore. Carbon black with grade N550 and vulcanizing agents were supported from Innovation group, Thailand.

2.2 Sample Preparation.

The rubber formulas used in this study are listed in Table 1. The rubber was orderly mixed with the ingredients in according to Table 1 using an internal mixer (MX500-D75L90) at 60°C and a rotor speed of 60 rpm. Then, the cure characteristic of the obtained compound was characterized using a moving die rheometer at 160°C. The sample was compression molded at 160°C.

IOP Conf. Series: Materials Science and Engineering **223** (2017) 012007 doi:10.1088/1757-899X/223/1/012007

Table 1. Composition of natural rubber and mixing step

Ingredients	Compound (phr)	
	1	**2**
Natural Rubber	100	100
SA:SBC	1:0/1:1/1:2/1:3*	1:1
GM	-	2.5
PEG	-	10
Carbon black	2	2
Zinc Oxide	3	3
Stearic acid	2	2
CBS	2	2
DPG	1	1
Sulfur	0.8	0.8

*The filler (SA:SBC) loading was kept constantly at 150 phr

2.3 Water Absorbent Properties Tests.

The rubber sample was cut into small pieces with a weight of about 1 g and a volume 2.0×2.0×0.2 cm^3 immersed in water for 30 days at room temperature. The weight and volume of swollen samples was checked frequently at the specific time intervals. The swelling by weightand swelling by volume were calculated according to Eq.1 and 2 respectively.

$$\text{Swelling by weight } (\%) = W_2\text{-}W_1 x100/W_1 \quad (1)$$

$$\text{Swelling by volume } (\%) = V_2\text{-}V_1 x100/V_1 \quad (2)$$

W_1, W_2 (unit in gram) is the weight of the sample before and after the swelling, respectively.

V_1, V_2 (unit in cm^3) is the volume (width x length x high) of the sample before and after the swelling, respectively.

2.4 Morphology of Water Swelling Rubber.

The effect of fillers GM and PEG on the interaction between water absorbent SA and rubber was studied. The fractures surfaces of the sample after swelling were investigated using a scanning electron microscope (JSM 5410LV/JEOL, Japan). Samples were quenched in liquid nitrogen and then broken into section and gold-sputtered onto the rough surface of them to increase the conductivity.

3. Results and Discussion

3.1 Effect of Water Absorbent SA: SBC Ratio on Swelling

Figure 1 shows the effect of SA:SBC ratio on the water adsorption of the rubber. It can be seen that the rubber filled with SA/SBC at the ratio of 1:0 showed the highest swelling rate. However, after few days the water swelling of the rubber filled with SA/SBC at the ratio of 1:0 clearly decreased. This was due to the falling out of SA from the rubber matrix. For the rubber filled with SA/SBC hybrid filler at the ratios of 1:1, 1:2 and 1:3, the water absorption increased with immersion time and then saturated around day 21. The rubber filled with SA/SBC hybrid filler at the ratio of 1:1 showed the highest water absorption. This was due to the presence of the highest SA content. The water absorption decreased with the increased SBC content. This is because the ability to absorb water of SBC is much lower that of SA. SBC is natural absorbent, it can absorb water 5-10 times from initial weight while SA is synthetic absorbent, it can absorb water about 100-500 times from

initial weight [9-10].

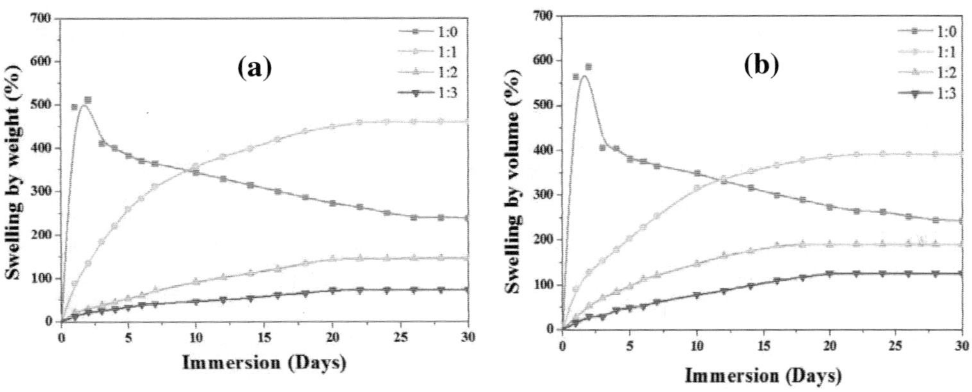

Figure 1. Water swelling by weight (a) and swelling by volume (b) of the rubber filled with various SA:SBC ratios when immersed in water for 30 days

3.2 Effect of GM and PEG.

Figure 2 shows the water swelling of the rubber filled with SA/SBC hybrid filler at the ratio of 1:1 with and without adding GM and PEG. It can be seen that the effect of GM on the water absorption, it was found that GM did enhance the water absorption and no SA falling out of the rubber was observed. Park JH and Kim D (2001) proposed that the increased interaction between SA and rubber by GM occurred via there action mechanism that the double bond in GM might react with the unsaturated groups innatural rubbers and the epoxidegroup with the carboxylic acid in SA[4].When considering the addition of PEG significantly increased the

IOP Conf. Series: Materials Science and Engineering **223** (2017) 012007 doi:10.1088/1757-899X/223/1/012007

water absorption. This is because PEG is hydrophilic material, thus increasing the hydrophilicity of the rubber. However, it was observed during the water absorption testing that the SA in the sample added PEG fell out of the rubber matrix acrossding to Figure 3(c). Therefore, to improve the property of the water swelling rubber, both PEG and GM are needed.Because this has capability to from stong hydrophilic network with rubber to generate highest water absoption [5,8]

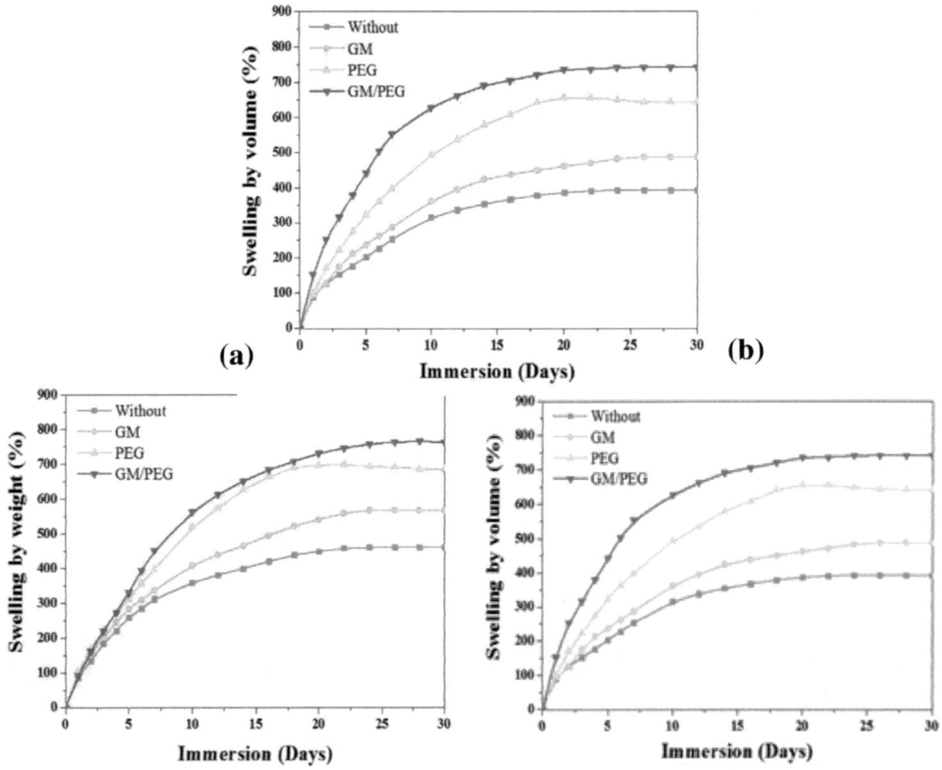

Figure 2. Water swelling by weight (a) and swelling by volume (b) of the rubber filled SA/SBC hybrid filler at the ratio of 1:1 with and without adding GM and PEG

Figure 3 showed SEM photographs of the fractured rubber filled with SA/SBC hybrid filler at the ratio of 1:1 without and with adding GM and PEG after swelling. The result showed that the swollen sample without adding fillers in Figure 3(a) showed a lot of voids due to the falling out of SA. However, when adding GM in Figure 3(b), the swollen sample had less voids. This can confirm that GM enhances the interaction between the SA and rubber, thus preventing the falling out of SA from the rubber.While adding PEG in Figure 3(c) showed a lot of void same without adding filler due to the falling out of SA. PEG is hydrophilic material, it can increase the hydrophilicity of the rubber but not enhance the interaction between the SA and rubber.

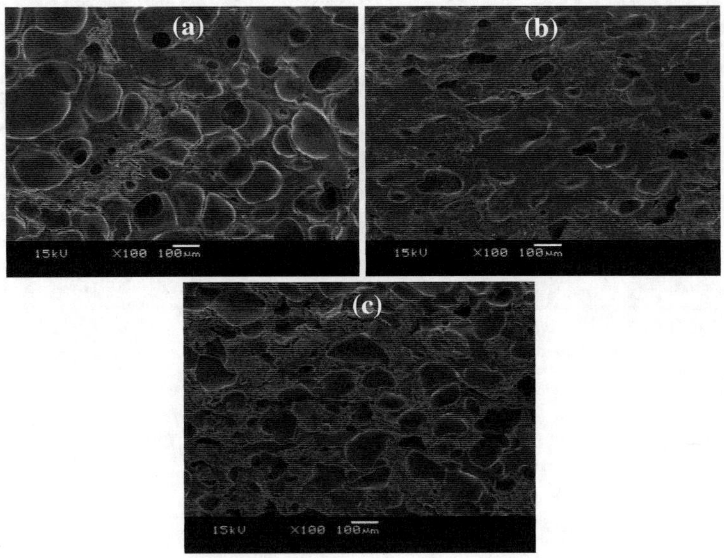

Figure 3. SEM photographs of the fractured rubber filled with SA/SBC hybrid filler at the ratio of 1:1 after swelling (a) without adding (b) with adding GM and (c) with adding PEG

4. Conclusion

Water swelling rubber was prepared from natural rubber filled SA/SBC hybrid filler. The suitable SA/SBC ratio was at 1:1. Although the higher SA content can lead to the higher water absorption, the SA was prone to fall out of the rubber. GM is necessary to enhance the interction between SA and rubber while PEG acts as the pathway of water and can strongly enhance the water swellability.

Acknowledgments

The study was supported by Special Task Force for Activating Research (STAR), Ratchadaphiseksomphot Endowment Fund and Chulalongkorn University.

IOP Conf. Series: Materials Science and Engineering **223** (2017) 012007 doi:10.1088/1757-899X/223/1/012007

References

[1] Saijun D *et al* 2009 *Songklanakarin J. Sci. Technol.* **31** 561.
[2] Wong G, Li M and Chen X 1998 *J. Appl. Polym. Sci*_.**68** 1219.
[3] Wong G, Li M and Chen X_1999 *J. Appl. Polym. Sci.***7** 577.
[4] Park JH and Kim D 2001 *J. Appl. Polym. Sci.* **80** 115.
[5] Nakason C *et al* 2013 *J. Euro. Polym.* **49** 1098.
[6] Zhang J *et al* 2007 *Mat*_. *Let.* **43** 316.
[7] Santiago F *et al* 2009 *J. Euro. Polym.* **31** 1.
[8] Wang C *et al* 2002 *J. Appl. Polym. Sci.* **86** 3120.
[9] Ozcan As and Ozcan A 2004 *J. Colloid*_. *Interface. Sci.*_**1** 276.
[10] Flory PJ and Rehner J 2001 *Chem. Phys* **11** 11.

IOP Conf. Series: Materials Science and Engineering **223** (2017) 012008 doi:10.1088/1757-899X/223/1/012008

Effect of carbon black composition with sludge palm oil on the curing characteristic and mechanical properties of natural rubber/styrene butadiene rubber compound

R Mohamed, N Mohd Nurazzi, M Huzaifah
GREEN Polymer Research Group, Faculty of Applied Sciences, Universiti Teknologi MARA (UiTM), 40450 Shah Alam, Selangor, Malaysia

E-mail : rahmahmd@salam.uitm.edu.my

Abstract. This study was conducted to investigate the possibility of utilizing sludge palm oil (SPO) as processing oil, with various amount of carbon black as its reinforcing filler, and its effects on the curing characteristics and mechanical properties of natural rubber/styrene butadiene rubber (NR/SBR) compound. Rubber compound with fixed 15 pphr of SPO loading, and different carbon black loading from 20 to 50 pphr, was prepared using two roll mills. The cure characteristics and mechanical tests that have been conducted are the scorch and cure time analysis, tensile strength and tear strength. Scorch time (t_{s5}) and cure time (t_{90}) of the compound increases with the increasing carbon black loading. The mechanical properties of NR/SBR compound viz. the tensile strength, modulus at 300% strain and tear strength were also improved by the increasing carbon black loading.

1. Introduction

Rubber and elastomers are some of the most common materials significantly known for its flexibility and their ability to absorb particle like carbon black, silica and clay in enhancing and obtaining the desired properties for specific applications wherever resistance to impact or toughness are desired. Fillers, chemicals, and other reinforcing materials usually are added to enhance the properties of the rubber. Oil is one of these substances which usually are added in the processing phase of the rubber compounding process, and acts as the processing aid in the rubber compounding. Usually the type oil that is used in the rubber compounding process are the highly aromatic (HA) oils. This type of oil was added to the rubber compound with other ingredients to improve the processability and performance of the compounds such as promoting wet grip and wear endurance.

Certain highly aromatic oils as such had been classified as carcinogenic as upon tested according to IP method called IP346,t showed an unhealthy index values exceeding 3% [1]. Numerous other studies had found that aromatic oil addition into rubber compounds, especially oils with high polycylic aromatic compounds (PAC) content which were identified as suspected carcinogens, might had contributed to increase cancer proliferation [2,3]. HA oils are often not actually bonded chemically to the rubber matrix, where they tend to leach from the mixture and migrate into their surroundings. For certain nations like United States and much of the European Union, the use of these aromatic oil in rubber compounding had been banned. Besides that, PAC containing aromatic amines such as 2-naphthylamine have long been associated with increased rates of urinary bladder cancer (UBC) in rubber workers [1]. Hence, the environmental problem caused by HA oil use shows the need for natural based processing oil. Studies done by Rahmah *et al.*, (2013) on epoxidised oil (EO) with NR/SBR compound found that EO oil can be an alternative sources to replace the aromatic oil (AO) as a plasticizer and processing aid [4, 5].

For tyre manufacturing, the common elastomers used are the NR/SBR mix. NR are usually mixed with synthetic rubber like SBR, commonly in the 60:40 ratio for rubber tyre and other compounded products. The SBR mixed often imparts suitable properties in meeting the performance needs of passenger car tyre treads [3,6,7]. The reinforcement of rubber properties by the incorporation of carbon black had been extensively studied. Recent studies with carbon black fillers and other fillers had been investigated by Ooi *et al.*, (2013) [8]. In their work, oil palm ash (OPA) as filler was compounded with NR using conventional laboratory-sized two roll mills. For comparison purposes, two commercial fillers (i.e. silica and carbon black (CB)) were also studied. The optimum loading of each type of filler was investigated and compared in terms of curing characteristics, tensile properties, rubber-filler interaction and dynamic mechanical analysis. Results showed that scorch time and cure time of the carbon black filled vulcanizates was the lowest compared to OPA and silica-filled NR vulcanizates, with the OPA compound offers the highest cure rate index. The carbon black filled vulcanizates had the highest tensile strength (26 MPa) due to the reinforcement that its filler imparted. Ulfah *et al.*, (2015) had also compounded NR with a blend ration of carbon black and silica fillers. They had found out that carbon black filler of the highest content ratio showed the highest tensile stress at 300% elongation for rubber compound with increasing amount of carbon black filler [9].

Waste cooking oil sludge sourced from waste oil palm cooling oil were used in this study; the waste oil used was collected from the oil trap as grease. The term "trap grease" technically refers only to kitchen waste, rather than septic grease. Sludge oil or trap grease are sometimes collected by rendering companies for purification and are often sold for non-edible applications such as burner fuel [7]. In the present study, sludge from palm oil (SPO) was used as an alternative source for aromatic oil. The effect of SPO loading on the cure characteristic and mechanical properties of NR/SBR compound had been investigated and reported.

2. Materials and Methods

2.1 Materials

NR of grade SMR 10 and SBR of grade Nipol 1502 were supplied by Rubber Research Institute Malaysia (RRIM). Nonox were supplied by Shanghai Rokem Industrial Co., Ltd and High Abrasion

Furnace Carbon black, HAF (N330) were supplied by Black Tiger Carbon Black Co. Ltd. Other chemicals such as zinc oxide and sulphur were supplied by Bayer (M) Sdn. Bhd. The SPO was obtained from Forest Research Institute Malaysia (FRIM) with its chemical compositions has been steam treated and filtered. Table 1 shows the specifications of SPO obtained from FRIM. Nonox was supplied by Shanghai Rokem Industrial Co., Ltd and high abrasion furnace carbon black, HAF (N330) was supplied by Black Tiger Carbon Black Co. Ltd and other chemicals such as zinc oxide and sulphur were supplied by Bayer (M) Sdn. Bhd.

Table 1. Specifications and compositions of SPO.

Specifications	Parameters
Code color	Black
Free Fatty Acid (FFA) *(Normal crude palm oil: FFA<5%)	42.77%
Moisture and Impurities (M&I)	2.47
Melting Point (2hours)	48.7°C
Flash Point	240°C
Total Fatty Matter	97.53
Density g/cc @70°	0.8615
Calorific Value	38.6MJ/kg

2.2 Compounding

Compounding of the ingredients as per Table 2 specifications were prepared in a two-step process; preparation of master-batch (sans sulphur and accelerator) was done using Banbury internal mixer (BR1600) at the RRIM under a set temperature of 150°C and mixing speed of 70rpm. The remaining ingredients were later added and mixed using a two-roll mill with 30rpm rolling speed. The sheeted rubber compound was conditioned under room temperature for 24 h. For this study, SPO content are fixed at 15 pphr while the carbon black (HAF N330) were varied to 50pphr (parts per hundred rubber).

Table 2. Formulation for rubber compounding for plasticizer effect.

Mix Number	1	2	3	4	5	6
NR (SMR 10)	60	60	60	60	60	60
SBR (NIPOL 1502)	40	40	40	40	40	40
Zinc Oxide	5	5	5	5	5	5
Stearic Acid	2	2	2	2	2	2
Nonox	3	3	3	3	3	3
HAF (N330)	**20**	**25**	**30**	**40**	**45**	**50**
SPO	15	15	15	15	15	15
Sulphur	2.5	2.5	2.5	2.5	2.5	2.5
CBS	0.6	0.6	0.6	0.6	0.6	0.6

Note: Natural Rubber (NR), Standard Malaysian Rubber (SMR), Styrene Butadiene Rubber (SBR), High Abrasion Furnace Carbon Black (HAF), Sludge Palm Oil (SPO), N-Cyclohexyl-2-Benzothiazole Sulphenamide (CBS)

2.3 Characterizations

According to the ISO 3417:1991 standard, a Monsanto oscillating disc rheometer (ODR) was used to determine the scorch (t_{s5}) and optimum cure time (t_{90}) at 150°C based on Eq. (1) and Eq. (2) [3].

Scorch time (t_{s5}) = Time to reach 5 unit increase in torque above minimum (at M_L + 5 units above it) (1)

Optimum cure time (t_{90}) = $0.9(M_H - M_L) + M_L$ (2)

Where, M_H = Maximum torque, M_L = Minimum torque

The tests for the tensile strength and tear strength properties of the rubber compounds were performed using an Instron universal machine (model Testometric MICR0500) at 500mm/min crosshead speed according to the ASTM D412 and ASTM D624 standards respectively.

3. Results and Discussion

3.1 Cure Characteristic

Figure 1 shows the effect of carbon black on both the cure and scorch time with fixed SPO loading (15 pphr). It can be seen that both characteristic increases with the increasing carbon black loading. From the result obtained, the maximum scorch time and cure time was achieved at 45 pphr of carbon black loading, which are 24 min and 65.85 min respectively. Both characteristic then experienced sudden decrease in the time recorded for 50 pphr of carbon black loading. This may due to the carbon black particle aggregating and did neither packed together at certain point nor does it distributed evenly. This process occurs mainly during the compounding (milling) process using the two roll mill machine.

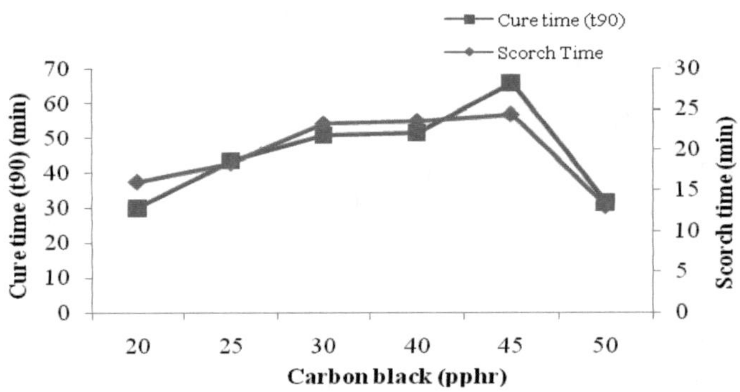

Figure 1. Effect of carbon black loading on scorch (min) and cure time (min).

The reduction of cure time upon carbon black loading above 45 pphr could be related to the excess carbon black, which indicates a presence of greater number of vulcanisation and crosslink site occurred with the carbon black and waste sludge oil applied onto the NR/SBR compounds. This phenomenon might be explained as the competition for cross-linking agents between the unsaturated rubber and thc excess carbon black, which resulted in the lowering of the vulcanization rate [10]. Other than the cross-linking linkages formation, the process of network formation had been explained by Karak *et al.,* (2000) [11]. As cure time decreases, the vulcanization process can theoretically be explained by the following processes. Both network formation and network degradation occur simultaneously, but the initial network formation rate was much lower than the degradation rate; subsequently, an optimum level can be reached before the rate of network structure degradation finally became much shorter for the NR vulcanizates as cure time decreases.

There were still scarcely enough works on the use and effect of waste or SPO and other vegetables oil forms for tyre tread compound apart from the limited number of works [12,13]. Rahmah *et al.,* (2013) had investigated the effect of AO and EO from vegetables oil on the scorch and curing time. It had shown that an EO addition up to 30pphr had reduced the cure time from 13 to 90% as compared to AO use [5].

3.2 Tensile Properties

Figure 2 shows the effect of carbon black (pphr) composition on the tensile strength of rubber compound. Tensile strength was observed to had maintained at values between 14MPa and 15MPa upon incorporation of 20 to 30 pphr of carbon black. The optimum strength was 19.32 MPa at 40 pphr carbon black loading. It shows that there is an interaction; the highest interaction at 40pphr of carbon black loading which occur between the carbon black and the rubber. However, the tensile strength was reduced again by about 10pphr for carbon black in excess of 40pphr. This may due to the overloading of carbon black in the compounding, or the possible use of the two roll mill machine which caused the carbon black particles not to be homogenized completely.

Besides that, the higher amount of carbon black had caused reinforcement between the carbon black particles themselves, fixed SPO and the rubber chains. This finally lead to the interfacial interaction reduction between the rubber chains due to agglomeration of carbon black particles [4]. The carbon black presence induces rigidity to the rubber, in which an increase in carbon black will increase the rigidity of the rubber compound. High rigidity is sometimes needed in rubber compound for tire manufacturing, which provides an improvement in strength.

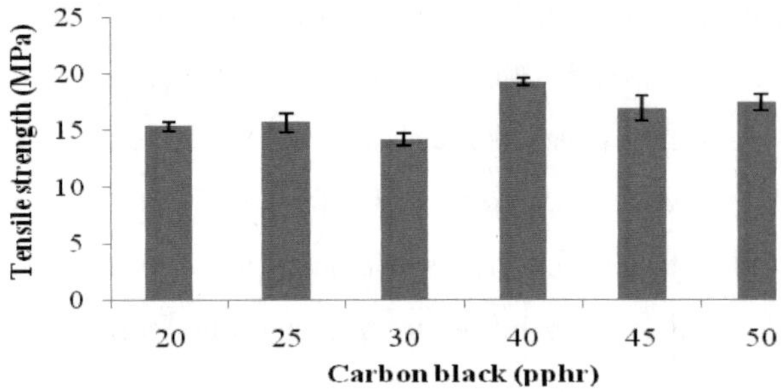

Figure 2. Effect of carbon black on tensile strength.

Figure 3 shows that there is an interaction between the carbon black and the rubber. The elongation at break reduces as the increase of carbon black between 20 to 45 pphr. This reduction may due to the carbon black reducing the elasticity. The sudden decrease of elongation at break at 50pphr carbon black may due to the uneven distribution of excessive carbon black in the rubber compound.

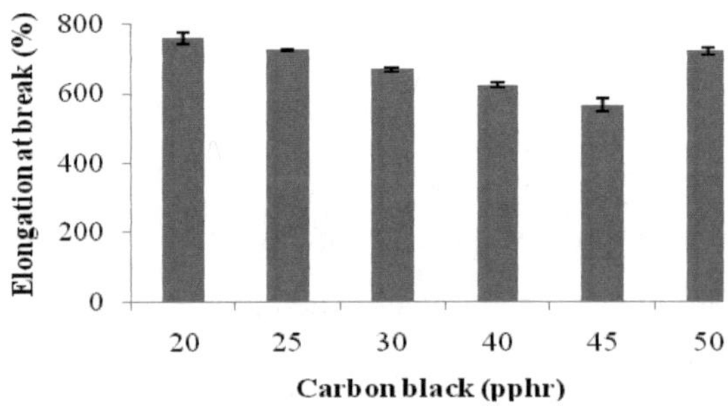

Figure 3. Effect of carbon black on elongation at break.

Figure 4 shows the 300% elongation results of the modulus at stress. Theoretically, the modulus is inversely proportional to the elongation at break, and it indicates the stiffness and rigidity of the materials tested. The 300% elongation results shows how the tensile modulus consistently increases with the carbon black content. The drop in modulus at 50 pphr may due to the effects of carbon black overloading. This tends to cause the uneven distribution of the carbon black present that tends to agglomerate at certain locations within the material.

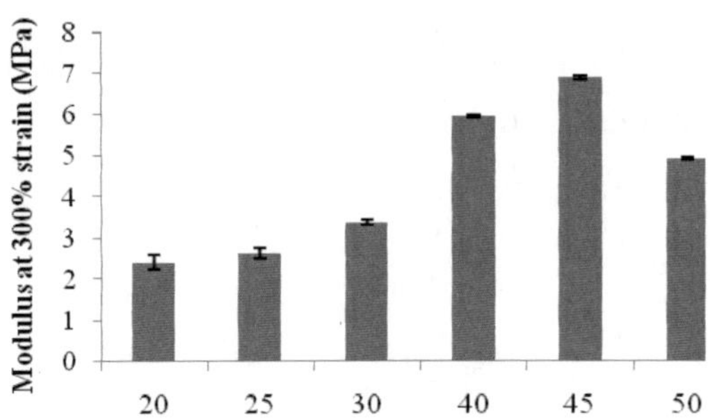

Figure 4. Tensile modulus at 300% strain.

3.3 Tear Strength

There are three types of sample for use in the tear tests, which are the trouser, angle or crescent samples. But for this test, the trousers sample was used. The tests were conducted by applying force to the material in a tensile direction, of which a value for the tear strength was calculated. From Figure **5** below, there were some small gradual observable increases in the tear strength when the carbon black content were increased from 20 to 40pphr. Upon reaching 45 pphr and 5 0pphr of carbon black a drastic increase of about four fold increase in the tear strength were observed.

Carbon black had long been thought to be a reinforcing agent but these higher reinforcing effects were only imparted upon higher amount of carbon black used. The high interaction of carbon black and rubber was apparent through this pronounced reinforcing filler effect.

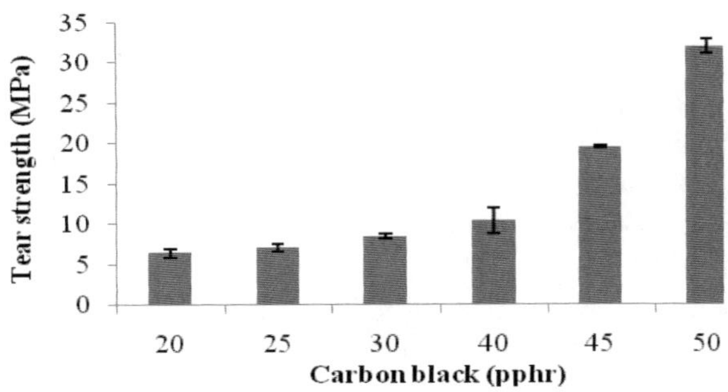

Figure 5. Effect of carbon black on tear strength.

This proves that the carbon black had acted as the reinforcing filler, hence caused better resistance to tear for rubber compound. Tear properties were markedly improved with a fourfold increase upon addition of 45pphr of carbon black. Fukahori *et al.*, (2013) stated that, during the tearing of rubber materials, the transition called elastic-viscous transition phenomenon exist; and this transition had occurred due to the crosslink density and visco-elastic energy dissipation [14].

4. Conclusions

In conclusion, the scorch time, cure time, tensile strength, and tear strength shall increase with corresponding increases in the amount of carbon black. Meanwhile, the tensile modulus at 300% strain had showed results that are inversely proportional to the elongation at break which due to the elasticity and stiffness properties of the compounds. An increase of carbon black up to 45pphr results in overloading effect, which causes an uneven distribution of the carbon black in rubber compounding process.

Acknowledgements

The authors wish to thank the Research Management Institute (RMI) of Universiti Teknologi MARA (UiTM), Malaysia for financial support rendered, and the Faculty of Applied Sciences (FSG), Universiti Teknologi MARA (UiTM) Shah Alam, Malaysia for providing the outstanding facilities which enables this study to be properly conducted.

References

[1] Talaska G *et al* 2012 *Toxicology letters* **213** 45-48.
[2] Dasgupta S *et al* 2007 *Polymer Testing* **26** 489-500.
[3] Azemi S 2009 *An Introduction to Polymer Science and Rubber Technology* (Pusat Penerbitan Universiti, Universiti Teknologi MARA) .
[4] Rahmah M *et al* 2013 *Adv. Mater. Res.* **812** 204-209.
[5] Rahmah M *et al* 2013 *Adv. Mater. Res.* **812** 138-144.
[6] Kukreja T R *et al* 2003 *J. Appl. Polym. Sci.* **87** 1574-1578.
[7] Kundu P and Kukreja T 2002 *J. Appl. Poly. Sci.* **84** 256-260.
[8] Ooi Z X, Ismail H and Bakar A A 2013 *Polymer Testing* **32** 625-630.
[9] Ulfah I M *et al* 2015 *Procedia Chemistry* **16** 258-264.
[10] Carli L N *et al* 2009 *Mater. Sci. Eng. C* **29** 383-386.
[11] Karak N and Gupta B R 2000 *Kautschuk Gummi Kunststoffe* **53** 30-34.
[12] Chung B *et al* 2002 *Rubber World* **227** 36-42.
[13] Wang M J *et al* 2001 *Rubber Chem. Technol.* **74** 124-137.
[14] Fukahori Y *et al* 2013 *Polymer* **54** 1905-1915.

IOP Conf. Series: Materials Science and Engineering **223** (2017) 012009 doi:10.1088/1757-899X/223/1/012009

The effect of carbon black loading and structure on tensile property of natural rubber composite

S Savetlana, Zulhendri, I Sukmana and F A Saputra
Mechanical Engineering Department, University of Lampung, Indonesia

E-mail: shirley.savetlana@eng.unila.ac.id

Abstract. Natural rubber composite has been continuously developed due to its advantages such as a good combination of strength and damping property. Most of carbon black (CB)/ Natural Rubber (NR) composite were used as material in tyre industry. The addition of CB in natural rubber is very important to enhance the strength of natural rubber. The particle loading and different structure of CB can affect the composite strength. The effects of CB particle loading of 20, 25 and 30 wt% and the effects of CB structures of N220, N330, N550 and N660 series on tensile property of composite were investigated. The result shows that the tensile strength and elastic modulus of natural rubber/CB composite was higher than pure natural rubber. From SEM observation the agglomeration of CB aggregate increases with particle loading. It leads to decrease of tensile strength of composite as more particle was added. High structure of CB particle i.e. N220 resulted in highest tensile stress. In fact, composite reinforced by N660 CB particle shown a comparable tensile strength and elastic modulus with N220 CB particle. SEM observation shows that agglomeration of CB aggregates of N330 and N550 results in lower stress of associate NR/CB composite.

1. Introduction

Natural rubber (NR) is widely used in today life. The important mechanical properties of NR are tensile property, tearing strength, wear resistance, etc. Carbon black can effectively reinforce NR because of the hydrophobic surface of CB filler matched to hydrophobic surface of NR [1,2]. In related to the particle size, the size of CB aggregate of 0.1-0.2 μm is also important in determining the utility of the CB reinforcement in NR. The aggregate then form agglomeration of 10-1000 μm in size. As the count of unit mass increases, aggregate interstitial spacing is reduced. Those properties of aggregates are called 'structure'. The aggregate that contained more particles were said has a higher structure [3]. The work on CB series of N990, N326, N351 and N358 reinforced NR shown that the frictional coefficient under abrasion test of SBR was increased as the structure of CB is higher [4]. The tensile strength and elastic modulus of CB reinforced EPDM was show the increase in tensile strength and elastic modulus as the surface area is increased, especially for N472 that has the largest surface area [5].

This work investigated the effect of CB loading on the tensile property of NR/CB composite. The effect of the CB structures i.e. CB N220, CB N330, CB N550 and CB N660 on tensile property also investigated. The tensile properties of NR without CB were investigated as comparison. SEM-

EDX images are used to explain the effect of the CB loading and structure on tensile property of NR/CB composite.

2. Materials and Methods

2.1. Materials

Materials use in this study was NR in form of latex. Latex was originated from NR cultivation in Lampung Province, Indonesia. Amoniac was use as anticoagulant of lateks during lateks transportation. Formic acid was used as coagulant after lateks was mixed with sulfur and carbon black. Carbon black N330 was used to investigate the effect of CB loading. Carbon black series of N220, N330, N550 and N660 were used to investigate the structure effect on the tensile property of NR/CB composite.

2.2. Composite Preparation

In preparation of composite, before mixing, all materials were weighted according to the percentage of its ingredients in Table 1. Table 1 show the composition of natural rubber with formic acid, refer as NRf1 and NRf2. NR/CB composite with different percentage of CB refer as NR/CB20, NR/CB25 and NR/CB30. Natural rubber with sulphur, refer as NRfS1, NRfS2, NRfS3 and natural rubber filled with carbon black structure, refer as NR/CB samples.

Table 1. Ingredients of NR and NR/CB composite

Sample	Natural Rubber	Formic acid	Sulphur	Carbon Black
NRf1	100	2	0	0
NRf2	100	3	0	0
NR/CB20	78	2.23	0	20
NR/CB25	73	2.23	0	25
NR/CB30	68	2.23	0	30
NRfS1	100	2	2	0
NRfS2	100	2	4	0
NRfS3	100	2	6	0
NR/CB	80	2	4	20

NRf : Natural rubber with formic acid.

NR/CB20: Natural rubber with addition of 20% carbon black. Carbon Black structure was N330.

NRfS1: Natural rubber with formic acid and sulphur.

NR/CB: Natural rubber with addition of different carbon black structure.

Preparations of the sample were as follow: firstly, latex was stirred for 2 min. For NRf sample, lateks was prepared for the second step. For NRfS sample, sulphur were added to latex and mixed. For NR/CB composite sample, the CB was added to latex and mixed. Secondly, each sample was mixed

IOP Conf. Series: Materials Science and Engineering **223** (2017) 012009 doi:10.1088/1757-899X/223/1/012009

with formic acid for 20 min before being place in the mould. Thirdly, each sample mixtures was poured into 50 x 50 x 5 mm moulds then let them sit for 30 min. Fourthly, each sample was taken out from the plastic mould and pressed under the load of 8 tons for 30 min. After pressing, each sample was taken out from the mould and place outside to dry. Finally, after 24 h drying, each sample was placed in the oven for curing at 150°C for 30 min. Sample then were cut to form dumble specimen.

2.3. Tensile Test and SEM Observation

Tensile test was prepared according to ASTM D412 Standard Test Method for Vulcanised Rubber and Thermoplastic Elastomer-Tension. The tensile rate was 500 mm/min. Temperature and humidity was 23°C and 52%, respectively. The SEM observation was performed on the cross section of the specimen under tensile load. The equipment for SEM observation and EDX analysis was JEOL JSM-6510LA and SEM-EDX EVO MA10.

3. Result and Discussion

Latex was susceptible to host of strong micro bacterial activity during the maturation because of the richness of the organic compound. Maturation is the time period between tree tapping and coagula processing [6]. For preservation of the latex, the latex was mixed with ammoniac before coagulation process. Many researchers used commercial dry rubber in making the NR/CB composite. Rarely, CB was added directly into fresh latex. In this research, latex was used as composite matrix after coagulated with formic acid and catalysed with sulphur. Therefore, to find appropriate percentage of formic acid and sulphur to get sufficient tensile strength, the tensile tests were also carried out to unfilled latex that have coagulated with formic acid and vulcanised with sulphur. For coagulation process, the NR was mixed with formic acid as coagulant, refer as NRf sample. In the range of formic acid loading tested in this study, the optimum tensile strength and strain of NRf was 2% as listed in Table 2.

Vulcanization is a process of compounding NR with different chemical. The chemicals such as sulphur, organic peroxide, silane, amino, phenolic resin, etc. Sulphur was widely used by rubber industries. Vulcanisation increases the elasticity of the rubber. The isoprene polymer chain was locked and limited the amount of chain slipping [7]. Tensile strength of natural rubber with sulfur which the samples refered as NRfS sample is shown in Table 2. Tensile stress and elastic modulus of NRfs were increases when sulphur loading up to 6 %, however the strain was decreases. It is suggest that four percent as the best loading of sulphur in refer to balance tensile stress and strain.

Table 2. Tensile Property of NR coagulated with formic acid and vulcanised with sulphur

Sample	Tensile Strength (MPa)	Elongation (%)	E (ε=0.05%) (MPa)
NRf1	0.417	517.3	4.47
NRf2	0.356	309.0	7.77
NRfS1	0.494	383.7	7.57
NRfS2	0.618	397.2	7.44
NRfS3	0.637	377.4	7.82

3.1. The Effect of CB Filler Loading on Tensile Property

Stress-strain curve of NR and NR/CB composite were showed in Figures 1(a)-(d). The mean stress-strain curve of NR and NR/CB composite was shown in Figure 1(e). The stress-strain curve of NR in Figure 1(a) shows the sample break at very high strain as compare to the NR/CB composites in figure 1(b)-(d). Obvious different in stress-strain curve between NR and NR/CB composite was observed in figure 1(e) that shown low stress and high strain of NR compare to NR/CB composite. As the CB loading increases, the composite failure strain decreases. It indicated the brittle behaviour of NR after filled with CB particle. In reverse, elastic modulus increases with CB loading.

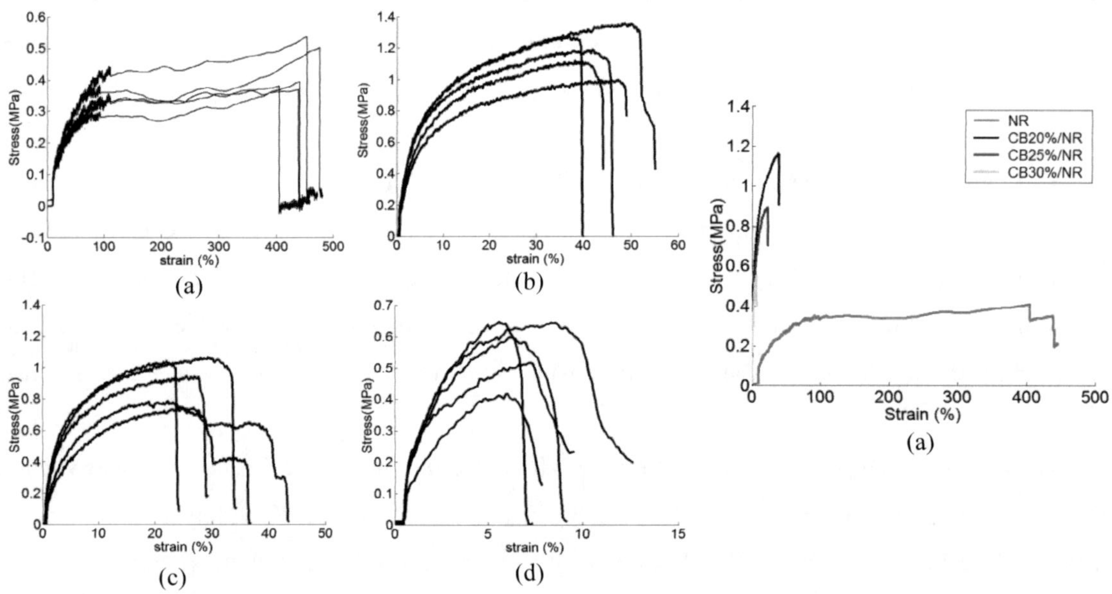

Figure 1. Stress Strain Curve of: (a) Natural Rubber (b) CB20%/NR Composite (c) CB25%/NR Composite and (d) CB30%/NR Composite (e) Mean value of NR and all composites

Composite of NR/CB20 show the highest tensile strength as listed in Table 3. This represents an effective reinforcement of CB in NR. The reinforcing potential is mainly come from formation of flexible filler network and strong polymer-filler coupling. The principal factors determined the

IOP Conf. Series: Materials Science and Engineering **223** (2017) 012009 doi:10.1088/1757-899X/223/1/012009

capability of reinforcement were (1) Van der walls force between CB and polymer, (2) the chemical cross link of polymer into the filler surface due to free radical reaction between carbon atoms in filler and polymer and (3) the mechanical interlocking of the polymer on to the filler surface.

Table 3. Tensile property of NR and NR/CB composite for different CB loading

Sample	Tensile Strength (MPa)	Elongation (%)	E (ε=0.05%) (MPa)
NR	0.437	442.2	2.48
NR/CB20	1.189	42.5	47.32
NR/CB25	0.916	25.2	34.17
NR/CB30	0.565	6.7	95.75

3.2. The Effect of Structure on the Tensile Property

The NR/CB tensile property as a function of series number was listed in Table 4. High CB structure exhibited a high amount of primary particle per aggregate. In reverse, the aggregate of low CB structure contained a low amount of primary particle. The number system of CB structure means the structure level for example N220 series structure was higher than N330. The sequence of CB structure for highest structure to the lowest was N220>N330>N550>N660. As expected, the tensile strength of NR/CB N220 was the highest among other. In other side, it was interesting that the NR/CB (660) composite show high value of elastic modulus and strain compares with other NR/CB (330) and NR/CB (550) composites. The high strain of NR/CB (660) composite might arise from the shape of aggregate. The aggregate formed agglomerates by weak van der walls force [8]. Like in filler loading that has an optimum value of CB loading on NR, the structure also has the limit which higher structure were no longer contribute to increasing the tensile property of the NR/CB composite. It is probably because the balance between particle loading and aggregate structure was achieved. Carbon black of N472 and N330 has the smaller aggregate size compare with N550 and N770. N472 and N330 have the higher BET specific surface area than N550 and N770 [5].

Table 4. Tensile property of NR and NR/CB composite with diferent structure of CB loading

Sample	Tensile Strength (MPa)	Elongation (%)	E (ε=0.05%) (MPa)
NR/CB (N220)	1.216	163.4	5.75
NR/CB (N330)	0.836	137.0	5.62
NR/CB (N550)	0.610	183.4	4.10
NR/CB (N660)	1.210	199.8	6.88

3.3. The Scanning Electron Microscope (SEM) Observation

SEM micrograph of NR and NR/CB composite that vary in CB loading is shown in figure 2. The homogeneous surface of NR can be observed in Figure 2(a) hence the strain of NR was the highest among others. The homogeneous mixture of 20% CB particle in NR is shown in figure 2(b). Coarse mixture of NR at 25% and 30% CB is shown in Figure 2(c) and (d). This explained the highest tensile strength of NR/CB20 compare to NR/CB25 and NR/CB30.

Figure 3 shows the SEM micrograph of NR/CB composite for different CB structure. The morphology of NR/CB (N220) and NR/CB (N660) in Figure 3(a) and (d) show fewer agglomerates of 2 and 11 μm in size, respectively, compare with NR/CB (N330) and NR/CB (N550) of 8 and 18 μm in size, respectively in Figure 3(b) and (c) . Hence, it result in higher tensile strength of NR/CB (N220) and NR/CB (N660) compare to NR/CB (N330) and NR/CB (N550). Theoretically, the lower structure i.e. NR/CB (N660) has the lowest tensile property. In this research, the less agglomeration of CB N660 is seen in Figure 3(d). It result in quite high tensile strength of NR/CB (N660) composite. Table 5 shows the EDX result of NR/CB (220) and NR/CB (550) composites. Both composite surfaces contain C, N, S and O elements. Both composite surfaces also contained almost the same amount of Carbon and Nitrogen.

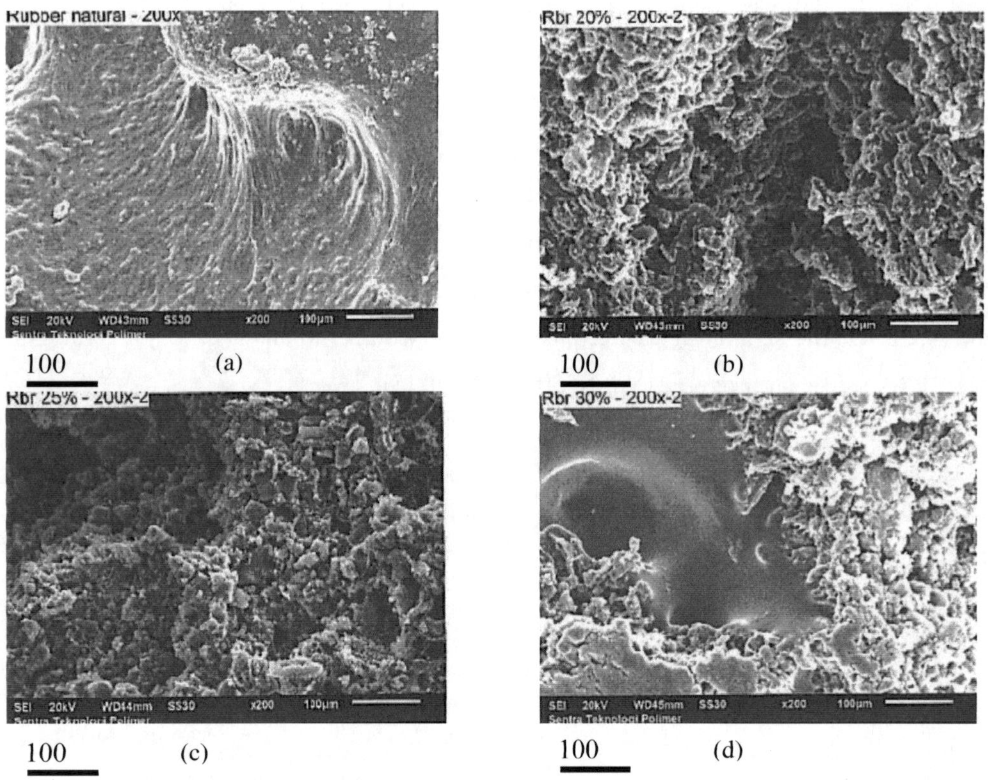

Figure 2. SEM Micrograph of NR/CB Composites: (a) NR (b) NR/CB20 (c) NR/CB25 (d) NR/CB30

The element in both composite was approximately the same. It indicated homogenity of both composite because the composition has the same composition. The different only in CB structure. The mapping of the element of C, N, S and O on the surface of each composite was depicted in Figure 4. It shown distribution of Sulphur and Carbon in composite. From Figure 4(a) and (b), it seems carbon evenly distributed in composite.

Table 5. EDX Result of NR/CB Composite with Different Structure of Carbon Black

Sample	Carbon (%)		Nitrogen (%)		Sulphur (%)		Oksigen (%)	
	wt.%	at.%	wt.%	at.%	wt.%	at.%	wt.%	at.%
NR/CB (N220)	81.40	84.61	14.72	13.12	1.98	0.77	1.91	1.49
NR/CB (N550)	81.59	84.65	13.95	12.41	3.07	2.39	1.39	0.54

Figure 3. SEM Micrograph of NR/CB Composites contained the Different Structure of CB (a) NR (N220) (b) NR/CB (N330) (c) NR/CB (N550) (d) NR/CB (N660)

Figure 4. Mapping of Element from EDX Result: (a) NR/CB (N220) (b) NR/CB (N550)

4. Conclusion

Mechanical properties of composite were tested using tensile test. Natural rubber was coagulated with formic acid. It shows that 2% volume fraction of formic acid gives sufficient tensile strength. For the rubber vulcanize with sulphur, the optimal tensile property was shown by 4% sulphur addition. The loading of 20% CB into the NR results in highest tensile strength. Further addition of CB decreases the tensile property of NR/CB composite. The strain decreases with CB loading. In reverse, elastic modulus increases with CB loading. From SEM observation, the surface of NR/CB20 showed the homogeneous CB dispersion in NR matrix. As the loading of CB increases the agglomeration of CB, it decreased the reinforcement of CB that results in low tensile strength.
The structure of CB affected the tensile stress of NR/CB Composite. From SEM observation, less agglomeration of NR/CB (N220) composite results in highest tensile strength. Large agglomeration size of CB N550 and N440 aggregates were clearly seen. This resulted in lower tensile strength of these composites. Although, theoretically, NR/CB N660 composite has lower structure among others. However, high tensile property was achieved because of limited amount of agglomerations of CB N660 agrregates.

From this work, the loading and structure of CB affected the tensile property of NR/CB composite. However, the formation of aggregates and agglomerations of CB also contributed. From EDX result, the surface of NR/CB (N330) and NR/CB (N550) composite consisted of C, S, N and O which matched with the composition of NR/CB composite.

Acknowledgment
The authors would like to thank the Ministry of Research, Technology and Higher Education of Indonesia for funding through PUPT research grant of University of Lampung. The authors also thank to 'UPT Laboratorium Terpadu dan Sentra Inovasi Teknologi' for SEM-EDX facility.

IOP Conf. Series: Materials Science and Engineering **223** (2017) 012009 doi:10.1088/1757-899X/223/1/012009

References

[1] Liu J *et al* 2008 *J. Macromol. Sci.* B **47** 689.
[2] Pasbakhsh P *et al* 2009 *J Appl. Polym. Sci.* **113** 1310.
[3] Thomas S *et al* 2014 *Composite and nano composite* (Cambridge: The Royal Society of Chemistry) Chapter 6 181-202.
[4] Hong C K *et al* 2007 *J. Mater. Sci* .**42** 8391-99.
[5] Li Z H *et al* 2008 *Express Polym. Lett.* **2** 695-704.
[6] Salomez M *et al* 2014 J. Appl. Microbiol. **117** 921-29.
[7] Verdejo R *et al* 2011 *In rubber-Clay Nanocomposites (Science technology and application)* ed Galimberti M (Hoboken: Jhon Wiley & Sons) 275-303.
[8] Fronlich J *et al* 2005 *Composites* A **36** 449.

IOP Conf. Series: Materials Science and Engineering **223** (2017) 012010 doi:10.1088/1757-899X/223/1/012010

Effect of urea deproteinization on catalytic hydrogenation of natural rubber latex

A Cifriadi[1], M Chalid[2], and S Puspitasari[1]
[1]Indonesian Rubber Research Institute, Jl. Salak No 1 Bogor 16151, West Java, Indonesia
[2]Departement of Metallurgy & Materials, Faculty of Engineering, University of Indonesia, Kampus UI Depok, Depok 16424, West Java, Indonesia

E-mail: puspitasari.santi@puslitkaret.co.id

Abstract. Natural rubber is unsaturated biopolymer which has low resistance to heat, oxygen, and ozone. Chemical modification of natural rubber by catalytic hydrogenation can improve its oxidative property. In this study, the catalytic hydrogenation of natural rubber was investigated in latex phase after reduction of protein content with urea. Hydrogenation of deproteinized natural rubber latex was performed by using diimide which generated insitu from hydrazine hydrate/hydrogen peroxide and catalyst (boric acid, cupric sulfate and cupric acetate) at 70°C for 5 h. The hydrogenation system was stabilized with sodium dodecyl sulphate. The hydrogenation of deproteinized natural rubber (HDPNR) was confirmed by FTIR analysis. The result indicated that cupric sulphate was extremely active catalyst which was showed by the elimination of C=C transmittance bands at 1660 cm^{-1} on HDPNR spectra and highest degree of hydrogenation. Furthermore, urea deproteinization increased possibility of side reactions during catalytic hydrogenation as seen on the reduction of gel content compared to undeproteinized natural rubber.

1. Introduction

The fresh natural rubber latex derived from *Hevea brasiliensis* (Willd.) Muell. Arg is composed of rubber particles and non-rubbers fraction mainly protein and phospholipid which are dispersed in water phase [1-3]. While on natural rubber particle itself is consist of 94% rubber hydrocarbon molecules of cis-1,4 polyisoprene surrounded by protein-phospholipid layer which maintain the stability of rubber particle [4]. Whereas long polyisoprene chain give advantages such good mechanical and dynamic properties to natural rubber characteristic [5]. The existence of reactive unsaturated C double bond within the polyisoprene backbone cause natural rubber is easily deteriorates when expose to oxygen, ozone, and sunlight [6]. This weakness limits the natural rubber application especially in the manufacture of outdoor rubber based product.

Thermal oxidative degradation of natural rubber can be prevented by physical technique such as blending or coating of natural rubber with another polymer which acts as a thermal oxidative stabilizer [7,8]. Another method is by chemical modification as hydrogenation of natural rubber in latex phase. Reduction of C double bond by hydrogenation without being followed by scissoring of the molecular chain will significantly improve the thermal oxidative resistance of natural rubber [9,10]. Catalytic

IOP Conf. Series: Materials Science and Engineering **223** (2017) 012010 doi:10.1088/1757-899X/223/1/012010

transfer hydrogenation is preferred than conventional. Transfer hydrogenation require an addition of non-H_2 compound and homogeneous catalys since they provide higher selectivity and simplicity mechanism owing to no involvement of diffusion mass transfer as occur in heterogeneous catalyst [9-12]. Diimide is highly reactive intermediate non-H_2compound which can be generated in situ during hydrogenation by oxidation of hydrazine hydrate by hydrogen peroxide, the decarboxylation of potassium azodicarboxylate, thermal decomposition of p-toluenesulfonyl hydrazine and the photochemical irradiation of 1-thia-3,4-diazolidine-2,5-dione [13-18]. The hydrogenated natural rubber (HNR) has similar molecular structure to ethylene-propylene alternating copolymer (EPDM) rubber [19,20]. Thus HNR is potential to substitute EPDM rubber to achieve better thermal oxidative stability on outdoor rubber based product [21].

The diimide hydrogenation is strongly depends on the ratio concentration of hydrazine hydrate/hydrogen peroxide, type of homogeneous catalysts, reaction condition and technique. Mahittikul *et al.,* (2007) found that cupric acetate was the most reactive catalyst for hydrogenation of high ammoniated natural rubber latex by semi-continous technique [22]. Furthermore, non-catalytic diimide hydrogenation by seedling technique was done by Veni and Ma'zam (2010) indicated that fresh field latex achieved highest hydrogenation efficiency compared to another type of the latex [23].

The use of hydrogen peroxide as strong oxidizing agent increase the possibility of side reaction such as depolymerization and gel formation by crosslink reaction of non rubber constituent which can compete to the diimide hydrogenation [6, 24]. Moreover, the protein contained in natural rubber are also suggested can reduce hydrogenation efficiency due to the protons of the amine group are electron donating species that may react with the diimide via metathesis hydrogenation [25]. This research used urea deproteinized natural rubber latex in order to reach high degree of catalytic diimide hydrogenation. Urea deproteinization offers simplicity on deproteinization method of natural rubber compared to enzymatic deproteinization [26,27]. The aim of the present research was to investigate the effect of urea deproteinization on the properties of hydrogenated rubber (HDPNR) synthesized by catalytic diimide hydrogenation of natural rubber latex by using various homogeneous catalysts.

2. Material and Method

Commercial high ammoniated natural rubber latex concentrated (NRL) (dry rubber content, DRC as 56%) was provided by Indonesian Rubber Research Institute (Bogor, Indonesia). Hydrazine hydrate (purity 80%), 30% aqueous hydrogen peroxide, boric acid, cupric sulfate, cupric acetate, urea, and sodium dodecyl sulfate (SDS) were purchased from Merck, Germany. Technical grade of silicon oil as antifoam agent, acetone and calcium chloride as coagulants were received from local supplier.

The latex deproteinization used Alfa Laval Centrifuge machine to separate the protein bound urea in serum fraction. While the hydrogenation was conducted in a 500 mL three necked round bottomed flask which was equipped with thermometer, reagent droplet vessel, and gaseous N_2 inlet tube. Heating and agitation were facilitated by Cimarex Thermolyne_hot plate magnetic stirrer. The qualitative analysis used Thermo Scientific Nicolet S5 ATR-FTIR. Thus the oxidative resistance of raw HNR was measured by using Wallace Rapid Plastimeter.

2.1. Urea Deproteinization of Natural Rubber Latex (DPNR)

Urea deproteinization of NRL procedure was referred to previous work [27]. NRL concentrate was added by 0.1% urea and dispersed in a SDS 1% solution until the DRC reached 30%. The NRL was incubated at room temperature for 60 min and low speed of agitation. The NRL incubation was followed by centrifugation at 15000 rpm. The cream fraction of NRL was redispersed in a distilled water to make 30% DRC and recentrifuged to wash residual protein in the NRL. The deproteinized natural rubber (DPNR) latex concentrate obtained was preserved with 0.2% ammonia and 0.2% SDS to ensure the stability before being used in diimide hydrogenation. DPNR was characterized the protein content by Kjeldahl Method.

2.2. Catalytic Diimide Hydrogenation of DPNR Latex (HDPNR)

The diimide hydrogenation of DPNR latex was run at various types of preservative and homogeneous catalyst. Semi-continous reaction technique was applied to the system. DPNR latex concentrate was fed into the flask, followed by addition of distillated water to make 30% DRC. Gaseous N_2 was purged into the DPNR latex under agitation for 15 min to remove dissolved oxygen. The temperature of the DPNR latex was raised gradually up to 70°C. Thus, 2 per hundred rubber (phr) of hydrazine hydrate and 0.05 phr of catalyst were added to the DPNR latex, respectively. 3 phr hydrogen peroxide was added dropwise at particular rate for 2 h in order to maintain an isothermic condition. The diimide hydrogenation was continued for 3 h after the last hydrogen peroxide droplets. At the end of the reaction, the hydrogenated deproteinized natural rubber (HDPNR) latex was coagulated by using the mixture of acetone and $CaCl_2$ 5% solution (1:1). The HDPNR coagulum was milled on creeper machine and dried at 50°C for 24 h before characterized.

IOP Conf. Series: Materials Science and Engineering **223** (2017) 012010 doi:10.1088/1757-899X/223/1/012010

2.3. Characterization of HNR

2.3.1. FTIR Spectroscopy Analysis.

The microstructure of HDPNR samples were determined by using *Thermo Scientific Nicolet S5 ATR-FTIR* spectroscopy. The FTIR spectra of HDPNR were scanned in transmittance mode at wavenumber within the range of 4000 to 500 cm^{-1}. Each HDPNR sampel was scanned twice before confirmed.

2.3.2. Iodine Number and Degree of Hydrogenation.

The conversion of unsaturated C double bond onto saturated C single bond on natural rubber molecular chain by diimide hydrogenation was determined by iodine number calculation referred to Wijs Method by following equation (1).

$$\text{Iodine Number} = N \times 12.69 \times \left(\frac{V_B - V_A}{m}\right) \quad (1)$$

Remarks :

N___: normality of sodium thiosulfate solution after standardization (N),

V_B___: volume sodium thiosulfate for blank titration (mL),

V_A___: volume of sodium thiosulfate for HDPNR titration (mL),

m___: mass of HDPNR (g).

Degree of hydrogenation was determined refered to the following equation (2)

$$\text{Degree of hydrogenation} = \frac{C_0 - C}{C_0} \times 100 \quad (2)$$

Remaks:

Co___: iodine number of DPNR

C___: iodine number of HDPNR

2.3.3. Original Plasticity (Po) and Plasticity Retention Index (PRI) Analysis.

The plasticity tests were designed to provide a rapid assessment of the susceptibility of raw rubber to thermal oxidative resistance. The test procedure for Po and PRI analysis were carried out according to the ASTM D 3194-04 standard. PRI was percentage ratio of Po before and after ageing at specified temperature and time (140°C for 30 min). Equation (3) described the PRI measurement.

$$PRI = \frac{Pa}{Po} \times 100\% \quad (3)$$

Remarks:

Pa___: plasticity after aging,

Po___: plasticity before aging

2.3.4. Gel Content Analysis.

IOP Conf. Series: Materials Science and Engineering **223** (2017) 012010 doi:10.1088/1757-899X/223/1/012010

Gel content of HDPNR was determined refers to following method: 0.125 g finely cut of HDPNR sample was diluted on 25 mL of chloroform and kept for 20 h. The swollen HDPNR then sieve through filter paper. The residue consist of macro gel together with the filter paper was dried at 100°C until definite weight was achieved. The gel content of HDPNR was calculated by equation (4).

$$\text{Gel Content} = \frac{\text{Gel weight}}{\text{HDPNR weight}} \times 100\% \qquad (4)$$

3. Results and Discussion

3.1. Preparation of DPNR Latex

Elimination of protein on NRL by urea deproteinization_is confirmed qualitatively by FTIR spectroscopy analysis as describes at Figure 1. NRL spectra shows the presence of protein layer indicated by_transmittance bands at 3285 cm^{-1} and 1544 cm^{-1} assigned to N-H stretching and N-C=O amide II vibration, respectively [6]. The bands are disappeared at DPNR spectra. The quantitative analysis also indicate that_the urea deproteiniation cause the protein content of NRL is significantly reduced from 1.37 to 0.44%. It was believed that protein attached_on the surface ofthe natural rubber particle by physical interactions had denatured by urea. Urea formed hydrogen bond to the moieties of the protein which was stronger than covalent binding between protein and rubber particle [28, 29]. Thus, the protein bound urea was dispersed on the serum phase of DPNR latex and being removed by surfactant washing during multistage centrifugation [30]. The function of protein in maintaining the rubber particles stabilization was replaced by ammonia as preservative or SDS surfactant as new stabilizer.

Natural rubber molecules on both NRL and DPNR spectras show distinct bands refer to cis-1,4 polyisoprene for the unsaturated C double bond stretching vibration and =C-H out of plane bending vibration for tri_substituted_olefinic at wavenumber around 1660 cm^{-1} and 833 cm^{-1}, respectively. Another important peaks are showed by transmittance bands at 738 cm^{-1}correlated to $-(CH_2)_3-$ species, 2959 and 1447 cm^{-1} for symmetric $C\text{-}H_3$ bonds stretching and bending vibrations, 2852 cm^{-1} and 1375 cm^{-1} for symmetric $C\text{-}H_2$ bonds stretching and bending vibration[10, 22, 31].

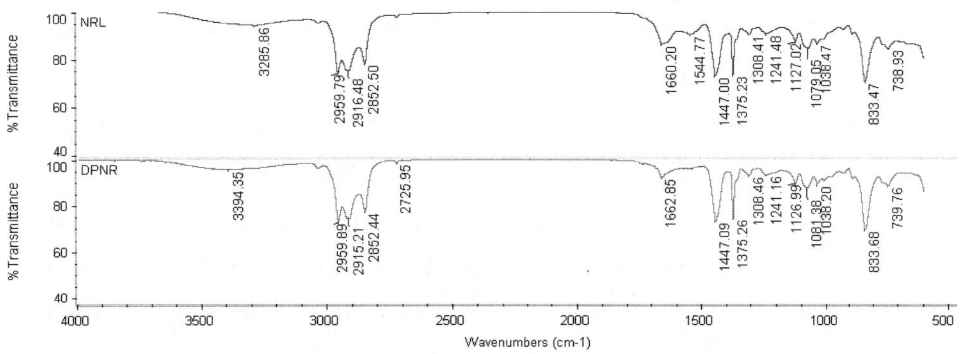

Figure 1. FTIR spectra of NRL and DPNR

3.2. Diimide Hydrogenation of DPNR Latex

FTIR spectras of all of the HDPNR show identical pattern as illustrated at Figure 2. After the hydrogenation process, the transmittance bands at 1660 cm^{-1} and 833 cm^{-1} decrease, followed by increasing band at wavenumber around 740 cm^{-1} assigned to $-(CH_2)_3-$ species due to the addition of hydrogen atom from diimide compound to C=C double bond on natural rubber molecular chain. The significant elimination of C=C double bond was occurred on cupric sulphate catalyst addition to hydrogenation of DPNR latex stabilized with SDS. The FTIR spectras also show peaks at wavenumber 3400 cm^{-1} and 1740 cm^{-1} representing OH and C=O bonds. Those transmittance bands indicated the occurance of side reaction such as depolymerisation during hydrogenation [32].

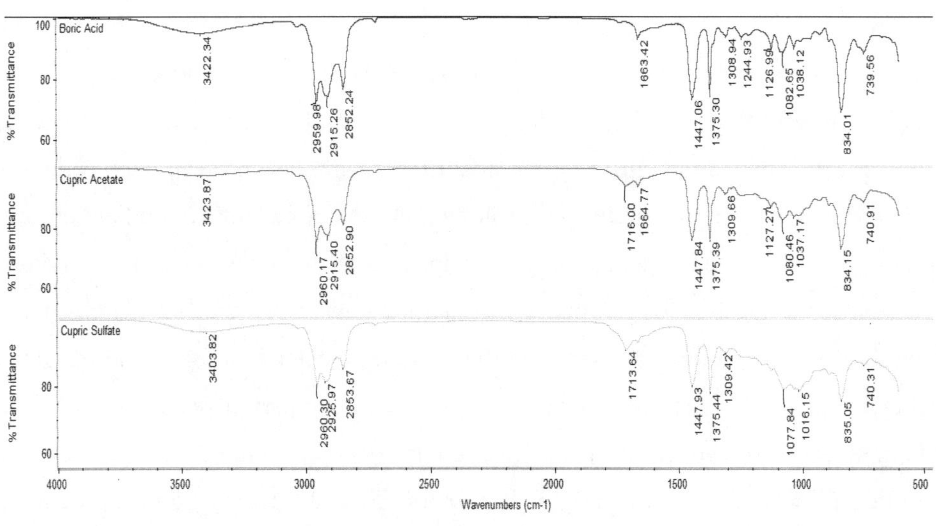

Figure 2. FTIR spectra of HDPNR prepared by various catalyst

Physical visualization of HDPNR as presented at Figure 3 can be used as primary evaluation to the hydrogenation efficiency since it characterizes side reaction. HDPNR prepared with addition of boric acid has yellowish colour and solid texture, whereas cupric ion catalysts produce brownish colour and soft sticky-paste like texture. The presence of cupric ion in hydrogen peroxide-latex hydrogenation system at certain temperature accelerated the oxidation of natural rubber molecule which caused chain scission or depolymerization. Cupric ion played as prooxidant compound, while hydrogen peroxide formed hydroxyl free radical. By the elimination of protein layer surrounded DPNR particle, the hydroxyl free radical and cupric ion became easier to attack the polyisoprene molecules. Hydroxyl free radical was believed to be more reactive than diimide compound, so at hydrogenation catalysed by cupric ion system, depolymerisation actually dominated. The similar mechanism could be found in the synthesis of liquid natural rubber.

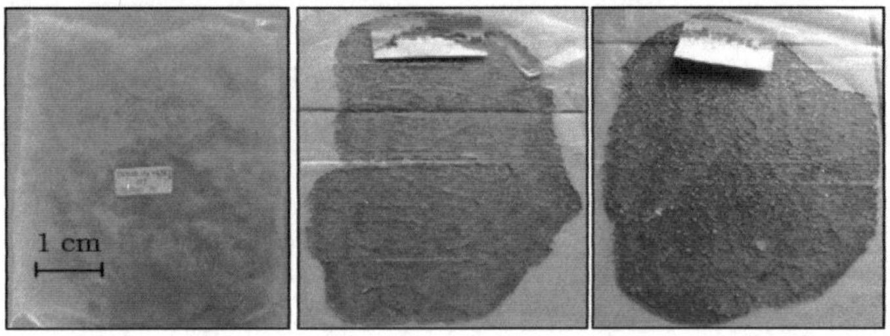

Figure 3. Physical visualization of HDPNR

The effect of ammonia and SDS in maintaining DPNR latex stabilization during diimide catalytic hydrogenation to the iodine number and degree of hydrogenation are ilustrated at Figure 4 and 5, respectively. The both figures demonstrate that SDS stabilizer produce higher efficiency of hydrogenation compared to ammonia preservative. On catalytic diimide transfer hydrogenation, SDS played dual functions, either as stabilizer of colloidal rubber particles and secondary catalyst due to the presence of sodium atom on the molecular structure. At the same latex stabilization systems, the effectiveness of catalysts in descending order are cupric sulphate > cupric acetate > boric acid. The cation and anion of the catalyst were strongly affected the degree of hydrogenation. Cupric ion was the most suitable metallic cation for catalyzing the diimide transfer hydrogenation of natural rubber. Sarkar et al claimed that the catalytic activity of cupric metal complex depended on the acidity of the ligand and the strength of the metal-ligand interaction [14]. The interaction between metal ion of cupric^{2+} and ligand such SO_4^{2-} was slightly higher than CH_3COO^-. Boric acid produced lowest degree of hydrogenation since it was more regarded as promoter than an catalyst on hydrogenation [22]. Veni and Ma'zam also purposed that between boric acid and hydrogen peroxide formed hydrogen bond which able to control hydrogen peroxide activity in the reaction medium [23]. The decrease on hydrogen peroxide activity inhibited the oxidation of hydrazine hydrate in generating diimide at the rubber particle surfaces.

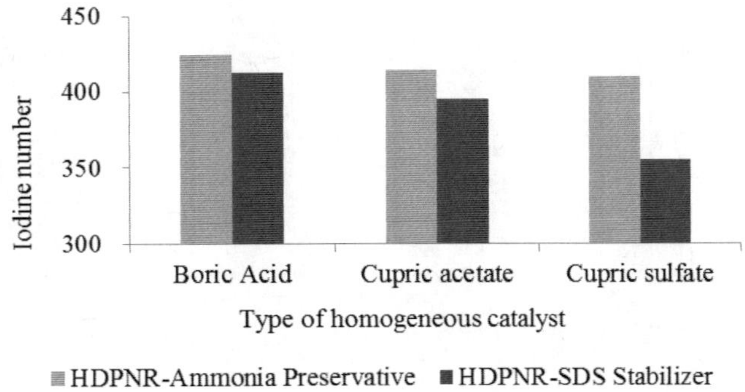

Figure 4. Iodine number of DPNR and HDPNR

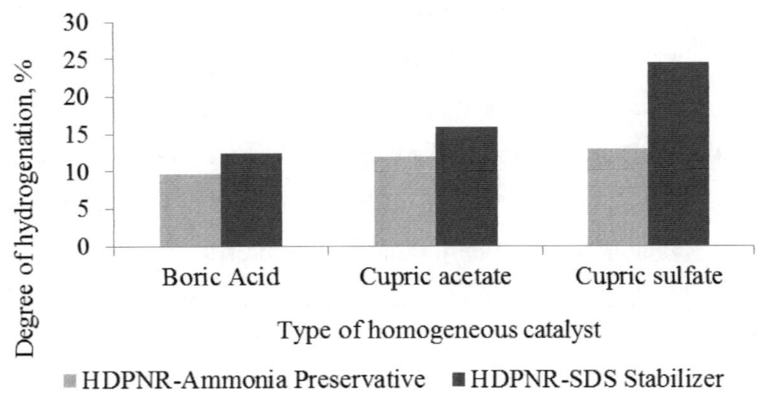

Figure 5. Degree of hydrogenation of the HDPNR

Test result of gel content_analysis of the HDPNR is described at Figure 6._All of the HDPNR samples have lower gel content compared to unmodified DPNR._The polyisoprene rubber molecular chains were composed of sol and gel fraction [33]. Gel fraction_could be used as indicator of the existence of branched or network structures in amongs rubber molecular chains [30]. Originally, gel fraction constributed by two mechanism such as interlinkage or crosslinking of rubber molecular chains and interaction of protein and phospholipid at the initiating ω-terminal and terminating α-terminal on rubber molecule [23,34,35]. In the study, the gel content formation only affected by the crosslink of rubber molecule since the protein-phospholipid had been elimainated during the deproteinization process. Blackley pointed out that the addition of small amount of polar substance in the latex system can greatly reduce the apparent gel content [36]. All reagents involved in the hydrogenation system were polar substances which could decomposed the gel fraction caused by intermolecular crosslink of natural rubber.

IOP Conf. Series: Materials Science and Engineering **223** (2017) 012010 doi:10.1088/1757-899X/223/1/012010

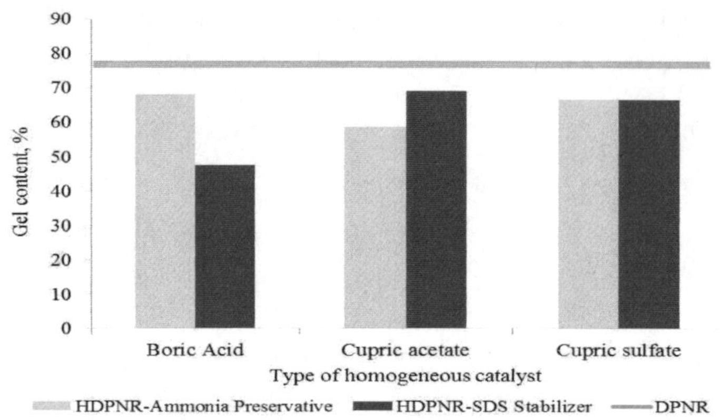

Figure 6. Gel content of HDPNR

High plasticity indicates better thermal oxidative resistance of natural rubber. Table 1 summarizes the plasticity test result of HDPNR. The undetected PRI value for HDPNR due to the samples were more not resistant to heat and oxygen exposure compared to unhydrogenated DPNR. At temperature of reaction as 70°C, hydrogen peroxide either react with hydrazine hydrate to generate diimide but also decomposed to form highly reactive free hydroxyl radical. The hydroxyl radical attacked unsaturated C double bond which ultimately led to chain scission. The dominations of this side reaction was characterized by tacky texture on HDPNR when expose to heat and oxygen.

Tabel 1. Po and PRI of HDPNR

Type of rubber	Po		PRI (%)	
	Ammonia	SDS	Ammonia	SDS
Unhydrogenated DPNR	54.5	54	27.8	27
HDPNR-boric acid	53	36	NA	NA
HDPNR- cupric acetate	NA	NA	NA	NA
HDPNR-cupric sulfate	NA	NA	NA	NA

Remarks: NA : the HDPNR was tacky along the PRI testing

4. Conclusion

The urea deproteinization of natural rubber latex was understood to be an ineffective procedure for preparing_diimide hydrogenation of natural rubber. The elimination of protein-phospholipid layer surrounded rubber particle increased the possibility of side reaction such as depolymerisation which also accelerated by the presence of cupric ion catalyst. It was showed by the elimination of C double bond at transmittance band around 1660 cm^{-1} on FTIR spectra and highest degree of hydrogenation. Furthermore, the reduction of gel content was also indicated the occurance of side reaction such decomposition of intermolecular rubber crosslink.

Acknowledgement

Authors gratefully acknowledge the support of Research, Technology, and High Education Ministry of Indonesian Republic for the Insentif_Riset SINAS 2015 (RT – 2015 – 0141) research funding.

References

[1] Rahman A Y A *et al* 2013 *BMC Genomic* **14** 75.
[2] Nawamawat K *et al* 2011 *Colloids and Surface A: Physicochemical and Engineering Aspects* **390** 157-166.
[3] Sansatsadeekul J *et al* 2011 *J. Bioscience and Bioengineering* **111(6)** 628-634.
[4] Norhanifah M Y *et al* 2015 *Procedia Chemistry* **15** 31-38.
[5] Hinchiranan N *et al* 2006 *J. Appl. Polym. Sci.* **100** 4499-4514.
[6] Mahittikul A *et al* L 2006 *J. Appl. Polym. Sci.* **1 00** 640-655.
[7] Piya-areetham P *et al* 2014 *Polymer Degradation and Stability* **102** 112-121.
[8] Ning N Y *et al* 2015 *e XPRESS Polymer Letters* **9(6)** 490-495.
[9] Piya-areetham P *et al* 2013 *European Polymer Journal* **49** 2584-2595.
[10] Mahittikul A *et al* 2009 *J. Molecular Catalysis A: Chemical* **297** 135-141.
[11] Wang D and Astruc D 2015 *Chemical Reviews* **115 (13)** 6621-6686
[12] Tangthongkul R *et al* 2005 *J. Appl. Polym. Sci.* **97** 2399-2406.
[13] Squillacote M *et al* 1993 *Tetrahedron Lettres* **34 (26)** 4137-4140.
[14] Sakar M D *et al* 2000 *Polymer* **41** 907-915.
[15] Samran J *et al* 2005 *J. Appl. Polym. Sci.* **95** 16-27.
[16] Smit C *et al* 2008 *J. Org. Chem* **73** (23) 9482-9485.
[17] Arayapranee W and Rempel G L 2009 *J. Appl. Polym. Sci.* **114** 4066-4075.
[18] Pieber B *et al* 2013 *Angew. Chem. Int. Ed* **52** 10241-10244.
[19] Inoue S and Nishio T 2007 *J. Appl. Polym. Sci.* **103** 3957-3963.
[20] Piya-areetham P *et al* 2013 *J. Molecular Catalysis A: Chemical* **372** 151-159.
[21] Pruttisirikul T 2010 *Reactive & Functional Polymers* **70** 674-683.
[22] Mahittikul A *et al* 2007 *Journal of Applied Polymer Science* **105** 1188-1199.
[23] Veni B N K and Ma'zam M S 2010 *J. Rubber Research* **13**(2) 103-109.
[24] Lin X *et al* 2004 *Applied Catalysis A: General* **276** 123-128.
[25] Kongparakul S *et al* 2012 *Top Catal* **55**: 524-529.
[26] Klinklai W *et al* 2004 *J. Appl. Polym. Sci.* **93** 555-559.
[27] Yamamoto Y *et al* 2008 *J. Appl. Polym. Sci.* **107** 2329-2332.
[28] Kawahara S and Chaikumpollert O 2012 *Transactions on GIGAKU* **1(1)** 1-8.
[29] Bennion B J and Daggett V 2003 *Proc Natl Acad Sci USA* **100** (9) 5142-5147.
[30] Nawamawat K *et al* 2010 *Macromolecular Symposia* **288** 95-103.
[31] Arayapranee W and Rempel G L 2013 *J. Polymer* 9p.
[32] Ibrahim S *et al* 2014 *Polymers* **6** 2928-2941.
[33] Nimpaiboon A *et al* 2014 *Advanced Material Research* **844** 101-104.
[34] Tangpakdee J and Tanaka Y 1997 *Rubber Chem Technol* **70** 707-713.
[35] Xie H Q, Li X D and Guo J S 2003 *Journal of Applied Polymer Science* **90** 1026-1031.
[36] Marinho J R and Tanaka Y 1999 *Journal of Rubber Research* **2 (4)** 231-238.

IOP Conf. Series: Materials Science and Engineering **223** (2017) 012011 doi:10.1088/1757-899X/223/1/012011

Effect of type and content of tackifier on adhesion of natural rubber and reclaimed natural rubber based sealant

P Raethong[1], K Boonkerd[2]
[1]Multidisciplinary Program in Petrochemistry & Polymer Science, Faculty of Science Chulalongkorn University, Thailand
[2]Department of Materials Science, Faculty of Science, Chulalongkorn University, Thailand

E-mail : parichat.rae@gmail.com

Abstract. Adhesion property of natural rubber (NR) and reclaimed natural rubber (RNR) based sealant with concrete was studied here. The effect of tackifier type and loading on the tensile properties of the rubber based sealant sandwished between two flat cements was evaluated. There are three different tackifiers including Coumarone-indene resin (CI), petro resin (PE) and gum rosin (GR). The result initially showed that at the 20 phr of tackifier both NR and RNR sealant mixed with CI consumed the highest force to separate the rubber based sealant from the concrete. This might be due to the highest compatiblility between CI and rubber. Regardless of the tackifier type, all NR based sealants showed the cohesive failure while all RNR based sealants only showed the adhesive failure. Moreover, the NR based sealant seemed to be stronger than the RNR based one. When considering the effect of CI loading on the adhesion, it was shown that for both NR and RNR based sealents, the highest stress was observed when the rubber based sealant loaded with the 20 phr of CI.

1. Introduction

Rubber adhesive and sealant are highly flexible material. They are used to join components or fill gaps between seams or on surfaces. Sealant features a reactive adhesion property adopted as a joint water-stop material for underground concrete structures, and used to prevent water leakage and protect reinforcing bars from centralizing cracks in mass concrete.

In general, both natural rubber based on polyisoprene and synthetic rubbers including silicone, polyurethane, polysulfide, styrene-butadiene rubber (SBR), butyl, acrylic or polyacrylate, isoprene, polyisobutylene, ethylene-vinyl acetate (EVA), vinyl and nitrile compounds are widely used in this application. Specifically for civil engineering, nowadays the rubber based sealantmade from the reclaimed butyl rubber has commercially come into wide use in Japan, largely replacing conventional PVC. Even though the reclaimed butylrubber based sealant has been commercially used for many years [1]. The excellent feature is that sealant adheres to green concrete during it's hardening process. This adhering mechanism is scientifically solved. It is based upon existence of active factors (carboxyl factor, etc.) generated during the production of sealant and while special treatment is given to butyl rubber. It was reportedthat metal oxide containing in the cement becomes metal water oxide by influence of water, and ion reaction is created together with active factor of sealant, thereby allowing chemical compound to be formed. Also, in addition to physical adhering mechanism of other anchoring effect, sealant adheres to green concrete through this feature [2]. The development of the rubber based sealant from other rubbers is recently concerned.

In the present study, the aim is to study the effect of tackifier type and loading on the adhesion between the rubber and concrete. Two kinds of natural rubber including virginnatural rubber and reclaimed natural rubber were used here. Coumarone-indene resin, petro resin and gum rosin were used as tackifier. A systematic study was carried out to understand the roll of tackifier type and content on the adhesion properties.

2. Experiment

2.1 Material

Natural rubber(NR) grade STR5L and reclaimed natural rubber (RNR) were purchased Sakdarungreaungkij Co., Ltd. Three tackifiers including coumarone-indene resin (CI), petro resin (PE) and gum resin (GR) were kindly supplied by Innovation Technology Co., Ltd.

2.2 Rubber Based Sealant Preparation.

Rubber based sealants were prepared by mixing 100 phr of rubber with 0, 10, 20 and 30 phr of tackifiers and other fillers in an internal mixer (MX500-D75L90) at a high temperature. The total mixing time in the internal mixer was 30 min. After discharging, the rubber compound was sheeted off to form the rubber sheet sample with a thickness of 4-5 mm using a two-roll mill.

2.3 Adhesion Testing.

To determine the adhesion between the rubber based sealant and concrete, a PVC pipe was as a mold. Firstly, the PVC pipe was placed on the sealant sheet. Then, the wet mixed cement was poured into the PVC pipe. The sample was left under air unitl the cement completely solidified. Secondly,the same procedure was repeated on the another side of rubber based sealant. The maximum load needed to separate the rubber based sealant from the concrete was determined

IOP Conf. Series: Materials Science and Engineering **223** (2017) 012011 doi:10.1088/1757-899X/223/1/012011

using an Instron model 5843 at a crosshead speed of 5 mm/min with the load cell of 1 kN.The failure behavior was observed and noted.

3 Result and Discussion

3.1 The Effect of Tackifier Type on the Adhesion.

The adhesion properties of the NR and RNR based sealant when loaded with the 20 phr of tackifier were shown in Table 1.

Table 1. Adhesion Propeties of the NR and RNR Based Sealant

Property	Samples					
	NR/CI	NR/PE	NR/GR	RNR/CI	RNR/PE	RNR/GR
Maximum load (N)	211.17	93.38	170.73	118.38	117.95	112.95
Strain at break (%)	3500	2043	1500	917	1875	428
Failure behavior	cohesive	cohesive	cohesive	adhesive	cohesive	adhesive

From the above results, in the case of NR based sealant, it can be clearly seen that the rubber based sealant loaded with PE required the lowest maximum load to separate the rubber from the concrete. Therefore, the NR based sealant loaded with PE tackifier was the weakest sealant. The PE tackifier is synthetic tackifier consisting of long aliphatic chains [3]. The reason why PE tacikifier gave the rubber sealant with the lowest adhesive strength might be attributed to the high chain entanglement between the rubber chains and the long molecule of PE, thus increasing the viscosity of the sealant and then inhibitting the sealant from wetting on the concrete. The NR based sealant loaded with CI and GR tackifier showed much higher adhesive strength than the one loaded with PE. This is due to the molecule of these two tackifiers are much smaller than the PE. However, the NR based sealant loaded with CI showed the highest adhesive strength. This might be due to the chemical structure of CI consisting less polar functional groups than that of GR, thus causing CI more compatible with the NR and then giving the NR sealant with higher adhesive strength.

The RNR used in this study is the cured rubber that was mechanically masticated to break down the crosslinking and shorten rubber chain, making the rubber be able to reprocess. However, the RNR still consists of crosslinking point, causing RNR in having higher viscosity than the NR. In this part, it was seen that the tackifier type hardly affected on the adhesive strength of the sealant [4]. This might be because the amount of tackifier used in this part was too low toloosen the viscosity of the RNR. Therefore, the RNR based sealant loaded with different tackifiers showed the similar adhesive strength [5].

When considering the type of rubber, it can be found that the NR based sealant showed the higher adhesive strength than the RNR based one. This might be due to the existence of partial crosslinking network inhibitting the wettability of the rubber onto the concrete, thus giving the lower adhesive strength.Most of the NR based sealant showed the cohesive failure,while all RNR based sealant except the one loaded with PE showed the adhesive failure. The failure behavior of the tested sample was shown in Figure 1. The cohesive failure indicated that the interaction between the rubber and concrete is stronger than the strength of the rubber. On the other hand, the adhesive failure indicated that the interaction between the rubber and concrete is poorer than the strength of the rubber.

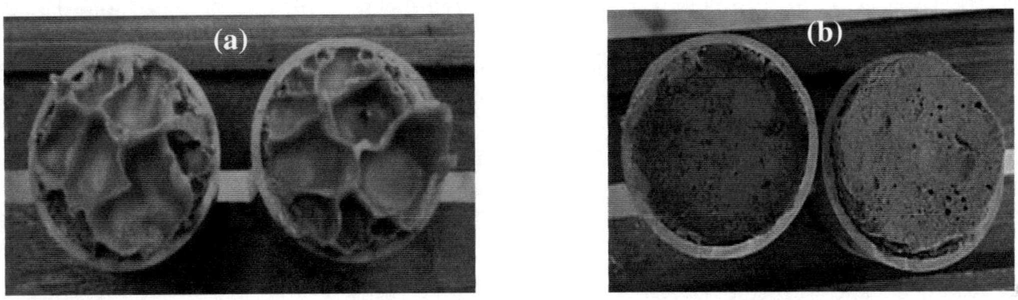

Figure 1. Failure Behavior of Sample (a) Cohesive failure (b) Adhesive Failure (diameter 40 mm)

3.2 The Effect of Tackifier Loading on the Adhesion.

In this part, only the rubber based sealant loading CI was focused. The loading of CI was varied from 0 to 10, 20 and 30 phr. The effect of CI loading on the adhesion properties was depicted in Table 2.

Table 2. Adhesion properties of NR and RNR adhesives with increasing CI from 0 to 30 phr

Property	Samples							
	NR/ CI0	NR/ CI10	NR/ CI20	NR/ CI30	RNR/ CI0	RNR/ CI10	RNR/ CI20	RNR/ CI30
Maximum load (N)	103.17	138.79	211.17	109.07	62.13	70.94	118.38	80.34
Strain at break (%)	1233	2250	3500	2750	425	1050	917	1400
Failure behavior	cohesive	cohesive	cohesive	cohesive	adhesive	adhesive	adhesive	adhesive

From the above result, for both the NR and RNR based sealant, it can be seen that the increase of CI content from 0 to 30 phr with the increment of 10 phr caused the maximum load or adhesive strength pass through the maximum. The presence of tackifier initially enhanced adhesion. However, when tackifier was overloaded, it caused the decrease of adhesion. This is generally due to the dilution effect of the rubber. The CI content gaving the highest adhesive strength was at 20 phr. At each CI content, the NR based sealant gave higher maximum load than the RNR one.

4. Conclusions

This study was focused on the development of the rubber formular for the rubber based sealant application. The best rubber formular can be considered from the highest adhesive strength between the rubber and the concrete. The result showed that amongst three different tackifiers including CI, PE and GR, CI gave the rubber based sealant with the highest adhesive strength. Although the NR is much more expensive than the RNR, the NR based sealant showed much higher adhesive strength than the RNR one. The rubber based sealant filled with CI exhibited the highest maximum tensile load at the 20 phr of tackifier. Therefore, from this study the best rubber formular for the sealant application is the NR filled with the 20 phr of CI.

Acknowledgments

The study was supported by HAYAKAWA RUBB
ER CO., LTD.

References

[1] Azura A R *et al* 2014 *J. Adh. Sci. Technol.* **28** 637.
[2] Poh B T and Yong A T 2010 *J. Appl. Polym. Sci.* **115** 1120.
[3] Poh B T and Yong A T 2009 *J. Appl. Polym. Sci.* **114** 3976.
[4] Poh B T and Yong A T 2009 *J. Adh.* **85** 435.
[5] Imran K and Poh B T 2012 *J. Adh.* **20** 132.
[6] Poh B T, Giam Y P, and Yeong F P 2010 *J. Adh.* **86** 846.
[7] Poh B T and The Y Y 2014 *J. Adh.* **90** 802.
[8] Soo K W *et al* 2016 *J. Adh.* **92** 135.

IOP Conf. Series: Materials Science and Engineering **223** (2017) 012012 doi:10.1088/1757-899X/223/1/012012

The effect of alkanolamide addition on cure and tensile properties of unfilled natural rubber compounds

I Surya[1], SF Siregar[1], H Ismail[2]
[1]Department of Chemical Engineering, University of Sumatera Utara, Medan, Indonesia
[2]School of Materials and Mineral Resources Engineering, Universiti Sains Malaysia, Penang, Malaysia

E-mail: indradanas@yahoo.com

Abstract. Effect of alkanolamide (ALK) addition on cure and tensile properties of unfilled natural rubber (NR) compounds was investigated. The ALK was synthesized from Refined Bleached Deodorized Palm Stearin (RBDPS) and diethanolamine, and incorporated into the unfilled-NR compound as a new rubber additive. The ALK loadings were 0.2, 0.4, 0.6, 0.8 and 1.0 phr. It was found that the addition of ALK exhibited shorter scorch and cure times and higher elongation at break of the unfilled NR compounds. The ALK also exhibited higher torque differences, tensile modulus and tensile strength of up to 0.6 phr of ALK loading and then decreased with further increases in the ALK loading. Scanning electron microscopy (SEM) proved that 0.6 phr loading of ALK in the NR compounds exhibited the greatest matrix tearing line and surface roughness which correlated to a higher tensile strength of the NR compound.

1. Introduction

In rubber technology, rubber recipe is the heart of rubber compounding which contains a group of base rubber and rubber additives [1]. Each of them has a specific function either in processing, curing, or end use of the rubber products. Rubber or blend of rubbers will provide rubbery behavior to the compound; rubber additives such as fillers can cheapen the cost of rubber products, modify modulus and processing properties; plasticisers are used to reduce viscosity, enhance deformability and alter properties; vulcanizing agents can crosslink the rubber chains; and accelerators are something that improve the action of vulcanizing agents to speed up the resultant crosslinking.

Rubber additives are classified into curative and non-curative additives. Curative additives play an important role in determining the cure characteristics of rubber compounds. They affect the rate and nature of the vulcanization process. In rubber technology, the popular curative additives are sulphur or other vulcanizing agents, zinc oxide, stearic acid and accelerators.

A lot of studies on curative additives have reported by some researchers. Coran studied on concentration effect of stearic acid, a palm oil fatty acid, on NR [2-3]. He found that the specific rate of vulcanization is decreased if the concentration of stearic acid is increased. Poh and Tang reported that the scorch time increases with the stearic acid content for all the rubbers investigated [4]. Barton and Hart reported that modulus values improve as the lauric acid content increases. The addition of lauric acid also improved the physical properties of the vulcanisates [5]. Ismail and Ruhaizat reported that the concentration of palm oil fatty acid affected not only the curing characteristics and but also the mechanical properties of the unfilled and calcium carbonate filled NR compounds [6-7]. The diamine salt of fatty acid, referred to as a multifunctional additive (MFA), can also be utilised as curative additive in filled NR compounds. The MFA improved the cure characteristics and mechanical properties of the silica-filled NR compounds [8] and white rice husk ash-filled NR compounds [9].

In this study, the effect of the addition of Alkanolamide (ALK) on the cure and tensile properties of the unfilled-NR compounds was investigated. The ALK is considered as a relatively new rubber additive, and it was prepared from Refined Bleached Deodorized Palm Stearin (RBDPS) and ethanolamine. Compared to others, this additive has unique nature of molecule which is structured by the combination of hydrophobic long chain hydrocarbon and hydrophilic amide as visualized in Figure 1.

2. Experimental Method

2.1. Materials

NR grade SMR L was used and obtained from Guthrie (M) Sdn. Bhd., Seremban, Malaysia. Other compounding ingredients such as sulphur, zinc oxide, stearic acid, N-isopropyl-N'-phenyl-p-phenylenediamine (IPPD) and Mercapto Benzothiazolyl disulfide (MBTS) were supplied by Bayer (M) Ltd., Sdn. Bhd., Petaling Jaya, Selangor, Malaysia. All materials were used as supplied. The ALK was synthesized in laboratory using Refined Bleached Deodorized Palm Stearin (RBDPS) and diethanolamine. Its molecular structure is shown in Figure 1.

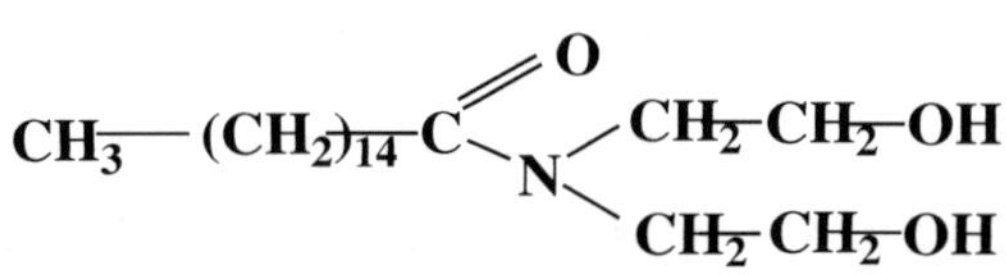

Figure 1. The molecular structure of alkanolamide

2.2. Compounding

A semi-efficient vulcanization system was used for the compounding. The recipe for the preparation of the NR compound is given in Table 1.

Table 1. The composition of the NR compound

Ingredients	Content (phr)*
NR (SMR L)	100.0
Zinc oxide	5.0
Stearic acid	2.0
IPPD	2.0
MBTS	1.5
Sulphur	1.5
ALK	0.0; 0.2; 0.4; 0.6; 0.8; 1.0

*parts per hundred parts of rubber

IOP Conf. Series: Materials Science and Engineering **223** (2017) 012012 doi:10.1088/1757-899X/223/1/012012

The compounding procedure was done in accordance with the American Society for Testing and Material (ASTM) – Designation D 3184 – 80. The compounding was done on a two-roll mill. Table 2 shows the designation and composition of the NR-based recipes used in this study.

Table 2. The designation and composition of the unfilled NR based recipes

Designation	Composition	Unfilled NR/ALK (phr)
A/0.0 (Control)	Unfilled NR	100/0.0
B/0.2	Unfilled NR/ALK	100/0.2
C/0.4	Unfilled NR/ALK	100/0.4
D/0.6	Unfilled NR/ALK	100/0.6
E/0.8	Unfilled NR/ALK	100/0.8
F/1.0	Unfilled NR/ALK	100/1.0

2.3. Cure Characteristics

The cure characteristics of the rubber compounds were obtained using a Monsanto Moving Die Rheometer (MDR 2000) which was employed to determine the scorch time (ts_2), cure time (t_{90}) and torque difference (M_H–M_L) value according to ISO 3417. Samples of the respective compounds were tested at 150°C. The rubber compounds were subsequently compression-moulded using a stainless steel mould at 150°C, with a pressure of 10 MPa and applying a laboratory hot-press based on respective curing times.

2.4. Tensile Properties

Dumbbell-shaped samples were cut from the moulded sheets according to ASTM D 412-93. Tensile tests were performed at a crosshead speed of 500 mm/min. Tensile tests were carried out with a universal tensile machine Instron 3366 to determine the tensile properties in terms of tensile strength (TS), stresses at 100% and 300% elongations (M100, M300) and elongation at break (EB).

2.5. Scanning Electron Microscopy (SEM)

The tensile fractured surfaces of the compounds were examined by using a Zeiss Supra-35VP scanning electron microscope (SEM). All the surfaces were examined after first sputter coating with gold to avoid electrostatic charging and poor image resolution.

3. Results and Discussion

3.1. Effect of Alkanolamide on Cure Characteristics and Crosslink Density

Effect of ALK on the cure characteristics of the unfilled NR compound is showed in Table 3. The scorch and cure times decreased with increased ALK loading. It was a cure enhancement which indicated ALK acted as a co-curing agent (secondary accelerator) in the curing. The amine content of the ALK caused the rubber compound more basic. Alkaline substances tend to improve the cure rate [10].

Table 3. The cure characteristics and tensile properties of unfilled NR compounds with and without ALK

Compounds	Cure characteristics			Tensile properties			
	ts_2, min.	t_{90}, min.	M_H−M_L (dN.m)	M100 (MPa)	M300 (MPa)	TS (MPa)	EB (%)
A/0.0	4.59	7.13	4.77	0.538	1.235	17.7	1070.0
B/0.2	4.52	7.05	4.85	0.544	1.236	19.3	1095.7
C/0.4	4.24	6.65	4.97	0.561	1.311	19.5	1116.5
D/0.6	3.96	6.45	5.21	0.586	1.364	20.1	1141.5
E/0.8	3.74	6.00	5.12	0.545	1.267	18.4	1154.0
F/1.0	3.73	5.83	4.89	0.516	1.231	16.5	1172.0

Note: M_L = Minimum torque; M_H = Maximum torque; M_H−M_L = Torque difference. A/0.0 = Unfilled NR without ALK B/0.2; C/0.4; D/0.6; E/0.8; F/1.0 = Unfilled NR with ALK.

The addition of up to 0.6 phr of ALK increased the torque difference (M_H−M_L). Further increases in the ALK loading decreased the torque difference. Torque difference represents shear dynamic modulus which indirectly related to the crosslink density of a rubber compound [11-14]. The addition of ALK affected the torque difference of the rubber compounds or the crosslink density of the rubber vulcanisate. It was due to the action of the ALK as a secondary accelerator which enhanced the state of curing process.

The reduction of crosslink density beyond 0.6 phr loading was due to the softening or lubricating effect of the excessive ALK which dissolved a part of the elemental sulphur and hence reduced crosslink density. This explanation was supported with the morphology study of the tensile fractured surface of the E/0.8 with excessive ALK. In Figure 2(D), it seemed smooth with a less matrix tearing and surface roughness. It indicated the more pronounced lubricating effect of the excessive ALK (15].

3.2. Effect of Alkanolamide on Tensile Properties

Effect of ALK loading on the tensile properties of the unfilled NR vulcanisates is shown in Table 3. The addition of 0.2 phr ALK produced vulcanisate B/0.2 with higher EB. Increasing the ALK loading caused further increases in the percentage of EB or the extensibility of the rubber vulcanisate. This was due to the function of ALK as an internal plasticiser which modified flexibility or distensibility of rubber vulcanisate.

The addition of ALK increased the M100 and M300 up to maximum level at 0.6 phr and decreased with further increases in the ALK loading. The results of tensile strength also exhibit similar trend. Tensile modulus and tensile strength are dependent only on degree of crosslink [8, 13]. The improvement of those properties was attributed to higher degree of crosslink (as indicated by a higher value of torque difference). The deterioration of those properties was due the excessive amount of ALK which dissolved other curatives and hence reduced degree of crosslink density (as indicated by a lower value of torque difference). As discussed earlier, crosslink density of a rubber compound is indicated by torque difference value. The greater the torque difference, the higher is the degree or total crosslink density [11, 13].

3.3. Scanning Electron Microscopy (SEM)

Figure 2 shows the SEM micrographs of the tensile fractured surfaces of the unfilled NR compounds with and without ALK. Compared to A/0.0 (Figure 2A), the fractured surfaces of C/0.4 and D/0.6 (Figure 2B and 2C, respectively) show many larger discontinuous slip tear lines with path deviation indicating a higher energy requirement to cause failure. As a consequence, it led to increased resistance to crack propagation and thus, caused an increase in crosslink density. Otherwise the fractured surface of E/0.8 (Figure 2D) exhibits the least matrix tearing and surface roughness which presumably correlated lower crosslink density due to the existence of the excessive ALK in the E/0.8 compound. The micrographs of the tensile fractured surfaces were in line with results obtained by Ismail et al. and Nabil *et al.,* (2013)[15]. They reported that an increase in crosslink density was responsible for the roughness and matrix tearing line of the fractured surface.

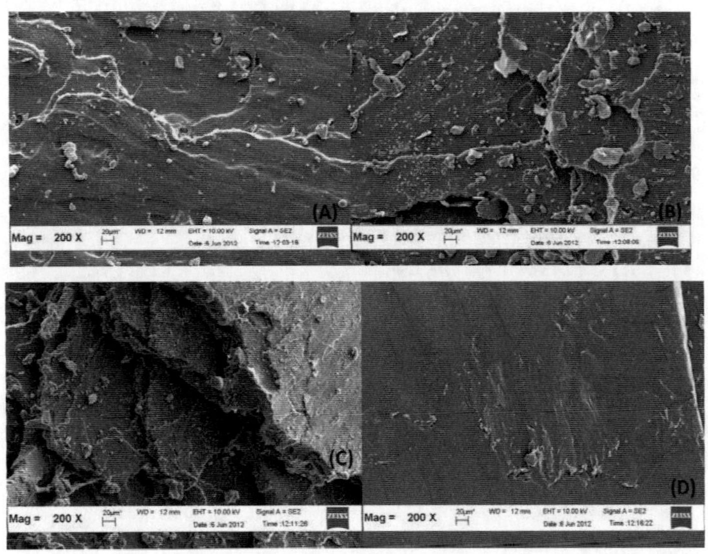

Figure 2. SEM micrographs of the failure fracture surface of unfilled NR compound at a magnification of 200 X, (A) A/0.0, (B) C/0.4, (C) D/0.6, and (D) E/0.8

It was demonstrated that ALK can be utilised as a new curative additive in unfilled NR compounds. ALK can be function as secondary accelerator which enhanced the cure rate. ALK also enhanced the torque difference, crosslink density, tensile modulus and tensile strength especially up to 0.6 phr. ALK also can be function as internal plasticiser which enhanced the EB

4. Conclusions

Alkanolamide can function as a curative additive in unfilled natural rubber compounds. Results indicate that alkanolamide can be utilized as a secondary accelerator and a plasticiser in the unfilled natural rubber compounds. The incorporation of alkanolamide enhanced the cure rate, reduced the scorch time and improved the flexibility by increasing the elongation at break. The incorporation of alkanolamide also enhanced tensile properties and crosslink density of the unfilled natural rubber vulcanisate especially up to 0.6 phr loading.

Acknowledgements

The authors would like to thank Universiti Sains Malaysia, for providing the research facilities for carrying out the experiment and for making this research work possible.

References

[1] Dick JS 2001 *Rubber Technology Compounding and testing for performance* Hamser Munich.
[2] Coran A 1964 *Rubber Chem Technol* **37** 3 689-697.
[3] Coran A 1965 *Rubber Chem Technol* **38** 1 1-14.
[4] Poh BT and Tang W 1995 *J Appl Polym Sci* **55** 3 537-542.
[5] Barton BC and Hart EJ 1953 *Rubber Chem Technol* **26** 3 510-521.
[6] Ismail H and Ruhaizat T 1998 *Polym-Plast Technol* **37** 4 483-494.
[7] Ismail H and Ruhaizat TA 1997 *Iran Polym J* **6** 2 97-104.
[8] Ismail H and Chia H 1998 *Polym Test* **17** 3 199-210.
[9] Ismail H *et al* 1999 *Eur Polym J* **35** 8 1429-1437.
[10] Surya I *et al* 2014 *Polym Test* **40** 24-32.
[11] Teh PL *et al* 2004 *Eur Polym J* **40** 11 2513-2521.
[12] Ismail H and Shaari SM 2010 *Polym Test* **29** 872-878.
[13] Surya I *et al* 2015 *Polym Test* **42** 208-214.
[14] Surya I and Ismail H 2016 *Polym Test* **50** 276-282.
[15] Surya I, Ismail H and ARashid A 2013 *Polym Test* **32** 1313-1321.

IOP Conf. Series: Materials Science and Engineering **223** (2017) 012013 doi:10.1088/1757-899X/223/1/012013

The influence of natural rubber – butadiene rubber and carbon black type on the mechanical properties of tread compound

L A Wisojodharmo, R Fidyaningsih, D A Fitriani, D K Arti, Indriasari, H Susanto
Center for Material Technology, Agency for the Assessment and Application of Technology, Building 224, Puspiptek Area, Serpong, Tangerang Selatan, Indonesia

E-mail: lieswisojo@hotmail.com

Abstract. The objective of this study is to optimize the performance of the tread compound, by mixing of specified ingredients, such as Natural Rubber (NR) and Butadiene Rubber (BR) at various compositions, and also the type of carbon black as filler. The composition of NR and BR are used as follows 100:0 ; 95:5 and 85:15. Other materials used as rubber additives are peptizer, ZnO, stearic acid, additives, processing oil, and curatives. Two types of carbon black namely N220 and N234 are used in this research. The un-vulcanized rubber from each formula was characterized of curing characteristics and Mooney viscosity. The vulcanized rubber was then characterized its mechanical and thermal properties. It is clear from the research that, the addition of BR into the NR compound increases the abrasion resistance, rebound resilience and hardness, yet the glass transition temperature of the rubber compounds decreases. On the other hand, it seems that the rubber compound reinforced N220 or N234 gives comparable results on both mechanical and thermal properties.

1. Introduction

Regarding its vast improvement in the last decades, tire industry has always won the first place considering the volume of the studies done on the properties of rubber. Many researchers, from all over the world, are trying to present new products with higher capabilities and efficiency.

From among these studies, the most has been dedicated to the tread of the tire for this is the part of the tire which has the greatest impact on the way and the type of use. The performance of tire tread was measured by combination of "magic triangle properties". They are rolling resistance, traction and wear. The balance of those properties increases the quality of tire tread [1].

Shiva and Atashi (2010) [2] have studied the optimization of failure properties of the Passenger Tire Tread Compound made of a compound of SBR with NR and BR, by altering curing condition and modeling rubber's behavior. Pham*et al.* studied the blends of two types of SBR with NR and the inherent properties of rubbers on the properties of the final compound [3]. The results showed an increase in Mooney viscosity and scorch time as the SBR increased in the compound. Kaushik *et al,* (2010) [4] studied the compound of NR/BR/HSR with the presence of different amounts of carbon black and found that the samples gained a very good abrasion resistance and in addition, as the size of carbon black particles decreased and became close to nano particle size, the strength of the samples increased.

IOP Conf. Series: Materials Science and Engineering **223** (2017) 012013 doi:10.1088/1757-899X/223/1/012013

Since tire is a composite structure made of substances with various properties, choosing each of these substances and the amount of each of them requires a rich knowledge. One of the substances which are added to the rubber compounds are reinforced fillers and have an important effects on the properties of the final product. Carbon black is one of the reinforcements that frequently used in tire industry and many researchers have studied its effect on mechanical and dynamic properties of tires. The size of carbon black particles and its specific surface are the most important factors affecting on tires' eventual properties [5].

Besides of reinforcing filler, rubber part has main role to the performance of tire tread. Cured BR imparts excellent abrasion resistance (good tread wear), and low rolling resistance (good fuel economy) due to its low glass transition temperature (Tg). In this study, optimization of BR composition in NR compounding formulation was studied to increase the abrasion resistance and tire retread performance. Instead of that, type of carbon black was also observed to proof the hypothesis that nano particle size carbon black would increase the strength of rubber compounding. Detail research on properties of compounding were performed without silica.

2. Experimental

All mixing ingredients were used as received. Natural rubber (NR) as Ribbed Smoked Sheet (RSS1) type with the density of 0.95 was provided by local supplier in Indonesia. Butadiene rubber (BR) with the Mooney viscosity 40 and the density of 0.92 was obtained from Goodyear Chemical. Carbon black N 220 and N234 were purchased from Cabot.

2.1 Preparation and Testing of Rubber Compound

A laboratory-sized kneader Moriyama DS3-10MWB-E was employed to prepare rubber compounds. The mixing was divided into two stages. In the first stage, the mixing condition was set at the temperature of 100°C and the velocity of 60 rpm. The mixing was started by mixing NR or NR-BR at various compositions along with plasticizers for about 5 min, and followed by adding zinc oxide, stearic acid, carbon black, other additives and a half of oil for about 6 min. The remaining oil was then added into the kneader for about 2 min. The compound was passed on the two roll mill at 70°C for 6 times. After the master batch was stored for a night to allow the rubber compound to rest, the batch then was mixed with curatives in the kneader. The temperature was set at 70 °C and the velocity of 60 rpm for about 2 min. At the end of the mixing, the batch again was passed on the two-roll mill, cooled and stored for another 24 h before it can be used for further analysis.

IOP Conf. Series: Materials Science and Engineering **223** (2017) 012013 doi:10.1088/1757-899X/223/1/012013

Six formulations were designed according to Table 1. In all formulations the total of rubber (NR and BR) was kept at 100 phr. Cure characteristics were determined using an MDR 3000 MonTech. Delta torque is obtained by subtracting of maximum torque (MH) and minimum torque (ML). Scorch time (ts_2) is the time to reach 2 units above minimum torque (ML) and optimum cure time (t_{90}) is the time to achieve 90% of delta torque above minimum. Sheets and test specimens were vulcanized by a compression molding Dr. Collin. The experimental curing press was set at the temperature of 140°C and the time of curing was based to t_{90} as shown in Table 2, which were obtained from curing testing.
Mooney viscosity was measured using a MV 2000 MonTech at 100°C, one min of preheating and four minutes of rotation ML (1+4)100 °C.

Table 1. Formulation design in phr

Component	FR1	F1	F2	FR2	F3	F4
NR	100	95	85	100	95	85
BR	-	5	15	-	5	15
Peptizer	3	3	3	3	3	3
ZnO	5	5	5	5	5	5
Stearic acid	3	3	3	3	3	3
N220	50	50	50	-	-	-
N234	-	-	-	50	50	50
Additives	8	8	8	8	8	8
Processing oil	6	6	6	6	6	6
Curatives	3	3	3	3	3	3

Table 2. The curing properties of rubber compounds

Compound name	ML (dNm)	MH (dNm)	Delta Torque (MH – ML)	ts_2 (min)	t_{90} (min)
FR1	2.65	15.66	13.01	3.23	9.47
F1	2.43	14.55	12.12	3.62	8.84
F2	2.70	15.25	12.55	3.58	8.65
FR2	2.48	14.38	11.91	3.87	9.10
F3	2.44	13.56	11.12	3.08	7.88
F4	2.47	14.99	12.52	3.81	8.74

2.2 Testing of Rubber Vulcanizates

Tensile strength was determined according to ASTM D412 by using Go Tech 7049 Universal Testing Machine 2kN, Taiwan. Abrasion resistance was performed with DIN abrasion tester machine AT150 as per DIN ISO 4649. Rebound resilience and hardness according to ASTM D2632 and ASTM D2240, respectively. The glass transition temperature (Tg) was measured using a Differential Scanning Calorimetry (DSC) Setaram, France according to ASTM D 3418-15.

IOP Conf. Series: Materials Science and Engineering **223** (2017) 012013 doi:10.1088/1757-899X/223/1/012013

3. Results and Discussion

The processing characteristics and physical mechanical properties of rubber compounds have been presented in graph or table below.

3.1 Unvulcanized Rubber Curing Characteristic and Mooney Viscocity

It can be seen from Figure 1 and Table 2 that the addition of BR into the NR compound does not have a significant effect on altering curing time, scorch time or delta torque. There is no significant difference of curing characteristics between rubber compound containing N220 and N234.

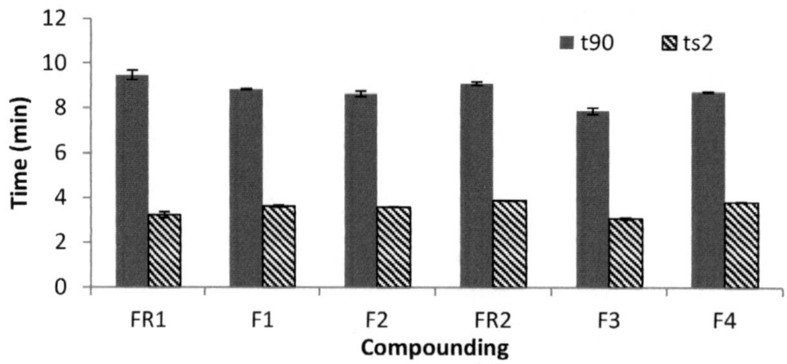

Figure 1. The optimum cure time and the scorch time of rubber compounds

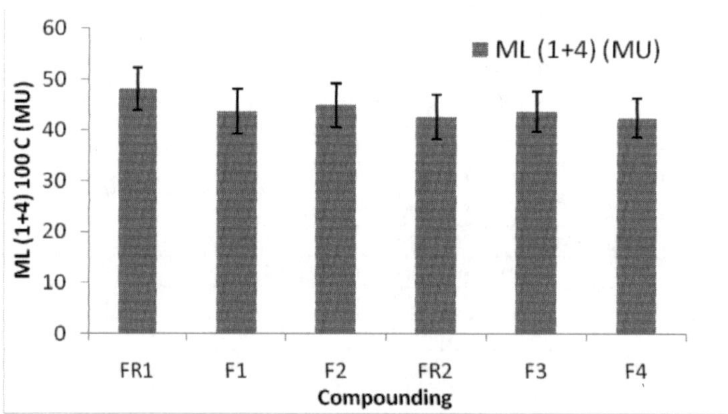

Figure 2. Mooney viscosity of rubber compounds

From Figure 2, the addition of BR into the rubber compound slightly lowering the Mooney viscosity (MV). It is because the MV of BR is lower than NR due to low molecular weight of BR compared to NR [6]. There is no significant difference of Mooney Viscosity between rubber compound reinforced N220 and rubber compound reinforced N234 since both types of carbon black has similar particle size and gave the same effect to the viscosity of compounding.

3.2 Characterization of Vulcanized Rubber

Hardness

Hardness may be defined as a material's resistance to permanent indentation. The Shore A Hardness Scale measures the hardness of flexible mold rubbers that range in hardness from very soft and flexible, to medium and somewhat flexible, to hard with almost no flexibility at all. It is expected that the addition of BR will not have a significant effect on altering the hardness of rubber compound. Yet in this research, the addition of BR decreased the hardness of rubber compound as presented in Figure 3. Hardness of rubber compound containing N220 is slightly higher than rubber compound containing N234 since lower particle size of carbon black could lead better dispersion level and increase the hardness.

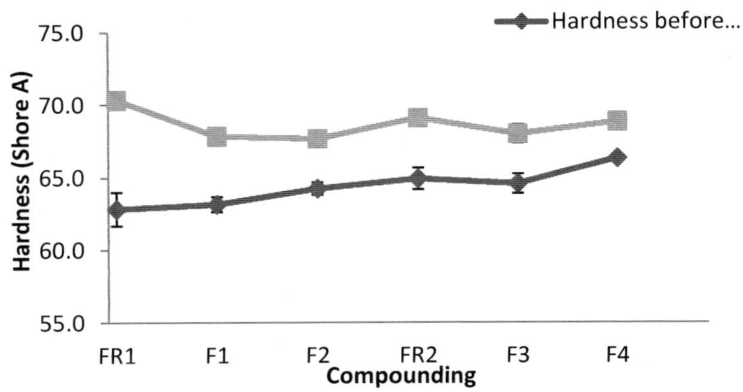

Figure 3. Hardness of rubber compounds

Tensile Strength

Tensile strength is the maximum tensile stress reached in stretching a test piece (either an O-ring or dumbbell). Tensile tests are used for controlling product quality and for determining the effect of chemical or thermal exposure or an elastomer. In the latter case, it is the retention of these physical properties, rather than the absolute values of the tensile stress, elongation or modulus, that is significant. Figure 4 shows the result of te4nsile strength measurement. The strength of rubber compound decreases significantly with the addition of 15 phr BR into the rubber compound. Compared to FR1 dan F1, F2 has lower tensile strength. So does F4 that showed significant low tensile strength after addition of BR. It is because unlike NR, BR does not have crystallization properties; therefore, the tensile strength of BR is very poor [7]. A not significant increasing point was observed in addition of N220 filler that gave higher tensile strength than N234.

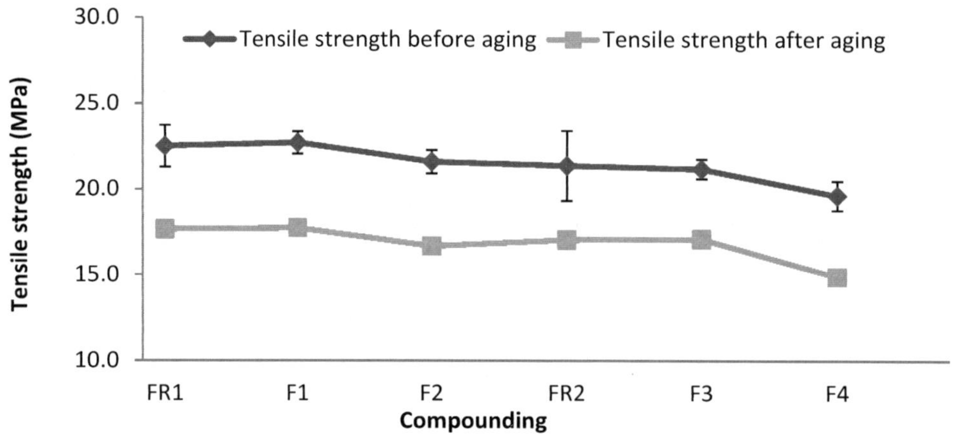

Figure 4. Tensile strength of rubber compounds

Abrasion Resistance

Abrasion resistance is the ability of a material to resist mechanical action such as rubbing, scraping, or erosion that tends progressively to remove material from its surface. When a product has abrasion resistance, it will resist erosion caused by scraping, rubbing, and other types of mechanical wear. This allows the material to retain its integrity and hold its form. This can be important when the form of a material is critical to its function, as seen when moving parts are carefully machined for maximum efficiency [8]. Abrasion resistant materials can be used for both moving and fixed parts in settings where wearing could become an issue.

The properties of a vulcanized rubber can be significantly influenced by details of the compounding. Practical materials will have, in addition to the base polymer, fillers, antioxidants, crosslinking agents, accelerators etc. All of these can have an influence on the physical and chemical stability of the finished material. For example, rubber abrasion resistance can be related quantitatively to the frictional force and the crack growth characteristic of the rubber [9].

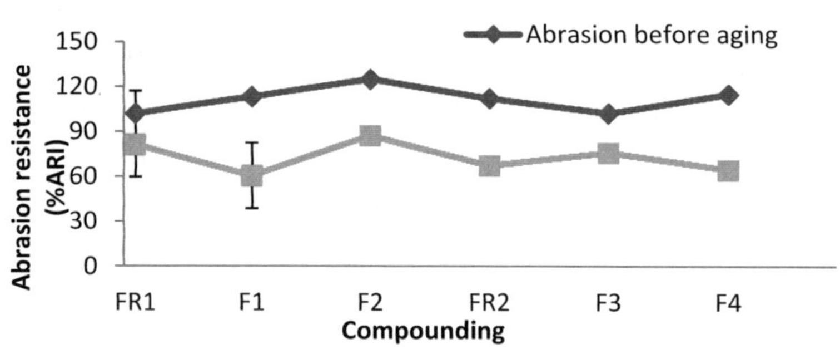

Figure 5. Abrasion resistance of rubber compounds

Abrasion behavior was presented by ARI% as shown in Figure 5. Along with the decreasing of hardness, BR substitution leads to higher abrasion resistance. Both rubber compounds containing N220 and N134 give a comparable value of abrasion resistance.

Rebound Resilience

Resilience of a rubber compound is a measurement of how elastic it is when exposed to various stresses. Resilience is the ratio of energy released in deformation recovery to the energy that caused the deformation. Measurement of rubber resilience can assist in choosing the right material for a given application, for example tread tyre.

The increasing of BR addition leads to resilience percentage improvement. It was in agreement with the decreasing of hardness and increasing of abrasion resistance [10]. Addition of N220 and N234 did not give any significant difference to the result of resilience testing.

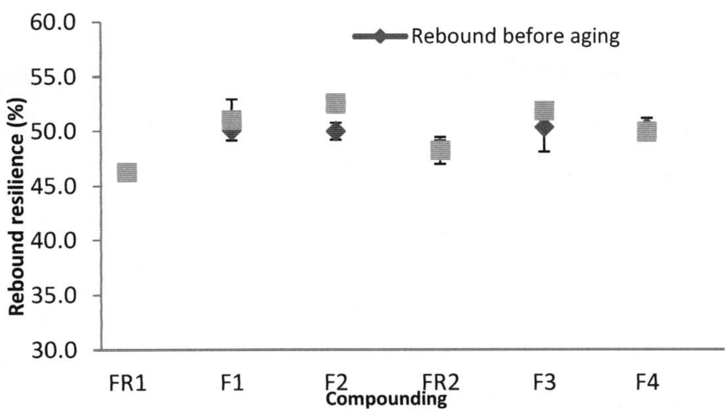

Figure 6. Rebound resilience of rubber compounds

The Glass Transition Temperature

Glass Transition is a method to characterize a property of a polymeric material. The glass transition is the temperature where the polymer goes from a hard, glass like state to a rubber like state [11]. Due to the result of mechanical properties above, the measurement of Tg were performed to F1, F2, F3, and F4 samples, which is consist of blended NR and BR. The result is shown in Table 3.

Table 3. Result of glass transition temperature (Tg)

Compounding	Tg (^{0}C)
F1	-58.89
F2	-59.44
F3	-58.95
F4	-59.53

IOP Conf. Series: Materials Science and Engineering **223** (2017) 012013 doi:10.1088/1757-899X/223/1/012013

Cured BR imparts excellent abrasion resistance (good tread wear), and low rolling resistance (good fuel economy) due to its low glass transition temperature (Tg). The low Tg, typically <–90C, is a result of the low "vinyl" content of BR. However, low Tg also leads to poor wet traction properties, so BR is usually blended with other elastomers like NR for tread compounds. It is expected that the addition of BR will have a significant effect on lowering the Tg of rubber compound [12]. Yet in this research, the addition of BR up to 15 phr only lowers the Tg by 1 degree compared to rubber compound containing 5 phr of BR. There is no significant difference on the Tg between rubber compound containing N220 and rubber compound containing N234 [13].

4. Conclusion

A blend of natural rubber (NR)/*cis*-butadiene rubber (BR) in the presence of carbon black filler showed excellent mechanical properties of a tread formulation. The mechanical properties such as abrasion resistance and rebound resilience have been improved with the presence of *cis*-butadiene rubber. Here the presence of carbon black N220 could give slightly better mechanical properties compared to carbon black N234 due to the better dispersion of N220 into the rubber compound. Due to good properties of these blending, it is possible to apply this formulation in tread tyre. Detail testing in order to fulfill the requirement of tread compound should be performed in advanced.

References

[1] Ghoreishy M H R *et al* 2010 *Iran J. Polym. Sci. Technol.* **23**(3) 247-254.
[2] Shiva M and Atashi H 2010 *Iran J. Polym. Sci. Technol.* **23**(3) 187-201.
[3] Pham T H *et al* 2001 *Polym Testing* **20** 539–544.
[4] Kaushik P *et al* 2010 *Mater and Design* **31** 1156-1164.
[5] Wang M J *et al* 1991 *Rub. Chem. Technol.* **64** 559-576.
[6] Wolf S and Wang M.J 1992 *Rub. Chem. Technol.* **65** 329-342.
[7] Bice J A E *et al* *Rub World.* **217** 58-67.
[8] Choi1S *et al* 2004 *Polym.. Adv. Technol.* **15** 122–127.
[9] Gatti L 2001 *Tire Technol. Int.* **1** 39-45.
[10] Rattanasom N *et al* 2007 *Polym. Test.* **26** 369-377.
[11] Brinke J W T *et al* 2003 *Compos. Sci. Technol.* **63** 1165-1174.
[12] Choi S-S 2002 *Polym. Test.* **21** 201- 208.
[13] Rattanasom N and Chaikumpollert O 2003 *J. Appl. Polym. Sci.* **90** 1793-1796.

IOP Conf. Series: Materials Science and Engineering **223** (2017) 012014 doi:10.1088/1757-899X/223/1/012014

Effect of maleated natural rubber on tensile strength and compatibility of natural rubber/coconut coir composite

O Ujianto[1], R Noviyanti[2], R Wijaya[1], B Ramadhoni[1]
[1]Center for Polymer Technology, Agency for Assessment and Application of Technology, Banten, Indonesia
[2]Chemistry Department, Islamic State University, Banten, Indonesia

E-mail: onny.ujianto@bppt.go.id

Abstract. Natural rubber (NR)/coconut coir (CF) composites were fabricated using co-rotating twin screw extruder with maleated NR (MNR) used as compatibilizer. The MNR was produced at three level of maleic anhydride (MA), and analyzed qualitative and quantitatively using FTIR and titration technique. Analysis on MNR using FTIR and titration methods showed that MA was grafted on NR chain at different percentage (0.76, 2.23, 4.79%) depended on MA concentration. Tensile strength data showed the best tensile strength was produced at 7 phr of MNR with 1 phr of MA level in MNR resulting 16.4 MPa. The improvement of compatibilized samples were more than 300% compare to uncompatibilized composite attributed to better interfacial bonding. The improvement on tensile strength was significantly influenced by MNR level and amount of MA added to produce MNR, as well as their interaction. The optimum conditions for producing NR-CF composite were predicted at 6.5 phr of MNR level with 1 phr of MA concentration added in MNR production, regardless screw rotation settings. Results from verification experiments confirm that developed model was capable of describing phenomena during composite preparation. Morphology analysis using scanning electron microscopy shows smooth covered fiber in compatibilized samples than that of without MNR. The morphology also showed less voids on compatibilized samples attributed to better interfacial bonding leading to tensile strength improvement.

1. Introduction

Natural fibers (NF) have been studied as reinforcement in composites due to its availability, environmental benign, and renewable. One of NF that widely applied in many applications is coconut coir [1-5]. It could be applied in range of application such as building, automotive, and packing materials [3] using wide range matrix variation ranged from thermoplastics [1, 2, 6, 7], thermosets [8-10], or rubbers [11-13]. It is also low density with high strength and elongation at break [14, 15]. This is beneficial to produce lightweight product with high performance. However, the coconut coir should be treated properly to improve mechanical [1, 8, 16, 17], thermal [17], and flammability [1] properties of composite.

In composite preparation, natural rubber (NR) is a material commonly used as matrix [11, 13, 18-20]. The combination between NR and natural fiber as coconut coir was good combination to produce strong but elastic material. The NR composites could be fabricated using solid processing techniques as roll mill or mixer [21, 22]. Despite its popularity, NR composite fabrication using roll mill had low dispersion of fiber, while batch processing was time consuming. On the other hand, extruder might produce better filler dispersion and improve mechanical properties [23, 24]. It also had higher production rate than mixer due to continuous process. However, there was limited study on the use of extruder for preparing NR composites. Therefore, it is of interest to better understand the use of extruder on NR composite preparation.

To improve interfacial bonding and mechanical properties, compatibilizer was commonly added. One of compatibilizer popularly produced was maleic anhydride (MA) grafted on polymer backbone. It was evidenced that the addition of MA based compatibilizer would increase mechanical properties. Maleated natural rubber produced in internal mixing improved tensile strength about more than 10% for natural rubber composite [25-28]. Higher improvement was obtained by grafting polymerization in torque rheometer [29], combining liquid natural rubber (LNR) and Polyethylene grafting MA to improve HDPE/NR/aramid composite [22], and modifying maleated natural rubber (MNR) with zinc acetate using solution technique [30]. The disadvantages of available techniques are time consuming and not environmental friendly due to solvent consumption. Therefore, our previous work studied the application of extruder to produce MNR [31]. However, there is limited information about amounts of compatibilizer and MA content, as well as their interactions in NR composite preparation using extruder. Moreover, as higher MA content produces higher gel content on compatibilizer thus decreases composite mechanical properties [32], it is beneficial to study the amount of MA in compatibilizer to produce the highest mechanical properties of NR composites.

This research aims to explore the composite of coconut coir and natural rubber with maleated NR (MNR) as the compatibilizer. The MNR was fabricated using twin screw extruder and characterized qualitative and quantitatively using FTIR and titration method. The composite samples were prepared according to three levels and three factors Box-Behnken experimental design. The effect of various level of MNR, amount of MA added in MNR preparation, and screw rotation on tensile strength were analyzed and optimized. The effects of MNR on sample morphology were observed using Scanning Electron Microscopy.

2. Materials and Methods

2.1. Materials

Natural rubber (NR), *Standard Indonesia Rubber* (SIR) 20, produced by Indonesian local company, Aneka Bumi Pratama, Palembang was used as matrix. Compatibilizer used in this research are maleated natural rubber (MNR) produced in our laboratory. Coconut fiber (CF) was obtained from local producer in Indonesia. Commercial grade additives were purchased from local supplier in South Tangerang, Indonesia. Butylated hydroxyl toluene (BHT) as antioxidant, zinc oxide (ZnO) as activator, N-cyclohexyl-2-benzothiazole sulphonamide (CBS) as accelerator, and sulphur (S) were used as vulcanization agent.

2.2. MNR Preparations

MNR was produced by grafting MA molecules on NR using shearing action in a twin screw extruder Collins ZK 25. Various MA concentrations namely 1, 8 and 15 phr were used to produce three different percentage of grafted MA. Chemical initiators was not used to minimize gel content and difficulties in characterization and processing [33]. Process parameters used in extruder are summarized in Table 1.

Table 1. Processing Conditions Applied in Twin Screw Extruder

T1	T2	T3	T4	T5	T6	T7	T8	Screw Rotation
40	160	165	170	175	180	185	190	40

2.3. Qualitative and Quantitative Analysis of MNR

Grafted MA in MNR was analyzed qualitative and quantitatively according to methods reported in our previous study [31]. The conversion of carboxylic acid concentration to grafted MA content was done as follows [34]:

$$MA\,(wt\%) = \frac{(V_0 - V_1)\,x\,N\,x\,Mr}{2W}\,x\,100\% \qquad (1)$$

where N is the concentration of NaOH (mol/L) in methanol. V_0 and V_1 are the volumes of NaOH used in blank test and the test with sample, respectively. Mr is the MA molecular weight (98 g/mol). W (g) is the weight of MNR sample.

Qualitative analysis of functional groups of MNR was conducted using FTIR Bruker Tensor 27 spectroscopy. The MNR samples were pressed and melted at 180 °C, 10 ton pressure, for 10 min. The FTIR spectra were recorded using ATR in the range of 4000 – 500 cm^{-1}.

2.4. Composite Preparations

The coconut fiber was immersed using NaOH at 5% concentration for 1 h before composite preparations. It was then washed using distillated water and dried at 100 °C for 1 h. The treated fiber was then cut and sieved with maximum 1 cm in length.

Mixing of NR and CF was done using a co-rotating twin screw extruder Collins ZK 25. The samples were prepared according to Box-Behnken experimental design (DoE) 15 combinations. Dependent variables used in this work were MNR level, MA concentration in MNR preparation, and screw rotation. 2 phr of BHT was added as antioxidant to avoid rubber degradation during processing. Variables used in this research are shown in Table 2.

Table 2. Experimental Conditions

No	Sample	MNR level (phr)	MA in MNR Preparation (phr)	Screw Rotation (rpm)	Fiber Content (phr)	BHT level (phr)
1	NR	-	-	70	0	2
2	NR-CF w/o MNR	-	-	70	30	2
3	NR-CF1	4	8	70	30	2
4	NR-CF2	1	8	100	30	2
5	NR-CF3	7	8	100	30	2
6	NR-CF4	1	1	70	30	2
7	NR-CF5	7	8	40	30	2
8	NR-CF6	1	8	40	30	2
9	NR-CF7	7	1	70	30	2
10	NR-CF8	4	8	70	30	2
11	NR-CF9	4	1	40	30	2
12	NR-CF10	7	15	70	30	2
13	NR-CF11	4	15	40	30	2
14	NR-CF12	4	15	100	30	2
15	NR-CF13	1	15	70	30	2
16	NR-CF14	4	8	70	30	2
17	NR-CF15	4	1	100	30	2

The extruded material was mixed with another additives using *Huicai* roll mill. The material was first rolled at 40 °C and 60 rpm for 10 min. After rolling, 5 phr of ZnO was added and rolled for 10 min. It was followed by mixing 2 phr of CBS to the material for 10 min. Vulcanization was done by mixing the rubber compound to 5 phr of sulphur for 30 min.

Composite samples were prepared from the rolled material using hot press Collins P300 P. The rolled compound was preheated at 150 °C without pressure for 5 min for softening the rubber. It was then pressed at 150 °C, 15 Bar for 10 min, and cooled at 25 °C for 25 min. All samples were conditioned for minimum 24 h before testing.

2.5. *Composite Characterizations*

Tensile tests were carried out to all samples and referred to ASTM D 412 type C. Tensile testing was carried out using universal testing machine Shimadzu AGS10 kNG at 500 mm/min of crosshead rate. Sample morphology was observed for untested selected samples. Analysis was performed using scanning electron microscope (SEM) model JEOL JSM-6510LA with 20 kV of accelerating voltage. Sample surfaces were platinum coated before analysis to minimize charging.

2.6. Modelling on Tensile Strength

Analysis on the effect of each variable on tensile strength was done according to statistical model using Minitab software. The model was developed from collected tensile strength data and regressed using polynomial equation as follows:

$$y = \beta_0 + \sum \beta_i x_i + \sum \beta_{ii} x_i^2 + \sum \beta_{ij} x_i x_j \quad (2)$$

Where y is the model response, x_i and x_j are independent variables, and βo, βi, βii, and βij are the coefficients for the constant, linear, quadratic terms, and interactions, respectively. The model was developed by only considering significance variables (p-value < 0.05) (Table 3). The less p-value suggests that the effect of term is more significant to influence tensile strength.

3. Results and Discussions

3.1. Characteristics of MNR

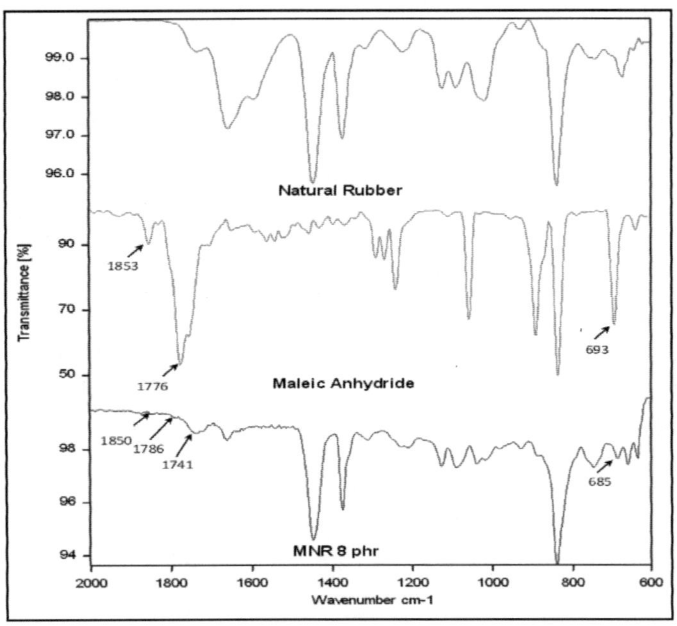

Figure 1. FTIR Spectra for MNR sample

Figure. 1 shows FTIR spectra for MNR with 8 phr of MA content. The MNR was selected because it was at middle position between the highest and lowest level of MA content. The spectra was focused in the range of 600 to 2000 cm^{-1} as reported in previous studies that the typical peak of MNR were observed in these ranges [28, 29, 34-36]. The absence of peak at 693 cm^{-1} suggested that MNR was successfully purified. The present of specific peak at 1785 cm^{-1} and 1850 cm^{-1} reflects symmetric (strong) and asymmetric (weak) C=O stretching vibration of succinic anhydride [28, 34] that suggests the presence of grafted MA onto NR.

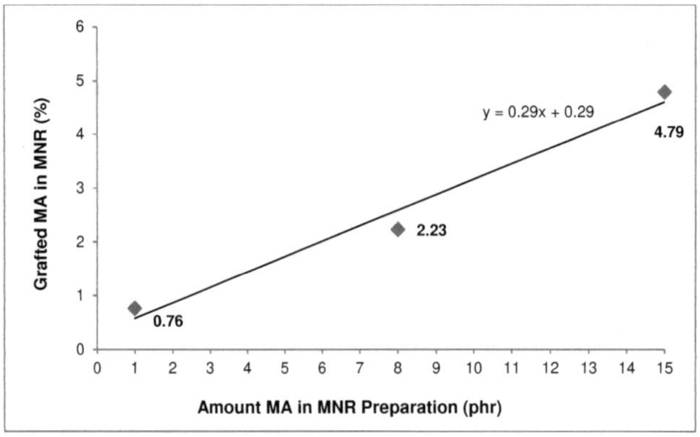

Figure 2. Percent Grafted MA in MNR at Different MA Concentration

The percentage of grafted MA on NR for each compatibilizer was shown in Figure 2. In general, grafted MA increased linearly as higher MA amount. The trend was similar to that of reported by another researchers [28]. This might be attributed to the absent of initiator during the MNR preparation that avoid to produce crosslink reaction among NR molecules. As a result, the interaction possibility of MA and NR chain through free radical reaction was higher. The lower percentage of grafted MA in this research than others [28, 37] might be attributed to different equipment used to produce MNR. The extruder used in this study was not able to facilitate longer mixing time, as it is influenced by screw rotation. On the other hand, mixing time can be controlled in mixer, so it facilitates longer reaction time.

3.2. Modelling on Tensile Strength

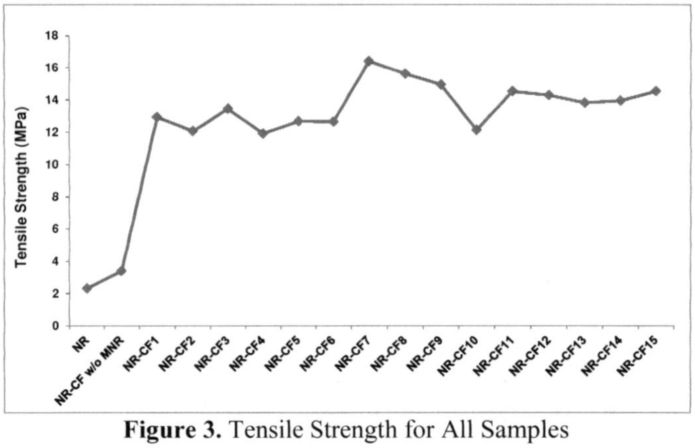

Figure 3. Tensile Strength for All Samples

IOP Conf. Series: Materials Science and Engineering **223** (2017) 012014 doi:10.1088/1757-899X/223/1/012014

. **Table 3.** P-value of Developed Model for Tensile Strength

Terms	P-Value	Note
Regression	0.006	Significant
Linear	0.003	Significant
MNR level	0.001	Significant
MA in MNR Preparation	0.021	Significant
Square	0.013	Significant
MNR level * MNR level	0.013	Significant
Interaction	0.003	Significant
MNR level * MA in MNR Preparation	0.003	Significant

The developed model to predict tensile strength was presented as follows:

$$\textit{Tensile strength} = 9.67 + 1.88 * \textit{MNR level} - 0.24 * \textit{MA in MNR} - 0.14 * \textit{MNR level} * \textit{MNR level} - 0.07 * \textit{MNR level} * \textit{MA in MNR} \quad (3)$$

The present of quadratic term for MNR level in the model suggested that the optimum value may be predicted in boundary ranges. On the other hand, the absence of quadratic term for MA concentration in MNR suggested that the effect of this factor was linier.

The absence of screw rotation in the model suggested that this factor was not significant to influence tensile strength. The insignificant effect of screw rotation might be due to uncontrollable mixing time in extruder. Low rpm produced longer residence time and facilitated longer reaction time. However, the low rpm would not produce shear as high as the high rpm. As a result, in low rpm, the reactive chain may be less. On the other hand, the high rpm would generate higher shear and produce more reactive chain, but it was not able to provide optimum reaction time.

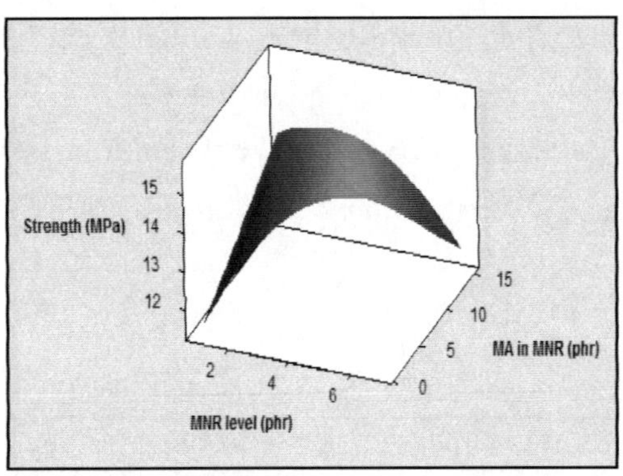

Figure 4. Surface Plot Effect of MNR Level and MA in MNR

IOP Conf. Series: Materials Science and Engineering **223** (2017) 012014 doi:10.1088/1757-899X/223/1/012014

Figure 4 describes effect of MNR level and MA in MNR interaction. At low MNR level (1 phr), the strength increased as higher MA concentration in MNR attributed to higher percentage of grafted MA (0.76 to 4.79%). On the other hand, the property decreased as higher MA in MNR (0.76 to 4.79%) at high MNR level (7 phr). This might be attributed to the migration of MNR to fiber, causing self-entanglement and resulting slippage [22]. These suggested that optimum percentage of grafted MA in final composite formulation should be considered as a crucial factor rather than single factor (MNR level or MA in MNR only).

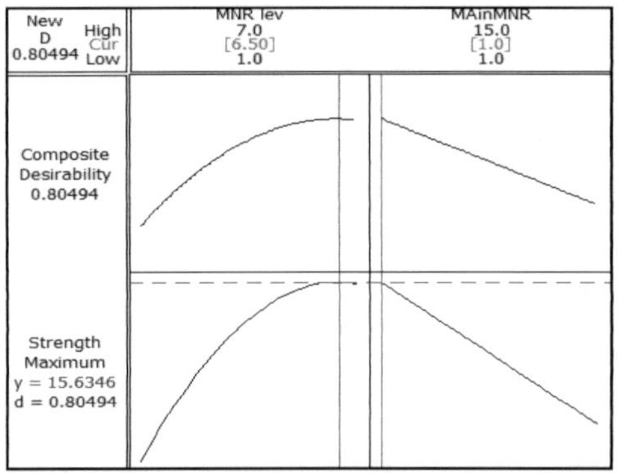

Figure 5. Response Optimizer Plot for Tensile Strength

The optimum conditions to produce tensile strength were predicted using response optimizer plot, shown in Figure 5. Screw rotation was not shown and optimized due to insignificant effect on strength as explained above. The plot suggests that the best tensile strength in applied boundaries are 6.5 phr of MNR level and 1 phr of MA concentration in MNR, at screw rotation 40 to 100 rpm, giving value of 15.6 MPa.

In order to verify the developed model, two additional experiments were performed. One experiment was carried out at optimum condition, while the other was done at random settings. 40 rpm was applied to produce sample at optimum condition (verification 1), because it needs less energy than at higher rpm. Random experiment (verification 2) was chosen in any condition within the boundary as proof of reproducibility.

Table 4. Settings and Tensile Strength of Confirmation Experiments

Sample	MNR level (phr)	MA in MNR (phr)	Screw Rotation (rpm)	Prediction (MPa)	Actual (MPa)	Error (%)
Verification 1	6.5	1	40	15.6	15.2	3%
Verification 2	7	1	100	15.5	15.8	2%

IOP Conf. Series: Materials Science and Engineering **223** (2017) 012014 doi:10.1088/1757-899X/223/1/012014

Table 4 shows settings applied to confirmation experiments. The results show that errors for both selected and randomly chosen experiments were less than 5% for all verification samples. These suggest that the model is capable of describing effects of proses and formula phenomena. Marginal difference in prediction of tensile strength (0.1 MPa) for optimized and verification 2 sample suggested that the effect of 0.5 MNR level was not significant to change tensile strength. This was also shown in Figure 5 that present flat region in the range of 6 to 7 phr MNR level. Moreover, the actual difference on tensile strength of verification samples shows 0.6 MPa. The value was lower than the standard deviation of mid-setting (NR-CF1, NR-CF8, NR-CF14) resulted 1.4 MPa. By considering insignificant effect of screw rotation (Section 3.2), this confirms result on modeling that shows insignificant difference between 6 and 7 of MNR level.

3.3. Compatibility of Natural Rubber / Coconut Coir

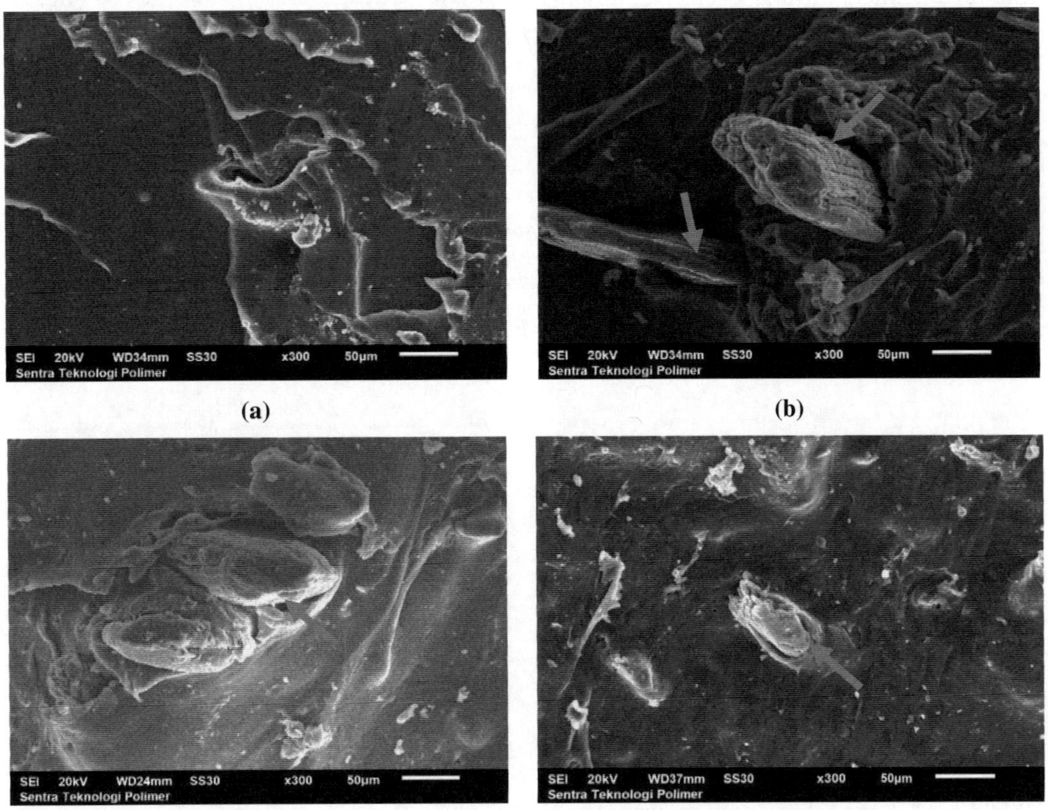

(a) **(b)**

(c) **(d)**

Figure 6. Sample Morphologies at 300x Magnification : (a) Rubber without filler, (b) NR-CF Composite without MNR, (c) Verification 1, (d) Verification 2

Morphology observations were done for selected samples, namely rubber, NR-CF composite without MNR, and verifications. The sample morphologies were shown in Figure 6. In general, the fiber morphologies of both verification samples (Figure 6 (c) and (d)) were covered smoothly than that of composite without MNR (Figure 6 (b) suggested improvement in interfacial bonding. The reason behind this is the compatibility between MNR and NR which has similar backbone. Meanwhile, the grafted MA might effectively react to hydroxyl functional group of bio fiber, as reported elsewhere [38, 39]. Moreover, by this chemical bonding, less voids was produced due to strong matrix fiber interaction for compatibilizer samples. As a result, stress could be transferred effectively from matrix to fiber leading to improvement in strength.

4. Conclusions

Natural rubber/coconut fiber composites were fabricated using twin screw extruder with MNR as compatibilizer. Qualitative and quantitative analysis on MNR samples show that maleic anhydride molecule was grafted on NR chain suggested that MNR was successfully produced. Analysis on collected tensile strength data showed that effect of MNR level and amount of MA in MNR were significant to improve tensile strength. On the other hand, there is no effect of screw rotation on tensile strength due to uncontrollable mixing time. Developed equation shows interaction between MNR level and amount of MA in MNR to improve tensile strength, with optimum conditions predicted at 6.5 phr of MNR level and 1 phr of MA concentration in MNR, regardless screw rotation settings. Analysis on the sample morphology showed that the improvement on tensile strength was influenced by the matrix fiber compatibility leading to better interfacial bonding.

Acknowledgment

Authors would like to thank to Center for Polymer Technology, Agency for Assessment and Application of Technology, Indonesia for financial support and Mr. Yogi A.S for conducting FTIR experiment.

References

[1] Ayrilmis N *et al* 2011 *Fibers and Polymers* **12** 919-926.
[2] Enriquez J *et al* 2010 *J. Thermoplastic Composite Materials* **23** 361-373.
[3] Mohammed L *et al* 2015 . *Appl. Polym. Sci.* **2015**.
[4] Naveen P and Raju T D 2013 *J. Nano Research* **24** 34-45.
[5] Oluwole O I *et al* 2015 *Acta Technica Corviniensis - Bulletin of Engineering* **8** 87-92.
[6] Azmi M A *et al* 2013 *IOP Conference Series: Materials Science and Engineering* **50**.
[7] Fernandes E M *et al* 2013 *Composites Sci. and Technol.* **78** 56-62.
[8] Hani A *et al* 2016 *Proceedings of the Institution of Mechanical Engineers, Part L: Journal of Materials Design and Applications* **230** 240-251.
[9] Narendar R *et al* 2014 *Materials & Design* **54** 644-651.
[10] Saw S K *et al* 2012 . *Appl. Polym. Sci.***125** 3038-3049.

[11] Geethamma V G *et al* 2005 *Composites Part A: Applied Science and Manufacturing*, **36** 1499-1506.
[12] Geethamma V G *et al* 2004 . *Appl. Polym. Sci.* **94** 96-104.
[13] Haseena A P *et al* 2005 *Progress in Rubber, Plastics and Recycling Technology* **21** 155-181.
[14] Alkbir M F M *et al* 2016 *Composite Structures* **148** 59-73.
[15] Faruk O *et al* 2012 *Progress in Polym. Sci.* **37** 1552-1596.
[16] Nunna S *et al* 2012 *Journal of Reinforced Plastics and Composites* **31** 759-769.
[17] Xu Z H and Kong Z N 2014 *Mechanics of Composite Materials* **50** 353-358.
[18] Martins M A *et al* 2008 *J. Mater. Sci.* **43** 6531-6538.
[19] Zaman H U *et al* 2010 *Journal of Reinforced Plastics and Composites* **29** 3064-3065.
[20] Xu C *et al* 2012 *J. Appl. Polym. Sci.* **125** 2449-2559.
[21] Pantamanatsopa P *et al* 2014 *Energy Procedia* **56** 641-647.
[22] Wong P Y*et al* 2006 *Polymers & Polymer Composites* **14** 731-739.
[23] Hietala M *et al* 2014 . *Appl. Polym. Sci.* **131**(2014).
[24] Ranganathan N *et al* 2016 *Polymers for Adv.Technol.* **27** 685-692.
[25] Cao X V *et al* 2012 *Polymer-Plastics Technology and Engineering* **51** 904-910.
[26] Hakim R N and Ismail H 2009 *Polymer-Plastics Technology and Engineering* **48** 910-918.
[27] Ismail H *et al* 2005 *Polymer Testing* **24** 856-862.
[28] Zeng Z *et al* 2009 *J. Polym. Res.* **17** 213-219.
[29] ong Z X *et al* 2013 *Chinese J. Polym. Sci.* **31** 1127-1138.
[30] Lopattananon N *et al* 2007 *J. Appl. Polym. Sci.* **105** 1144-1155.
[31] Ujianto O *et al* 2016 *Materials Science Forum* **864** 13-17.
[32] Sahakaro K and Beraheng S 2008 *J.Appl. Polym. Sci.* **109** 3839-3848.
[33] Nakason C *et al* 2002 *Polymer Testing* **21** 449-455.
[34] Nakason C *et al* 2004 *Polymer Testing* **23** 35-41.
[35] Afifi H and El-Wakil A A 2008 *Polymer-Plastics Technology and Engineering* **47** 1032-1039.
[36] Nakason C *et al* 2001 *J. Appl. Polym. Sci.* **81** 2008-2013.
[37] Nakason C *et al* 2006 *Polymer Testing* **25** 656-667.
[38] Anuar H and Zuraida A 2011 *Composites Part B: Engineering* **42** 462-465.
[39] Rozman H D *et al* 2000 *Eur. Polym. J.* **36** 1483-1494.

IOP Conf. Series: Materials Science and Engineering **223** (2017) 012015 doi:10.1088/1757-899X/223/1/012015

Utilization of amylopectin-grafted-poly(hexyl methacrylate) as bio-compatible agent for polypropylene/starch polymers blend

I S Purwaningsih and A S Handayani
Chemical Engineering Department,InstitutTeknologi Indonesia.

E-mail: aniek.handayani@iti.ac.id , anieksoemarto@gmail.com.

Abstract. PP film/starch blend at different ratio of composition and various amount of Amylopectin-grafted-Poly (Hexyl methacrylate) (Ap-g-PHMA) as bio compatible agent were prepared by blending method. In this study PP/starch blends were obtain 0.6% and 1% Ap-g-PHMA were processed in Rheomix. The structure characteristic, melt flow index (MFI) and thermal properties of the films product were characterized by Fourier Transform Infra-Red Spectroscopy (FTIR), Simultaneous Thermal Analysis (STA). The experimental result indicated that addition of Ap-g-PHMA to the PP/starch blend resulting in increased of MFI values more than three times, raised of starch content in PP matrix, and reduced crystallinity degree.

1. Introduction

Starch is one of the highest natural polymers that have some advantages such as edible, biodegradable, low cost, abundance in nature and renewable [1-3]. Starch has been widely used in various applications, including medicine, agriculture, textile, and especially for packaging. However, if it is compared to conventional synthetic plastic, starch has weakness, owing to its strong hydrophilic character, poor mechanical property and low thermal stability [3]. To improve its mechanical properties, starch has been incorporated with polymer synthetic such as polypropylene (PP). It has been known that PP is hydrophobic synthetic polymer with excellent film forming abilities, good mechanical and barrier properties, but poor adhesive and not biodegradable [4]. To overcome the disadvantages of starch and PP, several methods have been used. Blending PP to biodegradable polymer such as starch as low price material, reduce the final product price and raise the biodegradation rate [5]. Previous study has shown that blending of thermoplastic starch and Poly Lactic Acid (PLA) with no compatible agent lead to phase separation [6]. The addition of maleic anhydride (MA) as compatibilizer agent improves interfacial adhesion between starch and PLA resulting in improvement of thermal and mechanical properties of the blending [7]. Many compatible agents have been used in synthetic polymer productions or research. Among these materials, maleic anhydride-grafted-poly propylene (MAPP) is one of the most popular used. Utilization MAPP as compatible agent raised interface bonding strength, flexural modulus and other mechanical properties [8]. The objective of this research was developing PP film/starch matrix that compatible with Ap-g-

PHMA by melting compounding to improve its performance. This paper presents the impact of the Ap-g-PHMA as compatible agent and the influence of starch content in the PP film matrix. These composites specimen were characterized using Fourier Transform Infra-Red Spectroscopy (FTIR), thermal behavior by Simultaneous Thermal Analysis (STA) and analyze the Melt Flow Index (MFI).

2. Experiment

2.1. Materials

Natural starch, 2, 2'-Bipyridine (bpy) (with 99 % purity) and Dimethyl Suifo Oxide (DMSO) (with purity of 99.9%) were purchased from Sigma-Aldrich. *Ethyl α-bromoisobutyrate,* EB*i*B (with purity of 99 %), amylopectin (α-1, 4-glucan with α-1,6-cross-linking; one terminal group per 25 glucose units) and Hexyl Methacrylate (HMA) were also purchased from Sigma-Aldrich. CuCl (I) and CuCl (II) with purity 99.9% were purchased from Merck. Di Methyl Formamide (DMF) (at 99.9% purity) and Tetra Hydro Furan (THF with 99.9% purity) as solvent were purchased from Merck. Polypropylene film with MFI 10 g/10 min was supplied from Chandra Asri Petrochemical (CPAC-Indonesia)

2.2. Synthesis of Amylopectin Macro-Initiator (Ap-EBiB)

Amylopectin macro-initiator (Ap-EBiB) was prepared by combining amylopectin to EBiB as described in previous study [8].

2.3. Synthesis of Compatibilizer Amylopectin-Graft-PHMA (Ap-g-PHMA)

Copolymerization of Ap-g-PHMA was carried out as described in previous study [2]. Hence Ap-EBiB was reacted to [HMA] and (BPy/CuCl+$CuCl_2$) as ATRP catalyst system in mixed media reaction (DMF: H2O). Starch was dried in the oven at 105°C for 24 h to achieve its moisture content less than 5%. Then starch was compounded into a polypropylene film in the Minilab II Haake™ Rheomix CTW 5 conical that equipped with twin-screw extruder (Thermo Scientific Haake GmbH, Karlsruhe, Germany), a barrel temperature of 170 to 190°C from the feeding zone to the dead zone, respectively. Two levels of load was prepared for all materials. Starch was blend to PP film with a weight ratio of 40 and 50 % and was added to compatible agents at 0.6 and 1 phr respectively. Samples were compounded in specimens test by injection molded using a 5-ton Battlefield BA 200CD Plus machine, with a UNILOG 4000 control system (closed-loop control). A mold from Master mold Inc, having cavities for tensile specimen according to ASTM D638 was used.

2.4. Characterization

The Fourier transforms infrared (FTIR) analyses of Starch and PP/Starch/Ap-g-PHMA were performed using a Perkin-Elmer FTIR spectrometer (Spectrum 2000 series, resolution 2.0 cm, 100 scans) having diamond ATR device (attenuated total reflection) between 4000-500 cm^{-1}

Thermal analyses of composites were performed with a simultaneous thermal analysis. Sample (10.0 mg, dry weight) was loaded in aluminum pans and the moisture level was adjusted to 10% by adding deionized water. Samples were scanned from 30 to 500ºC, at heating rate of 10ºC/min, and a sealed empty pan was used as reference. Melt Flow Index (MFI) samples were measured using Zwick Melt Indexer with ASTM D.1238 as standard method.

3. Result and Discussion

Figure 1 shows the FTIR spectra of starch (a) that contain a broad band at 1007 cm^{-1}, which is related

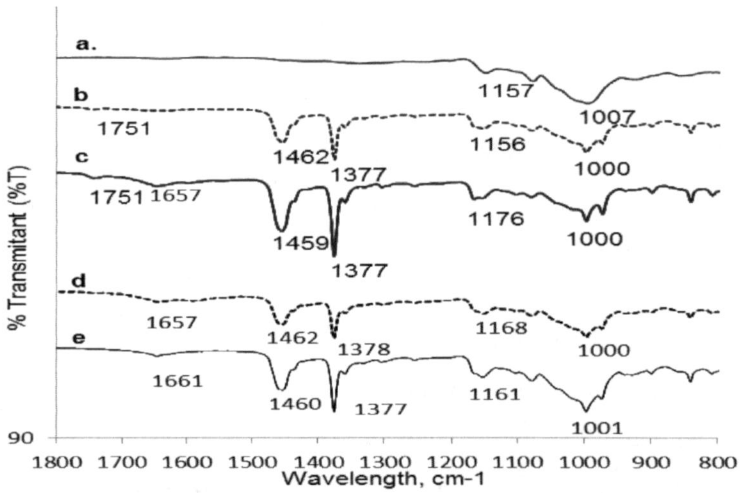

Figure 1. FTIR spectra of (a). Starch, (b).PP/Starch/Ap-g-PHMA (50/50, 0.6%) (c). PP/Starch/Ap-g-PHMA (50/50, 1%) (d), PP/Starch/Ap-g-PHMA (60/40, 0.6%), (e).PP/Starch/Ap-g-PHMA (60/40, 1%)

to CH_2OH groups, and band of 1157 cm^{-1} which is typical of C-O-C group. The FTIR spectrum of PP/Starch/Ap-g-PHMA film (b, c) demonstrated characteristic peaks of PP, starch and Ap-g-PHMA. The peak at 1751 cm^{-1} is the characteristic of Ap-g-PHMA carbonyl stretching band (C=O), also at peaks 1661cm^{-1} and 1657 cm^{-1} indicated of carbonyl stretching band. The spectrum that showed at peak 1462 cm^{-1} and 1377 cm^{-1}, respectively are attributed to C-OH stretching recorded as well as the change of starch. Other peaks at 1156 - 1176 cm^{-1} are typical of C-O-C group. The peaks at 1000-1007 cm^{-1} are the characteristic of CH_2OH [9].

The thermal properties, such as glass transition temperature (Tg), melting temperature (Tm) and heat of fusion ΔH_f are presented in the DSC curves in Figure 2 and 3. This figures show the DSC curves of

PP/starch/Ap-g-PHMA (60/40 at 0.6 and 1 phr respectively) and PP/starch/Ap-g-PHMA (50/50 at 0.6 and 1 phr respectively). The peak around 161.81°C corresponds to melting temperature of PP/starch/Ap-g-PHMA (60/40 at 0.6 phr), while peaks around 159.08°C, 160.58°C and 160.6°C correspond to melting point of PP/starch/Ap-g-PHMA ratio of 60/40 at 1 phr, 50/50 at 0.6 phr and 50/50 at 1 phr, respectively. The degradation temperature of pure starch is around 120 °C, while PP melting point is 173°C, so the melting point of the blends situated in between these two temperatures. The increase of starch weight in the blend, reduce the melting point of the components

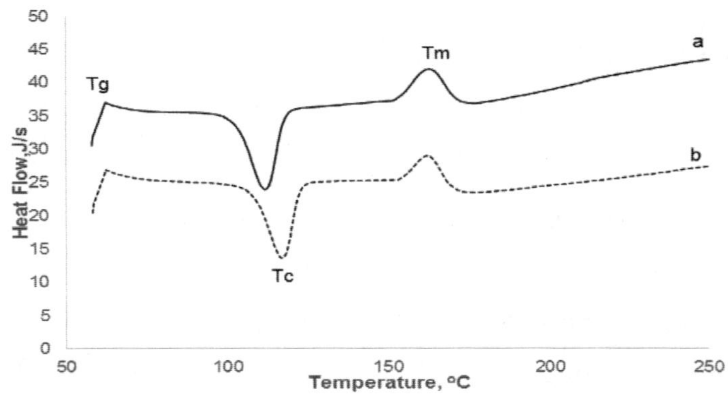

Figure 2. DSC- curves of PP/Starch/Ap-g-PHMA with (a).60/40/0.6 and (b) 60/40/1 ratios

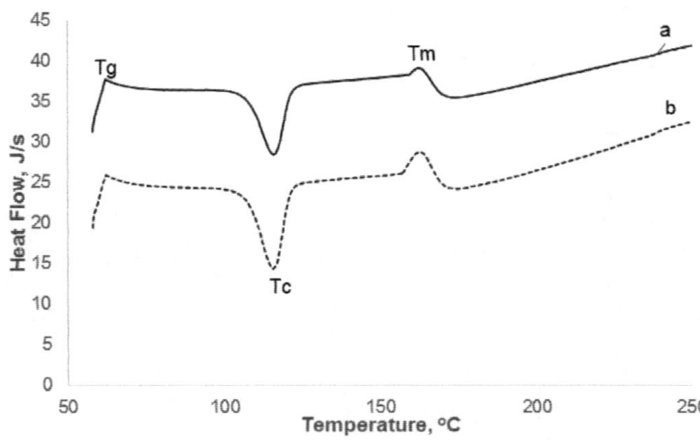

Figure 3. DSC-curves of PP/Starch/Ap-g-PHMA with (a).50/50/1 and (**b**) 50/50/0.6 ratios

Result of all composites thermal properties determination is summarized in Table 1. The higher level of coupling agent used, the lower of melting temperature (Tm) appeared. It is reduced to -0.9°C for ratio of 50/50, and -0.45°C for ratio of 60/40. In addition, the increased of fibers/PP ratio at lower levels of coupling agent causes a slight increase in melting temperature (+0.14°C at 0.6 phr). However, utilization of coupling agent at about 1%, slightly lower the melting temperature (-0.31°C). Usage of

Ap-g-PHMA as a coupling agent in PP/starch blends normally reduce the melting temperature (Tm) of pure PP. Since the melting temperature of starch is about 120°C and the Ap-g-PHMA melting temperature about 140.75°C, so that Tm composites definitely less than that of pure PP [2]. This indicates that Ap-g-PHMA compound influences the dispersion of starch in PP matrix, and enhance miscibility mixture as described by Simmons *et al.,* (2013) [10].

Table 1. Thermal Properties of PP/starch/Ap-g-PHMA blends.

Samples	Tg (°C)	Tm (°C)	Tc (°C)	ΔH_f (J/gr)	X_c
PP :100	-20.00	173.00	110.00	207.10	-
PP/starch/Ap-g-PHMA :50/50/0.6	63.94	160.58	117.1	1.81	17
PP/starch/Ap-g-PHMA : 50/50/1	64.37	160.60	117.3	1.89	18.25
PP/starch/Ap-g-PHMA : 60/40/0.6	64.70	161.81	112.5	2.53	22.77
PP/starch/Ap-g-PHMA : 60/40/1	65.14	159.08	118.3	3.08	24.80

From Table 1, it can be seen that PP blends with starch, increased glass transition temperature from -20°C to 64°C. The effect of Ap -g-PHMA as a coupling agent (at 0.6 and 1%) increase glass transition temperature around 0.67% at ratio of 50/50 and 0.68 % at ratio of 60/40. This result in accordance with the changes in molecular mobility which is the amorphous fraction is located between crystalline lamellae. This case normally is categorized as semi crystalline material relative to an amorphous situation. Hamma *et al.,* (2015) demonstrated that utilization of DSC on PP, at different degree of crystallinity, showed that almost all samples have Tg above Tg PE. This is indicated that the sample have semi crystalline properties. Crystallinity can be determined from calorimetric differential curve based on [11].

Table 1 shows also that with the increasing starch content at low level of Ap-g-PHMA, enhanced the crystallization temperature (T_c) from 112.5 to 117.1°C (4.08%). In contrast, at high level of Ap-g-PHMA, at the same condition of starch composition, the crystallization temperature decreased from 118.3 to 117.3°C (0.84%). Experimental result indicated that the higher of starch content in composites the higher crystallization temperature, but, reduced the degree of crystallinity. Based on Gonzales et.al., (2009) studied, it was confirmed that composites of PP/cassava starch below 30 phr increased degree of crystallinity, but crystallization temperature remain stable. According to them, the filler particles gave a nucleating effect that influences the growth of PP's spherullites. In this study, increasing starch content in PP matrix decreased degree of crystallinity, but increased crystallization temperature. Based on the thermal behavior, the composite in this experiment is categorized as semi crystalline as they have Tg, Tm and Tc [10].

Figure 4 shows the TGA curves of all samples. It is showed that an initial weight loss at 120 and 170°C corresponds to evaporation of water and dehydration of amylopectin [8]. While the weight loss step in the second range (266 – 320°C) was attributed to the degradation of starch compounds including degradation of glycoside ring [11]. The last degradation step is related to decomposition of PP matrix contained in Ap-g-PHMA at maximum temperature in the range of 400-460°C. Filler content in PP matrix lead reduction in chain mobility in the boundary layers [12]. The degradation process of the starch is highly complex. Starch structure need to be broken so that the amylose and amylopectin can be degraded. Starch is much easily degraded than PP, since it has carbon-oxygen bonds.

Figure 4 indicates also a residue that does not relate to PP. According to Gonzalez (2009), PP degrades through chain rupture, so it can be assumed that this residue relates to crosslink structure that is exist in starch granules. Due to highly hygroscopic in nature, it is imposible to eliminate all the water content in the blends, since water induces the increase in granule and swelling size [13-15].

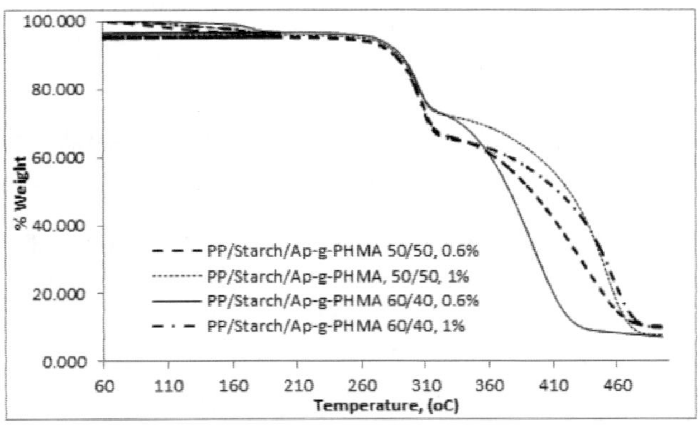

Figure 4. TGA- curves of PP/starch/Ap-g-PHMA 50/50/0.6 (— — —); 60/40/06 (——); 50/50/1 (•••••) and 60/40/1 (— · — ·)

The melt flow index (MFI) is a measure of viscosity inversion. Increase the MFI value by adding starch content, reduce its viscosity [16]. The MFI is used also as a measurement of the uniformity of flow rate of thermoplastic material. It is a practical method that widely used to relate the molecular weight and viscosity of polymer compounds.

Table 2. Melt Flow Index (MFI) of PP/Starch/Ap-g-PHMA blends

Samples	Ap-g-PHMA (phr)	MFI values g/10 min
PP film	-	10
PP/starch, 60/40	0.6	36±1.16
PP/starch, 60/40	1.0	63±1.62
PP/starch, 50/50	0.6	33±0.49
PP/starch, 50/50	1.0	57±0.47

IOP Conf. Series: Materials Science and Engineering **223** (2017) 012015 doi:10.1088/1757-899X/223/1/012015

Table 2 shows the MFI value at different compositions of PP /starch in polypropylene matrix. At all level of Ap-g-PHMA as compatible agent, the MFI value reduced with increasing the starch content. However, by adding the compatible agent content at same ratio of PP/starch, increase the MFI value until almost twice. In contrast, previous studied done by Suttivunarubet et al. (2016), utilized PP/starch/PP-g-MA at 30% starch content, demonstrated that compatible agent give low MFI. It was stated that compatibilization of starch in that mixture has poor dispersion, and incapable to widely disperse into the PP matrix [17, 18]

4. Conclusion

The addition of Ap-g-PHMA enhanced the MFI of PP/starch/Ap-g-PHMA blend. The MFI of pure PP film is 10 gr/10 min while its value of PP/starch/Ap-g-PHMA at 50/50 (0.6 phr) is 33±0.49 gr/10 min. It is recommended that these blends could be used as good interfacial adhesion and compatible material. The weight loss effect of Ap-g-PHMA is lower in PP/starch blends (60/40, 1 phr and 50/50, 0.6 phr) than that of in (60/40 0.6 phr and 50/50, 1phr). In all samples showed similar thermal properties pattern such as Tg, Tc and Tm. Hence indicated that PP/starch blends build semi crystalline material. Based on the TGA studies, starch produces a long-term effect on degradation process of PP, as the Ap-g-PHMA added as compatible agent, the decomposition temperature of the mixture raised.

Acknowledgement

The authors would like to thank the Ministry of Research Technology and Higher Education Republic of Indonesia, which has funded this research through Competitive Research Grant Simlintabmas Dikti 2016 under contract number: 807/K3/KM/SPK.LT/2016.

References

[1] Chalid M *et al* 2014 *Adv. Mater. Research* 90-94.
[2] Handayani A S *et al* 2015 *Mater. Sci. Forum* 306-310.
[3] Y. He *et al* 2013 *Mater. Chem. and Physics* **142** 1-11.
[4] Rhim J -W *et al* 2013 *Progress in Polymer Science* **38** 1629-1652.
[5] Shayan M *et al* 2015 *Carbohydr. Polym.* **124** 237-244.
[6] Sarazin P *et al* 2008 *Polymer* **49** 599-609.
[7] Leadprathom J *et al* 2010 *J. Metals, Materials and Minerals* **20** 87-90.
[8] Imre B, Pukánszky B 2013 *Eur. Polym. J.* **49** 1215-1233.
[9] Knitter M, Dobrzyńska-Mizera M 2015 *Mechanics of Composite Materials* **51** 245-252.
[10] Rodríguez-Llamazares S 2013 *J. Chilean Chemical Society* **58** 1643-1646.
[11] Hamma A *et al* 2015 J. Reinforced Plastics and Composites **34** 2045-2058.
[12] Challa G 1993 *Polymer Chemistry: an Introduction*, Prentice Hall.
[13] Gregorova A 2013 *Application of Differential Scanning Calorimetry to the Characterization of Biopolymers,* INTECH Open Access Publisher.
[14] Mali S *et al* 2005 *Carbohydr Polym* **60** 283-289.

[15] Rosa D *et al* 2007 *J. Mater. Sci.* **42** 551-557.
[16] Pedroso A G, Rosa D D S 2005 *Carbohydr. Polym.* **59** 1-9.
[17] Fechner P M *et al* 2005 *Carbohydr. Research* **340** 2563-2568.
[18] Waryat W *et al* 2013 *J. Agritech,* 33.

IOP Conf. Series: Materials Science and Engineering **223** (2017) 012016 doi:10.1088/1757-899X/223/1/012016

Effects of organoclay to miscibility, mechanical and thermal properties of poly(lactic acid) and propylene-ethylene copolymer blends

S Wacharawichanant, C Ounyai, P Rassamee
Department of Chemical Engineering, Faculty of Engineering and Industrial Technology, Silpakorn University, Nakhon Pathom 73000, Thailand

E-mail : sirirat.che@gmail.com

Abstract. The effects of propylene-ethylene copolymer (PEC or PEC3300) and clay surface modified with 25-30 wt% of trimethylstearyl ammonium (Clay-TSA) on morphology, thermal and mechanical properties of poly(lactic acid) (PLA) were investigated. The morphology analysis showed PLA/PEC3300 blends clearly demonstrated a two-phase separation of dispersed phase and the matrix phase and the addition of Clay-TSA could improve the miscibility of PLA and PEC3300 blends due to the decreased of the domain sizes of dispersed PEC3300 phase in the polymer matrix. From X-ray diffraction analysis showed the intercalation of PLA chains inside the Clay-TSA and this result implied that Clay-TSA platelets acted as an effective compatibilizer. The tensile properties showed the strain at break of PLA was improved after adding PEC3300 while Young's modulus, tensile strength and storage modulus decreased. The addition of Clay-TSA could improve Young's modulus of PLA/PEC3300 blends. The addition of Clay-TSA 7 phr showed the maximum of Young's modulus of PLA/PEC3300/Clay-TSA composites. The thermal properties found that the addition of PEC3300 and Clay-TSA did not change significantly on the glass transition temperature and melting point temperature of PLA. The percent of crystallinity of PLA decreased with increasing PEC content. The thermal stability of PLA improved after adding PEC3300.

1. Introduction

Poly(lactic acid) (PLA) is biodegradable polymer, an eco-friendly and biocompatible bioplastic. PLA can be processed by conventional processing methods such as extrusion, tubular blown film, injection molding, and thermoforming [1]. Another feature that makes PLA interesting is the fact that it can be processed similarly to polyolefins, and PLA has high strength, high modulus and transparency. However, the main drawbacks with PLA properties are too brittle for many applications, low heat distortion temperature (HDT) and low thermal stability compared to commercial polymers [2-5]. To extend the application field the improvement of mechanical and thermal properties is required. Therefore, it is interesting to study if the incorporation of reinforcements can improve the mechanical properties and thermal stability of PLA. Thus, some researchers have studied to improvethe toughness, mechanical and thermal properties, barrier properties, and degradation behavior of PLA by blending with plasticizers, rubber, thermoplastic starch (TPS) and other thermoplastic polymers [6-11]. For instance, Ferrarezi *et al.,* (2013) [9] prepared PLA/TPS blends by using poly(ethylene glycol) (PEG), a non-toxic polymer, as a compatibilizer. They found that the addition of

PEG resulted in the increase of PLA crystallization, due to its plasticizing effect, and improvement of the interfacial interaction between TPS and PLA matrix.

Nevertheless, the main problems with PLA toughening through plasticizers or rubber is the reduction in strength and modulus properties [6,8,11-12]. In the PLA/rubber blends, for example, reduction in stiffness mainly occurs due to the low modulus of the rubber component itself [12] and PLA/TPS showed the lower elastic modulus than pure PLA [9]. However, most of these polymers that are blended with PLA are partially immiscible and occur phase separation with poor mechanical performance because of a negligible entropy of mixing [13]. Thus, some authors have studied organoclay to improve the properties of polymer blendsand found that organoclay plays the role of both structural reinforcement and compatibilizer [14-16]. From some researchers have studied using the rubber and flexible polymer to improve the mechanical properties of PLA, and using organoclay to improve the miscibility and properties of PLA blends for more applications [12,17-18]. In stance, Baouz *et al.* (2013) [18] improved the toughening PLA with an ethylene-methyl acrylate-glycidyl methacrylate (E-MA-GMA) rubber by reactive blending in a twin screw extruder. The addition of E-MA-GMA rubber changed the brittle behavior of PLA to ductile and the E-MA-GMA rubber domain size increased with increasing rubber content in both the blends and nanocomposites.

The compatibilizing action of organoclay results in a decrease in interfacial tension and particle size and in a remarkable increase in mechanical and thermal properties of the immiscible blends. Thus, this work studied the influence of clay surface modified with 25-30 wt% of trimethylstearyl ammonium (Clay-TSA) on miscibility, mechanical and thermal properties of PLA/ propylene-ethylene copolymer (PEC) blends. PEC was used to improving the toughness of PLA and the Clay-TSA was used to improving the miscibility and modulus of PLA blends. The blends and composites with varying concentration of PEC3300 and Clay-TSA were prepared by a melt mixing technique in an internal mixer and molded by compression molding.

2. Experimental

2.1. Materials

PLA with the melt flow index (210°C, 2.16 kg) of 6.0 g/10 min and specific gravity of 1.24 g/mL was produced by NatureWorks LLC, USA, under the trade name of "Ingeo™ Biopolymer 2003D". The PEC was purchased from Chemical Innovation Co., Ltd, under the trade name of "VERSIFY™ 3300 Elastomer". After that this PEC was called PEC3300. The PEC3300 has the melt flow rate of 8 g/10 min, density of 0.867 g/mL and the ethylene content of 12 wt%, this work used the clay was surface modified by 25-30 wt% of trimethylstearyl ammonium (Clay-TSA), was produced by Aldrich Company.

2.2. Sample Preparation

Before blending, PLA was dried at 80°C for 4 h and Clay-TSA was dried at 80°C 24 h in an oven. PEC3300 was not dried before blending process. PLA/PEC3300 blends were prepared by melt blending in an internal mixer at 180°C and a rotor speed of 50 rpm for 15 min. The PEC3300 contents were 5, 10, 15 and 20 wt%. While PLA/PEC blends with Clay-TSA content 3, 5, and 7 phr were prepared in the same condition, and Clay-TSA was added after 5 min of mixing of the blends. After compounding, polymer blends without and with Clay-TSA were dried in an oven at 80°C for 1 h for reject moisture. Afterward the blends and composites were compression molded for testing.

2.3. Sample Characterization

The morphology of tensile fractured surfaces of PLA/PEC3300 blends without and with Clay-TSA was investigated by scanning electron microscope (SEM) instruments (JSM-6400 Scanning electron microscope, JEOL) at 15 kV accelerating voltage after gold coating. The tensile test was conducted according to ASTM D 638 with a universal tensile testing machine (LR 50k from Lloyd instruments, UK) at a crosshead speed of 50 mm/min. Each value obtained represented the average of five samples.

The thermomechanical properties (storage modulus) of PLA/PEC3300 blends without and with Clay-TSA was measured in bending film mode at a constant vibration frequency of 1 Hz, a temperature ranges from 30-140°C, and a heating rate of 5°C/min in a nitrogen atmosphere with a DMA (Pyris Diamond DMA, Perkin Elmer, USA). The glass transition temperature was detected by using the peak of tan δ from DMA test. DSC characterization (Pyris I, Perkin Elmer, USA) was performed to investigate the melting temperature of the blends and composites. The temperature was in the range of 50-200°C at a heating rate of 10°C/min in a nitrogen atmosphere for the heating scan then the analysis performed from the heating scan. The degradation temperature of PLA/PEC3300 blends without and with Clay-TSA was measured by TGA (SDT Q600, TA Instruments, UK). The temperature was in the range of 30-600°C at a heating rate of 10°C/min under nitrogen atmosphere.

3. Results and Discussion

3.1. Morphology

Figure 1 shows the SEM micrographs of the tensile fractured surfaces of pure PLA and PLA/Clay-TSA composites at various Clay-TSA contents. The tensile fractured surface of pure PLA was quite smooth which characteristic of a brittle semicrystalline polymer (Figure 1a). While tensile fractured surfaces of PLA/Clay-TSA composites in Figure 1b-d were relatively rough to the more ductile after adding Clay-TSA. Figure 2 shows the SEM micrographs of the tensile fractured surfaces of

PLA/PEC3300 blends at various PEC3300 contents. From PLA/PEC3300 blends clearly demonstrated a two-phase matrix-particle morphology and observed the minor phase of PEC3300 dispersed as spherical domains in PLA matrix. Moreover, the domain size of dispersed PEC3300 phase increased with increasing PEC3300 content. Additionally, it observes the microvoids surrounding the PEC3300 droplets indicates weak interfacial adhesion between PLA and PEC3300 due to the immiscible nature of the polymers. Because of the high interfacial tension between the two polymers, the spherical morphology is anticipated because of the minimization of the interfacial area [19].

Figures 3-4 show the SEM micrographs of the tensile fractured surfaces of PLA/PEC3300 blends with Clay-TSA content of 3 and 5 phr, respectively. Moreover, PLA/PEC3300 blends with Clay-TSA content of 7 phr showed similar morphology as Clay-TSA 5 phr. It is clear that the addition of Clay-TSA in PLA/PEC3300 blends, the blends morphology shows a significant reduction in domain size of dispersed PEC3300 phase that means a high level of mixing between two polymer matrices [20]. Moreover, as the amount of the Clay-TSA increased, domain size of the dispersed PEC33000 phase in the blend gradually decreased. This indicated that the compatibility between PLA and PEC3300 improved with the presence of Clay-TSA when compared with PLA/PEC3300 blends, and showed a good dispersion uniformly at the content of Clay-TAS 3, 5 and 7 phr. Thus, Clay-TSA played an important role in reducing domain size of the dispersed PEC3300 phase PLA/PEC3300 blends, and Clay-TSA acted as an effective compatibilizer and the obtained morphology affected on the mechanical properties of the composites.

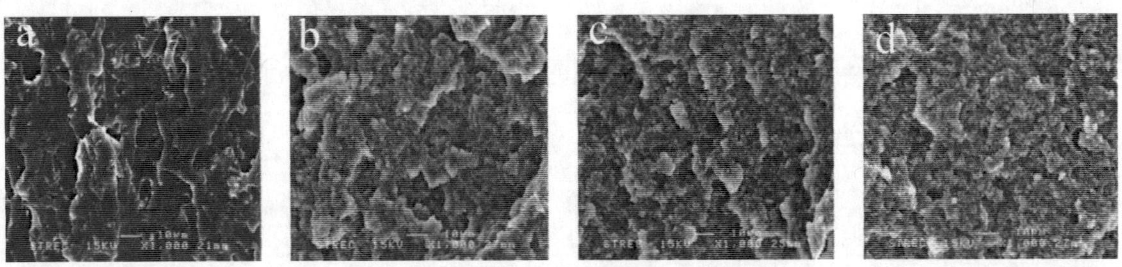

Figure 1. SEM Micrographs of the Tensile Fractured Surfaces of (a) Pure PLA, (b) PLA/Clay-TSA (100/3), (c) PLA/Clay-TSA (100/5) and (d) PLA/Clay-TSA (100/7)

3.2. XRD Analysis

The results from XRD patterns of Clay-TSA, PLA/PEC3300 (80/20) blends and PLA/PEC3300 blends with Clay-TSA showed that the Clay-TSA itself exhibited the characteristic peak at a 2θ of 3.67° correspond to the d-spacing of 2.41 nm. The peak position of clay-TSA was shifted to a lower 2θ region in the PLA/PEC3300/Clay-TSA composites indicated the intercalation of PLA chains inside the Clay-TSA. The relatively higher extent of intercalation in the PLA/PEC3300/Clay-TSA due to the different PEC3300 content (5, 10, 15 and 20 wt% of PEC3300 content). Interestingly, the characteristic peak of Clay-TSA in PLA/PEC3300 blends at PEC3300 content 5, 10, 15 and 20 wt% with Clay-TSA 5 phr were found at 2θ =2.44°, 2.48°, 2.41° and 2.48°, which corresponds to d-spacing of 3.61, 3.55, 3.65 and 3.55 nm, respectively. In the same time PLA/PEC3300 blends with Clay-TSA 3 and phr showed the same trend XRD results. This result indicated the dispersed clay was intercalated due to the intercalation of polymer chains usually increases the d-spacing of the clay platelets [17]. From this result implied that Clay-TSA platelets acted as an effective compatibilizer that improved the miscibility of PLA/PEC3300 blends, due to the location of intercalated clay silicate layers at the interface decreased the interfacial tension and dispersed-phase particle sizes in immiscible blends [21].

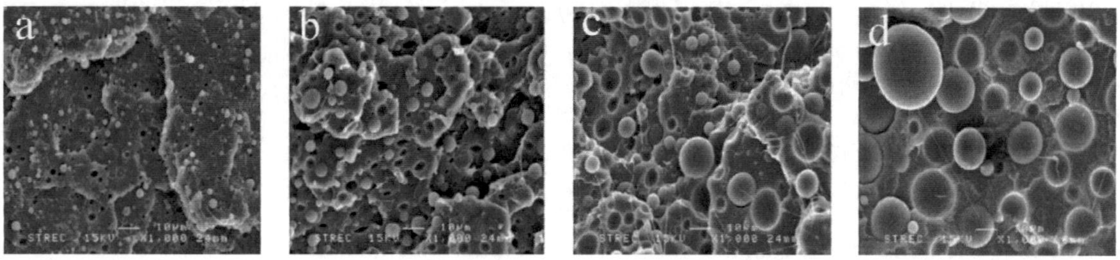

Figure 2. SEM micrographs of the Tensile Fractured Surfaces of PLA/PEC3300 Blends (a) 95/5, (b) 90/10, (c) 85/15 and (d) 80/20

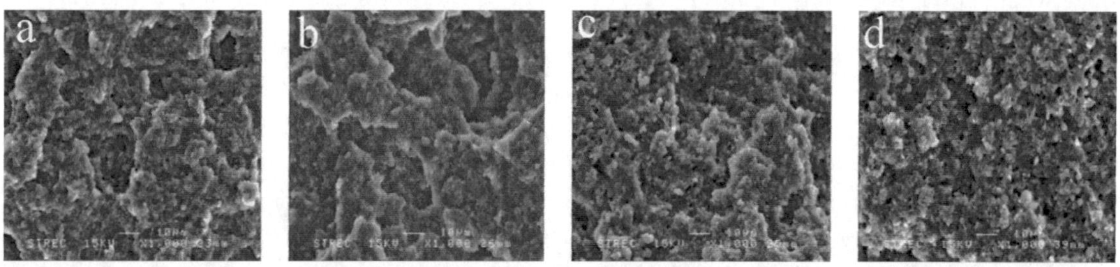

Figure 3. SEM micrographs of the Tensile Fractured Surfaces of PLA/PEC3300/Clay-TSA Composites (a) 95/5/3, (b) 90/10/3, (c) 85/15/3 and (d) 80/20/3

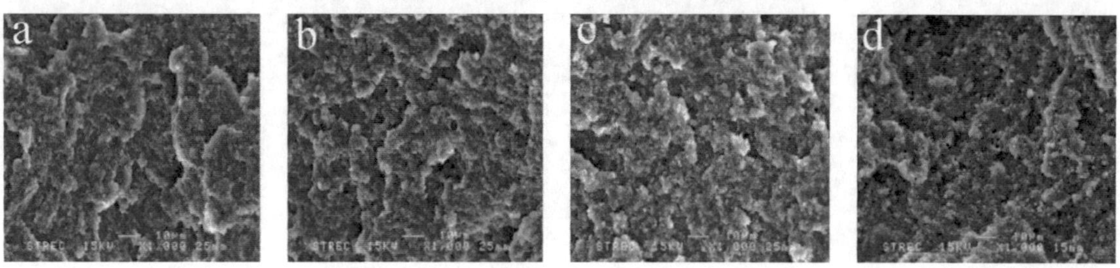

Figure 4. SEM micrographs of the Tensile Fractured Surfaces of PLA/PEC3300/Clay-TSA Composites (a) 95/5/5, (b) 90/10/5, (c) 85/15/5 and (d) 80/20/5.

3.3. Mechanical Properties

Figure 5 presents Young's modulus PLA/PEC3300 blends without and with Clay-TSA. The result found that Young's Modulus of PLA was 1,098.59 MPa and most of Young's modulus of PLA/PEC3300 blends was lower than pure PLA, and Young's modulus of PLA/PEC3300 blends decreased with increasing PEC3300 content because the polymer blends increased flexibility after adding PEC3300. The result found that Young's modulus of PLA/PEC3300/Clay-TSA composites was higher than that of PLA/PEC3300 blends, and most Young's modulus of PLA/PEC3300/Clay-TSA composites increasing PEC3300 content. So the incorporation of Clay-TSA significant improved Young's modulus of PLA/PEC3300 blends. The addition of Clay-TSA could improve Young's modulus of PLA/PEC3300 blends due to the reinforcement effect of the rigid inorganic Clay-TSA induced a substantial increase in the modulus of the composites owing to the stiffening effect of the Clay-TSA that induced chain immobilization [18].

The tensile strength of PLA/PEC3300 blends without and with Clay-TSA is presented in Figure 6. The result found that tensile strength of PLA was 53.11 MPa and tensile strength of PLA/PEC3300 blends was lower than pure PLA, and tensile strength of PLA/PEC3300 blends decreased with increasing PEC3300 content. The tensile strength of PLA/PEC3300 blends after adding Clay-TSA decreased with increasing Clay-TSA content. So, the addition of PEC3300 and Clay-TSA did not improve the tensile strength of PLA and PLA/PEC3300 blends.

The strain at break of PLA/PEC3300 blends without and with Clay-TSA is illustrated in Figure 7. It can be seen that the strain at break of PLA was 8.59% and strain at break of PLA/PEC3300 blends increased with increasing PEC3300 content. This implied that PEC3300 could improve the elongation and flexibility of PLA due to the nature elastomer characteristic of PEC3300. While the addition of Clay-TSA decreased strain at break of PLA/PEC3300 blends and decreased when the increase of Clay-TSA content. This result indicated that Clay-TSA decreased the movement of the polymer chain and reduced the elongation of PLA/PEC3300 blends.

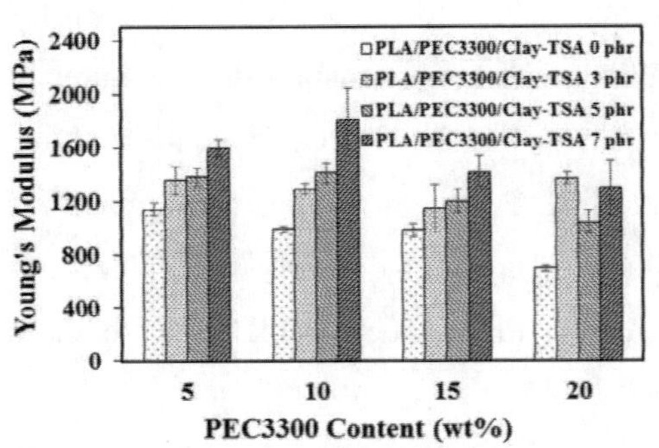

Figure 5. Young's Modulus of PLA/PEC3300 Blends without and with Clay-TSA.

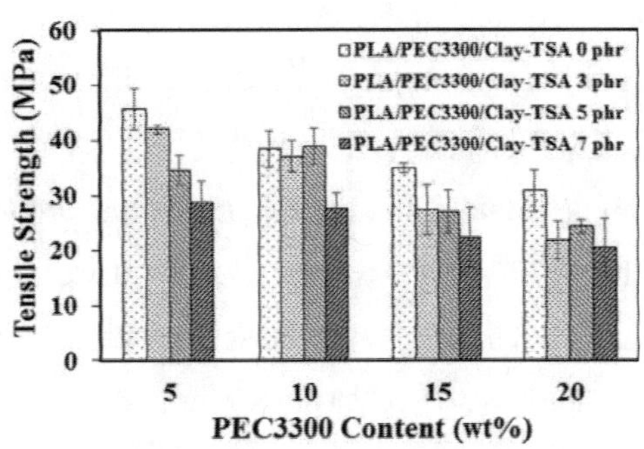

Figure 6. Tensile Strength of PLA/PEC3300 Blends without and with Clay-TSA.

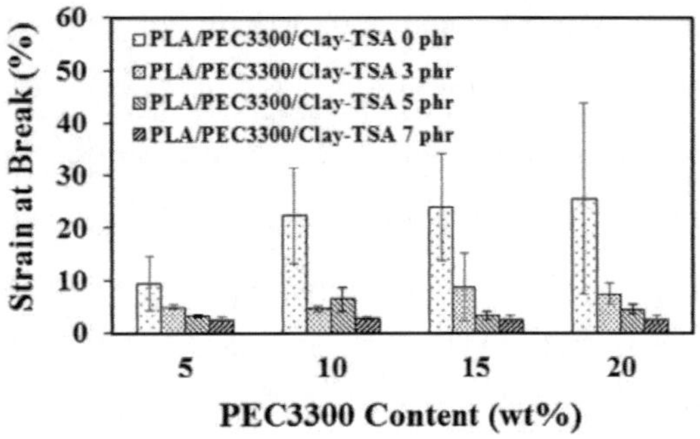

Figure 7. Strain at Break of PLA/PEC3300 Blends without and with Clay-TSA.

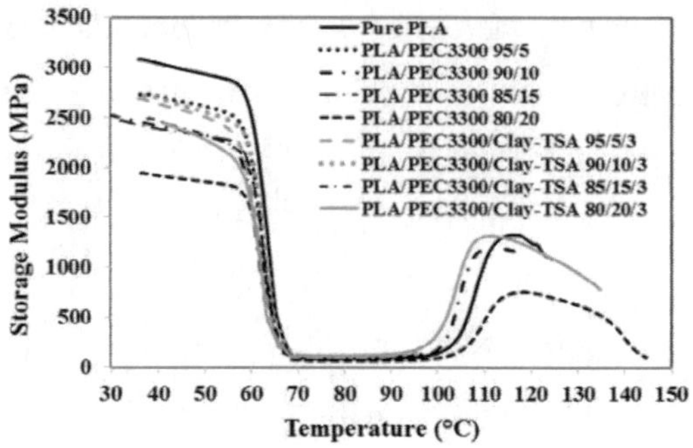

Figure 8. Storage Modulus of PLA/PEC3300 Blends without and with Clay-TSA 3 phr.

3.4. Thermomechanical Properties

The variation of storage modulus of PLA, PLA/PEC3300 blends and composites with Clay-TSA 3 phr is presented in Figure 8. The results found that the both the PLA/PEC3300 blends and pure PLA exhibited lower storage modulus as temperature increased. The sharply decrease was observed at the temperature higher than 63°C. The storage modulus of PLA/PEC3300 blends showed a similar trend to the pure PLA. The results found that the PLA/PEC3300 blends with Clay-TSA exhibited a higher storage modulus than the blends without Clay-TSA at a temperature lower than 63°C. The storage modulus sharply decreased was observed at the temperature higher than 63°C.

3.5. Thermal Properties

Table 1 shows the glass transition temperature, melting temperatures and percent crystallinity of pure PLA, PLA/Clay-TSA, PLA/PEC3300 blends and composites. The results revealed that the glass transition temperature of PLA was 65.5°C and glass transition temperatures of PLA/Clay-TSA at 3, 5 and 7 phr were 65.0, 64.7 and 61.1°C, respectively. Thus, that addition of Clay-TSA decreased the glass transition temperature of PLA at high Clay-TSA loading. The glass transition temperatures of PLA/PEC3300 blends at various PEC3300 content were in between 65.3 ± 1.0°C. Thus, the addition of PEC3300 had no effect on the glass transition temperature of PLA. The glass transition temperature of PLA/PEC3300/Clay-TSA composites at various PEC3300 and Clay-TSA contents were in between 64.7 ± 2.2°C. Therefore, the glass transition temperatures of the PLA/PEC3300 blends were slight changed when Clay-TSA was incorporated.

The melting temperatures of pure PLA, PLA/Clay-TSA, PLA/PEC3300 blends and composites are presented in Table 1. The melting temperature of PLA was 153.0°C and glass transition temperatures of PLA/Clay-TSA at 3, 5 and 7 phr were 154.2, 155.3 and 154.7°C, respectively. Thus, the incorporation of Clay-TSA slight increased the melting temperature of PLA. The melting temperatures of PLA/PEC3300 blends at various PEC3300 content were 152.8 ± 0.4°C. Thus, the addition of PEC3300 had no effect on the melting temperature of PLA. The melting temperatures of

IOP Conf. Series: Materials Science and Engineering **223** (2017) 012016 doi:10.1088/1757-899X/223/1/012016

PLA/PEC3300/Clay-TSA composites at various PEC3300 and Clay-TSA contents were 153.5 ± 0.4°C. This indicated that the addition of Clay-TSA had no effect on the melting temperature of the blends.

Table 1 shows the percent crystallinity of pure PLA, PLA/Clay-TSA, PLA/PEC3300 blends and composites, it was found that percent crystallinity of PLA was 25.2 and percent crystallinity of PLA dramatically decreased when added Clay-TSA. Moreover, PLA/PEC3300 blends showed the decrease of percent crystallinity with increasing PEC3300 content due to PEC3300 behaved as the amorphous polymer. However, the presence of Clay-TSA had a slight effect on the percent crystallinity of PLA/PEC3300/Clay-TSA composites.

The thermal stability of the PLA/PEC3300 blends without and with Clay-TSA was investigated by TGA. The degradation temperatures corresponding to the 10% weight loss (Td_{10}) were calculated and are shown in Table 1. The degradation temperatures of PLA and Clay-TSA were 320 and 294°C, respectively. The result showed the incorporation of Clay-TSA and PEC3300 improved the degradation temperature of PLA. This indicates that an improvement in the thermal stability in the PLA/PEC3300 blends due to PEC3300 had the higher degradation temperature. The degradation temperatures of PLA/PEC3300/Clay-TSA composites are illustrated in Table 1. The result showed that the addition of Clay-TSA did not improve the degradation temperature of PLA/PEC3300 blends and they had a slight influence on the thermal stability of the blends.

Table 1. Glass Transition Temperature (T_g), Melting Temperature (T_m), Percent Crystallinity and Degradation Temperature (T_d) of PLA/PEC3300 Blends without and with Clay-TSA

Sample	**T_g (°C)**	**T_m (°C)**	**%Crystallinity of PLA**	**T_d (°C)**
Pure PLA	65.5	153.0	25.2	320
PLA/PEC3300 (95/5)	66.1	152.6	23.8	334
PLA/PEC3300 (90/10)	64.0	152.8	22.2	334
PLA/PEC3300 (85/15)	65.3	152.4	24.0	326
PLA/PEC3300 (80/20)	66.1	153.4	19.0	332
PLA/Clay-TSA (100/3)	65.0	154.2	15.2	330
PLA/PEC3300/Clay-TSA (95/5/3)	66.3	154.1	24.1	332
PLA/PEC3300/Clay-TSA (90/10/3)	65.2	153.5	23.2	333
PLA/PEC3300/Clay-TSA (85/15/3)	65.2	153.5	20.3	330
PLA/PEC3300/Clay-TSA (80/20/3)	63.7	152.9	19.2	335
PLA/Clay-TSA (100/5)	64.7	155.3	7.9	328
PLA/PEC3300/Clay-TSA (95/5/5)	63.6	154.2	23.5	331
PLA/PEC3300/Clay-TSA (90/10/5)	65.0	153.4	21.8	334
PLA/PEC3300/Clay-TSA (85/15/5)	64.1	153.2	20.9	332
PLA/PEC3300/Clay-TSA (80/20/5)	64.4	153.2	19.4	320
PLA/Clay-TSA (100/7)	61.1	154.7	7.9	329
PLA/PEC3300/Clay-TSA (95/5/7)	68.4	153.9	22.2	332
PLA/PEC3300/Clay-TSA (90/10/7)	66.4	153.8	22.5	329
PLA/PEC3300/Clay-TSA (85/15/7)	59.2	152.9	20.5	329
PLA/PEC3300/Clay-TSA (80/20/7)	64.8	153.8	18.5	328

Figure 9-10 show the thermogravimetric curves of thermal degradation of PLA/PEC3300 blends without and with Clay-TSA 5 phr. It is evident that the thermal degradation of PLA showed only single step of weight loss, while the thermal degradation of the blends and composites proceeded by two degradation steps. It was found that the first step showed a large mass loss starting at around 300°C, this main region was thermal degradation of PLA. The second step showed a small mass loss starting at around 370°C, this region was thermal degradation of PEC3300. Moreover, this region showed the mass loss of PLA/PEC3300 blends at high PEC3300 content was lower than the blends at low PEC3300 content. This indicated that the addition of PEC3300 could improve the thermal stability of PLA blends. The thermogravimetric curves of PLA/PEC3300 blends with Clay-TSA 3 and 5 phr displayed the same trend curves.

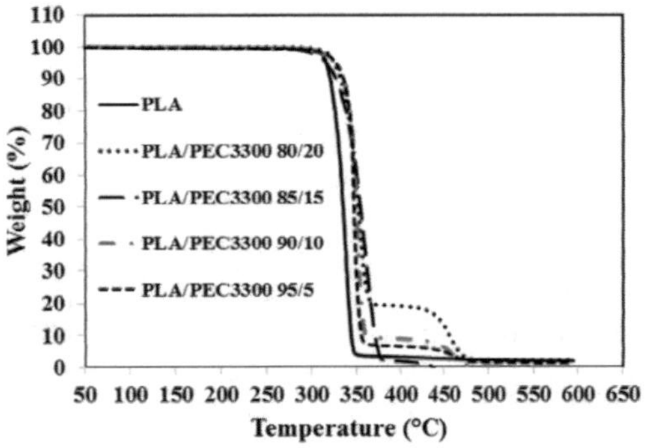

Figure 9. TGA thermograms of PLA/PEC3300 blends.

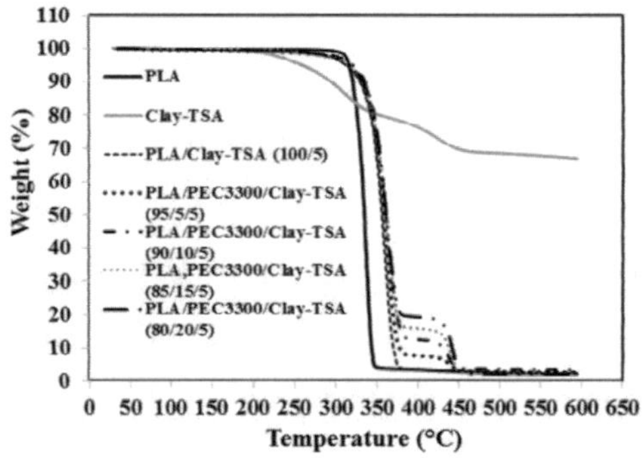

Figure 10. TGA thermograms of Clay-TSA, PLA, PLA/Clay-TSA 5phr and PLA/PEC3300 blends with Clay-TSA 5 phr.

4. Conclusions

The effects of Clay-TSA on miscibility, mechanical and thermal properties of PLA/PEC3300 blends were investigated. The morphology analysis showed PLA/PEC3300 blends clearly demonstrated a two-phase separation of dispersed phase and the matrix phase, and the addition of Clay-TSA could improve the miscibility of PLA and PEC3300 blends due to the decrease in interfacial tension and domain size of dispersed PEC3300 phase in PLA matrix. From XRD analysis showed the peak position of clay-TSA shifted to a lower 2θ region in the PLA/PEC3300/Clay-TSA composites indicated the intercalation of PLA chains inside the Clay-TSA. This result implied that Clay-TSA platelets acted as an effective compatibilizer due to the location of intercalated clay silicate layers at the interface decreased the interfacial tension and dispersed-phase particle sizes in PLA/PEC3300 blends. The tensile properties showed the strain at break of PLA was improved after adding PEC3300.

The addition of Clay-TSA could improve Young's modulus and storage modulus of PLA/PEC3300 blends. The addition of PEC3300 could improve the thermal stability of PLA.

Acknowledgment

The authors would like to thank Silpakorn University Research and Development Institute (SURDI) for the financial support of this project.

References

[1] Jaratrotkamjorn R *et al* 2012 *J. Appl. Polym. Sci.* **124** 5027.
[2] Halász K and Csóka L 2013 *Journal of Engineering* DOI: 10.1155/2013/329379: 1.
[3] Rawi N F M *et al* 2013 *J. Reinf. Plast. Comp.* **32** 1513.
[4] Tokoro R *et al* 2008 *J. Mater. Sci.* 43 775.
[5] Pilla S *et al* 2008 *Polym. Eng. Sci.* **48** 578.
[6] Juntuek P *et al* 2012 *J. Appl. Polym. Sci.* **125** 745.
[7] Tanrattanakul V and Bunkaew P 2014 *eXPRESS Polym. Lett.* **8** 387.
[8] Jaratrotkamjorn R *et al* 2012 *J. Appl. Polym. Sci.* **124** 5027.
[9] Ferrarezi M M F *et al* 2013 *J. Polym. Environ.* **21** 151.
[10] Liu J *et al* 2012 *J. Polym. Environ.* **20** 810.
[11] Bee S-T *et al* 2014 *Nucl. Instr. Meth. Phys. Res. B* **334** 18.
[12] Bijarimi M *et al* 2013 *J. Reinf. Plast. Comp.* **32** 1656.
[13] Choudhary P *et al* 2011 *J. Appl. Polym. Sci.* **121** 3223 .
[14] Yousfi M *et al* 2013 *Macromol. Mater. Eng.* **298** 757.
[15] Nazari T and Garmabi H 2012 *Polym. Compos.* **33** 1893.
[16] Nazari T *et al* 2012 *J. Appl. Polym. Sci.* **126** 1637.
[17] Hasook A *et al* 2006 *Polym. Eng. Sci.* **46** 1001.
[18] Baouz T *et al* 2013 *J. Appl. Polym. Sci.* **128** 3193.
[19] Wacharawichanant S *et al* 2015 *Polym. Plast. Technol. Eng.* **54** 1349 .
[20] Ray S S *et al* 2007 *Macromol. Mater. Eng.* **292** 729.
[21] Mallick S *et al* 2010 *J. Appl. Polym. Sci.* **116** 1010.

IOP Conf. Series: Materials Science and Engineering **223** (2017) 012017 doi:10.1088/1757-899X/223/1/012017

Effect of vulcanization temperature on curing characteristic, physical and mechanical properties of natural rubber/palygorskite composites

K C Lee[1], N A Md Yusoff [1], N Othman [1, 2], N A Mohamad Aini[1]
[1]Polymer Engineering Program, School of Materials and Mineral Resources Engneering, Universiti Sains Malaysia, Engineering Campus, 14300 Nibong Tebal, Penang, Malaysia.
[2]Cluster of Polymer Composite, Science Engineering Research Centre, Universiti Sains Malaysia, Engineering Campus, 14300 Nibong Tebal, Penang, Malaysia

Email : srnadras@usm.my

Abstract. This paper aims to determine the optimum vulcanization temperature on curing characteristic, mechanical and physical properties of natural rubber/palygorskite composites. Three variations of vulcanization temperature (140, 150 and 160°C) were conducted on the samples. Cation-exchanged method used to treat the palygorskite. Rheological measurements and mechanical testing (tear, tensile and fatique life) were conducted on the composites sample. Scanning electron microscopy (SEM) was used to reveal the dispersion of palygorskite in NR matrix. It was found that the scorch and cure time of NR/Palygorskite decrease with increasing of vulcanization temperature. The tensile strength and fatigue life was optimum at 140°C and slightly decreased with the increasing of vulcanization temperature. The SEM micrograph revealed the strength and weakness in the system. It can be concluded that the optimum properties of NR/Palygorskite composites were at 140°C of vulcanization temperature and most of the mechanical and physical properties were slightly decreased with the increasing of vulcanization temperature.

1. Introduction

Natural rubber (NR) is a unique biomass extracted from a tropical plant (cis-1,4-polyisoprene) and composed of carbon and hydrogen atoms only. The natural property, renewability which making it carbon neural, environmental friendly and biodegradable, has boost its demand around the world [1]. Nowadays, usage of natural rubber alone in rubber product manufacturing is not practical and economic anymore. Hence, the introduction of filler into the rubber compound has increased its mechanical properties while reducing manufacturing cost. One of the earliest and common nanofiller is carbon black that is relatively inert and stable materials [2]. However, with its current high price and its adverse effect toward environmental and human body, many researchers encouraged to find an ecofriendly alternative carbon black. Clay is identified as one of a good alternative for carbon black and many studies have been carried out to improve the performance of clay in a nanocomposite. Structurally, clays have different size and mineralogy compared to other fine-grained soils, which is the behavior, is more to plastic when contain water and become harder when dried it [3]. Palygorskite is a clay mineral type and a crystalline hydrated magnesium aluminium silicate mineral with needle shape result in unique colloidal properties and give high surface area and porosity particles when thermally activated [4-5]. The hydrophilic behavior of the palygorskite is incompatible and has a poor dispersion in the hydrophobic polymer phase [6]. Therefore, the idea is the organoclay surface need to undergo treatment such as cation exchange surfactant to improve the rubber-filler interaction when the polarity of pure natural clay is modifed [7]. Zheng *et al.*, (2004) [8] use trimethyloctadecylamine as cation exchange agent to treat the organoclay and compound it with the rubber matrix at 175°C while Rath *et al.*, (2012) [9], use palygorskite as filler and compared the properties of the treated and untreated nanocomposites at 120°C.

In the present study, the curing characteristic, physical and mechanical properties of the cation treated palygorskite of the natural rubber composite were investigated under different vulcanization temperature (140, 150 and 160°*C*). The properties of the NR/cation treated palygorskite composites were analyzed and compared with NR/untreated palygorskite composites. Then, the dispersion of the fillers in the rubber matrix was characterized using Field Emission Scanning Electron Microscopy (FESEM) and performed on the fracture surface of the composite.

2. Experimental

2.1. Material

Natural rubber SMR 20 (cis-polyisoprene- the glass transition is -70°C and density of 0.91 g/mL at room temperature), High Ammonia (HA) latex (60% total solid content and specific gravity is 0.95-0.96), zinc oxide, stearic acid, tetramethylthiuram disulphide (TMTD), zinc diethyldithiocarbamate (ZDEC) and sulphur was kindly provided by Zarm & Chemical Supplier Sdn Bhd. While antioxidant, anchoid, octdecylamine, potassium hydroxide (KOH) and calcium chloride was supplied by Sigma Aldrich (M) Sdn Bhd and used as received.

2.2. Methodology

2.2.1. Preparation of Palygorskite Clay.

20 g of clay was added into the 1000mL hot water (80°C) and stirred for 1h (300rpm). Later, mixed hot solution with mixture of 7.5 g of dedocyltriethylamine and 4.2 mL HCl in 500 mL of water and stirred at rpm 400 for another 1 h (80°C). The precipitate filtered and washed with hot distilled water until reached pH7 and dried at 60°C for 36 h before used. After that, the 6 pphr palygorskite clay was mixed with 3 pphr 10% KOH, 5 pphr anchoid and water and stirred for 2 h. Ball mill technique was applied to the dispersion mixture for 2 days to produce aqueous suspension.

IOP Conf. Series: Materials Science and Engineering **223** (2017) 012017 doi:10.1088/1757-899X/223/1/012017

2.2.2. Preparation of Master Batch.

200g of natural rubber latex was heated until it reached about 45 to 50°*C* and poured into the warm latex 20 g of modified palygorskite clay by cation ions exchanged and continuously stirred for 15minutes at 60°*C*. Then, 10% of CaCl solution was used to co-coagulate the latex compound and washed with excess distilled water until it reached pH 7. The coagulated latex was cut into small pieces and dried in an oven at 100°*C* for 2 h.

2.2.3. Mixing and Rubber Compounding

Two rubber compounds with different type of filler in the same amount of loading (6pphr) were prepared in semi efficient vulcanization (SEV) system based on summarized formulation (Table 1). After mixing, the compounds undergone compression moulding in order to produce sheet and button form of sample for testing with three different vulcanization temperatures.

Table 1. Formulation of rubber compounding

Ingredients	Pphr		
	Gum compound	Untreated Palygorskite Compound	Treated Palygorskite Compound
NR rubber	100	100	100
Stearic acid	2	2	2
ZnO	5	5	5
ZDEC	0.5	0.5	0.5
TMTD	1.5	1.5	1.5
Antioxidant	2	2	2
Sulphur	1.5	1.5	1.5
Palygorskite	-	6	6

2.2.4. Characterization

Curing characteristics of the rubber compounds were carried out using Monsanto rheometer (140, 150 and 160°C) for 30 min at 25 dNm range of the torque. Tensile test was carried out according to ASTM D412 while tear followed ASTM D624. Fatigue life test was carried out using a Monsanto fatigue to failure tester (FTFT) with dumbbell samples BS type E subjected to cyclic strain 100rpm. The fatigue life was calculated using Japanese Industrial Standard (JIS) average. Hardness test was performed according to ASTM D2240 at room temperature. Swelling test was performed by soaking the sample (30 x 5 x 2mm) in toluene until equilibrium for 72 h at room temperature. The rubber filler interaction, Qf/Qg was calculated using Lorenz and Parks equation [10]. Compression set test is measured the differences of original thickness and thickness after recovery expressed in percentage. FE-SEM was carried out using Zeiss Supra 35V SEM instrument. The samples from the surfaces of tensile testing were coated with a thin layer of gold palladium using a SEM sputter coater to prevent the occurrence of superfluous.

IOP Conf. Series: Materials Science and Engineering **223** (2017) 012017 doi:10.1088/1757-899X/223/1/012017

3. Result and Discussion

3.1. Curing Characteristic

The curing characteristics of unfilled and filled compound with untreated palygorskite and cation treated palygorskite obtained from rheometer graph are presented in Table 2. The results show the different cure characteristics due to different type of filler added into rubber composites.

Table 2. Cure characteristics of different composites at different vulcanization temperature

Compound	Temperature (°C)	M_L (dNm)	M_H (dNm)	t_{s2} (min)	t_{90} (min)	Cure Index (min^{-1})
Unfilled	140	0.14	6.32	3.75	7.14	29.5
	150	0.15	6.03	2.04	4.43	41.84
	160	0.11	5.65	1.28	1.99	140.85
Untreated palygorskite	140	0.30	5.47	4.38	7.40	33.11
	150	0.22	5.21	2.40	3.59	84.03
	160	0.21	4.89	1.45	2.02	175.44
Cation Treated palygorskite	140	0.27	5.53	2.40	3.81	70.92
	150	0.25	5.34	1.50	2.17	149.25
	160	0.23	5.11	1.06	1.51	222.22

Higher value of minimum torque (M_L) at temperature 140°C for all three compounds was recorded and decreasing with increasing of vulcanization temperature. Theoretically, when curing temperature arise, molecular chain has a tendency to absorbed the heat energy which produced during preheat stage and allowed the molecules to have higher mobility which led to lowering the rubber viscosity. Apart from that, the surface treatment also affected the minimum torque, M_L of rubber compound due to the present of hydroxyl group on the surface of untreated palygorskite tends to form agglomeration and led to increase in M_L of the compound [11]. The value maximum torque (M_H) of cation treated palygorskite/NR composites is higher compared to palygorskite/NR composites due to the formation of hydrogen bonding between amino group in the cation exchange agent (octadecylamine) and hydroxyl groups in the organoclay [12]. The presence of treated palygorskite reduced the scorch time and cure time (t_{90}) which accelerated the vulcanization prosess. This phenomenon is due to the cation exchange treatment which improve the compatibility of rubber-filler interaction by reduced silanol groups of palygorskite and able to homogenously distributed in the rubber matrix that leads the formation of crosslinking also increase the rate of vulcanization. In contrast, untreated palygorskite is incompatible with rubber matrices and have a high tendency to agglomerate, hence increase the scorch time when more torque needed to shear the agglomeration of filler in the matrix. The surface of hydrogen bond that absorbed basic accelerators and deactivate them is resulting in reducing rate of vulcanization process [13]. Cation treated palygorskite has the highest cure rate index at increased trend for 140, 150 and 160°C respectively and higher compare to the untreated palygorskite

IOP Conf. Series: Materials Science and Engineering **223** (2017) 012017 doi:10.1088/1757-899X/223/1/012017

composite. Hence from the viewpoint of cure rate index, it has obeyed the theory like aforementioned in cure time. Where the cation treated palygorskite has reduced silanol groups of palygorskite and hence increases the rate of vulcanization, while for untreated palygorskite composites the acidic hydroxyl, siloxane groups has absorbed the basic accelerators causing it to slow down the degree of vulcanization. The low viscosity of the component tends to form a continuous phase [14].

3.2. Tensile Properties

Figure 1 shows tensile strength and elongation at break of unfilled NR, NR/untreated palygorskite and NR/cation treated palygorskite composite at different vulcanization temperatures. The results show for treated palygorskite composite a descreasing trend while for unfilled NR and untreated palygorskite showed a quite same trend for 3 vulcanization temperatures. It can assume that, the high vulcanization temperature has deteriorated the mechanical properties of the rubber composite. Besides that, overall tensile strength of cation treated palygorskite is higher than untreated palygorskite rubber composite. This is because of the surface treatment was enhance the mechanical properties of the rubber composite by increase it compatibility and interfacial bonding with the rubber matrix at the same time causing the elongation at break become limited . This treatment lower down the surface energy and improves the wetting characteristic of the rubber matrix, and resulting in larger interlayer spacing. Additionally, the cation can provide functional group that can initiate the polymerization of monomers to improve the adhesion between the palygorskite and rubber matrix and the filler-rubber interaction and its mechanical properties.

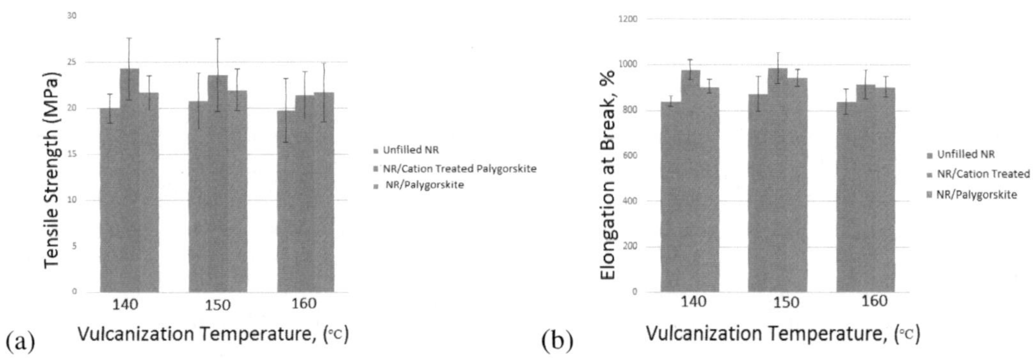

Figure 1. The effect of unfilled NR, NR/Cation Treated and Untreated Palygorskite/NR composite at different vulcanization temperature on (a) tensile strength (b) elongation at break.

The modulus showed higher value for treated palygorskite compared to untreated palygorskite in increasing trend with increasing temperature. According to Olad (2011) [15], when added rigid filler to the soft polymer matrix, it will carry the major ration of applied load to the polymer matrix under stress condition, if the interfacial interaction between the filler and matrix is adequate [16]. The good

IOP Conf. Series: Materials Science and Engineering **223** (2017) 012017 doi:10.1088/1757-899X/223/1/012017

compatibility and the present of amine group on the surface of palygorskite has led to better interaction of filler-matrix phase, resulting a higher stiffness. On the contrary untreated filler composite will have high tendency to slippage since the compound just need to break the crosslinkage, physical entanglement and matrix-matrix interaction compared to the treated palygorskite. The modulus properties reduced might be due to deterioration of the rubber chains when increasing the vulcanization temperature, resulting a phenomenon called chain scissoring.

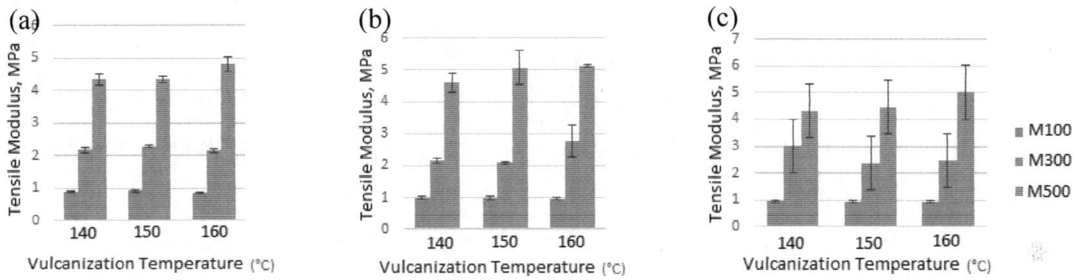

Figure 2. Tensile modulus of rubber composites at different vulcanization temperature, (a) Unfilled NR, (b)NR/ Cation Treated Palygorskite and (c) NR/untreated palygorskite

3.3. Tear Properties

Figure 3 shows the highest tear strength recorded at 150°C for NR/cation treated palygorskite. This could be explained by the ability of the crosslinks and rubber chains to crystalline upon the applied of tearing force which that turns the amorphous rubber into a semi crystalline material. This crystallite is highly oriented and act like filler crosslinks. Furthermore, due to the local reinforcement of strain-induced crystallites at highly strained crack tip, the crack growth resistance in natural rubber increased, resulting higher tear strength [17].

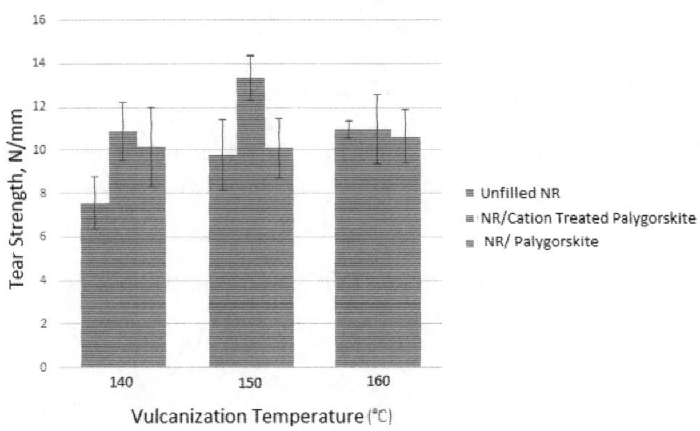

Figure 3. Tear strength of unfilled NR, NR/cation treated palygorskite, and NR/palygorskite composites at different vulcanization temperature

Based on the result, unfilled NR shows the lowest tear strength at 140°C followed 150°C and 160°C respectively. It because of low resistance in the gum rubber that prevent the rubber from resist the slippage that occurred due to the stress concentration compared to the composite filled with nano filler which able to reinforce the compound and form a surrounding layer of immobilized polymer.

3.4. Fatigue Life Properties

Figure 4 illustrates that fatigue life increase from 140°C (92.39 kcycles) to 150°C (100.64 kcycles), and decreased at temperature 160°C (86.17 kcycles). The decreasing of fatigue life properties from 150 to 160°C is due to the elevated temperature as stated by Gupta (2004) [18], where it was proven that higher curing temperature is not only decreased the curing time but also decreased the properties of the static strength and fatigue life. The fatigue to failure process involves initiation of microcracks and propagation of microscales and naturally occurring flaws due to combination of factors such as fillers, contaminants or void in the matrix, non-homogenous dispersed compound ingredients and surface flaws [19-20]. Besides, the main reason that affects fatigue life properties is crosslink density, which undergoes the reversion with increasing the vulcanization temperature. These crosslinks comprise two general types, which are monosulphide and disulphide crosslinks, and polysulphide crosslinks, which impart various properties to the rubber matrix. Polysulphide crosslink chain may break at higher vulcanization temperature and formed monosulphide crosslink chain.

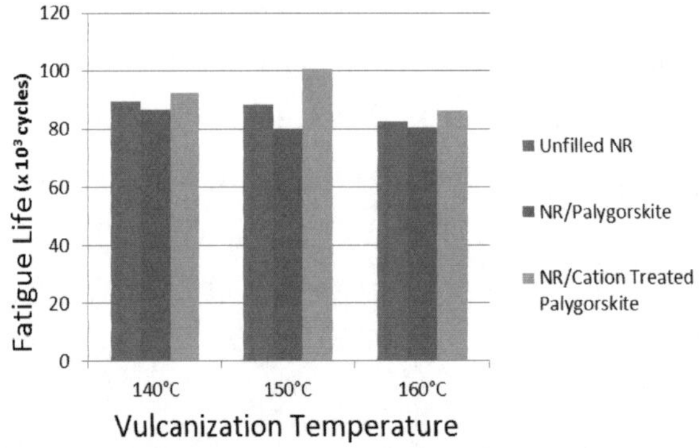

Figure 4. Fatigue life of Unfilled NR, NR/ untreated Palygorskite and NR/ Cation Treated Palygorskite composite at different vulcanization temperature

Figure 5 shows the rubber-filler interaction, Q_f/Q_g values on different composites at different vulcanization temperature. The higher values of the Q_f/Q_g, indicate the lower the rubber-filler interaction.

IOP Conf. Series: Materials Science and Engineering **223** (2017) 012017 doi:10.1088/1757-899X/223/1/012017

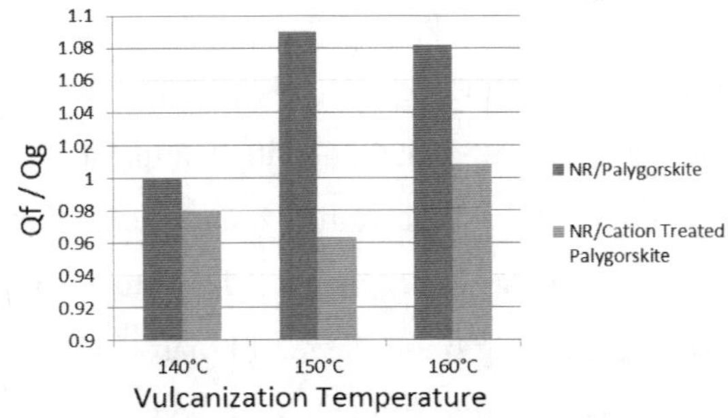

Figure 5. Q_f/Q_g of NR/ untreated Palygorskite and NR/ Cation Treated Palygorskite composite at different vulcanization temperature

The treated filler has better filler interaction than untreated filler composite due to the surface modification that replaced by some quartenary ammonium compound with a long hydrophobic tail enhances the compatibility between the filler and matrix. This method changes the behaviour of the filler from hydrophilic filler to hydrophobic filler.

3.5. Hardness Properties

Figure 6 shows at 140°C, unfilled NR shows the highest hardness compared to NR/ cation treated palygorskite and NR/ palygorskite, which reflects the lowest elongation at break 840% at the same temperature. Such result indicates that, at 140°C, the rubber composites achieved an optimum curing characteristic, where the results obtained was corresponded with one another. Besides temperature 140°C, cation treated palygorskite shows the highest hardness compared to the other two compounds. This is due to lower surface energy and larger interlayer spacing of cation treated palygorskite, which depends on presence of hydrophobic groups of surfactants, which enhanced the intercalate ability of palygorskite while improving physical and mechanical properties [20]. Elevated cure temperature produces a network with lower degree of crosslink density (thermal breakdown of monosulfide crosslink) and an increase in the extent of sulfidic main chain modification cause a reduction in modulus [21].

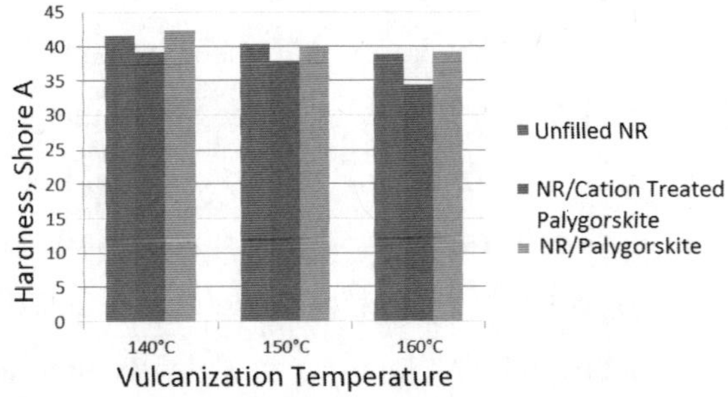

Figure 6. Hardness of unfilled NR, NR/ untreated palygorskite and NR/ cation treated palygorskite composite at different vulcanization temperature.

3.6. Swelling Properties

Figure 7 shows the crosslink density of the compounds at different vulcanization temperature. The crosslink density of the NR/ cation treated palygorskite composite samples was higher than NR/ palygorskite and unfilled NR composite at the same vulcanization temperature. This is because of there are many crosslinks produced in the rubber matrix when rubber matrices are swelled in toluene. The crosslink density showed a decreasing trend from 140 to 160°C. At vulcanization temperature of 140°C, highest values of crosslink density recorded which were 6.34 x 10^{-5} g/mol, 6.10 x 10^{-5} g/mol and 6.88 x 10^{-5} g/mol of the crosslink density for unfilled NR, NR/ untreated palygorskite and cation NR/ cation treated palygorskite composites respectively. Vulcanization temperature has significant effect on crosslink structure. Optimum properties are obtained when curing is done at the lowest possible temperature. However, to increase productivity, higher temperatures are frequently used.

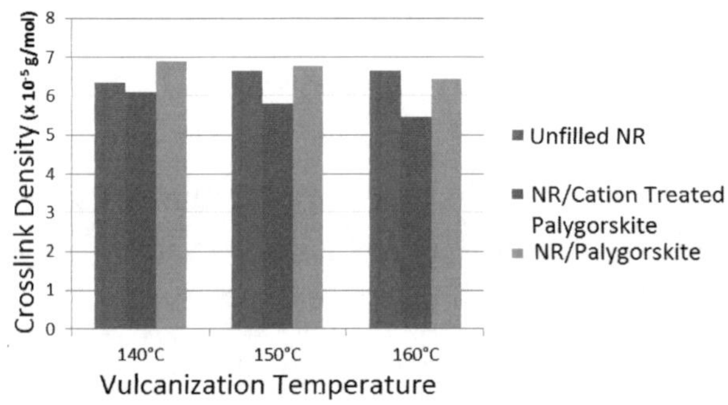

Figure 7. Crosslink density of the unfilled NR, NR/ untreated Palygorskite and NR/ Cation Treated Palygorskite composites at different vulcanization temperatures

3.7. Field Emission Scanning of Electron Microscope (FESEM)

Figure 8 shows the micrograph of tensile fracture surface of all three composite. Figure 8(a) shows the image surface of natural rubber composite without filler. It clearly shows that the surface has a small dot due to presence of additive like antioxidant, vulcanizing agent etc, which indicates that all ingredients homogenously distributed in the rubber matrices. Furthermore, Figure 8(b) shows the image of NR/palygorskite composite. The presence of needle shapes fillers shows homogenously dispersed and adhered perfectly to the matrix. This might be due to the good processing parameter and latex compounding method that allowed the palygorskite to reinforce the adjacent matrix and enhanced physical and mechanical properties. In contrast, Figure 8(c) shows rougher surface of NR/ cation treated palygorskite composite compare to Figure 8(a) and (b). It obviously shows the successfully effect of surface modification to improve the interaction with the presence of many tear lines on the fracture surface. It was proven that the composite needed a high energy to break the chain and it was designated the effect of better interaction between chains and has higher strength properties

[22]. Furthermore, the uniform tear lines in one direction presence on the NR/ cation treated palygorskite composite shows that the modified palygorskite undergo isotropic failure.

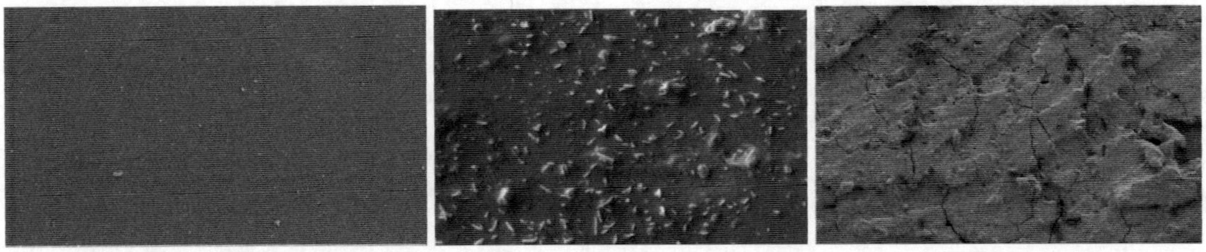

(a) (b) (c)

Figure 8. FESEM micrograph at 500 X magnification of (a) unfilled NR, 5000 X magnification (b)NR/ untreated palygorskite composite and (c) NR/ cation treated palygorskite composite

4. Conclusion

The result of NR/ cation treated palygorskite composite cured at 140°C demonstrated the optimum properties among others. These are corresponding with the expected outcome where the cation treated palygorskite have enhanced the compatibility and homogeneity in the rubber matrix. The improvement caused by the mobility of the rubber chains constrained. Furthermore, the investigation on morphology of cation treated palygorskite give strong evident and positive impact of the cation treated palygorskite to mechanical properties of rubber composite. Therefore, surface modification which using to modified surface of palygorskite clay was enhanced the mechanical properties of the composites. This method was improved filler – matrix interfacial interaction hence increases the mechanical properties.

References

[1] Kohjiya S and Ikeda Y 2014 Chemistry, Manufacture and Applications of Natural Rubber, *Woodhead Publishing Limited* 68-118.
[2] Dell L D *et al* 2006 *Journal of Occupational and Environmental Medicine* **48** 1219-1229.
[3] Stephen G and Martin R T 1995 *Clays and Clay Mineral* **43** 255-256.
[4] Haden J W 1963 Palygorskite: properties and uses *Clay Clays Mineral* **10** 284-290.
[5] Ke X B *et al* 2008 *B The Journal of Physical Chemistry* **112** 5000-5006.
[6] Gainnelis E 1996 Polymer layered silicate nanocomposites *Advanced Materials* **1** 29-35.
[7] Peeterbroeck S *et al* 2004 *Composites Science and Technology* **64** 2317-23 .
[8] Zhang L *et al* 2004 *Key Laboratory on Preparation and Processing of Novel Polymer Materials of Beijing,* Beijing University of Chemical Technology.
[9] Rath J P *et al* 2012 *Procedia Chemistry* **4** 131-137.
[10] Lorenz O and Parks C R 1961 *Journal of PolymerScience Part A Polymer Chemical* **50**(154) 299-312.
[11] Scurati A and Lin C J 2006 *Rubber Chemistry and Technology* **79** 170-197.
[12] Ge X *et al* 2015 *Applied Clay Science* **118** 265-275.
[13] Hosseini S M and Razzaghi-Kashani M 2014 *Polymer* **55** 6426-6434.

[14] Ismail H *et al* 2002 *Polymer Testing* **21** 565-569.
[15] Olad A 2011 *Advances in Diverse Industrial Applications of Nanocomposites* 113-138.
[16] Zhao S *et al* 2008 *Composites Science and Technology* **68** 2965-2975.
[17] Brüning K *et al* 2013 *Polymer* **54**(22) 6200-6205.
[18] Gupta M 2004 Processing and fabrication of advanced materials XIII *Stallion Press (S) Pte Ltd* 872.
[19] Nabil H *et al* 2011 *Key Engineering Materials* **471** *628-633.*
[20] Wu Y P *et al* 2006 *Macromolecular Materials and Engineering* **291**(8) 944-949.
[21] Singla P *et al* 2012 *Green and Sustainable Chemistry* **2** 21-25.
[22] Zhang H *et al* 2015 *Journal of Elastomers and Plastics* **48**(4) 331-339.

Physico-mechanical properties of silanized-montmorillonite reinforced chitosan-co-poly(maleic anhydride) composites

O A Saputra[1], A Fajrin[1], M Nauqinida[1], V Suryanti[1], E Pramono[1,2]
[1]Chemistry Department, Faculty of Mathematic and Natural Sciences, Sebelas Maret University, Jl. Ir Sutami 36A Surakarta, Indonesia.
[2]Inorganic Material Research Group, Faculty of Mathematic and Natural Sciences, Sebelas Maret University, Jl. Ir. Sutami 36A Surakarta, Indonesia.

E-mail: edi.pramono.uns@gmail.com

Abstract. To solve the problems of dependence on petroleum as starting material in the manufacturing of plastics in Indonesia, green plastic from biopolymer like chitosan to be one of promising options and alternative to replace the conventional plastics. However, to overcome the mechanical and physical properties of chitosan, the addition of reinforcement agent was introduced. In this study, silanized-montmorillonite (sMMt) has been prepared as a reinforcement agent in the chitosan-co-poly(maleic anhydride) (referred as Cs-MAH) matrix. Silanizing of montmorillonite is one of strategy to improve the interaction between montmorillonite and chitosan, consequently, the mechanical properties, tensile strength of composites contained 6 phr of sMMt improved 56.5% to chitosan. Moreover, the presence both MAH and sMMt on the comosites also reduced swelling degree and swelling area by 20.6% and 26.7%.

1. Introduction

World plastics production growth is increased every year [1]. Plastics to be one of the important materials in daily life. Many application such as automotive, packaging, aerospace, household tools, electrical devices and many other were replaced by plastics [2-4]. However, the availability of petroleum as a source of raw materials in plastic manufacturing will be depleting [5]. Moreover, almost all petroleum based plastics was difficult to degrade thus, the waste plastic to be environmental pollution. To overcome this problem, some conventional strategy was introduced, such as recycling, energy recovery and landfill disposal [6]. However, to provide an environmental friendly plastics for different types of applications in a sustainable manner, an alternative is needed. Therefore, in this study will be introduced bio-based plastic and it's modification as a candidate to replace conventional petroleum-based plastics.

Chitosan is a polysaccharide derived from deacetylated of chitin or known as poly(N-acetyl-D-glucopyranose). It composed mainly of β-(1,4)-linked 2-deoxy-2-amino-D-glucopyranose units. Chitosan and chitin, next to cellulose, are the second most abundant biopolymers [7]. Chitosan has

been extensively investigated for recent decades for food packaging, molecular separation, artificial skin, bone substitutes, water engineering and other, due to its good biocompatibility, biodegradability, multiple functional groups as well as its aqueous solubility [8]. However, its properties, such as water sensitivity, mechanical strength, thermal stability, rigidity and gas barrier properties are frequently not good enough for wide ranges of applications [9]. Recently, there has been a growing study in chemical modification of chitosan to improve its properties and widen its applications via graft copolymerization with maleic anhydride (MAH) which it is widely used as a grafting agent in various material [10, 11]. Incorporation of grafting material MAH in chitosan matrix increased the mechanical properties of the polymer, since the anhydride group could react with hydroxyls of chitosan [12, 13].
Various layered silicates from clay like montmorillonite (MMT) as a reinforcing material for polymers has received considerable research attention because of their potentially high aspect ratio and its unique characteristics [14]. It is clay type 2:1, which consists of the stacking a hydrated alumina-silicate layered clay made up of two silica tetrahedral sheets fused to an edge-shared octahedral sheet of aluminum hydroxide [15]. These superior properties could improve tensile strength and modulus as well as enhance thermal stability [9,16], reduce gas permeability and boost the flammability of matrix polymer [14].

Under natural condition, among layered of the MMT is likely to stack via the Van der Waals forces. Besides, clay are hydrophilic leading to incompatible with most hydrophobic polymeric materials. Thus, chemical modification on the clay or matrix polymer or both have been reached to enhance the interaction between the clay and the polymer [14]. The surface of the MMT is commonly modified via cation exchange technique or adding a compatibilizer materials to compatibilize the layered silicate with most hydrophobic polymer matrix. Recently, modification of the hydrophilic clay surface have been extended using silane coupling agent [16]. Wang *et al.* (2010) [17] developed GPTMS-modified zeolite for reinforcement of chitosan membranes, their study indicated that by using (3-glycidoxypropyl) trimethoxysilane (GPTMS) was able to achieve denser and stronger surface morphology. GPTMS is a crosslinking agent that has two functional groups, one is an organic functional group bonding with organic materials, and the other one is an epoxy group interacting with inorganic materials [18,19]. From some of these facts, so in this study examined the effect of modification of clay with a silane compound as a filler in the matrix of chitosan-co-poly (Maleic anhydrides) on mechanical and thermal properties. Moreover, the morphology, swelling degree and swelling area were evaluated.

2. Experimental

2.1. Materials

All material in this research was used without further furification. Chitosan (Cs) was purchased from PT. Biotech Surindo, Indonesia. Montmorillonite (MMt) and glycidiloxypropyl trimethoxysilane (GPTMS) were obtained from Sigma Aldrich, Singapore. Toluene, acetic acid (Act), maleic anhydride (MAH), benzoyl peroxide (BPO) and sodium hydroxide (NaOH) were purchased from Merck.

2.2. Preparation and Characterization of Silanized-Montmorillonite (sMMt)

About 10 g of montmorillonite was thermally treated by calcination method in the temperature of 400 °C for 12 h, namely treated-montmorillonite (t-MMt). The t-MMt (5 g) and GPTMS (16 mL) were sonically treated in 100 mL of toluene for 30 min. Then, the mixture was reflux and kept at boiling state for 24 h. The product was filtered and washed several times using dry toluene. The final product, referred as silanized-montmorillonite (sMMt) was dried overnight at room temperature and continuously dried again in oven at 80 °C for 24 h.

The characterization of the sMMt was then characterized using several instruments. Functional group analysis was conducted using FTIR (Fourier Transform Infra-Red) model Prestige21-Shimadzhu. The samples was prepared in fine powder by introducing KBr to form a pellet disc. The applied number of scan was 40 times with resolution 4 cm^{-1}. Crystallinity behaviour of samples was measured using XRD (X-Ray Diffraction) model Shimadzhu Lab-X 6000 by applied Cu Kα as light sources. The samples was prepared in fine powder before analysis. Morphology of sMMt was visualized using SEM (Scanning electron microsocpy)-EDAX model Phenom FEI.

2.3. Preparation of the Composites

The chitosan/clay composites were prepared via solution casting method. About 2 g of chitosan was dissolved in acetic acid solution (80 mL, 1.5%) at room temperature. Afterward, 10% wt of MAH was added into chitosan solution and stirred slowly. After 1 h reaction, amount of sMMt (0, 6, 9 and 12 phr) was added into the mixture solution. Mixing process was performed for 1 h at 70 °C. The mixture solution was poured into glass template and dried for 12 h at 50 °C. The film plastic was then washed using 0.05 M sodium hydroxide solution and washed again with distilled water for several times. Finally, the bioplastic sMMt reinforced chitosan-co-poly(maleic anhydride), referred as Cs-MAH/sMMt was dried at room temperature for two days.

2.4. Mechanical Testing

Mechanical testing was conducted using Mechanical Testing Machine Ray-Run M500-50CT. The sampel was prepared accoring to ASTM D638-02a method. The cross speed test was adjusted to 5 mm/min.

2.5. Swelling Testing

Swelling testing was used to determine the water adsorbed bility of composites which correlated with degree of humidity resistance. The composites was prepared into small pieces with 1 x 1 cm^2 area, approximately. The initial weight of samples is weighted. The samples were soaked into water up to 24 h and afterward, the final weight of samples were recorded.

3. Results and Discussion

3.1. Characterization of sMMt

In order to prove the surface modification of MMt with organo-silane compound, the characterization like FTIR and XRD has been conducted. Fourier transform infra-red is one of the most worthwhile instruments, which is used for chemically investigation. Figure 1 shows the FTIR spectrum of MMt and sMMt. Based on Figure 1, generally MMt has characteristic absorbance band at around 1060 cm^{-1} and 802 cm^{-1} corresponding to Si-O stretching vibration (indicated in Figure 1) and –OH deformation vibration from Al-OH group, respectively. After modification with GPTMS compound, the spectrum of sMMt is slightly different from their raw material. It was clearly see that at sMMt infra-red spectrum was found the –CH stretching vibration band at around 2937 cm^{-1} and 2877 cm^{-1}. The appearance of the bands are consequently of surface modification by organo-silane group which contained hydrocarbon backbone. Moreover, it is also observed a characteristic peak of –C-O-C- stretching vibration around 1211 cm^{-1} which slightly shift from the GPTMS spectrum. The presence of absorbance band at 905 cm^{-1} which is characteristic vibration of oxirene ring, is confirming allegation that MMt is successfully modified with GPTMS compound.

Chemically analysis, to prove the surface modification was conducted using XRD technique. Figure 2 showed that the XRD pattern of MMt and sMMt. The MMt have spesific 2θ peak at 8.66° with d_{001} basal spacing 10.2 Å. After modification, the 2θ is 8.74 (d_{001} = 10.1 Å). The decreasing of basal spacing of MMt after modification is due to the presence of siloxane network in MMt layer's surface, thus shorten the layer space. The surface modification of MMt by GPTMS compound is not change the MMt crystal structure. However, the crystallinity of MMt decreased after modification due to the formation of amorphous siloxane network in the MMt surface.

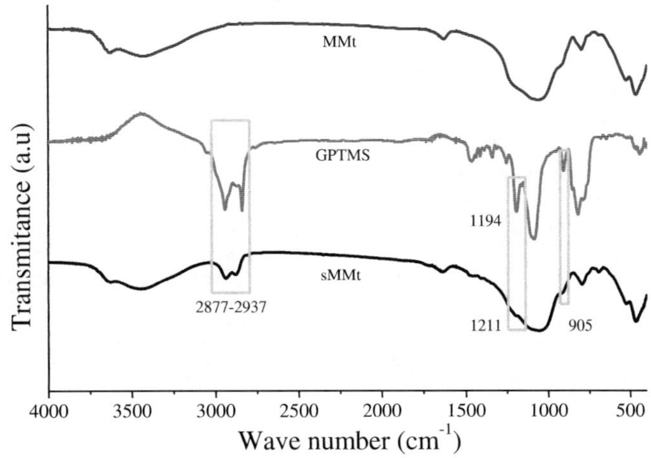

Figure 1. FTIR spectrum of MMt, GPTMS and sMMt.

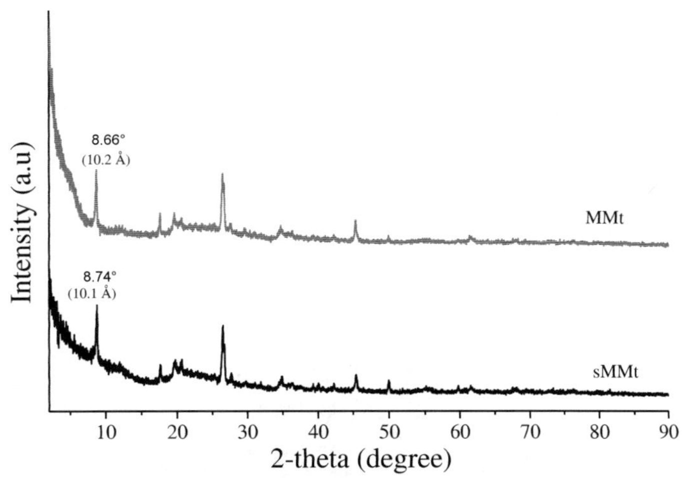

Figure 2. XRD pattern of MMt and sMMt.

The SEM image of both MMt and sMMt are presented in Figure 3 and 4, respectively. The MMt physical structure based on the SEM imgae is like bulk structure with fine surface. However, after surface modification, the MMt surface is more rough and it is conected with solid line as seen in Figure 4. The solid line which is pointed with yellow arrow, is probably the formation of siloxane netrowrk. Moreover, the presence of solid line is made the physical structure of sMMt to be porous in their surface.

IOP Conf. Series: Materials Science and Engineering **223** (2017) 012018 doi:10.1088/1757-899X/223/1/012018

Figure 3. SEM image of MMt.

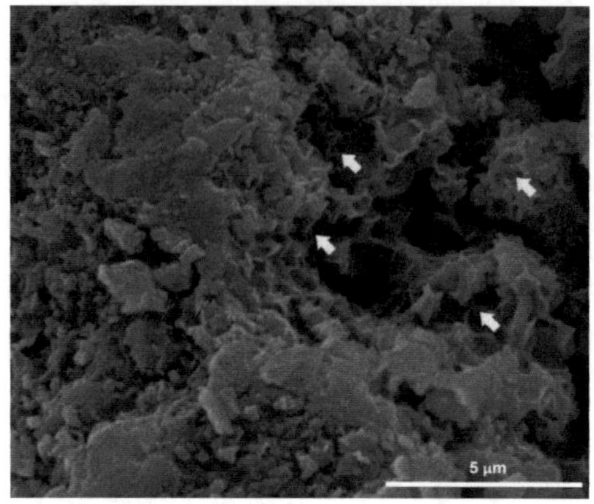

Figure 4. SEM image of sMMt.

Figure 5 showed the SEM-EDAX data of sMMt. This analysis is conducted to proved the surface modification of MMt by silane compound. Based on the data, it can be seen that the main elements of sMMt on the sufrace are C, O, Al, Si atoms, where the O, Al and Si atomic persentages are 36.84, 2.46 and 16.95%, respectively. Additionally, the C atomic persentation is reach 43.74%. The presence of the Carbon atom in the sMMt surface indicated the modification is successfully achieved due to the C atom is generated from silane compound. Moreover, the ratio of Al:Si is very high, which is 1:6.9. The improvement of Si atom is due to the effect of GPTMS addition, where each molecule of GPTMS contained 1 Si atom.

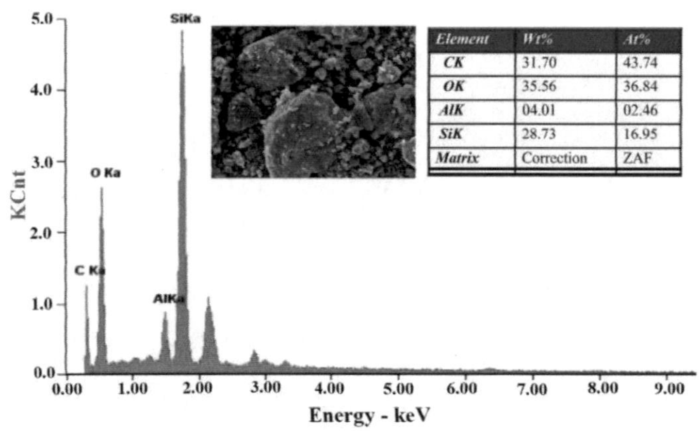

Figure 5. SEM-EDAX result of sMMt.

3.2. Mechanical Properties

Mechanical testing was conducted to determined the mechanical properties of composites. This testing was also used to determined the optimum sMMt addition on composites preparation. The mechanical properties are including tentile strength, elongation percentages and young's modulus. Tensile strength and elongation percentages of Cs-MAH/sMMt composites are presented in Figure 6. The tensile strength of chitosan-co-poly(maleic anhydrides) is 25 MPa. The addition of 3 phr of sMMt was able to improved the tensile strength of composites by 20% compared to chitosan-co-poly(maleic anhydrides). The increasing of the composites tensile strength was achieved until 6 phr addition of sMMt, which improved by 56.5% compared to chitosan. Afterward, the tensile strength of composites are decreased by 9.7 and 14.7% of Cs-MAH/sMMt composites contained 9 and 12 phr of sMMt respectively, compared to the composites contained 6 phr of sMMt. Based on this data, it can be concluded that the best condition regarding on tensile strength value is given at 6 phr addition of sMMt in chitosan matrix. The improvement of composites tensile strength by addition of clay is due to the clay minerals which contained alumina and silica layers can be distributed by intercalated or exfoliated [20]. Homogenous distribution of clay particles generated high mechanical properties of composites [21]. In this study, the presence of silane compound in MMt surface is able to act as compatibilizer agent [22]. GPTMS compounds which contained silane group and oxirene group are chemically interaction with both MMt surface to form siloxane network and amino group of chitosan [23]. Moreover, the MMt was known as reinforcement agent, thus it improved the mechanical strength and stiffness of materials [24]. Consequently, the elasticity of composites decreased by addition of reinforcement [25]. Elongation percentages of chitosan is 13.7% and it decreased by addition of sMMt. At optimum composition (6 phr of sMMt), the elongation percentages is 5.7% or decreased to 57.8% compared to chitosan. The more addition of sMMt into chitosan matrix was decreased the elasticity of composites.

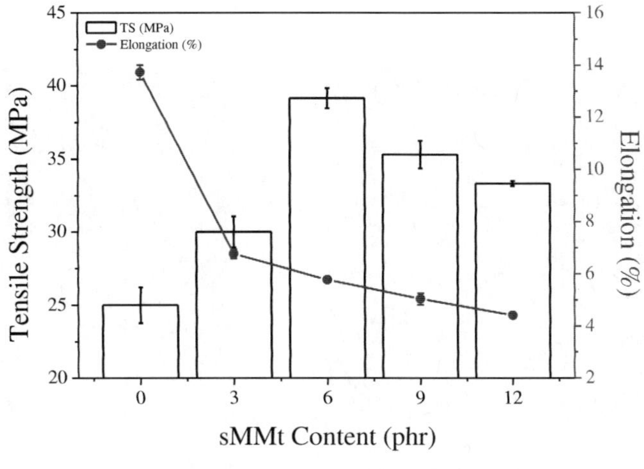

Figure 6. Elongation and tensile strength of Cs-MAH/sMMt.

Figure 7. is the young's modulus graph of chitosan/sMMt composites. This data explained the degree of composites britleness, where the higher of young's modulus value showed the composites tend to be brittle. The chitosan-co-poly(maleic anhydrides) has 182.4 MPa of young's modulus value.

However, the addition of high content of sMMt, which is 12 phr increased the young's modulus by 314%. At optimum condition, the young's modulus of composites improved to 271.9%.

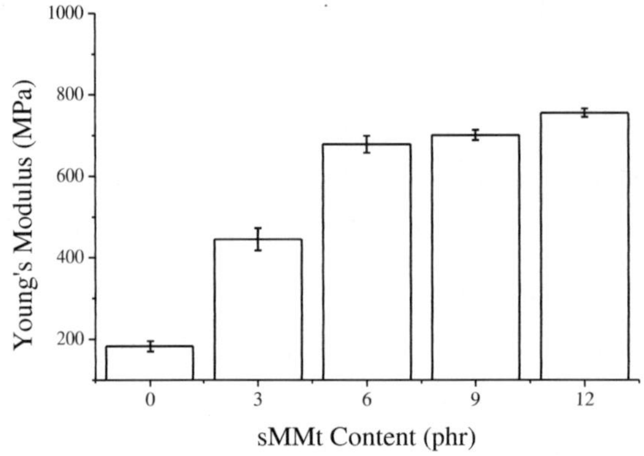

Figure 7. Young's modulus of Cs-MAH/sMMt.

3.3. Morphologycal Study

Morphology analysis using scanning electron microscopy was used to observe the fracture surface of sample after applied tensile testing. Figure 8 and 9 are the fracture surface morphology of optimum Cs-MAH/sMMt composites at different magnification. As seen in Figure 9, the fracture surface of composites after tensile testing showed strong interfacial bonding between chitosan and sMMt. It was indicated by the presence of cracking in the sample as result of high force distribution between matrix and reinforcing material [26]. However, at another spot is also found a flat fracture surface. This is due to the properties of clay particles tend to be brittle or the distribution of clay particles in chitosan matrix are unevenly well distributed.

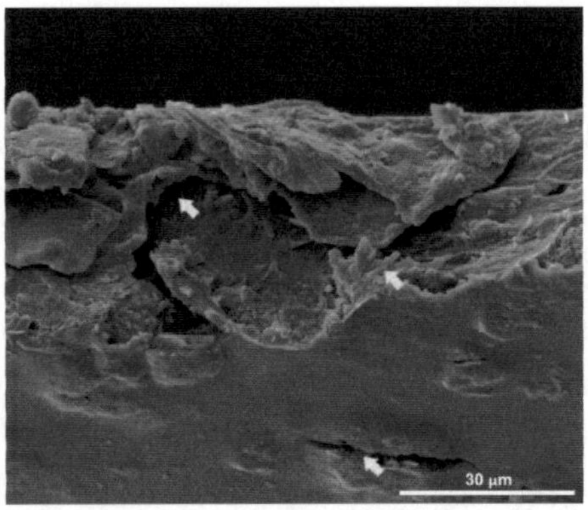

Figure 8. SEM image of Cs-MAH/sMMt cross section

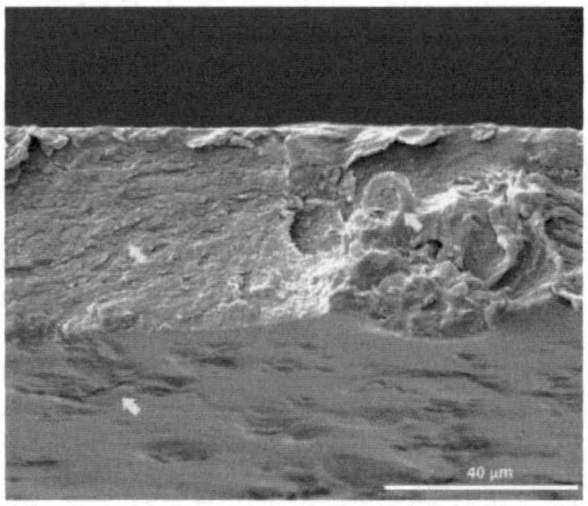

Figure 9. SEM image of Cs-MAH/sMMt cross section

3.4. Swelling Degree and Area of the Composites

Swelling properties is one of important testing to analyzed the humidity resistance of composites by soaking the sample in water. The plastic which is applied as packaging plastic should have high humidity resistance for long period application. To improved the humidity resistance of materials, one of strategy can be conducted by modification the materials surface using hydrophobic materials, such as silane compound [27]. Silane compound was able to form siloxane network, which have ability to limiting the contact of water and materials surface. It can be seen at Figure 10 that the addition of sMMt until 3 phr, decreased both swelling degree and area of composites by 20.6 and 26.7%, respectively. However, the higher of sMMt content on chitosan is turn both swelling degree and area to high value. It is due to the nature of clay mineral which swelled in water [28].

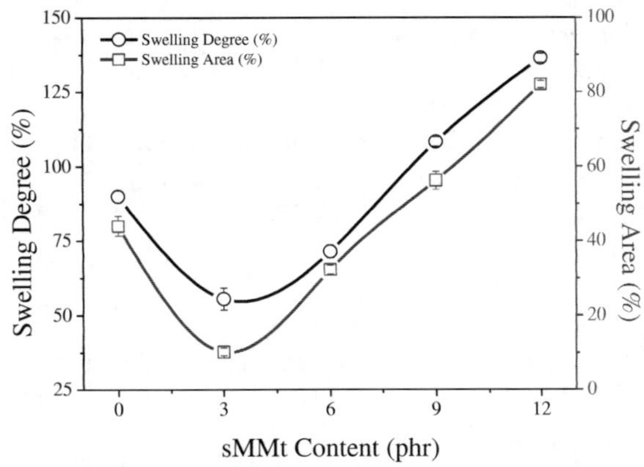

Figure 10. Swelling degree and area of Cs-MAH/sMMt composites.

4. Conclusion

Surface modification of montmorillonite clay using silane compound has been successfully achieved which proved by functional group and surface element analysis by FTIR and SEM EDAX, respectively. The presence of sMMt in chitosan was able to improve mechanical properties up to 56.5% at 6 phr sMMt addition. However, the elongation of the composites decreased by the addition of sMMt. By the mechanical testing, the composites contained 6 phr of sMMt to be the optimum condition in composites preparation. Based on swelling testing, the humidity resistance of composites contained 3 and 6 phr of sMMt are higher than chitosan. Thus, regarding to mechanical and physical properties, the chitosan/sMMt optimum composites can be used as green bio-based plastic to replaced the conventional plastic.

Acknowledgement

Author(s) would like to acknowledge to Indonesian General of Higher Education for financial support through PKM-Dikti grants 2016.

References

[1] Panda A K *et al* 2010 *Renewable Sustainable Energy Rev.* **14**(1) 233.
[2] Desai S D *et al* 2003 *J. Polym. Res.* **10** 141.
[3] Jeyanthi S and Rani J J 2012 *Int. J. Physc. Sci.* **7** 5765.
[4] Tall S *et al* 2001 *Polym. Adv. Technol.* **12** 279.
[5] Kreye O *et al* 2013 *Green Chem.* **15** 1431.
[6] Wu C and Williams P T 2010 *Fuel* **89** 3022.
[7] Fernandez J G and Donald E I 2014 *Macromol. Mater. Eng.* **299** 932.
[8] Jayakumar R *et al* 2005 *J. Carbohydrides. Polym.* **62** 142.
[9] Wang S F *et al* 2005 *Polym. Degrad. Stabil.* **90** 123.
[10] Kim S J *et al* 2001 *Polym.* **42** 4073.
[11] Pagnoulle C and Jerome R 2001 *Polym.* **42** 1893.
[12] Rocco A M *et al* 2001 *Polym.* **42** 5199.
[13] Techawinyutham L *et al* 2016 *Adv. Mechan. Eng.* **8** 1.
[14] Choi Y Y *et al* 2009 *Polym. Bull.* **63** 47.
[15] Xu Y, Ren X and Hanna M A 2006 *J. Appl. Polym. Sci.* **99** 1684.
[16] Chow W S and Neoh S S 2009 *J. Appl. Polym. Sci.* **114** 3967.
[17] Wang Y *et al* 2010 *Chem. Eng. Process.* **49** 278.
[18] Purwnto M *et al* 2016 *RSC Adv.* **6** 2314.
[19] Yanga C C *et al* 2011 *J. Power Sources.* **196** 4458.
[20] Chen B 2004 *British Ceramic Transactions.* **103**(6) 241.
[21] Pavildou S and Papaspyrides C D 2008 *Prog. Polym. Sci.* **33** 1119.
[22] Jiang S *et al* 2015 *Compos. Sci. Technol.* **110** 87.
[23] Ren L *et al* 2002 *Biomater* **23** 4765.
[24] Ismail H *et al* 2015 *J. Physic. Sci.* **26**(2) 89.
[25] Balakrishnan H *et al* 2012 *J. Thermoplas. Compos. Mater.* **27**(2) 233.
[26] Osman H *et al* 2010 *J. Compos. Mater.* **44**(12) 1477.
[27] Osman H *et al* 2011 *Polym. Compos.* **33**(4) 609.
[28] Saputra O A *et al* 2016 *J. Pure Appl. Chem. Res.* **5**(2) 85.

IOP Conf. Series: Materials Science and Engineering **223** (2017) 012019 doi:10.1088/1757-899X/223/1/012019

Novel route of synthesis for cellulose fiber- based hybrid polyurethane

F H Ikhwan, S Ilmiati, H Kurnia Adi, R Arumsari, M Chalid
Department of Metallurgy and Materials Engineering, Universitas Indonesia, UI Depok New Campus, Depok 16424, West Java – Indonesia

E-mail : fauziahanumikhwan@gmail.com

Abstract. Polyurethanes, obtained by the reaction of a diisocyanate compound with bifunctional or multifunctional reagent such as diols or polyols, have been studied intensively and well developed. The wide range modifier such as chemical structures and molecular weight to build polyurethanes led to designs of materials that may easily meet the functional product demand and to the extraordinary spreading of these materials in market. Properties of the obtained polymer are related to the chemical structure of polyurethane backbone. A number polyurethanes prepared from biomass-based monomers have been reported. Cellulose fiber, as a biomass material is containing abundant hydroxyl, promising material as chain extender for building hybrid polyurethanes. In previous researches, cellulose fiber was used as filler in synthesis of polyurethane composites. This paper reported a novel route of hybrid polyurethane synthesis, which a cellulose fiber was used as chain extender. The experiment performed by reacting 4,4'-Methylenebis (cyclohexyl isocyanate) (HMDI) and polyethylene glycol with variation of molecular weight to obtained pre-polyurethane, continued by adding micro fiber cellulose (MFC) with variation of type and composition in the mixture. The experiment was evaluated by NMR, FTIR, SEM and STA measurement. NMR and FTIR confirmed the reaction of the hybrid polyurethane. STA showed hybrid polyurethane has good thermal stability. SEM showed good distribution and dispersion of sorghum-based MFC.

1. Introduction

Cellulose, the main structural component of plant that confers strength and stability to the plant cell wall, is widely used in commercial material due to its properties such as biodegradability. Cellulose that provided from fibers has many hydroxyl functional groups which makes it potential as a chain extender in polyurethanes polymerization, but there has not been any study that proved it. In last decade, some efforts have converted to several forms like, MFC, microcrystalline (MMC) with several end use application [1]. The aim of this study, is to determine the availability of Sorghum bicolor plant cellulose as a chain extender in polyurethane polymerization.

Sorghum bicolor is a plant with a high level of photosynthesis. It takes only 3 – 5 months for the sweet sorghum to mature, and the species can be planted in very dry areas. The top grain, leaves and stalks can be used for forage, and sugars can be extracted from stalks for use in syrup and fermenting ethanol [2]. According to a study conducted by Bakeer and his colleagues in 2013, sorghum containing 32.30% of fiber with high crystallinity [3]. Sorghum bicolor and sorghum bagasse are only different in name, but the same sweet sorghum plant [4]. According to those study, sorghum bicolor worthy to be utilized as chain

IOP Conf. Series: Materials Science and Engineering **223** (2017) 012019 doi:10.1088/1757-899X/223/1/012019

extender because its fiber contain providing many hydroxyl groups.

Polyurethane (PU) chemistry has been studied intensively and is well developed. Typically, polyurethanes are obtained by the reaction of a diisocyanate with bifunctional or multifunctional reagent such as diols or polyols. Material design is achieved by varying the chemical structures and molecular weights of both the polyol and the diisocyanate compounds as the polymer precursors. A number of polyurethanes prepared from biomass-based monomers have been reported. So far, these polymers were synthesized by using several stages, such as pre-polymerization followed by chain extension, and have relatively low molecular weights. Also the direct polymerization of relatively high molecular weight monomers is a promising approach to obtain high molecular weight polyurethanes [5]. A study in 2013 explains that a lignin based polyurethane thermoplastics have a high tensile strength shown by its young's modulus. The study shows that by utilizing lignin's rigid cell properties, one can synthesize a high tensile strength polyurethane [6]. This study proves that sorghum bicolor can be a chain extender in this hybrid polyurethane (HPU) synthesis and serves a novelty route of hybrid polyurethane synthesis. The hybrid polyurethanes are obtained by pre-polymerization reaction of polyethylene glycol as the soft segment with the molecular weight of 4000 and 6000 and high molecular diisocyanate, 4,4′-Methylenebis (cyclohexyl isocyanate) (HMDI) as the hard segment, followed by copolymerization reaction which cellulose as a chain extender.

Over the past few years, several materials used by the automotive industries have been using PU – incorporated materials. The mentioned materials are especially referred to the materials needed for body kits, tires, and interiors. However, PU – incorporated materials usually do not have the ability to degrade naturally.

2. Materials and Methods

2.1 Materials

Polyurethanes were synthesized from mixtures of Polyethylene glycol (PEG) Mw:4000 g/mole and 6000 g/mole obtained from Sigma – Aldrich as the soft segment with the composition 0.0025 moles (10g) and 4,4′-Methylenebis (Cyclohexyl isocyanate) (HMDI) obtained from Sigma – Aldrich as the hard segment with the composition 0.01 moles (2.45 mL), Dibutyltin dilaurate (DD) obtained from Sigma - Aldrich as the polymerization catalyst with the composition 2 drops (0.1 mL), dimethyl formamide (DMF) as the solvent, Sorghum bicolor plant cellulose obtained from Chandra Asih as the chain extender with the composition ranging from 0.5– 2.0 g, and cellulose obtained from Sigma – Aldrich as the chain extender with the composition ranging from 0.5 – 2.0g.

IOP Conf. Series: Materials Science and Engineering **223** (2017) 012019 doi:10.1088/1757-899X/223/1/012019

2.2 Cellulose Preparation

Sorghum was crushed by using a crushing machine and sieved to get uniform size (50 mesh). 50g of fiber were soaked with 0.25 M of NaOH solution overnight, then washed 4 times with distilled water until the pH of fiber became 7. After that, the fiber trough bleaching stage, in this stage, fiber was soaked by using 5% NaClO solution for 4 h and then it was washed with distilled water until the pH became 7. Next stage was acid hydrolysis stage with 25% wt H_2SO_4 solution for 2 h and washed with distilled water. In all stages mentioned, fiber was immersed and stirred using hot plate magnetic stirrer. In the last step, cellulose obtained was dried by oven, at 60°C for 6 h.

2.3 Synthesis of Hybrid Polyurethane

The process started by preparation of schlenk tubes filled with DMF solvent and PEG was fluxed using vacuum suction machine and nitrogen gas three times. All of the flasks were then dipped into the oil bath and heated. The temperature is then raised to 75°C and stirring for 1h. After 1 h, HMDI were injected into the flasks in a nitrogen atmosphere. After that, 2 drops of Dibutyltin Dilaurate (DD) catalyst were added into each flask. Then stirring were conducted for 2 h at a constant temperature at 75°C [7].

After the solution became as white as milk, each of the schlenk tubes were filled by 0.5g chain extender (see each composition in Table 1 with the temperature raised to 80°C and stirring for 1h. The process was then terminated using 20 mL of water and was left intact for 24 h until a phase separation occurred, and the hybrid polyurethane will precipitate. After that, the precipitate will be separated using a filtration cloth and dried under room temperature for 3 days and will be continued with characterization afterwards.

2.4 Characterization

^{1}H-Nuclear Magnetic Resonance (^{1}H-NMR) and ^{13}C-NMR was used to analyze the structure of hybrid polyurethanes.^{1}H-NMR was used to identify the functional groups within the polyurethanes and to measure the amount of phenolic hydroxyl functional groups in the polyurethanes. Characterization procedure was conducted at DTMM FTUI using NMR Jeol 500MHz. Measurement condition are 23° C for the temperature and 6 Pa for the pressure (vacuum).

Fourier Transform – Infra Red (FTIR) to analyze and confirm the NMR data about the functional groups formed in the hybrid polyurethanes. Specifically, FTIR was used to identify the functional groups in the lignin. Characterization procedure was conducted at DTMM FTUI using FTIR PerkinElmer UATR Two Spectrum Two. The instrument has a limitation of range of wavelength from 450 – 4500 λ.

Simultaneous Thermal Analysis (STA) consisted of *Thermogravimetric Analysis (TGA)* and *Differential Scanning Chromatography (DSC)* to analyze physical changes such as Tg and Tm (thermal) and thermal stability.Characterization procedure was conducted at DTMM FTUI using STA PerkinElmer STA6000. The instrument has a limitation of rate of 5 min /°C and max temperature of 1000°C.

3. Results and Discussion

The reaction can be seen in Figure 1. The reaction was initiated by pre-polymerization process of polyethylene glycol with a diisocyanate (HMDI). The isocyanate groups are reactive towards active hydrogen groups such as amine (-NH), hydroxyl (-OH), water, carboxylic acid (-COOH), urea, and amide. Isocyanate will attack hydroxyl group forming urethane group. In the pre – PU forming stage, urethane groups are formed by reacting HMDI and PEG with DD (Dibutyltin dilaurate) as a catalyst and DMF (dimethyl formamide) as a solvent in nitrogen atmosphere. The reaction followed by copolymerization, which cellulose as chain extender to obtain polyurethanes.

In this stage, the isocyanate group from HMDI reacted with the hydroxyl group from PEG forming a pre – polymer which is an interspersed polyurethane copolymer. The pre – polymer in this stage still has isocyanate (N=C=O) group which will be eliminated after the addition of cellulose in the second step. The reaction of pre – polymer and cellulose occurs in the second step. Isocyanate group left on the end of the pre – polymer's chain attacked hydroxyl group in the cellulose forming a urethane chain. The reaction shows two groups of hydroxyl; aliphatic hydroxyl, and phenolic hydroxyl. The reaction rate for isocyanate and aliphatic hydroxyl is higher than the reaction rate for isocyanate and phenolic hydroxyl. The result of this process is the hybrid PU.

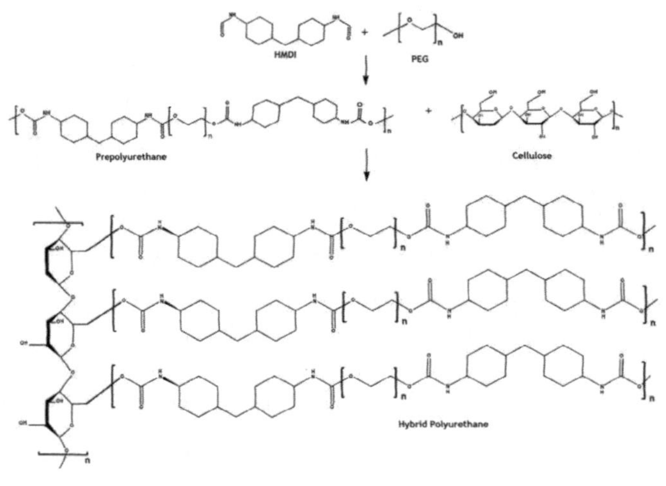

Figure 1. Purposed Synthesis Route of HPU

Figure 2 shows the FTIR spectra of each hybrid polyurethanes synthesized from PEG 4000 (a) and PEG 6000 (b) with addition of cellulose for each the macrodiol at the amount of 1.5 g for the top graph, 1 g for the middle graph, and 0.5 g for the bottom graph. The structure of the polymer was confirmed by FTIR spectrum. The presence of the amine bonds (NH_3) appears at 1280 – 1342 cm^{-1},C-O A*cyl* bond at 1242 cm^{-1}, and secondary amine at 2887 cm^{-1}. Secondary amine indicating that isocyanate bonds (N=C=O) opened and made urethane group bond with hydrogen. Urethane group bond in shows that hybrid polyurethane was obtained. According to the figure, fiber addition has influence on amount of polyurethane obtained, it is signed from high molecule vibration in 1242 cm^{-1}, 2887 cm^{-1} from polyurethane obtained and 1104 and 3361cm^{-1} from cellulose.

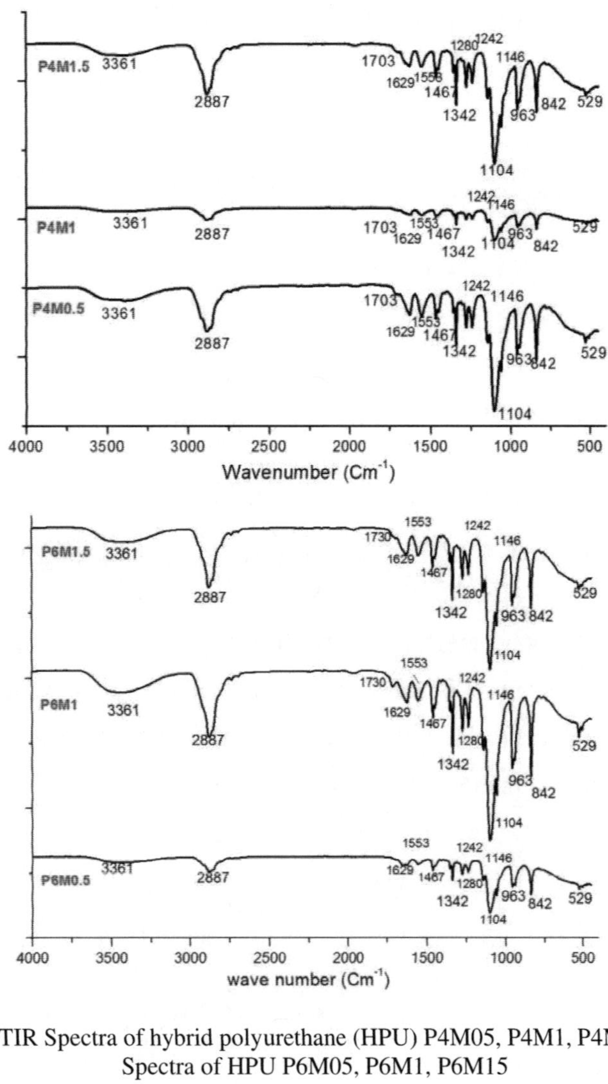

Figure 2. Top figure: FTIR Spectra of hybrid polyurethane (HPU) P4M05, P4M1, P4M15; Bottom figure: FTIR Spectra of HPU P6M05, P6M1, P6M15

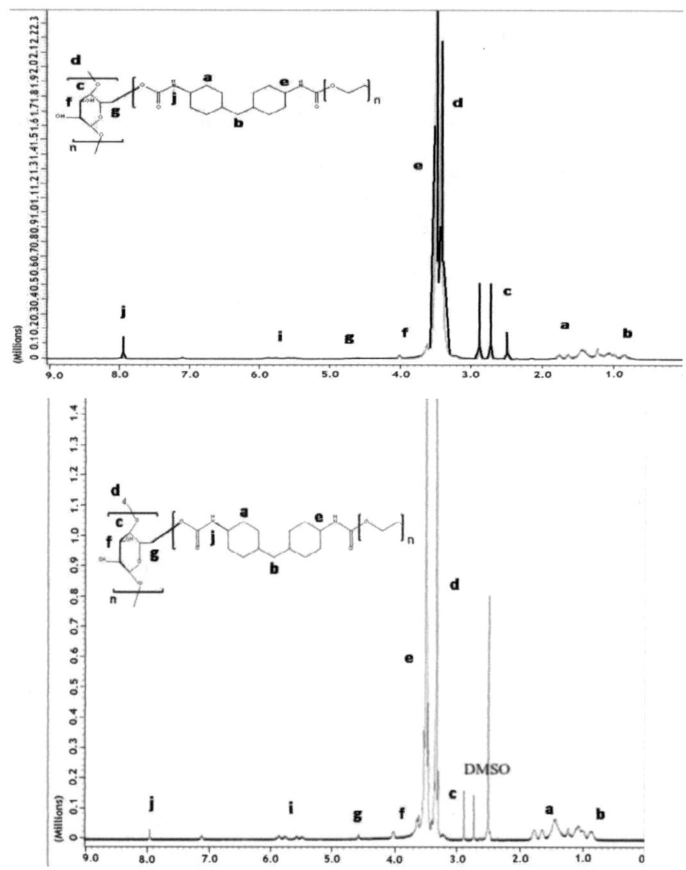

Figure 3. Top Figure:^{1}H-NMR Spectra of P4M1; Bottom Figure: ^{1}H-NMR Spectra of P4S1

Characterization of the polymers was performed by 1H- 13C-NMR, and FTIR. A representative ^{1}H and^{13}C-NMR spectrum for as one of the novel hybrid polyurethanes is shown in Figure 3. The urethane C=O group and amide groups appear in all of the samples, the former appears at about 0.865 until 1.075 ppm in ^{1}H-NMR and in 162.737 ppm in ^{13}C-NMR spectra. Meanwhile the chemical shift of the amide groups appear at about 8.2229 in spectrum for the NH.

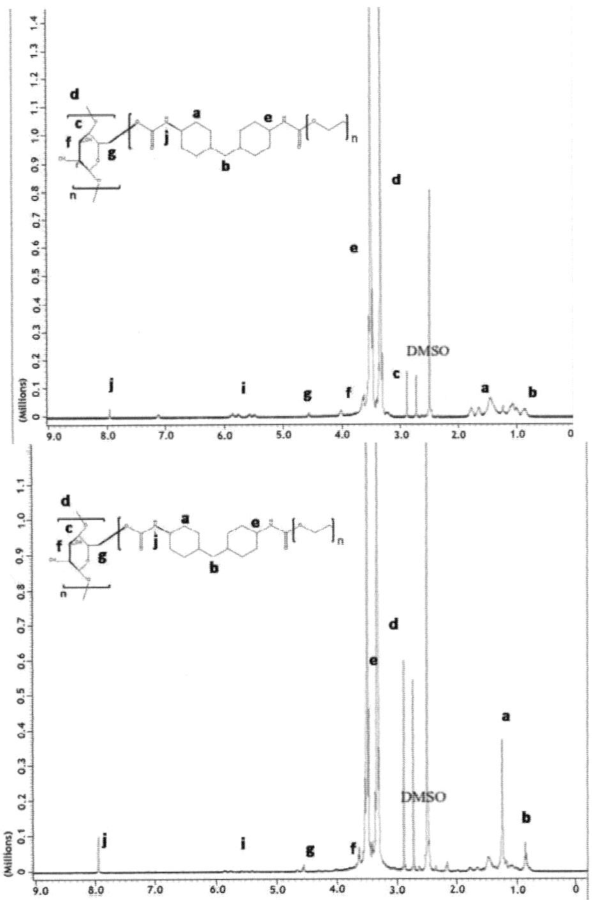

Figure 4. Top Figure: ^{1}H-NMR Spectra of P6M1; Bottom Figure: ^{1}H-NMR Spectra of P6S1

Based from the overall FTIR spectra from Figure 3 and 4, sample P4M1, P4S1, P6M1, P6S1 have a relatively similar absorbant bond. The structure of the hybrid polyurethane is also confirmed by FTIR spectrum, the presence of the amine bonds appears at 1342 and 1280 $cm^{-1,}$ while the urethane group and the amide group appear at 1703 cm^{-1} and 1553 cm^{-1}, respectively. According to the FTIR spectra, compared with P4M1 (Hybrid Polyurethane based from PEG – 4000 and 1 gr of cellulose), P4S1 (Hybrid Polyurethane based from PEG – 4000 and 1 gr of sorghum) has higher peak of OH that appears at 3400 cm^{-1}.

From the Figure 3 and 4, it was proven that peak increases at the addition of sorghum as a chain extender (P4S1 and P6S1) because there were more formed products (main and side product). The main product is the hybrid polyurethane while the side product is the $N\text{-}O_2$ stretch in 1392 cm^{-1} which only occurred in the sample with the sorghum addition.

IOP Conf. Series: Materials Science and Engineering **223** (2017) 012019 doi:10.1088/1757-899X/223/1/012019

3.1 Thermal Behavior of Hybrid Polyurethane

The thermal characteristics on the hybrid PU material was done by utilizing *Differential Scanning Calorimetry* (DSC). The result obtained from DSC was shown as heat flow vs. temperature graph (Figure 5 left). The graph doesn't show any heat flow decreasing peak, however it shows the increase in glass transition (Tg) and melting temperature (Tm). The increase in Tg and Tm was caused by the higher concentration of O-H in sorghum than in cellulose so more urethane chain was formed in the sample. Figure 5 right shows that the differences of thermal characteristics of a material were caused by the differences in utilizing PEG as a soft segment.

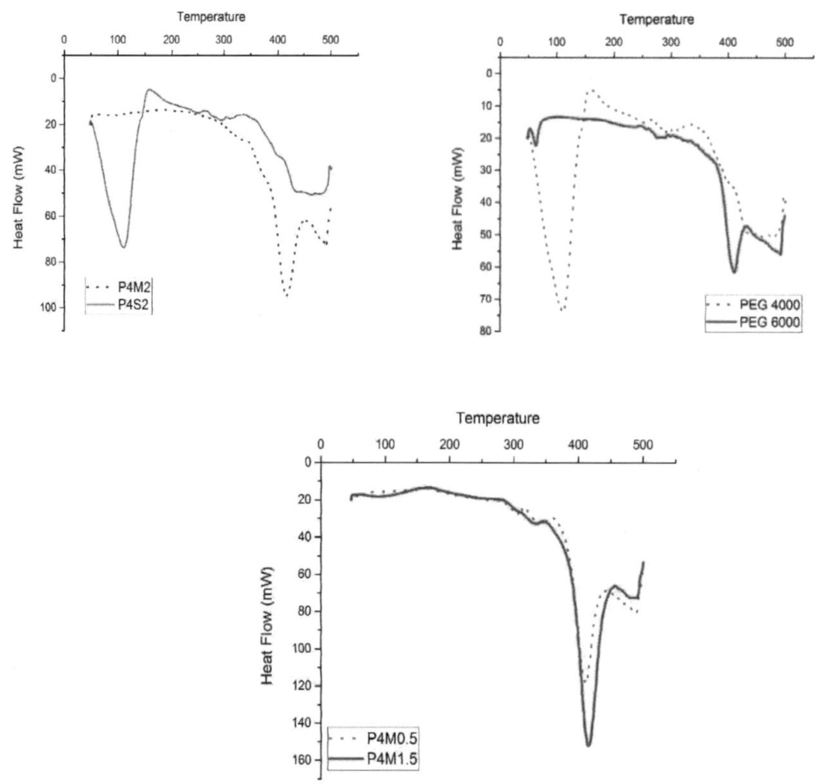

Figure 5. Left Figure: Sorghum and Cellulose Hybrid polyurethane P4M2, P4S2 DSC curve comparison; Right Figure: PEG 4000 and PEG 6000 DSC curve comparison; Bottom Figure: P4M0.5 and P4M1.5 DSC curve comparison.

The higher the molecular weight, the higher the melting temperature because the PEG – 6000 has more polyol chain than PEG – 4000. This causes PEG – 6000 have more cross linking with sorghum in forming a PU. Figure 5, Bottom Figure shows that the addition of cellulose as a cross linker in polyurethane affects greatly the thermal characteristics of a material. The greater the amount of cellulose added to a material, the higher its melting temperature will be because there are more cross linking happened in the process in forming a PU.

Table 1. DSC curve analysis results on hybrid polyurethane

Sample	Tg (°C)	Tm (°C)
P4M2	76.49	415.32
P4S2	110.60	431.78
PEG 4000	111.22	410.30
PEG 6000	63.70	435.00
P4M0.5	59.07	410.05
P4M1.5	104.20	415.07

3.2 Thermal Stability Analysis

Thermogravimetric Analysis was utilized to analyze thermal stability. The result of the test was shown as a graph that shows the reducing of polyurethane mass against temperature increase as shown in Figure 6.

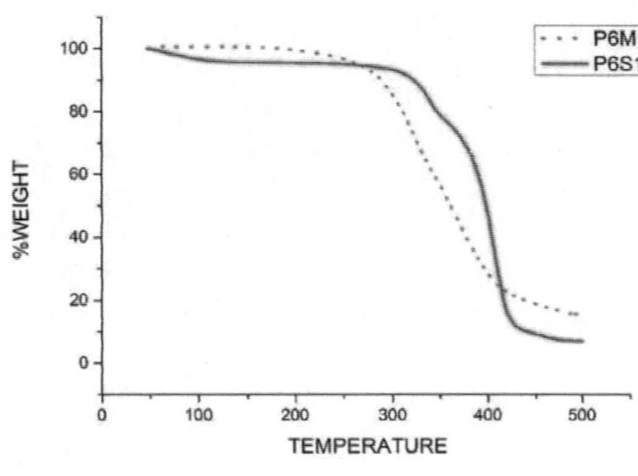

Figure 6. Polyurethane TGA Curve

According to the curve, it can be seen that, HPU cellulose obtained from sorghum bicolor addition (P6S1.5) has better thermal stability than HPU with pure cellulose addition (P6M1.5). It is predicted caused by the chemical strong bonds which come from the hydroxyl. The –OH functional group give a strong Van Der Walls intermolecular force on cellulose structure and provide the cellulose high mechanical and thermal stability. As this van der walls force increased from the –OH functional groups which has been bonded to the PU, it may arrange any ordered chain structure or strong attraction bonds with other functional groups which make it more difficult to break when the temperature is continuously increased. In this case, cellulose was obtained from sorghum predicted has more –OH functional than the pure cellulose, because cellulose from sorghum still containing hemicelluloses and lignin. This result was supported by Figure 7, where in the right of Figure 7 shows that there was bundles of cellulose obtained from sorghum in hybrid polyurethane sample, with diameter of cellulose about 16μm. While from the left Figure 7 it can be seen that surface of polyurethane with pure cellulose (P6M1.5) has no bundles like the right one does. It Indicates that the two type of cellulose has different size and diameter, so when it was mixed and reacted, the pure cellulose is able to be blended well with all of composition. While, cellulose form sorghum, with larger diameter and tend to congregate, gave a stronger influence on the thermal stability. The bundle was hard to break when the sample was heated with high temperature.

Figure 7. Left Figure: SEM Result of HPU with pure cellulose with 1000x magnifying; Right Figure: SEM Result of HPU with sorghum bicolor with 1000x magnifying.

4. Conclusion

The three composition of the base materials (PEG, HMDI, and Sorghum bicolor plant cellulose) have been successfully synthesized as hybrid polyurethane proven by the NMR and FTIR results in showing C=O, NH, amide, and C-O-H functional groups in all of the samples. Concentration variation of sorghum bicolor plant cellulose as a chain extender influences the thermal properties of the hybrid polyurethane proven by the DSC and TGA results in showing the glass transition and melting temperature comparison between 2 kinds of cellulose. Overall, the synthesis of hybrid polyurethane by using cellulose as the chain extender can be proven in this study.

References

[1] Yuanita E *et al* 2015 *Procedia Chemistry* **16** 608 – 615.
[2] Qi C 2013 *Industrial Crops and Products Elsevier.* 45:455-460.
[3] Bakeer B, *et al* 2013. *Ain Shams Eng J.* **4** 265–271.
[4] Widya F, Supriyanto, Iswanto A H 2015 *J. Eng. Technol. Sci.* 47 149 – 159.
[5] Chalid M 2012 *Levulinic Acidas a Renewable Source for Novel Polymers* RIJKSUNIVERSITEIT
[6] Saito Tomonori 2003 Naskar Amit K. 2013 *Development of Lignin – Based Polyurethane Thermoplastics. Royal Society of Chemistry*, RSC adv 3.21832.
[7] Firdaus D F *et al* 2015 *Indonesia J. Chem)* 130 – 140.

IOP Conf. Series: Materials Science and Engineering **223** (2017) 012020 doi:10.1088/1757-899X/223/1/012020

The effect of modified ijuk fibers to crystallinity of polypropylene composite

I Prabowo, J Nur Pratama, M Chalid
Department Metallurgy and Material Engineering, Faculty of Engineering, Universitas Indonesia, Kampus Baru UI Depok 16436, Indonesia

Email : chalid@metal.ui.ac.id

Abstract. Nowadays, plastics becomes concern associated with its degradation and environmental issues. It has led studies to develop an environmental-friendly material. To minimize the impact of those problems, recently the usage of natural fibers as a filler are introduced because of biodegradability and availability. The promising natural fiber is "ijuk" fiber from *Arenga pinnata* plant as a filler and polypropylene (PP) polymer as a matrix. Unfortunately, the natural fibers and polymers have the different properties on which polymers are polar while natural fibers are non-polar so that reducing the compatibility and resulting the poor crystallinity. To enhance the compatibility and crystallinity, ijuk fibers were prepared by multistage treatments including alkalinization with 5 and 10% sodium hydroxide (NaOH), oxidation with 3 and 6% sodium hypochlorite (NaClO) and hydrolysis with 20% sulphuric acid (H_2SO_4) in sequences. The purposes of multistage treatments are to remove the components such as lignin, wax, hemicellulose, to cause an oxidative fragmentation of remaining lignin and to annihilate the amorphous parts respectively. Fourier-Transform Infrared (FTIR) confirms the compatibility meanwhile Differential Scanning Calorimetry (DSC) reveals the crystallinity and Scanning Electron Microscope (SEM) displays surface morphology of polypropylene. The experiments were revealing that the effects of "ijuk" fibers by the multistage treatments of 5 and 10% NaOH resulting the crystallinity of polypropylene around 31.2 and 27.64% respectively compared to the crystallinity before adding the "ijuk" fibers for 16.8%. It indicates that the entire treatments increasing the compatibility and crystallinity of polypropylene. In addition, the use of 5% NaOH offers the better crystallinity than non-treated polypropylene. The experiments conclude that by adding alkalinized "ijuk" fibers of multistage treatments can increase the compatibility and crystallinity of polypropylene.

1. Introduction

Recently, composite materials where plastics as a matrix combined with fibers as a reinforcement are widely used in various applications including military, construction, packaging, household devices and automotive parts. The plastics are used commonly from petro-polymer because of low cost, low density, easily to process, have the mechanical properties compared to metals and ceramics while the fibers usually used carbon and glass fiber offering the high strength and high modulus elasticity [1]. But the problems emerged because the use of non-degradable materials such as plastics and glass or carbon fibers becomes concern regarding to its degradability. It had triggered the land pollution and had led studies to develop an environmental-friendly materials to reduce the impact of polluted environment. On the other hand, In several decades the use of natural fibers in composites has replaced inorganic materials in some applications due to the ability such as degradation and decomposition. Furthermore, the natural fibers have the properties for instance low cost, low density, and recyclable. They include many types of natural fiber such as flax, hemp, jute, kenaf, sisal, abaca, ramie, coir pineapple leaf fiber, bamboo, rice husk, oil palm and bagasse [2,3]. Another fiber still being developed is ijuk fiber or sugar palm fiber which has the advantages such as corrosion resistance, lower density, lower cost, low processing methods, and also available in the nature. It obtained from Arenga pinnata plant [4]. Bachtiar *et al.*, (2008) also found that ijuk fiber has good mechanical and elongation properties [5]. In some researches, ijuk fiber also merged with polymers such as epoxy and HIPS [6,7]. The problems, however, appear when the ijuk fibers are combined with polymers since they are hydrophobic whereas polymers are hydrophilic [8]. Consequently, the efforts to minimize the impact and to improve the compatibility between ijuk fibers and polymers consist of three different methods namely alkalization, oxidation and hydrolysis. Chalid *et al.*, (2015) found that alkalization method can increase the compatibility both in PLA polymers and ijuk fibers [9]. Other experiments involving the immersion of coconout fibers using $NaClO_2$ demonstrated the annihilation of lignin, impurities, wax and fatty substance from its surface [10]. Chalid *et al.*, (2015) also conducted the multistage treatments comprising alkalization, oxidation and hydrolysis by utilizing sodium hydroxide (NaOH), sodium hypochlorite (NaClO) and sulfuric acid (H_2SO_4) respectively to improve the compatibility and to enhance the fibrilation. They resulted that the fibers lost the amount of lignin and hemicellulose so the fibers reduced their hydrophobic property and were able to merge with polymers. In addition, the fibers had the high crystallinity as the hydrolysis treatment and it can be a potential candidate to use a nucleating agent while unifying with polymer as a matrix [11]. Another research conducted also by Chalid *et al.,* (2015) found that after ijuk fibers were treat by 0.25% of NaOH during alkalization treatment, the crystallinity of ijuk fibers rose due to exposure of micro-fibril cellulose [12].

Based on the explanations above, this study was performed to investigate the effect of modified ijuk fibers through multistage treatments consisting of alkalization, oxidation, and hydrolysis in sequence, in increasing crystallinity of the polypropylene composite. Furthermore, the experimental was performed to study the effects of alkaline concentration to crystallinity of polypropylene composite. The experimental results were evaluated by measuring FTIR spectra to characterize chemical content in composite, DSC is to measure the crystallinity and SEM to observe the morphological surface within composite material.

2. Materials and Methods

2.1 Materials

The polypropylene was supplied by the Chandra Asri Petrochemical Tbk, Indonesia. The “ijuk” fibers were purchased by the supplier on Tangerang, Indonesia. In addition, NaOH, acetic acid (CH_3COOH) and H_2SO_4 analytical grade were bought from Merck and NaClO analytical grade was obtained from Wako Chemical, Japan.

2.2 Material Preparations

Preparation of ijuk fibers consist of three stages namely, alkalization, oxidation and hydrolysis by NaOH, NaClO and H_2SO_4 in sequence. Alkalization was carried out to remove impurities, wax and dust on the fiber surface. It was carried out to annihilate the amount of lignin and hemicellulose in the fiber. Firstly, the fibers chopped using crusher machine and shieved to obtain uniform size (400-600 μm). Secondly, the fibers immersed into the NaOH solution with its concentration of 5 and 10%. Oxidation is aimed to reduce the amount of lignin and hemicellulose significantly and to make fibrilation of fibers. After alkalization process, the fiber was soaked by using sodium hypochlorite (NaClO) solution with concentration 3 and 6%. Hydrolysis is the process to enhance fibrilation and to increase the crystalline part of the fiber. The process is to immerse the fiber after oxidation process by using 20% of H_2SO_4 and the fibers were washed by distillated water until pH 7 afterwards the fibers were dried at the room temperature for one day. The final step is the hot melt mixing process with fraction volume of fibers and polypropylene is 10 and 90% respectively. The time to blend the materials is about six minutes and the temperature is 165°C. In this research, we compare between method A and B in which method A comprises alkalization 5% of NaOH, then immersing by 3 and 6% of NaClO, then by 20% of H_2SO_4 while B consists of alkalization process using 10% of NaOH, then continuing by 3 and 6% of NaClO, finally by adding 20% of H_2SO_4

2.3 Material Characterizations

2.3.1 Fourier Transmission Infra Red (FTIR)

FTIR spectra were recorded by using a Perkin-Elmer FTIR spectrometer. It was carried out to obtain an infrared spectrum of absorption of a solid, liquid or gas and to collect the resolution of high spectra on a widely range spectra data. It is utilized to characterize lignin, hemicellulose and cellulose fibers with or without multistage treatments and to show the compatibility of polypropylene after adding by the ‘ijuk” fibers. Each sample recording comprising 30 scans recorded from 400 to 4000 cm^{-1}

2.3.2 Differential Scanning Calorimetry (DSC)

DSC measurements were recorded by using a Perkin-Elmer 6000. It has range of temperature from 15 to 1000 °C and was carried out to measure the temperature and heat flows regarding transitional phase of materials. This technique conducts the experiment on which the material and reference are heated and chilled on certain condition. DSC also provides the information either quantitative or qualitative data related to the changes chemically and physically in endothermic and exothermic reactions.

2.3.3 Scanning Electron Microscope (SEM)

Surface morphology of polypropylene composite was observed by using Field Emission Scanning Electron Microscope FEI Inspect F50 to image the morphology of polypropylene composite with accelerating voltage of 30 kV, resolution and magnification are 1 nm and 500.000 times respectively. More specifically it was used to clearly investigate the compatibility between matrices and fibers.

3. Results and Discussions

3.1. Compatibility of Polypropylene Composite.

The observation on ijuk fibers shows some spectra either for untreated or for treated ijuk fibers. According to the Figure 1, the spectra on the peak 1.060 and 895 cm^{-1} indicate the specific spectra of lignin based on CO and CH stretch. The increase of intensity on 1.060 cm^{-1} also illustrates the increase of fiber crystallinity, whereas the peaks at 1043 and 1135 cm^{-1} show the decrease of percent transmission of FTIR spectra in which they are proportional with the decrease of amount of lignin components after hydrolysis treatment. Another peak on 1247 cm^{-1} shows the specific spectra of CO stretch describing ijuk fiber itself. Furthermore, the spectra on the peak 1653 and 1736 cm^{-1} represent the amount of hemicellulose. The peaks on range 2900 cm^{-1} and 3500-3300 cm^{-1} exhibit CH and OH stretch respectively [11].

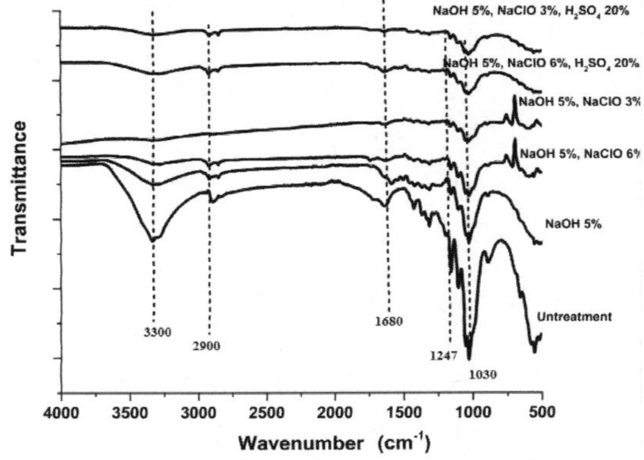

(a)

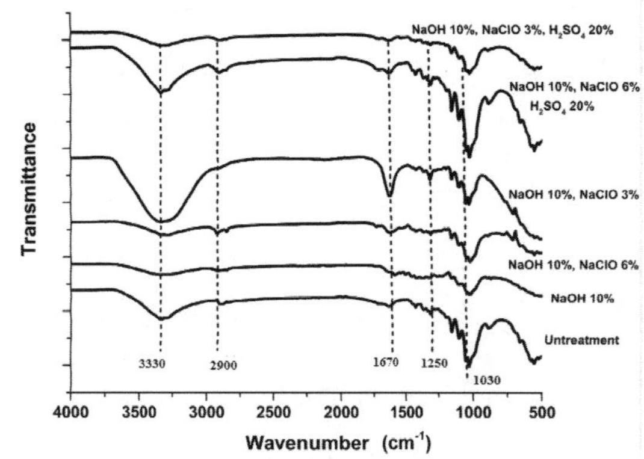

(b)

Figure 1. FTIR spectra of ijuk fibers (a) method A and (b) method B using untreated and treated fibers.

Meanwhile, the FTIR spectra of polypropylene are shown in the picture below.

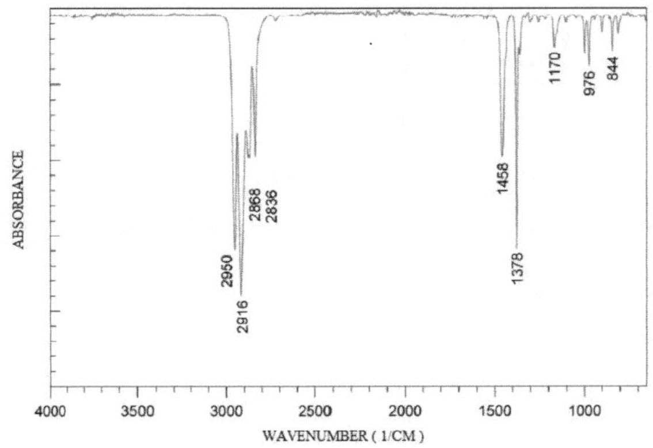

Figure 2. FTIR spectra of polypropylene.

The spectra on 2950, 2918, and 2836 cm^{-1} describe CH stretch while the peak on 1456 and 1376 cm^{-1} are specific spectra of polypropylene containing CH_2 deformation and symmetric CH_3 deformation. Whilst, the spectra on 998. 974 and 842 cm^{-1} show isotactic polypropylene band. Finally, the spectrum on the peak 1167 cm^{-1} indicates the C-C bending which is the backbone of polypropylene itself.

In order to investigate about the compatibility, the polypropylene composite was tested to see clearly whether there are bonds formed after mixing processes. The ijuk fiber itself is hydrophilic and polypropylene is hydrophobic, it will make the composite resulting the poor compatibility due to the moisture content of fibers [13].

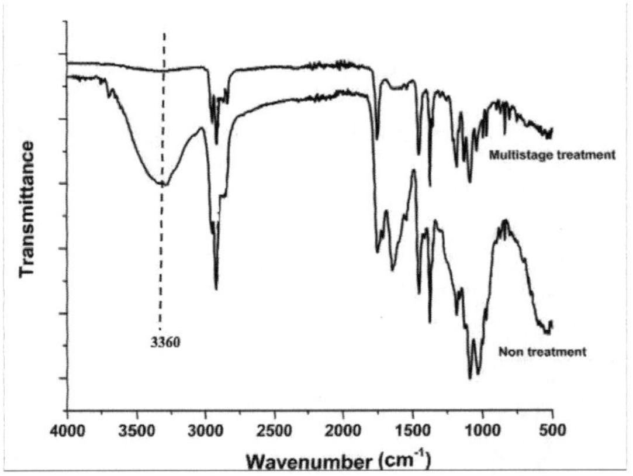

Figure 3. FTIR spectra of polypropylene composite using untreated and treated fibers

Based on the FTIR spectra above, after the mixing process with the ijuk fibers, the FTIR curve shows the new peak related to O-H stretch on 3561 – 3358 cm^{-1} and it also indicates the compatibility between ijuk fibers and polypropylene in composite caused by the bonding formation of polypropylene composite formed through the interaction between hydroxy alcohol and phenolic. The graph also shows that the number of functional groups has reduced in polypropylene composite with treated ijuk fibers compared with untreated ijuk fibers. Furthermore, it illustrates the compatibility has been achieved after multistage treatments.

Aside from investigating for the new bond formed by FTIR spectra, the compatibility also can be proven by monitoring the surface morphology between polypropylene and ijuk fibers as shown in the pictures below.

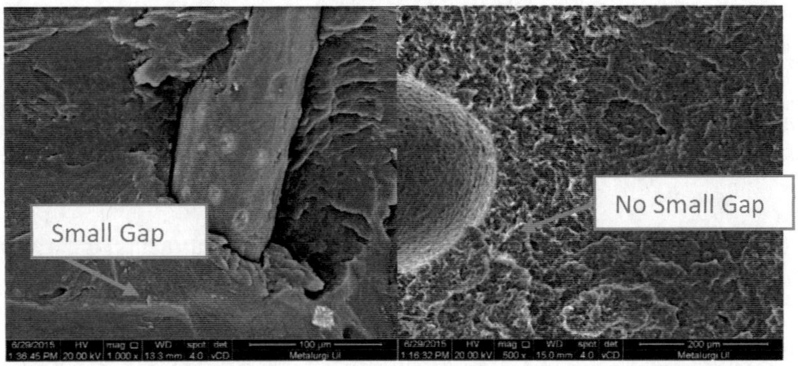

(a) (b)

Figure 4. The surface morphology of composite (a) Untreated fibers (b) Treated fibers.

Because ijuk fiber is non-polar meanwhile polypropylene is polar, the incompatibility will be formed after they are combined into composite so that the multistage treatments needed to enhance the compatibility. Multistage treatments can annihilate the hydrophilic component of ijuk fibers such as hemicellulose, lignin, and pectin so they will compatible with polypropylene. The Figure b above reveals that there is no gap between ijuk fiber and polypropylene which indicates good compatibility because the amount of hydrophilic components in fibers has been reduced by the multistage treatments. Meanwhile, The figure a exhibits that there is a small gap between fiber and matrix. It indicates poor compatibility because the fiber and polypropylene cannot wet each other due to the hydrophilic components of fibers [9,11]. Moreover, the untreated ijuk fibers can create the voids and holes when combined with the polypropylene matrix since during the mixing process they can release the volatile substances [14].

3.2. Crystallinity of Polypropylene Composite.

The crystallinity of ijuk fibers after multistage treatments increased compared to the untreated fibers based on the previous research [11]. In this steps, the multistage treatments are expected to remove the amorphous parts in ijuk fibers so the crystalline parts increase after treatments. The increase of crystalline parts is caused by hydronium ions can penetrate amorphous region and cause the hydrolytic cleavage on glycosidic bonds and sodium ions can dissipate hydrophilic components which are amorphous region within ijuk fibers by forming water vapor [9,15].

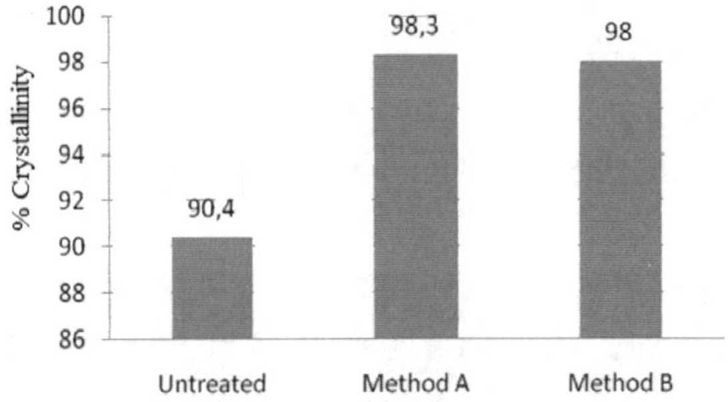

Figure 5. The crystallinity index of ijuk fibers in method A and method B.

As the crystallinity of ijuk fibers increased after multistage treatments, the crystallinity of polypropylene will also increase when combined with ijuk fibers.

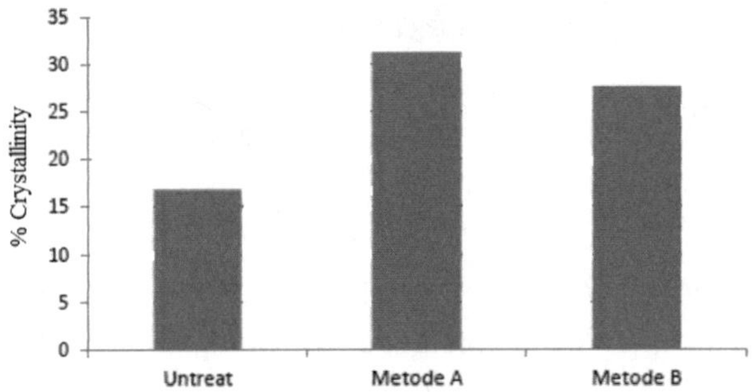

Figure 6. The crystallinity index of polypropylene composite after method A and B treatments.

The crystallinity index (CrI) above defines to examine the effect of the addition of ijuk fibers to polypropylene both in method A and B. The results display that after multistage treatments, ijuk fibers can increase the crystallinity of polypropylene composite compared with polypropylene with the untreated fibers. The crystallinity of polypropylene composite is proportional with the crystallinity of ijuk fibers when untreated ijuk fibers added to the polypropylene the crystallinity index was around 15% , but when using the treated fibers in method A and method B the crystallinity index was rising around 31 and 28% respectively. The crystallinity of polypropylene composite in method A (5% NaOH) is higher than in method B (10% NaOH) caused by there is a minimum concentration of sodium hydroxide used in alkalization treatment, in which the higher concentration causes the damage to the structure of fibers. The damage can reduce the amount of crystalline region so it does not maximize the alkalization treatments [16].

The increase of crystallinity of polypropylene composite also proportional with crystallization time which meant the time needed to crystalize during cooling process. According to DSC results, the increases of crystallization time to method A and B were 0.44 and 12 s obviously faster than polypropylene with untreated fibers which was 2.8 s.

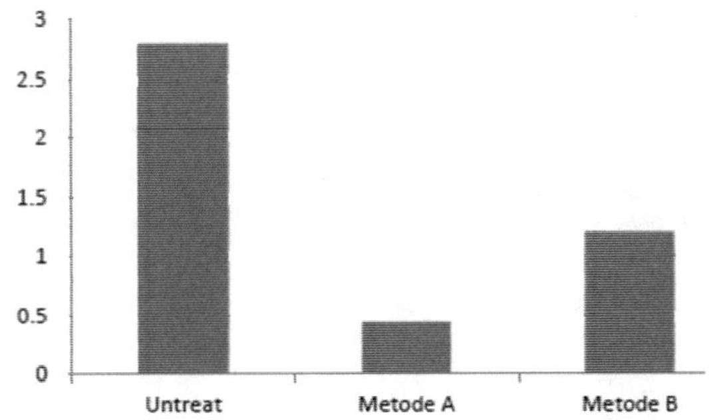

Figure 7. Crystallization time of polypropylene composite after method A and B treatments in seconds.

The crystallization time on method A is faster than on method B influenced by the higher crystallinity of ijuk fibers so it will initiate faster the polypropylene to crystallize vice versa method B results the slower crystallization time influenced also by lower crystallinity of ijuk fibers on method B. The crystallinity of method A (5 % NaOH) is higher because it has the lower concentration of sodium hydroxide whereas method B (10% NaOH) is higher concentration. The higher concentration will cause the damage of structure of ijuk fibers and reduce the crystalline region [12,16]. In other words, the more concentration we used on alkalization, the more crystalline regions are reduced. It will automatically affect crystallization time when the ijuk fibers combined with polypropylene. The ijuk fibers on method A has more crystalline regions than B so its crystallinity is higher than B and it will trigger faster crystallization time when combined with polypropylene.

4. Conclusions

According to our studies, the ijuk fibers prepared by multistage treatments (alkalization, oxidation, and hydrolysis) both in method A and method B reveal that the level of crystallinity increases due to the annihilation of amorphous and hydrophilic parts in fibers, method A can rise the crystallinity of polypropylene composite to 31% while method B rises to 28% more effective compared to untreated ijuk fibers. Aside from the percentage of crystallinity, time needed to crystallize when combined to polypropylene also speeds up which method A and method B can accelerate crystallinity to 0.44 and 1.2 seconds faster. Furthermore, these studies also probed that the good compatibility between ijuk fibers and polypropylene can be achieved by conducting fibers into

IOP Conf. Series: Materials Science and Engineering **223** (2017) 012020 doi:10.1088/1757-899X/223/1/012020

multistage treatments and it can be seen from the surface morphology of composite. In accordance with this study, the ijuk fibers were treated by multistage treatments can increase the crystallinity and good compatibility of polypropylene in material composite.

References

[1] Mohanty A K *et al* 2005. "*Natural fibres, biopolymers, and biocomposites: an introduction*," Boca Raton: CRC Press 1-36.
[2] Faruk O *et al* 2012 *J. Progress in Polym. Sci.* **37** 1552– 1596.
[3] Thiruchitrambalam M *et al* 2010 *J. Natural Fibers* **7** 307–323.
[4] Ishaka M R *et al* 2013 *J. Carbohydr. Polym.* **91** 699–710.
[5] Tserki V *et al* 2005 *Compos Part A – Appl. Sci. Manufac.* **36**(8) 1110–8.
[6] Bachtiar D *et al* 2008 *J. Mater. and Design* **29** 1285–1290.
[7] Bachtiar D *et al* 2010 *Smithers Rapra Technology.*
[8] Madsen B 2004. *Properties of plant fibre yarn polymer composites*. PhD thesis, BYGDTU, Technical University of Denmark ISBN **87**-7877-145-5.
[9] Chalid M and Prabowo Imam 2015 Int. J. Chemical, Nuclear, Material and Metalurgical Eng. **9**(2) 342-346.
[10] Muensri P *et al* 2011 *Composites* Part A **42** 173-179.
[11] Yuanita E *et al* 2015 Procedia Chemistry **16** 608-615.
[12] Chalid M *et al* 2015 *Mater. Sci. Forum* **827** 326-331.
[13] Li X *et al* 2007 *Polym. Environ.* **15**(1):25–33.
[14] Joseph P V *et al* 2003 *Compos Part A – Appl Sci Manuf,* **34**(3) 253–66.
[15] Johar N *et al* 2012 *Industrial Crops and Products* **37** 93-99.
[16] Bachtiar D *et al* 2008 *Mater. and Design* **29** 1285–1290.

IOP Conf. Series: Materials Science and Engineering **223** (2017) 012021 doi:10.1088/1757-899X/223/1/012021

Improvement of acoustical characteristics : wideband bamboo based polymer composite

M Farid, A Purniawan, A Rasyida, M Ramadhani, S Komariyah
Department of Materials and Metallurgical Engineering, Institut Teknologi Sepuluh Nopember (ITS), Surabaya, Indonesia.

E-mail : mofaredo@gmail.com

Abstract. Environmental friendly and comfortable materials are desirable for applications in the automobile interior. The objective of this research was to examine and develop bamboo based polymer composites applied to the sound absorption materials of automobile door panels.Morphological analysis of the polyurethane/bamboo powder composite materials was carried out using scanning electron microscope to reveal the microscopic material behavior and followed by the FTIR and TGA testing.The finding demonstrated that this acoustical polymer composite materials provided a potential wideband sound absorption material. The range of frequency can be controlled between 500 and 4000 Hz with an average of sound absorption coefficient around 0.411 and it met to the door panels criteria.

1. Introduction

The natural based composite material was widely used as sound absorption materials. They are used as noise absorption materials in offices, residential and industrial areas. Toyoda *et al.*, (2017) [1] investigated sound absorption enhancement with the wide-band frequency using material which was prepared by adding a filler to the cylindrical space microperforated panels absorbers (CMSAs) and rectangular panels space microperforated absorbers (RMSAs).

Mati-Bauoche *et al.,* (2016) [2] investigated the relationship between the sound absorption properties and dead-end pores or clusters. Park (2013)[3] investigated the acoustic resonator related to the micro-perforated panel absorber. Sanada and Tanaka [4] stated that the flexible panel was needed to expand the range of the sound absorption materials. Wideband sound absorption materials is one of three type of sound absorption materials. Bamboo have been developed by Koizumi and Farid as a potential natural fiber component of the sound absorption material. Koizumi *et al.* [5] has developed an acoustic material derived from polymer composite / bamboo fiber and reported that the fiber diameter reduction effect on the increase in sound absorption coefficient values. Farid *et al.* [6] studied sound absorption coefficient of composite polyester / bamboo fiber. The sound absorption coefficient reached 0.97 at a frequency of 1000 Hz.

There are various types of sound absorption materials, including the wideband type sound absorption material which is a challenge to develope it. This study focuses on obtaining a wideband type sound absorption materials by developing a composite materials based on natural fibers and polymers that meet product standards.

IOP Conf. Series: Materials Science and Engineering **223** (2017) 012021 doi:10.1088/1757-899X/223/1/012021

2. Materials and Method

The bamboo used was dendrocalamus asper (bambu betung). This bamboo was grown in East Java and the diameter was 10 cm. Based on the Dransfield' experimentation, dendrocalamus asper had a moisture content of 55%, air-dry moisture content of 15%, holoselulosa content of 53%, 19% pentosan, lignin 25% and ash 3%. NaOH were used to treat the bamboo fiber.The variable components consisting of bamboo powder and polyurethane composition. The first variable were 5% of bamboo powder, 95% of polyurethane (95% PU), the second variables were 10 % of bamboo powder, 90 % of polyurethane (90% PU) and 15 % of bamboo powder, 85% of polyurethane (85% PU). The specimen was characterized using scanning electron microscope. Sieving process needed 3 shifter tools which had a size of 450, 244 and 148 μm. This research used a fiber size less than 148 μm. Polyurtehane was prepared by mixing 70% of polyphenyl isocyanate and 30% of polypropylene glycol. Composite manufacturing process used blending method in which the polyurethane and bamboo powder had the same composition with total mass for absorption test 25 g. Finally, the ingredients were mixed and stirred for 60 s before it was poured into mold and deposited for 24 h. FTIR test was used to analyze functional groups. It offered quantitative and qualitative analysis for organic and inorganic samples. FTIR identifies chemical bond in a molecule by producing an infrared absorption spectrum and it was an effective analytical instrument for detecting functional groups and characterizing covalent bonding information. FTIR test was used to analyze functional groups.

Characterization of functional groups of composite were performed using the Thermo Scientific Nicolet iS 10 (Thermo Fisher Scientific.Inc, MA, USA) with a wavelength of 4000-400 cm^{-1}. Specification of FTIR instruments was as follows: Interface: PC USB 2.0. Laser: HeNe. Performance verification: ASTM E1421 to meet customer ISO/GLP requirements. Power supply: 100-240 V 50/60 Hz. Spectral range: 7800-350 cm^{-1} optimized, mid-infrared KBr beamsplitter 11000-375 cm^{-1} XT KBr extended range mid-infrared optics. Spectral resolution: Better than 0.4 wavenumer.

Meanwhile, surface morphology analysis used Scanning Electron Microscope (SEM) type FEI INSPECT S50. Sound absorption analysis using ASTM E1050 standard was used to determine the ability of the composite to absorb sound.

3. Results and Discussion

The synthesis process of polyurethane was made by mixing polyurethane A and Polyurethane B. Polyurethane A was used for isocyanate and polyurethane B was used for polyol. The result indicates that there was stretching vibrations bond O=C=N with wave lengtht of 2240 cm^{-1}
in Polyurethane and absorption bands with wavelength 1519 cm^{-1} for —C=C— bond and wave length of 1100 cm^{-1} for C—C—C bond. This is shown in Figure 1.

IOP Conf. Series: Materials Science and Engineering **223** (2017) 012021 doi:10.1088/1757-899X/223/1/012021

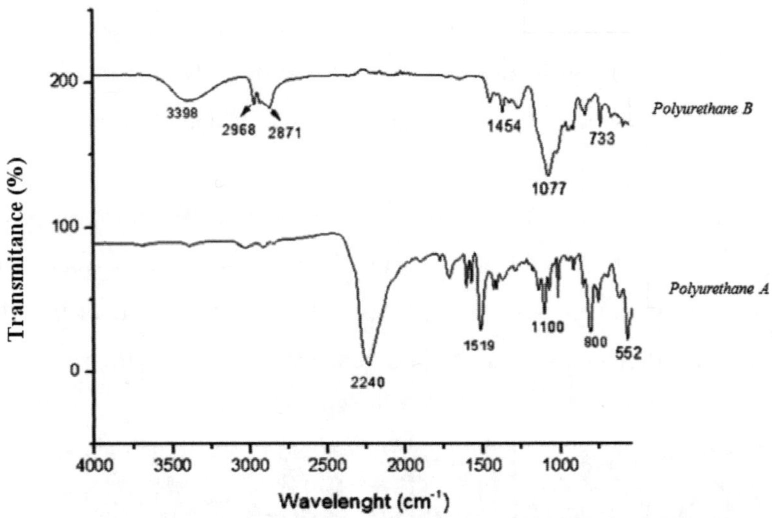

Figure 1. The result of FTIR analysis for polyurethane A and polyurethane B

This bond shows that polyurethane A was an *isocyanate*. Meanwhile, the absorption wave for double bond CH shows 800 cm^{-1} and stretching vibration with C=CH isocyanate bond shows wave length of 552 cm^{-1}. Therefore, we can conclude that *polyurethane* A was a type of polyphenyl Isocyanate. Polyurethane B graphic shows that there was stretching vibration for O-H bond with wave length of 3398 cm^{-1}. which means that polyurethane B was polyol. There was an obvious absorption energy for CH_3 on wave length of 1454 and 2968 cm^{-1}. The absorption peak for CH_2 appears on wave length of 733 and 2871 cm^{-1}. At the same time, the stretching absorption for CH_2OH was 1077 cm^{-1}; thus, polyol was known as polyprophylene glycol.

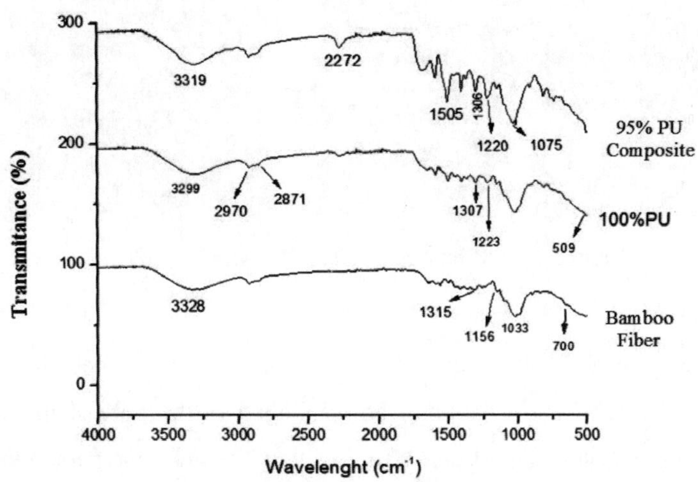

Figure 2. The Result of FTIR analysis for Bamboo Fiber, 100%PU and 95%PU composite.

FTIR test was performed to identify the presence of lignin, —OCH_3 groups is a part of the lignin and it had wavelength of 2860-3000 cm^{-1}. Figure 2 shows that there was no peak on wavelength 2860-3000 cm^{-1}; hence, there was no lignin on bamboo fiber after alkaline treatment. Composite graphic with 100% PU on wavelength 3299 cm^{-1} identified stretching vibration of N—H and on wavelength 1233 cm^{-1} which means there is abosoption of C—N.

Figure 2 shows the test results of FTIR bamboo fiber after alkali treatment. Alkali was intended to remove the lignin content of the bamboo; from this figure it was evaluated that there was an absent of the group -OCH_3 with wavelength 2860-3000 cm^{-1} which shows a part of the lignin. Moreover, the addition of 5% bamboo did not significantly change the structure of pristine polyurethane.

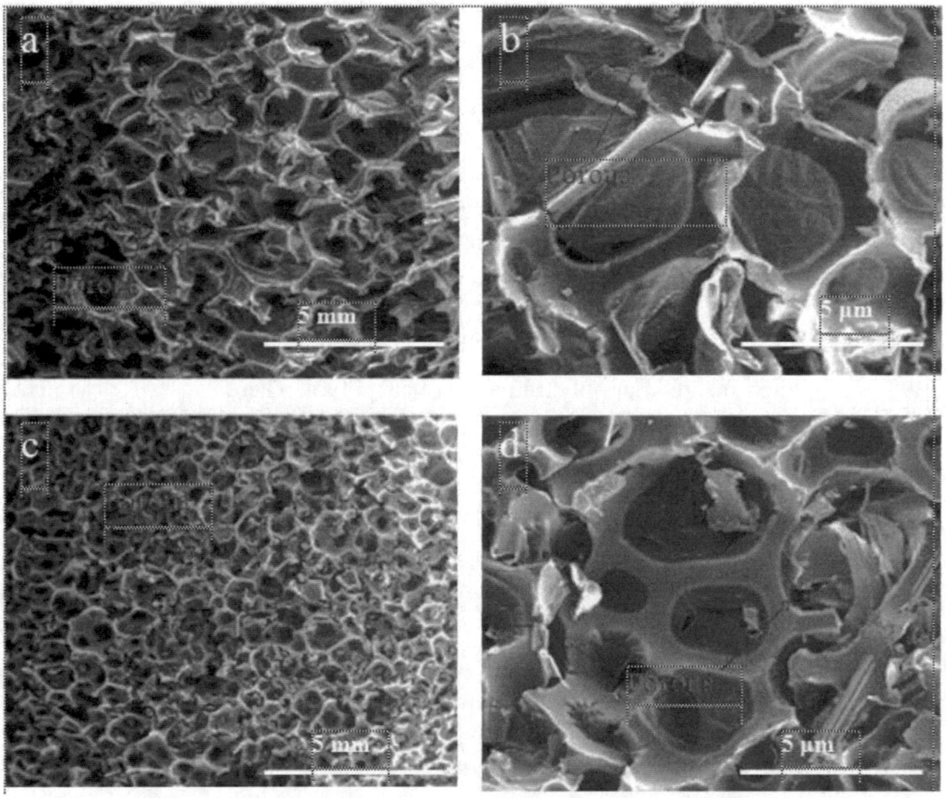

Figure 3. SEM images of 100% PU with 50x (a) and 250x (b) of magnifications and its composite (85% PU) with 50x (c) and 250x (d) of magnifications.

Figure 3 shows the SEM micrographs of the PU and its composites, revealing that there was interconnected porosity. The white part corresponded to the pore shape whereas the dark part was related to the open holes. This kind of porosity was important parameter to develop the acoustic behaviour [8]. However, we can evaluate that there were differences between the size and the distribution of the pore cell sizes. It was assumed that the addition of 15% bamboo fillers as reinforced can slightly alter the foam size distribution.

Typical porous cell diameter was obtained by inspecting visually the diameter of at least 20 porous cells on the surface of the foams [8]. Figure 3 (a) and (b) shows that the cell size became larger from 0.46 mm to 1.30 mm in diameter. Figure.3 (c) and (d) shows that the amount of porous cells more than 20 on the surface of the foams and it became larger from 0.48 to 1.66 mm in diameter.

The table shows mass of the specimen at a temperature of 250 °C. At this temperature, there was a little mass reduction. Even for pure PU, the mass increased to 100.5109. While the specimen with 5% of fiber, the mass decreased into 98.56787 and specimens with 10% of fiber, the mass degraded into 98.88169. Meanwhile, the specimens with 15% of fibre was 97.22951. The average working temperature stability of this material is in the range of 100 to 200 °C.

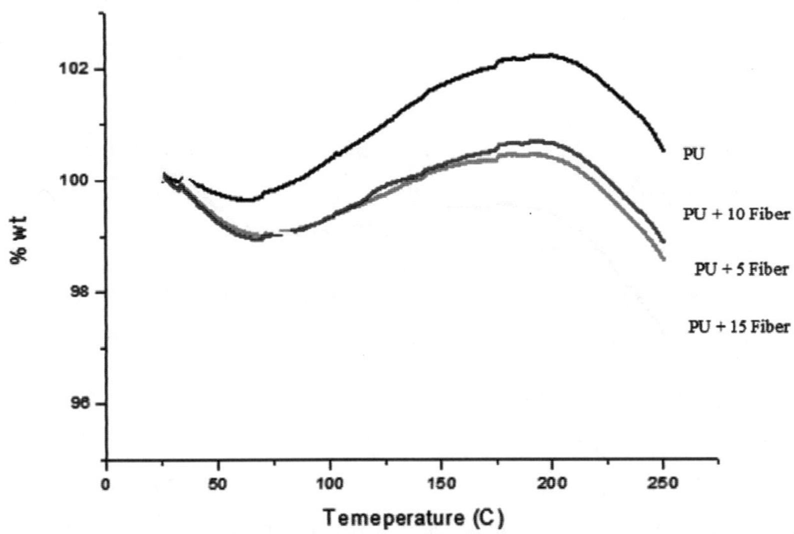

Figure 4. TGA of 100% PU, 95% PU, 90% PU, and 85% PU

The Figure 4 shows mass of the specimen at a temperature of 250 °C. At this temperature, there was a little mass reduction. Even for pure PU, the mass increased to 100.5109. While the specimen with 5% of fiber, the mass decreased into 98.56787 and specimens with 10% of fiber, the mass degraded into 98.88169. Meanwhile, the specimens with 15% of fibre was 97.22951. The average working temperature stability of this material is in the range of 100 to 200 °C.

IOP Conf. Series: Materials Science and Engineering **223** (2017) 012021 doi:10.1088/1757-899X/223/1/012021

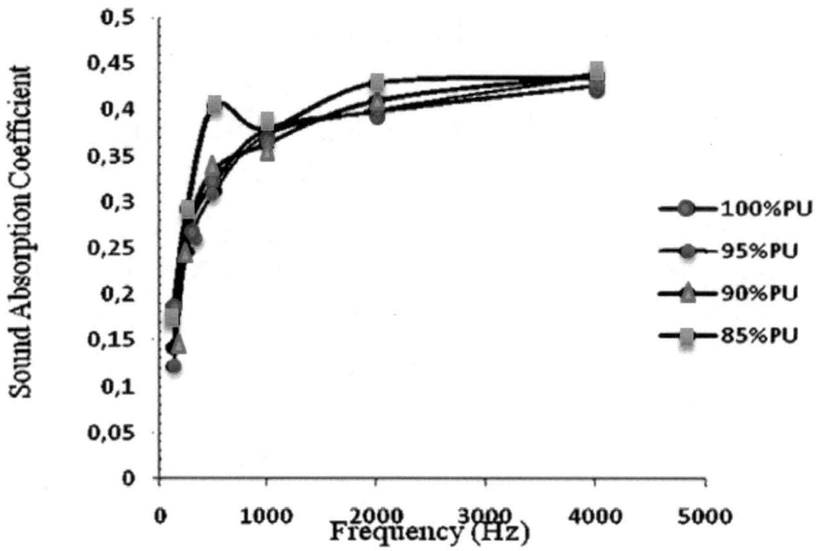

Figure 5. Sound absorption coefficient of 100% PU, 95% PU, 90% PU, and 85% PU

Figure 5 shows that the composite 85% PU had a sound absorption coefficient of 0.404, 0.379, 0.428, and 0.434 for each frequency of 500 Hz, 1000 Hz, 2000 Hz, and 4000 Hz. It was very interesting because the result was a relatively stable value. Thus, it can be concluded that with a composition of 85% of PU and 148-450 microns of bamboo powder it was obtained an average sound absorption coefficient stability 0.411. This value meets the standard criteria for door panel.

Referring to the research of Zhang (2012), Verdejo *et al.*, (2009) and Nordgren (2010), generally characteristic of polyurethane based acoustical composite material relied on aspects of cell size foam, vibroacoustic energy, the phenomenon of viscoelastic and viscoacoustic, air flow, sound pressure , kinetic energy, elasticity, permeability, flow-resistivity, etc. [8-10]. Generally, the sound absorption coefficient of the porous material affected by porous cell size and structure. Polyurethane foam having a larger cell size has a sound absorption coefficient greater than the PU which has a smaller cell size at a certain frequency. Vibro-acoustic energy occured both in the pores and solid frame of the material [8]. The large foam cell produced cell wall deformation larger, so more sound energy could be converted into kinetic energy [10].

4. Conclusion

The morphology, thermal stability, sound absorption coefficient, and functional groups of PU/bamboo fiber composite have been studied. Modification of long fibers reinforcement into a fine powder in the micron scale and the selection of the specific composition of polyurethane as a composite matrix provided a way to obtain wideband sound absorption materials. The addition of bamboo powder can slightly alter the foam size distribution. Formation of cells was closed cell. The addition of filler led to an increase in cell thickness. This contributed to the improvement of the sound absorption coefficient. The PU/Bamboo powder composite materials had a sound absorption coefficient of 0.411 and could be used in the frequency range from 500 to 4000 Hz; thus, this was in accordance with the criteria on the interior door panel of the car.

Acknowledgments

The Authors would like to express their gratitude to Department of Materials and Metallurgical Engineering ITS for the financial support.

References

[1] Toyoda M *et al* 2017 *Appl. Acoustics* **116** 311.
[2] Mati-Baouche N *et al* 2016 *Appl. Acoustics* **111** 179.
[3] Park H 2013 *J. Sound and Vibration* **132** 4895.
[4] Sanada A and Tanaka N. 2013 *J. Appl. Acoustics* **74** 509.
[5] Koizumi T *et al* 2002 The development of sound absorbing materials using natural bamboo fibers, high performance structures and composites (CA Brebbia and WP de Wilde).
[6] Farid M *et al* 2015 *Adv. Mater. Res.* **1112** 329.
[7] ASTM E-1050: Standard test method for impedance and absorption of acoustical materials using a tube, two microphones and a digital frequency analysis system.
[8] Zhang C 2012 *Materials & Design* **41** 319.
[9] Verdejo R *et al* 2009 *J. Compos. Sci. Technol.* **69** 1564.
[10] Nordgren E 2010 J. *Sound and Vibration* **329** 753.

IOP Conf. Series: Materials Science and Engineering **223** (2017) 012022 doi:10.1088/1757-899X/223/1/012022

Effect of lignin on morphology, biodegradability, mechanical and thermal properties of low linear density polyethylene/lignin biocomposites

M Ghozali[1], E Triwulandari[1], A Haryono[1] and E Yuanita[2]
[1]Research Center for Chemistry, Indonesian Institute of Sciences, Serpong, Indonesia
[2]Center for Chemical and Packaging Ministry of Industry, Jakarta, Indonesia

E-mail: muhammad.ghozali@lipi.go.id

Abstract. This research is purposed to study effects of lignin compositions on morphology, biodegradability, mechanical and thermal properties of low linear density polyethylene (LLDPE)/Lignin biocomposites. LLDPE/Lignin biocomposites has been manufactured by adding LLDPE, lignin and compatibilizer into rheomix at 200°C with a stirring speed of 70 rpm for 30 min. The composition of lignin added was 5, 10, 15, and 20 phr with compatibilizer 5 phr. LLDPE/lignin films has been made by using hydraulic hot press at 200-210°C with pressure of 6 bar for 20 min. Fourier Transform Infrared (FTIR) spectrum analysis was conducted to determine the functional groups of LLDPE/Lignin biocomposites. The surface morphology was observed by using Scanning Electron Microscope (SEM). The mechanical properties was measured as a tensile strength and thermal stability was measured by Thermogravimetric Analysis (TGA). In addition, biodegradation test was also conducted to determine the level of biodegradability. TGA results indicated that at 456°C LLDPE and lignin had similar thermal stability and the addition of lignin into LLDPE/lignin bicomposites can reduce the thermal stability up to temperature of 450-460°C. However, the thermal stability is increased at temperature over 460°C. The tensile strength and elongation at break of all LLDPE/Lignin biocomposites at various compositions is lower compared to those of LLDPE. The more lignin were added into LLDPE/Lignin biocomposites, the more the materials were biodegraded.

1. Introduction

Research on utilization and development of polymeric materials based on the use of renewable natural resources as raw material for the polymer industry has attracted the interest in the past few decades. Basically, the development of renewable materials for polymer comes from two main problems, which are the environmental problem associated with waste generated and the limitation of natural resources derived from fossil materials. Both of these problems has driven the efforts to find alternative materials based on renewable and biodegradable natural resources which can reduce the dependency on petroleum based material and simultaneously can solve the environmental problem caused by the generated waste.

Some of biomasses that can be used as raw material of polymers are starch, lignin, cellulose, protein, chitosan, and some vegetable oils. Among them, lignocellulose is considered as one of the most potential alternative material because it does not intersect with food crops. Lignocellulose is generally derived from wood containing cellulose, hemicellulose and lignin. Availability of lignin

IOP Conf. Series: Materials Science and Engineering **223** (2017) 012022 doi:10.1088/1757-899X/223/1/012022

reaches 30% of all non-fossil organic carbon on earth and commonly produced by paper industry in a significant quantity. The utilization of lignin as a new renewable biomaterial may be a perfect candidate for modification and chemical reactions due to its functional properties (rich in phenolic and aliphatic hydroxyl groups) for the development of new biomaterials. However, there is a major challenge in polymer applications associated with unclearly defined structure and diversity based on their origins, separations and fragmentation processes. Chemical modification of lignin can be classified into lignin fragmentation into phenolic or other aromatic compounds, synthesis of new chemical active sites and functionalization of hydroxyl groups [1].

Utilizations of lignin to produce biomaterial or biofuel have been previously performed [2-4]. Research on lignin in polyurethane applications have also been conducted [5-10,11-12]. Studies on chemical modification of lignin as biobased polymers [1,13,14], biodegradable [15], green composites [16] and green hydrogel [17] have also been reported. Various efforts have been reported to combine lignin into other natural polymers to form a full bio-based materials. There have been also reports on addition of lignin into starch, proteins, polycaprolactone, poly(hydroxyl-butyrate) [3,18], gelatin [19], polybutyleneadipate-co-terephthalate [20], chitosan [21], and polylactic acid [22-26]. The number of hydroxyl groups in the lignin causes a relative polar characteristic thereby providing better affinity to the polar polymer. However it also raises a problem when composited with nonpolar material because it is not compatible and will result in composites with poor stress transfer in terms of mechanical properties. Therefore, a compatibilizer to improve the inter-phases adhesion so that the material be compatible becomes necessary. Moreover, the compatibility of lignin can also be improved by reducing the content of hydroxyl, even simple esterification or alkylation of hydroxyl groups can favor the compatibility of lignin with non-polar polymer matrices [3]. Recent studies on utilization of lignin are more focused on the production of lignin-based thermoplastics. In addition, studies on performance of polymers and composites by mixing lignin or lignocellulosic material with thermoplastic resin has been reported [27-38]. Studies of the addition of the lignin into a non-biodegradable material are also intended to increase biodegradability of the resulted composites. Various studies on biodegradation of lignin have also been conducted [39-46].

In this research, LLDPE/Lignin biocomposites were manufactured by adding various composition of lignin into LLDPE to study the effects of lignin addition on the morphology, biodegradability, and mechanical and thermal properties of LLDPE/lignin biocomposites with polyethylene-grafted maleic anhydride as the compatibilizer. Lignin used in this research was a pure alkali lignin without modification.

IOP Conf. Series: Materials Science and Engineering **223** (2017) 012022 doi:10.1088/1757-899X/223/1/012022

2. Materials and Methods

2.1. Materials

The materials used in this study are a Linear Low Density Polyethylene (LLDPE) obtained from PT Chandra Asri, PE-g-MA as compatibilizer with viscosity of 500 cP (140ºC) (lit.) and density of 0.92 g/ml (25ºC) obtained from Sigma-Aldrich, and lignin with density of 1.3 g/mL (25ºC) also obtained from Sigma-Aldrich, potassium dihydrogen orthophosphate (KH_2PO_4), magnesium sulfate ($MgSO_4.7H_2O$), ammonium nitrate (NH_4NO_3), sodium chloride (NaCl), ferrouss ulfate ($FeSO_4.7H_2O$), zinc sulfate ($ZnSO_4.7H_2O$), manganous sulfate ($MnSO_4.H_2O$), potato dextrose agar, aspergillus niger, potassium monohydrogen orthophosphate (K_2HPO_4).

2.2. Composites and Films Preparation

LLDPE, lignin and PE-*g*-MA are mixed in advance to composite manufacturing. Those materials are stirred uniformly in Haake Rheomix PolyDrive. The composition of lignin added was 5, 10, 15 and 20 per hundred resin (phr) with the composition of compatibilizer was 5 phr. The mixture was processed in rheomix at 200°C with 70 rpm for 30 min. Chunks resulted from rheomix was then made into form of sheet using a Carver laboratory Hydraulic Hot Pressmodel 2702 at 210°C with a pressure of 6 bar for 20 min.

2.3. Characterizations

Analysis on functional groups of LLDPE, Lignin and LLDPE/Lignin bicomposite was conducted using FTIR at wave number of 4000-400 cm^{-1}. A total of 40 scans were accumulated in transmission mode with a resolution of 4 cm^{-1}. The morphological analysis of the LLDPE/Lignin biocomposite was investigated using a Scanning Electron Microscope (SEM) HITACHI SU3500. Thermal analysis was conducted using STA Linseis STAPT 1600. Biocomposites was heated up to 500°C with heating rate of 10°C/min in sample of small pieces with a weight of ± 20 mg. A universal testing machine (UTM) Strograph 50, Toyoseiki was used to evaluate the tensile strength dan elongation at break of LLDPE/Lignin biocomposites films according to ASTM D638. Biodegradation test was performed according to ASTM G 21-96. The medium for the fungi was prepared by dissolving KH_2PO_4 (0.7 g), $MgSO_4.7H_2O$ (0.7 g), NH_4NO_3 (1.0g), NaCl (0.005 g), $FeSO_4.7H_2O$ (0.002 g), $ZnSO_4.7H_2O$ (0.002 g), $MnSO_4.7H_2O$ (0.001 g) and agar (15 g) into 1L of water. Next step is the sterilization of agar medium by autoclaving at 121°C for 90 min. The sterilized agar medium is then poured into a sterile petri dish with depth of 3-6 mm and left until condensed. Samples biocomposites with size 2x2 cm^2 were planted on an agar medium, then added suspense fungus Aspergillus niger spores. Further, the samples were incubated for 68 days. The sample was then sterilized with alcohol 70%, then soaked in distilled water. Finally, the samples were dried at 50°C and then their dry weight were measured after biodegradation.

3. Results and Discussion

The composition of lignin added to LLDPE affects the color of LLDPE/Lignin biocomposites produced. The effect of lignin composition to the visual color biocomposite film is shown in Figure 1. The more lignin were added to the biocomposite then the darker the color of the produced LLDPE/Lignin biocomposites films. This is because the color of LLDPE used was translucent, whereas the lignin was dark brown. Therefore, the resultant composite tends to get darker with increasing lignin composition is added.

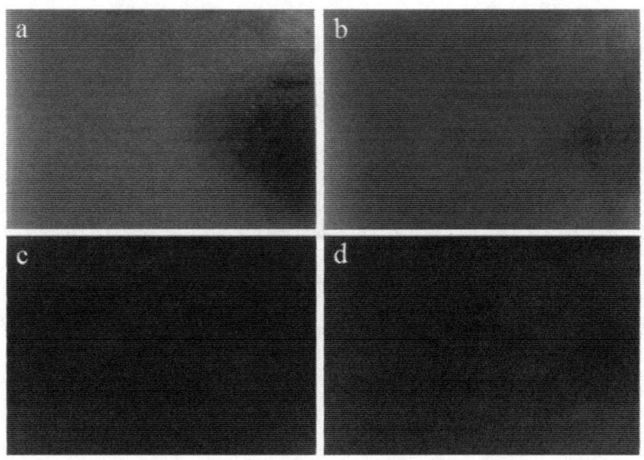

Figure 1. LLDPE/Lignin biocomposite films : (a) LLDPE/Lignin 5 phr; (b) LLDPE/Lignin 10phr; (c) LLDPE/Lignin 15 phr and (d) LLDPE/Lignin 20 phr

3.1. FTIR Analysis

The FTIR spectrum of LLDPE, Lignin and LLDPE/Lignin biocomposites are shown in Figure 2. From the FTIR spectrum, it is shown that between LLDPE, lignin and LLDPE/Lignin have different absorption peak. LLDPE spectrum (Figure 2a) consist of $–CH_2$- and $–CH_3$ in its structure which were shown by absorption peaks at wave numbers of 2953-2841 cm^{-1} which are antisym and sym CH stretching showing $-CH_3$ and $-CH_2$- groups in aliphatic compounds. Absorption peaks also appear at wave numbers of 1462 cm^{-1} and 1456 cm^{-1}, which are a scissor vibration of CH_2 and antisym deformation of CH_3 associated with the CH_2 and CH_3 in aliphatic compounds, respectively. The absorption peak at wave number of 1369 cm^{-1} shows C-C bonds while the absorption peak at wave number of 723 cm^{-1} is attributed to CH_2 in methylene chains which indicating the group $-(CH_2)_n-$ in hydrocarbons [47].

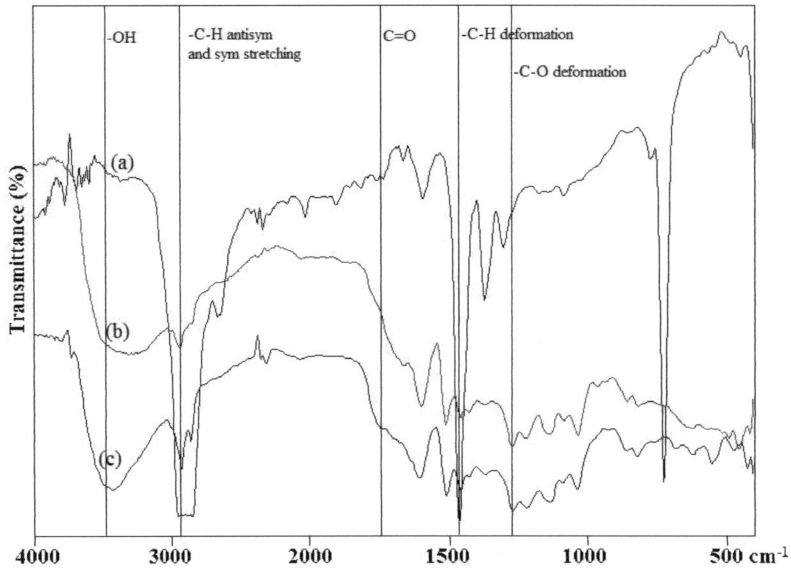

Figure 2. FTIR Spectrum of (a) LLDPE; (b) Lignin and (c) LLDPE/Lignin

Meanwhile, the FTIR spectrum of lignin (Figure 2b) has specific absorption peak at wave numbers of 3396-3203 cm^{-1} which is OH stretch indicating the presence of hydroxyl groups. The absorption peaks at wave numbers of 2953 cm^{-1} and 2843 cm^{-1} are CH stretch that show $-CH_3$ and $-CH_2-$ groups, respectively [22]. The absorption peaks at wave numbers of 1593 cm^{-1} and 1510 cm^{-1} are derived from C-C of aromatic skeleton vibrations, while the absorption peaks at wave numbers of 1458 cm^{-1} and 1425 cm^{-1} are attributed to the C-H deformation and C-H aromatic ring vibration [31]. The absorption peaks at wave numbers of 1269, 1134 and 1031 cm^{-1} are the deformation of C-O and C-H derived from syringyl and guaiacyl as lignin constituent monomers [22,25]. The FTIR spectrum of LLDPE/Lignin biocomposite (Figure 2c) showed has similar absorption peak with lignin, although the composition of lignin added into LLDPE is only maks 20 phr. It was indicated that absorption peak of LLDPE/Lignin biocomposite is mixture between absorption peak of LLDPE and lignin. Because of the $–CH_2-$ and $–CH_3$ absorption peak of LLDPE was overlapped with the CH_2- and $–CH_3$ absorption peak of lignin, so it seems the absorption peak of lignin was dominant in FTIR spectrum of LLDPE/lignin composite. Nevertheless, the FTIR spectrum of LLDPE/Lignin composite showed a new absorption peak at wave number of 1730 cm^{-1} [48] indicating the presence of C=O group derived from the compatibilizer and OH absorption peak shifting and becoming increasingly sharp.

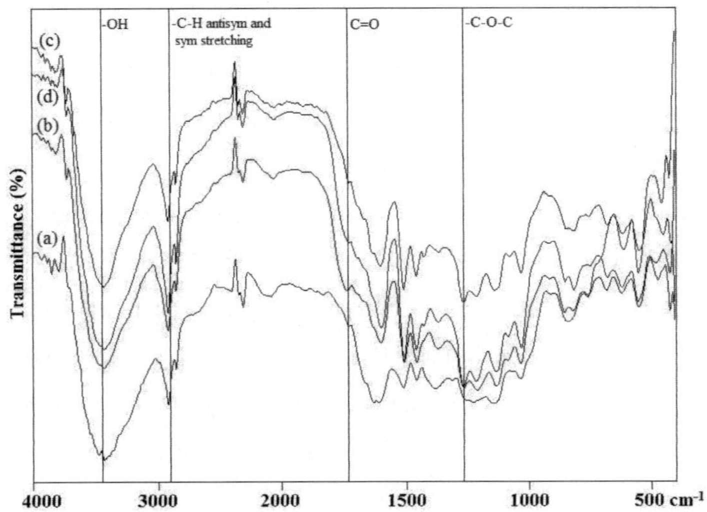

Figure 3. FTIR Spectrum of (a) LLDPE/Lignin 5 phr; (b) LLDPE/Lignin 10phr;(c) LLDPE/Lignin 15 phr and (d) LLDPE/Lignin 20 phr

Figure 3 shows the FTIR spectrum of LLDPE/lignin biocomposites at various lignin composition. All of the spectrums of LLDPE/Lignin were almost similar and showed a new absorption peak at wave number of 1730 cm^{-1} indicating the presence of C=O group derived from the compatibilizer and OH absorption peak shifting and becoming increasingly sharp. The absorption peaks of LLDPE/lignin biocomposites that indicate certain functional groups are presented in Table 1.

Table 1. Absorbance peaks ofLLDPE/Lignin biocomposite

Wavenumber (cm^{-1})	Functional Group	
3365	-OH	Lignin
2922	C – H	LLDPE
2852	$O-CH_3$	Lignin
1730	C = O	Compatibilizer
1508	C = C	Lignin
1458	$-CH_3$	LLDPE
1381	C – C	LLDPE
1267, 1217 and 1134	C – O – C	Lignin

It is shown in Table 1 that all LLDPE/lignin biocomposites show absorption peaks at wavenumber of 2922 cm^{-1} indicating the presence of C-H, wavenumber of 1458 cm^{-1} indicating the presence of $–CH_3$, and at wavenumber of 1381 cm^{-1} indicating the presence of C-C from LLDPE. Furthermore, all LLDPE/Lignin biocomposites show absorption peaks at 3365, 2852, and 1508 cm^{-1} indicating the presence of –OH, $O-CH_3$, and C=C, respectively. In addition, the absorption peaks at wavenumber of 1267, 1217 and 1134 cm^{-1} indicating the presence of C-O-C from lignin. It is also shown that the absorption peak at wavenumber of 1730 cm^{-1} related to C=O from the compatibilizer. The appearances of those absorption peaks shows that the LLDPE/lignin biocomposites was successfully manufactured.

3.2. Thermal Properties

The effect of addition of lignin composition on the thermal stability of LLDPE/lignin biocomposites was studied by thermogravimetric analysis (TGA). TGA curves of LLDPE, lignin and LLDPE/lignin biocomposites are shown in Figure 4.

LLDPE, lignin and LLDPE/lignin biocomposites have a single degradation step. LLDPE starts degrading at 250-300°C, while the lignin starts degrading at 40°C and a major degradation occurs at 300-500°C related to fragmentation of inter-linkage units [25]. At temperature below 450-460°C, LLDPE/lignin biocomposites are more susceptible to degrade compared to LLDPE, however, they are more resistant to thermal degradation when compared with lignin. On the contrary, at temperature over 460°C, LLDPE/Lignin biocomposites are more resistant to thermal degradation than LLDPE, however, more susceptible to thermal degradation compared to lignin.

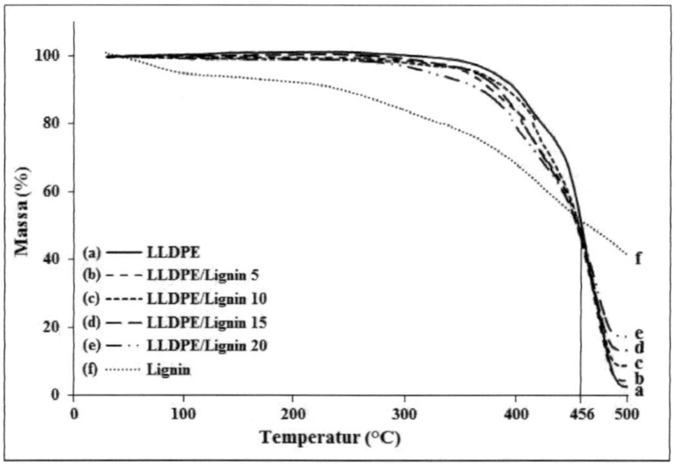

Figure 4. TGA thermograms of (a) LLDPE; (b) LLDPE/Lignin 5 phr; (c) LLDPE/Lignin 10phr;(d) LLDPE/Lignin 15 phr; (e) LLDPE/Lignin 20 phr and (f) Lignin

At 500°C the mass residu of LLDPE, LLDPE/lignin 5, LLDPE/Lignin 10, LLDPE/lignin 15, LLDPE/lignin 20 and Lignin are 2.58, 4.37, 8.73, 13.30, 17.33 and 41.39%, respectively. From the TGA curve of LLDPE and lignin, it is shown that at temperature over 450°C the thermal resistance of lignin was better than LLDPE which affects the thermal resistance of LLDPE/lignin biocomposites. The high thermal stability of lignin is caused by the presence of aromatic phenyl groups. In addition, the presence of hydroxyl groups also contributes to the thermal stability and improves the stability of aromatic structures in lignin and prevents damage due to temperature effects [22]. The increase in the residual mass is caused by the chemical structure of lignin, which shows a less contribution to the flammability of materials due to the high charring ability and the low heat release when burned [23,50]. This shows that the increase of lignin composition in the LLDPE/lignin biocomposites will increase the char content [24,34,51] and the residual mass and will increase its thermal resistance as well. TGA results indicate that the addition of lignin into composite LLDPE/lignin biocomposites can reduce the thermal stability up to temperature of 450-460°C compared to LLDPE. The thermal stability curve of intersection between LLDPE (Figure 4a) and lignin (Figure 4f) shown at 456°C. This intersection curve show that at temperatures below 456°C LLDPE is more resistant to degradation than lignin, whereas the temperature above 456°C LLDPE more susceptible than lignin. Therefore, the composite of LLDPE and lignin will result in LLDPE/lignin biocomposites with a thermal stability which is lower than LLDPE, however, higher than lignin. The degradation thermal stability of LLDPE/lignin biocomposites is also allegedly due to the empty space in the LLDPE/lignin biocomposites which is filled with oxygen or air trapped. Subsequently, it will initiate the degradation process resulting in lower thermal stability [49]. However, the thermal stability LLDPE/lignin biocomposites is increased at temperature over 460°C compared to LLDPE. The increase in thermal stability of LLDPE/lignin is due to the appearance of complex phenylpropanoid unit containing aromatic phenyl groups from lignin. In addition, the presence of multiple hydroxyl groups also contribute in the improvement of the stability of aromatic structures within the structure of lignin so that it will increase the mass of the remaining. Furthermore, the presence of a hydroxyl group also contributes to the thermal stability because it increases the stability of aromatic structures present in lignin. The increase in the residue is due to the chemical structure of lignin indicating a less contribution to the burn-ability due to the high ability to become charcoal and low heat release when burned to prevent it from degradation because of the influence of temperature [22,23,50].

3.3. Mechanical Properties

The effect of the addition of lignin composition to the mechanical properties of LLDPE/lignin biocomposites was studied by analyzing the tensile strength and elongation at break. The analysis results of tensile strength and elongation at break are shown in Table 2.

The tensile strengths of LLDPE, LLDPE/lignin 5, LLDPE/lignin 10, LLDPE/lignin 15 and LLDPE/lignin 20 are 218.4, 109.7, 86.4, 98.2 and 105.6 kgf/cm^2, respectively. While the elongations at break of LLDPE, LLDPE/lignin 5, LLDPE/lignin 10, LLDPE/lignin 15 and LLDPE/lignin 20 are 1251, 55.1, 32.3, 37.1 and 40%, respectively. The tensile strength and elongation at break of all LLDPE/Lignin biocomposites at various compositions is lower compared to those of LLDPE [35,36].

Table 2. Mechanical properties of LLDPE/Lignin biocomposite

	Tensile Strength (Kgf/cm^2)	Elongation at break (%)
LLDPE	218.4	1251
LLDPE/Lignin 5phr	109.7	55.1
LLDPE/Lignin 10phr	86.4	32.3
LLDPE/Lignin 15phr	98.2	37.1
LLDPE/Lignin 20phr	105.6	40

The decrease in tensile strength and elongation at break is considered as a result of non-uniformly distributed lignin in LLDPE/lignin biocomposite which causes an agglomeration affecting its stress distribution [26]. The elongation at break of LLDPE/lignin biocomposites was drastically decreased compared to LLDPE. This was probably caused by the presence of lignin in LLDPE/lignin biocomposites which results in fragility due to lack of interface adhesion. This lack of interface adhesion can be attributed to reduce the flexibility of LLDPE/lignin biocomposites. Moreover, the presence of lignin also increased the brittleness and affected the viscoelastic property of LLDPE/Lignin biocomposites. The hydrophilic characteristic nature of lignin results in incompatibility of lignin with LLDPE hydrophobic characteristic and tendency to form aggregates. This lignin's hydrophilic characteristic is also the cause that lignin is not tough to humidity which leads to high water absorption. Therefore, it causes the poor tensile properties of LLDPE/lignin biocomposites. In this case, a chemical modification is needed to improve lignin hydrophobic and decrease moisture absorption, leading to the enhancement of tensile properties of the composites [52]. Tensile strength and elongation at break shows decreasing tendency as the composition of lignin addition is increased. However, when the composition of lignin were added at 15 phr (LLDPE/lignin 15) and 20 phr (LLDPE/lignin 20), tensile strength and elongation at break increased in comparison with the addition of lignin by 5 phr (LLDPE/lignin 5) and 10 phr (LLDPE/lignin 15), but still lower than LLDPE. This is probably due to the lignin distribution in LLDPE/Lignin 15 and LLDPE/lignin 20 are more uniformly and less agglomerated compared to the lignin distribution in LLDPE/lignin 10, so that the tensile strength and elongation at break are increased.

3.4. Morphology

SEM micrographs of the fractured surface for LLDPE/lignin biocomposites films are shown in Figure 5. Figure 5(a), 5(b), 5(c), 5(d), and 5(e) show the SEM image of LLDPE, LLDPE/lignin 5, LLDPE/lignin 10, LLDPE/lignin 15, and LLDPE/lignin 20, respectively. Figure 5(a) shows relatively smooth and ductile surface of LLDPE. Figure 5(b) shows good lignin dispersion in the composite

IOP Conf. Series: Materials Science and Engineering **223** (2017) 012022 doi:10.1088/1757-899X/223/1/012022

surface. However SEM micrographs of agglomerated surface in the LLDPE/lignin biocomposite was shown in Figure 5(c) – 5(e). The addition of lignin from 10, 15 and 20 phr caused agglomeration of lignin. The agglomeration shown in the surface structure of bio-composites was caused by poor interfacial adhesion occurred between two phases [53]. Lignin was distributed non-uniformly in LLDPE/lignin biocomposites. LLDPE/lignin 15 (Figure 5d) and LLDPE/lignin 20 (Figure 5e) showed smoother surface than LLDPE/lignin 10 (Figure 5c). It was also probably the reason why the tensile strength and elongation at break LLDPE/lignin 15 and LLDPE/lignin 20 was higher than LLDPE/lignin 10. The addition of lignin into LLDPE made the surface structure of the resulted LLDPE/lignin biocomposites become rougher. This rough surface also possibly initiated faster biodegradation. The agglomeration can result in the lack of interface adhesion, thus, leads in a poor distribution of stress when pressured. Furthermore, it will result in the degradation of tensile strength and elongation at break. This phenomena was reported that the interfacial interaction between filler and polymer matrix in the composite affect the mechanical properties of the composite [54].

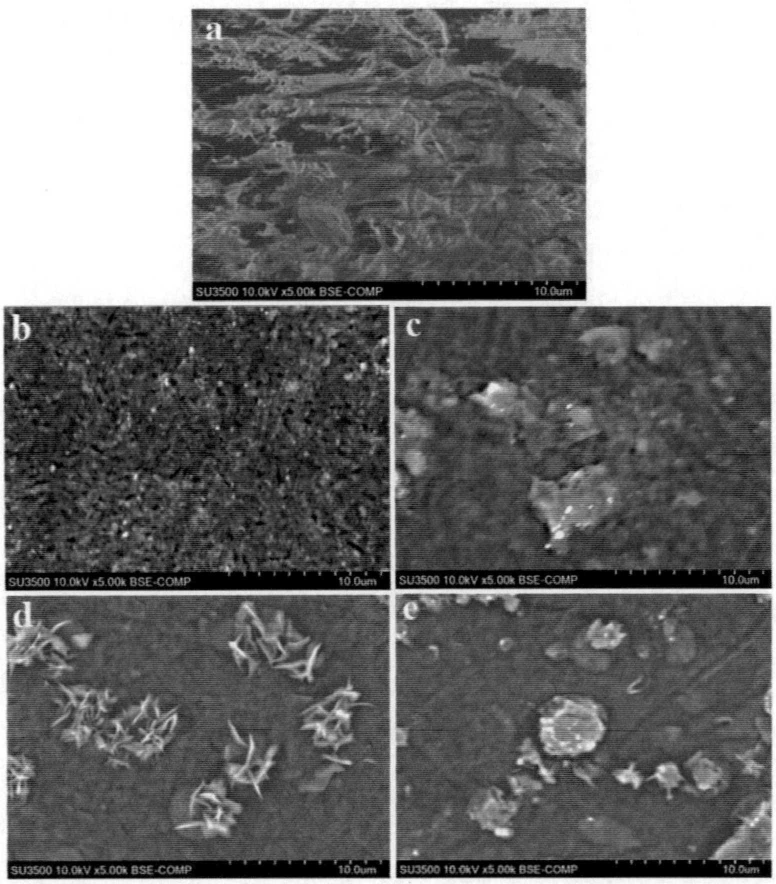

Figure 5. SEM micrograph of (a) LLDPE; (b) LLDPE/Lignin 5 phr; (c) LLDPE/Lignin 10phr; (d)LLDPE/Lignin 15 phr and (e) LLDPE/Lignin 20 phr

3.5. Biodegradability

The effect during degradation proses is greatly depended on hydrophilicity or hydrophobicity and particle dispersion. Moreover, the delay in biodegradability is caused by the increase in barrier properties of the material which can prevent water diffusion through the bulk [55]. The effect of lignin addition to the biodegradability of LLDPE/Lignin biocomposites is studied using ASTM G21-70. The results are shown in Table 3.

During 68 days of observation, the degradation rate of LDPE/Lignin 5, LLDPE/lignin 10, LLDPE/lignin 15 and LLDPE/lignin 20 are 0.1916, 0.2325, 0.3261 and 0.2864%, respectively. These LLDPE/lignin biocomposite degradations were associated with the presence of lignin which has many hydrophilic functional groups that promote biodegradation. The more lignin were added into LLDPE/lignin biocomposites, the more the materials were biodegraded. This is because as the number of lignin added increases, the biocomposites become more hydrophilic, and therefore, the material become more biodegradable.

Table 3. Biodegradability level of LLDPE/Lignin biocomposite

	Mass (gram)		**Biodegradation**
	Initial	**Final**	**(%)**
LLDPE/Lignin 5phr	0.3392	0.3386	0.1916
LLDPE/Lignin 10 phr	0.3227	0.3219	0.2325
LLDPE/Lignin 15 phr	0.2760	0.2751	0.3261
LLDPE/Lignin 20 phr	0.3318	0.3308	0.2864

However, at the lignin composition LDPE/lignin 20, the biodegradation result is lower than LLDPE/lignin 15. It was allegedly because of lignin has structural complexity and high molecular weight so the optimal biodegradability of fungi especially aspergillus niger to degrade the lignin is not more than 15 phr (LDPE/lignin 15). For lignin degradation requires specific enzyme, possibility some of the enzyme from aspergillus niger is non-spesific for lignin so its ability to degrade lignin is not optimal for the large composition [56].

4. Conclusion

The FTIR spectrum of all LLDPE/lignin biocomposite showed similar absorption peak with lignin and also show absorption peaks at wavenumber of 1730 cm^{-1} related to C=O from the compatibilizer. It was indicated that absorption peak of LLDPE/lignin biocomposite is mixture between absorption peak of LLDPE and lignin and shows that the LLDPE/lignin biocomposites was successfully manufactured. The composition of lignin added to LLDPE affects the color of LLDPE/lignin biocomposites produced. The degradation thermal stability of LLDPE/lignin biocomposites is allegedly due to the empty space in the LLDPE/lignin biocomposites which is filled with oxygen or air trapped. Subsequently, it will initiate the degradation process resulting in lower thermal stability. However, the thermal stability LLDPE/lignin biocomposites is increased at temperature over 460°C compared to LLDPE. The increase in thermal stability of LLDPE/lignin is due to the appearance of complex phenylpropanoid unit containing aromatic phenyl groups from lignin. In addition, the presence of multiple hydroxyl groups also contribute in the improvement of the stability of aromatic structures within the structure of lignin so that it will increase the mass of the remaining. The addition of lignin into LLDPE made the surface structure of the resulted LLDPE/lignin biocomposites become rougher and distributed non-uniformly in LLDPE/lignin biocomposites. This rough surface also possibly initiated faster biodegradation. The agglomeration can result in the lack of interface adhesion, thus, leads in a poor distribution of stress when pressured. Furthermore, it will result in the degradation of tensile strength and elongation at break. The addition of lignin results in enhancement of biodegradability properties of LLDPE/lignin biocomposites. These LLDPE/lignin biocomposite degradations were associated with the presence of lignin which has many hydrophilic functional groups that promote biodegradation.

Acknowledgements
This research was financially supported by "Kegiatan Mandiri" Research Center for Chemistry, Indonesian Institute of Sciences. The author would like to thank to Dr. Nino Rinaldi, Samsul Bhakri and Herlan Herdiawan for the help and support during this research.

References

[1] Laurichesse S, Avérous L 2014 *Progress in Polym. Sci.* **39** 1266–1290.
[2] Gordobil O *et al* 2016 *Industrial Crops and Products* **83** 155–165.
[3] Duval A, Lawoko M 2014 *Reactive & Functional Polymers* **85** 78–96.
[4] Mahmood N *et al* 2015 *Bioresource Technol.* **190** 416–419.
[5] Lee A, Deng Y 2015 *Eur. Polym. J.* **63** 67–73.
[6] Cinelli P *et al* 2013 *Eur. Polym. J.* **49** 1174–1184.
[7] Bernardini J *et al* 2015 *Eur. Polym. J.* **64**147–156.
[8] Zhang C *et al* 2015 *Polymer* **69** 52-57.
[9] Mahmood N *et al* 2015 *Eur. Polym. J.* **68**1–9.
[10] Cateto CA *et al* 2011 *Reactive & Functional Polym.* **71** 863–869.
[11] Huang J, Zhang L 2002 *Polymer* **43** 2287-2294.
[12] Chauhan M *et al* 2014 *Eur Polym J.* **52** 32–43.
[13] Norgren M, Edlund H 2014 *Current Opinion in Colloid & Interface Sci.* **19** 409–416.
[14] Stewart D 2008 *Industrial Crops and Products* **27** 202–207.
[15] Satyanarayana KG *et al* 2009 *Progress in Polym Sci.* **34** 982–1021.
[16] Wu RL *et al* 2009 *Bioresource Technol.* **100** 2569–2574.
[17] Thakur VK, Thakur MK 2015 *Int J. Biological Macromolecules* **72** 834–847.

[18] Mousavioun P *et al* 2013 *Industrial Crops and Products* **50** 270– 275.
[19] Shankar S *et al* 2015 *Int. J. Biological Macromolecules* **81** 267–273.
[20] Digabel FL, Avérous L 2006 *Carbohydr. Polym.* **66** 537–545.
[21] Nair V *et al* 2014 *Chemical Eng. J.* **254** 491–502.
[22] Gordobil O *et al* 2015 *Industrial Crops and Products* **72** 46–53.
[23] Yang W *et al* 2015 *Eur. Polym. J.* **71** 126–139.
[24] Spiridon I *et al* 2015 *Composites: Part B* **69** 342–349.
[25] Gordobil O *et al* 2014 *Polymer Degradation and Stability* **108** 330-338.
[26] Anwer MAS *et al* 2015 *Composites Part B* **82** 92-99.
[27] Morandim-Giannetti AA *et al* 2012 *Carbohydr Polym.* **87** 2563–2568.
[28] Canetti M, Bertini F 2007 *Composites Sci. and Technol.* **67** 3151–3157.
[29] Liu R *et al* 2014 *Composites Sci. and Technol.* **103** 1–7.
[30] Aradoaei S *et al* 2010 *J Non-Crystalline Solids* **356** 768–771.
[31] Yeo JS *et al* 2015 *J. Industrial and Engineering Chemistry* **31** 80–85.
[32] Fernandes EM *et al* 2014 *Composites: Part B* **67** 371–380.
[33] Klapiszewski L *et al* 2013 *Carbohydr Polym.* **94** 345– 355.
[34] Sailaja RRN, Deepthi MV 2010 *Materials and Design* **31** 4369–4379.
[35] Alexy P, KošíkováB, Podstránska G 2000 *Polymer* **41** 4901–4908.
[36] Cazacu G *et al* 2004 *Industrial Crops and Products* **20** 261–273.
[37] Xiong W *et al* 2015 *Industrial Crops and Products* **74** 285–292.
[38] Hilburg SL *et al* 2014 *Polymer* **55** 995-1003.
[39] Chen Y *et al* 2000 *Bioresource Technol.* **72** 169-183.
[41] Ko JJ, *et al* 2009 *Bioresource Technol.* **100** 1622–1627.
[42] Crestini C *et al* 1998 *Bioorganic & Medicinal Chemistry* **6** 967-973.
[43] Johansson E *et al* 2000 *Soil Biology & Biochemistry* **32** 1029-1032.
[44] Julinové M *et al* 2010 *Polymer Degradation and Stability* **95** 225-233.
[45] Machinet GE *et al* 2011 *Soil Biology & Biochemistry* **43** 1544-1552.
[46] Pham LTM *et al* 2014 *Enzyme and Microbial Technol.* **61–62** 48–54.
[47] Lambert J F *et al* *Organic structural spectroscopy, 1st ed* (Prentice-Hall)
[48] Valdes SS *et al* 1997 *J. Polymer Engineering* **17**(3) 257-268
[49] Dikobe D G, Luyt A S 2010 *eXPRESS Polymer Letters* **4(11)** 729–741.
[50] Detyothin S *et al* 2013 *Polymer Degradation and Stability* **98** (12) 2697–2708.
[51] Sahoo S *et al* 2011 *Composites Part A* **42**1710–8.
[52] Ku H Wang H *et al* 2011 *Composites: Part B* **42** 856–873.
[53] Panyasart K *et al* 2014 *Energy Procedia* **56** 406-413.
[54] Saha A K *et al* 1999 *J. Appl. Polym. Chem.* **71** 1505-1513.
[55] Lee WK *et al* 2002 *Langmuir* **18**(6) 2309–2312.
[56] Pe´rez J *et al* 2002 *J. Int. Microbiol.* **5** 53–63.

IOP Conf. Series: Materials Science and Engineering **223** (2017) 012023 doi:10.1088/1757-899X/223/1/012023

Structure and properties of organically modified poly(butylene adipate-co-terephthalate) based nanocomposites

A Rasyida[1,2], K Fukushima[2], M-C Yang[2]
[1]Department of Materials and Metallurgical Engineering, Faculty of Industrial Technology, Institut Teknologi Sepuluh Nopember (ITS) Surabaya, Indonesia
[2]Department of Materials Science and Engineering, National Taiwan University of Science and Technology, 43, Sec. 4, Keelung Rd, Taipei, Taiwan 106

E-mail :amaliyarasyida@gmail.com; amaliya@mat-eng.its.ac.id

Abstract. Poly (butylene adipate-co-terephthalate (PBAT) nanocomposites were prepared by melt blending PBAT with 5 wt.% of modified or unmodified montmorillonites (MMT). The effect of the presence of organic modifiers in MMT on the morphological, crystalline, thermal, and mechanical properties of PBAT nanocomposites was evaluated. The dispersion and distribution of the clays were studied by using wide angle X-ray analysis (WAXS), scanning electron microscopy (SEM), and transmission electron microscopy (TEM) analysis. Materials characterization techniques included: contact angle measurements, differential scanning calorimetry (DSC), thermogravimetry (TGA) and surface hardness analysis. As general results, nanocomposites exhibited different level of clay dispersion depending on the clay/organic modifier's chemical affinity with the polymer. Contact angle measurements show increases in the hydrophobicity level of PBAT based CLO30B, this could depict its high potential for packaging applications. In addition, Thermal analysis showed that clays partially hindered kinetics and extent of PBAT crystallization on cooling. In general, thermal properties of PBAT were improved by addition of clays, for a barrier effect of the nanoparticle towards polymer decomposition products ablation. In parallel, addition of clays led to enhancements in polymer hardness. These properties were found to be apparently influenced by clay dispersion level and chemical compatibility between the organic modifier and polymer matrix.

1. Introduction

In the last decade, biodegradable polymers have received much attention because of their wide range of promising applications in packaging, agriculture, and biomedical fields. The most often used biodegradable polymers have been polylactide (PLA), polyhydroxyalkanoate (PHA), polycaprolactone (PCL), and poly(butylene adipate terephthalate) (PBAT) [1]. Among these, PBAT has been widely studied due to its high commercial availability, biodegradability, and aromatic structure. The aromatic structure brings to this polymer an excellent pattern of physical properties such as: high performance properties and strong resistance to hydrolysis, and to bacterial and fungal attack [2]. Moreover, PBAT is flexible and has a higher elongation at break than most biodegradable polyesters, such as PLA to render them more suitable for packaging films [3]. The main limitations toward its wider industrial applications are its relatively insufficient thermal and mechanical resistance. However, it has been found that the above drawbacks can be overcome through filling techniques. In this way, the use of nano-sized fillers can potentially confer to polymer matrices multifunctional enabling properties (magnetic, optical, thermal, and mechanical properties) when compared to conventional polymer formulations [4–7].

Among the various nanoparticles studied, montmorillonites (MMT) have been one of the most studied layered silicates for the preparation of polymer nanocomposites due to their high availability, versatility and respectability towards the environment [8]. These nanoparticles are phyllosilicate clays of the 2:1 type. In their unmodified form usually contain Na^+ or K^+ ions [9], and they are only miscible with hydrophilic polymers [10]. Consequently, in order to increase their compatibility with polymers, it is sometimes necessary to convert the normally hydrophilic silicate interlayer environment into an organophilic one, for instance by ion-exchange reactions of the inter-gallery inorganic cations with primary, secondary, tertiary, and quaternary alkylammonium or alkylphosphonium cations [11], thus making possible high interfacial interactions between these organic modified clays and the polymer matrix.

Only few articles reported studies of PBAT/clay nanocomposites. Someya *et al.,* (2004,2005) [12,13] obtained certain level of clay intercalation and exfoliation in several compositions of PBAT/clay nanocomposites prepared by melt intercalation. Chivrac *et al.,* (2006, 2007) [14,15] have tested some commercial montmorillonites in PBAT compositions, revealing that higher intercalation degrees could be obtained by using particular preparation techniques and that the stiffness of PBAT matrix could increase with clay content due to the important interactions between PBAT and nanoparticles. In a recent work [16-18], we studied the morphology and properties of PBAT nanocomposites prepared by melt-mixing and by using different amount and types of nanoparticles including montmorillonite, fluoro-hectorite, and sepiolite.

The aim of this work is therefore to study whether the presence of organic modifiers can affect considerably the morphological, crystalline, thermal, and mechanical properties of PBAT. A comparative study of the properties of PBAT nanocomposites with and without any kind of clay modifier was performed in order to check whether the resulting nanocomposites could present potentially interesting properties for some industrial applications such as packaging applications.

2. Materials and Methods

PBAT (Mw = 140 kDa) with 44 mol% BT, was supplied by BASF AG (ECOFLEX* F BX 7011). Nanoparticles were supplied by Rockwood Additives Ltd. The characteristics of these clays are listed in Table 1. Nanocomposite preparation is based on our previous research [18]

Wide Angle X-ray Spectra (WAXS) were recorded at room temperature using Thermo ARL diffractometer X-tra 48 in the range 1-30° (2θ) (step size = 0.02°, scanning rate = 2 s/step) by using filtered Cu Kα radiation (λ = 1.54 Å).

IOP Conf. Series: Materials Science and Engineering **223** (2017) 012023 doi:10.1088/1757-899X/223/1/012023

Scanning electron microscopy (SEM) was carried out on the cryogenic fracture surface of the specimens using an SEM apparatus equipped with energy dispersive spectroscopy (EDS).

Transmission electron microscopy (TEM) was performed using a field emission transmission electron microscope (FE-TEM) (PhilipsTECNAI G^2F20) using an accelerator voltage of 120 kV. Ultrathin sections about 100 nm thick were cut with a Power TEOMEX microtome with a diamond knife and placed on a 200-mesh copper grid.

Table 1. Characteristics of nanoparticles

Type of nanoparticle	Commercial name	Modifier structure	Notation in text
Montmorillonite	NANOFIL 116	None	NANO
Montmorillonite[a]	CLOISITE 20A	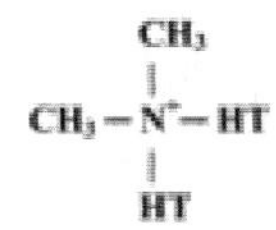	CLO20A
Montmorillonite[b]	CLOISITE 30B	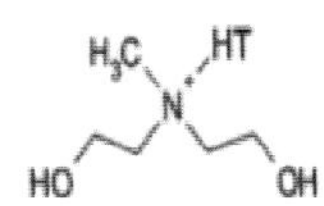	CLO30B

HT = Hydrogenated linear alkyl chains: C_{8-18}
a = Organic modifier content ca. 38 wt.% according to technical data sheet.
b = Organic modifier content ca. 30 wt.% according to technical data sheet.

Measurements of static water contact angle were performed according to the standard procedurein an air-conditioned laboratory at 25°C. A small drop of deionized water (~7 μL) was put onto the flat sample surfaces using a syringe such that the drop volume at the surface expands.

Differential scanning calorimetry (DSC) tests were carried out using a Perkin Elmer DSC 4000, under nitrogen atmosphere at a scanning rate of 10°C/min, sample size 5-6 mg in aluminium pans. Thermal history of samples was erased by a preliminary heating-cooling cycle at 10°C/min from 20 to 200°C.

The crystallinity (χ) of the polymer and nanocomposites was evaluated taking into account the amount of clay presents in nanocomposites by using the following expression [19]:

$$X = \left[\frac{\Delta H_m\ (t)}{\Delta H^o m\ x\ \left[\frac{\%\, filler}{100}\right]} \right] x\ 100$$

where $\Delta H_m(t)$ is the specific melting enthalpy of the sample, ΔH_m° is the melting enthalpy of the 100% crystalline PBAT (114 J/g) [14] and % filler is the weight percentage of the clay (5%). Thermogravimetric analysis (TGA) was carried out on 10 mg samples using a thermogravimeter (TGA Q500, TA) at 10°C/min from 50 to 800 °C under air flow.

IOP Conf. Series: Materials Science and Engineering **223** (2017) 012023 doi:10.1088/1757-899X/223/1/012023

The surface hardness was determined using a Durometer Type A (PTC instruments, model 306L, ASTM 2240-85). Three readings for each sample were taken for statistical purposes.

3. Results and Discussions

3.1. Morphology

3.1.1. *Wide Angle X-Ray Analysis (WAXS)*

The WAXS pattern of neat PBAT was characterized by five diffraction peaks (Figure 1) observed at 2θ = 16.1°, 17.3°, 20.2°, 23.1°, and 25.0°, corresponding to the interlayer distance of 0.55, 0.52, 0.44, 0.38, and 0.36 nm respectively, and indicating a crystalline structure for PBAT. These peaks were also observed at the same intensity for PBAT nanocomposites, indicating that no significant changes in the PBAT crystal structure was obtained by nanofiller incorporation. This behavior was in agreement with previously results obtained on PBAT based nanocomposites [14,16,17], as an unimportant transcrystallinity at the nanofillers/PBAT interface upon clay addition was observed.

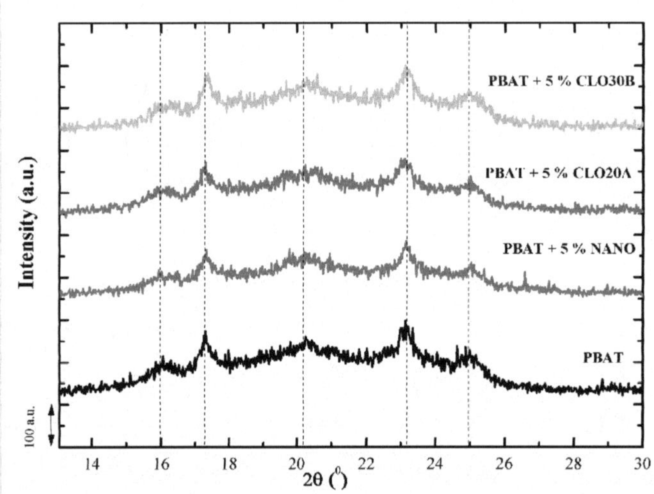

Figure 1. WAXS patterns of neat PBAT and nanocomposites

The WAXS patterns of NANO, CLO20A and CLO30B based nanocomposites in the lower angle range are presented in Figure 2. NANO was characterized by a single diffraction peak at 2θ = 7.0° corresponding to the basal reflection (001), and accounting for 1.3 nm interlayer distance (Figure 2a). On the other hand, CLO20A was characterized by two diffraction peaks at 2θ = 4.5° and 2.0° which corresponded to the interlayer distance 2.0 and 4.4 nm, respectively (Figure 2b) and related to the basal reflection (001) [17,19]. In parallel, CLO30B was characterized by a single diffraction peak at 2θ = 5.0°, corresponding to the interlayer distance of 1.8 nm (Figure 2c). Since the three clays had a similar chemical structure (Table 1), the higher interlayer distance found by pristine CLO20A and CLO30B as

IOP Conf. Series: Materials Science and Engineering **223** (2017) 012023 doi:10.1088/1757-899X/223/1/012023

compared to NANO could be attributed to the presence of the organic modifier between their clay platelets (Table 1).

NANO based materials showed a slight shift of the clay diffraction peak to lower angles corresponding to an increase of the interlayer distance of around 0.1 nm for nanocomposites, and indicating the obtainment of certain intercalation level in these nanocomposites without organo-modifying the nanofiller. Similar results were found by Chivrac*et et al.*, (2006) [14], obtaining intercalated-nanocomposites of PBAT with non-organo-modified nanofillers. Even more, it was also observed an almost complete disappearance of the peak for NANO based nanocomosites, suggesting extensive dispersion of the clay in the polymer matrix.

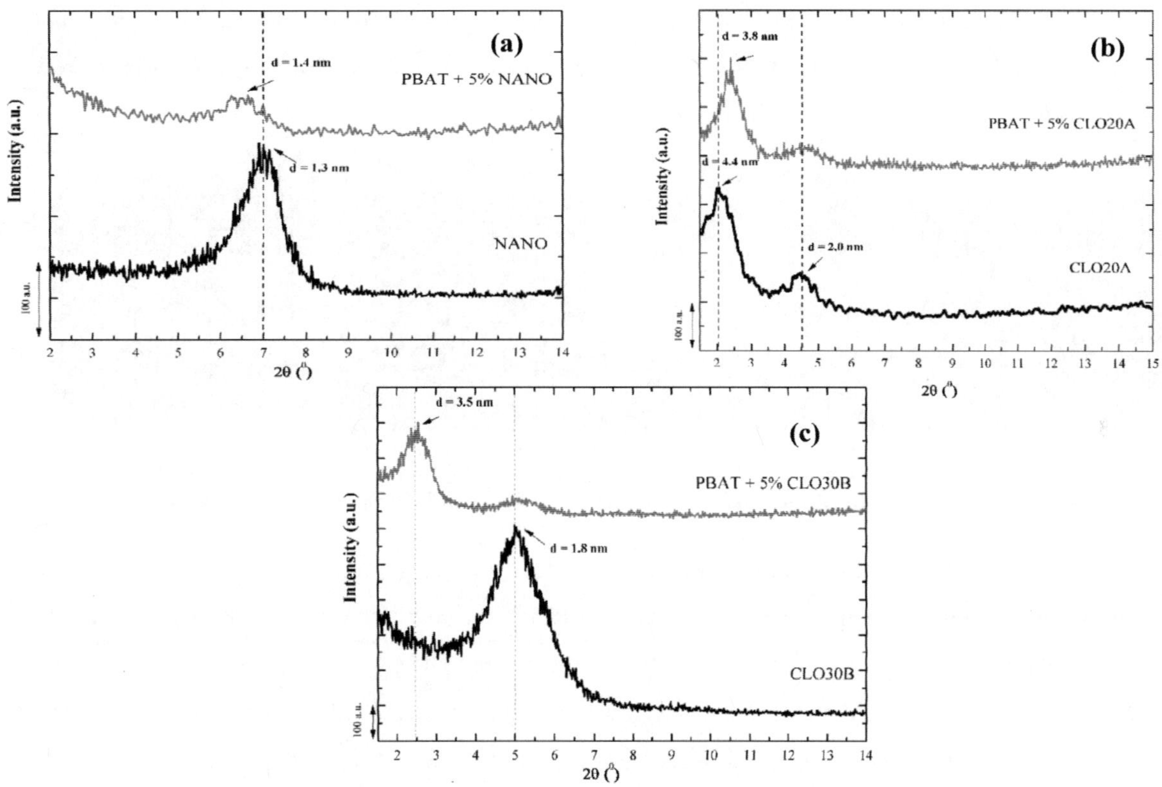

Figure 2. WAXS patterns of pristine clays and PBAT nanocomposites with 5% of (a) NANO, (b) CLO20A, and (c) CLO30B.

In the case of CLO20A based materials (Figure 2b), the main WAXS diffraction peak of CLO20A (*ca.* 0.6 nm) got lower after mixing with PBAT, suggesting less intercalation of the clay in PBAT. Another difference observed was a slight broadening of both WAXS peaks, indicating the possible occurrence of a disordered structure of CLO20A in PBAT.

IOP Conf. Series: Materials Science and Engineering **223** (2017) 012023 doi:10.1088/1757-899X/223/1/012023

On the other hand, CLO30B based nanocomposites (Figure 2c) exhibited a higher interaction/intercalation level with PBAT matrix. Indeed an important shift of the clay diffraction peak to lower angles was observed, corresponding to an increase of *ca.* 1.7 nm. Moreover, a significant decrease of the clay peak intensity was also observed in CLO30B based nanocomposites, which could account for the formation of an exfoliated/disordered structure.

3.1.2. *Transmisson Electron Microscopy (TEM)*

Figure 3a shows that PBAT/NANO nanocomposites exhibited certain level of intercalation together with the presence of clay micro-aggregates in agreement with XRD results. It was also possible to observe the presence of single clay layers in the polymer matrix (Figure 3b). When correlating these observations with the results obtained by WAXS analysis, it was possible to conclude that the significant decrease of the WAXS peak intensity of NANO could be associated to the attainment of a disordered structure in this system by intercalation and exfoliation.

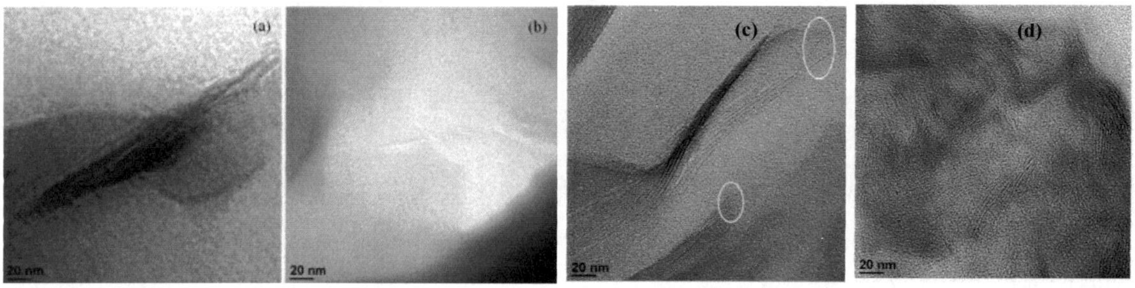

Figure 3.TEM patterns of PBAT + 5% NANO at 97 Kx of magnification: intercalated- and micro-structures (a) and exfoliated particles (b) PBAT/CLO20A (c) and PBAT/CLO30B (d) at 97Kx of magnification

Figure 3a shows that CLO20A formed micro-aggregation of the silicate layers in PBAT matrix, confirming non-extensive intercalation of this clay in PBAT as previously observed through WAXS analysis. Nevertheless, it is important to highlight that the presence of several single CLO20A layers in the polymer matrix (circles in Figure 3c) indicated the attainment of partially exfoliated nanocomposites, which could explain the broadening of the peak observed for these nanocomposites.

Figure 3d shows that PBAT/CLO30B nanocomposites exhibited a high concentration of single clay layers distributed along the polymer matrix as well as a high amount of intercalated structures of silicate particles in accordance with WAXS results.

3.1.3. *Scanning Electron Microscopy (SEM)*

Figure 4 shows the SEM micrographs of the fractured surfaces of PBAT and nanocomposites, revealing that all clays were highly distributed in the entire polymer matrix (see EDS images). PBAT/NANO nanocomposites (Figure 4a1 and 4a2) exhibited a good dispersion of clay particles, however some

IOP Conf. Series: Materials Science and Engineering **223** (2017) 012023 doi:10.1088/1757-899X/223/1/012023

microaggregates (*ca.* 5-10 μm) were still visible (circles in Fig. 4a1), in agreement with the clay aggregates observed by TEM.

In the case of PBAT/CLO20A nanocomposites, SEM images (Figure 4b) revealed the occurrence of micro-aggregation of clay particles (circles in Figure 4b1), confirming that the dispersion of this clay was less extensive in PBAT matrix, in accordance with XRD and TEM results.

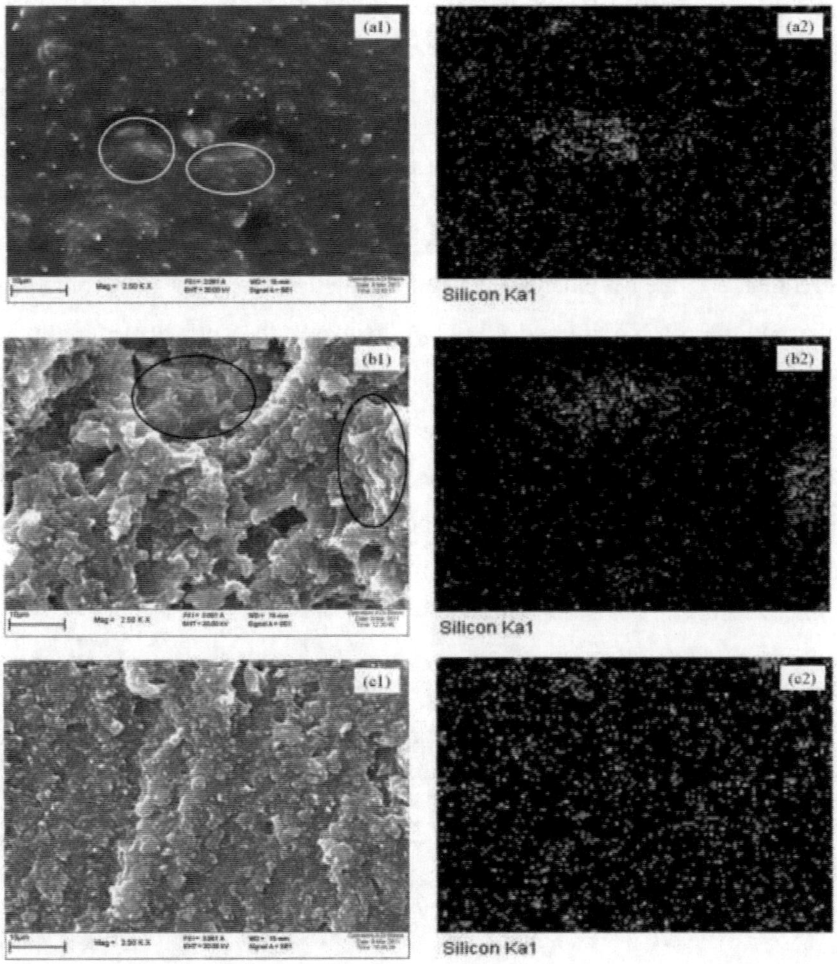

Figure 4. SEM and EDS images of fractured surface of PBAT/NANO (a1 and a2), PBAT/CLO20A (b1 and b2), and PBAT/CLO30B (c1 and c2).

In the case of PBAT/CLO20A nanocomposites, SEM images (Fig. 4b) revealed the occurrence of micro-aggregation of clay particles (circles in Figure 4b2), confirming that the dispersion of this clay was less extensive in PBAT matrix, in accordance with XRD and TEM results.

Figure 4c shows that PBAT/CLO30B nanocomposites exhibited no aggregates in the micron range (only very few aggregates of about 1-2 μm were observed in certain zones of the polymer matrix), in agreement with the high dispersion of these particles previously detected by WAXS and TEM analysis. The results from WAXS, TEM and SEM pointed out that PBAT/CLO30B nanocomposites presented higher degree of dispersion and intercalation than PBAT/NANO and PBAT/CLO20A. This was probably due to higher interactions between the carbonyl groups (>C=O) of PBAT and the surface hydroxyl groups of CLO30B layers and of its organic modifier [21]. Although NANO and CLO20A also had hydroxyl groups on the surface of their silicate layers, the absence of an organic modifier in NANO as well as the absence of hydroxyl groups in the organic modifier of CLO20A (see Table 1) could be able to hinder high interactions between these clays and PBAT, resulting thus in a lower clay dispersion level than that observed for PBAT/CLO30B.

3.2. Contact Angle Measurements

Wettability can be expressed by the contact angle of a liquid on a solid substrate. Figure 5 shows the contact angle photographs and measurements obtained from PBAT and its nanocomposites. These results show that the addition of NANO and CLO20A increases the surface wettability of PBAT, especially NANO. The higher wettability level observed for PBAT nanocomposites based on NANO and CLO20A can be attributed to the presence of surface hydroxyl groups on its clay layers, making the PBAT more hydrophilic. CLO30B also presents a high concentration of hydroxyl groups on its clay layers. However, its high level of dispersion in the polymer matrix could considerably decrease the wettability level of PBAT.

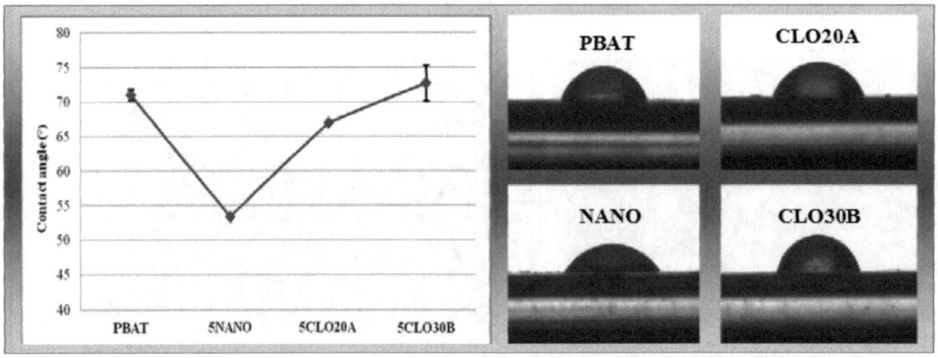

Figure 5. Contact angle measurements of PBAT and its nanocomposites

In this way, the presence of highly exfoliated and intercalated layers of CLO30B into the PBAT matrix, in agreement with morphological analyses, could act as a barrier against water through the bulk material, compensating the possible higher hydrophilicity effect of CLO30B particles on PBAT matrix, thus observing no increase in the hydrophilicity of PBAT upon addition of this clay [22-23].

IOP Conf. Series: Materials Science and Engineering **223** (2017) 012023 doi:10.1088/1757-899X/223/1/012023

3.3. Differential Scanning Calorimetry (DSC)

Table 2 reports the results obtained from the DSC thermograms of PBAT and nanocomposites during cooling and second heating. The results showed that the addition of 5% of layered silicate clays in PBAT matrix decreased the temperature of crystallization (T_c) by 6-11°C, enthalpy of crystallization (ΔH_c) by 5-7 J/g, melting temperature (T_m) by 1-5°C, enthalpy of melting (ΔH_m) by 9-11 J/g, and crystallinity (χ) of PBAT by 7-9%.

The decreases in T_c, ΔH_c, T_m, $\Delta H_m$and χ for PBAT nanocomposites indicated that the addition of 5% NANO, CLO20A and CLO30B could slightly hinder kinetics and extent of crystallization of PBAT on cooling, as well as promote the formation of largely imperfect crystals [14], which melted at lower temperatures than PBAT crystals. These last phenomena were probably caused by an increase in the polymer viscosity upon clay addition, which reduced the mobility of the macromolecular chains to the crystal growth front [16]. No particular differences in the trend of crystallization and melting were observed between nanocomposites.

Table 2. DSC data on PBAT and nanocomposites obtained from cooling and second heating

Sample	Cooling		Second heating		
	T_c* (°C)	ΔH_c (J/g)	T_m* (°C)	ΔH_m(J/g)	χ(%)
PBAT	109	31	133	32	28
PBAT + 5 % NANO	103	24	132	22	20
PBAT + 5% CLO20A	98	26	128	21	19
PBAT + 5% CLO30B	104	24	131	23	21

3.4. Thermo-Gravimetric Analysis (TGA)

Figure 6 reports the TGA curves for PBAT and nanocomposites in air. The thermal stability properties of PBAT were improved by the presence of the clays. Indeed, an increase of T_{max} of PBAT (T_{max} = 400°C) was observed, being these increases similar for all nanocomposites (*ca.* 3-10°C). This behaviour was attributed to a barrier effect of the clay towards polymer decomposition products ablation, which enhanced the overall thermal stability of the system, as well as assisted in the formation of char during thermal decomposition, thus tending to increase the maximum weight loss temperature of PBAT [1,16].

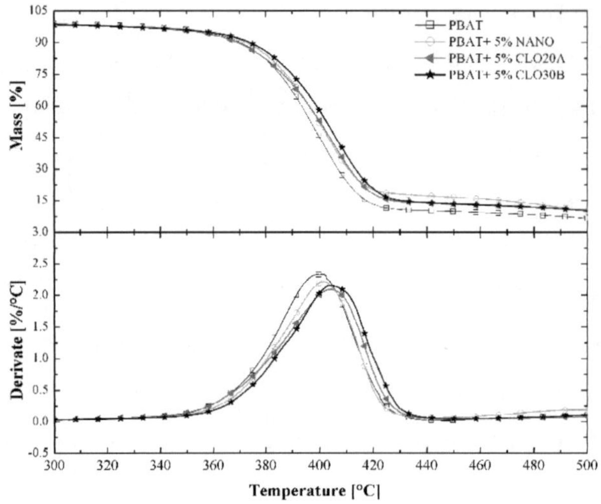

Figure 6. TGA mass loss curves for PBAT and nanocomposites in air

3.5. Surface Hardness

Figure 7 shows the hardness measurements for pristine PBAT and nanocomposites. Theaddition of nanoparticles brought some enhancements in the surface hardness of PBAT. These enhancements seemed to be independent of the clay type and dispersion level, observing similar increases upon addition of all nanofillers. These increases in hardness of PBAT matrix were associated to the inherent rigid nature of clays [24] and to the general good interaction/dispersion level between these clays and the polymer.Moreover, we have reported that the presence of clays increased thermo-mechanical properties as well as compressive strength and modulus elasticity of PBAT [18].

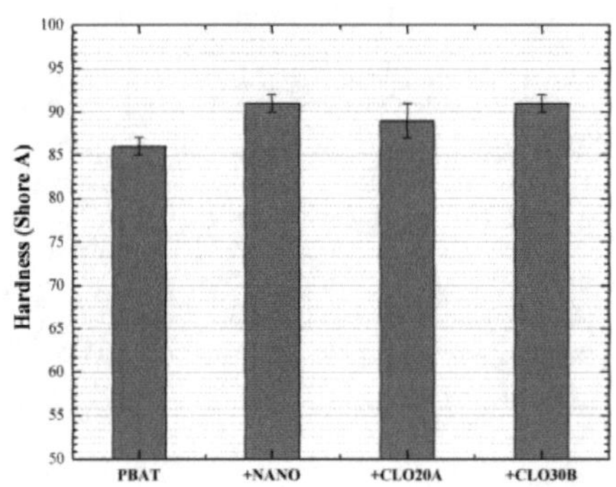

Figure 7. Hardness measurements for pristine PBAT and nanocomposites

4. Conclusions

All nanocomposites showed different levels of clay distribution and dispersion into PBAT matrix. According to WAXS, TEM, and SEM analysis, CLO30B exhibited higher dispersion and intercalation with PBAT than NANO and CLO20Arelated to higher interactions between the carbonyl groups of PBAT with the hydroxyl groups of CLO30B layers and its organo-modifier. Although NANO and CLO20A also had hydroxyl groups on the surface of their silicate layers, the absence of an organic modifier in NANO as well as the absence of hydroxyl groups in the organic modifier of CLO20A could elicit lower interactions between these clays and PBAT as compared to CLO30B.

Higher level of clay dispersion in the polymer matrix could considerably decrease the wettability level of PBAT. Concerning to industrial application, increases in the hydrophobicity level of PBAT based CLO30B could present a high potential for packaging applications, as a low hydrophilicity level is a requirement for materials with this final use.

Thermal analysis showed that there were no particular differences in the trend of crystallization and melting between nanocomposites, indicating that clay/organic modifier structure and clay dispersion level did not considerably affect the thermal properties of PBAT nanocomposites. Similar conclusions were obtained by TGA results, observing a similar improvement in the thermal stability of PBAT after addition of all nanoparticles.

The addition of all nanofillers also leaded to some enhancements in the surface hardness of PBAT which seemed to be independent of the type and dispersion level of clay, and associated to the inherent rigid nature of clays and to the general good interaction between clays and the polymer matrix.

Considering all results, it would be possible to predict a possible wide range of industrial applications for these PBAT based nanocomposites (such as packaging), however, it is important to consider that depending on the final use of these nanocomposites the organic modifier structure and clay dispersion level in the polymer matrix could play or not a key role in the desired properties to achieve for some determined used.

Acknowledgements

The authors would like to acknowledge Prof. S.P. Rwei from National Taipei University of Technology (NTUT), for making possible the preparation of the specimens by melt blending, as well as National Science Council (NSC) Grant Number NSC-100-2221-E-011-031 for funding.

References

[1] Sinha Ray S, Bousmina M 2005 *Progress in Mater. Sci.* **50** 962-1079.
[2] Li F *et al* 2007 *Polym. Degrad. Stab.* **92** 1053–1060.
[3] Bordes P *et al* 2009 *Prog. Polym. Sci.* **34** 125–155.
[4] Qin X Y *et al* 1999 *Nanostruct. Mater.* **11** 259–270.
[5] Ferrari M 2005 *Nat. Rev. Cancer* **5** 161–171.
[6] Vasir J K *et al* 2005 *Curr. Nanosci.* **1** 47–64.
[7] Webster T J *et al* 1999 *Biomaterials* **20** 1221–1227.

IOP Conf. Series: Materials Science and Engineering **223** (2017) 012023 doi:10.1088/1757-899X/223/1/012023

[8] Sinha Ray S, Okamoto M 2003 *Progress in Polym. Sci.* **28** 1539 - 1641.
[9] Krishnamoorti L *et al* 1996 *Chem. Mater.* **8** 1728.
[10] Webster T J *et al* 2000 *Biomaterials* **21** 1803–1810.
[11] Utracki L A *et al* 2007 *Polym. Adv. Technol.* **18** 1.
[12] Someya Y *et al* 2005 *J. Appl. Polym. Sci.* **95** 386–392.
[13] Someya Y *et al* 2004 *J. Appl. Polym. Sci.* **91** 1463–1475.
[14] Chivrac F *et al* 2006 *J. Polym. Environ.* **14** 393–401.
[15] Chivrac F *et al* 2007 *J. Polym. Sci. Polym. Phys.* **45** 1503–1510.
[16] Fukushima J *et al* 2012 *Mater. Sci. and Engineering:C* **32** 1331 - 1351.
[17] Fukushima K *et al* 2013 *J. Polym Res* **20** 302
[18] Fukushima K *et al* 2013 *Appl. Clay Science* **80-81** 291-298
[19] Fukushima K *et al* 2009 *Mater. Sci. and Eng.* **C 29** 1433-1441.
[20] Mohanty S, Nayak S 2009 *Int. J. Plastics Technol.* **13** 163-185.
[21] Mohanty S, Nayak S 2010 *Int. J. Plastics Technol.* **14** 192-212.
[22] Alexandre B *et al* 2009 *J. Membrane Sci.* **328** 186-204.
[23] Anadão P *et al* 2010 *Appl. Clay Sci.* **48** 127-32.
[24] Ramanand A J, Badal D 2009 *Reinforced Plastic and Composite* **28** 343.

IOP Conf. Series: Materials Science and Engineering **223** (2017) 012024 doi:10.1088/1757-899X/223/1/012024

Izod impact performance of hybrid carbon nanotube-alumina filled epoxy nanocomposites

M R Zakaria[1], H Md. Akil[1,2], M H Abdul Kudus[1] and N S A Ahmad Bakhtiar[1]
[1]School of Materials and Mineral Resources Engineering, UniversitiSains Malaysia, Engineering Campus, 14300 NibongTebal, Pulau Pinang, Malaysia
[2]Cluster for Polymer Composite (CPC), Science and Engineering Research Center, Engineering Campus, UniversitiSains Malaysia, 14300 NibongTebal, Pulau Pinang, Malaysia.

E-mail: hazizan@usm.my

Abstract. In this paper, multi-walled carbon nanotube (CNT) was hybridized with alumina (Al_2O_3) particles via chemical vapour deposition using nickel as a metal catalyst. The purpose of hybridization between CNT and Al_2O_3 particles is to prevent the agglomeration of the CNT due to van der Walls forces as well as to improve the impact performance of the epoxy nanocomposites. The Al_2O_3 particles work as "transport" for the CNT to homogenously disperse within the epoxy matrix. As a comparative study, the CNT-Al_2O_3 was also prepared through a physical mixing method. The assessment showed that the CNT–Al_2O_3 hybrid filler produced more homogeneous dispersion in the epoxy matrix with a higher Izod impact strength than the physically mixed CNT–Al_2O_3 filler. The izod impact strength of epoxy nanocomposites at 5 % wt of CNT–Al_2O_3 hybrid has increased up to 32 % compared to the neat epoxy.

1. Introduction

Epoxy resins belong to a major class of engineering polymeric materials, because of their outstanding mechanical strength, thermal stability, long pot life, low curing shrinkage and excellent adhesive strength [1]. The impressive performance characteristics of epoxy resin, together with their good processability and reasonable cost, are considered useful for intensively used diverse field applications [2]. To further improve their functionalities and expand their applications, epoxy resin has often been incorporated with various types of fillers; especially particulate filler, to produce composite materials [3].

Over the last decade, several research groups have focused their efforts on using carbon nanotube (CNT) as particulate fillers, due to their remarkable mechanical, thermal and electrical properties, as quoted by previous experimental and theoretical studies [4, 5]. However, CNT needs to be dispersed uniformly within the epoxy matrix, in order to reveal its extraordinary properties. The small size of CNT tends to aggregate or entangle with itself due to the van der Waals interaction [6, 7]. To overcome this dispersibility issue, the hybridization of CNT with inorganic filler, such as alumina (Al_2O_3), magnesium oxide (MgO), silicon dioxide (SiO_2), calcium oxide (CaO) and muscovite, has become a popular technique via several approaches [8-12]. The conventional hybridization method uses a ball milling approach to mix CNT and alumina. However, this method results in breakage and

IOP Conf. Series: Materials Science and Engineering **223** (2017) 012024 doi:10.1088/1757-899X/223/1/012024

damage of the CNT into smaller lengths, and is considered to be a major disadvantage. Therefore, the idea of the hybridization of CNT and Al_2O_3 via a chemical vapour deposition (CVD) method has been proposed by depositing the CNT onto the Al_2O_3 particles [13, 14]. In this concept, the Al_2O_3 particles work as a transport to help the CNT disperse homogenously into the epoxy matrix; as the dispersion of CNT is dependent on the dispersion of the Al_2O_3 particles.

In this paper, a CNT-Al_2O_3 hybrid compound was synthesised by growing the CNT on the Al_2O_3 particles under a methane atmosphere using nickel as the catalyst. The characteristics of the CNT-Al_2O_3 hybrid compound were examined using Field Emission Scanning Electron Microscopy (FESEM) and High Resolution Transmission Electron Microscopy (HRTEM) to observe the morphology of the CNT structure forming on the alumina particle. The aim of this study was to investigate the Izod impact performance of the CNT in the epoxy nanocomposite, which is relatively poor when CNT alone is added due to poor dispersion. The dispersion state of the CNT in the epoxy nanocomposites was also examined using FESEM. The possible explanations for the difference between the CNT-Al_2O_3 hybrid compound filled epoxy and the CNT-Al_2O_3 physically mixed filled epoxy are discussed.

2. Experimental

2.1. Production of CNT–Al_2O_3 Hybrid

A CNT–Al_2O_3 hybrid compound was synthesized using a chemical vapour deposition (CVD) method. The catalyst was prepared by precipitating $Ni(NO_3)_2.6H_2O$ on aluminium powder in a NaOH solution. The catalyst was then dried at 110°C for 2 h and calcined at 900°C to form a NiO-Al_2O_3 complex. The NiO-Al_2O_3 complex underwent a reduction process under hydrogen gas at 400°C for 2 h, followed by growth of CNT onto the Al_2O_3 under a methane and nitrogen gas atmosphere, at a gas flow ratio of 1:7 at 800°C for 30 min in a horizontal tube furnace. Physically-mixed CNT–Al_2O_3 was produced using a ball milling machine for 48 h at 20 rpm for comparison with the CNT–Al_2O_3hybrid compound. Pure CNT (supplied by SkySpring Nanomaterials Inc.) and Al_2O_3 were mixed at a weight percentage ratio of 12:100, based on the energy dispersive X-ray (EDX) analysis reported in our previous paper [12].

2.2. Characterization of Hybrid CNT– Al_2O_3 Filler

The morphologies of CNT–Al_2O_3 hybrid compound and CNT–Al_2O_3 physically mixed were analysed using a Leo Supra-35VP Field Emission Scanning Electron Microscope and a High Resolution Transmission Electron Microscope (Model Philip TECNAI 20 (200 kV)).

2.3. Preparation of CNT– Al_2O_3 Epoxy Nanocomposites

1.0, 3.0, and 5.0 wt% of CNT–Al_2O_3 hybrid compound and CNT–Al_2O_3physically mixed were dispersed in epoxy resin Diglycidyl Ether of Bisphenol A (DGEBA) using a QSonicasonicator machine at a frequency of 25 kHz for 30 min. A curing agent, trimethylhexame-thylenediamine (TMD) with mass ratio of 6:10 epoxy resin was then added to the mixture. The mixture was placed into a vacuum at 76 cm Hg pressure for 30 min to remove any trapped air and then poured into a silicon mould. Finally, the epoxy composites were cured at 120°C for 1 h. Table 1 shows the descriptions of the samples.

Table 1. Descriptions of the Samples.

Samples	Decriptions
HYB	CNT–Al_2O_3 hybrid compound
MIX	CNT–Al_2O_3 physically mix
Epoxy/HYB	Epoxy filled with CNT–Al_2O_3 hybrid compound
Epoxy/HYB1	Epoxy filled with 1 % wt CNT–Al_2O_3 hybrid compound
Epoxy/HYB3	Epoxy filled with 3 % wt CNT–Al_2O_3 hybrid compound
Epoxy/HYB5	Epoxy filled with 5 % wt CNT–Al_2O_3 hybrid compound
Epoxy/MIX	Epoxy filled with CNT–Al_2O_3 physically mix
Epoxy/MIX1	Epoxy filled with 1 % wt CNT–Al_2O_3 physically mix
Epoxy/MIX3	Epoxy filled with 3 % wt CNT–Al_2O_3 physically mix
Epoxy/MIX5	Epoxy filled with 5 % wt CNT–Al_2O_3 physically mix

2.4. Characterization of the Epoxy Nanocomposites

Izod impact tests of Epoxy/HYB and Epoxy/MIX were performed with a pendulum apparatus (Zwick Impact Testing Machine) using conventional V-notched specimens. Specimens were made in accordance with ASTM standard D256. At least five specimens were tested for each case to ensure reliability of the test results. The fracture surfaces of the Epoxy/HYB and Epoxy/MIX samples were examined by FESEM after coating with a 5–10 nm Au–Pd layer by sputtering.

3. Results and Discussion

Figure 1a-b show the SEM images of the HYB. From the observation, it can be seen that CNT was successfully grown on the alumina particles. The CNT was attached to the Al_2O_3 particles with a diameter of approximately 10 - 30 nm. The CNT was distributed homogenously on the Al_2O_3 particles and wrapped around the Al_2O_3 particles. The Al_2O_3 surface was fully covered by CNT due to the large amount of the CNT deposited. The dispersion of the CNT was dependent on the dispersion of the Al_2O_3 particles. Meanwhile, Figure 1c-d show the SEM images of the MIX, which illustrates that the CNT was poorly distributed and independent of the dispersion of the Al_2O_3 particles. In addition, CNT did not physically attach to the Al_2O_3 particles and tended to agglomerate due to the van der Waals interactions.

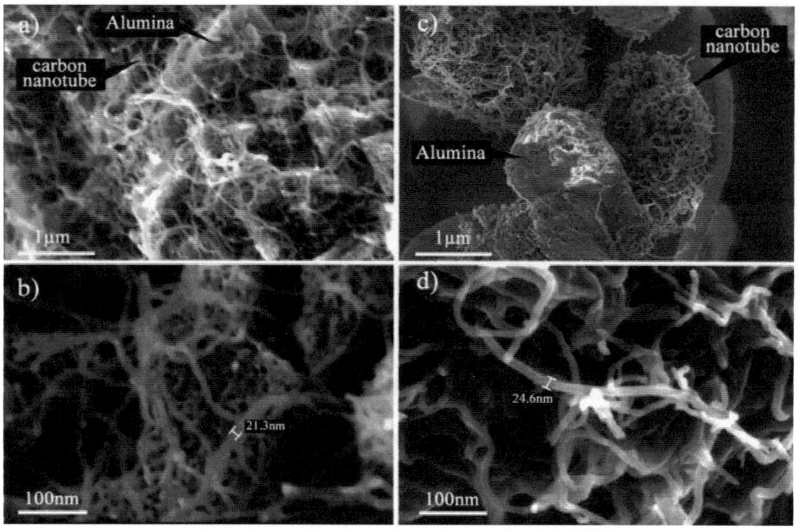

Figure 1: SEM Images of HYB with magnification of (a) 80,000X and (b) 200,000X and MIX with magnification of (c) 80,000X and (d) 200,000X.

Figure 2 shows the HRTEM images of the HYB. From the HRTEM images, the CNT was confirmed as multiwalled carbon nanotube (MWCNT) by the existence of hollow structures and multi-layered walls. The MWCNT contained several layers of graphene sheets along the longitudinal direction of the nanotube, approximately 15 - 25 walls, and an inner hollow with a diameter of 4 - 8 nm. The nickel catalyst can be seen clearly at the end of the MWCNT tip, which indicates the MWCNT grown by the tip growth mode. The size of the MWCNT is dependent on the size of the nickel catalyst as shown in Figure 2b.

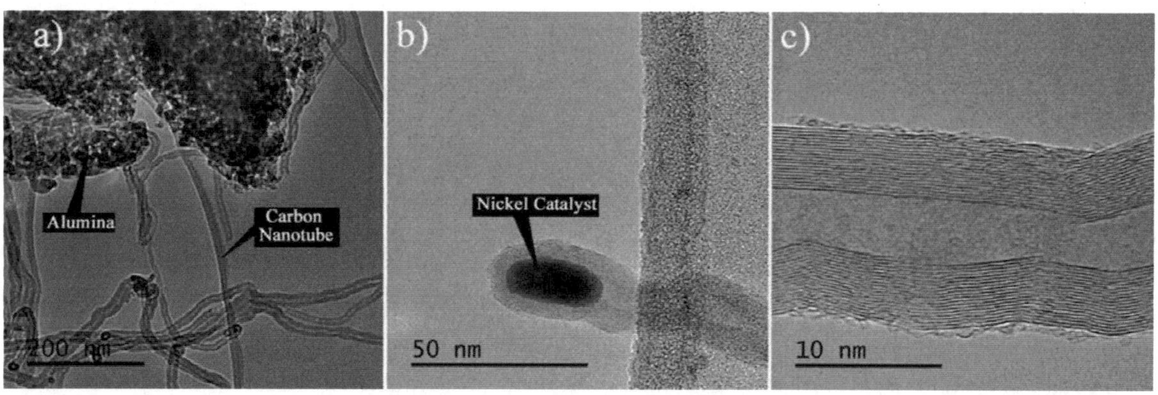

Figure 2. HRTEM Images of HYB.

Figure 3 shows the Izod impact strength of the Epoxy/HYB and Epoxy/MIX at various filler loadings. As shown in the figure, the Epoxy/HYB and Epoxy/MIX nanocomposites exhibited an improvement in Izod impact strength compared to the neat epoxy. The Izod impact strength increased gradually after the addition of the CNT-Al_2O_3 hybrid filler into the epoxy matrix. The enhancement of the Izod impact strength was due to the extraordinary properties of the CNT fillers in the epoxy matrix. From the graph, it can be seen that the Epoxy/HYB showed a higher Izod impact strength than the Epoxy/MIX. The highest Izod impact strength of the Epoxy/HYB was achieved at 5 wt% with 375 J/m, and exhibited an enhancement of up to 32% compared to the neat epoxy. Meanwhile, the Izod impact strength of the Epoxy/MIX3 was 335 J/m, which presented the highest Izod impact strength with an enhancement of up to 18% compared to the neat epoxy. The higher Izod impact strength of the Epoxy/HYB may be explained by the dispersion of CNT and alumina within the epoxy matrix being better than the Epoxy/MIX. The good dispersion of CNT in the Epoxy/HYB was due to the Al_2O_3 particles acting as a vehicle for the CNT to disperse within the epoxy matrix. As a result, this provided a uniform stress distribution and reduced the stress concentration in the epoxy composite. In the Epoxy/MIX, several agglomerations of CNT formed large particles, which consequently reduced the surface area of the CNT. Thus, the agglomeration of CNT led to a lower Izod impact strength.

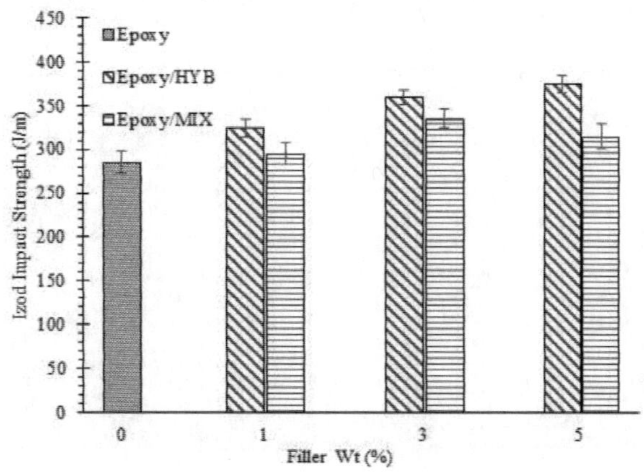

Figure 3. Izod Impact Strength of Neat Epoxy and the Epoxy Composites with 1%, 3% and 5% Weight Percentage of HYB and MIX.

In order to investigate the failure mechanism of epoxy nanocomposites, the fracture surfaces of the Epoxy/HYB and Epoxy/MIX specimens were comparatively examined usingFESEM. Figs. 4a-b show the surface of the Epoxy/HYB; which demonstrates that the HYB is distributed homogenously in the epoxy matrix. This HYB structure prevents the CNT from agglomerating with each other and ensures a large contact area between the CNT and the epoxy matrix. The bridging network, which is easily formed by HYB, enables it to transfer load from the matrix to the filler. Thus, enabling increased impact strength of the epoxy nanocomposites. Figure 4c–e show the surface of the Epoxy/MIX; which demonstrates that the MIX is not distributed homogenously in the epoxy matrix. As seen on the surface of the Epoxy/MIX, the CNT and Al_2O_3 particles were found to be dispersed separately and some CNT were detached from the alumina Al_2O_3 particle all together. This was probably caused by a weak interaction bonding between the CNT and the Al_2O_3 particles, as a result of the high frequency vibration sonication process. The agglomeration of the CNT and the Al_2O_3 particles in the epoxy matrix can be clearly observed from the close-up views shown in Fig. 4d and Fig. 4e, respectively. This agglomeration induces stress concentration and thus reduces the impact strength of the epoxy nanocomposites.

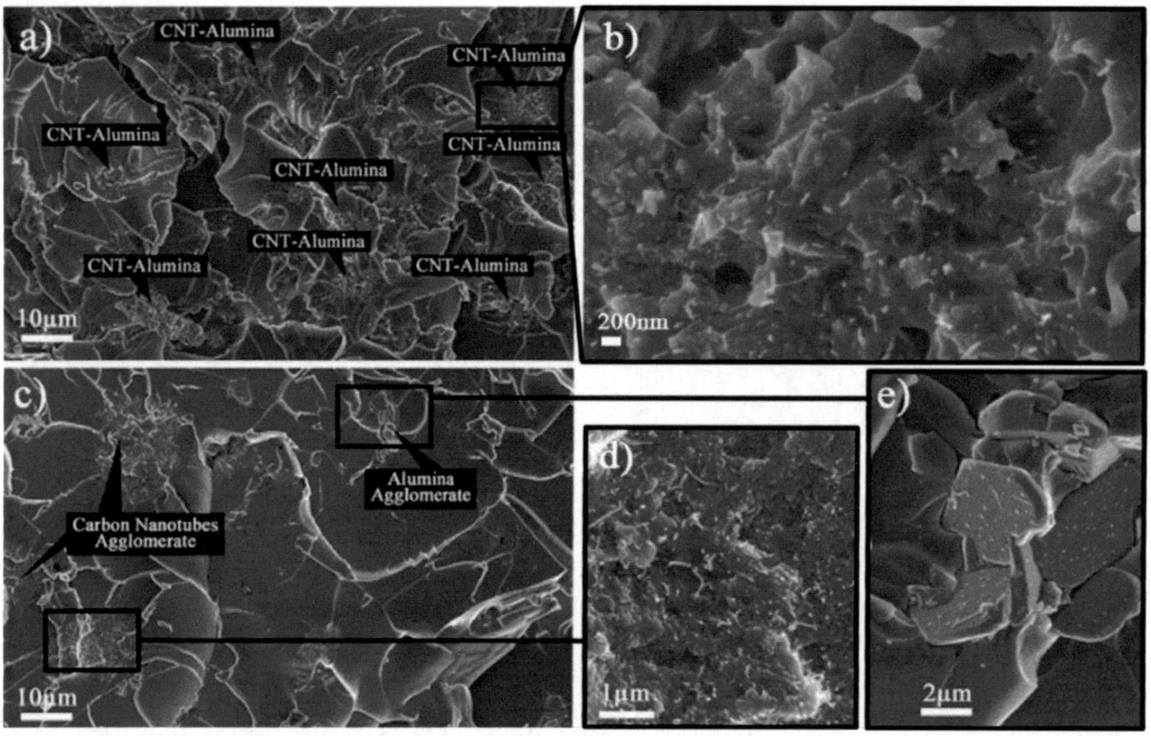

Figure 4. SEM images of the Epoxy/HYB at magnification of (a) 1,000X, (b) 20,000X and Epoxy/MIX at magnification of (c) 1,000X, (d) 10,000X and (e) 5,000X.

4. Conclusions

This study investigated the izod impact strength of CNT–Al_2O_3 filled epoxy composites. The CNT–Al_2O_3 hybrid compound was successfully synthesized via CVD to improve the dispersion of CNT in the epoxy matrix. The Epoxy/HYB composites showed higher izod impact strength than Epoxy/MIX. The izod impact strength of Epoxy/HYB was enhanced by approximately 32% at 5 % wt, whereas the izod impact strength of Epoxy/MIX was enhanced by approximately 18% at 5 % wt. In the Epoxy/HYB system, Al_2O_3 particles worked vehicles to disperse the CNT homogenously within the epoxy matrix. Homogenous dispersion of CNT–Al_2O_3 in Epoxy/HYB and some agglomeration of CNT–Al_2O_3 in Epoxy/MIX were observed using FESEM.

Acknowledgements

The authors would like to acknowledge Universiti Sains Malaysia (USM) FRGS 203/PBAHAN/6071337 for sponsoring and providing financial assistance during this research work.

References

[1] Kim M-G *et al* 2012 *Composites Part A: Appl. Sci. and Manufacturing.* **43**(9) 1620-7.
[2] Nam TH *et al* 2016 *Composites Part B:* Eng. **85** 15-23.
[3] Hong S-K *et al* 2015 *Composites Part A: Appl. Sci. and Manufacturing* **77** 142-6.
[4] Chechenin NG *et al* 2013 *Appl. Surface Sci.* **275** 217-21.
[5] Cha J *et al* 2016 Materials & Design **95** 1-8.
[6] Geng Y *et al* 2008 *Composites Part A: Applied Science and Manufacturing* **39**(12) 1876-83.
[7] Zakaria MR *et al* 2017 *Materials* **10**(3) 301.
[8] Zakaria MR *et al* 2014 *Composites Part A: Applied Science and Manufacturing* **66** 109-16.
[9] Zhu ZG *et al* 2013 *Sensors and Actuators, B: Chemical* **178** 586-92.
[10] Bakhtiar NSAA *et al* 2016 Materials & Design. **91** 46-52.
[11] Zakaria MR *et al* 2016 *Composites Part B: Eng.* **91** 235-42.
[12] Maghsoodi S *et al* 2010 *Appl. Surface Sci.* **256**(9) 2769-74.
[13] Li W *et al* 2014 *Composites Sci. and Technol.* **103** 36-43.
[14] Zakaria MR *et al* 2015 Composite Structures **132** 50–64.

IOP Conf. Series: Materials Science and Engineering **223** (2017) 012025 doi:10.1088/1757-899X/223/1/012025

Mechanical, thermal and morphological properties of poly(lactic acid)/ethylene-butyl acrylate copolymer nanocomposites

S Wacharawichanant, P Hoysang and S Ratchawong
Department of Chemical Engineering, Faculty of Engineering and Industrial Technology, Silpakorn University, NakhonPathom 73000, Thailand

E-mail: sirirat.che@gmail.com

Abstract. This paper reports a melt blend of poly(lactic acid) (PLA)/ethylene-butyl acrylate copolymer (EBAC) with organoclay content at 3 phr. The mechanical, thermal and morphological properties of PLA/EBAC blends and nanocomposites were investigated. The morphological analysis revealed EBAC phase dispersed as a spherical domain in PLA matrix and the domain size of EBAC dispersed phase increased with increasing EBAC content. The addition of organoclay could improve the miscibility of PLA/EBAC blends due to the decrease of domain size of EBAC dispersed phase. The mechanical properties indicated that the strain at break and impact strength of PLA increased when added EBAC, but Young's modulus and tensile strength decreased. Storage modulus increased with the addition of organoclay to the PLA/EBAC blends. The thermal properties found that the incorporation of organoclay in the PLA/EBAC blends did not effect on the glass transition temperature and melting temperature values relative to PLA. The degradation temperature of PLA improved with the addition of EBAC. This indicated that EBAC has more thermal stability and degradation temperature than PLA. From X-ray diffraction patterns displayed the characteristic peak in PLA/EBAC/organoclay nanocomposites appeared at the lower angle, which indicated the dispersed clay is intercalated in the polymer matrix. However, second-order diffraction peak appeared at the higher angle indicated that there was partially the conventional composite.

1. Introduction

Poly(lactic acid) (PLA) is a well-known biodegradable polymer and an eco-friendly material. PLA is linear aliphatic thermoplastic polyester, which can be synthesized by condensation of lactic acid or ring opening polymerization of lactide [1-2]. Another feature that makes PLA interesting is the fact that it can be processed similarly to polyolefins, and PLA has high strength, high modulus and transparency. However, the main drawbacks with PLA properties are too brittle, low-melt viscosity, softens at relative low temperatures, low-heat distortion temperature and low thermal stability compared to commercial polymers, thus PLA led to its restricted applications [3-5]. For more applications of PLA, it needed to improve the mechanical and thermal properties of PLA.

Some researchers have studied to improve the weak properties of PLA by blending with thermoplastic starch (TPS), natural rubber, plasticizers, and other more flexible polymers [4-7] and reinforcing materials [8-10]. However, the main problems with PLA toughening through plasticizers or rubber are the reduction in strength and modulus properties of the PLA blends can be observed [4,6,11-12]. Thus, the improvement of the reduced mechanical properties of PLA blends may be improved by the addition of clay due to clay has a high aspect ratio that it often exhibits remarkable improvement in material properties, resulting from the increased surface area between the polymer matrix and clay platelets [13]. Moreover, several studies have reported that organically modified clay or organoclay played the role as a compatibilizer for enhance the miscibility of immiscible polymer blends by effectively reducing the domain size of the dispersed phase [14-17]. For instance, Ray et al. [14] studied the effect of organoclay on the morphology and properties of poly(propylene) (PP) and poly[(butylene succinate)-co-adipate] (PBSA) blends. The results found that the presence of organoclay improved the tensile properties and thermal stability of virgin blends, and scanning electron microscopy (SEM) studyobserved the morphology of PP/PBSA blend was significant changed after adding organoclay. Mallick *et al.,* (2010) [15] studied the morphology and properties of poly(methyl methacrylate) (PMMA)/high-density polyethylene (HDPE) blends with organoclay. They reported that the average domain sizes of the dispersed HDPE phase in PMMA/HDPE (70/30 w/w) blends decreased in presence of organoclay. The incorporation organoclay increased the tensile strength and storage modulus of the PMMA/HDPE blends. Thus, the addition of organoclay to improve the miscibility and properties of PLA blends is interesting.

Some literatures have investigated the addition of organoclay to improve the properties of PLA blends for extend applications. For example, the study of Hasook *et al.*,(2006) [13] investigated the effect of organoclay on the mechanical properties of PLA/poly(ε-caprolactone) blends. They found that Young's modulus of PLA/organoclay composites was higher than that of neat PLA. The elongation of the PLA/PCL/organoclay composites also improved. Thus, the presence of organoclay can improve the strength and modulus of the PLA blends.

This work studied the influence of organoclay on mechanical, thermal and morphological properties of PLA/ethylene-butyl acrylate copolymer (EBAC)/organoclay composites. The composites with varying contents of organoclay were prepared by a melt mixing technique in an internal mixer and molded by compression molding.

2. Materials and Methods

2.1. Materials

PLA with the melt flow index (210°C/2.16 kg) of 6.0 g/10 min and specific gravity of 1.24 was produced by NatureWorks LLC, USA, under the trade name of "Ingeo™ Biopolymer 2003D". The ethylene-butyl acrylate copolymer (EBAC) has the melt flow index (190°C/2.16 kg) of 3.95 g/10 min and density of 0.926 g/mL. EBAC with 28.10 wt% of buthyl acrylate was produced by DuPont Company. The nanoclay surface modified with 25-30 wt% of octadecylamine (organoclay) was purchased from Aldrich Company.

IOP Conf. Series: Materials Science and Engineering **223** (2017) 012025 doi:10.1088/1757-899X/223/1/012025

2.2. Sample Preparation

Before blending, PLA was dried in an oven at 80°C for 4 h, and organoclay was dried at 80°C 24 h to eliminate moisture. EBAC was not dried before blending process. PLA/EBAC blends were prepared by melt blending in an internal mixer at 180°C and a rotor speed of 80 rpm for 15 min. The EBAC contents were 5, 10, 10 and 30 wt%. PLA/EBAC blends with organoclay content 3 phr were prepared in a same condition. PLA/EBAC first melted for 5 min and then organoclay was added and continuous mixed for 10 min. After compounding, PLA/EBAC blends without and with organoclay were dried in an oven at 80°C for 1 h for reject moisture. Afterward the blends and composites were compression molded at 180°C for testing.

2.3. Sample Characterization

The morphology of impact fractured surfaces of PLA/EBAC blends without and with organoclay was investigated by scanning electron microscope (SEM) instruments (Maxim 2000S, CamScan Analytical, UK) at 15 kV accelerating voltage after gold coating.

The impact strength of PLA/EBAC blends without and with organoclay was measured using a noted Izod impact tester (Zwick/material testing August-Nagelstr.11.D-89079 Ulm) at room temperature. The impact test was performed according to ASTM D 256. Tensile test was conducted according to ASTM D 638 with a universal tensile testing machine (LR 50k from Lloyd instruments, UK) at a crosshead speed of 50 mm/min. Each value obtained represented the average of five samples.

The thermomechanical properties (storage modulus) of PLA/EBAC blends without and with organoclay was measured in bending film mode at a constant vibration frequency of 1 Hz, a temperature ranges of 30-140°C, and a heating rate of 5°C/min in a nitrogen atmosphere with a dynamic mechanical analyzer (DMA) (Pyris Diamond DMA, Perkin Elmer, USA). The glass transition temperature was detected by using the peak of tan δ from DMA test.

Differential scanning calorimetry (DSC) characterization (Pyris I, Perkin Elmer, USA) was performed to investigate the glass transition temperature of the composites. The temperature was in the range of 50-200°C at a heating rate of 10°C/min in a nitrogen atmosphere for the heating scan then the analysis performed from the heating scan. The degradation temperature of PLA/EBAC blends without and with organoclay was measured by thermogravimetric analysis (TGA) (SDT Q600, TA Instruments, UK). The temperature was in the range of 30-600°C at a heating rate of 10°C/min under nitrogen atmosphere.

3. Results and Discussion

3.1. Morphology

Figure 1 shows the SEM micrographs of the impact fractured surfaces of pure PLA and PLA/organoclay (100/3) composites, respectively. Figure 1(a) shows the impact fractured surface of pure PLA presented a smooth surface, characteristic of a brittle semicrystalline polymer [6]. While impact fractured morphology of PLA/organoclay composites in Figure 1(b) was relatively rough.

Figure 2(a-d) shows SEM micrographs of impact fractured surface of PLA/EBAC blends with increasing EBAC contents from 5 to 20% (w/w). The micrographs of the blends were clearly demonstrated a two-phase separation of the dispersed phase and matrix phase. This indicates that PLA/EBAC blends are completely immiscible, where minor phase of EBAC disperses as spherical domains in PLA matrix. Moreover, the domain size of dispersed EBAC phase increased with increasing EBAC content. In general, in an immiscible blend system especially at high content of dispersed phase, polymers always coagulate individually due to their different chemical structures and high molecular weights [6]. Additionally, the microvoids surrounding the EBAC droplets indicate weak interfacial adhesion in the blends because of the immiscible nature of the polymers. Because of the high interfacial tension between the two polymers, the spherical morphology is anticipated because of the minimization of the interfacial area [18].

Figure 3 shows the SEM micrographs of the impact fractured surfaces of PLA/EBAC/organoclay composites. The addition of organoclay in the PLA/EBAC blend reduced the spherical domain sizes of dispersed EBAC phase significantly when compared with the blends absence organoclay. The spherical shape of dispersed EBAC phase was changed to elongated structure at high EBAC concentration. This indicated that the compatibility between PLA and EBAC was improved when organoclay was added. Thus, the organoclay plays an important role as compatibilizer in reducing the spherical domain of EBAC of PLA blends.

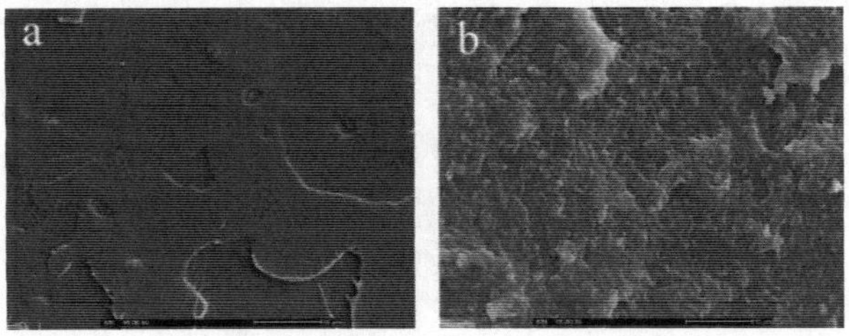

Figure 1. SEM micrographs of impact fractured surface of (a) PLA and (b) PLA/organoclay (100/3).

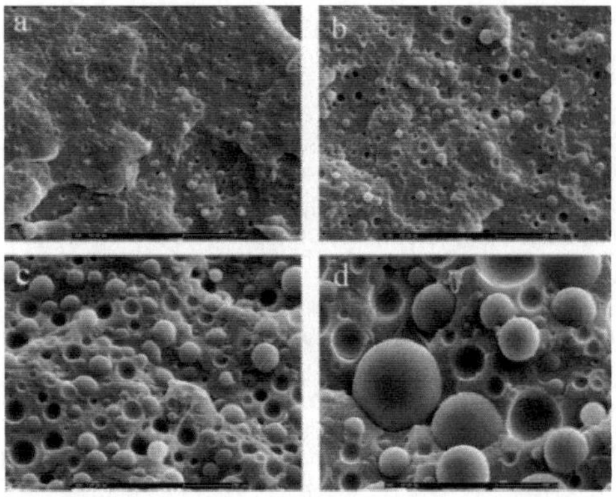

Figure 2 SEM micrographs of PLA/EBAC blends (a) 95/5, (b) 90/10, (c) 80/20, and (d) 70/30.

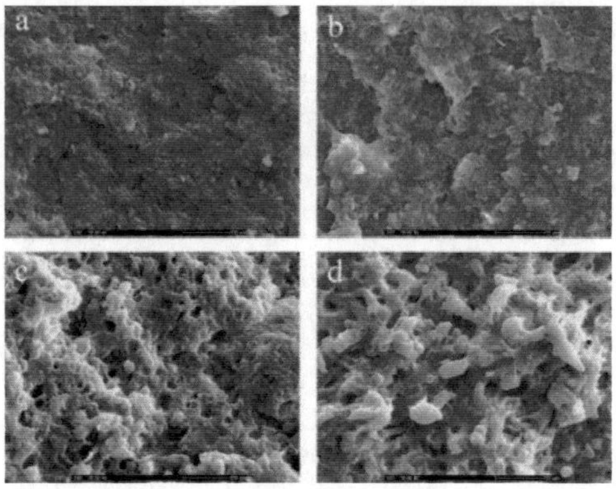

Figure 3 SEM micrographs of PLA/EBAC/organoclay composites (a) 95/5/3, (b) 90/10/3, (c) 80/20/3, and (d) 70/30/3.

3.2. Mechanical Properties

Figure 4 shows the impact strength of PLA/EBAC blends without and with organoclay at various EBAC contents. The impact strength of neat PLA was 2.77 mJ/mm^2. The impact strength of PLA/EBAC blends increased with increasing EBAC content up to 10% (w/w), this behavior was also observed in the PLA/natural rubber blends [4]. When the NR content reached 10% (w/w), impact strength of the PLA/EBAC blends was 6.57 mJ/mm^2, which is about two times higher than that of neat PLA. However, when EBAC content was increased to 20% and 30% (w/w), impact strength of PLA/EBAC blends was decreased to 5.32 and 3.60 mJ/mm^2, respectively, but it was still higher than that of neat PLA. This suggests that the optimum content of EBAC in PLA is 10% (w/w). The decrease in impact strength of PLA/EBAC blends at high EBAC content may be due to the enlargement of the dispersed EBAC caused by the coagulation of EBAC phase as shown in SEM micrographs. From the results indicated that EBAC could improve the impact strength of PLA. The impact strength of PLA/organoclay (100/3) was 2.71 mJ/mm^2 and the addition of organoclay decreased the impact strength of PLA/EBAC blends every concentration EBAC. However, the impact strength of PLA/EBAC/organoclay composites increased with increasing EBAC content up to 10% (w/w) same as the trend of PLA/EBAC blends, and it was still higher than that of neat PLA. This implied that the addition of organoclay did not improve the impact strength of PLA/EBAC blends.

Figure 5 shows Young's modulus of PLA/EBAC blends without and with organoclay at various EBAC contents. The Young's modulus of neat PLA was 1,152.08 MPa. Young's modulus of PLA increased when added EBAC content of 5% (w/w) and Young's modulus of PLA/EBAC blends decreased with increasing EBAC content when EBAC contents were 10, 20 and 30% (w/w). Thus, the reduction in Young's modulus was due to the result of the rubbery nature of EBAC and a weak interfacial adhesion between PLA and EBAC phases as previous shown in SEM micrographs [4]. The Young's modulus of PLA/organoclay (100/3) was 1,141.14 MPa and the addition of organoclay decreased the Young's modulus of PLA/EBAC blends every concentration EBAC and Young's modulus of PLA/EBAC/organoclay composites showed the trend like as Young's modulus of PLA/EBAC blends. This implied that the addition of organoclay did not improve Young's modulus of PLA/EBAC blends.

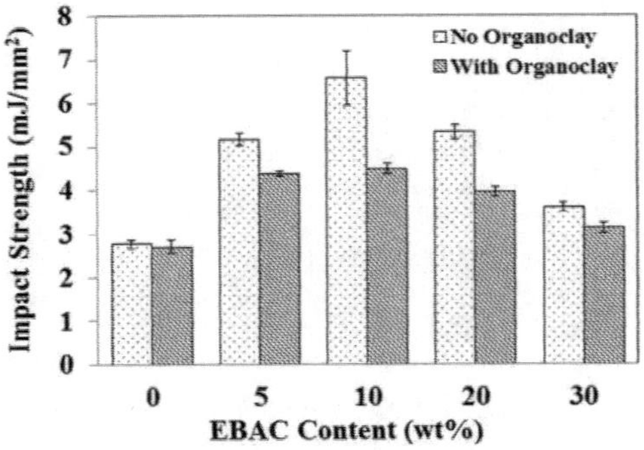

Figure 4. Impact strength of PLA/EBAC blends without and with organoclay 3 phr.

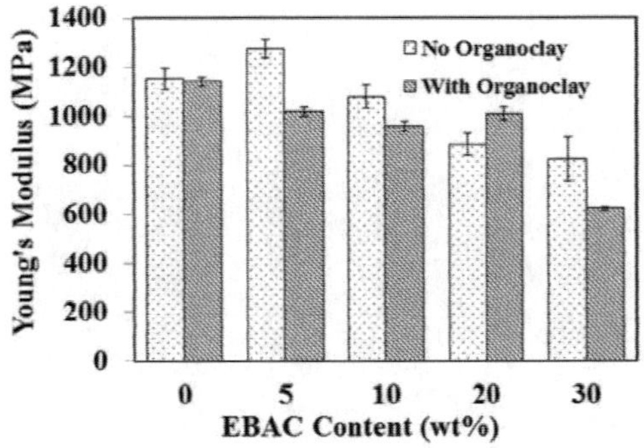

Figure 5. Young's modulus of PLA/EBAC blends without and with organoclay 3 phr.

Figure 6 shows the tensile strength of PLA/EBAC blends without and with organoclay at various EBAC contents. The tensile strength of neat PLA was 57.36 MPa. The tensile strength of PLA/EBAC blends decreased with increasing EBAC content. Thus, the reduction in tensile strength was due to the result of the rubbery nature of EBAC and a weak interfacial adhesion between PLA and EBAC phases as previous shown in SEM micrographs [4]. The tensile strength of PLA/organoclay (100/3) was 54.45 MPa and the addition of organoclay slight decreased the tensile strength of PLA/EBAC blends all concentration EBAC. This result indicated that the addition of organoclay did not improve the tensile strength of PLA/EBAC blends.

Figure 7 shows the stress at break of PLA/EBAC blends without and with organoclay at various EBAC contents. The stress at break of neat PLA was 57.36 MPa. The stress at break of PLA/EBAC blends decreased with increasing EBAC content. Thus, the reduction in stress at break was due to the result of nature of EBAC and a poor interfacial adhesion between PLA and EBAC phases [4]. The stress at break of PLA/organoclay (100/3) was 53.61 MPa and the addition of organoclay slight decreased the stress at break of PLA/EBAC blends at EBAC 5% (w/w), while the addition of organoclay increased the stress at break of PLA/EBAC blends in a range of 10-30% (w/w).

Figure 8 shows the strain at break of PLA/EBAC blends without and with organoclay at various EBAC contents. The strain at break of neat PLA and PLA/organoclay were 9.37% and 10.53%, respectively. The strain at break of PLA/EBAC blends increased with increasing EBAC content and the strain at break of the blends increased to 64.26% when EBAC content was 30% (w/w) [4]. The results indicated that EBAC could improve the strain at break of PLA. The incorporation of organoclay increased the strain at break of PLA/EBAC blends in a range of EBAC content 10%-20%

(w/w), while the other EBAC content showed the strain at break of PLA/EBAC/organoclay composites was small change.

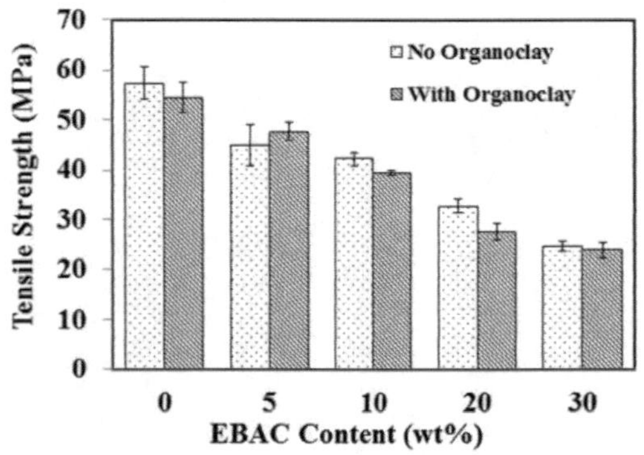

Figure 6. Tensile strength of PLA/EBAC blends without and with organoclay 3 phr.

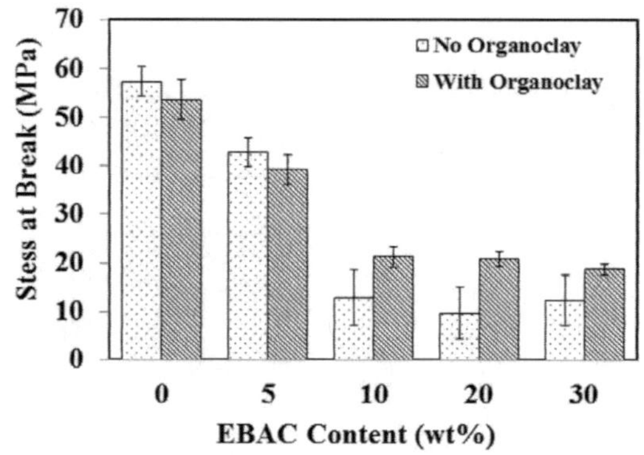

Figure 7. Stress at break of PLA/EBAC blends without and with organoclay 3 phr.

3.3. Thermomechanical Properties

The thermomechanical property of PLA/EBAC blends without and with organoclay was examined by DMA. The variation of storage modulus of PLA, PLA/EBAC blends and composites is presented in Figure 9. The storage modulus PLA/EBAC blends and PLA/EBAC/organoclay composites tended to decrease with increasing temperature. The storage modulus of pure PLA showed the highest value, and it was higher than PLA/EBAC blends and composites. The storage modulus of PLA/EBAC blends decreased with increasing EBAC content due to EBAC contain ethylene molecules which are more flexibility and an elongation than PLA.

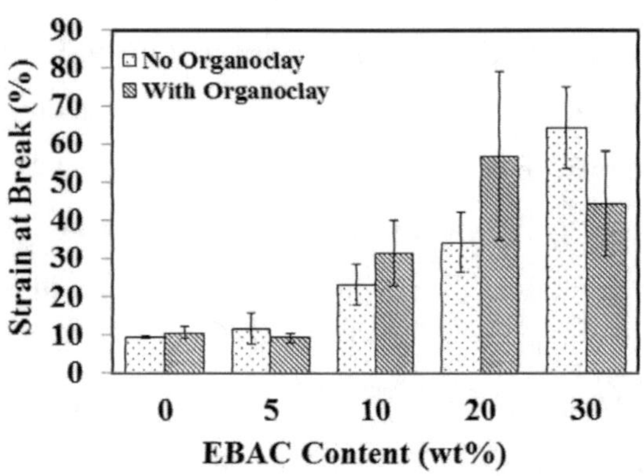

Figure 8. Strain at break of PLA/EBAC blends without and with organoclay 3 phr.

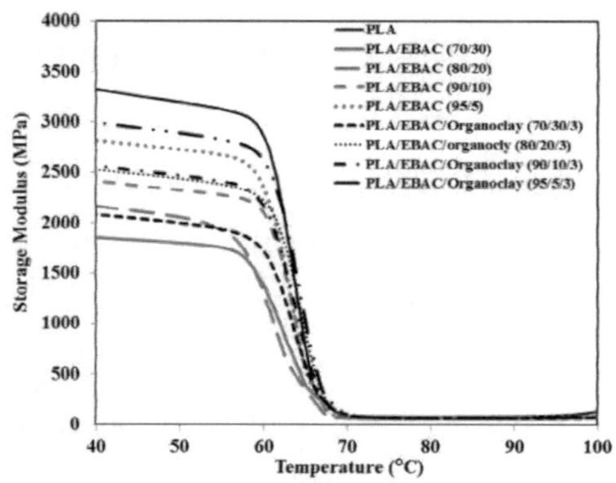

Figure 9. Storage modulus of PLA/EBAC blends without and with organoclay 3 phr.

The behavior of the decrease of modulus was also found in the PLA/natural rubber blends [4]. The presence of organoclay increased the storage modulus of PLA/EBAC blends in every concentration. Thus, organoclay could improve the storage modus of PLA/EBAC blends. This behavior was also observed in poly(methyl methacrylate) (PMMA)/high-density polyethylene (HDPE)/organoclay composites [15].

3.4. Thermal Properties

The results of thermal properties showed that the melting temperatures (T_m) of PLA and PLA/organoclay (100/3) were 151.3°C and 153.4°C, while T_m of blends and composites was 153.3 ± 1.3°C. Thus, the melting temperatures of PLA blends and composites were not significantly changed when EBAC and organoclay was incorporated in the polymer matrix. The glass transition temperatures of PLA and PLA/organoclay were 66.4 and 67.8°C, respectively. Generally, glass transition temperatures of PLA/EBAC blends were in between 67.1 ± 1.1°C. Thus, addition of EBAC had no effect on the glass transition temperature of PLA. The glass transition temperature of PLA/EBAC/organoclay composites were in between 67.7 ± 0.4°C. Thus, the glass transition temperatures of the PLA/EBAC blends were not significantly changed when organoclay was incorporated.

The degradation temperatures corresponding to the 10% weight loss were calculated. The degradation temperatures of PLA, EBAC and organoclay were 315.4, 412.1 and 366.2°C, respectively. This indicated the higher thermal stability of EBAC when compared with that of PLA. The result showed the incorporation of EBAC improved the degradation temperature of PLA. This indicates that an improvement in the thermal stability in the PLA/EBAC blends due to EBAC had the higher degradation temperature. The degradation temperatures of PLA/EBAC blends at 5, 10, 20 and 30 wt% of EBAC were 338.5, 339.8, 331.4 and 342.9°C respectively. While the degradation temperature of PLA blends composites was 340.3 ± 1.5°C. The result showed that the addition of organoclay did not improve the degradation temperature of PLA/EBAC blends.

Figures 10 and 11 show the thermogravimetric curves of thermal degradation of pure PLA, EBAC, organoclay and PLA/EBAC blends and composites. It is evident that the thermal degradation of PLA, EBAC, organoclay showed only single step of weight loss, while the thermal degradation of PLA/EBAC blends and composites proceeded by two degradation steps. It was found that the first step showed a large mass loss starting at around 300°C, this main region was thermal degradation of PLA. The second step showed a small mass loss starting at around 380°C, this region was thermal degradation of EBAC. Moreover, this region showed the mass loss of PLA/EBAC blends at high EBAC content was lower than the blends at low EBAC content.

IOP Conf. Series: Materials Science and Engineering **223** (2017) 012025 doi:10.1088/1757-899X/223/1/012025

3.5. XRD Analysis

From the XRD results, the dispersion state of the clay platelets in the polymer/clay nanocomposites can be identified, i.e., whether the dispersed clay is intercalated or exfoliated in the polymer matrix [13]. Figure 12 show the XRD pattern of organoclay, PLA, PLA/organoclay and PLA/EBAC blends without and with organoclay. The XRD results showed that there were no peaks in XRD data of neat PLA and PLA/EBAC blends, therefore, these peaks are attributed to the presence of clay. The organoclay exhibited peak at 2θ of 4.14°, which corresponds to d-spacing of 2.13 nm. The clay peak in PLA/organoclay (100/3) shifted to a lower 2θ region, indicating that the d-spacing increased to 3.20 nm. This implied that the distance between the clay platelets in the PLA/organoclay was larger than that in the neat organoclay, so the dispersed clay was intercalated in polymer matrix.

The characteristic peak of clay in PLA/EBAC blends at EBAC content 5, 10, 20 and 30 wt% with organoclay 3 phr were found at 2θ =2.76°, 2.72°, 2.73° and 2.74°, which corresponds to d-spacing of 3.20, 3.25, 3.23 and 3.22 nm, respectively. This result indicated the dispersed clay was intercalated due to the intercalation of polymer chains usually increases the d-spacing of the clay platelets [13]. However, second-order diffraction peak appeared at the higher region indicated that there was partially the conventional composite.

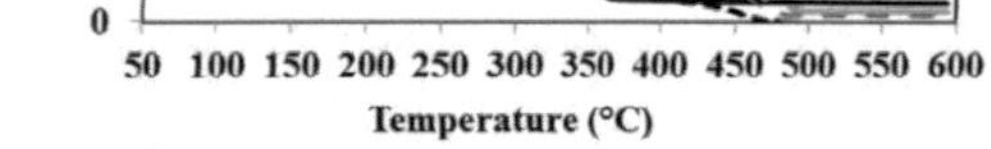

Figure 10. Thermogravimetric curves of PLA/EBAC blends.

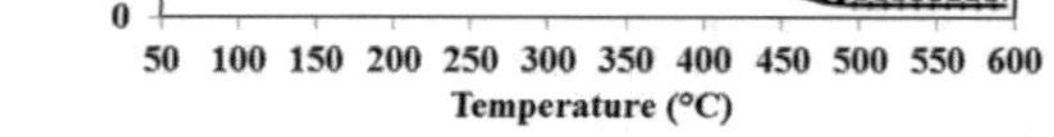

Figure 11. Thermogravimetric curves of PLA/EBAC/organoclay composites.

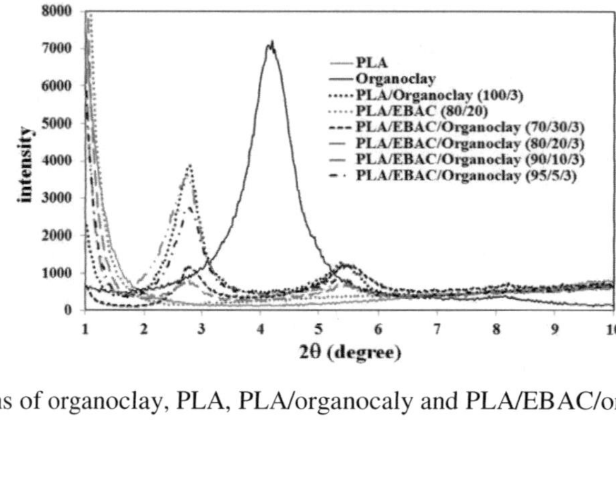

Figure 12. XRD patterns of organoclay, PLA, PLA/organocaly and PLA/EBAC/organoclay composites.

4. Conclusion

This paper reported the morphological, mechanical and thermal properties of PLA/EBAC blends and nanocomposites. The micrographs of PLA/EBAC blends demonstrated a two-phase separation of the dispersed phase and matrix phase, and the domain size of dispersed EBAC phase increased with increasing EBAC content. The addition of organoclay in the PLA/EBAC blend reduced the spherical domain sizes of dispersed EBAC phase significantly. The mechanical properties indicated that the strain at break and impact strength of PLA increased when added EBAC. The storage modulus increased with the addition of organoclay to the PLA/EBAC blends. The thermal properties found that the incorporation of EBAC in PLA and organoclay in PLA blends did not effect on the glass transition temperature and melting temperature of PLA. The degradation temperature of PLA was improved with the addition of EBAC. From X-ray diffraction patterns displayed the d-spacing of each PLA/organoclay and each PLA/EBAC/organoclay nanocomposites was larger than that of neat organoclay. These results imply that the spacing between clay platelets is greater in the PLA/organoclay and PLA/EBAC/organoclay nanocomposites than in the neat organoclay, this indicates the dispersed clay is intercalated in the polymer matrix.

Acknowledgments
The authors would like to thank Silpakorn University Research and Development Institute (SURDI) for the financial support of this project.

References

[1] Akos N I *et al* 2013 *Polym. Compos.* **34** 763.
[2] Spiridon I *et al* 2015 *Composites: Part B* **69** 342.
[3] Gumus S *et al* 2012 *J. Appl. Polym. Sci.* **123** 2837.
[4] Juntuek P *et al* 2012 *J. Appl. Polym. Sci.* **125** 745.
[5] Halász K, and Csóka L 2013 *J. Engineering* DOI: 10.1155/2013/329379: 1.
[6] Ferrarezi M M F *et al* 2013 *J. Polym. Environ.* **21** 151.
[7] Tanrattanakul V and Bunkaew P 2014 *eXPRESS Polym. Lett.* **8** 387.
[8] Jonoobi M *et al* 2012 *J. Polym. Environ.* **20** 991.
[9] Takatani M *et al* 2008 *Journal of Wood Science* **54** 54.
[10] Frone A N *et al* 2011 *Polym. Compos.* **32** 976.
[11] Bee S-T *et al* 2014 *Nucl. Instr. Meth. Phys. Res. B* **334** 18.
[12] Bijarimi M *et al* 2013 *J. Reinf. Plast. Comp.* **32** 1656.
[13] Hasook A *et al* 2006 *Polym. Eng. Sci.* **46** 1001.
[14] Ray S S *et al* 2007 *Macromol. Mater. Eng.* **292** 729.
[15] Mallick S *et al* 2010 *J. Appl. Polym. Sci.* **116** 1010.
[16] Das A K *et al* 2014 *Polym. Compos.* **35** 273.
[17] Yousfi M *et al* 2013 *Macromol. Mater. Eng.* **298** 757.
[18] Wacharawichanant S *et al* 2015 *Polym. Plast. Technol. Eng.* **54** 1349 .

IOP Conf. Series: Materials Science and Engineering **223** (2017) 012026 doi:10.1088/1757-899X/223/1/012026

Effects of number of ply, compression temperature, pressure and time on mechanical properties of prepreg kenaf-polypropilene composites

H S S Tomo[1], O Ujianto[1], R Rizal[2], Y Pratama[2]
[1]Center for Polymer Technology, Agency for Assessment and Application of Technology, Banten, Indonesia
[2]Chemical Engineering, Polytechnic STMI, Jakarta, Indonesia

Email : satsetijo@sentrapolimer.id

Abstract. Composite material thermoplastic was prepared from polypropilen granule as matrix, kenaf fiber as reinforcement and grafted polypropylene copolymer maleic anhydride as coupling agent. Composite products were produced as sandwich structures using compression molding. This research aimed to observe the influence of number of ply, temperature, pressure, and compression time using factorial design. Effects of variables on tensile and flexural strength were analyzed. Experimental results showed that tensile and flexural strength were influenced by degradation, fiber compaction, and matrix – fiber interaction mechanisms. Flexural strength was significantly affected by number of ply and its interaction to another process parameters (temperature, pressure, and compression time), but no significant effect of process parameters on tensile strength. The highest tensile strength (62.0 MPa) was produced at 3 ply, 210 °C, 50 Bar, and 3 min compression time (low, high, high, low), while the highest flexural strength (80.3 MPa) was produced at 3 ply, 190 °C, 50 Bar, and 3 min compression time (low, low, high, low).

1. Introduction

The development of composite technology and the use of natural fiber as reinforcerment come to better direction from year to year. The uses of natural fiber as reinforcement on composite material are cost as well as environmental reason. For example, the use of natural fibers as otomotif parts may save 60% more of energy consumption [1]. However, as different polarity between polymer matrix and natural fiber, coupling agent is needed to improve interfacial bonding between fiber and matrix [2-6]. Another aspect that commonly applied to improve fiber – polymer adhesion is alkali treatment as aqueous sodium hydroxide (NaOH). The purpose of this treatment is to remove cementing substance as lignin and hemicellulose [7-9].

It is common of using melt compounding technique to produce natural fiber thermoplastic composites [10-12]. However, this technique is not efficient to produce large product as building panel. Hence, prepeg method using compression molding may be useful to be used. In this technique, polymer sheets are prepared before composite processing. The sheets are then arranged as sandwich structure along with the fiber for reinforcement, and molten using compression molding. This technique was applied for some natural fibers composites as coconut husk [13], kenaf [14], and jute composites [15]. However, the study on the effect of number of ply as well as hot press conditions is still limited.

This research explored the effect of number of ply as process conditions to produce PP/kenaf fiber composites using compression molding. Experiment was run according to factorial design to achieve better understanding on the effects of variables on mechanical properties. Number of ply, temperature, pressure, and compression time were varied. Significant factors, main effects, as well as interaction influenced tensile and flexural strengths were determined and analyzed.

2. Experimental

2.1. Materials

Polypropilene (PP) used in this research was homopolymer PP HE 2.0 TF, with 2.4 g/10 min melt flow index (MFI), from PT. Chandra Asri, Banten, Indonesia. Kenaf fiber was obtained from local producer in Malang, East Java, Indonesia. Epolene G3003 maleated PP (PP-g-MA) from Eastman Chemical Industry was used as compatibilizers between PP and kenaf fiber.

2.2. Kenaf Fiber and PP Sheet Preparations

Kenaf fibers were chopped with length of ± 20 mm. The cut fibers were then immersed in alkali solution (NaOH 5%, 24 h) to remove lignin structure and improve matrix-fiber interaction [8]. After that, the fibers were strained and washed using demineralized water and dried using oven at 80 °C for 12 h.

Table 1. Processing parameters in extruder

Temperature (°C)										Screw Rotation
1	2	3	4	5	6	7	8	9	10	(rpm)
200	210	220	230	230	240	240	250	250	250	90

PP sheets were produced using Collin E 30 M single screw extruder equipped with callendering system. PP and 5 wt% of PP-g-MA were dry mixed prior to processing. The extruder setting parameters are shown in Table 1. Zones 1 to 7 were applied at extruder barrel, while 8 to 10 were at extrusion sheet die.

2.3. Composite Preparations

Composite samples were produced using Collin 300 P compression molding. Composite samples were prepared as sandwich structures (PP-fiber-PP or PP-fiber-PP-fiber-PP). The fiber content was set at 40 wt% as favourable amount to produce better mechanical properties [16]. Number of composite plies,

IOP Conf. Series: Materials Science and Engineering **223** (2017) 012026 doi:10.1088/1757-899X/223/1/012026

temperature, pressure, and compression time were varied according to Factorial Design as shown in Tabel 2. The conditions were set at low and high setting to explore effect of each parameter on tensile and flexural strength.

Table 2. Variables used in this research

Sample	Number of ply	Temp (°C)	Pressure (Bar)	Time (Min)
1	3	190	50	3
2	3	190	40	3
3	3	190	50	5
4	3	190	40	5
5	3	210	40	3
6	3	210	50	3
7	3	210	40	5
8	3	210	50	5
9	5	190	40	3
10	5	190	50	3
11	5	190	40	5
12	5	190	50	5
13	5	210	50	3
14	5	210	40	3
15	5	210	50	5
16	5	210	40	5

2.4. Composite Characterizations

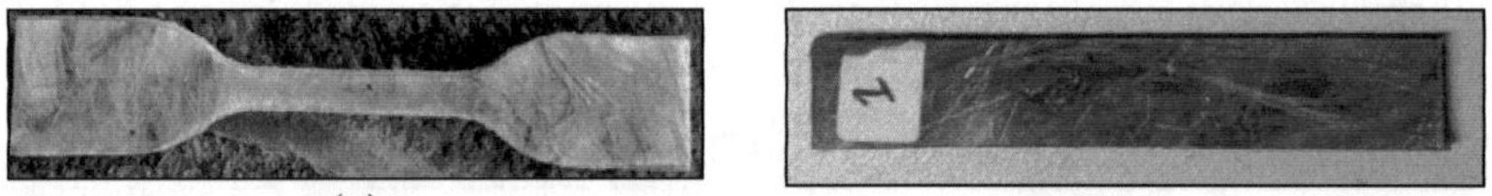

(a) (b)

Figure 1. Test specimens: (a) tensile test, (b) flexural test

Mechanical properties were analyzed from tensile and flexural strength. Tensile specimens (Figure 1) were prepared from compressed composites using puncher according to ASTM D638 type IV. Flexural specimens (Figure 2) were prepared according to ASTM D790. All specimens were conditioned at 23 ± 2 °C and 50% RH for 40 h. Both tensile and flexural tests were carried out using Universal Testing Machine (UTM) SHIMADZU type AGS-10kNG. Crosshead speed was applied at 5 mm/min for all tensile spesimens, but it varied from 0.8 to 1.8 mm/min, which was respected to each specimen thickness, for flexural specimens. The average of 5 (five) specimens were reported and analyzed using Minitab software.

IOP Conf. Series: Materials Science and Engineering **223** (2017) 012026 doi:10.1088/1757-899X/223/1/012026

3. Results and Discussions

Tabel 3 shows tensile and flexural strength data of PP/kenaf fiber composites. Tensile strength varied 19.1 to 62.0 MPa, while flexural varied from 43.1 to 80.3 MPa. The highest tensile strength was produced at 3 ply, 210 °C, 50 Bar, and 3 min compression time (low, high, high, low), attributed to better fiber – matrix interaction.

Table 3. Tensile and flexural strength data of PP/Kenaf Fiber Composites

Sample	Number of Ply	Temp (°C)	Pressure (Bar)	Time (Min)	Tensile Strength (MPa)	Flexural Strength (MPa)
1	3	190	50	3	39.3	80.3
2	3	190	40	3	34.3	72.5
3	3	190	50	5	19.1	77.2
4	3	190	40	5	43.7	45.0
5	3	210	40	3	33.4	43.1
6	3	210	50	3	62.0	64.3
7	3	210	40	5	26.5	70.5
8	3	210	50	5	39.5	77.4
9	5	190	40	3	53.0	54.9
10	5	190	50	3	46.7	54.4
11	5	190	40	5	38.4	64.3
12	5	190	50	5	46.7	50.3
13	5	210	50	3	47.8	44.2
14	5	210	40	3	34.3	68.1
15	5	210	50	5	31.3	53.9
16	5	210	40	5	25.5	52.5
Average					38.8 ± 11.4	60.8 ± 12.6

In these conditions, the matrix viscosity decrease as high temperature, with less degradation due to short compression time. This matrix, with high compression pressure, would able to penetrate the fiber structure and improve the fiber – matrix interaction. In contrast, the matrix difficulty to penetrate the fiber reduced the fiber – matrix interaction. This suggested why the lowest tensile strength was produced at 190 °C (low temperature). This condition might be worsen by matrix embrittlement on the composite layer due to high pressure and long compression time 50 Bar, and 5 min compression time (high, high) The highest flexural strength was produced at 3 ply, 190 °C, 50 Bar, and 3 min compression time (low, low, high, low), might be caused by the fiber compaction and void removal between fiber and matrix [17], while the lowest was produced at 3 ply, 210 °C, 40 Bar, and 3 min compression time (low, high, low, low) suggested some fiber decomposition at high temperature (210°C) [18] and fiber compaction at low pressure (40 Bar) [17].

Different conditions to produce highest and lowest tensile and flexural strength suggest variation effect of each parameter (number of ply, temperature, pressure, and compression time) and its interaction (for example: number of ply – pressure interaction, number of play – temperature – compression time interaction, etc) on the mechanical properties. It is of interest to explore effect of each variable on tensile and flexural strength. However, to give better explanation, significant effect of variables on desired properties should be observed to focus further analysis.

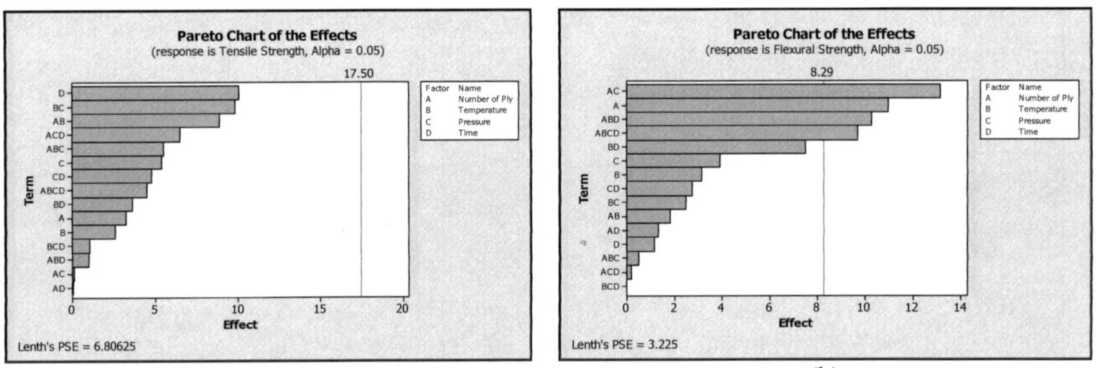

(a) (b)

Figure 2. Pareto chart of tensile and flexural strength of PP/kenaf fiber composites

Figure 2 shows significant factors to change tensile and flexural strength in 95% of confidence level. Variables significantly affected the tensile and flexural strength would have effect higher than minimal value (shown by vertical line on the graphs). There is insignificant effect of variables used in this research on tensile strength, in between boundaries at 40 wt% of kenaf. This suggested that there were complex interactions among process parameter that caused fiber compaction, degradation, and fiber – matrix interaction, etc. Unfortunately, as far as we know, it is still limited to find literature that explore the interaction effect of process parameters on tensile strength for kenaf/PP composite prepared by sandwich technique. Meanwhile, some factors as number of ply – pressure interaction, number of ply, and number of ply – temperature – time interaction, as well as all factor interaction have significant effect on flexural strength. Because of this, exploration on the effect of variables on flexural strength is more interesting than on tensile strength.

Figure 3 presents main effect plots of flexural strength of PP/kenaf fiber composites. The figure shows that flexural strength decreased as higher number of ply as well as temperature, but increased as higher pressure and compression time. The decrease on flexural might be caused by lower density with the higher number of ply as well as voids between fiber and matrix and fiber degradation on higher temperature [17, 18]. On the other hand, the increase on flexural strength might attribute to fiber compaction and matrix penetration which improved matrix – fiber interaction as well as composite density.

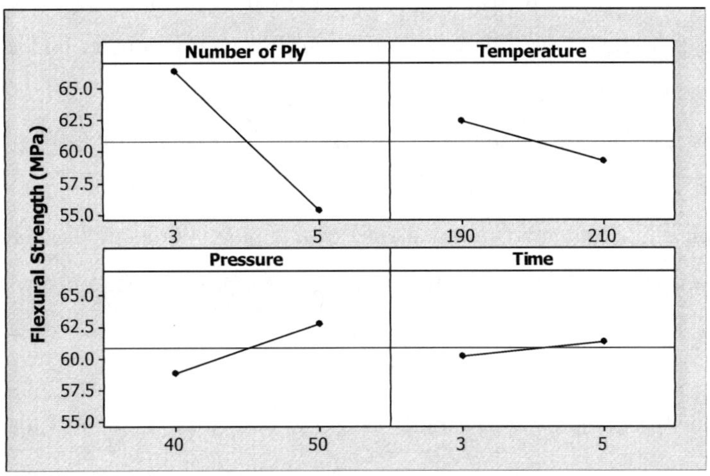

Figure 3. Main effects plot of flexural strength of PP/kenaf fiber composites

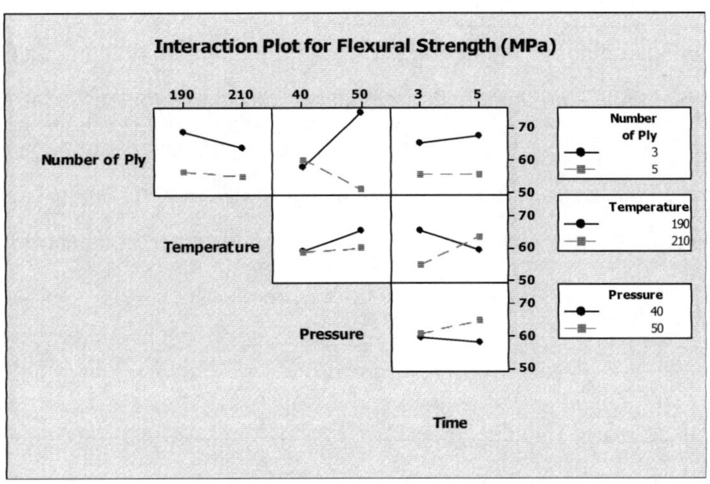

Figure 4. Interaction plot for flexural strength of PP/kenaf fiber composites

Main effect plots are useful to explain effect of individual variable on desired properties. However, as shown in Figure 2 (b), the flexural strength is significantly influenced by interaction between number of ply and other variables. For this reason, interaction plots should also be observed to analyze effect of interaction among process variables (number of ply, temperature, pressure, and compression time) on flexural strength.

Effects of interaction plots on flexural strength are shown in Figure 4. In general, higher flexural strength might be produced at less number of ply at any temperature, pressure, and compression time. This might be caused by the ease of molten polymer to penetrate the fibers and produce better interfacial bonding. In contrast, the figure also shows lower flexural strength at higher temperature suggested some degradation of kenaf fiber that started from 150 °C [18]. Exception at temperature and time interaction, the flexural increase might be caused by improvement interfacial bonding as the ease of polymer penetration. Further analysis on interaction plots shows higher result as pressure increase suggested the effect of fiber compaction and void removal on flexural [17]. While, flexural improvement as longer compression time was attributed to longer compaction process and interaction between matrix and fibers.

4. Conclusions

PP/kenaf fiber composites were prepared using prepreg technique. Compression molding was used to produce the composite samples. The results show that the highest tensile strength was produced at low number of ply, high temperature, high pressure, and low compression time, while the highest flexural strength was produced at low number of ply, low temperature, high pressure, and low compression time. However, analysis on the effects of number of ply, temperature, pressure, and compression time on tensile and flexural strength shows no significant effect of variables on tensile strength in between the boundaries, while number of ply, and its interaction to other parameters significantly influenced flexural strength. The decreased on flexural strength was atributed to degradation due to high temperature and less matrix – fiber interaction, while the increase was caused by fiber compaction and better interfacial bonding.

Acknowledgment

Authors would like to thank to Center for Polymer Technology, Agency for Assessment and Application of Technology, Indonesia for financial support.

References

[1] Holbery J and Houston D *The Journal of Minerals, Metals and Materials Society S.* **58** 80-86.
[2] Anuar H and Zuraida A 2011 *Composites Part B: Engineering* **42** 462-465.

[3] Feng D *et al* 2001 *Polym. Composites* **22** 506-517.
[4] Kim S -J *et al* 2008 *Polym. Testing* **27** 801-806.
[5] Noranizan I A and Ahmad I 2012 *Open J. Polym. Chem.* **02** 31-41.
[6] Saba N *et al* 2015 *Construction and Building Mater.* **76** 87-96.
[7] Agrawal R *et al* 2000 *Mater. Sci. and Eng.: A* **277** 77-82.
[8] Asumani O M L *et al* 2012 *Composites Part A: Appl. Sci.Manufac.* **43** 1431-1440.
[9] Chang W -P *et al* 2009 *Composite Interfaces* **16** 937-951.
[10] El-Shekeil Y A *et al* 2014 *Materials & Design* **58** 130-135.
[11] Han S O *et al* 2012 *Int. J. Polym. Sci.* **2012** 1-8 .
[12] Salleh F M *et al* 2014 *Composites Part B: Engineering* **58** 259-266.
[13] Zaman H U *et al* 2010 *J. Reinforced Plastics and Composites* **29** 3064-3065.
[14] Shibata S *et al* 2006 *Polym. Testing* **25** 142-148.
[15] Chandekar H and Chaudhari V 2016 *IOP Conference Series: Mater. Sci. Eng.* **149** (2016).
[16] Sallih N *et al* 2014 *Composites Part A: Appl. Sci. and Manufactur.* **61** 91-107.
[17] Alkbir M F M *et al* 2016 *Composite Structures* **148** 59-73.
[18] Bledzki A K *et al* 2015 *Industrial Crops and Products* **70** 91-99.

Synthesis and morphology of polyvinyl alcohol /zinc sulfide nanocomposite

M Sirait, Motlan
Department of Physics, Faculty of Mathematics and Natural Sciences, State University of Medan

E-mail : makmursirait@yahoo.com

Abstract. Polyvinyl alcohol (PVA) nanocomposites have been synthesized. The nanocomposites which prepared by sol-gel method were mixed to various concentrations of PVA using magnetic stirrer bar of 500 rpm at 80°C. After the solvent evaporates, the white suspensions were casted on to a flat metal and the film-like composites formed. Those samples were characterized including tensile for mechanical test follow ISO-527-2 and melting points for thermal analysis using DSC operated at 20°C/min with the temperature range of 30-300°C. The most homogeny and Young modulus samples were obtained at ratio 98:2 w/w % of PVA/ZnS, and the highest melting point were found at ratio 97:3% of PVA/ZnS.

1. Introduction

Nanotechnology describes the knowledge about the system as well as the proportion of equipment nanometers. One nanometer is equal to one millionth of a millimeter. Nanotechnology has been impact in science and engineering as well as each side of human life. Many people believe nanotechnology is able to cure the majority of medical diseases in humans. Indeed, most of the application of innovation in the field of nanotechnology is currently only speculative and theoretical, but already many are becoming practical applicative. For example, carbon nano-tubes, tubular carbon molecules are structurally unique and have the properties of electric currents better. Carbon nano-tubes has been applied to the high-resolution screen and reinforce the materials in the industry. Other practical applications of nanotechnology those are for packing, the health and automotive sector [1-3].

The nanoparticles used in the prevention of dirty clothes on the surface where the feathers attached with nano size so similar to taro leaf surface. Polymer nano sizes ranging from 10 nm to 100 nm is used for exterior wall paint, adhesives, paper coatings, upholstery fabrics, cosmetics as well as retaining UV rays. Retaining sunlight is also an example of the use of nanoparticles. It caused UV rays can be easily to disperse and adsorb to nanoparticles. The use of light barrier is very widespread in Australia to dominate the market by 60%. Nanoparticles of aluminium used to mix propellant (fuel) which is able to accelerate the burning of up to two times. Nanocopper mixed lubricating oil to prevent engine wear. Nanocalsium and phosphate composite is used as synthetic bone substitute human bone. The use of composites has advantages such as: lightweight, corrosion resistance, longer service life and has high elasticity properties. Nanoparticle technology, particularly in semiconductors has been expanding applications in biological and biomedical fields. Compound semiconductor nanoscale dimensions can be used as sensors to the cells of the human body, capable of detecting the cancer so that treatment would be more effective [4-8].

Polyvinyl alcohol (PVA) is one type of hydrophilic polymer that is widely used in various fields, especially chemistry, pharmaceuticals and health. PVA can be mixed with other materials to obtain a composite that better suit its usefulness [9]. Gea (2010) [10] has been mixing PVA with bacterial cellulose increase Young's modulus, Campos [11] mixing PVA with silver (Ag) to obtain a higher conductivity properties. Zhang *et al.,* (2011) [12] adds graphene oxide on PVA to produce a composite that has high tensile strength. The products produced as a result of this process generally has good physical properties, non-toxic and have the ability to absorb the relatively high water and biocompatible [13]. PVA hydrogel crosslinked is one of the PVA polymer modified. The hydrogel has a three-dimensional network structure which allows the inclusion of other substances into it. Therefore, the hydrogel matrix used for immobilization drugs, cells, enzymes, and polysaccharides [14].

PVA hydrogel is one type of hydrogel which in recent years have been developed for applications both for the purposes in the field of chemistry, health and biomedicine. It has unique properties, among others,PVA is sensitive to temperature and has a transition temperature of 58°C with partially hydrolyzed to 85°C with a degree of hydrolyzed. Because the melting point of PVA is relatively low, it is necessary to have an amplifier that can improve the physical properties of the material. So in the research, the addition of ZnS improved the mechanical and thermal properties of PVA/ZnS nanocomposite.

2. Materials and Methods

Materials used in this study are zinc asetat 99%, Thiourea 99%, Polyvynil Alcohol 98% each of Merck Darmstadt Germany. Tools used in this study are X-Ray Difraction (Philips Analitycal PW 1710 Based), Scanning Electron Microscopy (SEM) (Zeiss and JOEL Model), Universal Testing Mechanic (Laryee Universal Testing Mechine Wdw-10 model), Differential Scanning Calorimetry (Mettler Toledo type 821).

The method used for the manufacture of nanocomposite PVA/ZnS is a sol-gel method. PVA weighed 20 g dissolved in 200 mL of distilled water and stirred using a magnetic stirrer while heated with a hot plate until late at all. ZnS nanoparticles weighed 0.2 g (1% weight) dissolved in 200 mL of distilled water and stirred until dissolved. Both of these solutions incorporated into the three-neck flask and stirred using a magnetic stirrer with a speed of 500 rpm while heated with a hot plate to a temperature of 80°C. After distilled water evaporates and thickens like a gel solution, inserted into the glass mold that has been formed and allowed to dry and harden. The procedure is repeated with variations PVA:ZnS at ratio (100:0)%, (99:1)%, (98:2)%, (97:3)%, and (96:4)%. According to Mikrajuddin (2008) [15], percentage weight of nanoparticulate inserted very small about 0.5% to 5%. Samples were formed according to the sample test, tensile test to ISO 527-2 and DSC test with ASTM D3418-03. SEM used to determine the morphology of the sample surface.

3. Results and Discussion

Nanocomposite PVA/ZnS have been synthesized by sol-gel method. The product was formed in ISO 527-2 tensile testing with a length of 26 mm, 4 mm wide and 1.124 mm thickness. Every sample was made 5 pieces for each variation to obtain more accurate results and the resluts of measurements be averaged. Sample with only PVA seem translucent and transparent. Sample with a mixture of PVA and variations ZnS nanoparticles white colored slightly darker than the sample with only PVA. With only 1% mixture of ZnS nanoparticles are visible differences in terms of color samples, as well as for samples with a mixture of PVA and ZnS nanoparticles (98:2)%, (97:3)% and (96:4)% more white color and not transparent, this is due to the amount of nanoparticles are mixed more.

Tensile test was performed to determine the tensile strength, elongation at break, strain and Young's modulus. These samples were tested by tensile test equipment UCT with 5 variations mixture of PVA and ZnS nanoparticles. Recording the results of tensile tests nanocomposite PVA/ZnS nanoparticles to a mixture of PVA and ZnS.

Figure 1 shown Young's modulus represent of most reckoned by mechanic properties because it representing comparison of stress and strain, this properties show the delaying of a material. The Young's modulus of mixture PVA and nanoparticle ZnS at 98:2 % wt. The addition of 2% ZnS will multiply the number of cross-linking between the molecular chains that affect mechanical properties. It shown the largest Young's modulus obtained on the composition of PVA: ZnS (98:2)% which is 190.73 MPa. So the composition of the obtained ZnS 2% greater stress and smaller strain. PVA with a little amount of crosslinking will be relatively soft and flexible with the amount of the PVA crosslinking more. A reduction in compressive strength with the addition of PVA was also reported in previous research studies [16].

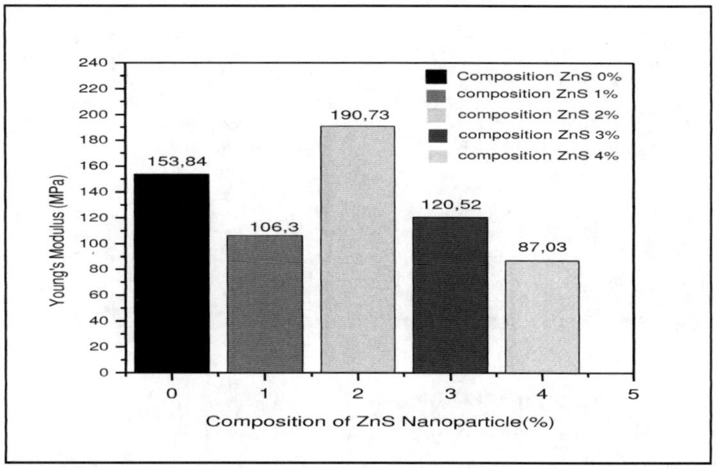

Figure 1. Bar charts Young's modulus of the composition of ZnS nanoparticles.

Thermal tests performed using diferential scanning calorimetry (DSC) to determine the melting point of the nanocomposite PVA/ZnS. DSC analysis with a mass of 6.8 mg using nitrogen gas with a flow rate of 20 mL / min and the heating rate was 20 °C/min. For the nanocomposite sample variation between PVA and ZnS nanoparticle mixture (0,1,2,3,4)% obtained thermogram results as Figure 2. It can be seen that in general the addition of nanoparticles gives rise from the melting point, but in the nanocomposite mixture of PVA/ZnS (98:2)% and (96:4) % are down. It can be caused by uneven mixture of nanoparticles and also the process of making that less than the maximum, in terms of mixing and drying. Greatest value of the melting temperature of the mixture contained in the nanocomposite PVA/ZnS in composition (97:3) % is 224.39 ^{0}C.

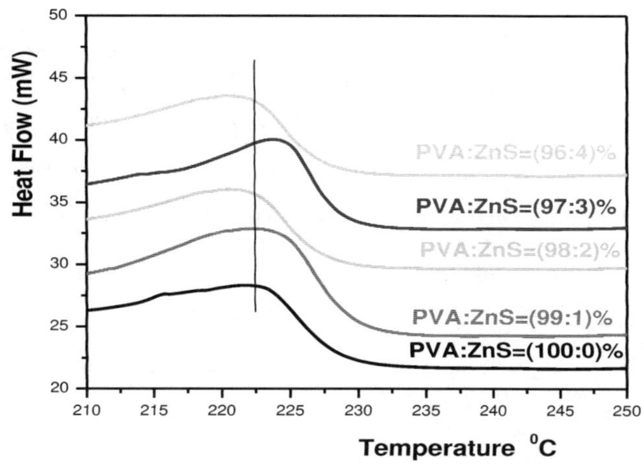

Figure 2. DSC thermograms of PVA/ZnS nanocomposite with ZnS composition of 0, 1, 2, 3, and 4%

IOP Conf. Series: Materials Science and Engineering **223** (2017) 012027 doi:10.1088/1757-899X/223/1/012027

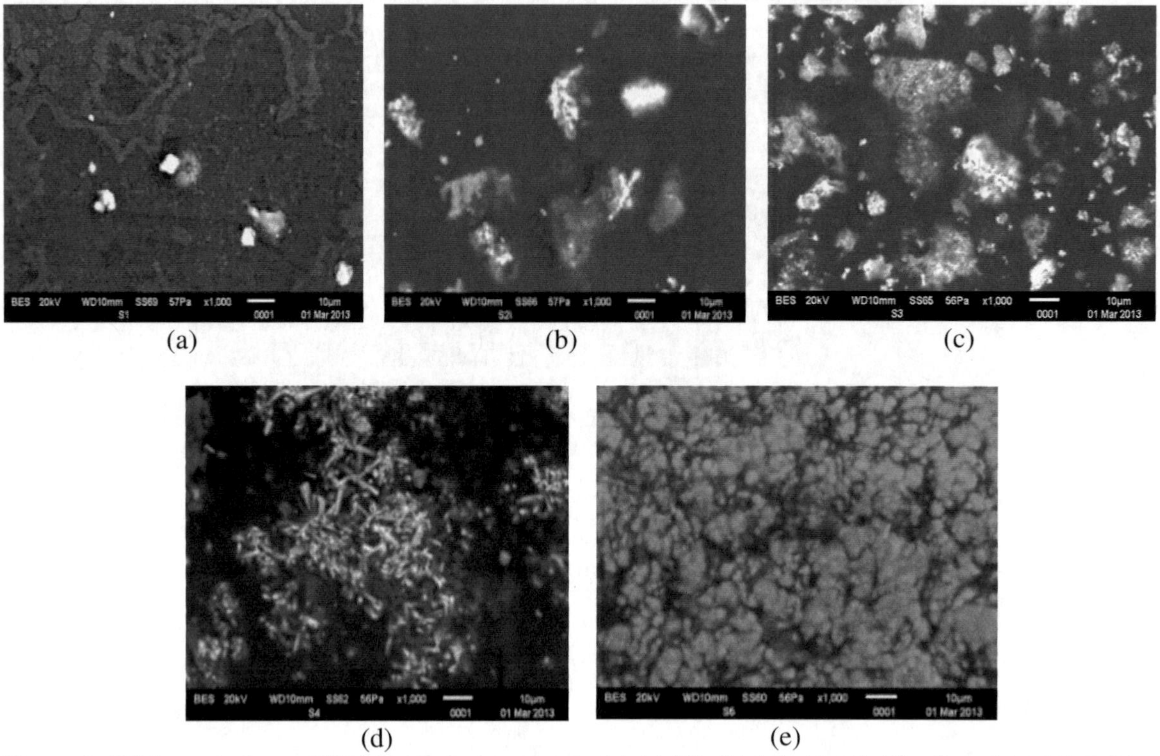

Figure 3. SEM morphology of PVA/ZnS nanocomposite with (a) 0%, (b) 1%, (c) 2%, (d) 3%, and (e) 4% ZnS in x1000 magnification.

The result of Figure 3c is better than other variations mixture made. The morphology of the surface has a uniform and homogeneous mixture. The results of mechanical tests also obtain better mechanical properties in this mixture. It is seen from Figure 3 that PVA/ZnS nanocomposite hydrogel with 2% ZnS contains lot of micro holes and the ZnS powder particles are uniformly distributed in the PVA matrix and well packed by the hydrogel. However, many micro holes minimize and even close because of the lower water content. For PVA/ZnS nanocomposite hydrogel with 2% ZnS, there exist some ZnS nanoparticle agglomerates. In other variations mixture have agglomeration of ZnS at some point. This occurs due to less than perfect stirring and heating unevenly. SEM results also linear with respect to the results of testing mechanical properties. This suggest the addition of nanoparticles causes an increase in its mechanical properties PVA/ZnS nanocomposite hydrogel with 2% ZnS contains). Although in some samples decreased upon the addition of nanoparticles, is seen on the addition of the composition of ZnS 1, 3, and 4%. This occurs due to the increase in areas that do not interact, so that the material is not increased mechanical strength [17].

4. Conclusion

Based on the results of research and discussion in the manufacture of nanocomposite Polyvinyl Alcohol (PVA) with a mixture of PVA and the composition of ZnS nanoparticles of 0, 1, 2, 3 and 4% is concluded as follows:

The results mechanical tests of PVA/ZnS nanocomposite to a mixture of PVA and the composition ZnS (98:2)% gained an average of the largest Young's modulus is 190.73 MPa. In this mixture also obtained morphology uniform and homogeneous. Thermal properties is better occur in the composition (97:3)%, obtained by the melting temperature of 224.39 °C with enthalpy 15.1036 J/g and the heat of 93.642 mJ.

References

[1] Mancini L H, dan Esposito C L 2008 *Nanocomposites : Preparation, Properties and Performance,* Nova Science Publisher, Inc, New York.
[2] Kumar A P *et al* 2009 *Progres in Polym Sci.* **34** 479-515.
[3] Ashby *et al* 1980 *Engineering Materials, An Introduction to their Properties and Applications*, edited by R.J. Brook, Pergamon Press, New York.
[4] Chang J Y *et al* 2000 *Biopolymers PVA Hydrogels Anionic Polymerisation Nanocomposites*, Springer-Verlag Berlin Heidelberg.
[5] Bielecki S *et al* 2005 *Bacterial cellulose, in: Polysaccharides and Polyamide in the Food industry.* Wiley VCH, Weinhein
[6] Bhushan B 2007 *Handbook of Nanotechnology*, Springer Science+Business Media, Inc. New York.
[7] Busnaina A 2007 *Nanomanufacturing Handbook*, CRC Press, New York.
[8] Mikrajuddin A *et al* 2008 *J. Nanosains & Nanoteknologi* **1**(2), 33- 57.
[9] Hossain K M A *et al* 2013 *Constructions and Building Mater.* **45** 20-29.
[10] Gea S 2010 *Innovative Bio-nanocomposites Based on Bacterial Cellulose*, Thesis Doctor, Queen Mary University of London.
[11] Campos J B *et al* 2012 *J of Nanomaterial.* **12**(10) 1 – 11.
[12] Zhang L *et al* 2011 *J. Mater. Chem.* **21** 10399-10406.
[13] Sun P *et al* 2009 *Pure and Appl. Chem.* **46** *533-540.*
[14] Erizal dan Rahayu 2009 *Indonesian J. Chem.* **9**(1) 19-27.
[15] Mikrajuddin A 2008 *Pengantar Nanosains* ITB Bandung.
[16] Makmur S *et al* 2014 *American J. Phys. Chem.* **3**(1): 5-8.
[17] Mikrajudin A 2012 *Pengantar Nanoteknologi,* ITB, Bandung .

IOP Conf. Series: Materials Science and Engineering **223** (2017) 012028 doi:10.1088/1757-899X/223/1/012028

Improvement of polypropylene (PP)-modified bitumen through lignin addition

E Yuanita[1], B E Hendrasetyawan[2], D F Firdaus[2], and M Chalid[2]
[1] Center for Chemical and Packaging, Ministry of Industry, Jakarta, Indonesia
[2] Department of Metallurgy and Material Faculty of Engineering Universitas Indonesia, Depok, Indonesia

E-mail: chalid@metal.ui.ac.id

Abstract. Polypropylene (PP) is usually added to bitumen to improve its mechanical properties, however, both of them have different chemical properties. To achieve best mechanical properties of the mixture, coupling agent such as lignin is importantly required. Lignin is an amorphous biopolymer, has bipolar characteristic due to its distinct chemical function which has carbonyl, carboxyl, hydroxyl and phenol chemical function. Otherwise, bitumen and PP have polar and non-polar characteristic, respectively. In the previous research, it is found that lignin is potential to be used as coupling agent. In order to confirm the potential of lignin as a coupling agent, there are various compounds of lignin on PP-bitumen mixtures used in this research. This experiment consists of several stages, ranging from sample preparation, characterization of raw materials, mixing, and characterization of the PP-Modified Bitumen. This experiment used hot melt mixing to mix lignin, PP, and bitumen. The result of this experimental was analyzed by using FTIR and FESEM. The addition of lignin make Polymer Modified Bitumen (PMB) getting better mixing and increase mechanical properties. Furthermore, FESEM characterization indicated that the addition of lignin gave better mixing of PP-Bitumen. FTIR showed a new chemical structure due to the addition of lignin. From this experiment, the addition of lignin can improve mixing between PP and Bitumen. So, we can use lignin as coupling agent.

1. Introduction

Bitumen is material which has been used for over century. The applications of bitumen as matrix material have a weakness such as plastic deformation when receiving pressure such as a vehicle. For countering that weakness, bitumen material should be added as filler. In order to obtain bitumen with better quality, an increasing number of research began to focus on bitumen modification.

Bitumen modification commonly used polymer materials. Polymer for bitumen modification include polyethylene waste (WPE), PP, tire crumb rubber [1-5]. From the research result, it shows a positive effect, which solves the waste problem in the beginning of this part. But when it is viewed chemically, bitumen has a hydrophilic characteristic that is polar and polymer is the opposite, hydrophobic and is non-polar, which make the two the exact opposite making them hard to interact with one another. What needed for the two of them to become one is the presence of coupling agent that acts as a bridge

IOP Conf. Series: Materials Science and Engineering **223** (2017) 012028 doi:10.1088/1757-899X/223/1/012028

between these two. The condition that this coupling agent need is that it needs to be both hydrophilic and hydrophobic.

Seeing lignin that has both of the conditions and have not been used optimally becomes a solution to this problem. Lignin is amorphous, aromatic biopolymer and the part of lignocellulose that is based on natural material such as wood and plants. Lignin also a phenolic polymeric compound that is the derivative of three hidroxynamel alcohol which is p-coumaryl alcohol, coniferyl alcohol, and synapylalcohol, has hydroxyl, carbonyl and carboxyl substituents [6-8]. Based on these compound lignin has unique characteristics, which is hydrophobic and hydrophilic. Because of it, we can use the characteristics as a coupling agent from different characteristics of other material. Previous research has shown the result that PP-modified bitumen with lignin have passed the penetration requirement for construction application [9]. In this work, we did research for further study of the compatibility level of lignin especially the composition of lignin to combine with PP plastic bag and bitumen from the chemistry bond and morphological characteristics.

2. Materials and Methods

2.1. Materials

In this experiment, bitumen with a penetration grade of 60/70 was used as base material for the modification. Details of properties are shown in Table 1. PP from plastic bag and lignin from paper mill waste were used as modifier and compatibilizer, respectively.

Table 1. Mechanical Properties of Bitumen (10)

Testing Item	Unit	Result	Specifications
Hardness at 25°C, 100g, 5 s	0.1mm	68	60-79
Softening Point	°C	49.9	48-58
Ignition Point COC	°C	310	Min. 232
Ductility 25°C, 5 cm/minute	cm	140	Min. 100
Density	g/cm^3	1.03	Min. 1.0
Solubility at CH_2Cl_3	%	99.09	Min. 99
Weight Loss TFOT	%	0.26	Max. 0.4
Hardness after TFOT	%	80.8	Min. 75
Ductility TFOT	cm	24.5	Min.50
Parafin Concentration	%	0.5	Max. 20

2.2. Methods

First, new PP plastic bag (purpose to minimize impurities) was crushed in to small part and prepared 4% w/w of bitumen. Secondly, lignin was prepared 0.1, 0.3 and 0.5% w/w of bitumen, respectively and then bitumen 60/70 was modified with 4% PP and lignin with those variables. Bitumen modification was perfomed at 180°C in 30 min using stirrer. Finally, PP-modified bitumen will go through for characterization, FE-SEM images were taken at an accelerated voltage of 20 kV, FTIR (Fourier Transform Infrared) were recorded using a Perkin-Elmer FTIR spectrometer, each sample recording consisted of 30 scans recorded from 500 to 4000 cm^{-1}.

3. Result and Discussion

3.1. Base Material Characteristic

Bitumen is the precursor material and the matrix of the mixture that has the color of black or dark brown and at room temperature has the shape of solid or semi- solid. Figure 1 indicates bitumen on FE-SEM testing to shows its base shape. Furthermore, Figure 2 shows the peak when bitumen is tested with FTIR. Peak 3349 cm^{-1} shows–OH functional group.

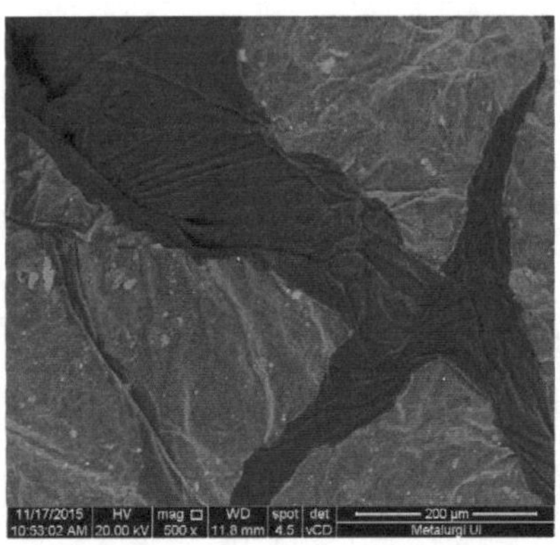

Figure 1. Bitumen Morphological Surface

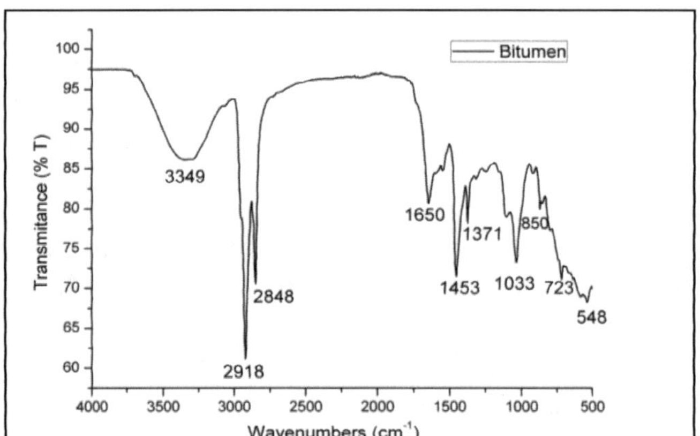

Figure 2. FTIR Spectra of Bitumen

Peak 2918 cm^{-1} and 2848 cm^{-1} shows C-H alkane chain. Peak 1650 cm^{-1} shows C=C alkane chain. Peak 1453 cm^{-1} and 1371 cm^{-1} shows CH_3 chain. Peak 1033 cm^{-1} shows functional group C=O carboxylic. Whereas, peak 850, 723, and 548 cm^{-1} shows functional group aromatic para, meta, and ortho. On the other hand, lignin is material which consist of hydroxyl (-OH), carboxylic (C=O), and aromatic group [6,

8] and shows peak 2292 cm^{-1}. Moreover, peak 2922 cm^{-1} is shown also in the FTIR chracterization of polypropylene (Figure 4). From this characterization, lignin has possibilty to become a bridge or link and approach to induce compatibility between bitumen and polypropylene.

3.2. Polymer Modified Bitumen (PMB) Morphology

Figure 3 is depicted morphology of PP, lignin, and bitumen mixtures. On Figure 3(a), it shows bitumen matrix and PP filler. PP filler has a border with matrix and coloured black. In Figure 3(b) shows the same bitumen matrix and PP filler, however, there's new white material. This white material is lignin that as can be seen in Figure 3(b). Furthermore, in the Figure 3(a) PP distribution on bitumen is not even and has poor dispersion because they form an agglomerate. In the other hand, in the Figure 3(b) PP and lignin distribution is better than in the Figure 3(a), however, the dispersion is still poor because they form agglomerate too. Agglomerate formed as a result of non-optimal mixing. Hence, to have a good mixing result, there are needs for raising the temperature and stirring the mixture during the mixing process. Future study is required so that the mixing can be optimized. This research will be done in the next project.

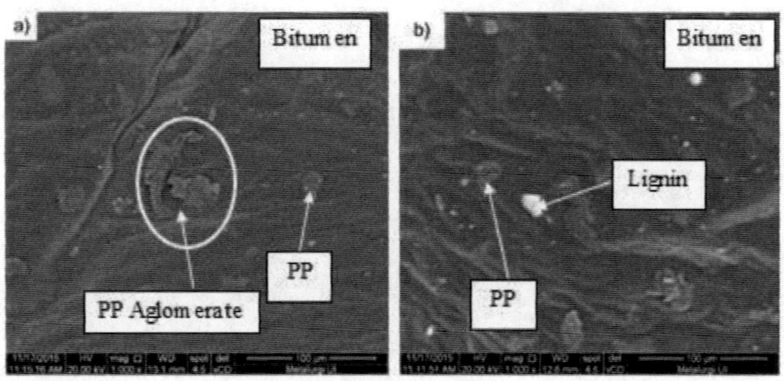

Figure 3. FE-SEM Image of PMB (a) without lignin, (b) with 0.3% lignin

3.3. Compatibility between Raw Materials in Polymer Modified Bitumen

Figure 4 indicates the different of functional groups possessed by non-lignin and lignin bitumen modification. On bitumen modification with lignin peak in 2921 and 2851 cm^{-1} shows the alkane C-H chain. Peak 1582 cm^{-1} shows the CH_2 chain. Peak 1466 and 1376 cm^{-1} showed CH_3 chain. Peak 1016 cm^{-1} shows the functional groups of ether C-O hydrogen bonding. Peak 863 cm^{-1} shows the aromatic functional groups. Peak 731 cm^{-1} shows the meta-aromatic functional groups. Assuming the existence of a hydrogen bond in the ether peak indicated a decline from 1055 into 1016. This, in line with the reports given by

IOP Conf. Series: Materials Science and Engineering **223** (2017) 012028 doi:10.1088/1757-899X/223/1/012028

Stéphanie Laurichesse [10]. Functional groups -OH on lignin PMB lost due -OH in the lignin and bitumen interact to bind together and bring H_2O gas. In this PMB too, appears a new group that is group C-O ether hydrogen bonds, this occurs due to the result of the reaction between the -OH lignin and bitumen. Possible reactions that create new groups can be seen in Figure 5. In Figure 4 spectra of PMB without lignin its look like spectra of PP, this condition because mixing between bitumen and polypropylene is not really good. In addition, PMB without lignin didn't show new chemical function group.

Possible reactions in Figure 5 shows that the reaction occurs in -OH groups. This is according to research conducted by Stéphanie Laurichesse who saw polar groups OH lignin potentially form the functional group ether [11,12]. On bitumen modification without lignin, peak formed only 2921, 2851, 1582, 1466 and 1376 cm^{-1}, so that the chains that were formed then only C-H alkanes, CH_2, and CH_3. In PMB without lignin is also -OH is reduced due to the interaction between bitumen with PP, but not forming ether. Peak FTIR that are formed from PMB without lignin also tend to be the same as the peak on the PP, so it is possible that mixing of PMB is not optimal. So, based on the result of mixing PMB without lignin is to confirm that with the addition of lignin resulted in the formation of a new cluster of ether. The cluster is indicated as an ether bond between bitumen with lignin. This shows that the lignin and bitumen have been compatible.

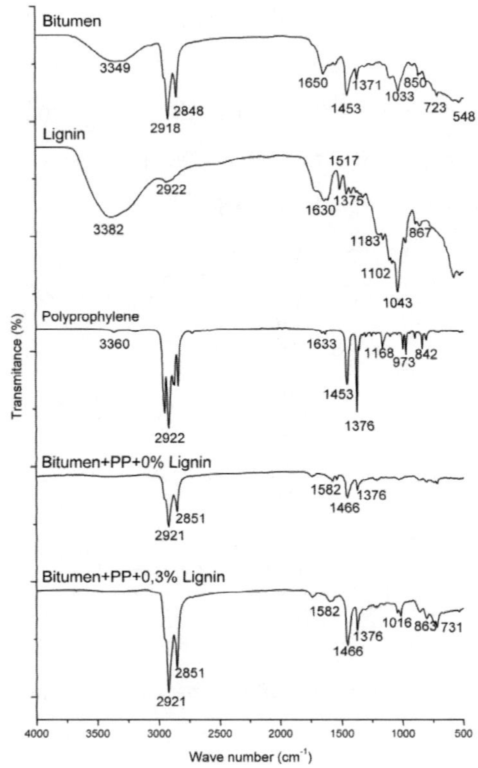

Figure 4. FTIR Spectra of Bitumen, Lignin, PP, PMB without and with Lignin

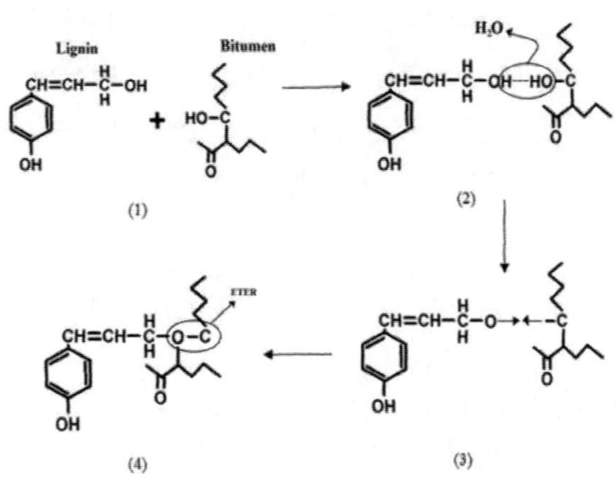

Figure 5. Possible Reaction (Interaction) between bitumen and lignin

Figure 6 shows the effect of the addition of lignin composition of the functional groups is formed. From this figure, it can be seen that the difference does not change the composition of the bond, but increase the intensity of the peak. With the addition of lignin in PMB, it can increase the intensity of the ether bond that is formed. Therefore, it can be assumed that the interaction between bitumen and PP increased hand in hand with the increase in the number of lignin in PMB.

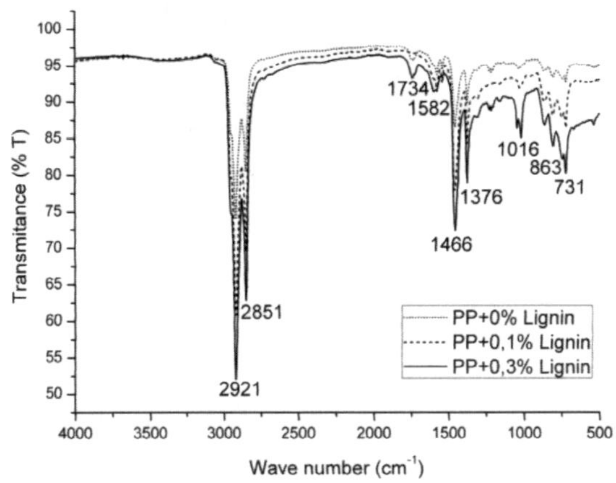

Figure 6. FTIR Spectra of PMB with Variaton of Lignin Composition

4. Conclusion

Lignin can be influential as a coupling agent between the PP plastic bag with bitumen which was indicated by the formation of ether peak and the loss of hydroxide peak. The increasing amount of lignin added to the mixture, increase the amount of ether produced. FE-SEM result indicated that with the addition of lignin, it can produce a better mixture between PP and bitumen. In addition, it also showed that the compatibility of the bitumen and PP with lignin was getting better.

References

[1] Fang C *et al* 2014 *Construction and Building Materials* **54** 320-5.
[2] Tapkın S 2008 *Building and Environment* **43**(6) 1065-71.
[3] Zhu J *et al* 2014 *J. Eur. Polym.* **54** 18-38.
[4] González V *et al* 2012 *Fuel Processing Technol.* **95** 137-43.
[5] Kishchynskyi S *et al* 2016 *Procedia Engineering* **143** 119-27.
[6] Vallejos ME *et al* 2011 *Bio Resources* 6(2) 1158-71.
[7] Bahl K *et al* 2014 *Polymer* **55(26)** 6754-63.
[8] Fernandes EM *et al* 2013 *Progress in Polym. Sci.* **38**(10) 1415-41.
[9] Putra WN *et al* 2017 *Macromolecular Symposia* **371**(1) 140-3.

[10] The Shell Bitumen Handbook 2015 Scotland: ICE P *Sci.* **39**(7) 1266-90.
[12] Le Digabel F, Avérous L 2006 *Carbohydr. Polym.* (4)537-45.

IOP Conf. Series: Materials Science and Engineering **223** (2017) 012029 doi:10.1088/1757-899X/223/1/012029

Thermal conductivity and tensile properties of tin oxide filled UPR/EPS composites with and without organic nanocrystal

R Mohamed, S A Syed Mustafa, Mohd N Norizan, L Suraya Amerudin
GREEN Polymer Research Group, Faculty of Applied Sciences, Universiti Teknologi MARA (UiTM), 40450 Shah Alam, Selangor, Malaysia

E-mail: ssyed_anas@hotmail.com

Abstract: Panel composite for building application made from waste expanded polystyrene (EPS) and unsaturated polyester (UPR) were fabricated and tested in this research for their tensile and thermal characteristics in determining its insulation properties. Research had been carried out to incorporate EPS into UPR with addition of two additives types which are tin oxide (SnO) stand alone, and tin oxide in combination with organic nanocrystals. The weight percentage of SnO had been varied from 0.1 to 2%, and the nanocrystals between 0.1 to 0.5%; both fabricated via casting and compression of the UPR/EPS composite. The testing performed are tensile properties by an Instron Universal Tensile Machine; and thermal conductivity test by using KD2 Pro hand-held thermal probe. It was found that, by using UPE/EPS, the additives affected the composites' thermal conductivity and tensile capabilities. For tensile strength, the control sample showed the highest tensile strength at 22.2 MPa. For thermal conductivity (k value), the lowest value was recorded by the 1% SnO stand alone at an approximate ~ 0.172 W/m.K. Via organic nanocrystal use fixed at 0.1%, the k value increases up to a certain point with increasing SnO. Addition of more SnO and nanocrystal incorporation had mostly increases its thermal conductivity and heat capacity. Heat capacity was also found to be at the lowest in the 1% SnO stand alone sample at 2.008 $MJ/m^3.K$.

1. Introduction

Composite materials has become one of the highly demanded construction material from industrial and commercial consumer worldwide. These composites are made from two or more constituent materials having different physical or chemical properties, that upon combination, produces a hybridized material with characteristics different from the individual components. Generally, the composite material is widely used in building construction. Besides that, the use of composite materials allows many new design possibilities such as orthopaedic biomaterials [1,2].

UPR is a widely utilized economical thermoset material due to its excellent processing ability and good cross-linking tendency as well as mechanical properties when cured; which lead it to be applied for boat construction and panelling manufacture [3]. EPS is a polymeric foam material that has certain desirable properties because of its structure. EPS is a rigid and tough, closed-cell, and often white in colour foam generated from polystyrene beads. Familiar uses include moulded sheets for building insulation and packing material for cushioning fragile items inside boxes [4].

IOP Conf. Series: Materials Science and Engineering **223** (2017) 012029 doi:10.1088/1757-899X/223/1/012029

Investigation regarding using EPS foam as continuous filler within UPR matrices had been investigated by numerous researchers [5–7]. These prior researches had focus on several distinct aspects of the EPS continuous filler filled UPR composites such as its thermal conductivity and its flame retarding ability.

This current research had fabricated EPS filled UPR matrix sheet with aerogel as an insulator and nanocrystals acting as an infrared reflector additive to impart better polymeric product life-cycle via lessening polymeric solar degradation. Kandare *et al.,* (2008) shown similar polymer matrix with zinc oxide (ZnO) and SnO acting as effective flame retardant synergist and smoke suppressants in several polymer systems. [8]

Granqvist, *et. al.* (2016) explained how the usage of infrared additives also played several functions based on its heat insulation properties. They also noted that SnO also had been used before as an infrared additives. In the production of certain film, SnO was used due to its high electrical conductivity and high in visible and near infrared region of solar radiation. [9,10]

Thermal conductivity measures the effectiveness of a material in conducting heat; and thus, acting reciprocally towards a material's thermal resistance and insulation. Thermal insulation is a material or material combinations, that, when properly applied, retards the rate of heat flow by conduction, convection, and radiation. Thermal insulation helps reduces a building's energy consumption through properly maintaining a building's internal environmental temperature, and preventing unnecessary heat flow between the walls of the building and its outside [11,12].

2. Materials and Methods

2.1 Materials

The selected thermo-set matrix material used for fabricating these composite systems is UPR system, Reversol consisting of vinyl ester oligomers having density of 1.12 g/cc, viscosity of 450-600cps with 41-44% styrene content. Methyl Ethyl Ketone Peroxide (MEKP) and Cobalt solution used as initiator and promoter each, were supplied together with the UPE resin by Revertex Sdn Bhd. These ingredients are similar to materials used by previous researches as codified by the research of Rashidan *et al.,* (2009) [13].

The EPS or Styrofoam filler was obtained from waste material; organic nanocrystal (ONC) was synthesised by Green Polymer Research Lab, UiTM Malaysia; and finally, SnO solutions were supplied by Merck KGaA.

2.2 Composite Fabrication and Testing

Recycled EPS, fixed at 10% parts by weight (%wt) of UPR resin, was blended together by utilizing a high speed agitation mixer until complete dissolution was attained. Gaseous contaminates present are eliminated from the mixture with vacuum suction, while solid contaminates are removed from the mixture after gravity settling for 24 h. The ONC utilized was independently synthesized by another researcher of the Green Polymer Research Lab, UiTM Malaysia from cellulose microfibers via proprietary acidolysis process, dried using vacuum drying, and was powderized using a ball mill machine.

The designated additives were then added to the mixture before the samples were prepared. The SnO-only set was varied between 0 to 2.0 wt.%; For the second set, SnO content was varied between 0 and 0.5 wt%, while ONC was set constant at 0.1 wt%. The MEKP and cobalt solution usage was set to 2.5%wt and 1 drops respectively for each sample. Compressive moulding process were done onto aluminium mould, where the mixture jellifies in the mould before any pressure was applied. Gelling time was set to be about 20 to 30 min. Once sufficient gel forms, low pressure press were applied for 1 h at 70°C to retain required shape, and complete curing were done by leaving the sample outside the mould for 24 h at room temperature. These steps are similar to the research process codified by Rashidan *et al.,* (2009) [13]

This paper was produced from a study conducted via qualitative analysis of the manufactured samples. The tensile properties of the samples were determined by an Instron Universal Machine (model Testometric MICR0500) which adheres to requirements of the ASTM D5083 standard. The samples were also tested via a thermal probe to determine their thermal conductivity, thermal diffusivity and volumetric heat capacity. The thermal utilized is the KD2 Pro Hand-held Thermal Probe which adheres to ASTM D5334-08 and IEEE Standard 422-03 as set by ASTM and IEEE.

3. Result and Discussion

3.1 Thermal Characteristics

The sample A and 1 are basically the same formulation despite produced as part of two separate set of test samples. This is shown by their collective data which are relatively similar and narrowly deviate from each other.

Thermal conductivity is a material property measuring the ease or difficulty of which heat energy is conducted through a specific type of material. The higher thermal conductivity, (k) value indicates the material is conducting heat faster as compared to more thermal resistant material.

Table 1. Thermal properties comparison of SnO-only and SnO-ONC samples.

Sample Code	Additives	Thermal Conductivity (W/m.K)	Volumetric Specific Heat ($MJ/m^3.K$)	Diffusivity (mm^2/s)
A	-	0.202	2.186	0.092
B	0.5 % SnO	0.190	2.030	0.094
C	1.0 % SnO	0.172	1.921	0.090
D	2.0 % SnO	0.280	3.134	0.089
1	-	0.211	2.371	0.089
2	0.1 % SnO - 0.1 % ONC	0.245	2.743	0.089
3	0.2 % SnO - 0.1 % ONC	0.331	3.699	0.089
4	0.3 % SnO - 0.1 % ONC	0.257	2.870	0.090
5	0.5 % SnO - 0.1 % ONC	0.258	2.890	0.089

The thermal conductivity indicated by Table 1 showed a minimum of 0.172 $Wm^{-1}K^{-1}$ as recorded by the 1.0wt% SnO sample, while the 0.2wt% SnO- 0.1wt% ONC sample recorded the highest at 0.331 $Wm^{-1}K^{-1}$. Hence, the data in Table 1 can show how increasing the amount of SnO affects thermal conductivity and volumetric specific heat in mixed way.

Sample D representing a probable excessive SnO addition reverses the conductivity decrease trend shown between samples A to C; which may indicate a limit which SnO may be integrated with the polymeric matrix. Adding ONC to the mix reports an opposite trend even at a lower concentration of tin oxide; which may indicates how ONC affecting and overloading the SnO integration capability not unlike of sample D.

As SnO and ONC are powdered materials, their heat transfer through radiation in the relatively porous media are probably affected by different parameters. According to Raed (2013) [14], this heat transference may depended on the optical properties or opacity of the solid material; e.g. carbon black addition will increases the extinction coefficient strongly, and thus decreases the radiative conductivity [15,16]. Other than that, the density/porosity could also influence the radiation absorption rate, and reduces their penetration through the porous medium, which will decrease the radiative thermal conductivity [17]. Existence of pore size within the composite, due to metal oxide and nanocrystal incorporation, will decreases the pore size which in turn reduces the overall radiative conductivity [18,19].

Volumetric specific heat indicates the amount of heat store within its material bulk and molecular bond for a given temperature change. This characteristic is important as it determines how the material reacts to heat exposure; whether it routes much of the heat into increasing its temperature due to its limited heat capacity, or storing said heat energy with little effect on its surface temperature [5,20]. As shown in the Table 1, it neatly follows the trend of the thermal conductivity in regards to the effects of additive integration. Thermal diffusivity, meanwhile, is one compound characteristic that is calculated by the ratio

IOP Conf. Series: Materials Science and Engineering **223** (2017) 012029 doi:10.1088/1757-899X/223/1/012029

of thermal conductivity to volumetric specific heat. This particular property measures the materials' thermal inertia; in which determines how fast the material changes temperature due to its thermal conductivity and specific heat, [5] or stated in another way, the speed of heat propagation via conduction during temperature changes with time. The heat propagation is hence faster for materials with higher thermal diffusivity [21].

Certain construction industry markets favours a material with low thermal diffusivity as it could withstand rapid changing environmental temperatures. Table 1 shows that the resulting thermal diffusivity of the samples are relatively low, fairly consistent, and parallels well the trends of both thermal conductivity and volumetric specific heat. These consistent ratios, as seen across these samples, thus may indicates that either additives had little impact on the resulted diffusivity values, and also that the changes present on both thermal conductivity and heat capacity are proportionally impacted by these additives.

Thermal diffusivity is often a quite problematic parameter to estimate than that of thermal conductivity. This is due to its strong correlation with the contact resistance, which in most cases, are relatively unknown. The error of estimation will be smaller when said thermal contact resistance is lower; errors for the probe parameters (i.e. power, heat capacity) are known; errors in timing accuracy as well as in measured temperature. Thermal conductivity is hence, in most cases, the best behaving parameter; whereas diffusivity and contact resistance are very sensitive to variations in heating power, probe heat capacity, and timing errors [21].

All in all, the inclusion of SnO lowers both thermal conductivity and specific heat until a certain threshold. This effect may occurs due to their metallic oxide nature, where the dispersed and trapped metal oxide particles might had created thermal barrier bubbles that acts similar to how air trapped in foams slowing heat conduction. Adding ONC might reduces this potential by replacing the metallic oxide integration.

3.2 Tensile Characteristics

Tensile test was conducted following the ASTM D5083 standard with the speed of crosshead at 5mm/minute. Based on Figure 1, the graphs shows how additives integration reduces the tensile strength of the composite slightly. Based on Figure 1, both SnO-only and ONC included samples shown a slight decreasing trend on the maximum strength; while their reported young's modulus are a mixed reaction to the additive integration. As maximum strength indicates how strong a force tensing upon the material are required to break it, the slight decrement between samples A to D and 1 to 5 had inferred that said additives used affects, although slightly, the materials' physical strength.

A possible reason for this observation is that the additives' particles may had agglomerated during the mixing and curing process, along with some molecular bonds dissociation between them and the polymer matrices. These lowers the reported strength by the increased sliding allowances created from this available dissociation.

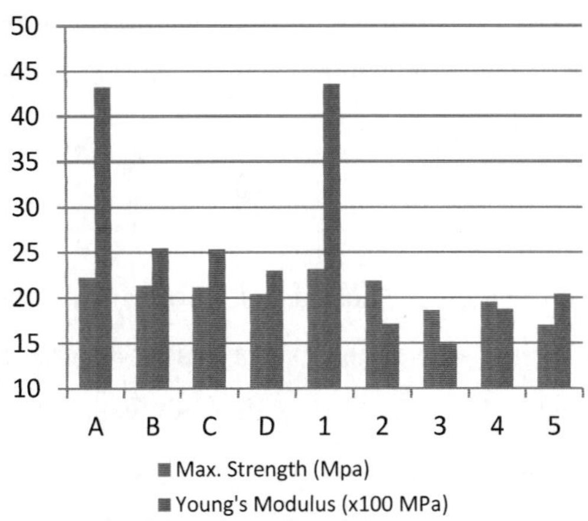

Figure 1. SnO-only and SnO-ONC samples tensile properties comparison.

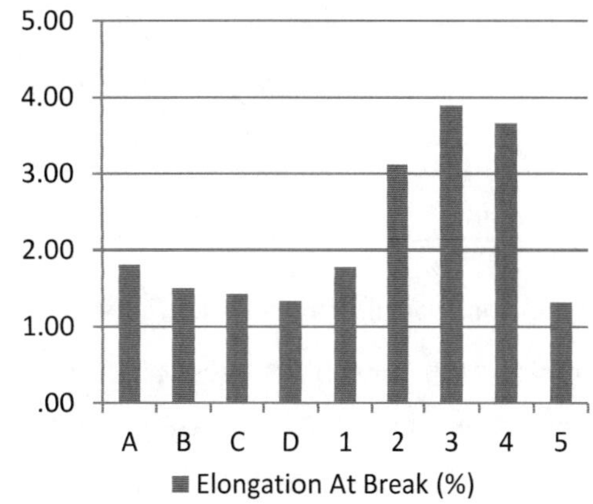

Figure 2. Elongation at Break of SnO-only and SnO-ONC.

The changes in their young's modulus were much pronounced than the changes in their material strength. Young's modulus indicates the stiffness of the material or the amount of forces required to deform it; a higher value signifies a greater strength need to change a materials shape, whereas any reduction meant a lesser force can now produce similar effects. The jarring difference between the starting sample A and 1 against their successor sample B and 2 are indicative on how the SnO and ONC additives had probably changed the molecular bond dynamics within the resin matrix material as to allow such gapping modulus reduction. The interaction between the bonds to resist tensile force are being affected by the presence of these metallic and organic additives, which hence resulted in less interaction between polymer matrix and EPS fillers.

Heterogeneous nucleating agents have been used industrially for polymers to reduce the process cycle times; via inducing crystallization at a lower degrees of supercooling (i.e., differences between melting temperature and crystallization temperature); and therefore producing translucent materials with smaller crystallite sizes. According to Wang *et al.*, (2004) [22] additives inclusions in a composite and its processing techniques are crucial basis to its properties. Rapid heating and cooling led to a distribution of particles;

leading to agglomeration of SnO and ZnO particles and resulted in low dispersion and lower tensile strength and modulus. Both ZnO and SnO can form faster nucleation formation, and creating noticeable UPR phase amorphocity difference and causes phase variation of composite; which introduces phase incompatibility.

A reason postulated to this effect was that the sample might have been poorly homogenized with poor resin wettability; thus SnO or ONC additions had disrupted the polymer cross-linking site. Improvement in wettability should enhances proper load transfer between the UPR/EPS matrix and the integrated additives. In addition, the probable presence of minute voids and bubbles may lead to ineffective stress transfer between the contents of the polymer matrix.

4. Conclusion

In conclusion, by utilising SnO and ONC, slight changes to the thermal and tensile properties of the composites were reported. Additional tin oxide alone reduces the studied thermal properties before an overloading of the material reverses the trend. Increasing addition of tin oxide with nanocrystal incorporation had increases both its thermal conductivity and heat capacity up to an upper limit as represented by a stable plateau. These observations could probably be attributed to tin oxide metallic nature and a speculated limit of its integration capability within the polymer matrix. In terms of the studied tensile characteristics, the materials' strength aren't directly affected by the additive integration, while the reported young's modulus were substantially affected. The main cause of these observations were postulated to be the effect of the cross-linking conditions between the matrix and its additive.

Acknowledgement

I would like to acknowledge the technicians, and staffs for their assistance during the sample fabrication and testing phases; and to my colleagues and lecturers for discussions related to this work. Special acknowledgement for Assoc. Prof. Rahmah Mohamed for her guidance and direction in producing this publication. I also wish to thank Universiti Teknologi MARA (UiTM) for their research funding under the Research Intensive Faculty (RIF) fund [File:600-RMI/DANA 5/3/RIF (64/2012)]

References

[1] Hull D 1981 *An introduction to composite materials* (London, United Kingdom: Cambridge University Press).
[2] Glodek T *et al* 2008 *Compos. Sci. Technol.* **68** 2994–3001.
[3] Smith W and Hashemi J 2009 *Foundations of Mater. Sci. and Eng.* (McGraw-Hill Education).
[4] Vaidya U *et al* 2000 *Compos. Part A Appl. Sci. Manuf.* **31** 761–72.

[5] Mohamed R *et al* 2013 *Adv. Mater. Res.* **664** 600–4.
[6] Ashari M A 2015 *The Properties of Unsaturated Polyester Resin (UPR) – Expanded Polystyrene (EPS) Composite* (Shah Alam, Selangor).
[7] Khosim R *et al* 2009 Physical properties and transmission of thermoset polyester composite filled expanded polystyrene.
[8] Kandare E *et al* 2008 *Polym. Degrad. Stab.* **93** 1996–2006.
[9] Jain G and Kumar R 2004 *Opt. Mater. (Amst).* **26** 27–31.
[10] Granqvist C G 2016 *Mater. Today Proc.* **3** S2–11.
[11] Al-Homoud M S 2005 *Build. Enviroment.* **40** 353–66.
[12] Mamunya Y P *et al* 2002 *Eur. Polym. J.* **38** 1887–97.
[13] Khosim R *et al* 2009 *Malaysian Polym. Int. Conf.*.
[14] Raed K 2013 *Investigation of Knudsen and Gas-Atmosphere Effects on Effective Thermal Conductivity of Porous Media* (Technische Universität Bergakademie).
[15] Lee D *et al* 1995 *J. Non. Cryst. Solids* **186** 285–90.
[16] Zeng S Q *et al* 1995 *J. Non. Cryst. Solids* **186** 271–7.
[17] Lu X *et al* 1995 *J. Non. Cryst. Solids* **188** 226–34.
[18] Smith D M *et al* 1998 *J. Non. Cryst. Solids* **225** 254–9.
[19] Glicksman L R 1994 *Low Density Cellular Plastics* ed N C Hilyarad and A Cunningham (London, United Kingdom: Chapman & Hall)
[20] Idicula M *et al* 2006 *Compos. Sci. Technol.* **66** 2719–25.
[21] Kukkonen I and Suppala I 1999 *Measurement of thermal conductivity and diffusivity in-situ: Literature survey and theoretical modelling of measurements* vol 3719.
[22] Wang Y C *et al* 2004 *Mater. Sci. Eng. A* **370** 41–9.

IOP Conf. Series: Materials Science and Engineering **223** (2017) 012030 doi:10.1088/1757-899X/223/1/012030

Effect of expanded organoclay by stearic acid to curing, mechanical and swelling properties of natural rubber nanocomposites

A Ramadhan[1], M Irfan Fathurrohman[1], A F Falaah[1], N Setyawan[2] and B Soegijono[3]

[1] Indonesian Rubber Research Institute, Bogor, Indonesia
[2] Indonesian Center for Agricultural Post Harvest Research and Development, Bogor, Indonesia
[3] University of Indonesia, Jakarta, Indonesia

E-mail: arifkaret@gmail.com

Abstract. The interlayer basal spacing of organoclay (OC) could be increased with stearic acid (SA) added, thus OC changed into expanded organoclay by SA (OCSA). The effect of various loadings of OCSA on the curing, mechanical and swelling properties of natural rubber (NR) nanocomposites were studied. The natural rubber/expanded organoclay (NR/OCSA) nanocomposites were prepared by melt intercalation using a laboratory open mill. The curing characteristics of NR compounds were determined using a Moving Die Rheometer (MDR). The X-ray Diffraction (XRD), Attenuated Total Reflectance Infrared (ATR-IR) Spectroscopy and Field Emission Scanning Electron Microscopy (FESEM) were used to study the dispersion of OCSA in the NR matrix. The mechanical properties of NR/OCSA nanocomposites such as tensile strength, elongation at break and hardness were determined using ISO standard and swelling of NR/OCSA nanocomposites in toluene were determined using ISO 1817. The results showed that the SA intercalated into the gallery of OC and reacted with the hydroxyl groups in OC. It was indicated with the shifting of the negative peak 1,700 to 1,723 cm^{-1} in the ATR-IR spectrum and increase the d-spacing of OC. The adding of various loadings of OCSA into NR could increase the torque and accelerate the curing of nanocomposites and it also could increase the mechanical and swelling properties of nanocomposites. The change in modulus at 100% elongation significantly increased with increasing the OCSA load until maximum loading at 10 phr. This trend was same with the hardness and modulus at 300% elongation. Meanwhile, the improvement of tensile strength and elongation at break was higher at 4 phr OCSA compared with the other loading. The increase of mechanical and swelling properties of NR/OCSA nanocomposites was due to intercalation/exfoliation of OCSA in NR matrix. It was revealed by appearing of the out-of-plane Si-O-(Al) stretch with peak value 1080 cm^{-1} in the ATR-IR spectrum and the peaks of OCSA in the XRD pattern was disappeared until the loading of OCSA 8 phr and the thickness of morphology of OCSA below 100 nm.

1. Introduction

The rubber filler is used to improve the mechanical properties and reducing cost. There are several rubber fillers, though the rubber industry usually uses carbon black as reinforcing filler. But carbon black is made from petroleum and causes of pollution. In the last two decades, research was focused to develop the other reinforcing filler. Clay has been used as a filler for many years. Unfortunately, the hydrophilic nature of the clay made these materials incompatible with the hydrophobic rubber matrix, with resulted non reinforcing effects in the rubber matrix. Recently, clay minerals could be used as a reinforcing rubber filler after modifying by intercalating quaternary ammonium compound into the interlayer space through ion exchange become organoclay (OC) [1].

The improvement of mechanical properties of rubber nanocomposite depends on the uniformly dispersed (exfoliated) and distribution of the clay in the rubber matrix. The fully exfoliated of clay within the rubber matrix is expected to enhance the mechanical properties of rubber composite [2]. The content of organoclay was below 6 wt% to get a completely exfoliated in the natural rubber (NR) matrix [3]. The high amount of clay due to the formation of agglomeration in rubber matrix as well as due to the cost factor. However, the small amount of organoclay did not improve the mechanical properties significantly. The tensile strength of NR-Styrene Butadiene Rubber (SBR) was not change in base rubber up to 2 wt% OC [4]. Therefore, the increase exfoliation of the OC with a sufficient amount in the rubber matrix is very important to improve mechanical properties.

The increase of exfoliation of OC in the rubber matrix can use expanded organoclay by using stearic acid (SA). From previous studies, SA had a beneficial effect on the intercalation/exfoliation on process of clay mineral in Nitrile rubber [5]. The adding OC 10 part hundred rubber (phr) and 10 phr SA could increase physical and mechanical properties of non polar rubber like NR, Ethylene Propylene Diene Monomer (EPDM), Butadiene Rubber (BR) and etc. [6]. The OC 10 phr and SA 2 phr could improve the tensile strength in polar and non-polar rubber [7]. SA intercalated in the gallery of OC, so that the d-spacing of OC increased with increasing SA concentration [8].

The present paper is aimed at studying the effect of various loadings of in-situ expanded organoclay by SA (OCSA) on the curing characteristic, mechanical and swelling properties of vulcanized NR/OCSA obtained by melt intercalation.

2. Experimental

Natural Rubber (NR) grade thin pale crepe was obtained from PTPN 8, West Java, Indonesia. Organoclay (OC) Cloisite® 20A was supplied by Southern Clay Products. Inc. Rockwood additives. Dimethyl dehydrogenated tallow quaternary ammonium was used as an organic modifier. The *d*-spacing of Cloisite® 20A was 2.42 nm. Zinc oxide (ZnO), stearic acid (SA), N-cyclohexylbenzhothiazole-2-sulfenamide (CBS), and sulfur were purchased from local suppliers.

OC and SA were mixed together in-situ by using porcelain mortar with the weight ratio of OC:SA (1:0.2). The mix was put in an oven at 100 °C for 1 h. After cooling, the obtained product is called expanded organoclay by SA (OCSA). The crystallographic spacing (*d*) of OCSA was measured using XRD Rigaku, Japan in transmission mode using CuKα radiation (λ = 0.1542 nm), X-ray diffractometer (40 kV, 30 A). The interaction between SA and OC was studied by ATR-IR Thermo

Scientific Nicole iS5. ATR-IR spectrum was measured between 500-4,000 cm^{-1} at a resolution of 4 cm^{-1}.

Natural rubber/expanded organoclay (NR/OCSA) nanocomposites were prepared by melt intercalation method in an open two roll mill with a speed of slow roll (24 rpm) and friction ratio (1.4) maintained at 65 ± 5 °C. The rubber compounds were conditioned at room temperature for minimum 16 h. The formulation compounds are described in Table 1. The loading of OC were varied (0, 2, 4, 6, 8, and 10). Because the weight ratio OC:SA = (1:0.2), the loading of OCSA also varied (0, 2.4, 4.8, 7.2, 9.6, and 12.0) phr. The total amount of SA was ajusted same at each formulation of compound (2 phr) as the common dosage of normal rubber compound. The deficit of SA was added at two roll mill (2.0, 1.6, 1.2, 0.8, 0.4, and 0). The curing characteristics of the rubber compounds were determined using an MDR at 150 °C according to ISO 6502. The compounds were then molded at electrically heated hydraulic press at 150°C with the respective cure time (t_{90}), previously determined from MDR. The sample's thickness of 2 mm sheets was molded.

Table 1. Formulation of natural rubber/expanded organoclay compounds

	Loadings (phr)					
	NROCSA-0	NROCSA-2	NROCSA-4	NROCSA-6	NROCSA-8	NROCSA-10
NR	100	100	100	100	100	100
OCSA	0	2.4	4.8	7.2	9.6	12.0
ZnO	5	5	5	5	5	5
SA	2	1.6	1.2	0.8	0.4	0
CBS	1.5	1.5	1.5	1.5	1.5	1.5
Sulfur	2.5	2.5	2.5	2.5	2.5	2.5

Tensile properties were measured according to ISO 37 by using universal testing machine Lloyd, with type 2 dumbbell test specimens. Hardness was measured by using Shore A durometer according to ISO 7619.

The improvement of mechanical properties of NR/OCSA nanocomposites with respect to the gum was calculated in the following [6]:

$$improvement = \frac{v(OCSA) - v(gum)}{v(gum)} x100 \quad (1)$$

The swelling of NR/OCSA nanocomposites in toluene was determined using ISO 1817 for 24 h at room temperature. The initial weight of the samples was taken, and then they were immersed into toluene. The samples were periodically removed from the test bottles, the surplus solvent was removed from the surface, and the samples were weighed immediately and then placed into the toluene again.

The swelling ratio, Q_t, was also determined from the weight of the sample in unswollen and swollen states. The equation as the following [9]:

$$Q_t = \frac{M_t - M_0}{M_0} x100 \qquad (2)$$

where Q_t is the swelling ratio, M_t the weight of the samples in the swollen states and M_0 the initial weight of the samples.

The intercalation/exfoliation of NR/OCSA nanocomposites was studied by XRD and ATR-IR. The morphology of NR/OCSA nanocomposites was characterized using FESEM INSPECT F50 (20 kV). Fractured surfaces of the vulcanizates NR/OCSA nanocomposites were obtained by tensile test. The surfaces were coated by gold/palladium to prevent electrostatic charging during examination.

3. Results and Discussion

3.1. Characterization of Expanded Organoclay (OCSA).

Figure 1 shows the XRD pattern of OC, SA and OCSA. The spectrum of OC presents two quite broad peaks. The first peak (d_{001}=2.52 nm) was related to intercalated ions with a paraffin-like structure [10], whereas the small second peak (d_{001} = 1.25 nm) was a portion of the clay layers has not been intercalated after chemical modification. The interlayer distance of OCSA was shifted from 2.52 nm of OC to 3.83 nm. It was indicated that the expanding space should be attributed to SA intercalation into a gallery of OC. The intercalation of SA were due to the van der Walls forces between SA and the alkylammonium ions [6] and the esterification between SA and OC [7]. The second peak (d_{001} = 2.01 nm) of OCSA shows that some SA molecules recrystallize during the cooling process. The second peak of OC was shifted from 1.25 to 1.33 nm. It was indicated that SA also intercalating into the clay layer which has not been intercalated by alkylammonium ions.

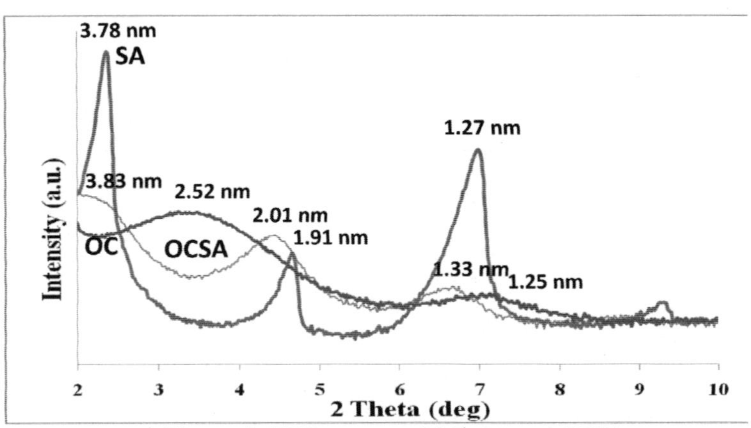

Figure 1. XRD pattern of OC, SA and OCSA

ATR-IR spectrums of OC, SA and OCSA are shown in Figure 2. There was no peak of carboxylic structure (1695.27 cm^{-1}) at spectrum ATR-IR of OC. Spectrum OCSA revealed that the peak at 1695.27 cm^{-1} is shifted to 1723.55 cm^{-1}, which means that carboxylic group were esterified by hydroxy groups of OC layers. The esterification could be the driving force for SA intercalating into OC and expand the spacing of OC [7].

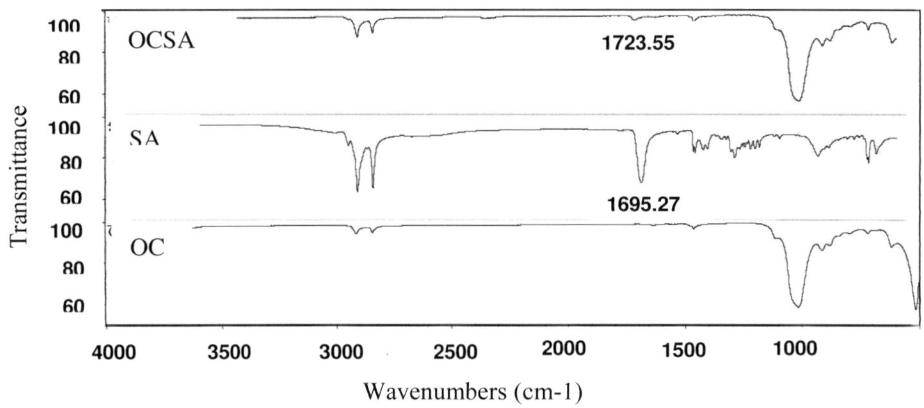

Figure 2. ATR-IR spectrums of OC, SA and OCSA

3.2. Curing Characteristic of NR/OCSA

The curing characteristics, expressed in terms of the scorch time (ts_2), optimum cure time (t_{90}), cure rate index (CRI) and torque value (S') of NR and NR/OCSA compounds are compiled in Table 2. The increase of loading of OCSA could decrease the scorch time and optimum cure time of NR compound and increased the cure rate index. From these results it can be assumed that OCSA can act as accelerators for the sulfur vulcanization system. The amine groups present in the OC facilitate the curing reaction of NR and it was a synergy with CBS accelerator that could accelerate the vulcanization [11].

Table 2. Curing characteristics of NR dan NR/OCSA compounds

Samples	S'_{max} (dN m)	S'_{min} (dN m)	ts_2 (min)	t_{90}(min)	ΔS' (dN m)	CRI[a] (min^{-1})
NROCSA-0	7.13	0.36	5.22	7.32	6.77	47.62
NROCSA-2	7.59	0.12	2.13	4.35	7.47	45.05
NROCSA-4	8.25	0.13	1.36	4.22	8.12	34.97
NROCSA-6	8.69	0.33	1.29	4.22	8.47	34.13
NROCSA-8	9.02	0.24	1.19	3.55	8.78	42.37
NROCSA-10	9.04	0.30	1.15	3.59	8.74	40.98

[a]CRI = Cure rate index = $100/(t_{90}-ts_2)$

The increase of OCSA loading of the NR compound could improve the value of torque maximum and delta torque ($\Delta S' = S'_{max} - S'_{min}$). The delta torque (ΔS') value was related to crosslink

IOP Conf. Series: Materials Science and Engineering **223** (2017) 012030 doi:10.1088/1757-899X/223/1/012030

density, it could be concluded that OCSA increases the crosslink density of NR. The increasing of ΔS' was attributed to the intercalation/exfoliation of OCAS in the NR matrix.

3.3. Mechanical and Swelling Properties of NR/OCSA

Figure 3 describes the improvement of mechanical properties of NR/OCSA nanocomposites. The improvement of all mechanical properties of all NR/OCSA nanocomposites always higher compared with NR containing only OC. The change in modulus at 100% elongation significantly increased with increasing the OCSA load. This trend was same with the hardness and modulus at 300% elongation, because the modulus at 100%, hardness and modulus at 300% was related to crosslink density of NR/OCSA compounds [12]. The improvement of tensile strength and elongation at break was higher at 4 phr OCSA compared with the other loading. The improvement of tensile strength in case of rubber/organoclay nanocomposites was given by some researcher [9,13,14]. Their studies suggested that the increase of tensile strength was related to the degree of dispersion (intercalation/exfoliation) of OC in the NR matrix. The exfoliation of OC causes an increase in surface area. Hence, the tensile strength of NR/OCSA at 4 phr OCSA recorded high tensile strength.

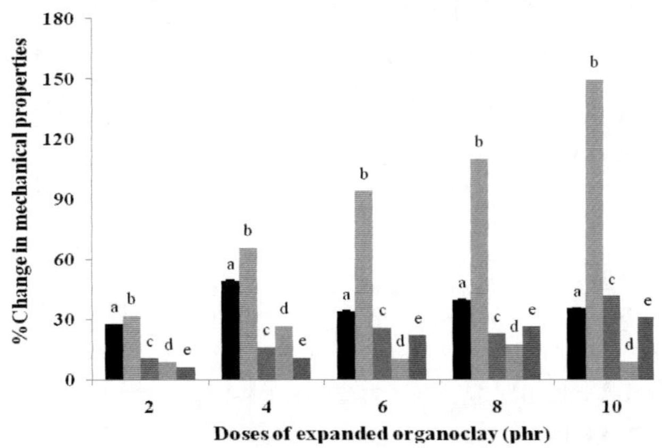

Figure 3. Improvement of mechanical properties: (a) tensile strength, (b) 100% modulus, (c) 300% modulus, (d) elongation at break, (e) hardness

The swelling curves of NR/OCSA nanocomposites with various loadings of OCSA are shown in Figure 4. The swelling curve was obtained by plotting between Q_t and time ($t^{1/2}$). Figure 4 shows that the swelling ratio decreased with increasing the loading of OCSA. It was due to the dispersion impermeable clay layers in the NR matrix which obtained the "Tortuous Path Model" [15]. Hence, the platelet structure of layered silicate in NR matrix could improve the swelling properties of NR vulcanizates.

IOP Conf. Series: Materials Science and Engineering **223** (2017) 012030 doi:10.1088/1757-899X/223/1/012030

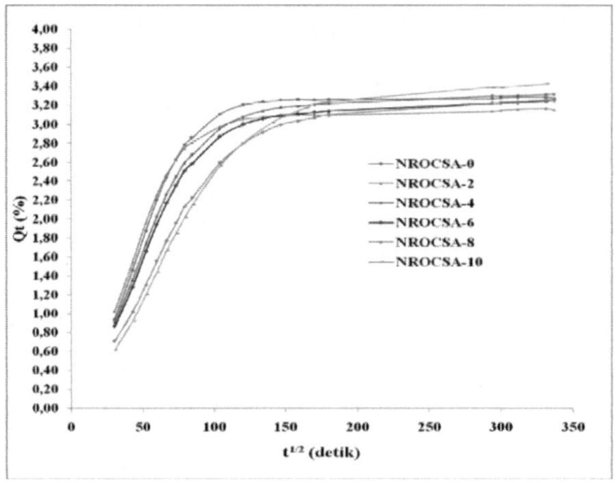

Figure 4. The swelling curves of NR and NR/OCSA nanocomposites with various loadings of OCSA

3.4. *Morphology Analysis of OCSA*

The morphology of OCSA was analyzed in order to study the intercalation/exfoliation of OCSA in the NR matrix. XRD patterns of NR/OCSA nanocomposites with various loadings of OCSA are shown in Figure 5. For NR without OCSA, there were no peaks in the NR vulcanizates XRD pattern. The XRD pattern of OCSA shows that the distance of spacing is 3.83 nm (2Theta = 2.3°). This peak disappeared after OCSA was added into NR compound. However, for 10 phr OCSA in the NR compound, the peak appeared and shifted to 3.99 nm (2Theta = 2.2°).

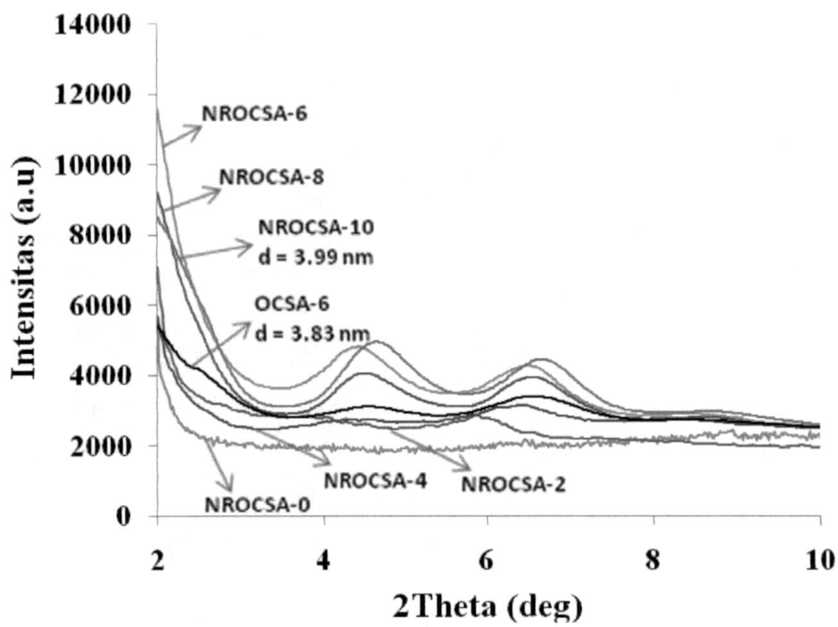

Figure 5. XRD patterns of NR and NR/OCSA vulcanizates

It means that the OCSA exfoliated in the NR matrix until 8 phr and for 10 phr of OCSA, there were partially intercalated and exfoliated of OCSA in NR matrix. The peak intensity of SA in the NR/OCSA vulcanizates at 2Theta 4.8° increased with increasing loading of OCSA in NR compounds. It was indicated that the SA molecules in the clay gallery were not reacting with zinc oxide and recrystallize after cooling process [16].

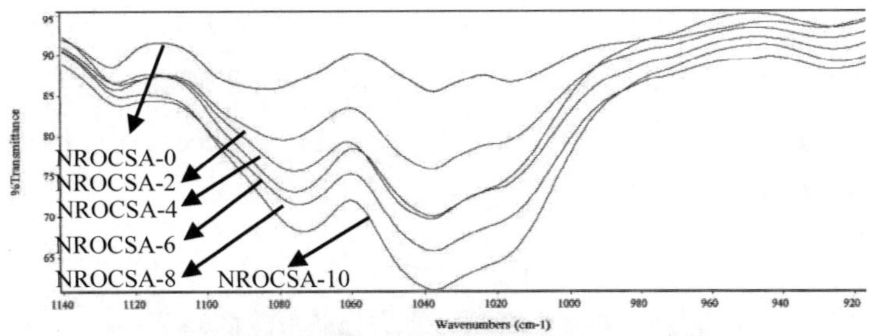

Figure 6. ATR-IR spectra of NR and NR/OCSA nanocomposites with various loadings of OCSA

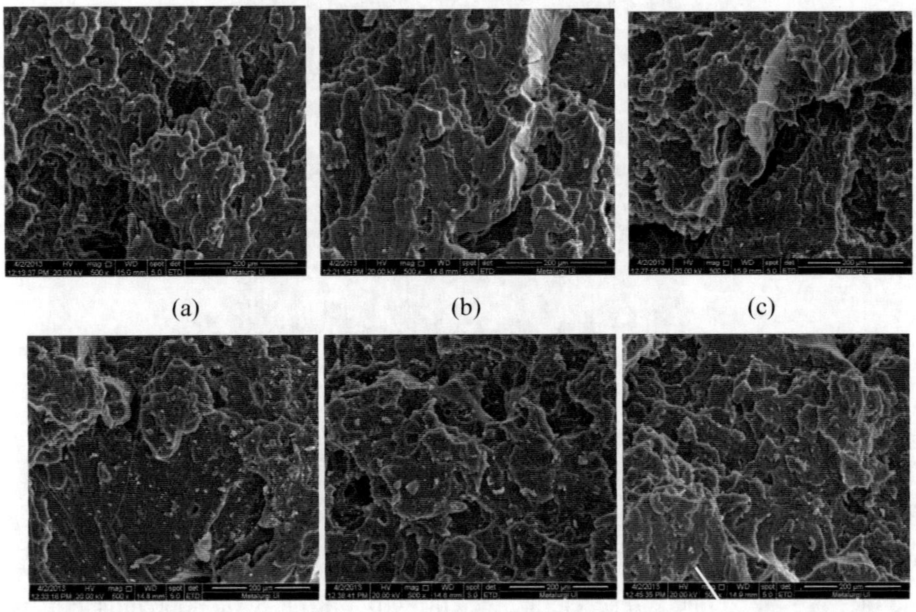

Figure 7. Morphology of NR/OCSA nanocomposite with various loadings of OCSA: (a) 0 phr; (b) 2 phr; (c) 4 phr; (d) 6 phr; (e) 8 phr and (f) 10 phr.

The degree of exfoliation of OC could be characterized by using ATR-IR spectra from 1140 to 900 cm^{-1}. Ijdo *et al.,* (2006) [17] studied the degree of OC delamination and orientation in nanocomposites materials using infrared and observed appeared the absorption band around 1080 cm^{-1}. The increase of this absorption band revealed the degree of exfoliation increased [18]. The spectrum ATR-IR of NR and NR/OCSA nanocomposites with various loadings of OCSA are shown in Figure 6. The transmission band of Si – O at 1000 – 1100 cm^{-1} increased with increasing the OCSA in the NR

IOP Conf. Series: Materials Science and Engineering **223** (2017) 012030 doi:10.1088/1757-899X/223/1/012030

compounds. This result was same with Diez *et al.,* (2011) used SBR nanocomposites [18]. The increasing of OCSA loading could increase the transmission band of Si – O – (Al) around 1080 cm^{-1}. It was revealed that the degree of exfoliation of OCSA in NR matrix increased with increasing the OCSA loading in NR compounds.

The morphology of NR and NR/OCSA nanocomposites are shown in Figure 7. The fractured surface of samples from tensile test revealed that there were some OCSA agglomerated particles appeared with increasing OCSA in the NR compound. The platelet structure of organoclay could be analyzed at NR/OCSA that containing 4 phr of OCSA as shows in Figure 8. The thickness of platelet structure was below 100 nm. We were difficult to analyze the thickness of others platelet because the platelet position on the fractured surfaces (Figure 8).

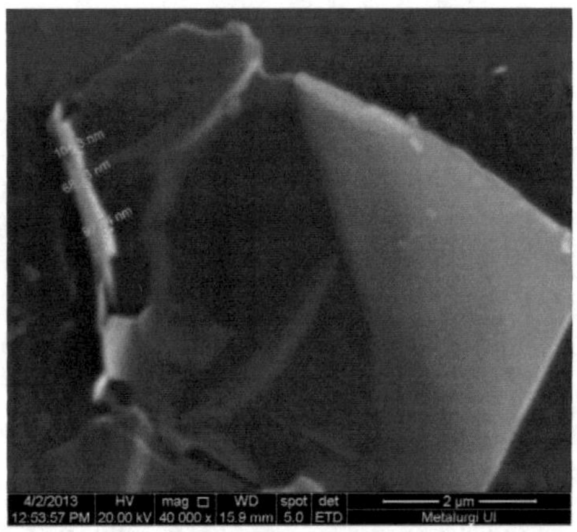

Figure 8. Morphology of NR/OCSA with containing 4 phr of OCSA

4. Conclusion

Curing characteristics, mechanical and swelling properties of NR/OCSA nanocomposites with various loadings of OCSA were studied. The results showed that the increasing of OCSA loading into natural rubber could increase the torque and accelerate the curing of NR/OCSA compounds and also increased the mechanical and swelling properties of nanocomposites. The change in modulus at 100% elongation significantly increased with increasing the OCSA load until maximum loading at 10 phr. This trend was same with the hardness and modulus at 300% elongation. Meanwhile, the improvement of tensile strength and elongation at break was higher at 4 phr OCSA compared with the other loading. The increase of delta torque, mechanical and swelling properties of NR/OCSA nanocomposites was due to better dispersion (intercalation/exfoliation) of OCSA in the NR matrix. The degree of intercalation/exfoliation of OCSA increased in the NR matrix was revealed by the band Si-O-(Al) with peak value 1080 cm^{-1} in the ATR-IR spectrum increased with increasing the loading of OCSA in the

NR compounds.The peaks of OCSA in the XRD pattern is disappeared until the loading of OCSA 8 phr and the thickness of morphology of platelet structure of OCSA below 100 nm.

References

[1] Jankovic L *et al* 2011 *Appl. Clay Sci.* **51** 438-44.
[2] Ramorino G *et al* 2009 *Comp. Sci. Tech.* **69** 1206-11.
[3] Gu Z *et al* 2010 *Appl. Clay Sci.* **50** 143-7.
[4] Mohan T P *et al* 2011 *J. Ind. Eng. Chem.* **17** 264-70.
[5] Das A *et al* 2008 *Macromol. Mater. Eng.* **293** 479-90.
[6] Das A *et al* 2011 *Appl. Clay Sci.* **51** 117-25.
[7] Ma Y *et al* 2008 *J. Appl. Polym. Sci.* **109** 1925-34.
[8] Fathurrohman M I *et al* 2013 *J. Mater. Sci. Eng. B*, **3** (9) 575-81.
[9] Li P *et al* 2002 *J. Appl. Polym. Sci.* **109** 3831-8.
[10] Le Puart L *et al* 2002 *J. Adhesion* **78** 645-62.
[11] Lopez-Manchado M A *et al* 2003 *Polym. Int.* **52** 1070-7.
[12] Ignatz-Hoover F and Brendan H T 2004 *Vulcanization Rubber Compounding* B. Rodgers (New York: Marcel Dekker Inc) pp 503-66.
[13] Varghese S *et al* 2003 *Polym.* **44** 3977-83.
[14] Madhusoodanan K N, and Varghese S 2006 *J. Appl. Polym. Sci.* **102** 2537-43.
[15] Xu B *et al* 2006 *Polym* **47** 2904-10.
[16] Ma Y *et al* 2007 *Polym. J.* **39** 1-7.
[17] Ijdo W L *et al* 2006 *Polym. Eng. Sci.* **46** 1031-9.
[18] Diez J *et al* 2011 *Polym. Eng. Sci.* **51** 1720-9.

Effects of natural zeolite and ferric oxide to electromagnetic and reflection loss properties of polyurethane nanocomposite

G Gultom[1,2], B Wirjosentono[3], M Ginting[4], K Sebayang[5]

[1] Polytechnic of Chemical Industry Technology, Medan, Indonesia
[2] Physics Post Graduate Department, University of Sumatera Utara, Medan, Indonesia
[3] Chemistry Department, University of Sumatera Utara, Medan, Indonesia
[4] Research Center for Physics, Indonesian Institute of Science, Serpong, Indonesia
[5] Physics Department, University of Sumatera Utara, Medan, Indonesia

E-mail : golfridg@gmail.com

Abstract. Microwave-absorptive polymeric composite materials are becoming important to protect interference of any communication systems due to increasing use of microwave-inducing devices. In this work, the microwave-absorptive polyurethane nanocomposites were prepared using natural zeolites of Sarulla North Sumatra and commercial ferric oxide as fillers. Weight ratios of the polyurethane to natural zeolite and ferric oxide were varied (90%:6%:4%; 80%:12%:8%; 70%:24%:6%) by weight. The fillers were prepared using ball milling technique and characterized for their particle size distributions using Particle Size Analyzer. The nanocomposites, prepared using in-situ reaction of polyethylene glycol, toluene diisocyanate and fillers. The complex permittivity (ε'and ε'') and complex permeability (μ' and μ'') as electromagnetic properties were calculated using NRW method after collecting real and imaginary S parameter using Vector Network Analyzer measurement at X band frequency. Results show ratio of the fillers will affect the permeability, permittivity and reflection loss of the materials. The best reflection loss was shown -40.588 dB (>99 % absorption) at ratio for polyurethane : nanozeolite : ferric oxide (80%:12%:8%) by weight observed at 10.92 GHz. According to the measurement and calculation was shown the polyurethane filled with natural nanozeolite and ferric oxide is a good electromagnetic wave attenuation material.

1. Introduction

The electromagnetic interference, a specific type of interference environmental pollution, is worsening due to the rapid development and utilization of wireless communications circuit devices and military applications. In the past decades, electromagnetic attenuation materials which comprise dielectric or magnetic fillers and polymer, have been commonly used to minimize the electromagnetic interference. The spinel ferrite have been utilized as the most frequent absorbing materials in various forms. The absorbing characteristics of materials depend on the frequency, layer thickness, complex permeability and complex permittivity. All the real and imaginary parameters of complex permeability and complex permittivity are found to increase with the increased of ferrite contents [1].

Zeolite mineral is a compound of aluminium silicate hydrate with alkali metal which is group of several types of minerals. Zeolite is aluminosilicate with a framework structure enclosing cavities occupied by a large ions and water molecules, both of which have considerable freedom of movement, permitting ion-exchange and reversible dehydration. Natural zeolite is natural mineral that composed of crystalline silica (SiO_2) and alumina (Al_2O_3), with cavities of metal ions, which is usually alkali and alkaline or earth metals, and water molecules. Unique characteristic, include very stable with very high adsorption capacity and selectivity and have large active pore structure (microporous) and has a high specific surface area. Natural resources have the potential to be further processed into products that can be used for broad applications, among others, as supporting the catalyst or catalysts, and slow release substances. Zeolite crystal structure of alumina silicate shaped frame (framework) three-dimensional, having cavities and channels as well as containing metal ions such as Na, K, Mg, Ca and Fe as well as water molecules. Sarulla natural zeolite chemical composition analysis using XRF showed the dominant chemical compounds are SiO_2 and Al_2O_3.

Usually, ferrite composites are used to achieve microwave absorption. The absorbing characteristics and electromagnetic characteristics could be varied by controlling the ferrite filler volume fraction in the composite materials [2]. Studies of either nanoparticles or core-shell nanopowders such as carbonaceous particles and polyurethane as matrix for the application in microwave absorbers were very interesting in the last decades [3,4,5]. However, studies of using nanozeolite as a filler for microwave-absorptive materials were very limited.

In this paper, with the aim finding high microwave absorption in GHz range filled with nanozeolite as an abundant local natural resources and ferric oxide with its strong magnetic characteristic were prepared. We investigate electromagnetic properties and the reflection loss properties due to nanofillers variation of weight. Our experimental results highest reflection loss was -40.588 dB at 10.92 GHz and as the result is polyurethane filled with nanozeolite and ferric oxide are promising for microwave-absorptive materials.

2. Materials and Methods

The fillers used for nanocomposite was Sarulla natural zeolite and Sigma Aldrich commercial ferric-oxide (α-Fe_2O_3) with 99.95% purity. The polyurethane matrix used was Sigma Aldrich commercial type which contains two-part monomers i.e. 60% wt polyethylene glycol has a density 1.01 g/mL and 40% wt toluene diisocyanate has a density 1.22 g/mL.

The polyurethane nanocomposites reinforced with natural nanozeolite and ferric-oxide were fabricated with in-situ reaction. The activated natural zeolite prepared using planetary ball mill for 28 h to complete nanozeolite with average particle size distribution $<$ 100 nm. Nanozeolite mixed for 4 h with ferric oxide using planetary ball mill to produce nanofillers. The homogen nanofillers divided into the same ratio and mixed with PPG and TDI for 1 min. Polyethylene glycol and toluene

IOP Conf. Series: Materials Science and Engineering **223** (2017) 012031 doi:10.1088/1757-899X/223/1/012031

diisocyanate with nanofillers of zeolite and ferric oxide are using in-situ reaction. Both of PPG and TDI with nanofiller mixed for 2 min with in-situ reaction at room temperature. The reaction of PPG and TDI and nanofillers polyurethane nanocomposites filled with nanozeolite and ferric oxide is hold and pressed for 30 min using iniversal testing machine with 5 mm sample thickness.

The crystal structure of the samples were determined by Shimadzu-700 X-ray Diffractometer (XRD) using Cu Kα radiation (λ = 1.54 Å) operated at 35 kV and 25 mA. The reflection loss of the samples was measured between 8 GHz and 12 GHz using Vector Network Analyzer Advantest type R 3770 and microstructure of the samples using SEM-EDX Carl Zeiss-Bruker Type MA10.

3. Characteristic of Product

Figure 1.a shows the X-Ray Diffraction spectrum of Sarulla natural zeolite. Observed from the XRD spectrum shows intensity at 2θ = 29.76; 27.67 and 27.92. Maximum peak occur at 29.76 with FWHM 0.124 which is at the aluminium silicate phase. Mordenite standard have a high intensity at 2θ = 27; 25.63 and 23. Compare with Sarulla zeolite, it shows that natural zeolite is mordenite, but another peak at natural zeolite at 29.76 shows that crystaline at Sarulla natural zeolit is not only modernit but also mixed with other crystalline and other impurities. Combination of filler varied of nanozeolit and ferric oxide with ratio (80% : 20%) and (60%:40%) by weight, observed from the XRD spectrum shows another high intensity at 2θ = 33.121 and 35.60, which is not occur at the nanozeolite XRD spectrum. The occurance of these intensity due to entrance of ferric oxide.

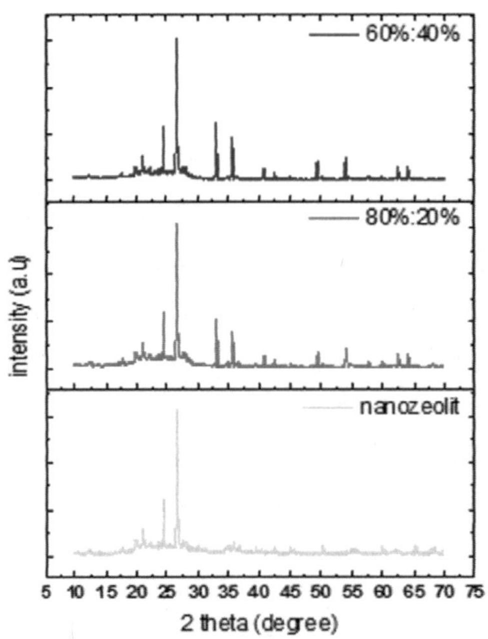

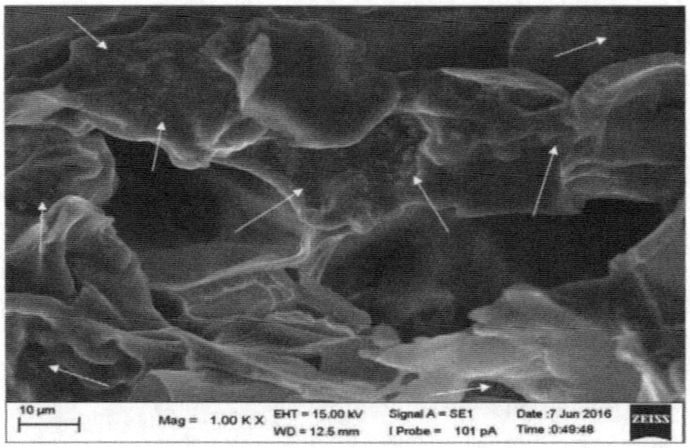

(b)

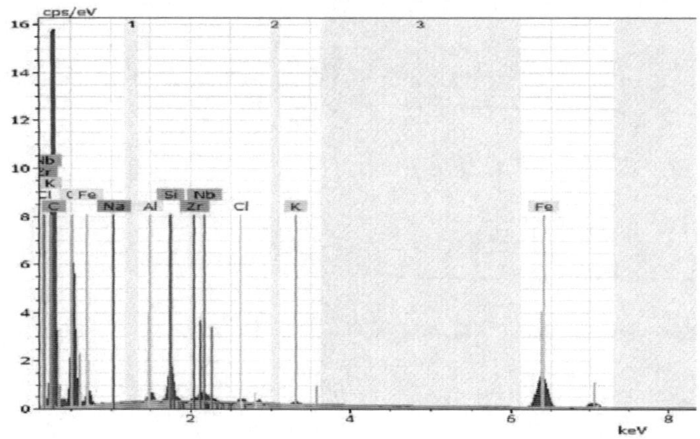

(c)

Figure 1. (a) XRD pattern of natural zeolite (b) SEM photograph of polyurethane nanocomposite (c) EDX polyurethane nanocomposite characterization

From Figure 1.b SEM photograph of polyurethane nanocomposite shows that dark colors are pores or cavities and bright color is polyurethane matrices and arrows show the occurance of agglomeration caused by nanozeolite and ferric oxide. Results of other studies suggest that the materials properties will be compatible with the polymer matrices and influenced by several factors such as filler particle size. The particle size will affect the bond of the filler and the matrix. The amount of surface area will be increased by the presence of a surface porous on the fillers surface as well as with the addition of natural nanozeolite and ferric oxide. Figure 1.c show EDX polyurethane nanocomposite characterization. According to the EDX characterization it was shown that the agglomeration is the effect of the fillers.

IOP Conf. Series: Materials Science and Engineering **223** (2017) 012031 doi:10.1088/1757-899X/223/1/012031

4. Results and Discussion

Characterization of electromagnetic wave absorption was carried out with a Vector Network Analyzer. From the scattering parameters S_{11} and S_{21} measured by the coaxial method using VNA in the frequency range of 8 - 12 GHz, the relative complex permeability μ_r (= μ_r' - jμ_r'') and the relative complex permittivity ε_r (= ε_r' - jε_r'') were calculated using Nicholson Ross Weir model. The electromagnetic properties which is dielectric and magnetic properties of polyurethane nanocomposites will characterize using NRW model.

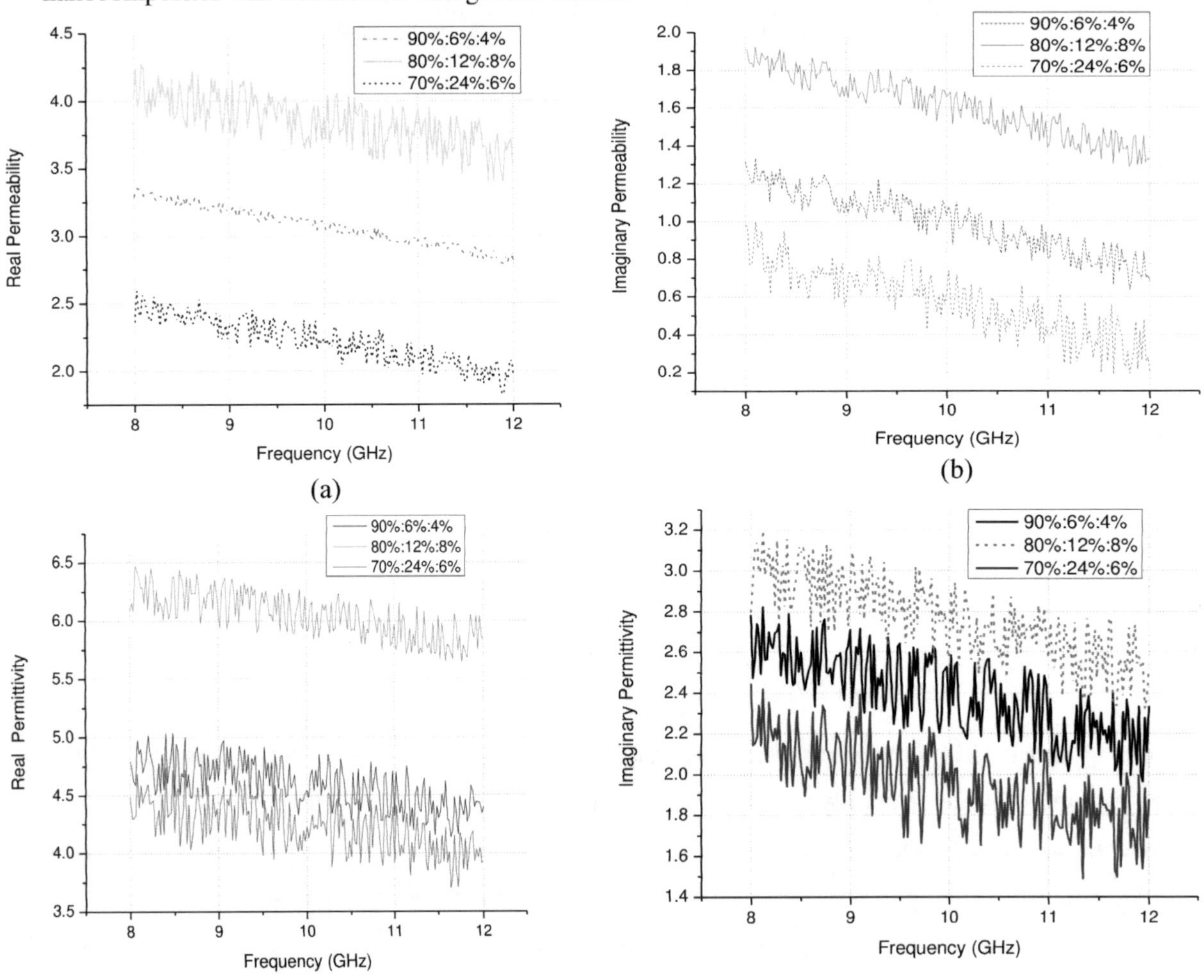

Figure 2. (a) Real permeability spectra (b) Imaginary permeability spectra (c) Real permittivity spectra (d) Imaginary permittivity spectra

From Figure 2 above show how variation of the filler affect the permittivity and permeability of the polyurethane nanocomposite. It is observed that in the lowest frequency region the polyurethane filled with nanozeolite and ferric oxide shows the larger value of real and imaginary for both permittivity and permeability. Dielectric constants depends on two factors, orientation and interfacial polarizations.

IOP Conf. Series: Materials Science and Engineering **223** (2017) 012031 doi:10.1088/1757-899X/223/1/012031

In case of polyurethane nanocomposite (80%:12%:8%), the orientation and interfacial polarizations should be high due to ferric oxide. The variation of both real and imaginary parts of permittivity with frequency may be happen due to space charge polarization. Polar-polar reaction between polyurethane and the fillers probably will affected the demagnetizing field. The demagnetizing field generated by the magnetic poles on the surface of magnetic fillers plays a very important and characteristics role in the permeability of material. Both μ' and μ'' shows decreasing trend with increasing of frequency, due to the lower resonance frequency range of the polyurethane nanocomposite. The fluctuation of the permittivity and permeability spectra caused by the pores of polyurethane nanocomposite. As a confirmation for the other research it was shown that trends of permittivity spectra was observed that in the lowest frequency region shows the larger value of real and imaginary permittivity [5].

The enhanced μ' and μ'' values will improve the microwave absorbing effect by transferring electromagnetic energy into heat energy. Magnetic losses, as the important phenomena in absorptive materials, are caused by the time lag of the magnetization vector behind the magnetic field vector [6]. The change of the magnetization vector is generally brought about by rotation of the magnetization or the domain wall displacement. These motions lag behind the change of the magnetic field and contribute to μ''. The smaller the particle size, the weaker the spins coupling at the surface of the nanozeolite and ferric oxide, which makes the magnetic relaxation behaviour more complex.

The magnetic and dielectric particles filled into the polyurethane matrix can act as a large number of charge domains and can contribute to increase the dielectric constant values due to interfacial polarization [7]. The dielectric loss mechanism occur due to the permanent and dominant polarization and their associated relaxation phenomena. The nanocomposite materials showed the microwave absorption properties according to their dielectric loss values. Polyurethane filled with nanozeolite and ferric oxide (80% : 12% : 8%) have better dielectric loss value than other nanocomposites and it achieved better microwave absorbing properties.

Reflection loss measurement using Vector Network Analyzer for polyurethane nanocomposites reinforcement with ratio of polyurethane : nanozeolite : ferric oxide = (90% : 6% : 4%); (80% : 12%: 8%) and (70% : 24% : 6%) by weight as a function of frequency in X-band (8-12 GHz) shows in Figure 3. Polyurethane nanocomposites with ratio (80% : 12% : 8%) have the optimum reflection loss of -40.588 dB at 10.92 GHz. The reflection loss of less than -40 dB means the microwave absorption more than 99% [8]. The microwave absorption being dependent on the ratio of SiO_2 and Al_2O_3 which is to dominant compounds of natural zeolite and ferric-oxide, the content of ferric-oxide in natural zeolite will modify the reflection loss and microwave absorption of the nanocomposite and will affect the complex permittivity and complex permeability of the nanocomposite. To satisfy the zero-reflection condition where maximum absorption would occur. Ferrite itself usually present electromagnetic characteristics of complex permeability less than complex permittivity. Minimum loss occurs when the thickness is about an odd multiple of one quarter of the wavelength of the incident frequency.

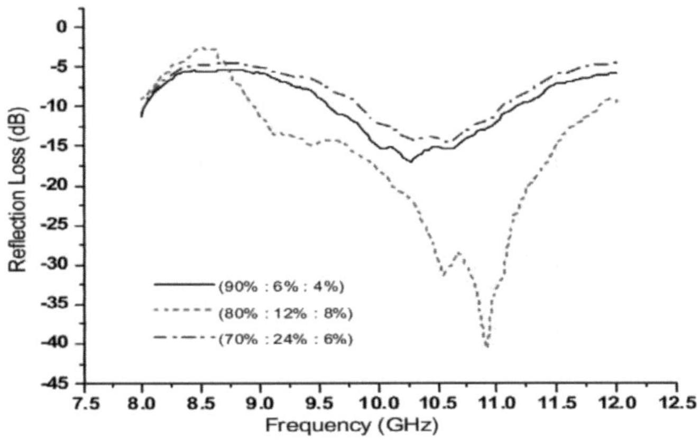

Figure 3. Reflection loss measurement of polyurethane nanozeolite and ferric oxide nanocomposites

5. Conclusion

Natural zeolite and ferric oxide ratio as a filler in polyurethane matrices will modify permeability and permittivity of the nanocomposite, by the addition of the the filler will cause a different effect in electromagnetic properties and microwave absorption due to material reflection loss. The nanocomposite absorb microwave at X band frequency for > 99 % absorption. This is due to the effect of aluminum silicate and ferric oxide binding with polar - polar reaction between filler and matrices. These reaction cause difference in input impedance and this will make a difference in the resonance at high frequency. Ferrite-oxide particle have a great microwave absorption characteristic, but in this case, the ratio between zeolite and ferric-oxide will affect the best microwave absorption. The unique properties of nanozeolite and polyurethane provide an important foundation for developing the materials with strong microwave absorption in wide applications prospects.

References

[1] Abbas SM *et al* 2006 Composites :Part A 37 2148-2154.
[2] Kong I *et al* 2010 *J Magnetism Magnetic Mat* **322** 3401-3409.
[3] Peng C H *et al* 2004 *J. Magnetic Mater* **284** 113-119.
[4] F. Qin and C. Brosseau 2012 *J. Appl. Phys* 111, 061301.
[5] Esfahani *et al* 012 *Preparation and Characterization of Foamed Polyurethane/Silicone Rubber/Graphite Nanocomposite as Radio Frequency Wave Absorbing Material : The Role of Interfacial Compatibilization,* J. Compscitech.
[6] Pawar R P *et al* 2013 *Proc. and Appl. Ceramics* **7** [1]: 1-7.
[7] Jamal E.M *et al* 2009 *Mater. Sci. and Eng. B* **156** 24-31.
[8] Kato Y *et al* 2002 *Materials Transactions M* **43** 406-409.

Improvement of thermal and mechanical properties of composite based on polylactic acid and microfibrillated cellulose through chemical modification

L Suryanegara[1], R A Nugraha[2], S S Achmadi[3]
[1)]Research Center for Biomaterials, LIPI, Cibinong, Indonesia
[2)]Department of Chemistry, Faculty of Mathematics and Natural Science, Bogor Agricultural University, Bogor, Indonesia
[3)]Department of Chemistry, Faculty of Mathematics and Natural Science, Bogor Agricultural University, Bogor, Indonesia

E-mail: l_suryanegara@yahoo.com

Abstract. Polylactic acid (PLA) is the most representative sustainable and bio-based polymer environmentally friendly that has a great potential to replace petroleum-based plastics. However, brittleness, low heat resistance, and slow crystallization limit the wide application of PLA. One of strategies to improve PLA properties is by reinforcing with microfibrillated cellulose (MFC). Unfortunately, the hydrophilic properties of MFC make it difficult to attain good dispersion in a hydrophobic PLA matrix. Therefore, modification of MFC was needed to increase its compatibility with PLA in the composite formation. In this experiment, MFC was modified with partial acetylation (degree of substitution: 1) and further grafted with lactide monomers through ring-opening polymerization using $Sn(Oct)_2$ catalyst. The result of acetylation and grafting were verified by infrared spectra. Composites were prepared by mixing PLA (molecular weight of 200,000) and the modified MFC at 9:1 ratio through organic solvent method. Followed by 8 min-kneading and hot pressing at 180°C, the resulted composites were evaluated for their mechanical and thermal properties. Thermal characterization carried out using differential scanning calorimetry measurements showed that the presence of modified MFC increased the temperature of glass transition and accelerated the crystallization of PLA. Mechanical properties measurement showed that the presence of modified MFC enhanced the elongation at break (1.1 to 1.8%), tensile strength (14.9 to 25.7 MPa), and modulus of elasticity (1.7 to 2.1 GPa). These results demonstrated that the modified MFC could extend the application of PLA in industry.

1. Introduction

In recent years, the concepts of green chemistry, green materials, and industrial ecology in the field of polymers have been developed. This concept aims to answer the environment problem and depletion of fossil oil resources. Polylactic acid (PLA) is one of the environmental-friendly polymers, which is versatile, made of renewable materials, energy and the production cost are low that can compete with other synthetic polymers. PLA is potential as an alternative substitution of petroleum-based polymers in the industrialized world. By its nature, PLA has good mechanical properties that are easily used in the manufacture of technical fiber and films, extrusion and thermoforming, and injection molding. However, it has some weakness including fragile, low stability and heat resistance, and the slow rate of crystallization. The weakness limits its industrial application [1]. Several studies reported that addition of microfibrillated cellulose (MFC) improved PLA properties [2][3].

In the application, MFC is used as filler in the PLA matrix. Its small size (10^{-9}-10^{-6} m) makes this nanofiber more effective as reinforcing compared to other fibers [4] and it is potential as a reinforcing material in the automotive industry, buildings, computers, and other products. Thus the formation of MFC-reinforced composite can give them the advantage in terms of the physical, thermal, and mechanical [5]. However, MFC has hydrophilic properties that inhibits its dispersion in the PLA matrix, forming a hydrophobic aggregation of the MFC and irreversible agglomeration due to strong hydrogen bonds. This makes composite formed with a low thermal stability [1]. Several researches were conducted to improve MFC properties either by exploring more sources of MFC [6] or by converting MFC to Nanofiber Cellulose (NFC) [7]. Therefore, some chemical modifications should be carried out with the aim of improving compatibility.

Partial acetylation in bacterial cellulose with a range of degree of substitution (DS) 0.04-2.77 had been reported by Kim *et al.*, (2002) [8]. The result showed that acetylation occurs on the surface of bacterial cellulose and does not damage the fiber morphology. Ifuku *et al.*, (2007) [9] had modified the bacterial surface nanocellulose acetylated fibers with various DS. Acetylation is able to improve various properties of fibers, effectively reduce the hygroscopicity of the bacterial nanocellulose, and enhance compatibility with other hydrophobic materials such as PLA matrix.

Another approach in reducing the hydrophilicity of MFC was reported by Goffin *et al.*, (2011) [1]. The surface of nanocellulose was grafted with polyester matrix using ring-opening polymerization of L-lactide, followed by mixing PLA to form a composite. Toluene was used as a solvent and $Sn(Oct)_2$ as a catalyst in the ring opening polymerization. Thermal property characterization of the composite indicated that such modification improves compatibility and also increases the strength of the composite. In preparing the PLA-MFC composite, Iwatake *et al.*, (2010) [10] used 2 methods, namely direct mixing and blending in organic solvent. Based on the mechanical properties of the products, it was suggested that mixing in organic solvent gave a better composite than that prepared through direct mixing.

This experiment aimed to improve thermal and mechanical properties of composite based on PLA and MFC through chemical modification to obtain a better homogeneity between the PLA and MFC. Partial acetylation allowed hydrophilic hidroxy groups at MFC structure partially substituted by acetyl groups that are hydrophobic, so that MFC could be compatible with hydrophobic materials such as PLA matrix. The acetylation product hereinafter was called MFC-acetate. The next process was grafting the MFC-acetate by employing lactide monomer. Lactide monomer was expected to grow at the backbone structure of MFC. The final step was preparing some composite which was expected to have some thermal and mechanical properties improved.

2. Materials and Methods

2.1 Materials

The materials used were MFC (Daicel Chemistry, Japan), lactide monomer (Sigma-Aldrich), $Sn(Oct)_2$ (Sigma-Aldrich) as polymerization catalyst, and PLA H-400 (Toyota).

2.2 Acetylation of MFC

Approximately 10 g wet MFC was squeezed using chamois until water seeps out. The sample was placed in a flask and 50 mL ethanol was added while stirring for 30 min. The suspension was then filtered using a suction means. The procedure was repeated by replacing ethanol with acetone and glacial acetic acid. After removing the free water, the sample was added dropwise an acid catalyst (CH_3COOH:H_2SO_4 95-97% (v/v) = 100:1) as much as 12 mL and allowed to stand for 1 min. Furthermore, some more acetic anhydride was slowly added (15 mL). The suspension was stirred by glass rod until thickened, allowed to stand for 90 min (started when acetic anhydride was added). The acetylation process was stopped by adding mixture of 3 mL of glacial acetic acid:distilled water (2:1) and allowed to stand for 30 min (stirring at medium speed in order to mix evenly). Samples that had been acetylated MFC was then dispersed into 500 mL of distilled water by vigorous stirring. The product was then vacuum-filtered by Büchner funnel and washed with $NaHCO_3$ 1 N until there was no more CO_2 bubbling out. The fiber was washed with distilled water until neutral (pH 6.5-7) and vacuum-filtered until water no longer dripping. Samples that had been acetylated were stored in tightly lid glass jar.

2.3 Grafting on MFC-acetate by Lactide Monomer through Ring Opening Polymerization

Samples of MFC-acetate that had been obtained previously (DS ± 1) were immersed in acetone and dried using toluene by centrifugation techniques. The samples were suspended in toluene in a three-necked reaction flask, stirred, and purged with nitrogen. The reaction flask was immersed in a water bath and heated to 80 °C. Then, 1 g of L-lactide monomer (in hot toluene) was added to the reaction flask containing the MFC-acetate suspension. $Sn(Oct)_2$ (2% of the monomer weight) was added through a sterilized needle to catalyzed the reaction. Polymerization was carried out for over 24 h, and stopped by adding a few drops of dilute HCl solution (0.1 M). Graft copolymerization products were obtained by precipitation in cold methanol and filtered.

IOP Conf. Series: Materials Science and Engineering **223** (2017) 012032 doi:10.1088/1757-899X/223/1/012032

2.4 Composite Preparation

The double modified MFC (10% weight of the intended composite) was suspended in methanol for 15 min and filtered. The ethanol was replaced by acetone and dichloromethane successively. Each inclusion process was repeated 3 times. In a separate beaker, PLA (90% of the intended composite weight) was dissolved in dichloromethane using a homogenizer to obtain a well-dispersed suspension. The suspension formed was stirred and the modified MFC was gradually added to the suspension at room temperature and kept stirring for 1 h. The blend of PLA/modified MFC in dichloromethane was spread out on a basin and evaporated in a fume hood at room temperature for 12 h followed by vacuum drying at 50 °C for 8 h. The composite layer was dried, weighed, and cut up (± 1 cm^2). In the next process, the composite was kneaded using laboplastomil at 160 °C, at 40 rpm speed for 8 min. Finally, the composite was molded into thin sheets using hot pressed at 180 °C for 14 min (preheating 6 min, pressing 3 min under 400 Pa, and cooling 5 min).

3. Results and Discussion

3.1 Microfibrillated Cellulose

MFC is cellulose which has been acid-hydrolysed [1] and mechanically processed in homogenized high pressure and a small aggregate [13]. According to Chakraborty *et al.*,(2006) [14], MFC has reached 0.1-1 m fiber diameter and minimum length of 5-50 μm. Based on morphological analysis using SEM (Figure 1) with a magnification of 1000× the fiber diameter was varied (111.7-157.9 nm), while the length is difficult to measure because each fiber is tangled in webs.

Cellulose microfibril consists of crystalline and amorphous regions in interspersed proportion depending on cellulose source obtainment. In addition, this compound has a regular structure in the form of a linear polymer repeat unit due to β-D-glucopyranose are linked to each other through β (1-4) bonds. Two hydroxy groups in position C2 and C3 (secondary) and that in C6 (primary) create intra- and intermolecular hydrogen bonds. Therefore, the cellulose fibers have high tensile strength and is insoluble in most solvents. Cellulose, and therefore, the MFC also shows strong hydrophilic character, as proven its moisture content as high as 90%. This strong hydrophilic character of MFC inhibits dispersion in the PLA matrix.

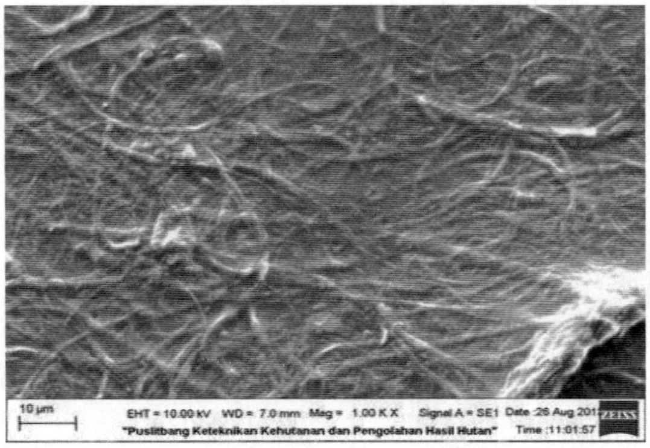

Figure 1. Morphology of MFC with 1000× magnification.

3.2 Modified MFC

Acetylation was meant to chemically modified MFC in order to alter certain properties based on their structure. In this study, the hydroxy groups on MFC was not completely acetylated, just sufficiently to render its compatibility with the hydrophobic PLA. Partial acetylation reaction in MFC-acetate can be evidenced by the degree of substitution (DS), as determined by titration techniques.

The DS of MFC-acetate was approximately 1. However, Ifuku *et al.*, (2007) [9] reported DS of 0.2 to 0.6 did not alter cellulose crystallinity region significantly, the extent of hydrogen bonds between cellulose fibers only slightly reduced. In this experiment, DS <1 was not obtained, presumably due to the use of sulfuric acid as a catalyst in acetylation causing structural deformation of the MFC. This was evidenced by the dissolution of the MFC when mixed with acetic anhydride and sulfuric acid. The deformation caused the acetylation of cellulose structure did not only occur on the surface of cellulose but all layers of cellulose fibers.

Analysis using FTIR spectrophotometer on neat-MFC, MFC-acetate, and MFC-acetic-lactic grafted (MFC-acetate-g-La) is shown in Figure 2 and Table 1. There is a typical absorption of cellulose at wavenumber 3236 cm^{-1} and wavenumber 1053 cm^{-1} belongs to –CO stretching on the cellulose ring. Furthermore, the intensity of the –OH absorption peak decreases and typical sharp absorption peak appears that belong to cellulose acetate, indicating –CO stretch (1759 cm^{-1}) and acetyl –CO bend (1234 cm^{-1}) [15]. The spectrum shows that the decrease in absorption intensity is due to partially substituted of the hydroxy groups by acetyl groups.

The second step of modification was graft copolymerization with lactide monomer by ring opening polymerization techniques. The graft copolymer can produce copolymer that is similar to the main chain and the grafted monomer, so that through graft copolymerization, the polymer properties can be modified in accordance with the purpose and utilization.

Table 1. Infrared spectra of neat MFC and modified MFC

MFC		MFC-asetat		MFC-asetat-g-La	
Wave number(cm^{-1})	Fungsional Group	Wave number(cm^{-1})	Fungsional Group	Wave number(cm^{-1})	Fungsional Group
3236	O-H stretching	3630	O-H stretching	3527	O-H stretching
2890	C-Hstretching	2947	C-Hstretching	2960	C-Hstretcing
1650	O-Hbend	1759	C=Ostretching	1752	C=Ostretching
1390	C-Hbend	1375	C-Hbend	1375	C-Hbend
1053	C-Ostretching	1234	C-Oacetate	1235	C-Oacetate
		1042	C-Ostretching	1047	C-Ostretching
		608	C-Hbend	604	C-Hbend

Covalent bonds between the main chain of the polymer grafts and ensure the permanence. Ring-opening polymerization technique is a good technique for polymerization using monomers such as cyclic lactone and lactide. Result of grafting on MFC-acetate is presented in Figure 2.

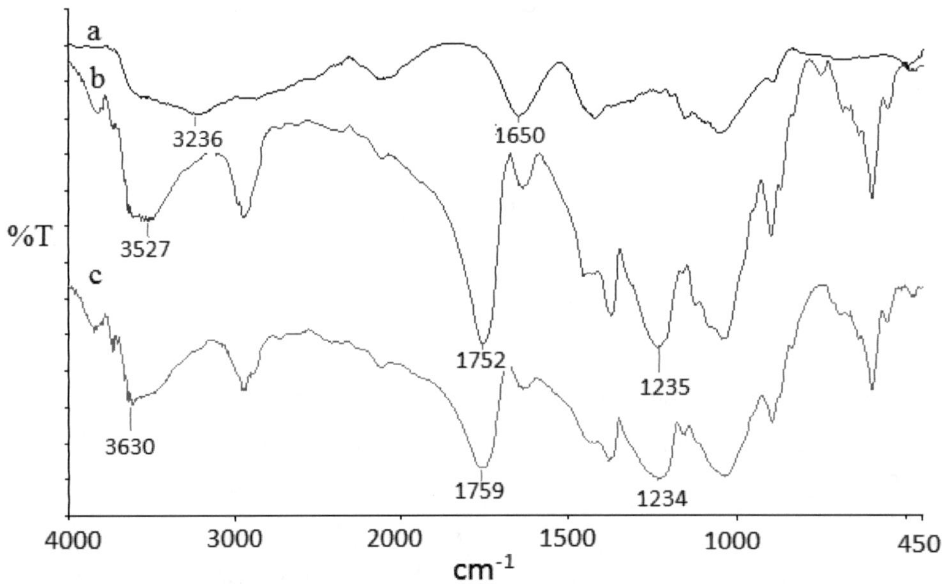

Figure 2. Spectrum of neat MFC (a), MFC-acetate (b) and MFC-acetate-g-La (c)

Figure 2 shows the absorption peaks are relatively similar to the spectrum of MFC-acetate, namely wavenumber 3527 cm^{-1} that belongs to ¬OH group, 1752 cm^{-1} belongs to ¬CO group, and 1235 cm^{-1} belongs to ¬CO acetyl group [15]. There is a trend of declining peak intensity of the free ¬OH and increasing peak intensity of –CO acetyl group, showing that the lactide has been grafted on the MFC-acetate backbone.

IOP Conf. Series: Materials Science and Engineering **223** (2017) 012032 doi:10.1088/1757-899X/223/1/012032

3.3 Analysis of Thermal Properties

Thermal properties of composites in this study were characterized using DSC instrument applying heat microscopy (scanning of low temperature to high temperature) and produced a thermogram curve. The thermogram curve displays the glass transition temperature (Tg), crystallization temperature (cold state) (Tc), and melting temperature (Tm) of the material being tested. Figure 3 displays the thermogram curve of neat-PLA, PLA composite/MFC, and composite PLA / MFC-acetate-g-La, respectively. Based on the analysis of the thermogram third curve (Table 2) shows that the Tg of PLA / MFC-acetate-g-LA (54.38°C) higher than the Tg of neat PLA (52.81 °C) and PLA/MFC (49.33°C). The shift of temperature indicated that the PLA reinforced with modified MFC can increase Tg. Increasing Tg fairly well because when the composite was subjected to heat, the rubber phase change to a higher temperature so that it was more heat-resistant than neat-PLA.

The increase was caused by when the PLA receives thermal energy, the polymer molecules is activated and induces vibrational and rotational motion. The presence of acetylated and lactylated MFC as filler reduced the flexibility of the molecular structure of PLA and inhibited the molecules vibration and rotation so that the high energy is required to achieve a Tg. In addition, Braun *et al.* [14] reported that the grafted-MFC formed percolation network on the filler particles that providing increase to Tg. Otherwise, the PLA/MFC (not modified) actually showed a reduction in the Tg. The phenomenon occur due to the unmodified MFC having incompatibility with PLA to form a weak interaction between the filler and the matrix, so that causing the Tg decline.

Based on crystallinity data, Tc PLA/MFC-acetate-g-La (93.20°C) and PLA/MFC (94.01 °C) is lower than Tc of neat- PLA (97.60°C) (Table 1). This shows that the decrease in Tc PLA reinforced with MFC was able to crystallize faster than neat- PLA. This is due to MFC action as a nucleating agent [12]. However, Tc of PLA/MFC-acetate-g-La is higher than PLA/MFC. The difference proves that the modifications to the MFC is able to improve the compatibility between the PLA. Thus, the effectiveness of the MFC as a nucleating agent can be further increased when dispersed evenly in the PLA matrix.

However, the melting data was not influenced by MFC as reinforcement in the PLA matrix. Based on data from Tm (Table 2), there was no significant decrease between Tm of PLA/MFC (168.29 °C) and Tm PLA/MFC-acetate-g-La (168.31°C) when compared with neat PLA Tm (168.35 °C). In the application, the Tm peak effect, fast or slow the process of getting molten form. The lower the temperature, it is better because it is faster to get a discharge considering PLA products have a narrow temperature range up degraded.

Table 2. Analysis of measurement results with DSC

Sampel	Tg (°C)	Tc (°C)	Tm (°C)
Neat-PLA	52.81	97.60	168.35
PLA/MFC	49.33	94.01	168.29
PLA/MFC-acetate-g-La	54.38	93.20	168.31

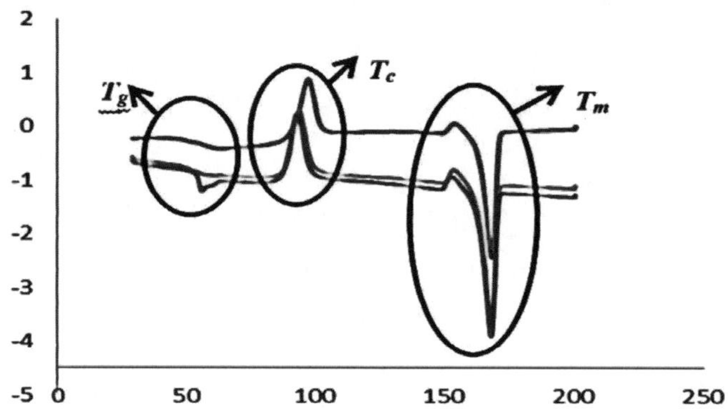

Figure 3. Thermogram of DSC where the upper line is neat PLA, the middle line is PLA/MFC, and the lower line is PLA/MFC-acetate-g-La

3.4 Analysis of Mechanical Properties

Reinforcing effect of MFC modified on the PLA matrix demonstrated through testing of mechanical properties: the elongation at break, tensile strength, and modulus of elasticity. Results of the analysis on the mechanical properties of neat-PLA, PLA/MFC, and PLA/MFC-acetate-g-La are shown in Table 3. It's known that the elongation break of modified or not modified sample showed no significant change, which is neat- PLA (1.18%), PLA/MFC (1.42%), and PLA/MFC-acetate-g-La (1.77%). However, it appeared that the presence of MFC as filler provides enhanced mechanical properties, although not real. This phenomenon occurs due to the interaction between MFC and PLA composite extend the point of deformation.

MFC reinforcement effect seen in rigidity and tensile strength of the resulting composites. All these facts proved that the MFC was able to increase the strength of the PLA. This increase was due to percolation network; cellulose fibers act as filler to strengthen the structure and were able to keep the stiffness matrix of PLA [3]. In addition, fillers restrict the movement of the molecular chain of the amorphous PLA [15].

Table 3. The result of mechanical strength test using UTM

Sample	Elongation at break (%)	Tensile strength (MPa)	Mod. Elasticity (GPa)
PLA	1.1±0.2	14.9±0.3	1.7±0.2
PLA/MFC	1.3±0.3	17.7±0.3	2.0±0.2
PLA/MFC-Acetate-g-La	1.8±0.3	25.7±2.2	2.1±0.1

However, dissolution of MFC on acetylation process was expected to affect strength of produced MFC-acetate and certainly have an impact on the composite strength of PLA/MFC-acetate-g-La. When MFC was soluble, it was suspected for structure deformation that makes hydrogen bonding between cellulose chains broken to form a space in the crystalline chain so that when the MFC-soluble is dispersed in water, the MFC that has not formed a tight crystalline structure as originally tough to prove it needs to be analyzed morphological crystalline using x-ray diffraction (XRD). The phenomenon was thought to be the cause of rigidity values PLA/MFC-acetate-g-La does not vary much with PLA/MFC.

PLA/MFC-acetate-g-La had a rigidity and a higher tensile strength compared with PLA/MFC. This difference proves that the MFC which had been modified (acetylated and lactylated) had a better compatibility with the PLA matrix than unmodified MFC. It is caused when a composite accept load (energy), then the energy will first be accepted by the matrix and then be transferred to the fibers as filler through the interface field. Therefore, the ability of the interface in the field of energy transfer from the matrix to the filler play an important role on the mechanical strength of a composite.

The good compatibility will form a large surface area of between matrix and filler so that the composite is able to transfer energy better. The presence of grafting in this study to increase the strength of a composite due to the formation of covalent bonds between the MFC and the grafted polymer chains [1] and will increase compatibility due to increased adhesion between the two bonds. In addition, the partial acetylation support grafting reaction in improving the strength of the composite produced. According to Braun *et al.,* (2013) [16], limiting the degree of grafting on the MFC through partial acetylation process successfully form a network thereby increasing the rigidity and strength of the composite.

4. Conclusions

The reinforcement of modified MFC improved the thermal and mechanical properties of PLA composites. Thermal characterization carried out using differential scanning calorimetry measurements revealed that the addition of modified MFC increased the temperature of glass transition and accelerated the crystallization of PLA. Furthermore, mechanical properties measurement showed that the addition of modified MFC enhanced the elongation at break from 1.1 to 1.8%, tensile strength from 14.9 to 25.7 MPa, and modulus of elasticity from 1.7 to 2.1 GPa. Thus the MFC modified (acetylated) can improve the properties of PLA composites.

Acknowledgement

This work was sponsored by the Ministry of Research and Technology Republic of Indonesia through Advanced Materials Competitive Program, LIPI and JICA through Satreps Project.

References

[1] Goffin L *et al* 2011 *Biomacromolecule* **12**(7) 2456–2465.
[2] Oksman K *et al* 2003 *Compos Sci Technol.* **9**(63) 1317–1324.
[3] Nakagaito A *et al* 2009 *Compos Sci. Technol.* **69**(7) 1293–1297.
[4] Angles M N, Dufresne A 2001 *Macromolecules* **34**(9) 2921–2931.
[5] Oksman K *et al* 2006 *Compos Sci Technol.* **66**(15) 2776–2784.
[6] Ramahdita G *et al* 2017 *Macromolecular Symp,* **371**(1) 69-74.
[7] Kitagawa K *et al* 2006 *Symp Japan Soc Poly Proc.* 171-172.
[8] Kim D Y *et al* 2002 *Cellulose* **9** (3) 361–367.
[9] Ifuku S *et al* 2007 *Biomacromolecules* **8**(6) 1973-1978.
[10] Iwatake A *et al* 2008 *Compos Sci Technol,* **68**(9) 2103–2106.
[11] Arifin B 2004 *Optimasi kondisi asetilasi selulosa bakteridari nata de coco* Fakultas Matematika dan Ilmu Pengetahuan Alam, Institut Pertanian Bogor.
[12] Suryanegara L *et al* 2009 *Compos Sci Technol.* **7**(69) 1187–1192.
[13] Carlmark A *et al* *Eur Polym J* **48**(10) 1646– 1659.
[14] Chakraborty *et al* 2006 *Holzforschung* **60**(1) 53-58.
[15] Silverstein M *et al* 2005 *Spectrometric Identification of Organic Compounds, 7th Ed,* Toronto: J Wiley & Sons.
[16] Braun J R *et al* 2012 *Biomacromolecules* **13(7)** 2013–2019.
[17] Siqueira G *et al* 2009 *Biomacromolecules,* **10**(2) 425-432.

Added value of lignin as lignin-based hybrid polyurethane for a compatibilizing agent

S Ilmiati, J Haris Mustafa, A Yaumal, F Hanum, M Chalid
Polymer Product Technology Laboratory, Department of Metallurgy and Materials Engineering, Universitas Indonesia, Depok, Indonesia

E-mail : chalid@metal.ui.ac.id

Abstract. As biomass-based material, lignin contains abundant hydroxyl groups promising to be used as chain extender in building hybrid polyurethanes. Consisting of polyehtylene glycol (PEG) content as hydrophobic part and lignin as hydrophilic part, the hybrid PU is expected to be as a novel compatibilizing agent in new materials production such as polyblends and composites. The hybrid PU was synthesized via two reaction stages, viz. pre-polyurethanization through reacting 4,4'-Methylenebis (Cyclohexyl Isocyanate) (HMDI) and PEG as polyol, and chain extention through adding lignin in the pre-polyurethanization system. The composition effect of lignin in hybrid PU syntehsis, to chemical structure corelated to hydrophobic to hydrophilic ratio, thermal and morphological properties, was evaluated by measuring NMR, FTIR, DSC, TGA and FE-SEM. The experiments showed that addition of lignin was able to extend the pre-polyurethane into hybrid polyurethane and to increase the lignin/polyol ratio in the hybrid polyurethanes, which were indicated by NMR and FTIR Analysis. And change of the ratio lead to increase the glass transition from 60.9 until 62.1°C and degradation temperature from 413.9 until 416.0°C. Observation of the morphology implied that addition of lignin gave more agglomerations. A Further investigation for this characterization study should be focused on a feasibility for this modified lignin as a novel compatibilizing agent.

1. Introduction

Lignin is one of the major components in biomass. Because of a high abundance, biomass is considered as the renewable main source for lignin [1]. In industry, lignin is obtained from the pulp industry, as waste [2]. Unfortunately, the lignin is still not optimally empowered yet. Therefore, study to enhance the added-value of lignin is an interesting issue. Rozman *et al.,* (2000) studied an using of lignin as compatibilizing agent between polypropylene and coconut fiber [3]. This study reported that lignin has a higher tendency to polarity than non-polarity, confirmed by literature [4]. Some chemical modifications have been performed, such as fragmentation of lignin into another phenolic hydroxyl or aromatic, a new active functional group synthesis to change the polarity of lignin, and functionalization of the hydroxyl group in order to enhance its reactivity [5]. Lignin was also modified as a main chain in synthesis of a polymer through grafting phenolic functional groups [6]. Further investigation to balance ratio of polar as hydrophilic part and non-polar as hydrophobic part, in lignin is a necessity.

As a versetile polymer with a wide range of product properties [7], polyurethanes are synthesized by reacting di-isocyanate compound with bifunctional reagents such as diols or polyols. Modification of polyurethanes through adding lignin therefore is an interesting study. Because of containing hydroxyl group as polar and major parts, lignin is a potential chain extender agent in modifiying polyurethane to a hybrid polyurethane, which is synthesized by reacting di-isocyanate and polyol such polyethylene glycol (PEG) as a non-polar part.

Consisting of polyehtylene glycol (PEG) content as hydrophobic part and lignin as hydrophilic part, the hybrid polyurethanes are expected to be as a novel compatibilizing agent in new materials production such as polyblends and composites. The hybrid PU was synthesized via two reaction stages, viz. pre-polyurethanization through reacting 4,4'-Methylenebis (Cyclohexyl Isocyanate) (HMDI) and PEG as polyol, and chain extention through adding lignin in the pre-polyurethanization system. The composition effect of lignin in hybrid PU syntehsis, to chemical structure corelated to hydrophobic to hydrophilic ratio, thermal and morphological properties, was evaluated by measuring NMR, FTIR, DSC, TGA and FE-SEM.

2. Materials and Methods

2.1. Materials

All raw materials were analytic grade chemicals, viz., lignin from Technical Chemicals Indonesia, 4,4-methylenebis (cyclohexyl isocyanate) (HMDI), polyethylene glycol (PEG) with MW 6000 g/mol, and dibutyltin dilaurate (DD) as catalyst purchased from Sigma Aldrich.

2.2. Methods

The process started by dissolving 15 g of PEG-6000 with 25 mL of DMF in the reactor schlenk tube inert nitrogen gas at room temperature. The solution was fluxed three times by using vacuum suction machine and nitrogen gas. And the tube was dipped into the oil bath heated up to 80°C under stirring about 1 h. After HMDI injection, 2 drops of dibutyltin dilaurate (DD) catalyst were added into the flasks with nitrogen atmosphere. After stirring for 2 h at 80 °C, the mixture was added by lignin dissolved in 5 mL of DMF. This reaction was performed under stirring for 1 h, and was terminated by adding some drops of water into the tube. Composition of the chemicals during the experiments is shown in Table 1. The hybrid polyurethane products were characterized by measuring Nuclear Magnetic Resonance (NMR) and Fourier Transform Infra-Red Spectroscopy (FTIR Spectroscopy) to analyse the chemical structure of the products, and measuring Simultaneous Thermal Analysis (STA) and Scanning Electron Microscopy (SEM) to evaluate thermal properties and morphology of the products. The integration heights from prominent hydrogen peaks on NMR spectra, representing lignin as the polar (hydrophilic) part and PEG as the nonplar (hydrophobic) part, were used to calculate ratio of the polar (p) and the non polar (np) as shown in equation (1) [8]:

$$\frac{p}{np} = \frac{\frac{Total\ integration\ of\ spectrum\ (H-a)}{3\ atoms\ H}}{\frac{Total\ integration\ of\ spectrum\ (H-e)}{4\ atoms\ H}} \qquad (1)$$

Table 1. Materials Composition in Experimental Conditions

Sample	PEG-6000, g	HMDI, mL	Lignin, g
H-06	15	0.6	0.5
L-05	15	2.5	0.5
L-2	15	2.5	2
L-4	15	2.5	4

3. Result and Discussion

3.1. Synthesis of The Hybrid Polyurethanes

This synthesis was performed by pre-polymerizing PEG with certain compositions of a diisocyanate in a dried solution, followed adding lignin as chain extender to synthesize the lignin based-hybrid polyurethanes with the presence of dibutyltin dilaurate as the catalyst in both stages. Those reaction stages are systematically depicted in Figure 1.

This reaction was expected to obtain the hybrid polyurethanes with functional groups of urethane and amide as shown in Figure 1. Presence of these functional groups is indicated by analysing the products with ^{1}H- ^{13}C-NMR and FTIR. Corresponding FTIR spectrum to Lignin and the novel hybrid polyurethanes is shown in Figure 2.

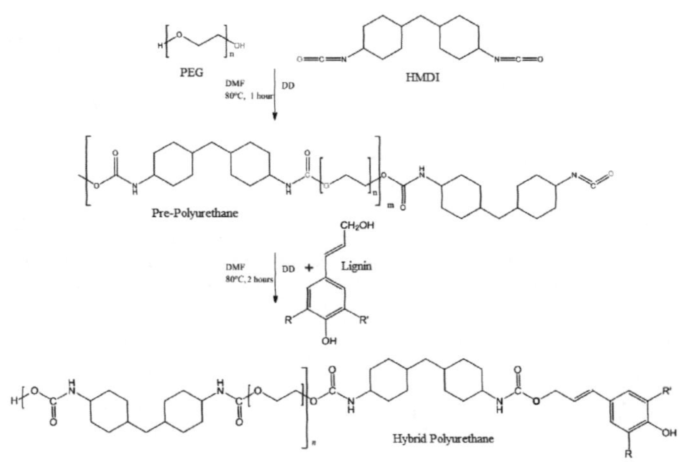

Figure 1. Purposed synthesis reactions of lignin based-hybrid polyurethanes

FTIR spectra for lignin based on p-coumaryl alcohol, indicates some peaks at some wave numbers, such as 3363 cm^{-1} for hydroxyl groups, 1630 cm^{-1} for C=C alkenes chain, 1517 cm^{-1} and 1375 cm^{-1} for CH_2 chain, and 1183 cm^{-1} and 1104 cm^{-1} for sulfone chain, and 867 cm^{-1} for para-aromatic groups. The FTIR analysis is confirmed by Lisperguer *et al.,* (2009) reported that lignin has hydroxyl and methoxyl groups, aliphatic chain, and aromatic [9].

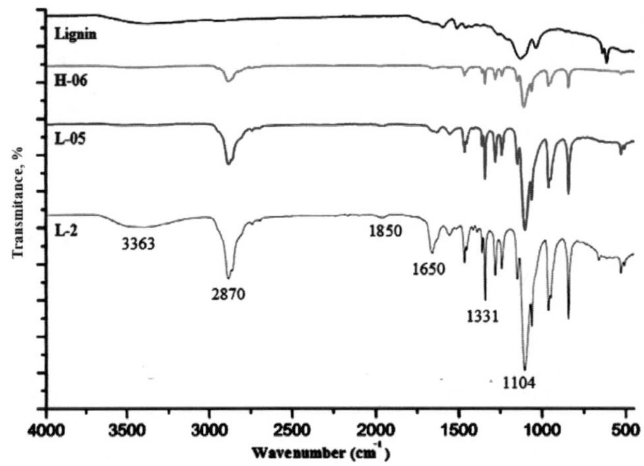

Figure 2. FTIR spectra for lignin and the hybrid polyurethanes

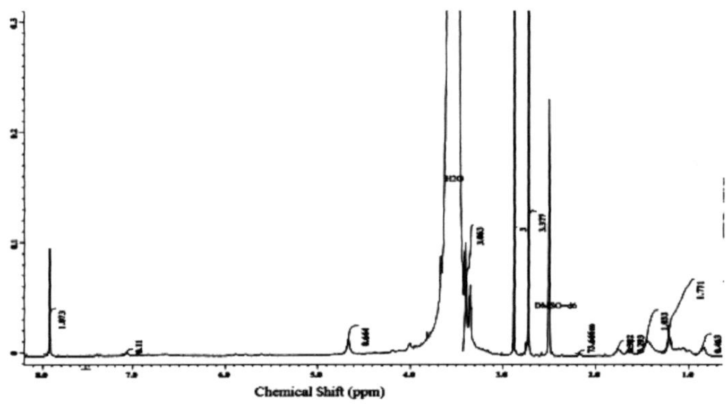

Figure 3. H-NMR spectra of L-2, as representative of the hybride polyurethane products

Table 2. Chemical structure and chemical shifts on NMR spectra for L-2

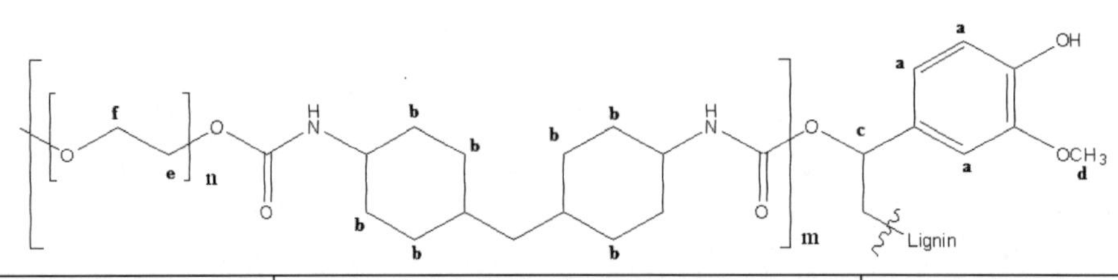

Hydrogen Position	Sample / chemical shift (ppm)			Remarks
	H-05	L-05	L-2	
H-a	7.0	7.05	7.06	Benzene
H-b	0.8; 1.18; 1.86	0.79; 1.18; 1.5	0.8; 1.2; 1.4	Cyclic HMDI
H-c	2.1; 2.32	2.13; 2.37	2.1; 2.36	Benzyl
H-d	4.13	4.13	3.96	ArO-CH_3
H-e	4.63	4.63	4.66	PEG
H-f	3-4	3-4	3-4	PEG
H-g	8.1	7.8	7.9	N-H

Furthermore, FTIR spectrum for the hybrid polyurethane products as H-06, L-05 and L-2 confirms presences of the urethane and amide functional groups 1850 at and 1650 cm^{-1}, respectively. The presences of the amine bonds in the hybride polyurethane appears at 2870 cm^{-1}. Chalid *et al.,* (2011, 2012, 2013) reported that the important group of the polyols can be observed at about 3363 cm^{-1} for hydroxyl groups and 1104 cm^{-1} for ether groups. Isocyanate groups from HMDI have important group at about 2250 cm^{-1}[10-13]. On the FTIR spectra of the hybrid polyurethanes (in Figure 3), the peak at 2250 cm^{-1} is not visible at all. This indicates that the product does not have active cyanate groups any more. Overall, the peaks in FTIR spectrum for product H-05, L-05 and L-2 are approximately similar. The FTIR analysis about the hybrid polyurethane above was confirmed by characterization of the products with NMR measurements, as shown in Figure 3. The analysis of the NMR spectra was perfomed by referring to simulation using *Molecular Modeling Pro Plus* Software for *Windows* released by *ChemSW, Inc.* and some literatures. Chalid *et al.,* (2011, 2012, 2013) [10-13] reported important spectrum for HMDI based-polyurethanes on H-NMR are at 6.0-8.5 ppm for benzene; 0.8-2.14 ppm for cyclic HMDI and 4.6 ppm for N-C-H from urethane group. In details, analysis on NMR spectra for the hybrid polyurethane product is presented in Table 2. The chemical structure analysis conludes that synthesis reactions of lignin based-hybrid polyurethanes are appropriate as the proposed reactions in Figure 1.

The hybrid polyurethanes obtained from this study are a potential finding to be empowered as a comptibilizing agent due to consisting of PEG as hydrophobic part and lignin as hydrophilic part in their chemical structures [14-15]. The next discussion is therefore concerned about ratio of polar to non polar parts.

3.2. The Ratio of Polar and Non Polar Parts

One of approaches to study the hybrid polyurethanes feasibility as a compatibilizing agent is comparison of hydrophobic and hydrophilic part amount [16]. The hydrophobic (non polar) part in the hybrid polyurethane structure is represented by chemical shifts (δ) of hydrogen peak in the outer PEG-6000 in H-NMR spectra, while the hydrophilic (polar) part is represented by chemical shifts (δ) of hydrogen peak in aromatic lignin. Based on this, comparison of polar (p) and nonpolar (np) parts, as shown in Table 3, can be calculated by the equation[8].

Table 3. Values of p/np for the products

Sample	p/np
H-05	0.14
L-05	0.18
L-2	2.38

Table 3 implies that the ratio of polar to non polar parts is proportional to lignin and HMDI amounts. The results indicate that HMDI has a contribution to bind lignin molecules with prepolyurethane molecules (*see* Figure 1). Furthermore increasing of lignin amounts leads to increasing of the polar parts in the hybrid polyurethane products. Finally modification of lignin through adding prepolyurethanes may be able to enhance polarity portion in the lignin, as a compatibiling agent.

3.3. Thermal Properties

Modification of chemical structure on the polyurethane through adding lignin to obtain the hybrid polyurethanes, of course leads to an interesting change in their properties such as thermal properties.

Figure 4 and Table 4 show some interesting thermal properties obtained from DSC measurements. Increasing of lignin amounts added to prepolyurethane gives an effect to enhancement of the glass transition and degradation temperatures from 102.9 - 105.1°C and from 401.5 - 4016.0°C, respectively. Enhancement of the glass transition temperatures may be caused by increasing entanglement behavior of the molecular products due to presences of lignin part as chain extender for prepolyurethanes, indicated the p/np ratio, leading to irregularity and molecular weight of the molecular products. Beside that chemical structure of lignin, which is rigid has a contribution to enhancement of glass transition [17].

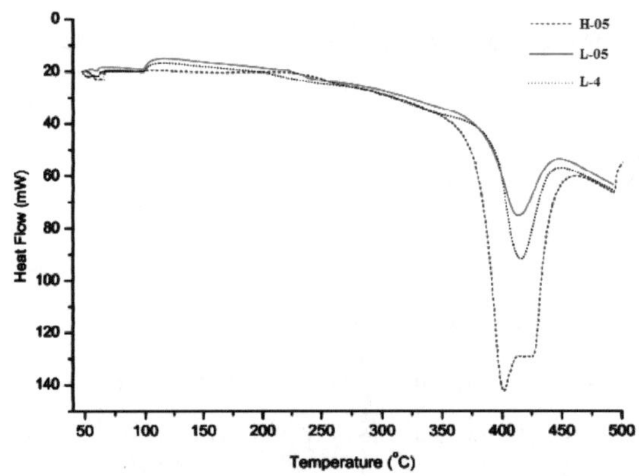

Figure 4. DSC curves for lignin and the hybrid polyurethanes as H-05, L-2 and L-4

Table 4. Tg and Td extracted data from Figure 4

Sample	Tg, °C	Td, °C
Lignin	-	420.0
H-05	102.9	401.5
L-2	103.9	413.9
L-4	105.1	416.0

Furthermore presences of lignin also lead to enhancement of the rigid compoenents in the hybrid produt. This component has an effect to increase thermal stability, indicated by increasing the degradation temperature of the products due to increasing lignin amounts. This analysis is confirmed by comparing pure lignin to the hybrid products, as shown in Figure 4 and Table 4. This is also in accordance with the p/np ratio and the literatures [17-18].

3.4. Morphology

Addition of lignin into prepolyurethane structure and using of HMD in excess amount to bind between prepolyurethane and lignin molecules enhance rigd segments in the products. These segments tend to agglomerate with a certain distribution, in their morphology. Furthermore this behavior was evaluated by observing FE-SEM, as shown in Figure 5.

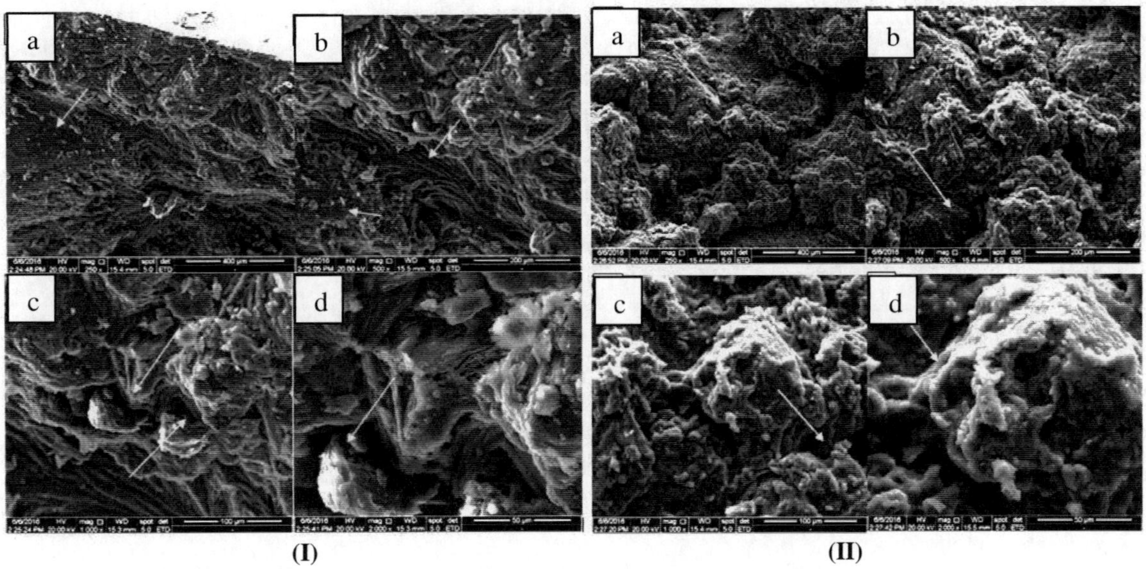

Figure 5. The morphology of the sample surface (I) H-05 and (II) L-05.
Bar scale magnifications are a) 400 μm b) 200 μm c) 100 μm and d) 50 μm

The morphology on the hybrid product obtained from adding HMDI in excess amounts (*see* Figure 5(II)), appears rough and sketch presence along certain direction which is the phase of delicate segment of polyurethanes. Meanwhile, in Figure 5(I) there are light designated aglomerates. These behaviours being suitable with analysis of the ratio of polar to non polar parts and thermal properties, indicate that the phase separation occurs on polyurethane parts and lignin parts.

4. Conclusions

Polyurethane formation reaction of lignin occur in the whole sample obtained to the functional groups of urethane and amide in the hybrid polyurethane structures, through analyzing NMR and FTIR spectrum. Compositions variation on the reagents such as HMDI and lignin leads to enhancement of the rigid segments, indicated by values of the ratio of hydrophilic as polar to hydropobic as non polar parts. This enhancement is suitable with tendency in the glass transition and degradation temperature enhcement, confirmed by observing morphology for the hybrid polyurethane samples. The findings may be able to promote potency of the hybrid polyurethanes as comptibilizing agent.

References

[1] Heitner C, Dimmel D 2010 *Lignin and Lignans: Advances in Chemistry.* Florida: CRC Press.
[2] Tuck C O *et al* 2012 *Science* **337** 695−699.
[3] Rozman H D *et al* 2000 *Eur. Polym. J.* **37** (1) 1483–1494.
[4] Glasser W G, Sarkanen S 1989 *ACS Symposium Series* **398** American Chemical Society, Washington, DC.
[5] Putra W N *et al* 2017 *Macromol. Symposia* **371** 140-143.
[6] Laurichesse S, Avérous L 2014 *Progress in Polym. Sci.* **39** 1266–1290.
[7] Zia K M *et al* 2007 *Reactive & Functional Polym.* **67**(8) 675-692.
[8] Izunobi J U, Higginbotham C L 2011 *J. Chem. Educ.* **88** 1098–1104.
[9] Lisperguer J *et al* 2009 *J. Chil. Chem. Soc.* **54** 460-463.
[10] Chalid M *et al* 2012 *Appl. Mech. Mater.* **229** 297–302.
[11] Chalid M *2011 Adv. Mater. Res.* **277** *112-119.*
[12] Chalid M *et al* 2013 *Adv. Mater. Res.* **789** 274-278.
[13] Chalid M *et al* 2015 *Polym. Plastics Technol. Eng.* **54** 234–245.
[14] Markham R L 1990 *Adv. Pol. Tech.* **10** (3) 231-236.
[15] Imre B, Pukanszky B 2013 *Eur. Polym, J.* **49**(6) 1215–1233.
[16] Yu C *et al* 2016 *Mater. and Design* **107** 171–177.
[17] Hatakeyama H *et al* 2013 *J Therm Anal Calorim* **114** 1075–1082.
[18] Ciobanu C *et al* 2004 *Industrial Crops and Products* **20** 231–241.

IOP Conf. Series: Materials Science and Engineering **223** (2017) 012034 doi:10.1088/1757-899X/223/1/012034

The effect of ratio chitosan-bentonite and processing time on the characterization of chitosan-bentonite composite

E Savitri, A Budhyantoro
Chemical Engineering, University of Surabaya

E-mail : savitri_ma@staff.ubaya.ac.id

Abstract. Chitosan/bentonite composite was synthesized to modify and enhance its characteristics to adsorp both anionic and cationic substances. The influence of several factors on synthesis of the chitosan/bentonite composite were investigated. The study focused on obtaining the best operating conditions that would produce homogenous chitosan/bentonite composite. The effect of weight ratio of bentonite to chitosan, reaction times and pre-treatment were investigated. The composite were characterized by Fourier transform infrared spectroscopy, X-ray diffraction and Scanning electron microscopy.The optimum chitosan/bentonite composite was obtained at weight ratio of bentonite to chitosan 1 : 2 and 12 hours reaction. On these conditions, more chitosan can interact with the matrix of bentonite to form chitosan – bentonite composite.

1. Introduction

Chitosan is the second largest biopolymer after cellulose. Chitosan is interesting because of its low cost, availability of amino and hydroxyl groups in the structure, which are potential to react [1]. However, the chitosan has limitation in their application due to its low surface area, and mechanic resistance [2]. In order to overcome the limitation, several works have been conducted on the development of chitosan based materials such as membranes [3], fibers [4], films [5] and scaffolds [2]. Chitosan film have prepared by Dotto *et al.,* (2013) to remove dyes from aqueous media. The film showed high adsorption capacities, good mechanical properties and easy for separation. Esquerdo et al. have developed chitosan scaffolds which has good characteristics for adsorption, including porosity of 92% and specific surface area of 1135 m^2g^{-1}. Chitosan based materials with good mechanical and functional characteristics, able to remove anionic and cationic dyes, are scarce. An alternative is the development of chitosan/bentonite composite, once chitosan is polycationic and bentonite has anionic groups in its structure. Another study showed that Gunister have prepared and characterized chitosan-montmorillonite composite [6]. They used chitosan which has an average molecular weight of 400,000. The composite was prepared by mixing chitosan solution in different concentration into montmorillonite suspension. chitosan is attached on the surface of clay particles. XRD analyses displayed that organic molecules did not enter enough sufficiently into the layers of clay structures. The effects of polymers on the montmorrilonite dispersions are widely studied [7-10]. Because of that, in this study we focused on the bentonite pre-treatment. Bentonite used in this study were both

IOP Conf. Series: Materials Science and Engineering **223** (2017) 012034 doi:10.1088/1757-899X/223/1/012034

intercalated and without treatment. By intercalating process, the interlayer of bentonite might be wider and the chitosan could enter into the layers of bentonite structuce. This work aimed to synthesized and characterized a composite based on chitosan and bentonite. Chitosan/bentonite composite was prepared and characterized by potential test of the suspension, Fourier transform infrared spectroscopy (FTIR), and X–ray diffraction (XRD).

2. Experimental

2.1. Materials

Natural bentonite was collected from Pacitan, East Java, Indonesia. Bentonite was grinded and sieved through 140 mesh. After sieving, Bentonite was intercalated by using HDTMA-Cl 1% as surfactant. Chitosan was produced by deacetylation of chitin in sodium hydroxide solution at 125°C for 3 h process. After deacetylation process, chitosan was washed by water and then dried.

2.2. Preparation of Chitosan – Bentonite

A definite amount of bentonite was dispersed in water and then chitosan in 1% acetic acid solution was added into the dispersion. The dispersion was mechanically stirred for 6 – 24 h at room temperatures. The composites having weight ratio of bentonite to chitosan 2 : 1; 1 : 1; and 1 : 2 were prepared. After processing, the composite was filtrated and dried in the oven.

2.3. Characterization of Chitosan-Bentonite Composite

The potential measurements were carried out using potential Instruments to identify the potential of ions in the suspension system. Bentonite has been identified as mainly smectite group minerals using X-ray diffraction (XRD) and FTIR analysis methods. X-ray diffraction (XRD) measurements were performed by a PANalytical PW 3373/00 X'Pert X-ray diffractometer, Netherlands which used CuKa, 1.54 Å radiation at 40 kV and 30 mA. Fourier transform infrared spectroscopy (FTIR 8400S Shimadzu,Japan) was performed using the KBr pellet with a concentration of 1 % in the range of 4000 - 400 cm^{-1}. All of the spectra were recorded at room temperature with a resolution of 4 cm^{-1} for 45 scans. Spectral outputs were recorded in transmittance mode as a function of wave number.

3. Results and Discussions

3.1 The Effect of Bentonite : Chitosan Ratio on the Composite

The Bentonite which is used to synthesize the composite were both fresh bentonite and also bentonite that has been intercalated with HDTMA-Cl.To study the effect of ratio bentonite to chitosan, the samples were characterized on the weight ratio of bentonite to chitosan 2:1, 1:1 and 1:2. The potential test was conducted on samples for bentonite suspensions, chitosan solutions and chitosan-bentonite composite suspension system, separately. The potential test is performed to determine the ions changes in the suspension system before and after the composite formed. The potential test data was shown in the Table 1.

Table 1. The potential data for bentonite suspension, chitosan solution, bentonite chitosan before and after composite processing

Bentonite – Chitosan ratio, g/g	Potential, mV				
	Bentonite suspension	Chitosan solution	Before processing	After processing	Δ Potential
B:C = 1:2	244	318	562	349	213
B:C = 1:1	260	318	578	359	219
B:C = 2:1	263	319	582	357	225
B-HDTMA:C = 1:2	246	318	564	354	210

Table 1 showed that the potential of bentonite suspension increased when the amount of bentonite added increased. Otherwise, because the amount of chitosan solution used was unchanged in number so that the potential of chitosan solution remained constant. Theoritically, the number of charge in the suspension system increase because of both materials. But, Table 1 showed that the potential of the system after process was lower than that of system before process. It meant that there is phenomena that the charges of both materials combine (negative charge attributed to bentonite and positive charge attributed to chitosan) so that the potential system after process reduced. The reducing charge because of the charge combination between bentonite and chitosan was described by potential difference (Δ Potential). Potential difference in this study might explain the amount of chitosan was attached on to bentonite in composite system. The higher the potential difference, the higher the composite developed. The potential difference (Δ Potential) of the composites increase with increasing the amount of fresh bentonite added into the composite system. The maximum potential difference was obtained at the weight ratio of bentonite to chitosan 2:1. Although the value of the highest potential difference was obtained by the weight ratio bentonite to chitosan 2:1, but the increase of potential difference is not proportional to the amount of added bentonite. Increasing bentonite as much as four time just gives insignificant increasing in potential difference of composite. In comparison, the ratio of bentonite to chitosan is 1:2 has been obtained from a composite with a potential difference is quite good value. Otherwise, the intercalated bentonite gives the lowest value of potential difference on the composite. The presence of HDTMA on the bentonite gives the limitation of chitosan molecule to interact with the bentonite system. HDTMA was attached in the multilayer of bentonite at pretreatment process and blocked the chitosan molecules to enter the bentonite multilayer system.

Figure 1 shows that chitosan peaks look similar to bentonite and chitosan–bentonite composite peaks in the range of 3400 - 3600 cm^{-1}. It is the characteristic of the -OH vibrating stretching asymmetric -OH bond at 3447.20 cm^{-1}.

On the FTIR spectrum, the peak in the range 500-700 cm^{-1} is Si-O bending vibration. The peak is attributed to bentonite that changed in the absorption before and after the process. The change in the peak indicated the evolution in the condition of the bentonite pores. The bentonite pores were more open than the pores of the initial bentonite. This is due to the inclusion of the chitosan molecules into the layer structure of bentonite.

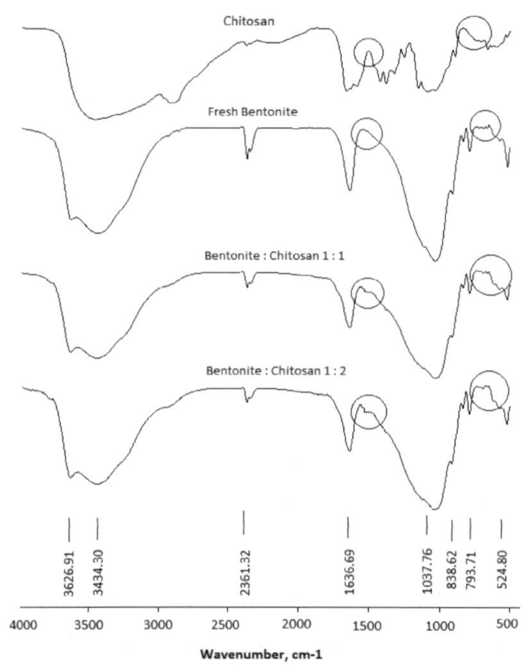

Figure 1. FTIR spectra of chitosan, fresh bentonite and chitosan-bentonite composite at different ratio for 6 h processing

Figure 1 also gives the FTIR spectra of material based chitosan-bentonite for different ratio of bentonite to chitosan. According the FTIR spectra, the ratio of bentonite:chitosan 1:2 was the optimum ratio rather than other weight ratio. It looks at the peak in the range 1500-1550 cm^{-1}. The peak at this wavenumber is the highest absorbance compared to the peak of other weight ratio of bentonite-chitosan. It referred to the higher amount of chitosan in the composite system. Eventhough, the weight ratio bentonite to chitosan 1:2 was optimum result compared to other system, but this result was not as expected. It caused that the amount of bentonite at that ratio is the smallest so that the number of bentonite pores provided is also limited. The limitation of the bentonite pore will limit the number of chitosan molecule to occupy the bentonite space. The weak peak in the range 1500-1550 cm^{-1} is a bending vibration of N-H bond of the amine group in chitosan. This weak peak indicates that the number of chitosan moleculesthat enter into the bentonite pore is still limited. It was probably caused by the high molecular weight of chitosan used and did not suitable to the bentonite pore size. The peaks in the range 850-1450 cm^{-1} are sharper than that of chitosan-bentonite composite. It might be the occurance of the binding system of the chitosan and bentonite.

3.2 Effect of The Processing Time on The Composite

To study the effect of the processing time on the composite, the samples were characterized on the ratio of bentonite to chitosan 1:2 for 3 to 24 h processing. The potential test datas are shown in the Table 2 and the FTIR spectras of material are shown in Figure 2

Table 2. The potential data for bentonite suspension, chitosan solution, bentonite chitosan before and after composite processing at different processing times

Processing Time, h	Potential, mV				
	Bentonite suspension	Chitosan solution	Before processing	After processing	Δ Potential
3	246	319	565	351	214
6	244	318	562	334	228
12	246	319	565	340	225
24	246	318	564	346	218

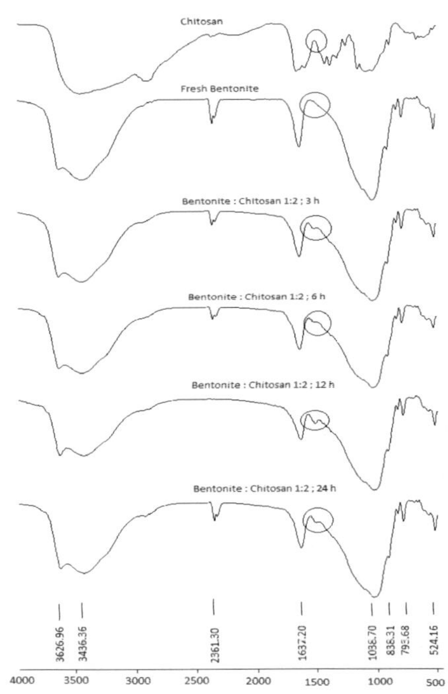

Figure 2. FTIR spectra of chitosan, fresh bentonite and chitosan-bentonite composite at different procesing times for bentonite : chitosan ratio 1:2

IOP Conf. Series: Materials Science and Engineering **223** (2017) 012034 doi:10.1088/1757-899X/223/1/012034

Figure 2 shows that the optimum condition of composites processing are for 12 h processing. The result was supported by the potential test datas that showed that processing time gave a high potential difference of the composite. It can be seen on Figure 2 at peak in 1500-1550 cm^{-1} that indicating the vibration of N-H bonds of the chitosan.

The data indicated that chitosan has interacted with the bentonite better than that of other processing time. At the time of 3-6 h, the chitosan probably have not attacthed strongly with the bentonite, while the 24 h processing time, the chitosan might occure desorption due to the prolong of processing time.

3.3. The Characterization of Chitosan-Bentonite Composite by XRD

In this section, the focus of the study was to determine how the effect of the variations on the bentonite-chitosan ratio and the length of processing time on the main peak shifting at 2θ = 5.828. The shifting peaks can be used to determine whether the chitosan molecule entered into the bentonite multilayers structure.

Table 3. XRD difraction of the fresh bentonite and chitosan-bentonite composite

Samples	2θ, degree	D_{001} spacing, °A	I-count
Fresh Bentonite	5.83	15.15	695
Composite B:C =1:2; 6 h	5.91	14.93	485
Composite B:C =1:1; 6 h	5.92	14.91	347
Composite B:C =2:1; 6 h	5.86	15.05	476
Composite B:C =1:2; 12 h	6.03	14.64	407

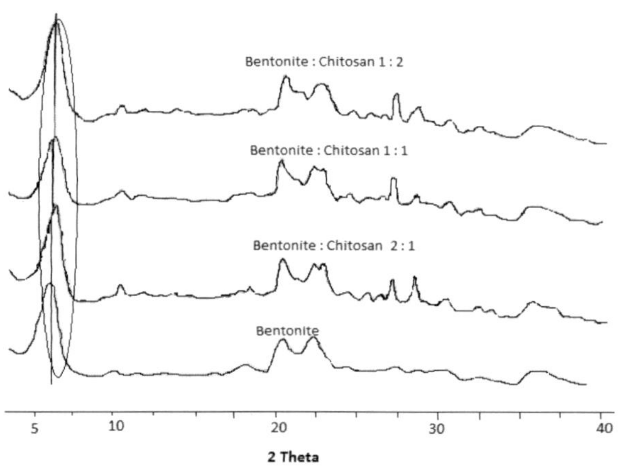

Figure 3. X-ray Diffractogram of fresh bentonite and chitosan-bentonite composite at different bentonite-chitosan ratio for 6 hours processing

The main peak shifting and d_{001} spacing changing are shown in Figure 3. It can estimate how large the bentonite pore size formation. The XRD Diffractogram of the samples showed the shifting of the main peak of the composite slightly to the right that showed the distance between the bentonite layers were getting smaller. The shifting peak was not significant, this indicated that only a small number of chitosan molecules entered into the pores of bentonite and was unable to change the bentonite interlayer distance. The chitosan molecules did not enter into the bentonite structure due to a high molecular size of chitosan molecules. However, the data FTIR has supported that chitosan has interacted with the bentonite eventhough the interaction might be on the outer surface of bentonite.

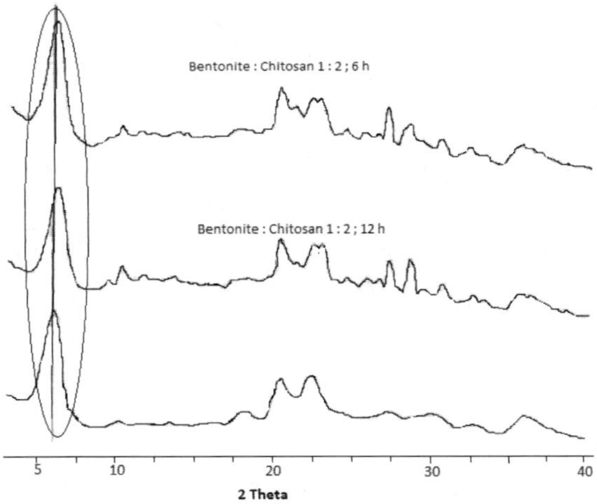

Figure 4. X-ray Diffractogram of fresh bentonite and chitosan-bentonite composite at different processing times at bentonite – chitosan ratio 1:2

According to the Figure 4, it can be observed that the composite results of both samples (6 and 12 hours process) shifted the main peak slightly to the right compared with fresh bentonite. The main peak shifting was not significant indicates that the number of entering chitosan was very low. Chitosan molecules did not change the bentonite interlayer distance. Both of samples looked like similar result. The main peak shifting was not significant.

4. Conclusions

In this research, the optimum conditions of the synthesis of composite chitosan - bentonite were at weight ratio of bentonite / chitosan 1: 2 and processing time of 12 h. On these conditions, more chitosan can interact with the matrix of bentonite to form chitosan-bentonite composite. The chitosan molecules probably located on the outer surface of the bentonite and not enter in the layer structure of bentonite.

References

[1] Wan Ngah W S *et al* 2011 *Carbohydr. Polym.* **83** 1446–1456.
[2] Esquerdo V M *et al* 2004 *J. Colloid Interface Sci.* **424** 7–15.
[3] Aliabadi M *et al* 2013 *Chem. Eng. J.* **220** 237–243.
[4] Mirmohseni A *et al* 2012 *Bioresour. Technol.* **121** 212–220.

[5] Dotto G L *et al* 2013 *Chem. Eng. J.* **214** 8–16.
[6] Gunister E *et al* 2007 *Carbohydr.Polym.* **67** 358-365.
[7] Alemdar A *et al* 2005 *J. Mater. Sci.* **40** 171–177.
[8] Alemdar A *et al* 2005 *Colloids and Surfaces A-Physicochem. Eng. Aspects* **252** 95–98.
[9] Billingham J *et al* 1997 *Vibrational Spectroscopy* **14** 19–34.
[10] Chang S H *et al* 1992 *J. Rheol.* **36** 273–287.

IOP Conf. Series: Materials Science and Engineering **223** (2017) 012035 doi:10.1088/1757-899X/223/1/012035

Unsaturated polyester/expanded polystyrene composite : thermal characteristics and flame retardancy effects

R Mohamed, S A Syed Mustafa, Mohd N Norizan, L S Amerudin
GREEN Polymer Research Group, Faculty of Applied Sciences, Universiti Teknologi MARA (UiTM), 40450 Shah Alam, Selangor, Malaysia

E-mail: ssyed_anas@hotmail.com

Abstract. Panels for energy efficient buildings has to meet certain requirements such as low thermal conductivity and inherent flame retardancy characteristics, before being eligible for buildings and construction applications. Expanded Polystyrene (EPS) as waste material had been incorporated as filler in Unsaturated Polyester Resin (UPR) composites. The composite are fabricated as flat panel window or glazing to replace glass. In this study, different EPS content incorporated was found to affect flammability and thermal characteristics. Core additives such as Flame Retardant (FR) and Antioxidant (AO) were added to the composite for imparting flame retardancy and prevent aging of the composite. The result obtained via the comparison of the various composite systems studied had revealed that organic and metal oxide flame retardant (FR) additives imparts higher flame retardancy levels than others, but each type of additives had interacted with the polymeric matrixes differently. The thermal conductivity, k value, as measured from a handheld thermal probe had showed a minimum of ~0.124 W/m.K for the 1%wt zinc oxide sample, while the highest k value of ~0.280 W/m.K was exhibited by the 2%wt tin oxide sample. The first 1%wt of either metal oxide FR initially decreases both the thermal conductivity, k value; and volumetric specific heat, $C_{p,v}$ of the samples. At 2%wt, increases in k value were obtained . The flammability was reduced with the use of organic Phosphate Ester FR, which had reduced the flame speed to about 37.1% of the original flame speed. With the equivalent mixture of all three organic FR system, the flammability reduced less than 30%, while the metal oxide FR additive doesn't reduces the flammability.

1. Introduction

All products applied in exterior glazing units must able to withstand the climate exposure during several decades of use; this includes the use of glass panelling or polymer sheets [1]. Aiding the conservation of energy usage are now a prerequisite for modern green materials. High insulation materials and daylight harvesting, for example, are some of the energy management techniques that reduces overhead lighting via utilization of the ambient (natural & artificial) light present in a space; flame retardancy characteristic is another prerequisite [2]. Green Technology can contribute to the glazing industry by the development of new materials using recycled plastic, which could also impart some insulation characteristic, while fulfilling the building requirement for flame retardancy.

Sustainable building is the practice of designing, constructing, operating, maintaining, and removing buildings in ways that conserve natural resources and reduce pollution. Rating systems

IOP Conf. Series: Materials Science and Engineering **223** (2017) 012035 doi:10.1088/1757-899X/223/1/012035

provide an effective framework for assessing most buildings' environmental performance. A good Green Building Rating System (GBRS) should cover key indicators reflecting a building's characteristics while keeping their performance in balance [3].

Composite employing Unsaturated Polyester Resin (UPR) matrix have been used for many years in a broad technology fields such as naval construction, offshore applications, waterlines, and building construction. UPR is the chosen thermoset material due to its excellent processing ability; good cross-linking tendency, as well have good mechanical and weathering properties. Studies had also been done to improve its thermal properties by incorporation of Expanded Polystyrene (EPS) into UPR with relevant diluents [4,5].

In the light for an alternative building material, EPS had been used as a filler for thermoset insulating panels; and tests were performed for thermal insulation studies and flame retardancy. Thus, the preparation and characterization of recycled EPS filled UPR composite systems had systematically been investigated. Core additives such as flame retardant (FR) agents, metal oxides and antioxidant (AO) for preventing aging were added to the UPR-EPS composite for imparting selected properties.

Flame retardant functions to increase the resistance of a material to ignition and, once ignited, reduces the rate of flame spread via combustion suppression by acting either through the vapour phase or the condensed phase by chemical and/or physical mechanisms [6,7]. The use of a flame-retardant additive may prevent a small fire from becoming a catastrophe, to which most fires involving polymer composites often quickly evolved as it consume the volatile combustible material generated from the thermal degradation of polymers. Flame ignition itself depends on numerous variables such as oxygen availability, temperature, physical and chemical properties of polymer [8,9].

The fire resistance of polyester based resins could improved via blending with flame retardant additives such as triphenyl phosphate and antimony trioxide; such of which are generally utilized for production of fume hoods, electrical equipments, building panels and navy boats as noted by Nazan, 2016 [5].

Flame retardants can be incorporated into polymeric materials either as additives or as reactive materials. Additive types are widely used by blending directly into the polymeric material. Problems such as poor compatibility, leaching and reduced mechanical properties are often the commonly encountered drawbacks of utilizing additive type FR. Reactive type FR overcome these shortcomings by directly co-polymering the FR with the polymer itself, thus making the flame retardancy quality inherent to the material itself. The main concern and limitation however of this method is the toxicity due to use of halogenated monomers [8–12].

IOP Conf. Series: Materials Science and Engineering **223** (2017) 012035 doi:10.1088/1757-899X/223/1/012035

Thus, this paper shall emphasise on the flammability and thermal aspects of a UPR/EPS composite system with varying organic and metal oxide FR additives utilized both with and without antioxidant inclusion.

2. EXPERIMENTAL

2.1 Materials and Method

The selected thermoset matrix material used for fabricating these composite systems is the Reversol UPR system, consisting of vinyl ester oligomers having a density of 1.12 g/cc, a viscosity of 450-600 cps with 41-44% styrene content. Methyl Ethyl Ketone Peroxide (MEKP) and Cobalt solution used as initiator and promoter each, were supplied together with the UPE resin by Revertex Sdn Bhd. These ingredients are similar to materials used by Rashidan, *et al.* [8], which was the basis for this study.

The EPS or Styrofoam filler was obtained from waste material; Irganox, an antioxidant was supplied by Ciba Geigy; Phosphate Ester, Silesquioxane, and Melamine as flame retardants were supplied by Tina Organics (P) Ltd. Plasticizers & Allied Chemicals; Zinc Oxide and Tin Oxide solutions were supplied by Merck KGaA.

2.2 Composite Fabrication and Testing

Recycled EPS, fixed at 10% parts by weight (%wt) of UPR resin, was blended together with the resin by utilizing a high speed agitation mixer until complete mixture dissolution was attained. The gaseous contaminates present were eliminated from the mixture via vacuum suction, while any solid contaminates were removed from the mixture after a gravity settling phase of 24 h.

Designated additives were then added to the mixture before the samples were cured. The FR content was varied between 0 to 2.5%wt. Each FR are designated a code for distinction. The organic-based FRs of phosphate ester, silesquioxane, and melamine were designated as FRA, FRB, and FRC respectively. Tin oxide and zinc oxide representing metal oxide-based FRs were respectively designated as FRD and FRE.

The AO content was set to 0.5 g for samples impregnated with organic FR additives, which was absent for the samples impregnated with metal oxide FR additives. The MEKP and cobalt solution usage was limited to just 2.5%wt and 1 drops respectively for each sample.

The samples were shaped using compressive moulding process onto an aluminium mould, where the mixture made to partially jellifies in the mould before any pressure was applied upon it. The gelling time was set to be about 20 to 30 min. Once sufficient gel forms, low pressure press were applied for 1 h at 70°C to retain the required shape, and complete curing was done by leaving the

IOP Conf. Series: Materials Science and Engineering **223** (2017) 012035 doi:10.1088/1757-899X/223/1/012035

sample outside the mould for 24 h at room temperature. These steps are also similar to the process used by the study of Rashidan, *et al.,* (2009) [8].

This paper was then written upon the conduction of sufficient qualitative analysis of the manufactured samples. The flammability testing of the sample was subjected to the procedures underlined under the ASTM D635 standard. This particular standard utilizes prepared specimen size of 125 mm x 20 mm x 3 mm. A fixed distance of 75cm for combustion was used to complete flame propagation, with the sample being placed to burn in a horizontal position. Flame initiation was performed by using a candle flame at room temperature. The linear burning rate were assessed from time taken to reach the specified distance. The flame speed as defined by its equation:

$$\text{Linear Burning Rate, } V(\text{mm/s}) = L/t$$

Where: L = the burn length between two reference marks, in millimetres (The designated length is 75mm)
t = the time(s) (The flame's elapsed time between the two reference mark)

The samples with metal oxide FR additives were also tested via a thermal probe to determine their thermal conductivity, thermal diffusivity and volumetric heat capacity. The thermal properties of samples were determined by use of a KD2 Pro Hand-held Thermal Probe which adheres to ASTM Standard D5334-08 and IEEE Standard 422-03.

3. Result and Discussion

3.1. Thermal Characteristics

Owing to unavoidable circumstances, only the latterly produced metal oxide FR samples of FRD (tin oxide) and FRE (zinc oxide) were available for thermal characteristic testing via a thermal probe in determining their thermal conductivity, thermal diffusivity and volumetric heat capacity. These thermal characteristics are oft-considered qualities in green materials especially those involved in the construction industry. A low thermal conductivity, and the resulting thermal diffusivity, helps retain heat from dissipating against the end-users' needs.

Table 1. Thermal properties comparison of metal oxide FR.

Samples Code	Additive in UPE/EPS Composite	Conductivity W/m K	Diffusivity mm^2/s	Volumetric Specific Heat $MJ/m^3 K$
FRD 00	UPR-EPS Mix.	0.193	0.096	2.008
FRD 05	0.5%wt FRD	0.190	0.094	2.030
FRD 10	1.0%wt FRD	0.172	0.089	1.921
FRD 20	2.0%wt FRD	0.280	0.089	3.134
FRE 05	0.5%wt FRE	0.181	0.090	2.006
FRE 10	1.0%wt FRE	0.124	0.091	1.368
FRE 20	2.0%wt FRE	0.147	0.098	1.505

The thermal conductivity indicated by Table 1 showed a minimum of 0.124 $Wm^{-1}K^{-1}$ as recorded by the FRE 10 sample, while the FRD 20 sample recorded the highest at 0.280 $Wm^{-1}K^{-1}$. Figure 3 below shows how increasing the amount of FRD/E decreases both thermal conductivity and volumetric specific heat before an excessive additives content instead reverses the trend.

The initial decrement could be postulated as an effect of the additives agglomerating caused by its insufficient amount [13], which may lead to the resin having larger contact on the probe, and thus, a larger impact on the heat flow. The latter reversal could be achieved due to the inherent metallic nature of the metal oxide FR used. Increasing its amount facilitates the heat flow which hence recorded as an increase in the latter samples.

In contrast, their thermal diffusivity is a compound characteristic obtained from the ratio of thermal conductivity to volumetric specific heat. This property measure thermal inertia in that how fast the material changes temperature due to its thermal conductivity and specific heat. In the construction industry, a material with low thermal diffusivity are preferable as the quality may indicates a material which could withstand noticeable environmental temperatures changes.

As such, FRD 10 had shown characteristic of a good green building material due to its relatively low thermal conductivity and specific heat give birth to its low diffusivity. This sample might be suitable for tropical climate such as in Malaysia.

Surprisingly, FRD 20 also sports a similar low diffusivity of FRD 10 despite its conductivity and specific heat are the highest of the set. This hypothetically makes FRD 20 a suitable alternative to FRD 10 for use in other type of climates where the temperature changes are faster or with more extreme temperature ranges.

All in all, the inclusion of metal oxide FR improves the thermal resistance quality of the polymeric system by lowering both thermal conductivity and specific heat until a certain threshold. This reduction effect may occurs due to the trapped and initially agglomerating metal oxide particles might had created thermal barrier bubbles acting similar to the air trapped in foams slows heat conduction. Further study are required to determine the validity of this hypothesis.

3.2 Flammability Test

Table 2. shows the time it took to burn the sample for a fixed distance.

Table 2. Linear burn rate comparison of different FR additive composition.

Sample Code	Additive in UPE/EPS Composite	Elapsed time(s)	Linear Burning (mm/s)	Linear Burning (mm/min)
F00	0.5%wt AO	510.6	0.147	8.813
FRA	2.5%wt FRA - 0.5%wt AO	812.4	0.092	5.527
FRB	2.5%wt FRB - 0.5%wt AO	709.8	0.106	6.340

FRC	2.5%wt FRC - 0.5%wt AO	627.6	0.120	7.170
FRABC	2.5%wt FRABC - 0.5%wt AO	805.2	0.093	5.589
FRD 00	Similar to F00; AO absent	408.3	0.184	11.021
FRD 05	0.5%wt FRD	198.1	0.379	22.716
FRD 10	1.0%wt FRD	178.5	0.420	25.210
FRD 20	2.0%wt FRD	152.3	0.492	29.547
FRE 05	0.5%wt FRE	184.7	0.406	24.364
FRE 10	1.0%wt FRE	165.1	0.454	27.256
FRE 20	2.0%wt FRE	100.4	0.747	44.821

Based on the Linear Burning formulation, the linear burning rate decreases as a longer elapsed time is recorded. As Figure 1a indicated, the burning time taken for the three types of organic FR shows how each FR affects the flame speed/burn rate either individually or in tandem.

FRA, Phosphate Ester, shows the best performance as a flame retardant additive for the system at a 37.1% flame speed reduction; even an equivalent mixture of all three in the FRABC system was second best in relation to it. FRC or melamine, on the other hand, was the least efficient FR additive for this polymeric system, although it does boast a fair 18.6% reduction in flame speed.

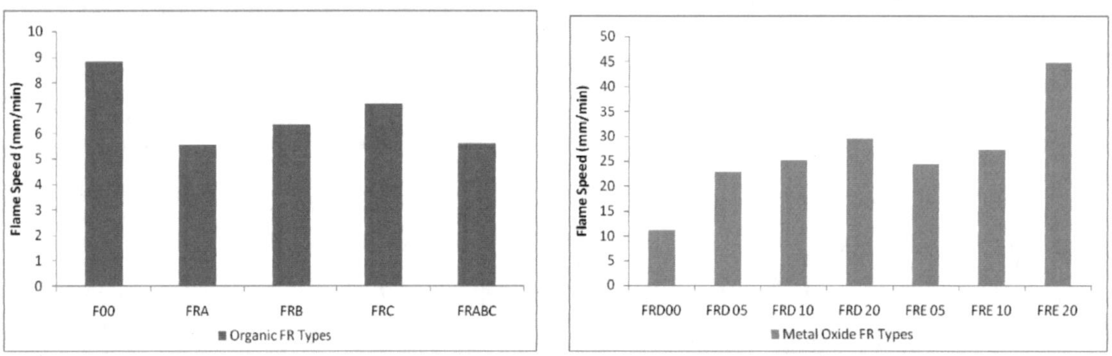

a b

Figure 1. Linear flame speed comparison of a) Organic FR types, and, b) Metal Oxide FR levels.

As shown in Figure 1b, the use of FRD and FRE (Tin and Zinc Oxide respectively) instead showed some failure in achieving any flame retardancy effects for the polymeric system. Although the absence of AO may enable an easier ignition and a faster flame rate as seen in the difference of F00 and FRD00 in Table 2 above, each increment of either material increases the flame speed. The introduction of either into the system triggers a doubling in flame speed as compared to the organic FR, although subsequent amount only adds a small flame speed boost except the result as shown by the outlier of FRE 20 that significantly adds flame speed.

In regards to the effect of the presence or lack of AO in the systems upon material ignition and flame speed, previous studies had implied that synergism in the gas phase reaction between organic

FR and AO could have existed. The synergism mechanism may involved evolved gases, which could be intercepted by AO and causes retardancy in smoke formation and flame suppressants; which may be ascribed to the inhibiting radical formation at gas phase level [9,14,15].

In comparison, metal oxide employed by Feng *et al.,* (2016) [16] showed that thermoplastic incorporated with CeO_2 and FR resulted in better flame retardancy behaviour. The CeO_2 played a pivotal role in the increase of LOI value of the intumescent flame retardant polypropylene composites, and presents excellent synergism. The interaction between CeO_2 and IFR leads to the formation of numerous closed cells without holes and formation of a compact and homogeneous char layer during burning was revealed. Similar findings are also reported by Kandare *et al.,* (2006); Katsoulis *et al.,* (2011); and Naik *et al.,* (2013) [17–19].

A probable explanation to the flame speed disparity between organic and metal oxide FR might be in the manner which both FR types were integrated into the polymeric systems. Unlike the organic FR which often covalently bonded directly into the matrix, the metal oxide FR requires a suspension system to enable through mixing of the particle into the polymeric matrix. These suspensions usually utilizes relatively flammable solutions such as toluene which are counter-productive to the task expected of the additive. Until further study to determine this hypothesis, this shall be best reasoning to explain this data disparity.

4. Conclusion

In conclusion, by utilising suitable FR that easily bond with the polymeric system; the flammability, flame speed, and thus its fire hazard maybe reduced up to nearly 40%. Otherwise the effects desired may not be obtained and may even backfired due to the low capability of the additive to the system. Thus, FR compatibility should be a concern for its use.

In terms of the studied thermal characteristics, the inclusion of metal oxide FR had improved the thermal resistance quality of the polymeric system by lowering both thermal conductivity and specific heat until an observed limit. A study on how this quality improvement actually occurs within the polymeric matrix is required.

Acknowledgement

I would like to acknowledge the technicians, and staffs for their assistance during the composite fabrication and testing phases; and to my colleagues and lecturers for discussions related to this work. Special acknowledgement for Assoc. Prof. Rahmah Mohamed for her guidance and direction in producing this publication. I also wish to thank Universiti Teknologi MARA (UiTM) for their research funding under the Research Intensive Faculty (RIF) fund [File:600-RMI/DANA 5/3/RIF (64/2012)]

References

[1] Jelle B P *et al* 2012 *Sol. Energy Mater. Sol. Cells* **96** 1–28.
[2] Eames P C 2008 *Vacuum* **82** 717–22.
[3] Nilashi M *et al* 2015 *Knowledge-Based Syst.* **86** 194–209.
[4] Mohamed R *et al* 2013 *Adv. Mater. Res.* **664** 600–4.
[5] Avcioglukalebek N 2016 *Int. J. Adv. Res. Sci. Eng. Technol.* **3** 1355–61.
[6] Lu S Y and Hamerton I 2002 *Prog. Polym. Sci.* **27** 1661–712.
[7] Laoutid F *et al* 2009 *Mater. Sci. Eng. R Reports* **63** 100–25.
[8] Khosim R *et al* 2009 *Malaysian Polym. Int. Conf.*
[9] Burchill P J and Pearce P J 1996 *Polym. Mater. Encycl.* 2204–2210.
[10] Kicko-Walczak E 1999 *Polym. Degrad. Stab.* **64** 439–42.
[11] Pereira C M C *et al* 2009 *Polym. Degrad. Stab.* **94** 939–46.
[12] Glodek T *et al* 2008 *Compos. Sci. Technol.* **68** 2994–3001.
[13] Fadhil M M, Abood A N and Fahan M M 2015 Mechanical Properties of TiO 2 Reinforced .Polystyrene- Modified Unsaturated Polyester **7** 109–19.
[14] Wilkie C A 2000 *Synerg. Adjuv. Antagon. FR Syst.* **4.**
[15] Kandare E *et al* 2010 *Fire Mater.* **34** 21–38.
[16] Feng C *et al* 2016 *Thermochim. Acta* **627–629** 83–90.
[17] Kandare E *et al* 2006 *Polym. Degrad. Stab.* **91** 1209–18.
[18] Katsoulis C *et al* 2011 *J. Fire Sci.* **29** 361–83.
[19] Naik A D *et al* 2013 *Polym. Degrad. Stab.* **98** 2653–62.

IOP Conf. Series: Materials Science and Engineering **223** (2017) 012036 doi:10.1088/1757-899X/223/1/012036

Utilization of bagasse and coconut fibers waste as fillers of sandwich composite for bridge railway sleepers

K A Soehardjo, A Basuki
Center for Material and Technical Product Researcher, Bandung, Indonesia

E-mail : koentariadisoehardjo@gmail.com

Abstract. The bridge railway sleepers is an essential component in the construction of railways, as the foundation of the rail support in order to withstand the load a train that runs above it. Sleepers used in bridge construction are expected to have a longer service life, lighter weight and durable so that can be used more efficient. This research was carried out to create a model of bridges railway sleepers made of sandwich structured composite from fiber glass, epoxy resin with fillers waste of bagasse (sugar cane pulp mill) or coconut fiberboard (copra industry) that using polyurethane as an adhesive. The process of making was conditioned for small and medium industrial applications. Railway sleepers' specifications adapted to meet the requirements of end user. The process steps in this research include; lay-up fiberglass combined with bagasse/coconut fiberboard (as fillers), gluing with epoxy resin, molded it with pressure to be solid, curing after solidification process. The specimens of composite, bagasse and coconut fiber board were tested for tensile and compressive strength.The prototype were tested of mechanical test: flexural moment test to the stand rail, flexural moment test to the middle of the sleepers and tensile strength test on one side of the sleepers, in accordance to SNI 11-3388-1994 Method testing of single block concrete sleepers and bearing single rail fastening systems. The results of mechanical testing all variations meet the technical specifications of end user such as test results for flexural moment on all prototypes, after load test, there is no visible crack. While in the tensile strength test, it seem the prototype with coconut fiberboard filler, shows better performance than bagasse fiberboard filler, the decisions is just depended on techno economic and lifetime

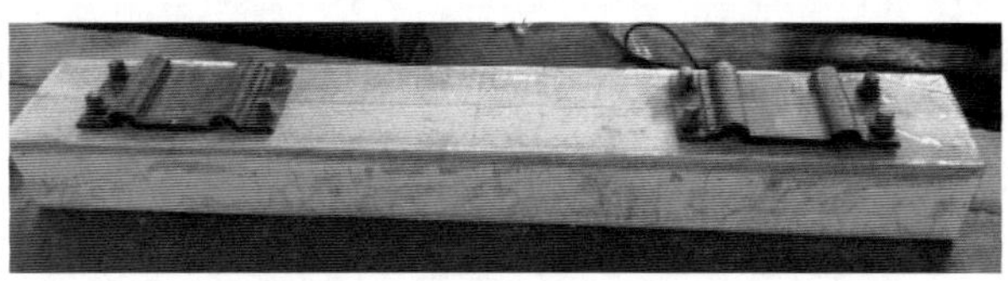

Figure. Bridge railway sleepers

1. Introduction

Sleepers of the railway bridge made of hefty wooden. In indonesia, the demand of wooden sleepers is 12.371 unit per year, it came from Java and Sumatra region. The source of timber is restricted nowadays, it effects the price of wooden sleepers and difficult to get, especially for size of sleepers 18 by 24 by 180 cm or 22 by 24 by 200 cm. This situation increases act of illegal logging in Indonesia. This research purpose is to find a solution for these matter, a substitute material which can easily reproduced. Composite material which combined in shape of sandwich structure is one solution from several alternative [1]. It has an improvement aspect, such as ; weather resistance, better compresive, tensile and flexural strength, lighter compared to concrete sleepers. This sandwich structure can consist of glass fiber, filler component (bagasse or coconut fibres, in shapes as a fiberboard) combined with epoxy resin as a matrix. Coconut fibers is natural resource which placed Indonesia as 2nd coconut producer in the worlds. Bagasse is a side-product of sugar mill which haven't optimized in Indonesia [2]. In consideration of manufacturing technology is appropriate technology, as well as the acquisition of raw material is quite easy and cheaper. The research that has been done is bridge railway sleepers from sandwich structure composite. Sandwich structure made up of a minimum of three parts, namely laminate top; core and laminate bottom. As it has been known that a plate, when the thicker plate is bent required the greater force compared to thinner plates. This happens because they relate to the value of the moment of mass inertia. Moment of mass inertia of thick plates is greater than the thin plate [3]. In the sandwich structure, the core is useful to strengthen the structure by means of insert between the two laminate that is a laminate top and bottom laminate. In addition to increasing the flexural strength with a layered arrangement of the mechanical structure is more resistant to damage rate. Another advantage gained by using a sandwich structure when compared to the solid structure (full laminate) is flexural rigidity sandwich will increase 8 times higher than the flexural rigidity solid [4]. Laminate structure composed of sheets of glass fiber piled such that the dough is mixed with an epoxy resin [5], hardener by way of lay-ups. At preparation, to achieve the optimal of tensile strength, compressive strength and flexural strength, fiber orientation (0°, 45°, 90°) is set properly. Preparation of fiber orientation, the number of layers of fiber, the type of glass fiber construction, arrangement and the pile fiber volume fraction between the fiber and the matrix, pressure is needed to regulate and control the fiber volume fraction [6,7]. It is no less important is the pressure and lay up process must be equitable to reduce the air or gas which may be trapped within the composite cause voids in the composite due to trapped air or gas that would weaken the structure. The void will be the initial cracks or delaminating of the structure receiving cyclic loading. Void will also cause stress concentration at the time of receiving the structure of static load. An epoxy resin matrix is best for all kinds of fibers in composites manufacture; it is there are several reasons including: the epoxy resin is attached very well with many kinds of fillers, reinforcing agent and the substrate. The chemical structure of epoxy resin is the result of reaction between diepoxy and diamine joint together and going crosslink to form large molecules and a substance that is hard and very strong [8,11].

2.Material and Methods.

2.1 Material and Equipment.

Materials are epoxy resins, epoxy hardener, epoxy thinner, catalyst, glass fiber S.B (Stitch Bonded), Woven Roving (WR) 200, WR400, WR600 and WR800) [7], fiberboard bagasse / coconut fibers, polyurethane thinner cleanser The equipment used is mold and pressure regulating device to sleepers of the railway bridge in the form of test specimens and prototype.

2.2 *Methods.*

The First Preliminary Study.

Experiment that have been done in autoclave to set the condition of process, used variations of fabric construction, orientation, number of layers, the specimens size 10x300x400mm with specimen code : 1.7SB (room), 2.6SB (60°C), 3.40WR200 (room temperature , 4. 20WR400 (room temperature), 5.20WR400 (30°C), 6.20WR400 (60°C), 7.8WR200 + 14WR400 (30°C), 8.16WR200 + 6SB (room temperature), 9. 2WR200 + 7SB (room temperature), 10.4WR200 + 2SB + 6WR400 (room), 11. 4WR200 + 2SB + 10WR400 (room temperature), and then Mechanical Properties testing: Hardness, tensile strength, flexural strength, evaluation and analysis of data test and Conclusions.

The Second Preliminary Study.

1. Specimens Size 25x300x400mm with orientation fabric 0^0, 90^0.
   a.100WR200, b. 67WR400, c.44WR600, d. 33WR800, e. 20SB
2. Specimens size 10x300x400mm with orientation fabric 0^0, 90^0
   a.40WR, b. 27WR400, c.18WR600, d. 13WR800, e. 20SB,

The Third Preliminary Study

1. 2 WR200 + 4 Stitch Bonded+ 2WR200, lay up epoxy resin, with total thickness of 6 mm

2 .1 WR400 + 4 Stitch Bonded+ 1 WR400, lay up epoxy resin with a total thickness of 6 mm

3 .3 WR200 + 2 Stitch Bonded+ 3 WR 200, lay up epoxy resin with a total thickness of 6 mm

Making of Bagasse Fiberboard

Bagasse filtration to separate the fine fiber and coarse fiber

Every type of bagasse fibers blended with 13% polyurethane and 10% MC catalyst and then press 2 minutes using mold, next make sandwich : layer I: fine fiber bagasse dough (15%), layer II: coarse fiber

IOP Conf. Series: Materials Science and Engineering **223** (2017) 012036 doi:10.1088/1757-899X/223/1/012036

bagasse dough (70%) and layer III: Fine fiber bagasse (15%), density 250kg / m^3 Press time 3 min, steam 170° C, release, cutting & coating (size 20x140x1800 mm)

Making of Coconut Fiberboard

Coconut fiber is described to be separated between the long fibers, short fibers and coco pith, then weighed 90%, was added 10% poly urethane, 10% MC catalyst and then mixing in mold arrangement with density 200 kg/m^3 , pressed 2 min , steam 170°C, curing with blower 10-26°C 2 days, press about 2 min, release, cutting and coating (size 20x140x1800 mm).

Making of Bridge Railway Sleepers from Composite Sandwich Structure with Filler Sandwich of Bagasse or Coconut Fiberboard

Fiberglass lay-up process layer by layer in the mold for composite filling left and right sides 4WR200, 28 SB,4WR200 (Size 40X140x1800mm), laminate bottom 2WR20, 14 SB, 2 WR200 (size 20x240x 1800mm), inserting 5 pieces bagasse or coconut fiberboard as core/ filler sandwich (size 20x140 x1800mm), Fiberglass lay-up process layer by layer in the mold for laminate top (size 20x240x1800mm), variation using (2WR20, 14SB, 2 WR200), pressed, curing at room temperature one day/ curing using autoclave at 60°C 1 h, finishing, nut and bolt installation. Bridge railway sleepers mechanical testing.

3. Results and Discussion.

3.1 The First Preliminary Study.

Result of Fiberglass Composite Mechanical Testing in the First Preliminary study shown in table 1. From the evaluation and analysis of test data can discuss things as follows: Table 1 shows the relationship between the number of layers, temperature and time of curing on the results of the testing of tensile and flexural strength and hardness by grouping variety of fabrics of glass fiber which is stitch bonded, woven roving 200 (WR200) and woven roving 400 (WR400). Can be explained that the characteristics of the fabric is determined by the fabric construction, namely: Stitch bonded a fabric of glass fiber yarns coated with long fibers were shown irregularly, intended to improve the physical properties of the fabric. Woven roving is a fabric of glass fiber yarn, which in WR200 has a finer cross-section of a cross than WR400, WR200 so widely used as the outer layer of the composite to produce a smooth surface [6,8] .

Test result from Table 1, codes 3, 4, 5 and 6 which has a less number of layer composite 20 WR400 but have a good result compare to composite 40WR200, especially for hardness, tensile and flexural strength

IOP Conf. Series: Materials Science and Engineering **223** (2017) 012036 doi:10.1088/1757-899X/223/1/012036

value. This condition is caused by a larger cross-section of WR 400 compared to WR200. WR 400 consist of pile of fabric which bulkier, contain more pores of microfibril and thicker than WR200 (0.4 mm compare to 0.2 mm). During impregnation process, the epoxy resin will enter and disperse into pores of microfibril fiber.

Table 1. Mechanical Testing Result of Fiberglass Composite at the First Preliminary study

Code	Type	Tempature & Time	Orientation of fabric (°)	Tensile strength (kg/cm^2)	Flexural strength (kg/cm^2)	Hardness (HRr)
1	7 SB	room 1 day	0/90	1524.97	3122.3	120.0
2	6 SB	60°C 1 h	0/90/45	1893.99	**3216.1**	112.7
3	40 WR200	room 1day	0/90	1198.78	2374.1	116.7
4	20 WR400	room 1day	0/90/45	1763.51	2982.7	117.2
5	20 WR400	30°C 1 h	0/90	1320.08	2801.2	117.7
6	20 WR400	60°C 1 day	0/90/45	1612.64	2877.7	121.0
7	8 WR200 14 WR400	30°C 1 day	0/90/45	1620.80	2324.2	118.7
8	16 WR200 6 SB	room 1 day	0/90/45	1996.94	2252.8	105.0
9	2 WR200 7 SB	room 1 day	0/90/45	**2064.22**	2079.5	**123,0**
10	4 WR200 2 SB 6 WR400	Room 1 day	0/90/45	1997.96	2909.3	104.3
11	4 WR200 2 SB 10 WR400	Room 1 day	0/90/45	1569.83	1932.7	103.5

This will form better polymerization so it produces better bond between polymer epoxy resin and glass fiber. The composite will have solid and stronger bond which increase the value of hardness, tensile and flexural strength. WR 200 has smaller cross section, more dense and smaller pores between microfibril. During impregnation process the epoxy resin will be more difficult to enter and cause fewer chemical bonding between epoxy resin and glass fiber. This will form the composite less solid and less robust. The composite of WR 200 produces hardness, tensile and flexural strength values less than composite of WR400. Table 1 code 4, 5 and 6 shown that polymerization of the epoxy resin and glass fiber using 20WR400 at room temperature need curing time up to 1 day . The shorter curing time due to the higher temperature, is needed to improve production efficiency. This experiment was done by varying the temperature of 30 and 60°C. At temperature curing of 60°C for 1 h, the polymerization process will be more perfect than curing at room temperature as those in the code 6, and code 5. Consequently, testing

IOP Conf. Series: Materials Science and Engineering **223** (2017) 012036 doi:10.1088/1757-899X/223/1/012036

results of hardness, tensile strength and flexibility are high enough. In terms of tensile and flexural test, when compared between code 5 with of orientation $0^o/90^o$ and code 4 and 6 with orientation of $0^0/90^0/45^0$, the overall orientation of fabric greatly affect the results of physical tests where the orientation of the combined $0^o/90^o/45^o$ better than the orientation of $0^o/90^o$. The composite of epoxy resin and glass fiber are composite types of stitch bonded, WR200 and WR400. Test results are shown in Table 1 of the code 7, code 8, code 9, code 10 and code 11. The amount of the stitch bonded layer which gives more hardness and higher tensile test, can be seen in code 9 (2WR200 7SB) with the highest grades of hardness and tensile. It is because stitch bonded fabric consists of coated glass fiber (long fiber) were shown irregular resulting in such fabrics have excellent strength in diagonal direction, especially composites with $0^o/90^o/45^o$ orientation. Additionally, stitch bonded more bulky than WR200 and WR400, so that when the epoxy resin is impregnated it can be more evenly distributed among the pores of the fiber micro fibrils well. Therefor during the curing process epoxy resin will polymerize well so as to have a good physical test results. In this experiment epoxy resin used is type 3, namely the type of diglycidyl ether of bisphenol A (Bakelite EPR 174) in the presence of a catalyst (curing agent) types polyaminoamide (Bakelite EPH 340) will be formed polymerizes to form composite [11]. The composite quality was also affected by temperature and time of curing; epoxy resin type 3 can be polymerized at room temperature up to about 60^oC. Provision of heat at the time of polymerization allows the voids (air trapped in the composite). The Void is an initial crack at the time of testing physical/mechanical. Voids can also occur during the impregnation into glass fiber, coating overcrowding will cause epoxy resins difficult to enter the pores of the glass fiber micro fibrils. Hardness test results on code 10, 8 and 11 had low values. This may be due to the variation that a layer number is high enough, so that when the epoxy resin impregnation difficult to enter the pores of the resin means micro fibrils epoxy is not well dispersed into the glass fiber, which resulted in a void that is the initial crack, so that by the time their hardness testing void will make the test result is lower than that void. Thus the hardness and tensile test results have a higher value because more dominant fiber strength is supported with a combination of fiber orientation $0^o/90^o/45^o$ [10].

3.2 The Second Preliminary Study.

The effect of variety of construction and the number of layers of glass fiber on tensile strength (kg/mm^2), flexural strength (kgf/mm^2), compressive strength (kg/mm^2) and hardness (R-HRr) test result can be seen at Table 2 [6,8].

IOP Conf. Series: Materials Science and Engineering **223** (2017) 012036 doi:10.1088/1757-899X/223/1/012036

Table 2. Mechanical Testing Result of Fiberglass Composite at the Second Preliminary Study

Code	Tensile Strength (kg/cm^2)	Flexural Strength (kgf/cm^2)	Compressive Strength (kg/cm^2)	Hardness (R-HRr)
WR200	1883	**3227**	**3208**	112
WR400	2395	2558	1790	**118**
WR600	2017	1935	2162	**118**
WR800	1757	1773	2004	105
SB	**3324**	3222	1671	96.5

The highest tensile strength as achieved on code SB of 3324 kg/cm^2, while the highest flexural strength test results is achieved in code WR200 of 3227 kgf/cm^2. The highest compressive is achieved on code WR200 of 3208 kg/cm^2 and the highest hardness achieved on WR400 and WR600 118 HRr. This can be explained as follows: Characteristics of composites generated from these experiments is determined by a two-phase forming composite:

First phase strong (reinforcement), in this experiment used glass fiber in the form of fabric with construction of WR 200, WR 400, WR600, WR800 and Stitch bonded [11] and the second phase binder (matrix) used epoxy resin. It is also affected by temperature and curing time, epoxy resins can be polymerized at room temperature up to about 60°C [12]. Curing at room temperature necessary drying time up to 24 h with elevated curing temperature curing time is expected to be shorter to improve production efficiency. The experiments have been done at 60°C. Provision of heat at the time of polymerization allows the voids (air trapped in the composite). The Void is an initial crack at the time of testing physical/ mechanical [13]. Voids can also occur during the impregnation into glass fiber, coating overcrowding will cause epoxy resins difficult to enter the pores of the glass fiber micro fibrils. From the above data value tensile strength and flexural strength highest in the testing code SB because It is because stitch bonded consists of a cloth coated glass fiber (long fiber) were sown irregular resulting in such fabrics have strength diagonal direction is excellent, especially composites this orientation has $0^0/90^0/45^0$. Additionally stitch bonded more bulky than woven roving, so that when impregnated epoxy resin can be evenly distributed among the micro-porous fiber fibrils well, during the curing process will polymerize well so as to have the results of tensile strength and flexural strength are good. Hardness values are closely related to the compressive strength which in this experiment the best value achieved in the code WR200 use WR200 has a thick fabric that is thinner so the use of more fiber, specimens were produced having comparisons % of the fibers is higher than % resin, this causes the composite generated more solid, so as to have hardness and compressive strength is higher.

3.3 The Third Preliminary Study

Mechanical Testing Result of Fiberglass Composite in the Third Preliminary Study show in table3

Table 3. Mechanical Testing Result of Fiberglass Composite at the Third Preliminary Study

Code	Tensile Strength (kg/cm^2)	Flexural strength (kgf/cm^2)	Compressive Strength (kg/cm^2)
1	2487.6	5527.9	**3654.75**
2	1922.9	**6080.0**	2848.3
3	**2612.8**	**6080.0**	1909.99

The tensile strength, the flexural strength and the compressive strength test results do not really give a significant difference this means all variations gave no effect on the tensile strength, flexural test compressive strength of the composite of glass fiber and epoxy resin, it is because the number of layers of glass fiber in an amount equal only distinct variation amount 2 layer WR 200 more give the tensile, flexural and compressive strength is slightly higher. Therefore for making bridge railway sleepers, use code 1. From this research obtained information that WR 200 will provide high tensile strength properties and provide a smooth surface material, while Stitch Bonded will give flexural strength properties and high compressive strength. Therefore, the manufacture of products Railway Bridge sleepers WR200 and SB use of glass fiber and epoxy resin. The mechanical properties of coconut fiberboard have an higher the mechanical properties of bagasse fiberboard.

3.4 Mechanical Testing Result of Coconut and Bagasse Fiberboard at the Third Preliminary Study

Table 4. Mechanical Testing Result of Coconut and Bagasse Fiber board
at the Third Preliminary Study

code	Tensile strength (kg/cm^2)		Flexural strength (kgf/cm^2)		Compressive strength (kg/cm^2)	
	Coconut fiber	Bagasse fiber	Coconut fiber	Bagasse fiber	Coconut fiber	Bagasse fiber
1	10.42	7.81	11.34	8.51	3.11	2.33
2	9.70	7.28	11.85	8.89	3.20	2.40
3	9.88	7.41	10.15	7.61	3.07	2.30
4	8.32	6.24	9.97	7.48	3.17	2.38
5	8.78	6.59	10.65	7.99	3.40	2.55

In Table 4 seen that bagasse fiberboard have lower mechanical properties than coconut fiberboard on tensile strength, flexural strength and compressive strength , therefore if the bagasse or coconut used as a sandwich filler for Bridge Railway Sleepers, the mechanical properties of coconut fiberboard will have an higher mechanical properties than bagasse fiberboard, it is because coconut fiber have a longer fiber and higher tensile strength than bagasse, and also bagasse fiber more brittle than coconut fibers.

Mechanical Testing Result of Bridge Railway Sleepers from Glass Fiber and Epoxy Resin Material Composite with Sandwich Filler of Bagasse or Coconut Fiberboard.

The Static Load Bridge Railway Sleepers

Table 5. Testing Result of the static load on the Bridge Railway Sleepers

No	Type Testing	Design Moment (kg.m)	Design load (kg)	Description	Requirements
1	Seated vertical load rail (Type I)	1500	13139	no structural cracks	fulfilled
2	Seated vertical load rail (Type II)	1500	13139	no structural cracks	fulfilled
3	Moment at midspan (Type I)	930	3750	no structural cracks	fulfilled
4	Moment at midspan (Type II)	930	3750	no structural cracks	fulfilled

Static load test performed on Bridge Railway Sleepers for each type I is Bridge Railway Sleepers using bagasse fiberboard as filler and type II is bridge railway sleepers using coconut fiberboard as filler Results of testing the static load on the Bridge Railway Sleepers can be seen on table 5 shows that the bridge railway sleepers that were tested did not experience structural cracks during the design moment and when the moment of the design work for 3 minutes, so that the the bridge railway sleepers are eligible

Lift Test Rail Fastening Systems the Bridge Railway Sleepers

lift Test rail fastening systems carried on one of the stand bearing the number 2 for each types (type I and II) shows Table 6 Testing result of the lift Test rail fastening system

Table 6. Testing Results of Lift Test Rail Fastening Systems of the Bridge Railway Sleepers

Item	Weight	Specimen Type I	Specimen Type II
Load measured at separation (P)	kg	850.00	1031.25
heavy railway sleepers	kg	126.90	126.90
heavy framework	kg	22.24	22.24
load test (Pt)	kg	999.14	1180.39
1.5 Pt	kg	1498.71	1770.59

IOP Conf. Series: Materials Science and Engineering **223** (2017) 012036 doi:10.1088/1757-899X/223/1/012036

The Bridge Railway Sleepers burdened with a load of 1.5 Pt. For the test specimen type I load magnitude 1.5 Pt amounted to 1498.71 kg while for the test object of type II of 1770.59 kg. From visual observation after b the Bridge Railway Sleepers burdened with a load of 1.5 Pt for each specimen, it can be seen that: no component of the fastening system is broken, cracked or collapsed (good conditioned), rail cannot be separated, no loose or torn inserts. From the results of this test it can be concluded that the bearing under test meets the requirements of the test lift rail fastening systems.

Rail fastening system endurance test against longitudinal load

Test specimen rail fastening system is set up complete with a rail. Longitudinal load given in stages and each increment load recorded magnitude of displacement in the longitudinal direction. At the load of 10.9 kN shift in the longitudinal direction shall not exceed 5 mm. Testing of rail fastening system resistance to longitudinal loads, said that a shift by 5.5 mm, occurred at the time of the rail burdened with a load of 677.5 kg (6.65 kN). From this test it can be seen that the specimen (rail fastening apparatus / fastening) does not meet the requirements specified longitudinal strength. This may be due to rail clip used is not suitable.

4. Conclusions

Conclusions from preliminary study are the highest tensile strength test results achieved on code SB of 3324 kg/cm^2, the highest flexural strength test results achieved in code WR200 of 3227 kgf/cm^2, The highest compressive strength test achieve on code WR200 of 3208 kg/cm^2 and the highest hardness achieved on WR400 and WR600 of 118 HRr. From this research obtained information that Wr 200 will provide high tensile strength properties and provide a smooth surface material, while Stitch Bonded will give flexural strength properties and high compressive strength. Therefore, the manufacture of products Railway Bridge sleepers WR200 and SB use of glass fiber and epoxy resin. The mechanical properties of coconut fiberboard have an higher the mechanical properties of bagasse fiberboard

The better results of mechanical testing of the Railway Bridge sleepers all variations meet the technical specifications of end user, just look at the techno economic factors and wear time. The mechanical properties of the Railway Bridge sleepers using coconut fiberboard higher than using bagasse fiberboard as sandwich filler

Ackknowledgements

The authors would like to thanks to Mr. Sulaefi Nasserie as an expertise of composite sandwich structure and Mr Suryadi Rachmat as an expertise of bridge railway sleepers for assistance in manufacturing process technology

References

[1] Manalo A *et al* 2010 *Composites Structures* **92** 603-611.
[2] Charles E. Knox 2001 *Fiber Glass Reinforcement* Technical Director, Uniglass Industries New York.
[3] Ashori A *et al* 2012 *Construction and Building Materials* **27** (1) 126-129.
[4] Klang E C and Kuo T M 2004 *Impact Responses of 3-D, Woven and Laminated Composite Material* North Caroline State University.
[5] http://plsc.ws/macrog/epoxy.htm 2008 *Epoxy Resins : Polymer Science Learning Center, Department of Polymer Science* The University of Southern Mississippi.
[6] Suhardjo K A 2012 *Journal of Industrial Research* **VI** (3) 203-289.
[7] ITW overcoat, 2010: "Fiber Glass-Ever coat, Choosing The Right Resin- Epoxy Resin" a Division of Illinois Tool Works Inc.
[8] Suhardjo K A *et al* 2011 *Journal of Industrial Research* **V** (2) 99-112.
[9] Jacqueline I, Kroschwitz 2003 *Epoxy Resins, Concise Encyclopedia of Polymer Science and Engineering* John Wiley & Son.
[10] Ozawa S 2005 *Polymer J. Japan* **19** 199.
[11] ITW overcoat 2010 Fiber Glass-Ever coat, Choosing The Right Resin- Epoxy Resin" a Division of Illinois Tool Works Inc.
[12] Lynn S Penn and Chiao T T 2002 *Epoxy Resins* Ciba Geigy Corp Ardsley New York and Lawrence Livermore labolatory, University of California, Livermore California.
[13] Mahmoood M Shokrieh, Rahmat M 2006 *Composite Structures* **76** (4) 326-37.
[14] Sadeghi J, Barati P 2012 *J. Structure and Infrastructure Engineering* **8** (112) 1151-59.

IOP Conf. Series: Materials Science and Engineering **223** (2017) 012037 doi:10.1088/1757-899X/223/1/012037

Preparation of catalyst coated membrane by modified decal transfer method for proton exchange membrane fuel cell

Indriyati[1], **Y Irmawati**[2] **and B Prihandoko**[2]
[1] Research Unit for Clean Technology, Indonesian Institute of Sciences (LIPI), Jl. Sangkuriang Kompleks LIPI Gedung 50, Bandung, Indonesia 40135
[2] Research Center for Physics, Indonesian Institute of Sciences (LIPI), Kawasan PUSPIPTEK Serpong, Tangerang Selatan, Indonesia 15314

E-mail: indriyati@lipi.go.id

Abstract. A new catalyst coated membrane (CCM) was prepared by modified decal transfer method. A structure of ionomer/catalyst/carbon/substrate was used to facilitate the transfer of catalyst layer from decal substrate to the membrane at quite low hot-pressing temperature (120 °C) for 8 min. Several decal substrates were tested to select a proper substrate, namely PTFE cloth, PTFE film, aluminium foil, and OHP transparent sheet. The transfer degree of catalyst layer was estimated. Elemental analysis and SEM-mapping were performed to evaluate the residue, whereas contact angle measurement was conducted to characterize the hydrophobicity of decal substrates. The results showed that PTFE cloth and PFTE film transferred approximately 90% of catalyst layer onto the membrane, while the other two substrates were around 70%. Furthermore, the elemental analysis of the residue on the substrate revealed that it was mainly composed of carbon and fluorine for PTFE cloth and PTFE film. This result supports other findings that PTFE cloth and PTFE film are suitable as decal substrate at low temperature hot pressing for fabricating CCM.

1. Introduction

Proton exchange membrane fuel cells (PEMFCs) have been already commercialized for stationary application and become one of potential candidates for automobile applications. PEMFCs show high energy efficiency and low temperature operation compared to conventional combustion engine. As the heart of PEMFC, preparation of membrane electrode assembly (MEA) directly influences the cell performance. Many studies have been made to improve MEA preparation [1-4]. Typically, there are two categories of MEA fabrication, namely catalyst-coated on electrode (CCE) and catalyst coated on membrane (CCM). The performance and durability of CCM are usually higher than that of MEA prepared by the CCE method due to better interfacial contact between membrane and catalyst and better catalyst utilization [5-7].

Instead of direct coating the catalyst onto the membrane that might undergo swelling, catalyst layer can be deposited onto the membrane by decal method. Catalyst inks are coated over a substrate to form a catalyst layer, and then transferred onto the membrane by hot pressing under certain temperature and pressure condition [7,8]. Low Pt loading and low catalyst losses during MEA fabrication can be obtained by this way [9]. Furthermore, to facilitate the transfer of catalyst layer

IOP Conf. Series: Materials Science and Engineering **223** (2017) 012037 doi:10.1088/1757-899X/223/1/012037

from the decal substrate to a membrane, Park *et al.*, (2008) [10] introduced a layer composed of carbon powder and Nafion ionomer prior to coating the catalyst.

Conventionally, decal transfer is performed by hot pressing at 150-200°C [7,11,12]. As a consequence, the H^+ form of Nafion membrane should be converted to other cationic forms such as Na^+ or K^+ to give more mechanical strength. Subsequence re-protonation by boiling the MEA in sulfuric acid solution is also required. This process seems to be more complex and the temperature of hot pressing is too high. Recently, Cho *et al.*, (2009) [4] have reported low temperature decal transfer for direct methanol fuel cell (DMFC) by adding carbon layer and Nafion ionomer layer prior to hot pressing at 140 °C.

Among many aspects that has been studied regarding decal technique, a choice of the materials as a decal substrate have been a challenging task to get a complete transfer of catalyst layer onto membrane [4,8,13]. The choice of decal substrate depends on several parameters, such as chemical and mechanical stability, hydrophobicity, adhesive strength with catalyst layers [14] as well as hot press condition. Cho *et al., (2009)* [4] have chosen Teflon film as decal substrate, while Xie *et al.*, (2004) [8] preferred Kapton film than Teflon film to avoid the problem of ionomer segregation.

In this work, the catalyst layer was transferred onto the membrane by decal method as previously report [4], but at lower temperature. Various substrates such as Teflon cloth, PTFE film, Al foil and OHP transparency sheet were studied by measuring the degree of catalyst transfer in order to find the best substrate for hot pressing at 120 °C for 8 min. SEM and elemental analysis were performed on the residue remained on the substrate to investigate whether it is catalyst particles or carbon powder.

2. Experiment

In all methods, Nafion® 117 membrane (DuPont Inc.) and 5wt.% Nafion ionomer solution (DuPont Inc.) were used. Catalyst Pt/C (20wt.% Platinum on Vulcan XC 72) and Carbon Vulcan XC 72 for carbon layer were purchased from Fuel Cell Earth LLC. All chemicals such as hydrogen peroxide (H_2O_2), sulfuric acid (H_2SO_4), ethanol, and isopropyl alcohol (IPA) were in analytical grade and purchased from Merck. Four substrates were used, such as virgin electrical grade film of Teflon PTFE purchased from Fuel Cell Earth LLC (denoted as PTFE film), commercial grade of woven glass fiber coated with PTFE (denoted as PTFE cloth), commercial grade of aluminium foil (denoted as Al foil), and commercial grade of OHP transparency sheet (denoted as OHP sheet).

The membranes were pre-treated to eliminate any organic or inorganic impurities prior to use in MEA preparation. The treatment process was accomplished by successively soaking the membranes in 3% H_2O_2, deionized water, 0.5M H_2SO_4, and then deionized water again at 70-80 °C for 1 h each step

[15]. Carbon ink was made of carbon Vulcan XC-72 and ethanol, while catalyst ink was prepared by mixing Pt/C, Nafion ionomer solution, and IPA ultrasonically for 60 min. The weight ratio between Pt and Nafion in catalyst ink was set at 65:35.

The detail procedure of the catalyst coated membrane prepared through the decal transfer method is schematically represented in Figure 1. Firstly, a thin carbon layer was applied over each substrate, and this was followed by addition of catalyst layer. An additional layer of Nafion ionomer was then sprayed on the coated catalyst layer. The prepared multi-layer substrates (outer ionomer/catalyst/carbon/substrate) were then hot pressed on both sides of the membrane at 120 °C for 8 min to form CCM. The hot-pressing conditions used for this work have been optimized in our lab. The geometric active area of all the MEAs was 3 cm x 3 cm.

The residue remained on the substrate after decal transfer was examined using SEM-EDS (Hitachi SU 3500) to verify a complete transfer of the catalyst. SEM images obtained by BSE detector was done at 15 kV accelerating voltage. Prior to measurement, the samples were coated using gold sputtering. For evaluating the hydrophobic properties of the substrates, contact angles between water and the substrates were measured using sessile drop method. For each measurement, a water droplet was placed onto the surface of the substrate and images were captured with a digital camera. Furthermore, the contact angles were estimated by measuring the angle between water drop and the substrate regarding to drop shape analysis.

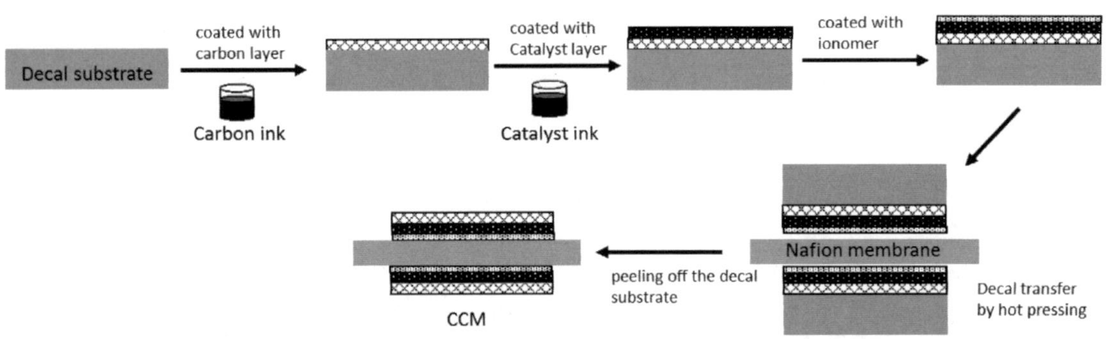

Figure 1. Schematic Diagram of Low-Temperature Decal Method.

3. Results and Discussion

A schematic procedure of low temperature decal transfer is shown in Figure 1. The advantage of this method is significantly lower hot pressing temperature, even if it is compared with previously report by Cho *et al.*, (2009) [4]. By using this low temperature for decal process, H^+ membrane can be used instead of Na^+ membrane, which is used for high temperature decal process [16]. Therefore, re-protonation of membrane by boiling MEA in sulfuric acid solution is not necessary, hence it shortened the process. This is another advantage of this method, which is the simplicity in the process scheme.

To facilitate a good transfer of catalyst layer from a decal substrate to a membrane, carbon layer was applied onto the substrates prior to depositing catalyst layer as previously report [4]. The composition of materials in this carbon layer is equated to that used in microporous layer (MPL) deposited on the gas diffusion layer (GDL). Thereby, it is expected that no negative effect on MEA performance.

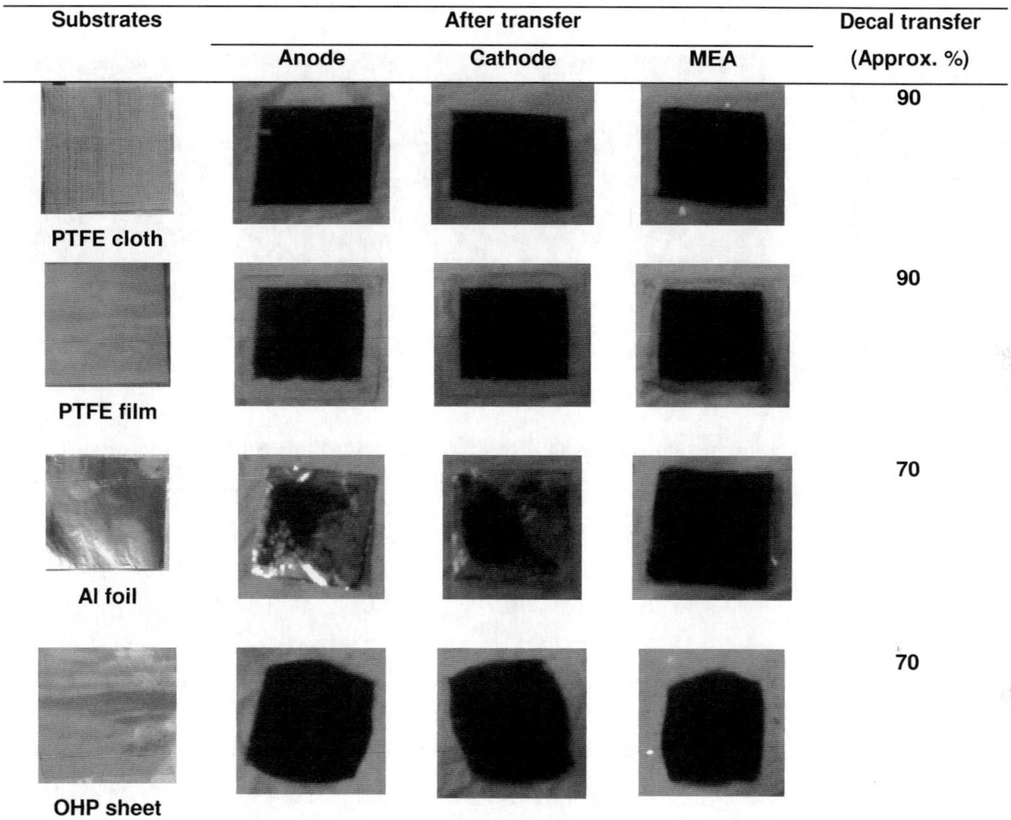

Figure 2. Photograph of Decal Substrates before Coating Process and After Catalyst Layers were Transferred to The Membranes with Active Area of 3 cm x 3 cm

Photographs of decal substrates after hot-pressing are given in Figure 2. It is obvious that the transfer degree of the catalyst layer depends on the type of substrate. As shown in figure 2, PTFE cloth and PTFE film have a fairly good transfer, even though the surface of PTFE cloth and PTFE film appears slightly dark after being peeled off from the catalyst layer. A black residue also remained on the surface of OHP sheet and Al foil, but they appear thicker than that of on PTFE cloth and PTFE film. It is assumed that the black residue was carbon layer, and it will be checked by EDS and described in the later session. Furthermore, Al foil shows not suitable as a substrate for hot pressing condition at 120 °C for 8 min. This is probably due to low mechanical stability of Al foil, which is too thin for this hot pressing condition. OHP transparent is also demonstrated to be inadequate as a substrate for this decal transfer method, since it underwent dimensional changes of MEA. This could be explained that hot pressing temperature of 120°C is above glass transition temperature (Tg) of OHP film, which is made of polyethylene terephthalate (PET) (Tg between 65-80 °C). Above its glass transition temperature, the film becomes soft and changes the dimension due to pressing. Mehmood and Ha [9] also experienced changes in geometric dimension of MEA when using Teflon film as decal substrate after flat-press at 150°C under pressure of 8 MPa for 8 min.

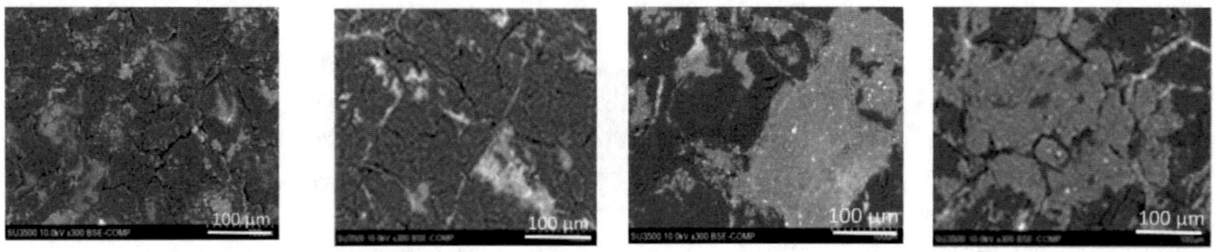

(a) (b) (c) (d)

Figure 3. SEM Images of Residual Surface of (a) PTFE cloth, (b) PTFE film, (c) Al foil, and (d) OHP sheet.

Table 1. Elemental Analysis of Residue on the Substrate Surface in Atomic Percentage.

Element	PTFE cloth	PTFE Film	Al foil	OHP Sheet
C	78.34	85.84	87.22	80.49
O	2.57	3.05	3.90	8.48
F	18.78	10.77	8.05	10.47
S	0.26	0.30	0.41	0.32
Pt	0.06	0.04	0.42	0.25

In order to identify the dark material (residue), small pieces were cut from the middle of the substrates that experienced hot-pressing at 120°C and then were subjected to SEM-EDS analysis. Figure 3 presents SEM images of the residue remains on the substrates obtained by BSE detector. BSE is beam electron that is reflected from the samples by elastic scattering. BSE images can provide information about the distribution of different elements in the sample because BSE signal is strongly related to the atomic number (Z) of the specimen. In Figure 3, bright contrast represent the present of Pt particles. Analyzing the residue using EDS (Table 1) reveals that the residues are composed of carbon and fluorine. Only small amount of Pt left on the PTFE cloth and PTFE film. This indicates that catalyst layers are completely transferred for the case of PTFE cloth and PTFE film. On the other hand, the residue on Al foil and OHP sheet contained quite high Pt. This can be explained that the residue was not only carbon layer, but also catalyst layer which is contained 0.45 mg/cm^2 of Pt.

In term of measuring the hydrophobicity of substrates, the sessile drop method was used to measure static contact angles. The sessile drop method requires a flat surface upon which a liquid drop can be placed and those substrates fulfill this requirement. As can be seen in Figure 4, all substrate exhibit almost the same contact angle with Al foil and OHP sheet were slightly lower than the others. This result indicates that high degree of catalyst transfer onto PTFE cloth and PTFE film was not because those substrates have slightly higher hydrophobicity, but because they have good mechanical stability for press process at 120°C. Mechanical stability of materials at high temperature is closely related to thermal properties especially with glass transition temperature. It should be ensured that hot pressing temperature is below the glass transition temperature of the substrates.

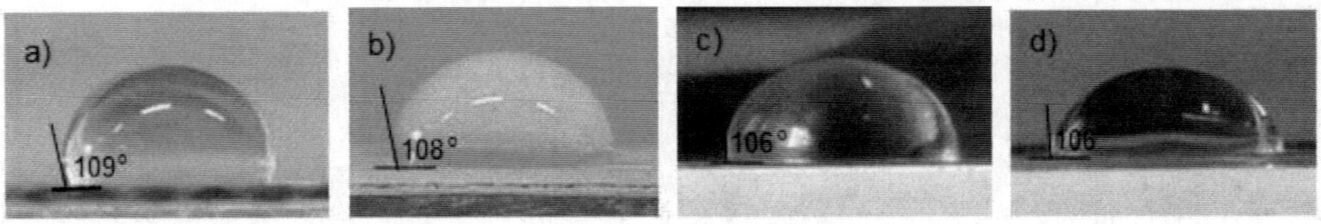

Figure 4. Photograph of a Drop of Aquades on Surface of a) PTFE Cloth, b) PTFE Film, c) Al Foil, and d) OHP Sheet.

4. Conclusion

The MEA prepared by decal transfer method has been developed at low temperature hot pressing of 120 °C. By using this low temperature, this method can provide shorter process because it is not necessary to re-protonation of Nafion membrane since the H^+ form of Nafion membrane can be directly used instead of using Na^+ form of membrane as in conventional decal method. Among several types of substrates, PTFE cloth and PTFE film showed the most suitable substrates for this method. OHP sheet was not strong enough to be processed at temperature of 120°C due to low glass transition temperature. Moreover, evaluation on residue left on the substrates proved a complete catalyst transfer to the membrane for PTFE cloth and PTFE film, while there were some Pt catalyst remained on the surface of Al foil and OHP sheet based on elemental analysis. Based on the results, mechanical stability at this hot pressing condition was the main reason for different degree of transfer rather than hydrophobicity of substrates.

Acknowledgement

Authors sincerely acknowledge Indonesian Institute of Sciences under Riset Unggulan for financial support. We also thank to fuel cell team for valuable discussion and Testing Laboratory of Research Center for Physics which has facilitated the characterization of SEM images and elemental analysis.

IOP Conf. Series: Materials Science and Engineering **223** (2017) 012037 doi:10.1088/1757-899X/223/1/012037

References

[1]. Towne S *et al* 2007 *J. Power Sources* **171** 575-84.
[2]. Prasanna M *et al* 2008 *Electrochim. Acta* **53** 5434-41.
[3]. Sun L *et al* 2008 *Solid State Ionics* **179** 960-5.
[4]. Cho JH *et al* 2009 *J. Power Sources* **187** 378-86.
[5]. Lindermeir A *et al* 2004 *J. Power Sources* **129** 180–7.
[6]. Tang H *et al* 2007 *Electrochim. Acta* **52** 3714-8.
[7]. Song SQ *et al* 2005 *J. Power Sources* **145** 495-501.
[8]. Xie J *et al* 2004 *J. Electrochem. Soc.* **151** A1084-93.
[9]. Mehmood A and Ha HY 2012 *Int. J. Hydrogen Energy* **37** 18463-70.
[10]. Park HS *et al* 2008 *J. Electrochem. Soc.* **155** B455-60.
[11]. Wilson MS *et al* 1995 *Electrochim. Acta* **40** 355-63.
[12]. Song S *et al* 2005 *J. Power Sources* **140** 103-10.
[13]. Yoon YJ *et al* 2011 *J. Power Sources* **196** 9800-9.
[14]. Mehmood A and Ha HY 2013 *Int. J. Hydrogen Energy* **38** 12427-37.
[15]. Scott K *et al* 1999 *J. Power Sources* **79** 43-59.
[16]. Saha MS *et al* 2010 *Electrochem. Comm.* **12** 410-3.

IOP Conf. Series: Materials Science and Engineering **223** (2017) 012038 doi:10.1088/1757-899X/223/1/012038

Effects of pore forming agents on chitosan-graft-poly(N-vinylpyrrolidone) hydrogel properties for use as a matrix for floating drug delivery

E Budianto, M F Al-Shidqi and A H Cahyana
Department of Chemistry, University of Indonesia, Depok 16424 Indonesia

E-mail : emilb@ui.ac.id

Abstract. Eradicating *H. pylori*-based infection by using conventional oral dosage form of amoxicillin trihydrate finds difficulties to overcome rapid gastric retention time. Encapsulating amoxicillin trihydrate in floating drug delivery system may solve the problem. In this research, the floating drug delivery system of amoxicillin trihydrate encapsulated in floating chitosan-graft-poly(N-vinyl pyrrolidone) hydrogels containing $CaCO_3$ and $NaHCO_3$ as pore forming agents has been successfully prepared. Pore forming agents used was varied with the ratio of 10 to 25% pore forming agents to total mass of the used materials. The hydrogel were characterizedusing FTIR spectrophotometer and stereo microscope. As pore forming agents compositions increased, the porosity (%) and floating properties increased but followed by decrease in drug entrapment efficiency. Most of the floating hydrogels possessed floating ability longer than 180 min and the highest porosity was found in hydrogel containing 25% $NaHCO_3$. Hydrogel containing $CaCO_3$ showed sustained drug release profile than hydrogel containing $NaHCO_3$. However, the optimum formulation was achieved at composition of 10% $NaHCO_3$ with 57% of drug entrapped within the hydrogel and 43% drug released. The results of these studies show that $NaHCO_3$ is an effective pore forming agents for chitosan-graft-poly(*N*-vinyl pyrrolidone) hydrogel preparation as compare to $CaCO_3$.

1. Introduction

Heliobacter pylori (*H. pylori*) is one of the most frequent and persistent bacterial infections worldwide [1]. *H. pylori* is associated with peptic ulcer disease, gastric ulcers, mucosa-associated lymphoid tissue lymphoma and gastric cancer [2]. As stated in Indonesian Pharmacopoeia (2013), Amoxicillin is administered orally as trihydrate for the eradication of *H. pylori* in forms of tablets, capsules, or oral suspensions. Amoxicillin is known to have short plasma half-lives of 1 to 1.5 h [3]. Amoxicillin limitation of having short plasma half-lives cause amoxicillin should be administered by multiple dosing at regular interval. This conventional route of administration leads to a rapid rise and fall in drug concentrations within plasma which may fall outside the therapeutic range for significant time periods [4]. Several approaches have been recommended to overcome these limitation, one of them is the controlled drug delivery system. Nevertheless, conventional controlled drug delivery system still faces some kind of difficulties. One of such difficulties is the short gastric emptying time [5]. The short transit time means the drugs within oral controlled release dosage fromare not completely released. This results in ineffective administered dose which is then accompanied by lesser bioavailability [6].

IOP Conf. Series: Materials Science and Engineering **223** (2017) 012038 doi:10.1088/1757-899X/223/1/012038

Approaches to increase the gastric residence time of drug formulation have been developed in recent years, including gastroretentive drug delivery system (GDDS). Floating drug delivery system (FDDS) is one of GDDS and it may be used for various drugs that are locally active in the stomach, possess narrow absorption window in gastrointestinal tract, degrade in colon, or are insoluble in alkaline pH [7-9]. In FDDS, while thesystem is floating on the gastric contents, the drug is released at a desired rate from the system [8]. By using amoxicillin in FDDS dosage form, the administered dose of amoxicillin can overcome the limited gastric transit time, thus increasing its bioavailability.

During the last two decades, natural and renewable polymers have attracted interest in researches related to drug delivery. Among various renewable polymers, chitosan has been evaluated as a potential polymer for drug administrated orally [10]. Chitosan is natural, biocompatible, and biodegradable polymer prepared from partial deacetylation of chitin, the second prevalent form of polymerized carbon in nature [11,12]. In this research, chitosan-graft-poly(N-vinyl pyrrolidone) has been prepared and used as floating drug delivery system matrix with amoxicillin trihydrate as a model drug. The floating drug delivery system was prepared using two different pore forming agents, i.e. calcium carbonate ($CaCO_3$) and sodium bicarbonate ($NaHCO_3$). The effects of different pore forming agents and its efficacy toward the hydrogel matrix properties were examined.

2. Materials and Methods

2.1 Materials

Amoxicillin trihydrate API was kindly provided by PT. Dankos Laboratories Tbk, Chitosan (94.13 % deacetylation degrees, 25.20 cps viscosity) was purchased from PT. Biotech Surindo,N-vinyl pirrolidone(analytical grade) was purchased from Sigma-Aldrich), calcium bicarbonate ($CaCO_3$) and sodium bicarbonate ($NaHCO_3$)(analytical grade) were purchased from Merck, acetic acid (glacial, analytical grade) purchased from Merck, N,N-dimethylbisacrilamide (NNMBA, analytical grade) was purchased from Sigma-Aldrich

2.2 Methods

Preparation of chitosan-graft-poly(N-vinyl pirrolidone) hydrogel and chitosan-graft-poly(N-vinyl pirrolidone) floating hydrogel matrix.An exact amount of 0.2 g dry chitosan was first dissolved in 15 mL of 2% acetic acid using a round bottom flask. Then, the flask was placed in a thermostated bath to initiate the polymerization at 60° C and followed by the addition of 0.1 g ammonium peroxide as initiator. A slow stream of nitrogen gas was passed continously during the polymerization reaction and the reaction mixture was stirred for 10 min. 0.8 mL of N-vinyl pyrrolidone monomer and NNMBA crosslinker (2%) was added into the reaction mixture. The reaction was done after 180 min and stopped by letting air into the flask. The reaction mixture was poured in a beaker and the grafted product and was neutralized to pH 8.0 by addition of 1 N NaOH. Then, 50 mL methanol was added to the hydrogel product while stirring. After complete dewatering for 8 h, the product was filtered, washed with fresh methanol, and dried at 50°C. To prepare floating hydrogel matrix, calcium carbonate with various composition were added to the hydrogel product when it reach pH 5. The same procedure was followed to prepare floating hydrogel with sodium bicarbonate. Table 1 list the various composition of pore forming agents added during hydrogel and floating matrix synthesis.The amoxicillin encapsulation was performed by using in-situ method. Drug solution was prepared by dissolving 500 mg amoxicillin trihydrate in 5 mL distilled water. The drug into the reaction mixture was added before the addition of crosslinker. The same variation of pore forming agents composition as indicated in Table 1 were used in the synthesis of amoxicillin trihydrate loaded floating hydrogel.

Table 1. Synthesis condition of chitosan-graft-poly(N-vinyl pirrolidone) hydrogel and chitosan-graft-poly(N-vinyl pirrolidone) floating hydrogel matrix containing $CaCO_3$ and $NaHCO_3$.

Chitosan (g)	NVP (g)	APS (g)	MBA (g)	$CaCO_3$/ $NaHCO_3$ (g)	% PFA
				0	0
				0.107	10
0.2	0.834	0.1	0.0167	0.1808	15
				0.214	20
				0.2675	25

Hydrogel physical characterization: porosity measurement of hydrogel and invitro buoyancy study. Wet hydrogels after being immersed in pH 1.2 solution were weighed after blotted. The dried hydrogels were also weighed before the hydrogels being immersed. The porosity was calculated based on the following equation:

$$(1)\ \frac{W_w - W_d}{\rho V}$$

Where W_w and W_d are mass of wet and dry hydrogel respectively; is density of pH 1.2 solution, and V is volume of the hydrogel [13].

Floating properties of hydrogel were evaluated as floating lag time and floating time. Floating lag time can be defined as the time required for hydrogels to rise to the surface and float, while floating time is the length of time when hydrogel completely float on the surface of 5 mL pH 1.2 medium. The measurement was carried out in 37 ± 0.1 °C and was measured by visual observation.

Amoxicillin trihydrate entrapment efficiency. The weighed hydrogels were powdered and dissolved in 5 mL pH 1.2 solution, sonicated for 1 h, diluted further and analysed spectrophotometrically at 272 nm using Shimadzu-2450 UVV spectrophotometer. In vitro release study. The loaded hydrogels were immersed in beaker containing 20 mL pH 1.2 solution and placed in a shaking waterbath at 37 ± 0.1 °C. At 10, 20, 30, 60, 120, and 180 min time interval, 1 mL of the sample was withdrawn, diluted, and analysed spectrophotometrically at 272 nm. Equal amount of fresh medium was replaced immediately after withdrawal of the test sample.

Hydrogel characterizations were conducted using Fourier transform infrared spectroscopy (FTIR) and Olympus SZX16 stereo microscope. FTIR analysis was performed by grounding and mixing hydrogel samples with KBr powder with ratio 1:9. The measurements were carried out within the range of 4000-400 cm^{-1}. Morphology of hydrogels was studied using stereomicroscope with 1x magnification and 10x zoom.

3. Results & Discussion

3.1 FTIR Characterization

Figure 1(a) and (b) illustrate the FTIR spectra of blank hydrogel and hydrogels containing $CaCO_3$ and $NaHCO_3$, respectively. The broad band showed at 3200-3650 cm^{-1} is due to stretching vibration of -OH groups in chitosan while the band at 1598 cm^{-1} is due to bending vibration of -NH group of chitosan. The presence of C=O stretching vibration band at 1674 cm^{-1} and increase in band area of stretching vibration of C-N group at 1078 cm^{-1} indicates that poly(N-vinyl pyrrolidone) chain has been successfully grafted to chitosan backbone. Addition of $NaHCO_3$ and $CaCO_3$ into chitosan-graft-poly(N-vinyl pyrrolidone) does not alter the existing functional group within the hydrogel matrix, for all characteristic bands of blank hydrogel are present in hydrogel containing both pore forming agents. The new bands appeared at 828 cm^{-1} in $CaCO_3$ and 877 cm^{-1} in $NaHCO_3$-containing hydrogels attributed to the carbonate peak which indicates pore forming agent present in hydrogel network. Meanwhile, another new band appeared at 2514 and 1793 cm^{-1} in $CaCO_3$ containing hydrogel. These bands does not appeared in the $NaHCO_3$-containing hydrogels. The appearance of the new bands indicating that there is chemical interaction between the hydrogel matrix and $CaCO_3$. It is taught that the Calcium ion from $CaCO_3$ bind to the polymer structure of chitosan-graft-poly(N-vinyl pyrrolidone) hydrogel resulting in polymer-metal complex. The formation of complex through chelation of divalent cations with poly(N-vinyl pyrrolidone) may be through oxygen of poly(N-vinyl pyrrolidone) [16]. Figure 1(c) illustrates the FTIR spectra of blank hydrogel and floating hydrogel matrix loaded with amoxicillin trihydrate. The FTIR spectra shows one of the characteristic band of amoxicillin at 1515 cm^{-1} indicating the bending vibration of N-H group. The band appeared at blank hydrogel, hydrogel containing $NaHCO_3$, and hydrogel containing $CaCO_3$ at 1520, 1517, and 1513 cm^{-1}, respectively. The bands show no significant shift from the original position in the spectrum of pure amoxicillin trihydrate at 1515 cm^{-1} [17]. This shows no evidence for any chemical interaction between amoxicillin and hydrogel matrix and the pore forming agents [18]. The loaded amoxicillin trihyrate drug are binded to the hydrogel matrix though physical interactions between the functional groups of drug and the hydrogel itself.

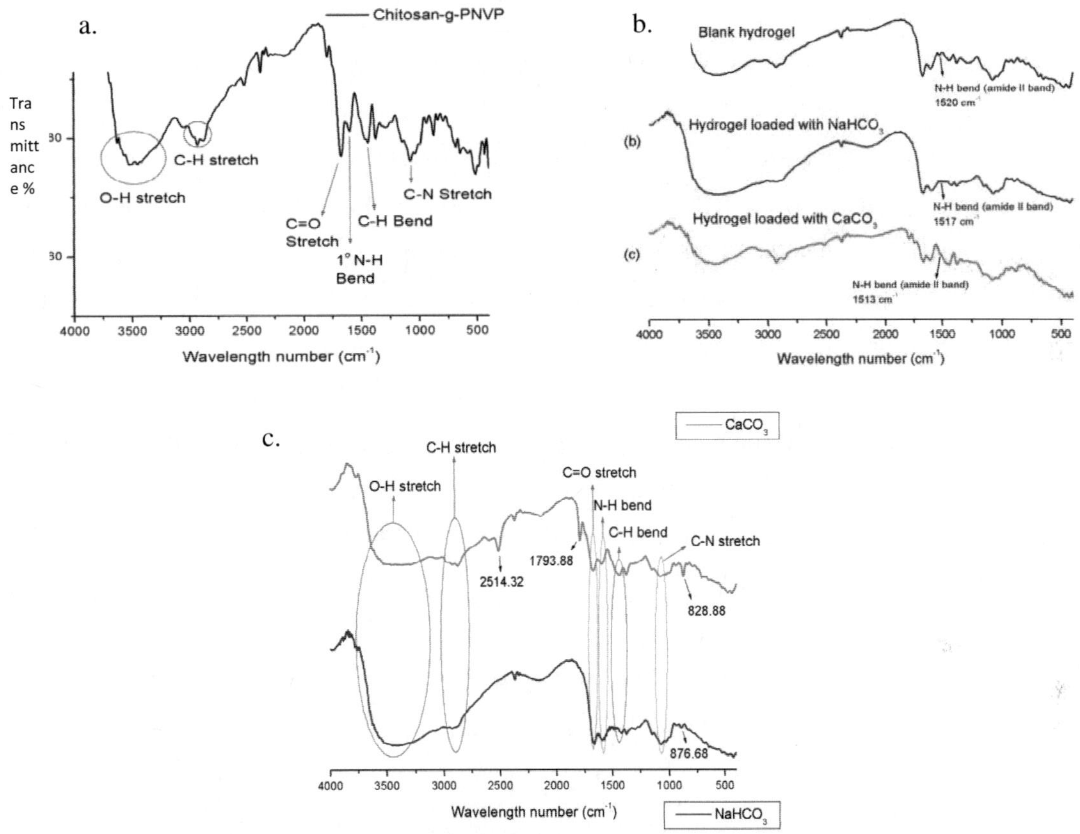

Figure 1. FTIR spectrum ofchitosan-graft-poly(N-vinyl pyrrolidone) hydrogel (a), spectra of hydrogels loaded with $NaHCO_3$ and $CaCO_3$ (b), and spectra of floating hydrogels loaded with amoxicillin trihydrate (c).

3.2 Porosity Evaluation of Floating Hydrogel

During the immersion of hydrogel matrix within acidic medium, carbonate salts of pore forming agent react with the acid to produce carbon dioxide. The evolving gas permeates throgh the hydrogel leaving gas bubble or pores [19].

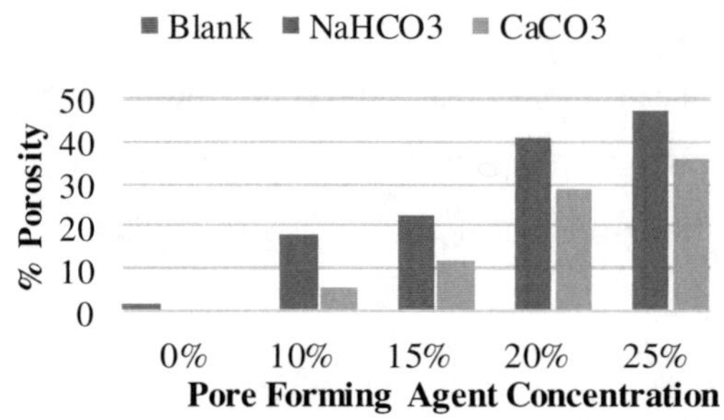

Figure 2. Porosity of floating hydrogels as a function of pore forming agents concentration

Porosity of floating hydrogels study was conducted to determine the effects of pore forming agents on pore structure of hydrogels. The results of the porosity study are shown in Figure 2. Blank chitosan-graft-poly(N-vinyl pyrrolidone) hydrogels possess very low porosity (1.4%). This shows that the hydrogel that is free from pore forming agent has a very compact and rigid internal structure. It was observed that the porosity of hydrogels increased as the concentration of pore forming agents increase from 10 to 25%. Hydrogels with 10% concentration of $NaHCO_3$ and $CaCO_3$ show low porosity which are 18 and 5%, respectively. At the highest concentration of pore forming agents, hydrogels with 25% of $NaHCO_3$ and 25% of $CaCO_3$ show higher porosity, where 47% and 36% for $NaHCO_3$and $CaCO_3$ hydrogels, respectively. The results clearly shows that $NaHCO_3$ produced higher porosity of the hydrogels than $CaCO_3$.

Table 2. The results of *in vitro* buoyancy study of chitosan-graft-poly(N-vinyl pyrrolidone) floating hydrogels

Pore forming agent		Floating lag time (min)	Floating time (min)
Control (0%)		Not floating	Not floating
$NaHCO_3$	10%	34	>180
	15%	20	>180
	20%	6	<20
	25%	5	<20
$CaCO_3$	10%	Not floating	Not floating
	15%	53	>180
	20%	15	>180
	25%	13	>180

Table 2 shows the floating ability of the hydrogel matrix in pH 1.2 solution at 37 °C that stimulates the gastric fluid condition. Hydrogels containing pore forming agents shows the ability of floating. In contrast, hydrogels without pore forming agents could not float because their lower porosity ,thus higher density, hindered their ability of float. The floating lag time of hydrogels increases with the addition of pore forming agents. This is due to the evolving of CO_2 gas produced from the effervescent reaction in hydrogel matrix which causes the hydrogel to float on the pH 1.2 solution in such faster time. When comparing $CaCO_3$ and $NaHCO_3$, floating lag time for hydrogels containing $NaHCO_3$ are

IOP Conf. Series: Materials Science and Engineering **223** (2017) 012038 doi:10.1088/1757-899X/223/1/012038

the shortest. It takes 5 min for hydrogels containing to 25% $NaHCO_3$ to reach the surface whereas for hydrogels containing 25% $CaCO_3$ are 13 min. Meanwhile, hydrogels containing 20% and 25% $NaHCO_3$ are only able to float less than 20 min and later dissolve. This is due to increase of pore forming agents which lead to decrease of the hydrogel strength as the hydrogels become highly porous and fragile at the higher concentration of pore forming agents [13,18].

3.3 Drug Entrapment Efficiency

The drug entrapment efficiency of The results show that as the concentration of pore forming agent increases from 10% to 25%, a decreased in drug entrapment efficiency are observed 57 to 23% for $NaHCO_3$ and 49 to 12% for $CaCO_3$. Hydrogels that do not contain any pore forming agents show high entrapment efficiency as compared to other hydrogels that loaded with pore forming agents. This may be due to thefact that hydrogels without pore forming agent have an extremely compact internal structure that are able to retain the drugs within its matrix [19].

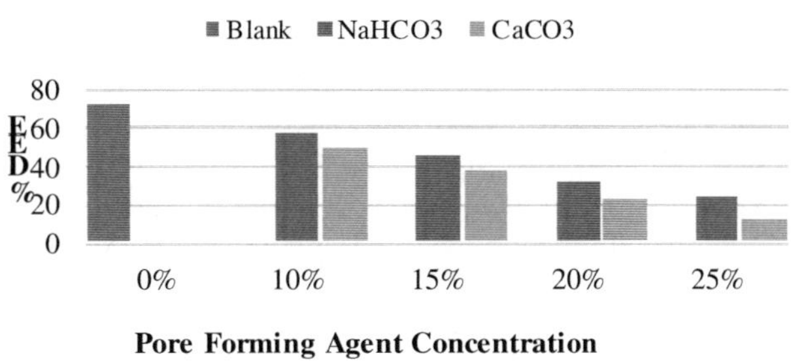

Figure 3. Drug entrapment efficiency of floating hydrogels wit different concentration of sodium bicarbonate and calcium carbonate

On the other side, the decrease in drug entrapment observed in the addition of concentration of pore forming agents is possibly due to the higher porosity of the hydrogels. The less dense internal structure of the hydrogels which contain high amount of pore forming agents causing less physical interaction between the matrix and the drug resulting in decreased drug entrapment. Addition of pore forming agent tends to make hydrogels possess lower drug entrapment efficiency (<50%) and $CaCO_3$ tends to decrease the drug entrapment efficiency than $NaHCO_3$.

3.4 In Vitro Release Profile of Amoxicillin Trihydrate

The in vitro release profile of amoxicillin trihydrate from floating hydrogel matrixes which contain various concentration of $NaHCO_3$ and $CaCO_3$ are shown in Figure 4(a) and 4(b). The amount of amoxicillin relase from hydrogels loaded with pore forming agents are higher than blank hydrogels. Lower drug release profile of blank hydrogels is due to the less porous nature of hydrogel network that retain drugs more effectively. The rate of drug release is found to increase as the amount of $CaCO_3$ and $NaHCO_3$ increases.

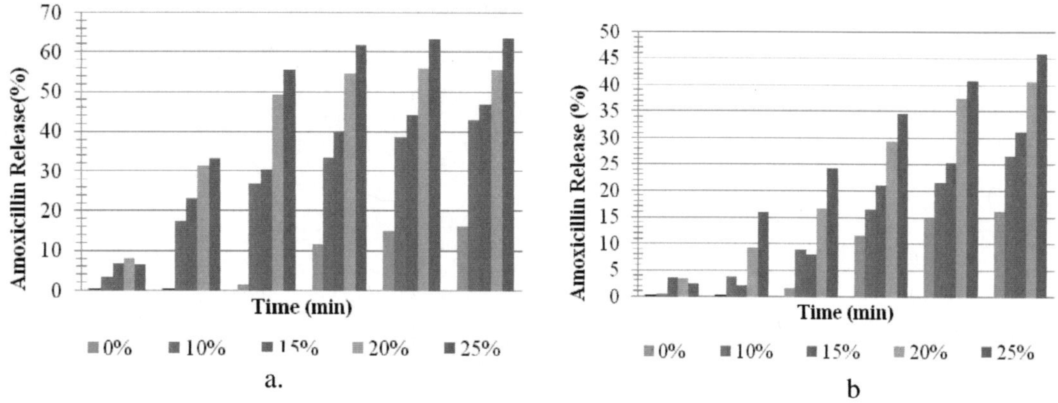

Figure 4. The *in vitro* release of amoxicillin trihydrate from hydrogel containing $CaCO_3$(a) and $NaHCO_3$(b)

It is observed that hydrogels containing $NaHCO_3$ as pore forming agent showed higher drug release than hydrogels containing $CaCO_3$. At first 10 min, the release of amoxicillin from 25% of $CaCO_3$ and $NaHCO_3$ filled hydrogel are 3 and 7%, respectively. At both pore forming agents, increase in pore forming agents' concentration is followed by gradual increases in amount of drug release. At t = 180 min, drug release for 25% $CaCO_3$ and 25% $NaHCO_3$ are 46 and 64%, respectively. It is also observed that different pore forming agent produces different pattern of drug release where hydrogels containing $CaCO_3$ shows slower and more controlled release than $NaHCO_3$. This is possibly due to the presence of smaller pore size in hydrogels containing $CaCO_3$ that able to control the drugs diffusion process from hydrogel matrix [10]. Meanwhile, hydrogels containing $NaHCO_3$ show fast drug release pattern at higher concentration (20 and 25%) and much slower and controlled release pattern at lower concentration of $NaHCO_3$ (10 and 15%). The fast release pattern at higher concentration is marked with sudden increase of the amount released drug between t = 10 min and t = 30 min. The fast release pattern is expected due to higher porosity of hydrogels causing drug diffusing quickly from the floating hydrogel matrix [20].

3.5 Morphology of Hydrogels

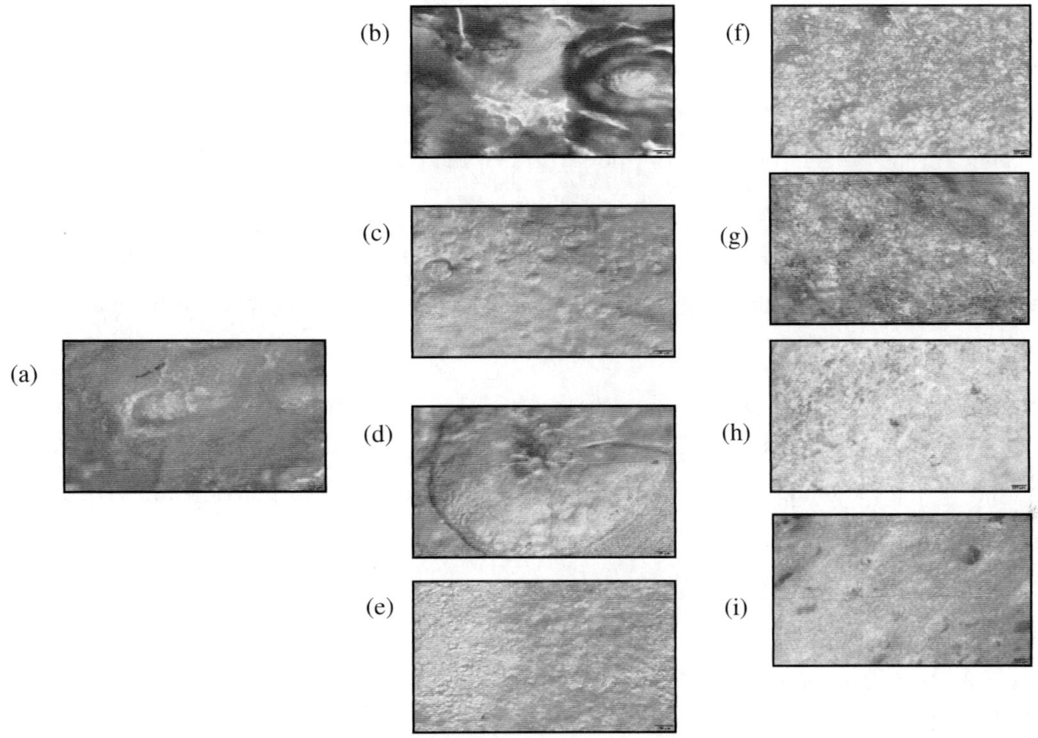

Figure 5. Morphology of blank hydrogel (a), hydrogel containing (b) 10% (c) 15% (d) 15% (e) 20% of $NaHCO_3$ and hydrogel containing (f) 10% (g) 15% (h) 20% (i) 25% of $CaCO_3$

The morphology study of floating hydrogels was studied using stereo microscope and the results are illustrated by Figure 5. The hydrogels were studied after being used in drug release test. In general, the study shows that the hydrogel surfaces are eroded. The acidic condition of pH 1.2 solution causes protonation of amine groups in chitosan. The protonation leads to repulsion in the polymer chains and dissociation of intramolecular hydrogen bonding [21]. Hydrogels with $NaHCO_3$ as pore forming agent are more susceptible to surface erosion, shown by more translucent surface seen at Figure 5(a)-5(b). This may due at higher concentration of $NaHCO_3$, the high porosity and fragile nature of hydrogels facilitate the erosion process. The more intact structure of $CaCO_3$ containing hydrogels as can bee seen at Figure 5(f)-5(g) can be attributed to the lower porosity of $CaCO_3$ containing hydrogels.In addition, change in chemical properties of floating $CaCO_3$-containing hydrogels as explained in Section 3.1 may also explained the more resistant nature of hydrogels towards surface erosion in pH 1.2 solution.

4. Conclusion

At the same amount of pore forming agent concentrations, $NaHCO_3$ and $CaCO_3$ affects the properties of floating chitosan-graft-poly(N-vinyl pyrrolidone) hydrogels to different extent. $CaCO_3$ can alter the physical properties of hydrogel more significantly than $NaHCO_3$ as seen from the morphology study results and the chemical properties. Addition of pore forming agentat different concentrations increase the buoyancy property and porosity of the hydrogel matrixes but lowers the drug entrapment efficiency. Overall, on the buoyancy study, drug entrapment efficiency, and drug release pattern, it was demonstrated that $NaHCO_3$ at 10% concentration is the most effective pore forming agent for floating drug delivery system employing chitosan-graft-poly(N-vinyl pyrrolidone) as the matrix.

Acknowledgment

We would like to thank the University of Indonesia for supporting this study through Hibah Publikasi Internasional Terindeks untuk Tugas Akhir Mahasiswa No. 2011/UN2.R12/HKP.005.00/2016 and the Department of Chemistry University of Indonesia for the additional support.

References

[1] Tonkic A 2012 *Helicobacter* **17** 1-8.
[2] Bardonnet PL *et al* 2006 *J. Control. Release* **111** 11–18.
[3] Martindale 2009 *The Complete Drug* Reference ed Sweetman SC 36 (London: Pharmaceutical Press).
[4] Siepmann 2012 *J Fundamentals and Applications of Controlled Release Drug Delivery* ed (New York City: Springer) 21.
[5] Arora S *et al* 2005 *AAPS Pharm. Sci.Tech.* **6** 372–90.
[6] Patil J M *et al* 2006 *J. Sci. Ind. Res. (India)* **65** 11-21.
[7] Narang N 2010 Int. J. App. Pharm. **3** 1-7.
[8] Singh B N and Kim K H 1999 *J. Control. Release* 63 235-259.
[9] Pahwa R *et al* 2012 *Expert Opin. Drug Deliv.* **9** 525-539.
[10] Sokker H H *et al* 2008 *Carbohydr. Polym.* **75** 222-229.
[11] Zargar V *et al* 2015 *Chem. Bio. Eng. Rev* 2 1-24.
[12] Bhattarai N *et al* 2010 *Adv. Drug Deliv. Rev* **62** 83-99.
[13] Selvakumaran S *et al* 2016 *Carbohydr. Polym.***135** 207-214.
[14] Sahasathian T *et al* 2010 *Arch. Pharm. Res.* **33** 889-899.
[15] Srivastava A *et al* 2010 *Catbohydr. Polym.* **80** 790-798.
[16] Anasuya K V *et al* 2014 *Indian J. Adv. Chem. Sci.* **2** 12-15.
[17] Songsurang K *et al* 2011 *AAPS Pharm. Sci. Tech.* **12** 35-45.
[18] Choi B Y *et al* 2002 *Int. J. Pharm.* **239** 81-91.
[19] Jassem N A and Rajab N A 2012 *Karbala J. Pharm. Sci.* **4** 166-176.
[20] Krishnan V *et al* 2010 *Trends Biomater. Artif. Org.* **24** (3) 139-145.
[21] Mirzaei E B *et al* 2013 *Polym-Plast Technol.* **52** 1147-53.

IOP Conf. Series: Materials Science and Engineering **223** (2017) 012039 doi:10.1088/1757-899X/223/1/012039

Synthesis and characterization of PVA blended $LiClO_4$ as electrolyte material for battery Li-ion

I Gunawan, Deswita, B. Sugeng, Sudaryanto
Center of Advanced Materials Science and Technology, National Nuclear Energy Agency, South Tangerang, Indonesia

Email : irgun.IG@gmail.com

Abstract. It have been synthesized the materials for Li ion battery electrolytes, namely PVA with the addition of $LiClO_4$ salt were varied 0, 5, 10, 15 and 20% by weight respectively. The objective of this study is to control the ionic conductivity in traditional polymer electrolytes, to improve ionic conductivity with the addition of lithium perchlorat ($LiClO_4$). These electrolyte materials prepared by PVA powder was dissolved into distilled water and added $LiClO_4$ salt were varied. After drying the solution, PVA sheet blended $LiClO_4$ salt as electrolyte material for Li ion battery obtained. PVA blended $LiClO_4$ salt crystallite form was confirmed using X-Ray Difraction (XRD) equipment. Observation of the morphology done by using Scanning Electron Microscope (SEM). While the electrical conductivity of the material is measured using LCR meter. The results of XRD pattern of $LiClO_4$ shows intense peaks at angles $2\theta = 23.2$, 32.99, and 36.58°, which represent the crystalline nature of the salt. Particles morphology of the sample revealed by scanning electron microscopy are irregular in shape and agglomerated, with mean size 200-300 nm. It can be concluded that polycrystalline particles are composed of large number of crystallites. The study of conductivity by using LCR meter shows that all the graphs represent the DC and AC conductivity phenomena.

1. Introduction

Safety concerns have limited the full utilization of lithium-ion batteries. Extensive efforts have been made to formulate an electrolyte that is nonflammable and still works well. It has been pointed out that there is a trade-off between an electrolyte's flammability and its performance in a cell. One strategy is to employ gelled polymer electrolytes as alternatives to the currently used organic carbonate electrolytes. The polymer electrolytes have high thermal stability, but their lithium-ion conductivity is low, either due to high viscosity or due to low mobility ion-conducting mechanisms. Another popular means is to add flame-retardants to increase thermal stability and to decrease flammability. The reported flame retardant additives include organicphosphates, phosphites, triazine, organic halogens, biphenyls, and a combination of halogens and phosphates [1].

All-solid-state cells using a solid electrolyte instead of a liquid electrolyte have safety advantages such as nonleakage and nonflammability [2]. An all-solid-state cell with sulfur itself as an active material could hardly be charged and discharged. An all-solid-state cell using sulfur-based materials as a positive electrode and $Li_2S–P_2S_5$ glass–ceramics or $Li_2S–SiS_2$ glasses as a solid electrolyte exhibited excellent cycling performance with a large capacity at room temperature [3].

Solid polymer electrolytes (SPE) show a lower ionic conductivity than the liquid electrolyte; however, they are less reactive with lithium, increasing the safety of battery. SPE, such as PEO-Li–salt complex, can be introduced to restrain the dissolution of polysulfides into the electrolyte. However, the linear PEO-Li–salt polymer electrolytes generally have some problems in lithium polymer battery. The problems are active reactivity of lithium anode at high temperature and low ionic conductivity at low temperature. Some researchers have tried to overcome the problems by the use of ceramic filler such as Al_2O_3, SiO_2, and TiO_2 in PEO based polymer electrolyte. It was reported that the additive improved the conductivity of PEO-base electrolytes as well as their interfacial resistance between the electrolyte and the lithium electrode. Moreover, the additions of ceramic fillers are accompanied with amorphous PEO structure. The amorphous PEO electrolyte also improves ionic conductivity [4].

Polymer electrolytes (PEs) are ionically conducting materials that may be used in the fabrication of solid-state electrochemical devices, particularly rechargeable batteries, electrochromic displays, capacitors and sensors. During the last decades different systems have been extensively studied and most of them were based on poly(ethylene oxide). More recently new types of electrolytes based on natural polymers (like cellulose derivatives, chitosan, starch or natural rubber) have been proposed due to their biodegradability, low production cost, good physical and chemical properties and good performance as SPEs (solvent-free polymer electrolytes) [5].

Several studies have been done in order to increase the conductivity of PVA. Kim et. al has studied PVA containing the salt lithium trifluoromethane sulfonate ($LiCF_3SO_3$, LiTf) to overcome the low ionic conductivity PVA. It was found that the ionic conductivity of the PVA-based SPE increased with increasing salt concentration [6]. However, the relationship between the structure and properties of PVA based SPE is still needed to be explored.

In this study, we prepared PVA containing lithium perchlorat ($LiClO_4$) with various salt content to investigate the relationship between the bulk structure and electrical properties of SPE. The bulk structure is studied by scanning electron microscope (SEM) and X-ray diffraction (XRD). While the electrical properties is studied by electrochemical impedance spectroscopy (EIS). The objective of this study is to control the ionic conductivity in traditional polymer electrolytes, to improve ionic conductivity with the addition of lithium perchlorat ($LiClO_4$). This study is part of research material covering the battery anode, cathode and electrolyte materials in PSTBM, BATAN [7-10].

IOP Conf. Series: Materials Science and Engineering **223** (2017) 012039 doi:10.1088/1757-899X/223/1/012039

2. Materials and Methods

The powder of Poly Vinyl Alcohol, PVA (Aldricth, pro analysis, Germany) was dissolved into distilled water and added lithium chlorate salt $LiClO_4$ (Aldrich, pro analysis, Germany) were varied of 0, 5, 10, 15, and 20 % (w/w) respectively. After the drying process of the solution in an oven at a temperature of 70 ^oC, PVA sheet with addition of salt $LiClO_4$ obtained as an electrolyte materials for Li ion batteries. The PVA sheet with addition of $LiClO_4$ salt were then characterized by using X-Ray Difraction (XRD, Shimadzu XD 610, Jepang), Scanning Electron Microscope (SEM, JEOL JSM 6510 LA, Jepang) and conductivity measurement by using LCR meter HIOKI 3532-50, China. The X-ray diffraction (XRD) measurement was carried out using a Cu Kα radiation source (λ=1.5406 Å), with a step size of 4° min^{-1} from 5 to 90°.

3. Result and Discussion

PVA–$LiClO_4$ blended polymer electrolytes have been prepared by a solvent casting technique. XRD study revealed the occurrence of complexation between PVA and $LiClO_4$ in the composite matrix polymer. Figure 1 shows the XRD patterns of PVA with various weight ratio of $LiClO_4$. The XRD pattern of $LiClO_4$ shows intense peaks at angles $2\theta \approx 20$, and 41° marked by bar sign in the Figure 1. The typical XRD pattern of PVA with 20 % $LiClO_4$ listed in Table 1 below shows intense peaks at angles $2\theta = 19.85$, 19.90, 41.03 and 41.135^o, which represent the crystalline nature of the salt. As shown in Figure 1b, 1c, and 1d, the XRD pattern of PVA with various $LiClO_4$/PVA weight ratio show the incorporation of $LiClO_4$ into PVA decrease the crystallinity of PVA electrolyte. The calculated degree of crystallinity listed in Table 2. Figure 1 also shows the XRD patterns of Li-based PVA, most of the peaks pertaining to $LiClO_4$ disappeared in the Li-based PVA complexes, this indicates the complete dissolution of the $LiClO_4$ in the polymeric matrix. It is evident from Figure 1c and 1d that the amorphous nature is predominant in the Li-based PVA complexed system. It can be observed from the semi-crystalline peak of the PVA ($\theta \approx 20^o$) about the semi-crystal behavior of the PVA. The addition of $LiClO_4$ causes the decrease in the degree of the crystallinity and consequently the increase in the amorphicity of the material and its conductivity. The diffraction spectra of Li-based PVA showed the characteristic peaks of $LiClO_4$ predominantly, indicating that the crystal structure of $LiClO_4$ remained unaltered and no new phases were formed during the processing.

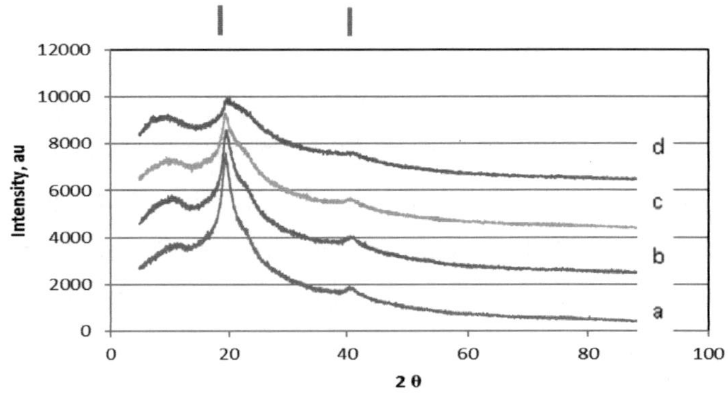

Figure 1. The XRD patterns of PVA with various weight ratio of $LiClO_4$, (a) 0% $LiClO_4$, (b) 5% $LiClO_4$, (c) 10 % $LiClO_4$ and (d) 20 % $LiClO_4$.

Figure 2a shows particle morphology of the sample revealed by SEM. The figure as seen is clear because there are no binder added yet. Meanwhile in Figure 2b, 2c and 2d correspond to the addition of $LiClO_4$ 5, 10 and 20 % into PVA matrix visible presence of particles in the matrix PVA. The particles are irregular in shape and agglomerated, with size about 200-1000 nm.

Table 1. The XRD pattern of PVA with 20 % $LiClO_4$.

Pos. [°2Th.]	Height [cts]	Rel. Int. [%]
19.8532	659.65	100.00
19.9030	329.83	50.00
41.0285	158.91	24.09
41.1350	79.46	12.05

Table 2. The calculated degree of crystallinity of composites prepared

Weigth percent of $LiClO_4$ in the PVA	degree of crystallinity %
0	14.03
5	13.11
10	10.96
20	10.79

It can be concluded that polycrystalline particles are composed of large number of crystallites. Due to pronounced particle agglomeration, grain boundaries are not quite clear, and within a relatively broad grain size distribution. On the thin particle edges a thickness of several nanometers can be observed. We believe that this layer is responsible for improved electrical conductivity. As more pores appear, the movement of electrons is resisted; and as the pore size increases, the charge-discharge efficiency decreases.

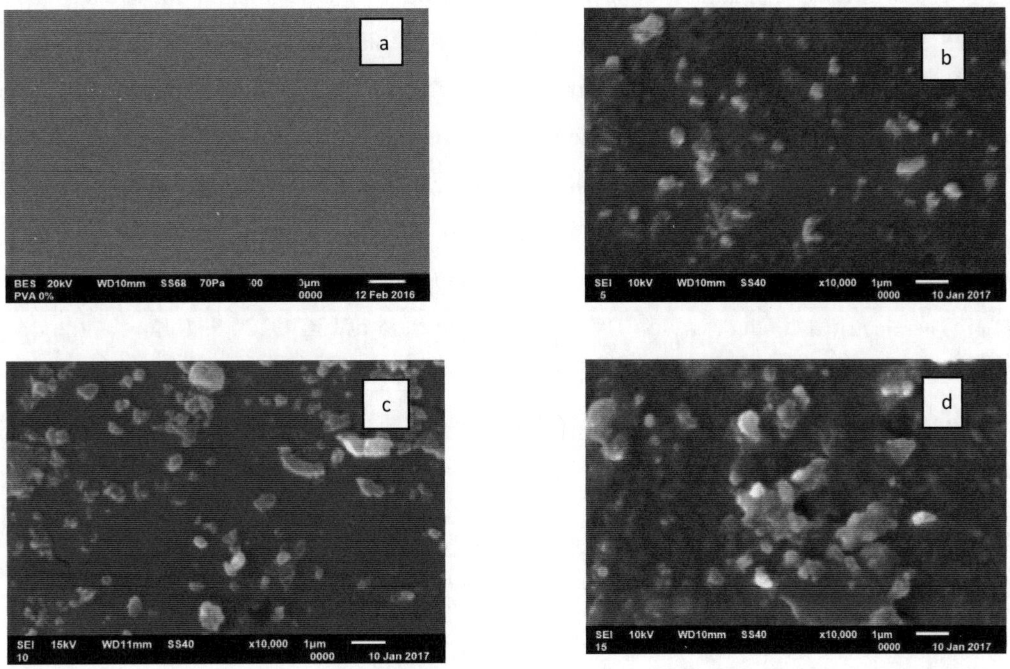

Figure 2. Morphology of the PVA–$LiClO_4$ blended polymer electrolytes at various weight ratio of $LiClO_4$, (a) 0% $LiClO_4$, (b) 5% $LiClO_4$, (c) 10 % $LiClO_4$ and (d) 20 % $LiClO_4$ revealed by scanning electron microscopy.

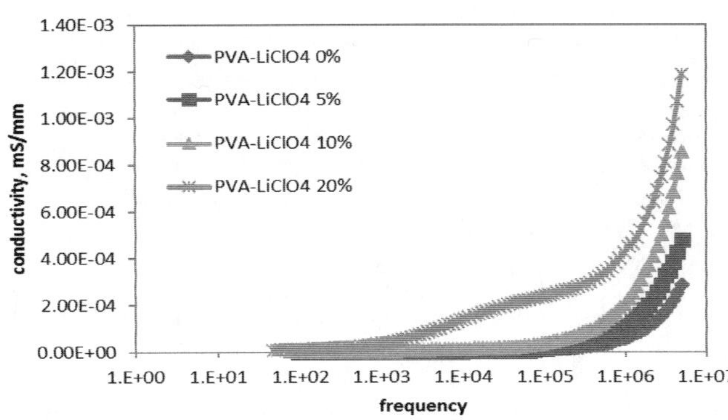

Figure 3. The electrical conductivity of PVA with various weight ratio of $LiClO_4$ as function of frequency.

The electrical conductivity of the electrolyte was calculated by equation, $\sigma = \lambda/(R_b X A)$ where R_b is the bulk resistance from AC impedance, λ the film thickness and A the surface area of electrode [11].

Figure 3 shows the electrical conductivity of PVA with various weight ratio of $LiClO_4$. The conductivity of the material as a whole is a function of frequency, the graph shows the electrical conductivity of PVA with weight ratio of $LiClO_4$ 0, 5, and 10 % have the same shape of straigth lines at frequencies up to 800 kHz, and then increased with increasing frequency. At low frequencies up to 800 kHz the electrical conductivity is measured following the phenomenon of DC conductivity, and then increased with increasing frequency following the phenomenon of AC conductivity. DC conductivity is frequency independent conductivity typically assigned to the hopping conduction and AC conductivity has a power law behavior in terms of frequency. There are three effects that contribute to the AC conductivity: (i) electrode effects (active at low frequencies), (ii) DC plateau (at intermediate frequencies) and (iii) defect process [12]. The electrical conductivity of PVA with weight ratio of $LiClO_4$ 20 % follow a straigth lines at frequencies up to 1 kHz, makes a curve at frequencies up to 800 kHz, and then increased gradually with increasing frequency. Table 1 and Figure 4 show the electrical conductivity of PVA with various weight ratio of $LiClO_4$ measured at frequency of 750 kHz.

Table 3. The Electrical conductivity of PVA with various weight ratio of $LiClO_4$ measured at frequency of 750 kHz.

$LiClO_4$ in the PVA (w%)	Electrical conductivity (750 kHz) mS/mm
0	9.48E-05
5	1.78E-04
10	1.86E-04
20	4.00E-04

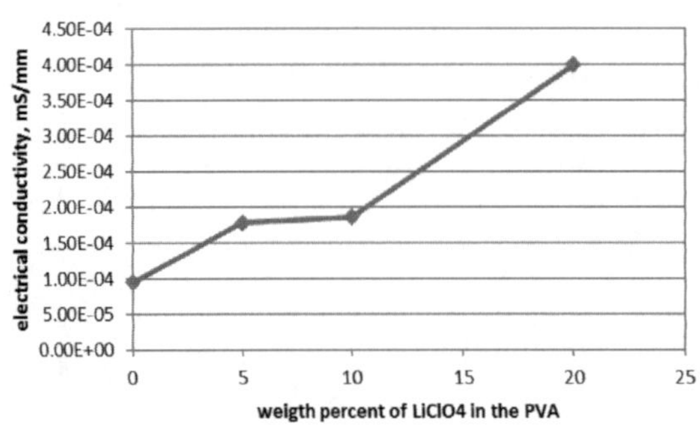

Figure 4. The electrical conductivity of PVA with various weight ratio of $LiClO_4$ measured at frequency of 750 kHz

It can seen in Table 3 and Figure 4 that the effect of $LiClO_4$ addition into PVA tends to increase the conductivity of the material, especially in the addition of 0 to 5% and from 10 to 20%. Meanwhile, in addition $LiClO_4$ as much as 5 to 10% the value of conductivity is almost the same. PVA is an insulating material, with the addition of a conductive salt $LiClO_4$ it is clear that the more salt $LiClO_4$ added the conductivity of the composite increases.

4. Conclusion

The XRD patterns of Li-based PVA, most of the peaks pertaining to $LiClO_4$ disappeared in the Li-based PVA complexes, this indicates the complete dissolution of the $LiClO_4$ in the polymeric matrix. The diffraction spectra of Li-based PVA showed the characteristic peaks of $LiClO_4$ predominantly, indicating that the crystal structure of $LiClO_4$ remained unaltered and no new phases were formed during the processing. For next study suggested to conduct exploration on the addition of 15-20% $LiClO_4$. Particle morphology of Li-based PVA revealed by scanning electron microscopy showed the particles are irregular in shape and agglomerated, with size under micron. It can be concluded that polycrystalline particles are composed of large number of crystallites. The electrical conductivity of PVA with weight ratio of $LiClO_4$ 0, 5, 10 and 20% measured at frequency of 750 kHz are 9.48E-05, 1.78E-04, 1.86E-04 and 4.00E-04 mS/mm respectively.

Acknowledgement

Authors would like to thank to Drs. Gunawan, MSc as Head of Center of Advanced Materials Science and Technology and DR. Abu Khalid Rivai as Head of Advanced Materials Science Division, National Nuclear Energy Agency.

References

[1] Li J, Daniel C, and Wood D 2011 *J. Power Sources* **196**(5) 2452–2460.
[2] Hayashi A *et al* 2003 *Electrochem. commun.* **5**(8) 701–705.
[3] Nagao M *et al* 2011 *Electrochim. Acta*, **56**(17) 6055–6059.
[4] Jeong S S *et al* 2007 *J. Power Sources* **174**(2)745–750.
[5] Leones R *et al* 2012 *Express Polym. Lett.* **6**(12) 1007–1016.
[6] Kim Y *et al* 2012 *J Nanosci Nanotechnol.* **12**(4) 3529–2533.
[7] Gunawan I and Wagiyo H 2015 *Macromol. Symp.* **353** (1) 225–230.
[8] Gunawan I *et al* 2011 *Proceedings of The International Conference on Innovation in Polymer Science and Technology* 53–57.
[9] Honggowiranto W *et al* 2014 *Proc.The International Conference on Materials Science and Technology* 497–504.
[10] Kartini E *et al* 2014 *14th Asian Conf. Solid State Ionics* **2** 978–981.
[11] Wang J *et al* 2003 *Electrochim. Acta* **48** (13) 1861–1867.
[12] Sadhukhan S 2011, *Preparation and Characterization of Polymer Electrolyte*, (National Institute of Technology, Rourkela, India).

IOP Conf. Series: Materials Science and Engineering **223** (2017) 012040 doi:10.1088/1757-899X/223/1/012040

Provision of micro-nano bacterial cellulose as bio plastic filler by sonication method

Maryam[1], D. Rahmad [1], Yunizurwan[1] , A. Kasim [2], Novelina[2], Emriadi[2]
[1]Polytechnic ATI Padang, Tabing, Padang, West Sumatra, Indonesia, 25171
[2]Andalas University, Pauh, Padang, West Sumatra, Indonesia, 25163

E-mail : maryam.atip@gmail.com

Abstract. Research and development of bioplastic has increased recently as a solution for substitution of conventional plastic which have many negative impacts to environment. However, physical properties and mechanical properties of its still lower than conventional plastic. An alternative solution for that problem is by using fillers that can increase the strength. Bacterial cellulose is considered as potential source for filler, but still need to be explored more. The privileges of bacterial cellulose are easy to get and does not have lignin, pectin, and hemicelluloses which are impurities in other celluloses. This research focused on gaining bacterial cellulose in micro-nano particle form and its impact on increasing the strength of bio plastic. Ultrasonication has been used as method to form micro-nano particle from bacterial cellulose. The result showed this method may form the particle size of bacterial cellulose approximately ± 3μm. Next step, after getting ± 3μm particle of bacterial cellulose, is making bio plastic with casting method by adding 1% of bacterial cellulose, from the total material in making bio plastic. Physical characteristic of the bio plastic which are tensile strength 11.85 MPa, modulus young 3.13 MPa, elongation 4.11% and density 0.42 g/cm^3. The numbers of physical properties showwthat, by adding 1% of bacterial cellulose, the strength of bio plastic was significantly increase, even value of tensile strength has complied the international standard for bio plastic.

1. Introduction

Research and development of bio plastic is increased recently, however in comparison bio plastic industry is stagnant. Bio plastic from starch takes about 50% of the market in bio plastic. Unfortunately, physical properties, mechanical properties and thermal properties of bio plastic from starch are lower than conventional plastic. Moreover, the strength, flexibility, and stability are considered as most important characteristic on producing bio plastic. Using filler which may increase those properties is an alternative solution for that problem [1].

Nanocellulosa particle is a new kind of cellulose form which is characterized by an increase in crystallinity, aspect ratio, surface area, and increase in the ability of dispersion and biodegradation. With those new characteristics, nanoselulosa particles can be used in a lot of fields, such as filler to reinforcing polymers, additives for biodegredable products, strengthening the membrane, thickener for dispersion, and media for carrying drug and implant [2]. Bacterial cellulose has several advantages as a source of cellulose compared to wood or non-wood resources, among others, it can be harvested in a shorter time, easily cultivated, and reduce the exploitation of forests to meet the needs of cellulose for industry. Furthermore, not like other cellulose, bacterial cellulose has no lignin, pectin, and hemicelluloses which are impurities in cellulose which means fewer steps in production. More studies on bacterial cellulose need to be done inorder knowing the use of bacterial cellulose for various areas of application not only limited on food product as we know recently.

Nanomaterial is a very attractive material compare to its macroscopic form due to significantly different characteristic. There is quantum attractive phenomena which appear due to particle size reduction into nano form. With nanotechnology every material can be lighter, more stable and increase in functionality. Nanotechnology offers many advantages and opputunities in the field of product innovation and development of bio plastic.

Nano material can be made by top-down method such as ball milling and ultasonication. Ultrasonic is sound vibration with higher frequency than human, which are approximately more that 20 kHz [3]. Ultrasonication is the most effective technique in mixing, reaction process, and breaking materials that use high energy input [4]. Up level of ultrasonic range for gas is 5 MHz and 500 MHz for liquid and solid material [5]. The wide range used of ultrasonic can be break into two part. First part is sound with low amplitude (high frequency). Usually use for analyzing speed and absorption of wave in the range 2 to 10 MHz. Second is high energy wave that located in 20 to 100 kHz. This wave can be used as cleaner, in the area of plastic composting, and organic or non organic materials modification [5].

High intensity ultrasonicationcan induce physical and chemical properties of material. One of physical effects of high-intensity ultrasonication is emulsification. This effect can be used in Some applications such as dispersion of filler in the polymer base, emulsification of inorganic particles inpolymer base, as well as the formation and cutting of plastic [6]. Chemical effects of ultrasonication can causes increasing interaction between molecules in materials that lead into changing of chemical properties of materials.The change is caused by ultrasonic wavelength, which is higher than molecules wave length.The wave interaction occurs through a liquid medium. The wave produced by the electric power is transmitted by the liquid media to the intended field through acoustic cavitations phenomena. This phenomenon makesa local pressure in the fluid and rising the temperature [7].Ultrasonication on liquids has various parameters such as frequency, pressure, temperature, viscosity, and concentration of sample. Ultrasonication affect the degradation of polymer [7].

2. Material and Method

2.1 Materials and Tools

Coconut water, urea, sugar, acetic acid, sodium hydroxide, sodium hypochlorite, sago starch, and glycerin were purchased from a local retailer. Distilled water was produced in own laboratorium.

Tools are hotpplate stirrer, beaker glass, beaker, 41 Whatman filter papers, plastic mould, Elmasonic S 300 H ultrasonic batch and Buchi B 290 spray dryer.

2.2 Methodology

This research was conducted in three stages: first is production of bacterial cellulose, second is making of micro-nano from bacterial cellulose that is produced in first stage, and final stage is manufacturing bioplastics. The method of making micro-nano cellulose bacteria is ultrasonication. The layer of bacterial cellulose was dryed then tested tensile. Bacterial cellulose was heated in NaOH 5%, 100°C for 4 h. After that, washed and dryed and then grinding. Cellulose fiber,water content, water absorbtion in the raw materials are calculated. Next step is ultrasonication of bacterial cellulose at 55°C for 12 h. This process is done in two times.

Producing nano bacterial cellulose was started by sieving it in 100 mesh sieve (±149 μm). Furthermore, this 100 mesh bacterial cellulose, is ultrasonicated with 20 kHz-10 MHz frequency. Ultrasonic wave would generate acoustic cavitations when put in liquid medium. In the process ultrasonic wave was beamed into liquid medium to produce cavitations bubble.

During the cavitations process, bubble collapse would appear which the break of wave by sound is. This phenomenon made a high energy hotspot. This hotspot is local heat around 5000 K with 1000 ATM pressure and heating rate approximately 1010 K/s which could change the bacterial cellulose into nano form. The final step was producing nano bacterial cellulose into powder. This process was done by drying nano bacterial cellulose with spray dryer. Casting method was used to manufacture the bioplastics. 1% nanoselulosa added into 100 mL of distilled water plus 3 mL of glycerol and 10 g of sago starch and then stirred for 1 h at 80-85°C. After 1 h, the solution then printed in the mould.

Analysis of the results was performed tensile test, SEM analysis, FTIR and density.

Scanning Electron Microscope Analysis

Determination of particle size and morphological surface with SEM Analysis. The morphology of micro-nano bacterial cellulose was observed by using a field emission scanning electron microscope (FE-SEM, S-3400N, Hitachi, Japan) at an accelerating voltage of 10 kV. Powdered micro-nano bacterial cellulose were spread onto the copper grids coated with a carbon supported film and a thin Au–Pd conductive coating was then used to reduce the charge effect. Photos is taken with magnification desirable. SEM analyzes were performed on the raw materials and the results of micro-nano bacterial cellulose.

Fourier Transform Infra Red Analysis

FTIR Analysis is an analytical technique used to identify organic, polymeric, and in some cases, inorganic materials. The FTIR analysis method uses infrared light to scan test samples and observe chemical properties.

Tensile Testing of Thin Plastic Sheeting (Film) ASTM D882

Tensile tests measure the force required to break a specimen and the extent to which the specimen stretches or elongates to that breaking point. Tensile tests produce a stress-strain diagram, which is used to determine tensile modulus. Specimens are placed in the grips of the Universal Tester and pulled until failure. For ASTM D882 the test speed and grip separation are based on the elongation to break of the material. Elongation and tensile modulus can be calculated from crosshead displacement, or with an extensometer. Tensile testing is performed on the raw material (bacterial cellulose) and bioplastics.

3. Result and Discussion

3.1 Bacterial Cellulose

Cellulose, which is produced with fermentation, is kind of microbial polysaccharide that formed by cellulose fibres from *Acetobacter xylinum*. This cellulose is famous as bacterial cellulose (BC). One of the food products made from coconut water are classified as food dessert that can be used as raw material for bioplastics is bacterial cellulose (bioselulosa) because base on the chemically classified cellulose. BC is a very unique material for cellulose is produced lignin free, has high mechanical properties and does not damage the environment (biodegradable) that can replace synthetic polymer that is currently widely used, both in the food industry and non-food. Bacterial cellulose is a polymer that has supramolecular and higher in crystallinity. Physical properties are yellowish white in colour, transparent, and formed in slippy and elastic layers.

Purification processes performed by heating in 5% NaOH. Purification is intended to eliminate the components of non-cellulose and residual bacteria that remain. The components of non-cellulose is expected to block the hydrogen bonding between cellulose molecular chains that lead to a lowering of the mechanical strength properties of cellulose.

Bacterial cellulose has 0.9-1.0% of cellulose fibers. Base on tensile testing analysis of bacterial cellulose has 380.56 Mpa in tensile strength test. The more layers formed the higher tensile strength number. Modulus young approximately 18 Gpa, elongation 11.3 % and density around 1.15 g/cm^3. Bacterial cellulose has high porosity, hydrophilic and 700% water absorption.

Figure 1. Bacterial cellulose (BC)

Bacterial cellulose is raw material in this research can be seen in Figure 1. The high water content in the bacterial cellulose (wet form) to be a problem in the drying process so that the necessary reduced in size, and then dried (dry form). Bacterial cellulose is reduced in size aims to remove the water that is bound in the material so that it will speed up the drying process. Dried bacterial cellulose is milled with a mill grinding to be a powder. The test results SEM the layer of bacterial cellulose can be seen in Figure 2. SEM analysis showed that the morphology of the fiber of bacterial cellulose is still composed and bonded with regularly.

Figure 2. SEM analysis bacterial cellulose

3.2 Micro-nano Bacterial Cellulose

Ultrasonication process has been successful in reducing the size of the bacterial cellulose particle to ± 3 μm. The test results SEM the particle of bacterial cellulose can be seen in Figure 3. SEM analysis results indicate that the particle size varies between 2.61 until 4.02 μm. It does not look homogenous. The overall condition of the particles seen from SEM images tend to clump together. Ultrasonication duration of the process may lead to particles that have been split to merge back so particles clump together from the SEM image. Clumps occur due to user does not use surfactants. Sonication treatment resulted in a decrease in the molecular weight of the bacterial cellulose which ultrasonic waves break the chemical bonds of bacterial cellulose. Because the bonds are broken, the size

IOP Conf. Series: Materials Science and Engineering **223** (2017) 012040 doi:10.1088/1757-899X/223/1/012040

becomes smaller. The longer the time giving ultrasonic waves to the solution of bacterial cellulose, cellulose chemical chain truncation process also continues to run. Jin Li *et al.*, (2008) claimed the fastest ultrasonic degradation occurs in polymers with the largest molecules. In this study, the degradation process is slow due to the size of the bacterial cellulose molecules are very small [8]. Hydrolysis process using hydrochloric acid (HCl) with 3.5 M concentration conditions, 55°C temperature for 6 h. This process has been successful in reducing the size of the bacterial cellulose fibers to ± 7 μm. SEM testing results against bacterial cellulose can be seen in Figure 3b [9].

Figure 3. SEM analysis results

Preparations micro-nano bacterial cellulose stored in liquid form can be seen in Figure 4(a), dry (b) and in packaged form (**c**). It can be applied as reinforcement in the manufacture of bioplastics.

(a) (b) (c)

Figure 4. Nano bacterial cellulose form liquid (a), dried (b), in package (c)

3.3 Physical Characteristic of Bioplastics

The physical characteristics of bio plastics which produced by adding 1% of nano bacterial cellulose are tensile strength of 11.85 MPa, Young modulus of 3.13 MPa, elongation of 4.11% and density 0.42 g/cm^3. The characteristic of tensile strength is lower than bioplastics from breadfruit starch-chitosan-sorbitol with a value of 16.34 MPa [10]. There is a possibility to increase the tensile strength by increasing the percentage of nanobacterial cellulose. Nevertheless, the elasticity is higher when compared to the starch bioplastics breadfruit-chitosan-sorbitol with a value of 2.72 MPa [10].

According to Bourtoom (2008) [11], to increase the elasticity of the bioplastic can be done by the addition of plasticizers. The tensile strength value fulfilled the standard for tensile strength values which is 10-100 MPa, but lower when compare to the polypropylene (conventional plastics) 24.7 MPa [12]. Test FTIR as an analysis conducted to determine the functional groups on the polymer. FTIR spectra generated from testing bioplastics are presented in Figure 5. Based on the image seen on the identification of functional groups in bioplastic films formed on IR absorption spectrum is a combination of the specific functional group contained in each of the constituent components of bioplastics. So it can be seen clearly that the bioplastics obtained a bioplastic produced through a process of mixing (blending), it can be seen because no new functional groups. In addition, it can be concluded also that the plastic is formed while still having hydrophilic properties such as its constituent components. The existence of functional groups such as CH (frequency 2900-3010 cm^{-1}), $C\text{-}NO_2$ (frequency 1300-1500 cm^{-1}), -N = N (frequency from 1575 to 1650 cm^{-1}) and - (CH_2) n (frequency 750 cm^{-1}), shows the plastic film can be degraded by either ground. Because nanotechnology is still relatively expensive, the recommendations are using bioplastics for intelligent/smart packaging.

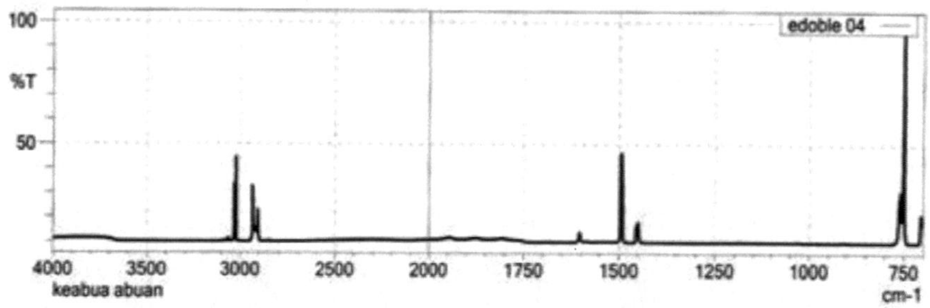

Figure 5. FTIR results

4. Conclusion

The particle size of micro-nano bacterial cellulose which was produced is ± 3 μm. The physical characteristics of bioplastics by adding 1% of micro-nano microbial cellulose which are tensile strength of 11.85 MPa, Young modulus of 3.13 MPa, elongation of 4.11% and density of 0.42 g / cm^3.

Acknowledgement

This work is supported by centre of training and education of the industrial ministry of Indonesia. Analysis was done at Polytechnic ATI Padang.

IOP Conf. Series: Materials Science and Engineering **223** (2017) 012040 doi:10.1088/1757-899X/223/1/012040

References

[1] Wicaksono R *et al* 2013 *J Tek Ind Pertanian.* **23** (1) 38-45.
[2] Ioelovich M 2012 *J. Nanosci. Nanotechnol* **2** (2) 9-13.
[3] Tipler PA 1998 *Fisika untuk Sains dan Teknik*. Prasetyo L & Adi RW. Jakarta: Erlangga.
[4] Pirrung M C 2007 *The Synthetic Organic Chemist's Companion*. New Jersey:John Wiley & Sons Inc.
[5] Mason T J, Lorimer J P 2002 *Applied Sonochemistry : The Uses of Power Ultrasound in Chemistry and Processing*. Verlag: Whiley-VCH.
[6] Suslick K S and Price G J 1999 *Annu. Rev. Mater. Sci.* **29** 295-326.
[7] Wardiyati S 2004 Prosiding Pertemuan Ilmiah IPTEK Bahan; Serpong, 7 Sep 2004. Serpong: P3IB Batan 419-424.
[8] Jin Li *et al* 2008 *J. Appl. Polym. Sci.* **109** 2417-2425.
[9] Maryam *et al* 2016 Laporan Penelitian. Politeknik ATI Padang.
[10] Setiani *et al* 2013 Valensi **3** (2) 100-109.
[11] Bourtoom T 2008 Songklanakarin *J. Sci. Technol.* **30** 149-165.
[12] Krochta J M, De Mulder-Johnston, C 1997 *J. Food Technol.* **51**(2), 61–74.

IOP Conf. Series: Materials Science and Engineering **223** (2017) 012041 doi:10.1088/1757-899X/223/1/012041

Calcium modified edible Canna (*Canna edulis* L) starch for controlled released matrix

A P Putri, M Ridwan, T A Darmawan, F Darusman and A Gadri
Department of Pharmacy, Bandung Islamic University, Bandung, Indonesia

E-mail : arlina.prima.p@unisba.ac.id

Abstract Canna edulis L starch was modified with calcium chloride in order to form controlled released matrix. Present study aim to analyze modified starch characteristic. Four different formulation of ondansetron granules was used to provide dissolution profile of controlled released, two formula consisted of 15% and 30% modified starch, one formula utilized matrix reference standards and the last granules was negative control. Methocel-hydroxypropyl methyl cellulose was used as controlled released matrix reference standards in the third formula. Calcium starch was synthesized in the presence of sodium hydroxide to form gelatinized mass and calcium chloride as the cross linking agent. Physicochemical and dissolution properties of modified starch for controlled released application were investigated. Modified starch has higher swelling index, water solubility and compressibility index. Three of four different formulation of granules provide dissolution profile of controlled released. The profiles indicate granules which employed calcium Canna edulis L starch as matrix are able to resemble controlled drug released profile of matrix reference, however their bigger detain ability lead to lower bioavailability.

1. Introduction

Canna indica L or known as achira is a perennial plant from South America, which it cultivation has expanded to Asia. Apart from being known as top 25 source of starch, considerable studies have suggested that achira rhizomes have immense potential not only for functional ingredient in food industry but also source of biopolymer for pharmaceutical industry [1]. Canna have been proven as promising biomaterial for industrial application based on their characteristic similarities with cassava starch that common for this utilization [2]. Starch is a familiar term for complex carbohydrate, $(C_6H_{10}O_5)_x$, that can be found in the seed, bulbs and tuber of plants, composed of plentiful of atoms which corresponding to value of x, as given formula [3]. Native starch denotes for untreated starch, that can be customize through physical, chemical and enzymatic methods to enhance their functional properties as called modified starch.

Starch polymer chain are commonly cross-linked with two- or poly-function compounds such as adipic acid, citric acid or phosphoric acid, to increase polymer molecular weight, stabilized granule structure, high temperature, acidity or shear resistant paste, whereas the result depend on material and method [4,5]. Calcium chloride are also a cross-linked agent, that usually use to modified alginate or pectin, adopt to bring the same effect for *Canna edulis* L starch and studies their properties and function as granule matrix utilize it gelling properties. Studies about utilization of calcium chloride as starch cross linking agent and it influence to starch physicochemical properties are still limited. Chowdary and Sundari (2008) was studied the utilization of calcium chloride as cross linking agent for potato starch and using the modified as matrix for diclofenac granules, but the information about characteristic of modified starch are still limited [6]. Ondansetron, for preventing postoperative nausea and vomiting and also prescribe for preventing nausea caused by pregnancy [7]. Controlled released formulation of ondansetron, aimed to tailoring drug released rate for increase patient compliance by reducing drug bioavailability and intake frequency [8].

In this study three of four formulation of ondansetron granules containing matrix granules were prepared, two formulas employing two different concentration of calcium *Canna edulis* L starch, and methocel-hydroxypropyl methyl cellulose at the third as controlled released matrix reference standards. The fourth formula set as negative controlled which only consisted drug and diluents. All formulas were evaluated for controlled release ability by analyzed drug dissolution profiles. The influence of calcium chloride as cross linking agent for *Canna edulis* L starch will be observed by physicochemical properties determination, included amylose content analysis, swelling index, water solubility, compressibility index, Hausner ratio and powder flow rate. The results shown that calcium modification on *Canna edulis* L starch influence it characteristic and dissolution profiles indicate that the modified starch regulated drug release in dissolution media.

2. Materials and Methods

2.1. Materials

The initial raw material, native *Canna edulis* L starch was isolated from achira rhizomes from Cibogo, West Java, Indonesia. Sodium hydroxide, calcium chloride, ethanol, acetic acid, iodine, hydrochloric acid, lactose monohydrate and amylum (soluble starch) were analytical grade reagents from Merck®. Methocel-hydroxypropyl methyl cellulose (Methocel K15M) also pure chemical from Sigma-Aldrich, Ondansetron hydrochloride di-hydrate was a give from PT. Combiphar Indonesia. Distilled water was used throughout the work.

2.2. Preparation of Cross-linked Starch

Cross-linked starch was prepared according to the method of Chowdary and Sundari (2008) [6]. Starch slurry from 5 part of *Canna edulis* L starch was dispersed in 50 part of distilled water, added with 10 % w/v NaOH. Solution was mixing for 30 min or continued until thick gelatinized mass was formed. A 300 mL $CaCl_2$ (20 % w/v) was added vessel contained with gelatinized mass while stirring at 1000 rpm stirrer. The stirring was remained for 1 h to precipitate calcium starch. The modified starch was collected by vacuum filtration, washed repeatedly with distilled water and dried at 80°C. Dried starch samples were then grounded in a mortar and sieved (mesh No. 80).

2.3. Physicochemical Analysis

2.3.1. Starch Amylose Content *Amylose Content Determination.*

was determined using calorimetric method [9]. Complex solution was obtained from 100 mg sample added with 1 mL 95 % ethanol and 9 mL 1 N NaOH, placed in 200 mL volumetric flask. The sample was heated for 10 min in water bath to gelatinized the starch, cooled and add with distilled water to up volume. A 5 mL aliquot was transferred to 100 mL volumetric flask, to which was added 1 mL 1 N CH_3COOH and 2 mL iodine solution, and the volume was made up to 100 mL with distilled water. Solution was homogenized by shaker and color intensity was determined by spectrophotometric method, that based on Beer's law, correspond to amylose content of samples, using a UV/VIS spectrometer (Shimadzu UVmini-1240).

2.3.2. Swelling Index and Water Solubility.

An aqueous solution of sample starch was made from 1 g of starch per 100 g of solution. The suspension was steady shaken for 30 min at 90°C in thermostatic water bath (Memmert). Subsequently it was centrifuged for 10 min using 80-2 Centrifuge at 3000 rpm. In separated supernatant, dry powder was determined gravimetrically, and the precipitate remaining in the centrifuge tubes was weighed [10].

2.3.3. Compressibility Index and Hausner Ratio.

Bulk and tapped density were resolute using 100 g sample. The bulk volume was measured after manually tapping the cylinder two times on flat table top surface. The tapped volume was amount with the Electrolab ETD-1020 Tapped Density Tester after tapping in increments of 10 and 500 taps. The bulk and tapped densities were used to calculated the Carr's compressibility index (Eq. 1) and the Hausner ratio (Eq. 2) to provide a measure of the flow properties and compressibility [11].

$$\text{CI} = \frac{\rho_{tap} - \rho_{bulk}}{\rho_{tap}} \, x \, 100 \qquad (1)$$

$$\text{HR} = \frac{\rho_{tap}}{\rho_{bulk}} \qquad (2)$$

Where ρ_{tab} is the tab density and ρ_{bulk} is the bulk density.

2.3.4. Flow Rate

Sample flow ability was defined by angle formed by the horizontal base of the bench surface and the edge of a cone-like pile of granules known as angle of repose. Stainless steel funnel with 15 mm orifice and 120 mm tall was used, placed 16 cm above the surface. After a cone from 100 g of samples was build, height (h) and radius (r) of the granules forming of cone of the base were measured. The angle of repose (θ) was calculated as follows:

$$\theta = \tan^{-1}\left(\frac{h}{r}\right) \tag{3}$$

2.4. *Granule Preparation*

Each pack of 600 mg granules contains of 16 mg of ondansetron hydrochloride di-hydrate were prepared using calcium starch and methocel K15M as matrix in different proportion, accept for control formula. The required quantities of drug matrix materials were mixed thoroughly in a mortar by following geometric dilution technique. Mucilago amylum was employing as binder and lactose as filler to form a dough mass and dried until mass water content less than 3 % afterward it passed through sieve. The mass was dried at 50°C and later it passed to mesh No. 16 to obtain granules.

2.5. *Ondansetron Determination*

Ondansetron was estimated by spectrophotometric (Shimadzu UV mini 1240) method based on measurement absorbance at 310 nm. The method obeyed Beer's law in the concentration 6-18 μg/mL.

2.6. *Drug Release Study*

Drug release from matrix granules was studied using 6 station dissolution tester (LID-6, Vanguard) paddle type. A 500 mL HCl 1 N was use as dissolution media, employing a paddle stirrer at 50 rpm and temperature set at 37±0.5°C. For different time intervals over period of 8 h, filtered samples of 5 mL of each station were withdrawn. For comparison, ondansetron release from methocel K15M and control, granules without matrix, are also studied.

2.7. *Statistical Analysis*

All experiments were performed in triplicate and Statistical Package for the Social Science (SPSS, Version 20.0) was used to statistically analyzed the result. The results were express as the mean ± SD and Paired T-test analysis express with alphabet notation (a and b) to indicate data significant differences ($P>0.05$).

3. Results and Discussion

3.1. *Calcium Starch Characteristic*

The modified *Canna edulis* L starch was successfully synthesize resulting off white odorless powder, which when dissolved in water giving alkaline pH. Physicochemical determination of modified starch was conduct to learn starch characteristic change that caused by forming of cross liked in starch molecules.

3.1.1. Amylose Content.

Blue color solution is formed form long branch chains amylopectin, like amylose, that bind with iodine configure a single helical complex during titration. The amount of amylose founded in calcium starch was about 26 percent lower than native (Table 1). Richard and William study found that the swelling behavior of starch is base on the property of its amylopectin content, and amylose acts as both a diluents and an inhibitor of swelling [12]. The maximal swelling are also related to the molecular weight and the shape of amylopectin [13]. The data shown in Table 1 indicate that most of swelling and water solubility of *Canna indica* L starch mostly influent by the shape of amylopectin, since higher amylose and lipid content can strongly inhibited swelling at all temperatures.

Table 1. Starch physicochemical characteristic before and after modification.

Physicochemical	Native starch	Calcium starch
Amylose content (%)	57.28±1.40^a	31.04±0.20^b
Swelling index (w/w)	2.13±0.10^a	3.44±0.11^b
Water solubility (%)	2.44±0.84^a	17.47±0.60^b
Compressibility index (%)	16.89±0.83^a	23.81±0.61^b
Hausner ratio	1.20±0.01^a	1.29±0.03^b
Angle of repose (θ)	Do not flow^a	Do not flow^a

All values are means of triplicate determination ± standard deviation

3.1.2. Swelling Index and Water Solubility.

Cross linking modification starch characteristic will be effected by chemicals and raw material [5]. Cross linked starch using different chemical and native starch tends to increasing starch structure crystallinity [14], lowering solubility [15] likewise the swelling index [16]. Other studies generate water insoluble characteristic [3] and higher degree of substitution will increased the effect [17]. Table 1 shown both swelling index and water solubility of native to modified starch have increasing value. Results presented the impact of cross linked agent gives unlikely outcome, the lower amylose content of calcium starch exhibit higher swelling index as well as water solubility. This findings can be affected by branch chain lengths distribution of amylopectin and X-ray type of starch which contribute to heterogeneous effect [22]. Characteristic of *Canna indica* L amylopectin allows starch to interact more with water molecules. Modification of calcium chloride provide the change of amylose and lipid solubility behavior in the granule, which allow the amylopectin to stretch and start starch gelatinization led to increase swelling volume and higher water intake. Starch swelling index can be used to foreseen gelatinization capability that will be exploited at controlled release mechanism. Increasing of molecular structure enlargement proficiency in water environment expected to prolong the drug-matrix binding, consequently time of release.

3.1.3. Flow Rate

Powder flow is a key requirement for pharmaceutical manufacturing process to understanding mixing, packaging and transportation. The compendium methods include several measurements among others are angle of repose, compressibility index and Hausner ratio [11]. The cross linked produced by calcium chloride upsurge starch compressibility index and Hausner ratio. Increasing value of both characteristics will aggravated powder flow rate. Both native and modified starch angle of repose cannot be identified because they were not able to form the cone in the base surface, both were not surge through the V-funnel. All three test indicate that modification enlarge starch particles size so it will takes higher force when compressing granules into tablet, raise the Hausner ratio and lowering the powder flow ability. This inability indicate that both starch will be suitable for wet granulation method in tabletization [18].

3.2. Drug Analysis

3.2.1. Ondansetron Determination

Amount of ondansetron in granules was estimated by UV spectrometric method based on the measurement of absorbance at 310 nm [19]. Series of concentration of reference ondansetron hydrochloride di-hydrate solution obtained valid linearity, y=0.0407x + 0.0138.

3.2.2. Release Profile

Dissolution profile of a granule or tablet depends on dissolution condition, composition, surface and hardness, among other factors [21]. Concentration of ondansetron that release from granules studied using dissolution apparatus with hydrochloric acid as dissolution liquid to mimicking digestive system. All four formulations of ondansetron, diluents and matrix was shown in Figure 1. Total drug that release from F1 was peaking immediately after 5 min and descent in line with increasing of time. Other formulation produces different outline where after the same onset drug release in constant rate.

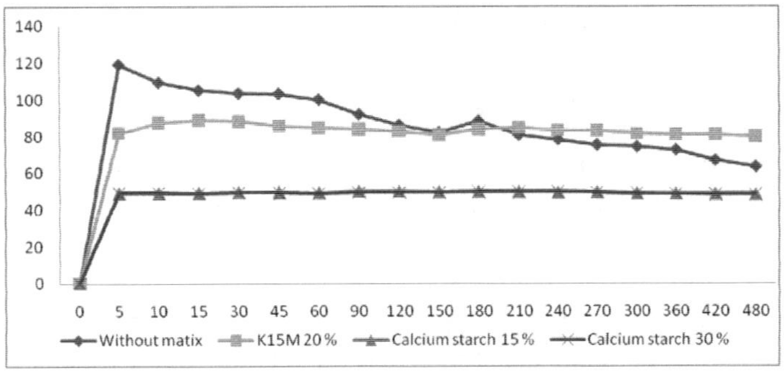

Figure 1. Ondansetron release from granule from four different formula shown that modified starch have controlled release profile.

The second formulation where metochel K15M employing as matrix give the appropriate profile as controlled released, more that 80 % of drug was in dissolution fluid. Although the profile of modified starch shown linier controlled profile but since the proportion of drug in liquid was less that level of requirement the formula optimization is still needed.

4. Conclusion

Calcium starch was synthesized from achira rhizome employing new cross linking agent produced off white and odorless powder with homogeneous particle size obtain by sieve with mess No. 80. The cross linked starch possesses a lower portion of amylose than native. Physicochemical study of the modified starch indicates that it suitable for controlled release matrix, with higher swelling index and water solubility. Based on the contradiction between amylose content and swelling behavior, indicate that in *Canna indica* L starch characteristic more affected by the shape of amylopectin.
Ondansetron granules are made with four different formulations, where two of it utilizing calcium starch as controlled released matrix. Drug release study was obtaining by observe granules dissolution profiles. Most of the water molecules interact with amylopectin structure, the entrapping mechanism inhibit by the strength of cross linked effect at the granules. The dissolution profiles indicate granules which employed calcium Canna edulis L starch as matrix are able to resemble controlled drug released profile of matrix reference, however their bigger detain ability lead to lower bioavailability.

References

[1] Andrade-Mahecha M M *et al* 2015 *Carborhydates* **123** 406–15.
[2] Valencia G A *et al* 2012 *J. Polym. Eng.* **32** 531–37.
[3] Jubril I, Muazu J and Mohammed G T 2012 *J. of App. Pharm. Sci.* **2**(2) 32–6.
[4] Wattanachant S *et al* 2003 *Food Chem.* **80**(4) 463–71.
[5] Kapelko M *et al* 2015 *Food Chem.* **167** 124–30.
[6] Chowdary K P R and Sundari P T 2008 *Int. J. Chem. Sci.* **6**(3) 1189–95.
[7] Tramèr M R *et al* 1997 *Anesthesiology* **87** 1227–89.
[8] Bhowmik D *et al* 2012 *The Pharma Innovation* **1** 24–32.
[9] William P C *et al* 1970 *Cereal Chem.* **47** 411–21.
[10] Kapelko M *et al* 2012 *Food Chem.* **135** 1494–1504.
[11] Rakhi B S *et al* 2008 *AAPS Pharm. Sci. Tech.* **9**(1) 250–8.
[12] Tester R F and Morrison W R 1990 *Cereal Chem.* **67(6)** 551–7.
[13] Alcázar-Alay S C, Meireles M A A *Food Sci. Technol.* **35(2)** 215-36.
[14] Atichokudomchai N and Varavinit S 2003 *Carbohydrate Polymers* **53(3)** 263–70.
[15] Yeh A I and Yeh S L 1993 *Cereal Chem.* **70(5)** 596–601.
[16] Hoover R and Sosulski F 1986 *Starch* **38(5)** 149–55.
[17] Koo S H *et al* 2010 *Food Hydrocolloids* **24** 619–25.
[18] Visavarungroj N *et al* 1990 *Int. J. of Pharmaceutics* **59(1)** 73–8.
[19] Dibbern H W 2002 UV and IR spectra: pharmaceutical substances (UV and IR) and pharmaceutical and cosmetic excipients (IR) (Aulendorf Germany).

IOP Conf. Series: Materials Science and Engineering **223** (2017) 012041 doi:10.1088/1757-899X/223/1/012041

[20] Smith A A *et al* 2009 *J. of Pharm. Sci. and Research* **1(4)** 48–54.
[21] Blanco M *et al* 2006 *J. Pharm. Sci.* **95(10)** 2137–44.
[22] Jane J *et al* 1999 *Cereal Chem.* **76(5)** 629–37.

IOP Conf. Series: Materials Science and Engineering **223** (2017) 012042 doi:10.1088/1757-899X/223/1/012042

Preparation of micro-fibrillated cellulose based on sugar palm ijuk (*Arenga pinnata)* fibres through partial acid hydrolysis

A Saputro, I Verawati, G Ramahdita, M Chalid
Department of Metallurgy and Materials Engineering, Universitas Indonesia, Jalan Kampus Baru UI, Depok, Jawa Barat, Indonesia

E-mail: chalid@metal.ui.ac.id

Abstract. The aim of this study was to isolate and characterized micro-fibrillated cellulose (MFC) from sugar palm/ijuk fibre (*Arenga pinnata*) by partial sulfuric acid hydrolysis. Cellulose fibre was prepared by repeated treatments with 5 wt% sodium hydroxide 2 h at 80 ^{o}C, followed by bleaching with 1.7 wt% sodium chlorite for 2 h at 80°C in acidic environment under stirring. MFC was prepared by partial hydrolysis with sulfuric acid in various concentrations (30, 40, 50, and 60 % for 45 min at 45 ^{o}C) under stirring. Fourier Transform Infrared, Field Emission Scanning Electron Microscope, Thermo Gravimetric Analyzer and X-ray Diffraction characterized cellulose fibre and MFC. FTIR measurements showed that alkaline and bleaching treatments were effective to remove non-cellulosic constituents such as wax, lignin and hemicellulose. FESEM observation revealed conversion into more clear surface and defibrillation of cellulosic fibre after pre-treatments. XRD measurement revealed increase in crystallinity after pre-treatments and acid hydrolysis from 54.4 to 87.8%. Thermal analysis showed that increasing acid concentration reduced thermal stability.

1. Introduction

Cellulose is the main constituent component responsible for the strength and stability of the plant cell wall. Cellulose has a linier chain of ringed glucose molecules created by repeating link of β-D-glucose [1]. This is confirmed by presence of hydroxyl groups (-OH) are participated the formation of strong inter- and intra-molecular bond [2, 3]. Cellulose has a wide range of potential application due to, low cost, renewable and biodegradability. It can be used as, reinforcing agent, biomaterials (drug delivery), energy harvesters, energy storage, and nanostructure via templating with nanocellulose [4-8]. Since cellulose has such big potential applications, several developmental efforts to convert cellulose into various forms such as microfibrillated cellulose (MFC), microcrystalline cellulose (MCC), nanofibrillated cellulose (NFC), and nanocrystalline cellulose (NCC) is still being conducted. There are two routes for conversion of cellulose into micro/nanocellulose, chemical route and mechanical route. Since the energy consumption of the mechanical route is rather high, chemical route seem a promising research topic [9].

Sugarcane, coconut, rice husk, pineapple, bamboo and bagasse are mostly used resource in Indonesia. Widely applied in Indonesia is also a kind of natural fibres isolated form sugar palm fibre *(Arenga pinnata)* or locally knows as ijuk. Ijuk fibres can be used as a broom, binders, filters and

IOP Conf. Series: Materials Science and Engineering **223** (2017) 012042 doi:10.1088/1757-899X/223/1/012042

rooftop. Ishak *et al.,* (2013), previously reported chemical and physical properties of ijuk fibre [10]. Mechanical properties were also reported such as; tensile strength (276.6 MPa), tensile modulus (5.9 GPa), and elongation (22.3 %) [10]. With superior mechanical properties, sugar palm fibre is expected to replace the use of synthetic fibres. Sahari *et al.,* (2013) made bio-composite of the sugar palm fibre reinforced with starch [11]. Bachtiar *et al.,* (2008) also reported effect of alkalinisation on mechanical properties sugar palm fibre reinforced epoxy [12]. Chalid *et al.,* (2015) reported that sugar palm fibre could be reinforced on PLA and act as a nucleating agent of for making polypropylene composite [2,13,14]. However, effects of partial acid treatment on cellulose ijuk fibre have not been exploited before. The aims of this paper are to analyse the effect of chemical treatment (pre-treatment and various concentration acid hydrolysis treatments) on cellulose ijuk fibre followed by characterization of the properties of the isolated MFC.

2. Materials and Methods

2.1. *Materials*

Ijuk *(Arenga pinnata)* fibre was purchased from PD Markum Jaya, South Tangerang, and Indonesia. Analytical grade sodium hydroxide (NaOH) pellet, 99% acetic Acid (CH_3COOH), 96.1% sulfuric acid (H_2SO_4), 25% sodium chlorite ($NaClO_2$) were purchased from Merck, German.

2.2. *Preparation of Micro-Fibrillated Cellulose from Ijuk Fibre*

Raw ijuk fibres were previously cut and crushed into small pieces around 30-40 mesh were washed thoroughly with tap water and dried at room temperature for 24 h. The fibres were treated with 5 w% sodium hydroxide solution at 80°C for 2 h in a solid to liquid ratio of 1:20 (w/v) to remove non-cellulosic components. This alkali treatment was repeated four times. After treatment, alkali-treated ijuk fibres were recovered by filtration and washed with deionized water until neutral pH. Afterwards, ijuk fibres were treated with acidified 1.7 wt % sodium chlorite solution at 80°C for 2 h in a solid to liquid ratio of 1:20 (w/v). After addition of sodium chlorite for four times, bleached ijuk fibres were recovered as described above. The bleached ijuk fibres were then partially hydrolysed with aqueous sulfuric acid solutions having four different concentrations 30, 40, 50 and 60 w% at 45°C for 45 min under stirring. Then, each suspensions was diluted for 5 fold with cold water. Materials remained unhydrolyzed were separated by decantation, and pH of the remained suspension was adjusted to 2-3. The samples were filtered; the filtrate was washed with 0.1 M sodium hydroxide until pH reaches 6-7. Samples were dried at 60 °C for 24 h. The samples abbreviation are presented in Table 1.

Table 1. Samples code

Sample Code	Treatments
Native ijuk	Without chemical treatment
Alkali-treated ijuk	Treated with 5 w% sodium hydroxide for 2 h, 80 °C, repeated 4x
Bleached ijuk	Treated with 1.7 w% sodium chlorite for 2 h, 80°C, repeated 4x, pH 4-4.5
Hdr 30	Treated with 30 w% sulfuric acid for 45 min at 45 °C
Hdr 40	Treated with 40 w% sulfuric acid for 45 min at 45 °C
Hdr 50	Treated with 50 w% sulfuric acid for 45 min at 45 °C
Hdr 60	Treated with 60 w% sulfuric acid for 45 min at 45 °C

2.3. *Characterizations*

Fourier transform infrared (FTIR) spectroscopy was used for functional group analysis of ijuk fibres. FTIR spectra at 450-4000 were recorded using a PerkinElmer UATR Two. Thermo Gravimetric Analyzer using Perkin Elmer STA 6000 under an argon atmosphere with gas flow of 10 mL/min and heating rate 10 °C/min from room temperature until 500 °C, analyzed the thermal behavior. The crystallinity index of the ijuk fibres were estimated by X-Ray Diffraction (Phillips *X ray Diffractometer*) under the condition with monochromatic Cu K-α radiation of λ= 15.418 nm; 40 kV, 40 mA with step size 0.02° and time/step around 20 s from 10° to 40° (2θ). The crystallinity index (CrI) was determined by an empirical method using the following equation by Segal, L *et al.*, (1959) [15];

$$CrI\ (\%) = \frac{(I_{002} - I_{am})}{I_{002}} x100\% \qquad (1)$$

Where I_{002} is the maximum intensity of the (002) lattice diffraction at $2\theta = 22.5°$ and I_{am} is the intensity of the scattered diffraction by the amorphous part of the sample at $2\theta = 18°$. The morphology of ijuk fibre was observed by INSPECT F50 Field Emission Scanning Electron Microscope (FE-SEM) (FEI Co., part of Thermo Scientific, Oregon, USA) after sputter coated with AuPd.

3. Results and Discussion

3.1. *Compounds Composition*

Figure 1 shows the FTIR spectra of the ijuk fibre samples recorded for (a) pre-treatments and (b) acid hydrolysis with sulfuric acid. The result shows that the peak at 3500-3390 cm $^{-1}$ wavenumber that represent C-H and O-H occur in every spectra, so do ~1635 and 2990-2880 cm^{-1} wavenumber also occur in every spectra. These functional groups are responsible to the water absorption and C-H stretch respectively [16, 17]. These three main peaks are hydrogen and hydroxyl bond, which occur in cellulose structure. This shows that the pre-treatment (alkali and bleached) and hydrolysis did not

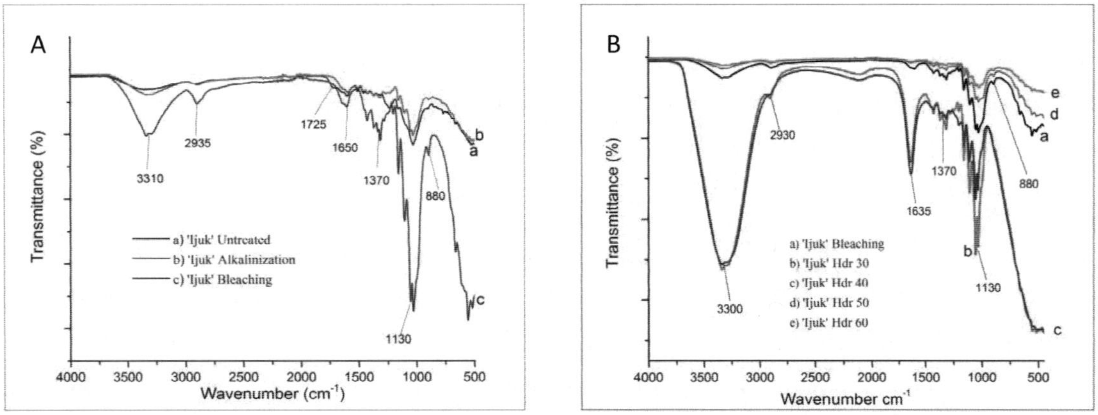

Figure 1. FTIR Spectra of ijuk fibre after (A) pre-treatments (B) various acid hydrolysis concentrations treatment

eliminate the cellulose from ijuk fibre [2,16,17]. Furthermore, 1700-1735 cm^{-1} wavenumber occur at untreated ijuk fibre. The wavenumber represent C=O functional group from acetyl and uranic ester group from hemicelluloses or ester linkage of the carboxylic group of lignin/hemicellulose [16,17]. As we can see, the prominent peak did not see at pre-treated fibre. It shows that pre-treatment was successfully eliminated lignin and hemicelluloses at certain amount. Same result reported by Kegarzadeh *et al.,* (2012) for kenaf fibre untreated and Johar *et al.,* (2012) for rice husk fibre untreated. Another peak at 1450-1330 cm^{-1} that occurs in every spectra show symmetric bending of C-H, also C-O and C-H from aromatic function group in polysaccharides [18]. The 1200-1030 cm^{-1} and ~880 cm^{-1} wavenumber peaks represent C-O and C-H function group from pyranose ring skeleton [19]. There are no significant differences between spectra and concentration in sulfuric acid except the intense of the absorption. This means there is no structure transformation of cellulose molecule after acid hydrolysis treatment [2,16,17].

3.2. Thermal Properties

Thermo Gravimetric Analysis (TGA) observed effect of multistage treatment in MFC preparation to thermal behavior of ijuk fibre. Figure 2 show TGA curve from untreated, alkalinization treated, bleach treated, and various acid hydrolysis concentrations treated of ijuk fibre. The result shows weight loss about 2-7 % happened at below 100 °C in almost every sample. It is occurred by vaporize of moisture content from every sample. At higher temperature, the weight loss increased because the fibre were starting to degrade. Compared to untreated and alkalinization treated fibres (Figure 2), bleached ijuk fibre has highest stability of thermal behavior. Bleached ijuk fibre is start to degrade at ~225°C while untreated and alkalinization fibre start to degrade at ~100 °C. This degradation temperature difference because effect of the hemicelluloses and pectin elimination.

Hemicelluloses and pectin have lower degradation temperature. So do the alkalinization, it has lower degradation temperature due to partial removal of non-cellulosic ijuk fibre [16]. Afterward, the effect of various sulfuric acid concentrations to thermal behavior of the ijuk fibre was observed. The results show that hydrolysis treatment with 60 % sulfuric acid has the lowest stability of thermal behavior, around ~140°C. It is caused due to the residual of sulfuric function group. Sulfuric function group gives lower the degradation temperature of ijuk fibre. Same result reported by Fahma *et al.,* (2011) that shows nanowhiskers cellulose from coconut husk, hydrolysis treated, has degradation temperature ~135°C with 65 % acid concentration. It also shows that the highest stability of thermal behavior occur with 30 % acid concentration [20]. The 30 % acid concentration has highest thermal behavior stability due to successfully removing sulfuric function group on MFC by sodium hydroxide. At high temperature, >350 °C, residual weight of native and alkalize fibre are higher than bleached, hydrolysis at 50-60 % due to non-cellulosic composition is higher in untreated, and alkalinization treated fibre than bleached, and hydrolysis treated. The pyrolysis of non-cellulosic produces higher residual weight than pyrolysis of cellulose [20].

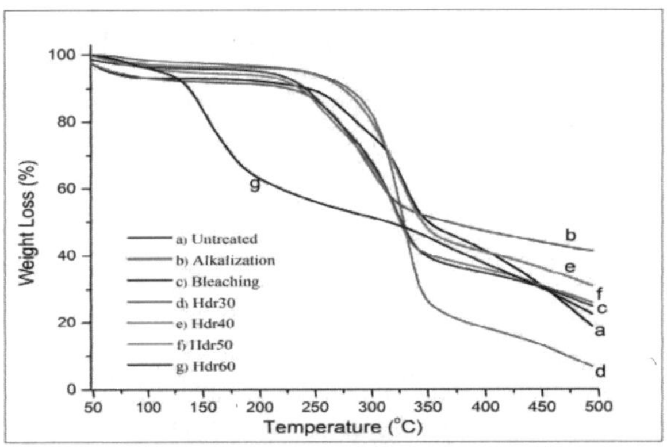

Figure 2 Thermal behaviour of ijuk fibre with various

3.3. *Crystallinity*

Observation of X-ray result to every chemical treatment and crystallinity index (CrI) which calculated using Segal equation (1) could be seen in Figure 3a-b. Figure 3a show the effect of pre-treatments to the crystallinity of ijuk fibre. The results show a significant enhancement of crystallinity compared to the crystallinity of untreted ijuk fibre. Based on Figure 3b, crystallinity of native ijuk fibre is 57.1 %, increased to 63.1 % by alkalinization, and increased to 85.7 % by bleaching treatment. It shows that the pre-treatment was effectively eliminating the non-cellulose compounds (pectin, hemicelluloses and lignin) which have amorphous structure due to the complexity of the molecule structure. In Figure 3a, all diffraction has the same pattern, and pre-treatment of ijuk fibre increase the sharpness of the peak, which indicates the increase of crystallinity.

Its peak can see the typical of cellulose. At peak about 2θ = 18° indicates amorphous zone and 2θ = 22.5°, 34-35° indicates crystalline zone. The increase of crystallinity by pre-treatment is consistent with our previous observation before [2]. Effect of hydrolysis concentration on crystallinity of ijuk fibre can be seen in Figure 3a-b. Based on calculation, sulfuric acid with 60 % concentration has highest crystallinity index compared to 30, 40, and 50% concentration. Hydronium ion from sulfuric acid was expected to penetrate amorphous region of cellulose, causing hydrolytic cleavage of glycoside bond and the release of individual crystallite [21]. Hydrolysis with 30 and 40%, concentration do not cause any significant change to the crystallinity of ijuk fibre [16,17,20,22].

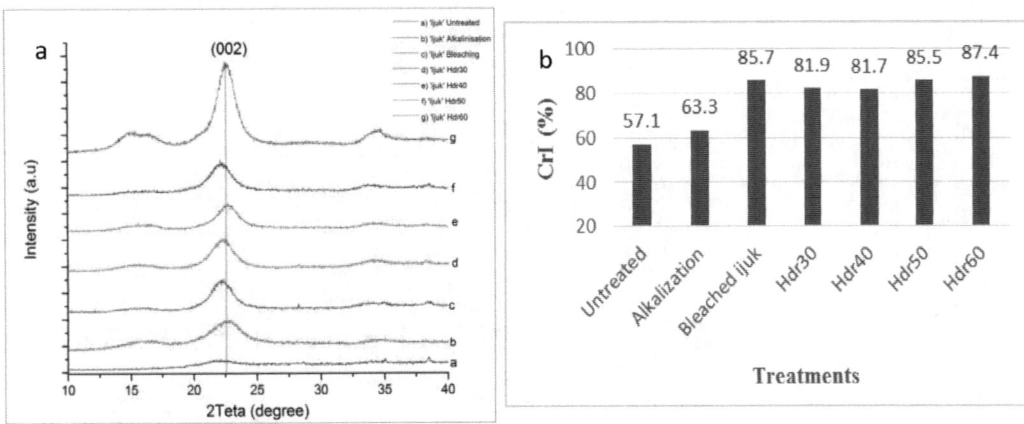

Figure 3. a) X-ray diffactogram b) Crystallinity Index calculations of ijuk fibre with various treatments

3.4. *Morphology*

The microscopic evolution of the ijuk fibre after each stage of chemical treatment is present in Figure 4a-g. The images show that pre-treatment of ijuk fibre clearly modified the fibre surface (Figure 4a-c). Figure 4a is the image of native ijuk fibre. Native ijuk fibres still contain impurities and wax layer. The diameter size of untreated ijuk fibre is around 50-80 μm. Alkalinization-treated ijuk fibres produce an impurities-free surface (Figure 4b). Alkalinization treatment also promotes the defibrillation phenomena. Defibrillation phenomena decrease the diameter size of ijuk fibre caused the size of the fibre decreased from 100 μm to 10-20 μm but not entirely. Cleaner surface of ijuk fibre increase the effectiveness of chlorite to oxidation hemicelluloses and lignin. Figure 4c shows ijuk fibre after oxidation process. It shows clear image of ijuk fibre, and as we can see, the diameter size reduce to <20 μm entirely. Pre-treatment process has reduced the ijuk fibre size to MFC. The size reduction of ijuk diameter is also consistent with our previous report before [2]. The diameter size is decrease from native cause the separation of the fibre primary cell wall due to removal of non-cellulosic contents [23].

The effect of different hydrolysis concentration can be seen in Figure 4d-g. Increasing sulfuric acid concentration leads to decreasing diameter of ijuk fibre. Figure 4d-g shows the size of ijuk with 30, 40, 50, 60 % of hydrolysis concentrations are ~15 ; 10 ; 2.5 ; 2 μm respectively. This shows an increase in the concentration of sulfuric acid, has also increased the content of hydronium ion (H^+), which responsible for penetrating amorphous area compared crystalline structure during 60% acid hydrolysis

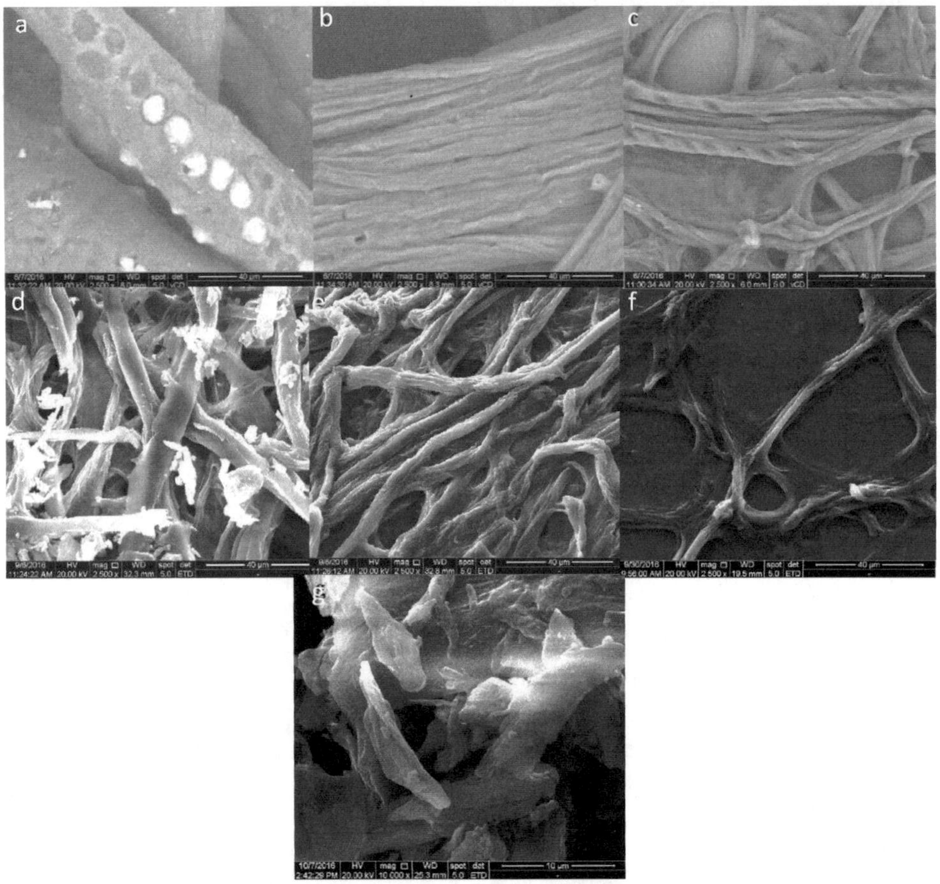

Figure 4. FESEM images of ijuk, a) untreated, b) alkalinization, c) bleaching, d) hdr30, e) hdr40, f) hdr50 and g) hdr60

concentration [24]. However, there is no significant difference to the morphological behaviour of ijuk fibre. Nevertheless, the change is which occur only in the fibre hydrolysis 60% (Figure 4g). This expected to the start of the nanocellulose formation due increasing amount hydronium ion (H^+) promotes hydrolytic cleavage of the amorphous region, but not enough to separate fibre into nanosize

[24]. The addition of temperature, time, and a higher concentration suggested capable to form nanocellulose fibre.

4. Conclusion

Cellulose fibres were extracted form ijuk fibre by alkalinisation followed bleaching treatment. Cellulose characterized by FTIR spectroscopy, thermal behaviour, X-Ray diffraction and Electron microscope examination. The results show pre-treatments has successfully removed non-cellulosic contents, increased thermal stability, enhanced crystallinity and caused defibrillation of cellulose. The effect of different concentration of sulfuric acid causes size reduction of ijuk fibre and increase crystallinity around 87.1%. In contrast, sulfuric acid (60%) reduce thermal stability of ijuk due the present of sulphate group. Electron microscope examination showed hydrolysis with a concentration of 60% has not enough to reduce microfibrillated cellulose into nano size.

Acknowledgment

The authors would like to thank to Ministry of Research and Higher Education Republic of Indonesia for funding this research through Program Kreativitas Mahasiswa (PKM-P) 2016.

References

[1] Moon R J *et al* 2011 *Chem. Soc. Rev* **40** 3941-94 .
[2] Yuanita E *et al* 2015 *Procedia Chem.* **16** 608-15.
[3] Kabir M M *et al* 2012 *Compos: Part B* **43** 2883-92.
[4] Kim J H *et al* 2015 *Int. J. Precis. Eng Manuf. - GT* **2** 197-213.
[5] Ng H M *et al* 2015 *Compos: Part B* **75** 176-200.
[6] Peng B L *et al* 2011 *Can. J. Chem. Eng.* **9999** 1-16.
[7] Roman M and Dong S 2007 *J. Am. Chem. Soc.* **129** 13810-811.
[8] Roman M *et al* 2009 *J. Am. Chem. Soc.* **1017** 81-91.
[9] Khalil H A *et al* 2014 *Carbohydr. Polym.* **99** 649-65.
[10] Ishak M *et al* 2013 *Carbohydr. Polym.* **91** 699-710.
[11] Sahari J *et al* 2013 *Mater. Des.* **49** 285-89.
[12] Bachtiar D *et al* 2008 *Mater. Des.* **29** 1285-90.
[13] M. Chalid *et al* 2015 *Mater. Sci.* **827** 326-31.
[14] Chalid M and Prabowo I 2015 *Int. J. Chem. Mol. Nuclr. Mater. Metal. Eng.* **9** 342-46.
[15] Segal L *et al* 1959 *Text Res J* **29** 786-94.
[16] Johar N *et al* 2012 *Ind. Crops Prod.* **37** 93-9.
[17] Kargarzadeh H *et al* 2012 *Cellulose* **19** 855-66.

IOP Conf. Series: Materials Science and Engineering **223** (2017) 012042 doi:10.1088/1757-899X/223/1/012042

[18] Janoobi M *et al* 2009 *Bioresour.* **4** 626-39.
[19] Alemdar A and Sain M 2008 *Compos. Sci. Technol.* **68** 557-565.
[20] Fahma F *et al* 2011 *Cellulose* **18** 443-50 .
[21] Lima M d S and Borsali R 2004 *Macromol. Rapid. Comm.* **25** 771-87.
[22] Lu P and Hsieh Y 2012 *Carbohydr. Polym.* **87** 564-73.
[23] Maheswari C *et al* 2012 *Biomas. Bioenergy.* **46** 555-63.
[24] Chena Y W *et al* 2016 *Carbohydr. Polym.* **151** 1210–19.

IOP Conf. Series: Materials Science and Engineering **223** (2017) 012043 doi:10.1088/1757-899X/223/1/012043

Formulation and synthesis of hydrogels having lower critical solution temperature near body temperature

A Z Abidin, H P R Graha and D A Trirahayu
Department of Chemical Engineering, Faculty of Industrial Technology
Institut Teknologi Bandung
Jl. Ganesa 10 Bandung, Indonesia 40132

Email: zainal@che.itb.ac.id

Abstract. Copolymerization between bacterial cellulose nanocrystal (CN) and methyl cellulose (MC) was carried out using UV light to produce a biocompatible hydrogel at body temperature and liquid at room temperature. Viscosity and salt effect of the MC and copolymer solution at room temperature and its Lower Critical Solution Temperature (LCST) were evaluated. The analysis showed that the higher concentration of methyl cellulose and salt content in the solution produced lower LCST and higher solution viscosity. All samples of polymer solution with MC concentrations of 1 and 2% have a viscosity less than 5000 cP at room temperature. The solutions with MC concentration of 1, 2, and 3% have respectively LCST of 59, 58, and 57°C, while its copolymer solutions with CN concentration of 0.1, 0.3, and 0.5% have respectively LCST of 55, 51, and 41°C. The salt addition to the solution of MC-CN copolymer with concentrations of 1x and 1.5x Phosphat Buffered Saline (PBS) produces respectively LCST of 47 and 38°C. The results suggest that the copolymer solution of MC-CN could produce a lower LCST and the addition of salt could amplify the effect of LCST decrease that can be used to produce a biocompatible hydrogel with LCST as close as body temperature.

1. Introduction

Bone damaged such as bone fractures take a long time to recover. In fact it is possible that the patients will suffer permanent disability. Recent solution to overcome this problem is using polymer based biomaterial, a non-living material used in the medical field and is intended to interact with biological systems [1]. In 2005 the worldwide biomaterials market reached $300 billion with a projection to continuesly grow at a rate of 20% per year [2]. New challenges in human health problems answered by the development of a revolutionary biotechnology, such as the development of artificial organs, tissue engineering, and controlled drug delivery system that cannot be separated from the role of biomaterials, especially biocompatible polymer hydrogel.

Hydrogel is an absorbtive material made from polymers with a hydrophilic group bonded (crosslinked) physically or chemically so that it can absorb and hold water in three-dimensional structure without solute [3]. Some hydrogels show physical properties changing due to environmental condition changes and called as smart hydrogel [1]. The hydrogel type that is suitable for bone tissue engineering is called temperature responsive hydrogel [4]. Some polymer solutions have Lower Critical Solution Temperature (LCST), a temperature below which polymer solution remains in solution form (one phase) and above which the polymer solution will separate into two phases, i.e.: liquid phase that contains no polymer and solid phase that riches with polymer [5]. These changes can take place reversibly. In the bone healing issue, LCST should be as close as possible to the physiological body temperature of about 37°C.

The common polymer used in smart hydrogel research is poly N-Isopropyl Acrylamide (P-NIPAAm) because it has LCST close to the physiological body temperature. Unfortunately, the N-Isopropyl Acrylamide as the monomer of P-NIPAAm is very expensive. Therefore, in the present research, other alternative polymeric materials which are cheaper and have a LCST as that of P-NIPAAm is used.The alternative material is modified methyl cellulose (MC) polymer. MC polymer has a LCST much higher than body temperature without any modification.

Many Researchers have worked in evaluating the temperature-sensitive properties of MC. Li (2002) conducted a study on the characteristics of methyl cellulose solution. This study revealed that methyl cellulose had LCST temperature of about 60°C [6]. Modification of MC thorough combination of MC with carboxy methyl cellulose, polyethylene glycol and chitosan had also been done and showed a potential to be an injectable material for post-surgery glue [7]. Methyl cellulose can also be combined with agarose for injectable property and can quickly turn into a solid gel at room temperature [8]. The combination of chitosan and methyl cellulose in various salt solutions were also made [9]. Another research on modifying properties of MC is by chemical crosslinking either with a cross linker glutaraldehyde or with gamma radiation [10].

In the previous works by the authors, the experiment have shown that the composite of bacterial cellulose and polyvinyl alcohol (PVA) showed potential uses in bone healing [11]. Cellulose is biocompatible [12] and has a fairly high tensile strength, the ability to store water in large quantities, high crystallinity, and easy to set up (high mouldability) [13]. The main issue in using this composite is that it is in solid form at room temperature and requires surgery to install. Our research here is focusing on formulation and synthesis of thermosensitive hydrogel from bacterial cellulose, that has liquid phase at room temperature so it is injectable and has solid phase at body temperature so the hydrogel solidifies.

2. Materials and Methods

2.1 Materials

Materials used in this study are commercial nata de coco as bacterial cellulose source, technical NaOH, Methyl cellulose SM4000, NaCl, KCl, Na_2HPO_4, KH_2PO_4, HCl 95%, Phosphat buffered saline

IOP Conf. Series: Materials Science and Engineering **223** (2017) 012043 doi:10.1088/1757-899X/223/1/012043

(PBS) solution, benzophenone, citric acid, and ethanol 95%. Home-made UV chamber was used for copolymerization process. The chamber has four 9W UV lamps.

2.2 Method

2.2.1. Synthesis of Cellulose Nanocrystals

The synthesis of cellulose nanocrystals consist of 2 steps: First pretreatment, nata de coco was soaked in NaOH 1 M for 2 h at 80°C then washed several times. Next, hydrolysis of cellulose was performed with 65% sulfuric acid for 1 h at 45°C. Then filtered using Whatman 42 filter paper several times until the solution become clear that indicated that most cellulose nanocrystals are on the filter paper and then dried.

2.2.2. Making MC Solution

The solvent used to make MC solution here were distilled water, NaCl solution, and PBS solution. MC solution was made by 2 step dissolution: The first used hot dissolution at 80°C and then continued with cold dissolution at room temperature.

2.2.3. Preparation of MC-CNCopolymer

The copolymer was made from MC and CN monomers. The CN monomer was prepared by sonication of CN in 30 mL of distilled water for 2 h. CN suspensions then stirred with a magnetic stirrer at a speed of 1500 rpm for 15 min. Benzophenone (0.01% g/v solution) was dissolved in 95% ethanol in 1 mL. Benzophenone solution and citric acid (at 0.05% g/v solution) was added to CN solution. The solution then combined with MC solution to achieve desired composition and stored in refrigerator for minimal 4 h. The mixture solution of MC and CN was poured into a petri dish and placed into UV-Chamber. UV exposure had been given for 30 min and then the copolymer product was stored in polyethylene plastic vials.

2.2.4. Characterization

CN size was characterized using a Particle Size Analyzer (PSA) machine. The viscosity of the polymer solution was measured using Brookfield RV viscometer at room temperature. FTIR characterization was conducted to analyze the copolymerization result by comparing the FTIR spectra of precursors with that of copolymer product. There are two methods of LCST measurement used in this study. The first is spectrophotometric method, used for polymer solutions with salinity variations without CN. For each polymer solution, transmittance of wavelength 500 nm was measured at temperatures varied from 30 to 70°C with 2°C interval. Then its LCST was determined at transmittance of 0.5. Spectrophotometric method could not be applied to copolymer samples because the samples were cloudy so the measurement would not be accurate. Instead the LCST was determined using Differential Scanning Calorimetry (DSC) tests by adding water as solvent and varying temperature from 30 to 58°C, then the results verified by visual observation.

IOP Conf. Series: Materials Science and Engineering **223** (2017) 012043 doi:10.1088/1757-899X/223/1/012043

2.2.5. Experimental Variation

Sample variation in the experiment shown in Table 1. Each sample is given a special name to make easy the analysis.

Table 1. Experimental Variations and Sampel Name

No	Variables	Variations	Sampel Name	No	Variables	Variations	Sampel Name
1	Methyl Cellulose and NaCl Concentrations (12 sampels)	MC 1% - NaCl 0%	MC1	3	Methyl Cellulose and PBS Concentration (9 sampels)	MC 1% - PBS 0.5x	MCa
		MC 1% - NaCl 0.5%	MC2			MC 1% - PBS 1x	MCb
		MC 1% - NaCl 0.7%	MC3			MC 1% - PBS 1.5x	MCc
		MC 1% - NaCl 0.9%	MC4			MC 2% - PBS 0.5x	MCd
		MC 2% - NaCl 0%	MC5/A			MC 2% - PBS 1x	MCe
		MC 2% - NaCl 0.5%	MC6			MC 2% - PBS 1.5x	MCf
		MC 2% - NaCl 0.7%	MC7			MC 3% - PBS 0.5x	MCg
		MC 2% - NaCl 0.9%	MC8			MC 3% - PBS 1x	MCh
		MC 3% - NaCl 0%	MC9			MC 3% - PBS 1.5x	MCi
		MC 3% - NaCl 0.5%	MC10	4	PBS Concentration (2 sampel)	PBS 1x	F
		MC 3% - NaCl 0.7%	MC11			PBS 1.5x	G
		MC 3% - NaCl 0.9%	MC12				
2	Nanocrystal Cellulose (NC) Concentration (4 sampels)	NC 0%	B				
		NC 0.1%	C				
		NC 0.3%	D				
		NC 0.5%	E				

3. Results and Discussion

3.1. Celulose Nanocrystal

Cellulose particle size distribution is shown in Figure 1. Assuming all particle size are homogeneous, the smallest size of cellulose nanocrystal is 127 nm and the average size is 171 nm. Based on this distribution it can be stated that the cellulose nanocrystal has the right size distribution.

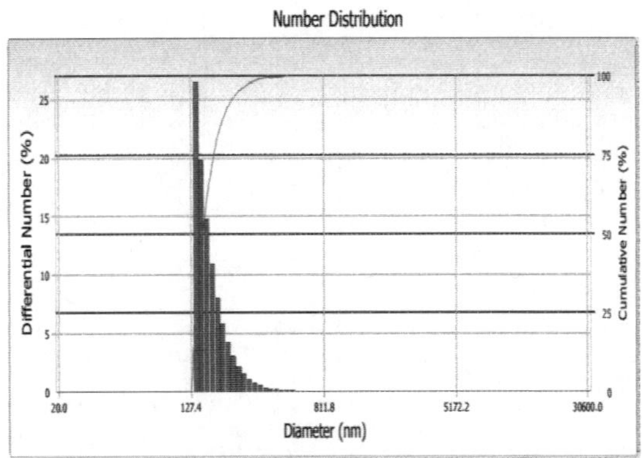

Figure 1. Cellulose Nanocrystal Particle Size Distribution

3.2. Copolymerisation

To evaluate the copolymerization of methyl cellulose and cellulose nanocrystal, the IR spectrum of the copolymer is compared to the spectrum of methyl cellulose (Figure 2). The copolymerization is indicated by weakening the -OH peak at wave number around 1315 cm^{-1} that found in all samples. The peak weakening is caused by some -OH groups of CN react to form –C-O-C- groups. Another peak representing-OH groups with hydrogen bonds is indicated by the wave number 3500-3200 cm^{-1}, which was found in all samples, including the reference sample (A). The peak is weaker in all samples containing CN (C, D, E, F, and G), though the different may not be significant.

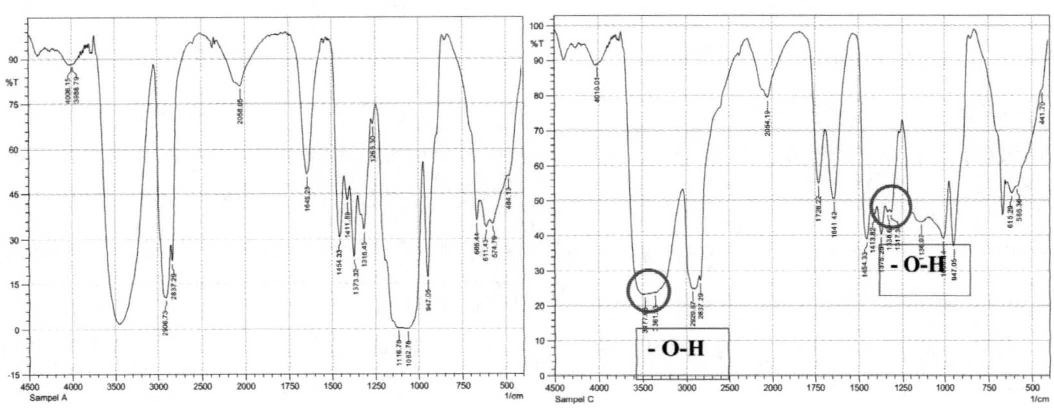

(a) Sample A (b) Sample C

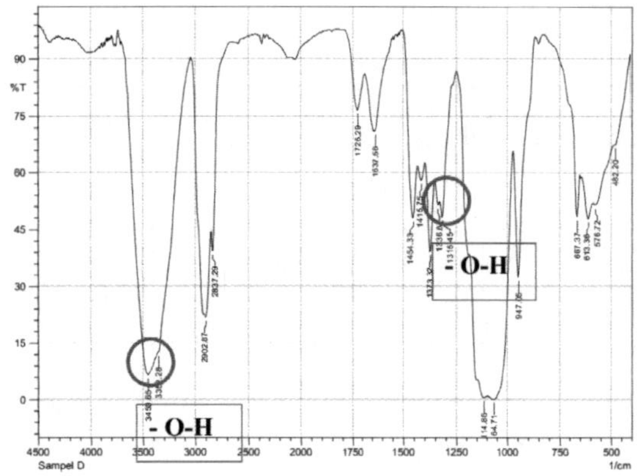

(c) Sample D

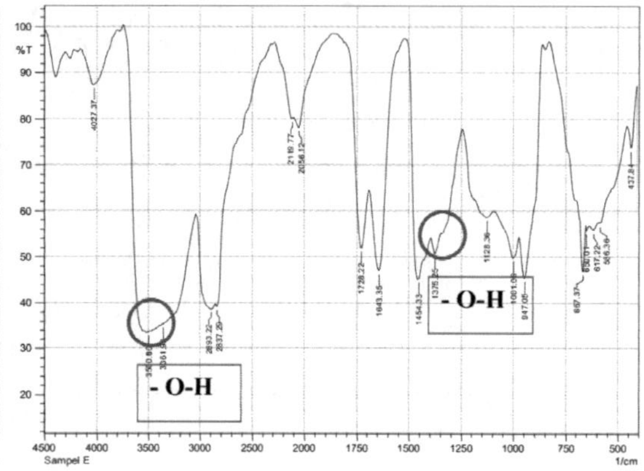

(d) Sample D

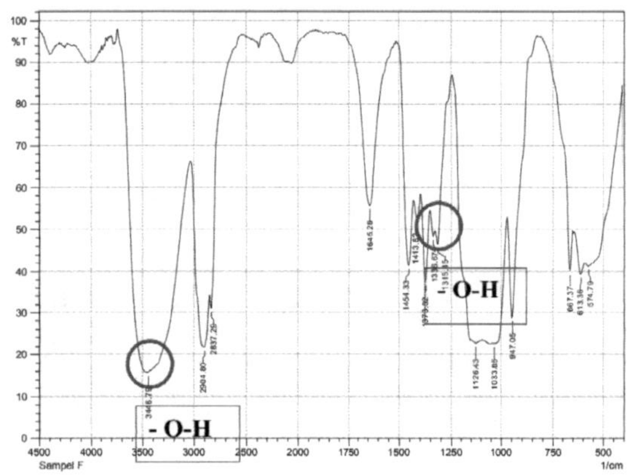

(e) Sample F

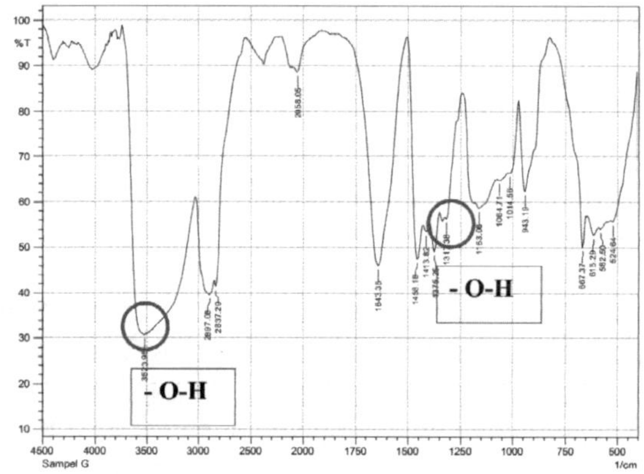

(f) Sample G

Figure 2. Spectrum Comparison of The Copolymer (B), (C), (D), (E), (F) and Methyl Cellulose (A)

3.3. Efect of Salt Concentration

Viscosity of a fluid is a measure of its resistance to gradual deformation by shear stress or tensile stress. It is an important factor to understand the durability of the polymer solution. The results of viscosity measurements of polymer solution samples MC1-MC4 are presented in Figure 3. At concentration of 1% MC, the viscosity decreases when the RPM increases. The RPM increment is proportional to shear rate increment in the polymer solution. This means the MC solution has tension thinning properties. It is interesting to note here that increasing concentration of NaCl makes solution viscosity going up.

Viscosity measurements of polymer solution samples MC5-MC8 and MC9-MC12 is presented in Figure 4 and Figure 5. The addition of NaCl has almost no effect on the viscosity of the solution. As in concentrations of 1% and 2%, the 3% MC is also thinning tension. The higher the concentration of NaCl, the more viscous the solution will be..

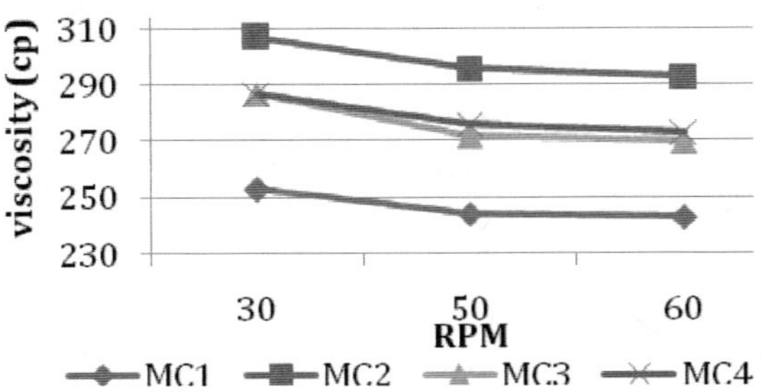

Figure 3. Viscosity Profile of Sample MC1-MC4

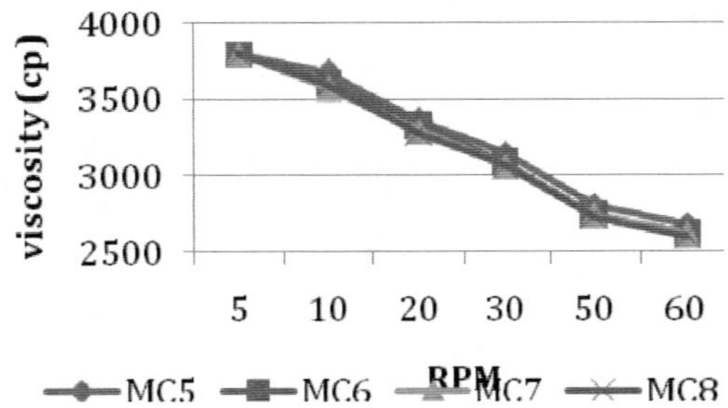

Figure 4. Viscosity Profile of Sample MC5-MC8

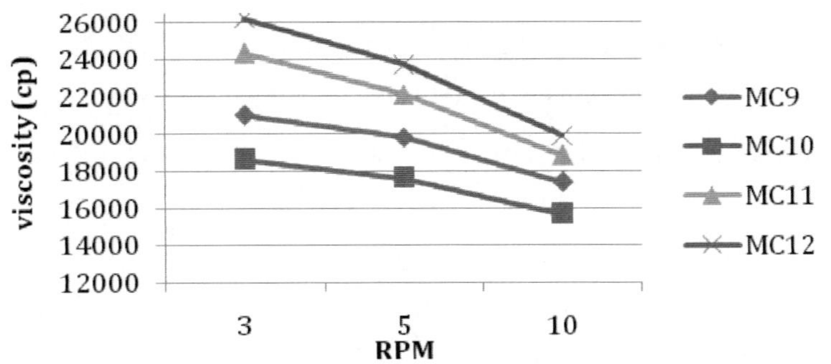

Figure 5. Viscosity Profile Sample MC9-MC12

3.4. Effect of PBS Concentration

Polymer solution viscosity of samples MCa-MCc, MCd-MCf, and MCg-MCi are presented in Figure 6, Figure 7, and Figure 8. Viscosity of 1% MC solution in PBS is lower compared to that of solution with the same concentration in NaCl. 1x PBS is equivalent to the concentration of NaCl 0.9%. The same phenomenon was also observed at MC concentration of 2 and 3%. The PBS concentration give significant impact in sample MCd-MCf dan MCg-MCi, but not in sample MCa dan MCc.

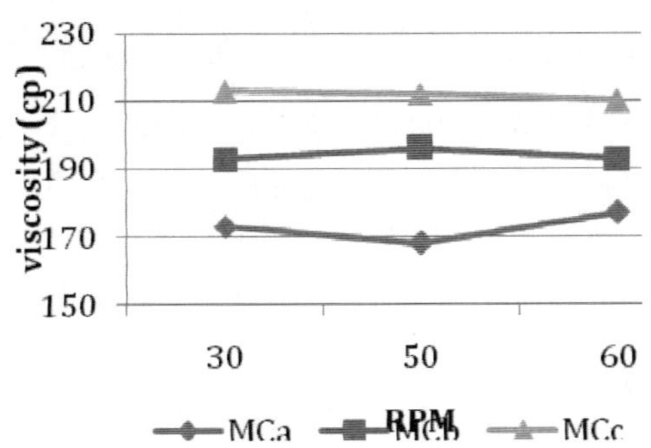

Figure 6. Viscosity Profile of Sample MCa-MCc

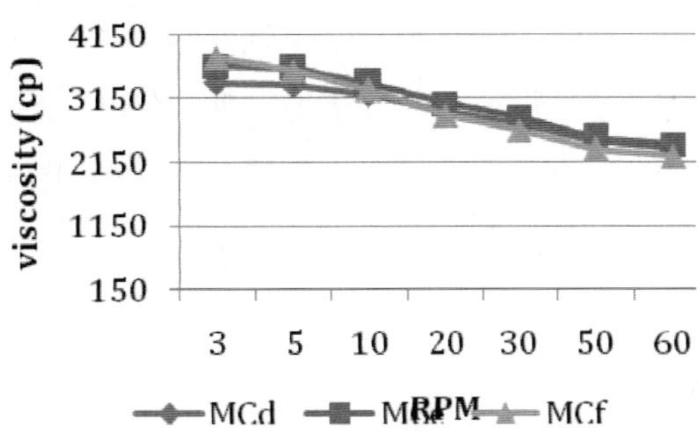

Figure 7. Viscosity Profile of Sample MCd-MCf

IOP Conf. Series: Materials Science and Engineering **223** (2017) 012043 doi:10.1088/1757-899X/223/1/012043

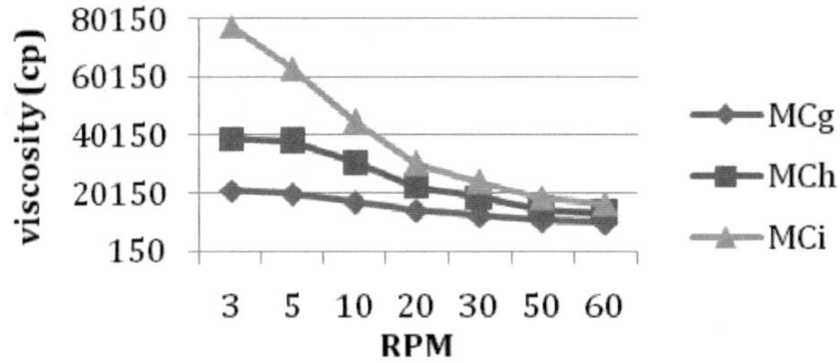

Figure 8. Viscosity Profile of Sample MCg-Mci

3.5. Effect of CN Concentration

Samples B-E viscosities are presented in Figure 9 and Figure 10. Higher CN concentration produces more viscous copolymer solution. However, the viscosity values are lower than the viscosity of 2% MC solution. This lower viscosity is desirable and an advantage that makes the fluid injected easier .

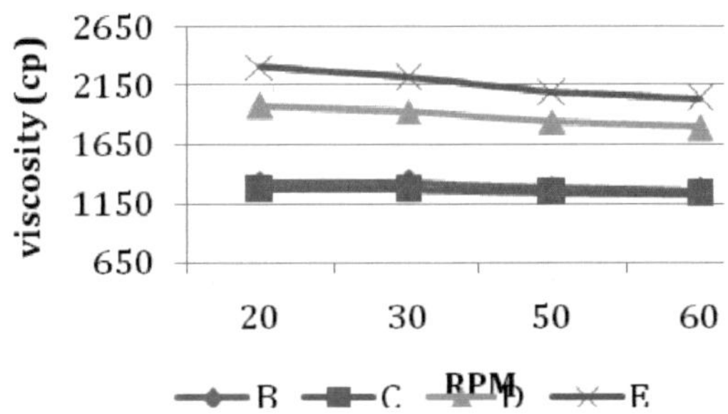

Figure 9. Viscosity Profile of Sample B-E

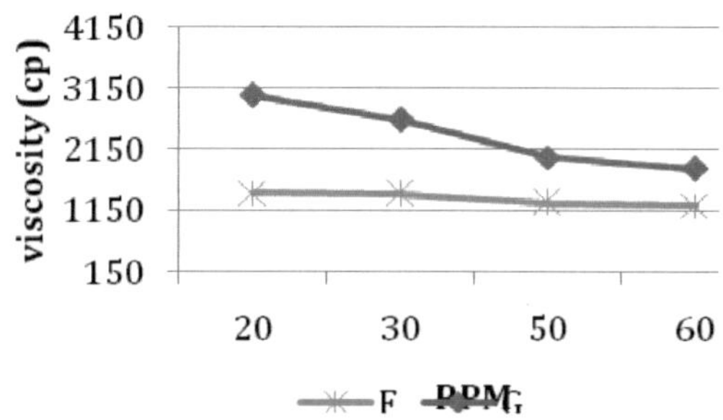

Figure 10. Viscosity Profile of Sample F-G

3.6. LCST (Sol-Gel Temperature)

Results of LCST measurement using spectrophotometer are presented in Figure 11. While results of LCST measurement using DSC and visual observations are shown in Figure 12. From Figure 11 we can see that sample G compared to the other samples has LCST closest to body temperature (38°C).

IOP Conf. Series: Materials Science and Engineering **223** (2017) 012043 doi:10.1088/1757-899X/223/1/012043

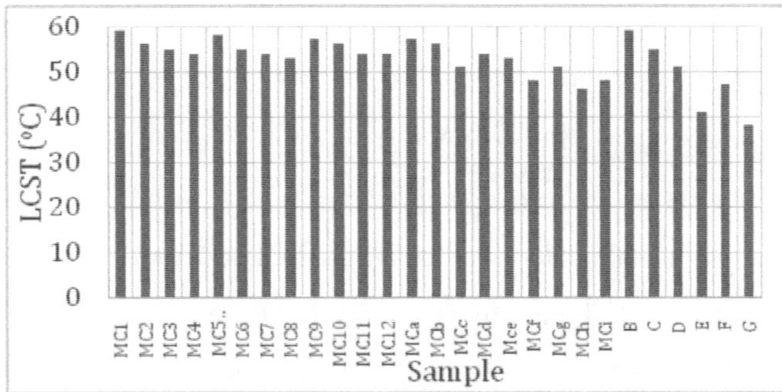

Figure 11. LCST of Samples MC1-MC12, MCa-MCi, and B-G

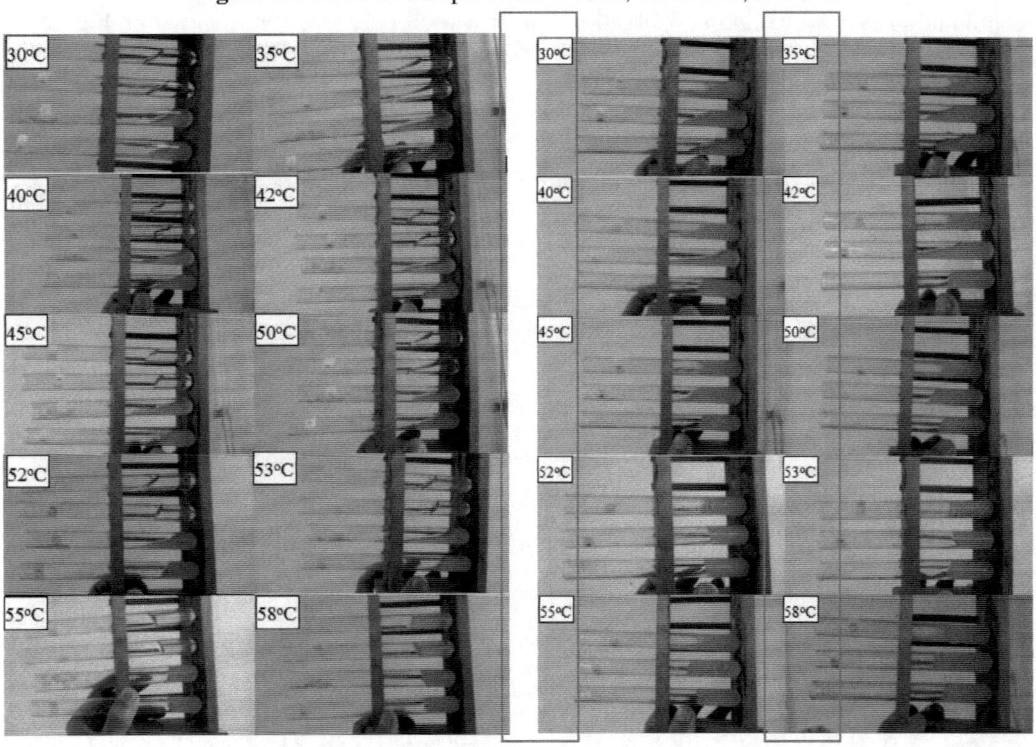

a b

Figure 12. Visual Observation of Samples a) A-D (from top to bottom), b) E-G (from top to bottom)

Addition and grafting process in copolymerisation of CN and MC give effects on decreasing LCST. The higher CN concentration results in lower LCST as shown by data of sample C-E. This is due to MC and NC that merged into the copolymer. The existence of CN which is not soluble in water increases hydrophobic effect that appears when the temperature raises above the sol-gel temperature.

The presence of salt will amplify the effect of CN on lowering LCST. The lowest LCST observed in sample G which contain 2% MC and 0.3% CN in 1.5x PBS solution. Ions from dissociation of the salt is the main factor affecting the decrease of LCST. Cations contained in the PBS is Na^+ and K^+, while the anion-anion is Cl^-, $H_2PO_4^-$and $H_2PO_4^-$. Anion plays a major role in lowering the sol-gel temperature. Hofmeister series gives an idea of how strong anion can make the process of sol-gel phase changes. Hofmeister series ions actually shows ability to precipitate proteins but the same phenomenon is also observed in the solution of MC[14].

Anions of Hofmeister series are: citric^{3-}> tartric^{2-}= SO4^{2-}> HPO4^{2-}> CrO4^{2-}> acetate^{-}> HCO^{3-}> Cl^{-}> NO3^{-}> CLO3^{-}[14]. Anions are located on the right called cosmotrof and have hygroscopic nature. In PBS solution, $H_2PO_4^-$has most dominant effect in lowering the LCST either MC or MC-CN copolymer solution. The effects of these ions are also observed in the results of viscosity measurements. Increasing the salt concentration increases the viscosity. Ions from salt will disturb hydrogen bonding in water and raises solvation. Ion solvation phenomenon is shown in Figure 12 [15].

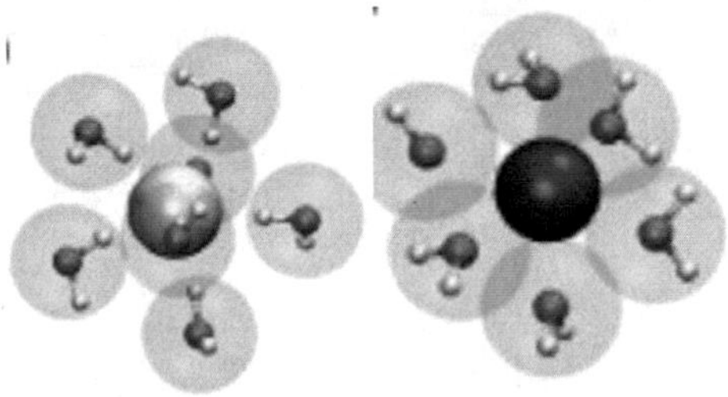

Figure 13. Solvation Process. Anion (left) and Kation (right)

4. Conclusion

Formulation and synthesis of a hydrogel having LCST near body temperature had been made from copolymers of methyl cellulose (MC) and cellulose nanocrystals (CN) that were prepared by the grafting method using UV light. Higher concentration of MC and CN in the solution will be lowering LCST and increasing solution viscosity. Higher concentration of NaCl and PBS in the solution will also be lowering LCST and increasing solution viscosity. All samples of the polymer solution with concentrations of 1% and 2% MC have viscosities less than 5000 cp at room temperature. Copolymer solution with CN concentration of 0.1, 0.3, and 0.5% have LCST of 55°C, 51°C, and 41°C, respectively. Copolymer solution of MC-CN with PBS concentration 1x and 1.5x have LCST of 47°C and 38°C, respectively. Best formulation obtained in this study to produce LCST as close as possible to body temperature is at 2% MC, 0.3% CN, and 1.5x PBS solution. It would be possible to get a better formula having LCST below body temperature by further optimization and modification of the present result.

References

[1] Jun L X 2009 *Smart Biomaterials* Dissertation National University of Singapore.
[2] Chu P K and Liu X (Ed) 2008 *Biomaterials Fabrication and Processing HANDBOOK* (CRC Press, Boca Raton).
[3] Amin M *et al* 2012 *Carbohydrate Polym.* **88** 465-473.

[4] Liao H T *et al* 2011 *Tissue Engineering: Part C* **17** (11) 1139 – 1149.
[5] Kopecek J and Yang J 2007 *Polymer Int.* **56** 1078–1098.
[6] Li L *et al* 2002 *Langmuir* **18** 7291-7298.
[7] Zhang Y *et al* 2014 *Carbohydr. Polym.* **101** 171– 178.
[8] Martin B C *et al* 2008 *J. Neural Eng.* **5** 221–231.
[9] Tang Y *et al* 2010 *Carbohydr. Polym.* **82** 833–841.
[10] Rimdusit S *et al* 2012 *Engineering J.* **16**(4).
[11] Abidin A Z and Graha H P R 2014 *Adv. Mater. Res.* **950** 24-28.
[12] Esposito A *et al* 2005 *J. Biomater.* **26** 4101–4110.
[13] Svensson A *et al* 2005 *Biomaterials* **26**(4) 419-31.
[14] Bain M K *et al* 2012 *Int. J. Biological Macromol.* **51** 831–836.
[15] Rosler E 2013 *The Hofmeister series a review of monatomic ions near the hydrophobic/water interface* Thesis. University of Amsterdam.

IOP Conf. Series: Materials Science and Engineering **223** (2017) 012044 doi:10.1088/1757-899X/223/1/012044

Study of the influence of ZnO addition on the properties of chitosan-banana starch bioplastics

L Sapei, K S Padmawijaya, O Sijayanti and P J Wardhana
University of Surabaya, Surabaya 60293, Indonesia

E-mail: lanny.sapei@staff.ubaya.ac.id

Abstract. Plastics have been widely used in our daily life due to their relatively low cost and practical uses. However, plastics are hardly degraded and thus creating lots of environmental problems. Bioplastics have been developed in order to replace the conventional plastics since they are easily degraded and environmentally friendly. In this research, chitosan-banana starch bioplastics were made with the ratio of 70:30 in the presence of 30% glycerol as the plasticizer. ZnO with varying concentrations of 1, 3, and 5% were added into the biopolymer mixtures before casting in order to improve their mechanical properties. It turned out that 3% ZnO significantly increased the bioplastics tensile strength up to ~36MPa. Vice versa, elongation and swelling percentage were decreased as ZnO concentrations increased. Bioplastics prepared with 3% ZnO were degraded within relatively short time in 90 min. These chitosan-banana starch bioplastics reinforced by ZnO seems quite promising to substitute some commercial conventional plastics.

1. introduction

Plastic production has been rapidly growing since the beginning of 20th century. Plastics have been very useful for our daily life and become an integral part of human life. It has been widely used for many applications due to its versatilepractical use, lightness, convenience and relatively low-cost. However it has been creating an accumulated environmental problem since plastics have been hardly degraded in the nature. Due to an increasing demand of eco-friendly packaging materials, renewable and biobased plastics such as chitosan/ protein/ starch based bioplastics have been increasingly developed [1-4]. However, the composite bioplastics were generally still lacking of the mechanical properties, thus some fillers such as cellulosic fiber [5] or particles i.e. clay [6], bentonite [7], SiO_2 [8], ZnO [5, 9, 10] were added to improve their mechanical properties. Over recent years, ZnO has become a considerable research interest due to its low cost, easy bioavailability, biocompatibility and possibility of performing surface modifications with different functional groups [11]. It is a non-toxic, environmentally friendly material and shows an antimicrobial activity [11-13]. It is also ideal to be used as reinforcing fillers in polymer composites since it possesses exceptional mechanical properties, low coefficient of thermal expansion, and high thermal conductivity [11].

According to our previous investigation, bioplastics derived from chitosan and banana starch (70:30) with the addition of 30% glycerol as the plasticizer showed a relatively high elongation of about 28% but still possessed a relatively low tensile strength of about 2.5 MPa [1]. The ZnO addition

IOP Conf. Series: Materials Science and Engineering **223** (2017) 012044 doi:10.1088/1757-899X/223/1/012044

was thus expected to be able to significantly enhance the tensile strength of the plasticized chitosan-banana starch bioplastics. The objective of this study was to investigate the effect of varying ZnO concentrations on the overall mechanical properties, swelling behaviour, as well as biodegradability process of bioplastics.

2. Materials and Methods

2.1. Materials

Banana starch was derived from green saba banana ('pisang kepok') which contain amylose and amylopectin with the ratio of 37:63 (w/w). The moisture content of banana starch ~8%. Glycerol purchased from PT. Brataco (Surabaya, Indonesia) was used as plasticizer. Chitosan flakes were purchased from Institut Pertanian Bogor (Bogor, Indonesia). It was derived from shrimp shell and was pale yellow in color. Its de-acetylation degree was 88.5% and viscosity was 20 cps. Technical ZnO powder and technical NaOH flakes was purchased from PT. Brataco (Surabaya, Indonesia).Glacial acetic acid was purchased from PT. Merck Tbk (Indonesia). EM4 (Effective Microorganisms 4) solution containing mixed microbial cultures (Lactobacillus 8.7 x 10^5; phosphate solubilising bacteria 7.5 x 10^6; yeast 8.5 x 10^6 some actinomycetes and photosynthetic bacteria in mineral containing solution) was purchased from PT. Songgolangit Persada (Jakarta, Indonesia). Deionised water was used during the experiments.

2.2. Methods

2.2.1. Preparation of chitosan-starch bioplastics

Chitosan-starch bioplastics were prepared with chitosan:starch ratio of 70:30 (v/v). ZnO particles were added to improve the mechanical properties of the bioplastics. Chitosan solution was prepared by adding 2.8 g chitosan into 70 mL 1% (v/v) acetic acid solution. The emerging gas bubbles had to be removed out prior to mixing with the other ingredients. Starch slurry was prepared by suspending 1.2 g banana starch into 20 mL 1% (v/v) acetic acid solution. The slurry was then heated to its initial gelation temperature of 55°C while stirring. ZnO as an additive was weighed depending on its concentration (1, 3, 5%) in the mixture. ZnO particles were added into 10 mL of 1% (v/v) acetic acid solution and then mixed until they were homogenously dispersed in the solution. Afterwards, chitosan, starch and ZnO in aqueous solution were mixed together and 30 mL of glycerol was added in order to have chitosan-starch-ZnO:glycerol ratio of 100:30 (v/v) or 30% (v/v) glycerol relative to the polymer mixtures. Finally, casting of bioplastics was conducted by pouring the final mixture into a petri dish and then stored at 70°C. Having been dried, bioplastics were soaked with 4% (b/v) NaOH solution for several minutes and carefully removed out. The bioplastics were then rinsed under the flowing water and dried at room temperature prior to analysis.

2.2.2. Mechanical Properties Measurement

The mechanical properties of bioplastics in terms of tensile strength and elongation percentage were measured using an Autograph AG-10TE (Shimadzu, Japan). Bioplastics specimen with the dimension of 7 cm x 2 cm was prepared. The specimen thickness was measured using a thickness gauge instrument (Mitutoyo, Japan). Tensile strength was performed by elongating the specimen and the load carried by the specimen before breaking was measured. Tensile strength was calculated by dividing the maximum load by the original cross-sectional area of the specimen.

$$\text{tensile strength (MPa)} = \frac{\text{F (kgF)}}{\text{A (cm}^2\text{)}} \quad (1)$$

where F was measured load before breaking and A was cross-sectional area of the specimen (width x thickness).

Elongation was calculated as the percentage increase in length that occurred before the specimen was broken under tension.

$$\%\ \text{elongation} = \frac{L - L_0}{L_0} \times 100\% \quad (2)$$

Where L was the final length of specimen before breaking and L_0 was the initial length of specimen.

2.2.3. Swelling Behavior Analysis

A specimen of bioplastics was immersed in water up to 10 min and the weight was monitored every 2 min. Bioplastics were dried up prior to weighing. Swelling percentage could be calculated as

$$\%\ \text{swelling} = \frac{W - W_0}{W_0} \times 100\% \quad (3)$$

Where W_0 = initial sample weight and W = final sample weight.

2.2.4 Biodegradability Test

Bioplastics were cut into 2 cm x 2 cm and put into a petri dish. Afterwards, 20 mL microbial solution (EM4) which was previously diluted for 5 times was added. The physical changes of bioplastics such as changes in size, thickness, and color during the degradation processes were observed every 30 min.

3. Results and Discussion

Chitosan-starch (70:30) bioplastics containing 30% glycerol was reinforced with varying concentration of ZnO (1, 3, and 5%) in order to increase the mechanical properties. The influence of ZnO concentration on the tensile strength and elongation could be seen in Figure 1.

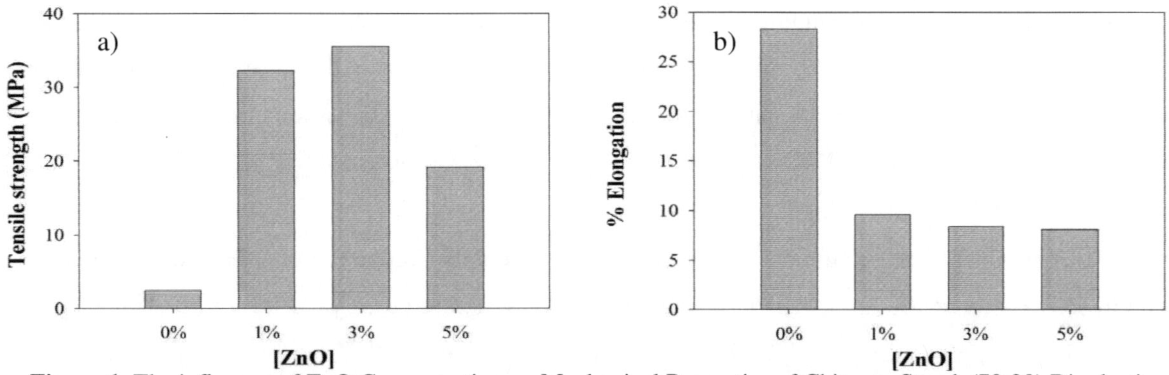

Figure 1. The influence of ZnO Concentration on Mechanical Properties of Chitosan-Starch (70:30) Bioplastics containing 30% Glycerol. a) Tensile Strength; b) % elongation.

It was obvious that tensile strength tended to increase with the increase of ZnO concentration up to 3% and slightly decreased when ZnO concentration was increased to 5%. There was a significant increase of about 15 times when bioplastics 3% ZnO was incorporated in contrast to that without ZnO addition. The highest tensile strength achieved upon the addition of 3% ZnO was ~36MPa which was much higher in comparison to the highest tensile strength of cassava based starch bioplastics of about 32 MPa after being reinforced by 9% ZnO [5]. The addition of nano-ZnO up to 4%wt into glycerol plasticized-pea starch only increased the tensile yield strength from 3.94 to 10.80 MPa [10]. The tremendous increase in tensile strength of ZnO reinforced bioplastics in our experiment could be caused by the use of chitosan as the backbone of the matrix which inherently possess higher tensile strength. Tensile strength was increased as the chitosan ratio relative to the banana starch was increased [1].

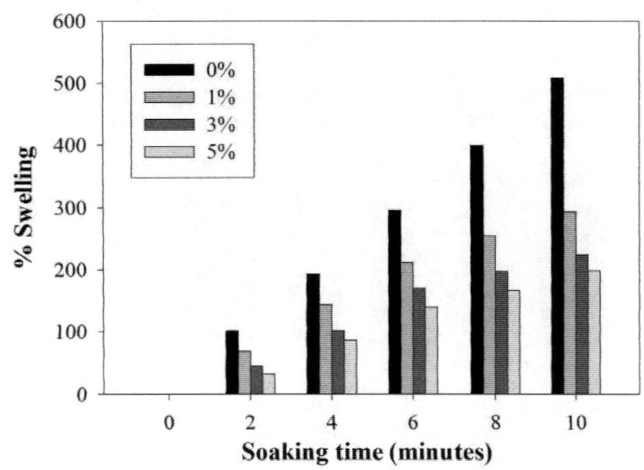

Figure 2. Swelling behaviour of chitosan-starch (70:30) bioplastics containing 30% glycerol upon the addition of varying concentrations of ZnO.

Vice versa, the elongation was decreasing as the ZnO concentration was increasing. The elongation was dropped from ~28 to ~8% when 3% ZnO was incorporated into the bioplastics. There were no significant changes in elongation when ZnO concentrations were varied within the ranges of 1 to 5%. Concentration ZnO of 3% was then taken as the optimum due to its significant effect on the improvement of the tensile strength.

The water uptake ability of bioplastics was significantly decreased upon the addition of ZnO. The higher the concentration of ZnO, the lower was the swelling percentage (Figure 2). Bioplastics without ZnO addition underwent swelling quite easily mainly due to the presence of glycerol. ZnO tended to retard the swelling process of the bioplastics since it was hardly soluble in water. This finding was in line with the previous research incorporating ZnO into bioplastics [9-10]. Furthermore, swelling percentage was increasing with time in all cases.

Table 1. Biodegradability tests of chitosan-starch bioplastics.

[ZnO]	**Degradation time (minutes)**	**Remaining specimen (%)**
0%	> 180	± 20
1%	120	0
3%	90	0
5%	90	0

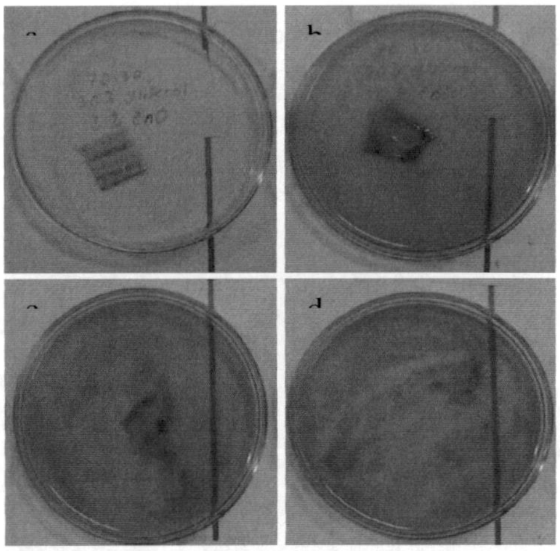

Figure 3. Biodegradability test of chitosan-starch (70:30) bioplastics containing 30% glycerol upon the addition of 3% ZnO after certain time. a) 0 min; b) 30 min; c) 60 min; d) 90 min.

Finally, the biodegradability of chitosan-starch bioplastics reinforced by ZnO was tested by immersing the bioplastics into mixed microbial EM4 solution. From Table 1, it turned out that the bioplastics incorporated with ZnO particles were all completely biodegraded within 120 min in

IOP Conf. Series: Materials Science and Engineering **223** (2017) 012044 doi:10.1088/1757-899X/223/1/012044

contrast to that without ZnO addition. The biodegradability of the bioplastics tended to increase with the increase of ZnO concentration. The biodegradability process of bioplastics reinforced by 3% ZnO was completely vanished within 90 min as depicted in Figure 3.

ZnO has been known as an antimicrobial agent [11-13] so that it was supposed that ZnO was able to retard the biodegradation process of bioplastics. However, ZnO could act as a catalyst in the hydrolytic degradation of biopolymers [14]. The incorporation of ZnO into the plasticized chitosan-starch bioplastics not only increased the tensile strength but also helped accelerate the biodegradation process of bioplastics immersed in the microbial solution. The specific role of ZnO on the biodegradability mechanisms needs to be further investigated. This research showed the immense potential of ZnO as an additive for enhancing the mechanical properties of eco-friendly bioplastics. This bioplastics could replace the commercial conventional packaging plastics.

4. Conclusions

The incorporation of ZnO particles into the plasticized chitosan-starch bioplastics (70:30) has improved the mechanical properties of the biocomposites. The optimum ZnO concentration of 3% was found to increase the bioplastics tensile strength of about 15 times from ~2.5 to ~36 MPa. The elongation percentage as well as swelling percentage of bioplastics was decreasing with the increase of ZnO concentration. Furthermore ZnO addition helped accelerate the biodegradability process of bioplastics. These results suggested that ZnO could have become a potential filler to be used for reinforcing bioplastics while also supporting the bioplastics recycling ability towards the environmentally-friendly waste management leading to the sustainability living of the next generations.

Acknowledgements

We wish to thank Ms. Dyah Ayu Ambarsari for the technical assistance during the experiments conducted at Polymer and Membrane Laboratory, University of Surabaya, Surabaya, East Java, Indonesia

References

[1] Sapei L *et al* 2015 *J. Chem. Pharm. Res.***7** 101-5.
[2] Liu H *et al* 2013 *J. Food Engineering* **116** 588-97.
[3] Gonzalez-Gutierrez J *et al* 2010 *Bioresource Technol.* **101** 2007-13.
[4] González-Gutiérrez J *et al* 2011 *Carbohydr. Polym.* **84** 308-15.
[5] Amni C *et al* 2015 *Jurnal Litbang Industri* **591**-9.

IOP Conf. Series: Materials Science and Engineering **223** (2017) 012044 doi:10.1088/1757-899X/223/1/012044

[6] Mbey J A *et al* 2012 *Carbohydr. Polym.* **88** 213-22.
[7] Sarifuddin N and Ismail H 2013 *Bioresources* **8** 4238-57.
[8] Torabi Z and Nafchi M A 2013 J. *Chemical Health Risks* **3**33-42.
[9] Alebooyeh R *et al* 2012 *J. Chemical Health Risks* **2**13-6.
[10] Ma X, Chang P R, Yang J and Yu J 2009 *Carbohydr. Polym.* **75** 472-8.
[11] Díez-Pascual A M and Díez-Vicente A L 2014 *ACS applied materials & interfaces* **6** 9822-34.
[12] Xie Y *et al* *Appl. Environment microbial.* **77**2325-31.
[13] Sirelkhatim A *et al* 2015 N*ano-Micro Letters* **7** 219-42.
[14] Lizundia E *et al* 2016 *RSC Advances* **6** 15660-9.

IOP Conf. Series: Materials Science and Engineering **223** (2017) 012045 doi:10.1088/1757-899X/223/1/012045

Water absorbency of chitosan grafted acrylic acid hydrogels

N Astrini[1], L Anah[1], A Haryono[2]
[1] Research Unit for Clean Technology, Indonesian Institute of Sciences, Bandung 40135, Indonesia
[2] Research Center for Chemistry, Indonesian Institute of Sciences, South Tangerang 15314, Indonesia

E-mail: n_astrin@yahoo.com

Abstract. Acrylic acid (AA) monomer was directly grafted onto chitosan (CTS) using potassium persulfate (KPS) as an initiator and methylenebisacrylamide (MBA) as a crosslinking agent under an inert atmosphere. One factor affecting the swelling capacity of the obtained hydrogel, KPS concentration, were studied. The hydrogel products were characterized using Fourier Transform Infrared spectroscopy (FTIR) for chemical structure and scanning electron microscopy (SEM) for morphology. Swelling of the hydrogel samples in distilled water and saline solution (9% NaCl) was examined. Swelling capacity of the CTS-g-PAA hydrogels in distilled water (88.53 g/g) was higher than in NaCl solution (29.94 g/g) The highest swelling capacity value was obtained when the grafted reaction was carried out using 2.5wt% initiator

1. Introduction

With growing worldwide ecological awareness, there has been a trend to produce low cost, biodegradable materials with improved properties. Lately, there has been a great interest to produce thermoplastic composites reinforced with natural products [1]. The chemical modification of synthetic and natural polymers is a convenient route to obtain materials with different chemical and physical properties. Considerable interest has been focused on the modification by grafting synthetic polymers onto the most abundant naturally occurring polysaccharides such as cellulose, starch, and alginates. Many actual as well as potential applications for products obtained from natural and chemically modified polysaccharides exist as, for example, products with enhanced hydrophilic characteristics. For instance, modified polysaccharides containing carboxyl functions as anionic groups are adequate for the preparation of ionotropic gels. Ionotropic gels are products capable of inter-acting with polyvalent metal cations. These are water insoluble but they swell considerably with the addition of water or aqueous metal salt solutions. Much recent research deals with this topic and its metal chelating capacity [2]. This type of material, with a large capacity for water retention, is known as a hydrogel. In particular, chitosan (CTS) is becoming popular because it has lower cost, lower density and exhibit better processing ability. Nowadays, CTS is no longer a waste by-product from the seafood processing industry because currently this material is being used by biomedical and pharmaceutical industry [3,4].

CTS, a high molecular weight polysaccharide from chitin, is the second most abundant natural polymers after cellulose in the world, and is excellent biodegradable material and can be degraded into nontoxic products in vivo. Reactive $-NH_2$ and $-OH$ groups present in CTS repeating unit are convenient for graft polymerization of hydrophilic vinyl monomers, making it an efficient way to acquire hydrogels with novel properties (chemical structure is provided below in Figure 1) [5–7]. Grafting polymerization of hydrophilic monomers onto CTS could enhance hydrophilicity and water uptake capacity of the backbone [8].

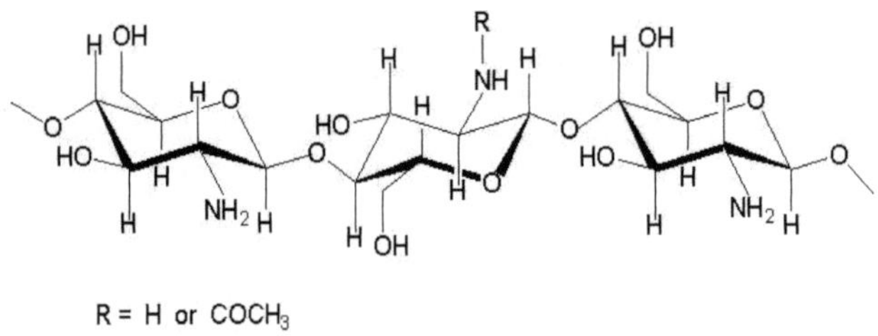

Figure 1. Structure of chitosan

Due to its excellent biocompatibility and biodegradability, CTS and its derivatives were widely applied in the fabrication of biomedical materials, enzyme and cell immobilization, especially for drug delivery. Since CTS is easily soluble in acidic solutions, crosslinking of CTS to form a network is the only way to prepare chitosan hydrogels. When an anionic monomer such as acrylic acid is grafted onto CTS (in the presence of a crosslinking agent monomer), an ampholytic hydrogel containing both cationic and anionic charges is prepared. Therefore, by introducing anionic charges ($-COO^-$) onto CTS, a hydrogel with swelling ability at various pHs is prepared [10].

Graft copolymerization of vinyl monomers onto polysaccharides is an efficient route for the preparation of hydrogels.. Because of the presence of certain functional groups along the polymer chains, hydrogels are often sensitive to the conditions of the surrounding environment, which are referred to as "intelligent materials" or "smart materials". For example, the water uptake of these materials may be sensitive to temperature, pH, or ionic strength of the swelling solutions [11,12]. The aims of this work are to synthesize graft copolymer of CTS onto acrylic acid using KPS as initiator in aqueous medium and under nitrogen atmosphere. The resulting CTS grafted acrylic acid was characterized by using Fourier-Transform Infrared Spectroscopy and Scanning Electron Microscopy.

2. Materials and Methods

2.1. Materials

Acrylic Acid (AA) and potassium persulfate (KPS) from E-Merck, chitosan (CTS) and *N,N*-methylenebisacrylamide (MBA) from Sigma-Aldrich

IOP Conf. Series: Materials Science and Engineering **223** (2017) 012045 doi:10.1088/1757-899X/223/1/012045

2.2. Instrument and Apparatus

FTIR spectra of the hydrogel samples were recorded on a FTIR spectrophotometer (IR Prestige-21 Shimadzu) in the range of 4000–400 cm^{-1}, using KBr pellets. SEM studies were carried out in a SEM instrument (JSM-6360LA, JEOL, Ltd) after coating the sample with gold film.

2.3. Preparation of CTS-g-PAA

A series of hydrogel from CTS and AA, were synthesized according to the following procedure. Appropriate amount of CTS was dissolved in 50 mL acetic acid solution (1%) in a 1L four-neck flask, equipped with a mechanical stirrer, a reflux condenser, a funnel and a nitrogen line. The reactor was placed in a water bath preset at 60°C. After being purged with nitrogen for 30 min to remove the oxygen dissolved from the system, and certain amount of KPS was introduced to initiate CTS to generate radicals. Ten minutes later, the mixed solution of 7.2 g AA, certain amount of MBA and 90 mL water was added. The water bath was kept at 80 °C for 3 h to complete the polymerization, and then the resulting product was poured into (6 N) NaOH solution to pH 7.0 by neutralization. The swollen product was dehydrated with aceton and dried at 70°C to a constant weight.

2.4. Swelling Measurements

CTS-g-poly(acrylic acid) sample (0.10 g) was put into a weighed teabag and immersed in 100 mL distilled water and allowed to soak for 24 h at room temperature [13]. The equilibrated swollen gel was allowed to drain by removing the teabag from water (~10 min). The bag was then weighed to determine the weight of the swollen gel. The absorbency (equilibrium swelling) was calculated using the following equation:

$$\text{Absorbency } (Q_e) = (W_s - W_d)/W_d \quad (1)$$

Where, W_s and W_d are the weights of the samples swollen in water and in dry state, respectively. So, absorbency was calculated as grams of water per gram of resin (g/g) [14].

2.5. Measurement of Swelling Rate

The swelling rate of the samples was measured according to the following process. Sample (0.10 g) was immersed amount of distilled water and 9 wt% NaCl solution in several 500 ml beakers. At certain time intervals, the water absorbency (Qt) of the sample was measured. The measurement condition is the same as that for equilibrium water absorbency [4].

3. Results and Discussion

3.1. Characterization of CTS-g-PAA

Chemical modifications are promising ways to obtain various utilizations of polysaccharides, such as CTS in this study. Of possible modifications, graft copolymerization is expected to be quite promising for developing sophisticated functions [15]. Cross-linked CTS-g-poly(AA) was formed through freeradical polymerization mechanism. The –OH groups present in the backbone polymer are the active sites for the formation of CTS-g-poly(AA) and takes place through the following mechanism as described in Scheme 1. The persulfate was decomposed under heating and produced sulfate anion-radicals that remove hydrogen from –OH groups of CTS backbones. Therefore, this persulfate-saccharide redox system results in active centers capable of radically initiating polymerization of AA leading to graft copolymer. One of the most important features of graft polymerization is unwanted formation of homopolymer, homopoly(AA), that is not chemically bonded to the substrate poly (CTS). Homopolymer can result if the initiator used is one that produces free radicals in solution (in the presence of vinyl monomer AA initiating homopolymerization) before creating the macroradicals.
Once a grafted chain has been initiated and begins to propagate, chain transfer from the growing grafted chain end can occur with some species in the medium to yield free radicals that could initiate the growth of homopoly(AA) chains. Since the CTS is an anionic polysaccharide [16] the cross-linking of the MBA takes place through secondary bonding forces with the backbone as well as through multiarmed cross-linking of MBA with CTS [17,18].

The infrared spectra of CTS and CTS-g-PAA are depicted in Figure 2. The FTIR spectrum of CTS has strong peak around 3448 cm^{-1} due to the stretching vibration of O–H, the extension vibration of N–H, and inter-molecular hydrogen bonds of the polysaccharide. In addition, the typical absorption bands of CTS at 1654, 1594 and 1328 $cm^{-1}$are assigned to amide I, amide II and amide III, respectively. In the region of 1000–1200 cm^{-1}, CTS presents a broad absorption band centered at 1091cm^{-1}, corresponding to stretching of C–O bond of secondary –OH. The band of 1034 cm^{-1} is derived from C–O stretching of primary –OH. The asymmetric stretching of C–O–C appears at 1155 cm^{-1} [19]. After the reaction, these characteristic absorption bands get weaken or masked with other absorption bands, implying that reactive functional groups of CTS, such as $–NH_2$ and –OH, have took part in the copolymerization reaction. The absorption bands in the spectrum of CTS-g-PAA at 1722, 1556, 1454 and 1407 cm^{-1} arearisen from PAA and assigned respectively to stretching C-O vibration, asymmetric –COO– stretching vibration, bending vibration of C–H and symmetric –COO– stretching vibration. The occurrence of these bands in CTS-g-PAA spectrum is a powerful evidence of PAA existence. From above analysis, it is no doubtful that during the reaction, PAA has been grafted on the backbone of CTS [15].

IOP Conf. Series: Materials Science and Engineering **223** (2017) 012045 doi:10.1088/1757-899X/223/1/012045

Initiation

$^{-}O_3$-S-O-O-SO_3^{-}	$\longrightarrow$	2 $SO_4^{\cdot -}$	(1)
$SO_4^{\cdot}$ + H_2O	$\longrightarrow$	$H_2SO_4^{-}$ + $^{\cdot}OH$	(2)
CTS-OH + $SO_4^{\cdot}$	$\longrightarrow$	CTS-$O^{\cdot}$ + $H_2S_4^{-}$	(3)
CTS-OH + $^{\cdot}OH$	$\longrightarrow$	CTS-$O^{\cdot}$ + H_2O	(4)
M + $^{\cdot}OH$	$\longrightarrow$	$^{\cdot}M$-OH	(5)
M + $SO_4^{\cdot -}$	$\longrightarrow$	$^{\cdot}M$-SO_4	(6)

Propagation

CTS-OH + $^{\cdot}M$-OH	$\longrightarrow$	CTS-O-$M^{\cdot}$ + H_2O	(7)
CTS-O-$M^{\cdot}$ + nM	$\longrightarrow$	CTS-O-$(M)_n$-$M^{\cdot}$	(8)
CTS-$O^{\cdot}$ + nM	$\longrightarrow$	CTS-O-$(M)_{n-1}M^{\cdot}$	(9)
$M^{\cdot}$-OH + nM	$\longrightarrow$	HO-$(M)_n$-$M^{\cdot}$	(10)

Termination

CTS-O-$(M)_n$-$M^{\cdot}$ + $M^{\cdot}$-$(M)_n$-O-CMC	$\longrightarrow$	CTS-O-$(M)_n$-M_2-$(M)_n$-O-CTS (*Graft copolymer*)	(11)
CTS-O-$(M)_{n-1}$-$M^{\cdot}$ + $M^{\cdot}$-$(M)_{n-1}$-O-CMC	$\longrightarrow$	CTS-O-$(M)_{n-1}$-M_2-$(M)_{n-1}$-O-CTS (*Graft copolymer*)	(12)
CTS-O-$(M)_n$-$M^{\cdot}$ + $^{\cdot}OH$	$\longrightarrow$	CTS-O-$(M)_{n+1}$-OH	(13)
HO-$(M)_n$-$M^{\cdot}$ + $^{\cdot}M$-$(M)_n$-OH	$\longrightarrow$	HO-$(M)_n$-$(M_2$-$(M)_n$-OH (*Homopolymer*)	(14)

$M^{\cdot}$ = monomer free radical   CTS-$O^{\cdot}$ = Backbone free radical

Scheme 1. Mechanism of the graft copolymerization reactions

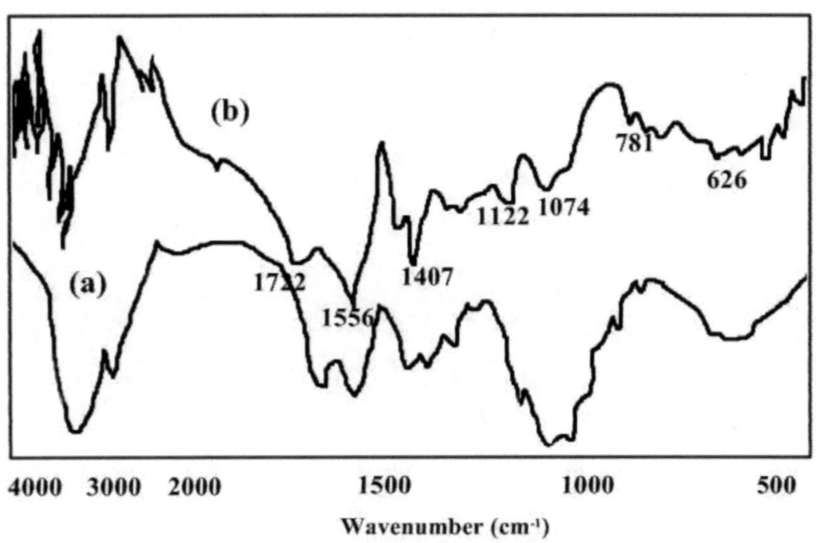

Figure 2. FTIR spectra of (a) CTS and (b) CTS-g-PAA

One of the most important properties that must be considered is hydrogel microstructure morphologies. The surface morphology of the samples was investigated by scanning electron microscopy. Figure 3 shows an SEM micrograph of the polymeric hydrogels (CTS-g-PAA) obtained from the fracture surface. The hydrogel has a porous structure. It is supposed that these pores are the regions of water permeation and interaction sites of external stimuli with the hydrophilic groups of the graft copolymers [14].

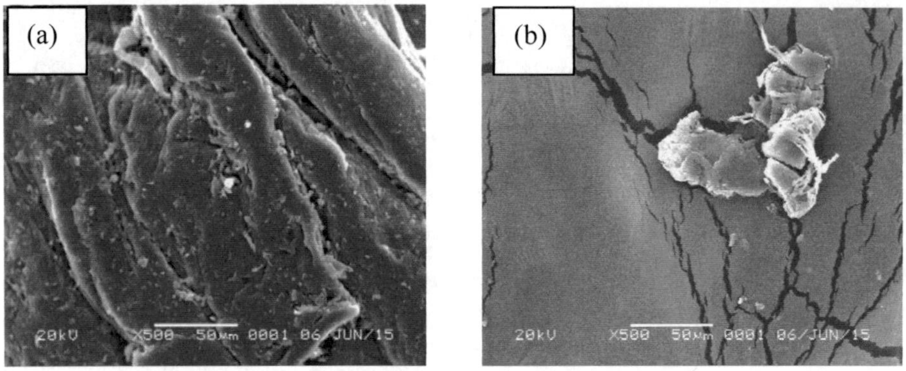

Figure 3. SEM micrographs of (a) CTS and (b) CTS-g-PAA

3.2. Effect of Initiator Content on Water Absorbency

The effect of initiator content on water absorbency for CTS-g-PAA hydrogel is shown in Figure 4. As can be seen, the water absorbency in distilled water increases with increasing KPS content from 1.0 to 2.5 wt%, and then decreases with further increase in the content of KPS. When the content of KPS reaches 2.5 wt%, the water absorbency in distilled water is 88.53 g g^{-1}. The curve of water absorbency did not change in 9 wt% NaCl solution and the water absorbency is in the range of 26.8 to 29.9 g g^{-1}. The increase in water absorbency with increasing the ratio of initiator may be ascribed to the increase in the active sites on the backbone of CTS, arising from the attack of sulfate anion-radical. The increase in the number of active sites on the CTS led to an increase in PAA grafting onto CTS backbones [4]. As the KPS content increased, a large amount of radical was produced, which in turn created more macroradicals in the main chain of CTS. However, if the massratio is increase further, the water absorbency is decreased. This was attributed to the fact that more radicals would enhance the terminating step via bimolecular collision, and the molecular weights of the branched chains are thus decreased [20, 21].

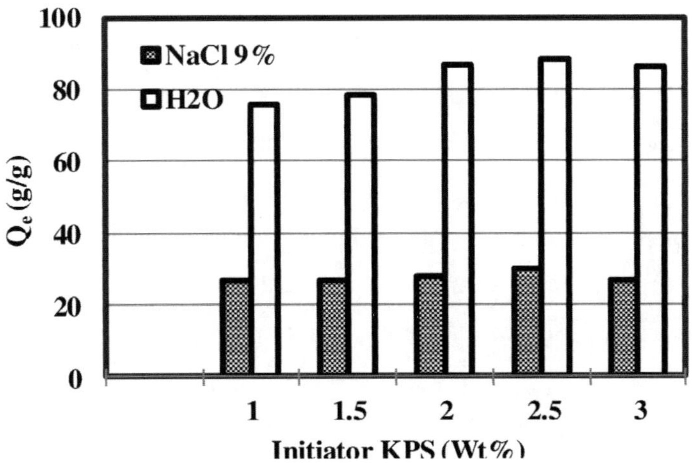

Figure 4. Variation of water absorbency in distilled water and in 9 wt% NaCl solution for the CTS-g-PAA hydrogel with KPS contents

Figure 5. displays the swelling rate of hydrogel in distilled water and 9 wt% NaCl solution. It was found that the swelling rate of the hydrogel is higher in 0–20 min, and the time of reaching swelling equilibrium is about 2.0 h. The ratios of water absorbency in 1 min to that of swelling equilibrium of CTS-g-PAA with 2.5 wt% KPS, are 91.47% and 76.48% in water and 9 wt% NaCl respectively.

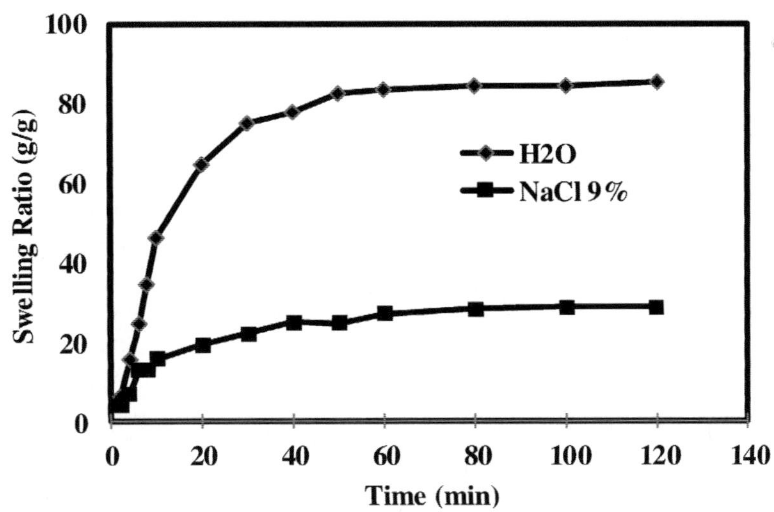

Figure 5. Swelling rate in distilled water and in 9 wt% NaCl solution for CTS-g-PAA with 2.5 wt% of KPS

4. Conclusions

Superabsorbent hydrogels, CST-g-PAA, were synthesized through grafting of AA onto CTS. Swelling capacity of the hydrogels was found to be affected by initiator concentrations. The water absorbancy both in distillate water and in 9 wt% NaCl solution increased with increase in initiator content. Swelling capacity of the CST-g-PAA hydrogels in distilled water (88.53 g/g) was higher than in NaCl solution (29.94 g/g) The highest swelling capacity value was obtained when the grafted reaction was carried out using 2.5 wt% initiator The changes of chemical structure were characterized by FTIR spectroscopy which showed a new band at 1722 cm^{-1} attributed to the carbonyl group of acrylic acid.

References

[1] Tsao C T *et al* 2010 *Carbohydr. Res.* **345** 1774–1780.
[2] H. Minamisawa *et al* 1999 *Anal.Chim. Acta.* **378** (1–3) 279.
[3] Liu Z *et al* 2006. *J.Appl Polym Sci.* **101** (1) 799-804.
[4] Flores-Ramirez *et al* 2008. *J. Biomat Sci, Polym Edn.*,**19(2)** 259-273.
[5] Mahdavinia G R *et al* 2004 *Eur Polym J.* **40** 1399.
[6] Dergunov S A *et al* 2004 *Physicochem Eng Asp.* **238** 13.
[7] Pourjavadi A, Mahdavinia GR. 2006 *Polym Polym Compos.* **14** 203
[8] Mahdavinia G R *et al* 2006 *J Appl Polym Sci.* **99** 1615.
[9] Yuntao Xie and Aiqin Wang, 2009 *J Polym Res.* **16** 143–150.
[10] Pourjavadi A, Mahdavinia GR 2006 *Turk J Chem.* **30** 595 – 608.
[11] Hosseinzadeh H *et al* 2006 *J. Biomater. Sci. Polym.Edn.***15** 1499-1511.
[12] Hitush K and Shin K 1997 *J. Appl. Polym. Sci.* **64** 2259-2263.
[13] Zohuriaan-Mehr M J and Pourjavadi A, 2003. *J. Polym. Mater.* **20** 113-120.
[14] Sadeghi M and Yarahmadi M 2011 *African J Biotech.* **10** (57) 12265- 12275.
[15] Yian Zhenga *et al* 2011 *Analytica Chimica Acta.* **687** 193–200.
[16] Mishra A et al 2003 *Colloid Polym. Sci.* **281** 187-189.
[17] Chen Q *et al* 2001 *Chinese Chem. Lett.* **12** 1029-1032.
[18] Balbir Singh Kaith and Kiran Kumar 2007 *Iran. Polym. J.* **16**(8) 529-538.
[19] Zheng Y, Wang A 2009 *J. Macromol. Sci. A.*, **47** 33–38.
[20] Chen H Tan 2006 *Carbohydrate Research.*, **341** 887-896.
[21] Pourjavadi A *et al* 2004 *Eur. Polym. J.* **40** 1336-1370.

Effect of glycerol and zinc oxide addition on antibacterial activity of biodegradable bioplastics from chitosan-kepok banana peel starch

Y E Agustin and K S Padmawijaya
Chemical Engineering Department, University of Surabaya, Surabaya, Indonesia

E-mail : yuana@staff.ubaya.ac.id

Abstract. Bioplastic is a biopolymer plastic that can be degraded easily by microorganisms so it can be used as alternative replaced commercial plastic. This research aims to study the effects of additive (glycerol and zinc oxide) addition in the characteristic of antimicrobial activity and biodegradability bioplastic from chitosan and Kepok banana peel starch. In this research, bioplastics were synthesized by chitosan as the backbone and antimicrobial, Kepok banana peel starch as filler, glycerol as plasticizer, also ZnO as an amplifier. Bioplastics were characterized their antimicrobial activity using agar diffusion method (zone inhibition assay) and biodegradability test using microbe (EM4). The result showed the optimum composition of bioplastic is kitosan 4 - 30% starch – 5 mL glycerol – 5% ZnO gives the good antimicrobial activity towards gram positive and gram negative bacteria, and this bioplastic will be degraded within an hour and 12 min. Thus, this bioplastics may have potential to be use for food packaging by having biodegradable properties and also inhibit bacterial growth.

1. Introduction

In the last decades there has been an increasing interest in the development of renewable materials with biodegradable properties in attempt to contribute to the sustainable development and to reduce the environmental impact of non- biodegradable petroleum based plastics. Bioplastics that are made from renewable resources which are easily broken down by microorganism represent a new generation of plastics that can reduces the dependence on fossil fuels also at the same way can reduces the impact to the environment. In order to solve the problem, several studies have been reported the use of biodegradable starches from different sources to prepare films and coating with different properties. Starch is the most important polysaccharide polymer used to develop biodegradable films, as it has potential to form a continuous matrix. Starch exhibits some disadvantages such as strong hydrophilic character and poor mechanical properties as compared to conventional synthetic polymers, which make it an adequate for some packaging purposes [1, 2].

Nowadays, interest in antimicrobial polymer packaging was growing rapidly, especially in field of medicine, healthcare, and food industry. Antimicrobial polymer packaging is one of the most promising active packaging systems that used to protect surfaces from microorganism including parasites, bacteria, germs and fungi. Chitosan has been proved to be nontoxic, biodegradable, biofunctional, biocompatible and possess antimicrobial characteristics. Therefore chitosan have been used as a packaging material for the quality preservation of a variety of food [3]. Starch based film is considered as an economical material for antimicrobial packaging when we blended it with chitosan as antimicrobial agents. Blended between chitosan with renewable and abundant agro-resource based such as polysaccharides, particularly starch can be utilized to reduce the manufacturing cost, improved moisture resistance and accelerated degradation [4].

The aim of the present work was to obtain bioplastics that have antimicrobial ability to restrain or inhibit the growth of spoilage and or pathogenic microorganisms. Blends between banana peel starch and chitosan using glycerol as a plasticizer will produce biodegradable plastics that can have antimicrobial activity towards microorganisms. The antimicrobial effect was tested on *Staphylococcus aureus* (gram positive bacteria) and *Escherichia coli* (gram negative bacteria) using zone of inhibition test on solid media.

2. Materials and Methods

2.1 Materials

Banana peel starch was isolated from white kepok banana that plenty found in Indonesia using water as a solvent extraction. Chitosan flakes with 85% de-acetylation degree were used as the backbone of bioplastics and also antimicrobial agents. Glycerol was used as plasticizer and purchased from PT. Brataco (Surabaya, Indonesia). NaOH flake technical was purchased from PT. Brataco (Surabaya, Indonesia). Glacial acetic acid was purchased from PT. Merck (Indonesia). EM4 (Effective Microorganism 4) solution containing mixed microbial cultures was purchased from PT. Songgolangit Persada (Indonesia). Deionised water was used during the experiments. Zinc oxide was used as the filler to strength the bioplastics film.

2.2. Preparation of Chitosan- Starch Bioplastic Film

Starch slurry was prepared by suspending kepok banana peel into water with ratio 1 : 2 (%w/v) for 10-12 h. Separation between slurry and filtrate was used by filtration method. Slurry was dried at 40°C for 12 h. Chitosan- starch bioplastics were prepared with 4% (w/v) chitosan and various ratio of starch (0, 10, 20, 30, 40, 50%) (w/w). Both mixtures were prepared by suspending into acetic acid glacial as the solvent, process was conducted at 55°C. Afterwards, 30 mL of glycerol was added in order to have chitosan- starch- glycerol ratio of 100 : 30 (v/v). Various additive glycerol (0, 5, 10 mL) and zinc oxide (1, 3, 5%) (w/v) was added to increase the biodegradability and antimicrobial activity of bioplastics. At the end, bioplastics casting were conducted by pouring the final mixtures into a petri dish and dried at 70°C. After dried, bioplastics were soaked with 4% (w/v) NaOH solution until it can

IOP Conf. Series: Materials Science and Engineering **223** (2017) 012046 doi:10.1088/1757-899X/223/1/012046

be easily remove out. Bioplastics were then rinsed under flowing water and dried at room temperature prior to analysis.

2.3. *Biodegradability Test*

Bioplastics were cut into 2 cm x 2 cm and then were put into petri dish. Afterwards, 20 mL microbial solution (EM4) which was previously diluted for 5 times was added. The degradation process such as color and dimension changing were monitored every 30 min.

2.4. *Antimicrobial Activity Test*

There are several methods to evaluate the antimicrobial activity of the bioplastics. This research was used disc diffusion (Kirby Baeur) test as antimicrobial activity test for bioplastics film. Bioplastics with 1 cm diameter were placed on solid nutritive media (nutrient agar) which contain a suspension of gram positive bacteria (*Staphylococcus aureus*) or gram negative bacteria (*Escherichia coli*) swabbed evenly on its surface. Afterwards, petridish was incubated for 18- 24 h. The incubation condition depends on the culture that was used on the test. After the incubation period, the inhibitory zone (clear zone) has to be determined around the bioplastics samples on the agar plate [5].

3. Results and Discussion

3.1. *Optimization of Starch Composition to Biodegradable Ability of Banana Peel Starch- Chitosan Blended Bioplastics*

Biodegradable test was also analyzed for commercial plastics as a comparison study. The result of biodegradability test towards banana peel starch- chitosan bioplastics was provided inside the Table 1 and Figure 1 as follows. Table 1 show the increasing rate of degradable ability of bioplastic when concentration of starch was increased in bioplastic and the visual degradation process of banana peel starch- chitosan bioplastic could be seen in Figure 1.

Table 1. Biodegradability test of chitosan- starch bioplastics at various composition of starch (4% Chitosan)

Starch composition (%)	Degradation time
0	2 h 16 min
10	1 h 57 min
20	1 h 42 min
30	1 h 32 min
40	1 h
50	34 min
Commercial plastic	>2 weeks

IOP Conf. Series: Materials Science and Engineering **223** (2017) 012046 doi:10.1088/1757-899X/223/1/012046

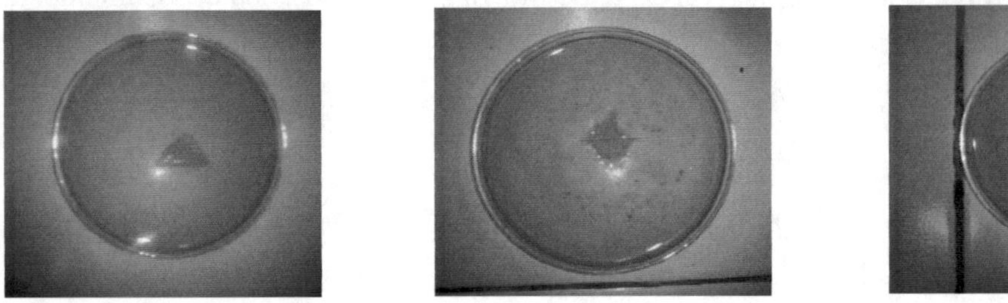

(a) (b) (c)

Figure 1. Biodegradability Test for 30% Banana Peel Starch- 4% Chitosan- 5 mL glycerol- 5% ZnO Bioplastic (a) at the beginning (b) after 1 h (c) after 1 h and 30 min

3.2. Effect of Glycerol to Biodegradable Ability of Banana Peel Starch- Chitosan Blended Bioplastics

The effect of glycerol on the biodegradability of banana peel starch- chitosan based bioplastics was shown in Table 2.

Table 2. Biodegradability Test of Chitosan- Starch Bioplastics (4% Chitosan- 30 % Starch) with Addition of Glycerol

Glycerol (mL)	Degradation time
0	1 h 32 min
5	1 h 7 min
10	31 min

When glycerol was added in the 30% banana peel starch-chitosan bioplastics, the degradablility of bioplastics film also increased. Glycerol can reduce intermolecular forces in starch or chitosan and thus decreased the strength of the films significantly [6]. The result indicated that addition of glycerol which acted as a plasticizer decreased the strength of banana peel starch- chitosan composite films.

3.3. Effect of Zinc Oxide to Biodegradable Ability of Banana Peel Starch- Chitosan Blended Bioplastics

Zinc oxide was loaded into banana peel starch- chitosan bioplastics as filler. The illustration of interaction between zinc oxide with starch- chitosan bioplastics was shown in Figure 2.

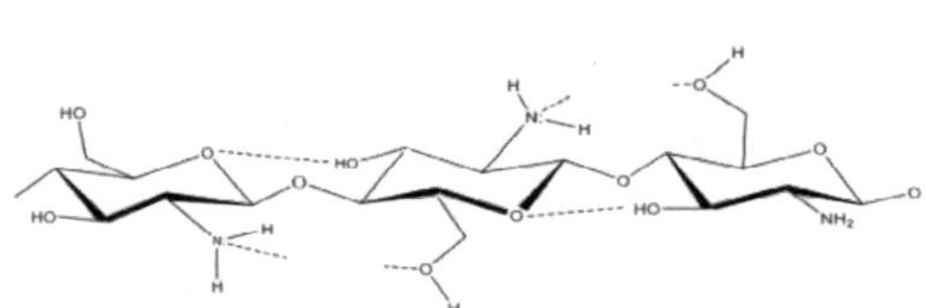

Figure 2. Interaction between Zinc Metal with Chitosan

The interaction between zinc oxide with bioplastics was a complex metal- chitosan. Zn^{2+} ion will become a bridge and substitute for the lost intra and intermolecular hydrogen bond. Biodegradability rates decreased as the increased of zinc oxide concentration in banana peel starch-chitosan bioplastics as could be references are clearly legible on the track of IOP Author Guidelines. Check in composite bioplastics will inhibit bacterial growing and can also decrease the degradation rate of bioplastics.

Table 3. Biodegradability Test of Chitosan- Starch Bioplastics

Zinc oxide (%)	Degradation time
1	35 min
3	42 min
5	1 h 12 min

3.4. Antimicrobial Activity of Banana Peel Starch- Chitosan Bioplastics

The results after 24 h incubation time was shown in Table 4 and Figure 3. The inhibitory effect was measured based on clear zone surrounding circular film disc. Measurement of clear zone diameter was started from the outside diameter of film bioplastics.

Table 4. Antimicrobial Activity Test of Chitosan- Starch Bioplastics (4% Chitosan)

Starch composition (%)	Clear zone diameter (mm)	
	Gram positive	Gram negative
0	6.25	8.875
10	4.125	8.75
20	3.75	7.375
30	3.25	6.625
40	1.125	5.375
50	0.25	2.75
Biodegradable plastic	0	0
PE plastic	0	0

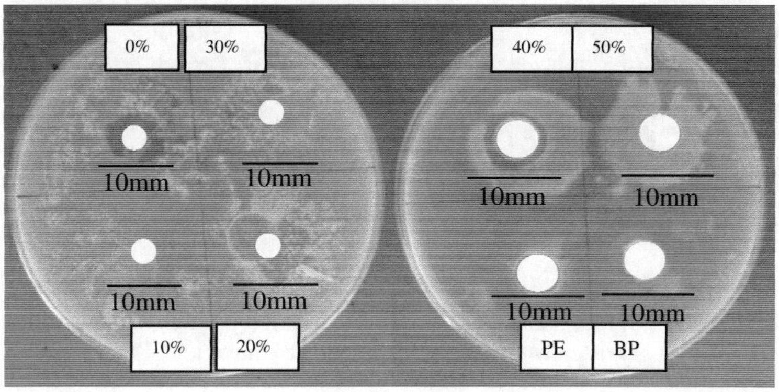

(a) (b)

Figure 3. Antimicrobial Activity Test for Banana Peel Starch- Chitosan Bioplastics against Gram Positive Bacteria (*S. aureus*) (a) 0-30% Starch Compositions (b) 40-50% Starch Compositions, Biodegradable Plastic (BP), and Polyethylene Plastic (PE)

If there is no clear zone surrounding, it means that there is no inhibitory zone, and the diameter was valued as zero. Films bioplastics showed fair antibacterial activity against gram positive (*S. aureus*) and gram negative (*E. coli*) bacteria. Commercial biodegradable plastics and polyethylene (PE) plastic was used as control for antimicrobial activity test. In general as shown on Table 4, chitosan bioplastics without starch addition will given the biggest clear zone against gram positive and gram negative bacteria [3]. Antimicrobial activity will decreased as the increased of starch addition on banana peel starch- chitosan bioplastics. The results were showed that banana peel starch- chitosan bioplastics have bigger clear zone against *E.coli* compare to *S. aureus*. It means banana peel starch- chitosan bioplastics have more effective antimicrobial activity against gram negative (*E. coli*) bacteria than gram positive (*S. aureus*) bacteria. Commercial biodegradable plastic and polyethylene (PE) plastic were not showed any antimicrobial activity.

Table 5. Antimicrobial activity test of chitosan- starch bioplastics (4% chitosan- 30% starch) with additive

Additive	Composition	Clear zone diameter (mm)	
		Gram positive	Gram negative
	0 mL	3.25	7.375
Glycerol	5 mL	0.5	7
	10 mL	0	4.5
Zinc oxide	1%	4.5	7.5
(5 mL glycerol)	3%	8.5	10.375
	5%	8.75	10.625

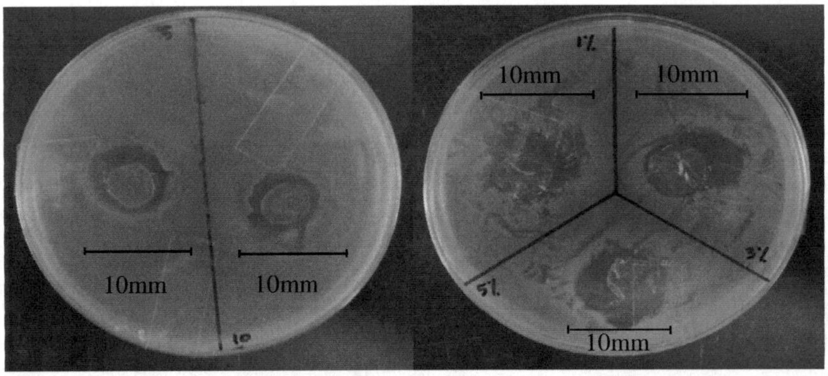

(a) (b)

Figure 4. antimicrobial activity test for banana peel starch- chitosan bioplastics (a) for addition 5 and 10 mL of glycerol (b) for addition 1,3,5% of zinc oxide

Addition of glycerol as a plasticizer into chitosan bioplastics was decreased the antimicrobial activity of bioplastics because glycerol filled the space between hydrogen bond and will weakness the chitosan strength that can affect chitosan antimicrobial activity. Zinc oxid (ZnO) is an efficient antimicrobial agent since zinc oxide was able to exhibit an antimicrobial effect by diffusion of soluble species into agar medium [7]. Decrease of microbial population with respect to zinc oxide concentration on banana peel starch- chitosan bioplastics. The antimicrobial activity analysis also tested for banana peel starch- citosan bioplastics with addition of glycerol and zinc oxide as shown on Table 5 and Figure 4.

4. Conclusion

Biodegradability increased with the increase of banana peel starch concentration in starch-chitosan bioplastics. On contrary their antimicrobial activity will decreased. The starch- chitosan based bioplastics demonstrates more effective antimicrobial activity against gram negative (*E. coli*) bacteria than gram positive (*S. aureus*) bacteria. Addition of glycerol which acted as plasticizer will increase the biodegradability of bioplastics but at the same time will reduced their antimicrobial activity. Decreased of microbial growth will respected to zinc oxide concentration because of their antimicrobial ability. The banana peel starch- chitosan bioplastics may have potential applications for antimicrobial packaging by their fast biodegradable rate and antimicrobial activity that can inhibit bacterial growth to improve the food safety.

References

[1] Priya B *et al* 2014 *Carbohydr. Polym.* **109** 171-179.
[2] Mollah M Z I et al 2016 *J Organic Polym. Mater.* **6** 11-24.
[3] Tripathi S et al 2008 *e-Polymers* **093.**
[4] Salleh E and Muhamad I I 2010 *AIP Conference Proc.* **1217** 432.
[5] Kun E and Marossy K 2013 *Mater. Sci. Forum* **729** 430-435.
[6] Tonny W *et al* 2014 *J. Chem. Eng. Chem. Res.* **1** 343-352.
[7] Pasquet J *et al* 2014 *Physicochem. and Eng. Aspects* **457** 263-274.

IOP Conf. Series: Materials Science and Engineering **223** (2017) 012047 doi:10.1088/1757-899X/223/1/012047

Sengon wood (*Paraserianthes falcataria (L.) Nielsen*) carbon as supporting material for electrochemical double layer capasitor

Wulandhari, N Syarif, I. Waruwu, M. Ridho Prayogo
University of Sriwijaya, Palembang South Sumatera, Indonesia

E-mail: nnsyarif@gmail.com

Abstract. A microstructure carbon electrodes was potentially developed into energy storage device, i.e electrochemical double layer capacitor (EDLC) or super capacitor. The structure has a large surface that able to increased the capacitance of electrodes. Sengon Wood Carbon (SWC) was one of microstructure carbon which mainly has honeycomb structure. SWC was prepared hydrothermally along with microwave heating. SWC has 202 m^2g^{-1} of surface area, 14,4 $S.cm^{-1}$ of conductivity and crystalline carbon peak at 29.55° with honeycomb structure. Mixtures of honeycomb sengon carbon and graphite were casted it into thin layer electrode (TLE). The electrodes are fabricated into EDLC along with aluminum foil, and their performance are tested by using Galvanostatic and capacitance meter. TLE had 2.984 to 3.547 μF/g of capacitance, initial voltage of EDLC ranged from 0.67 to 0.42 V. Capacitance of 3 cm x 4 cm EDLC ranged from 30.6 to 60 μF. Galvanostatic Charge-Discharge (GCD) indicated that SWC suitable for EDLC application .

1. Introduction

Advance technology in electrical equipment has influenced the increasing of energy storages demand. Energy storage such as battery, fuel cell and super capacitor or Electrical Double Layer capacitor (EDLC) has became a mass production to fulfill needs. EDLC has some excellent things compare to other energy storage i.e: high energy density, high energy storage, simple principal and easy construction and also environmental friendly as well as free corrosive content. One of essential factor for EDLC is electrodes performance.

Several research have previously made carbon electrodes from various nature derived carbons, i.e gelam wood bark that contain 40-45 % of carbon, 200-1200 m^2g^{-1} of surface area, 0.001-1.5 Scm^{-1} of conductivity, 0.01-28 Fg^{-1} of specific capacitance and contained micro-nano-structure [1]. Cui *et. al.,* (2013) [2] reported that microstructure of carbon i.e honeycomb structure exhibited lower sheet resistance and showed a record-high fill factor of 72% as a power conversion efficiency. Liu *et al.,* (2007) [3] reported extruding of a mixture of bituminous coal and organic additive for making activated carbon honeycomb (ACH) monoliths. It has significant effects on pore size distribution, lower values of the BET surface area and total pore volume. In general, it was concluded that microstructure showed a good properties for carbon electrode application.

IOP Conf. Series: Materials Science and Engineering **223** (2017) 012047 doi:10.1088/1757-899X/223/1/012047

A microstructure carbon was potentially obtained from sengon wood which abundantly planted in forestry. This research mainly discussed about development of sengon carbon as electrode to be applicated in EDLC. The high performance of EDLC was indicated by its capacitance and charge-discharge curve.

2. Materials and Methods

2.1 Materials

Sengon wood purchased from local forestry, distilled water, graphite, aluminum, stainless steel net sheet at 206 mm x 36 mm and thick 1 mm, gel polymer PVA (Merck), PVAc, K_2CO_3(Merck),Na_2CO_3(Merck), $Na_2B_4O_7.10H_2O$, $C_8H_{10}NNaS_8.3H_2O$, aluminum foil, polyurethane.

2.2 Preparation of SWC

Sengon wood was sieved of sengon wood sized of 20 mm in length and 2 mm in diameter into chip, dried for 2-3 days under sunlight. The chip was powdered by using disk mill from JIMO FF-15. 40 g of sengon wood powder was blend with 100 mL distilled water, then stirred for 3 min. The sengon wood powder were heated hydrothermally as mentioned in Titiricci *et al.*, (2007) [4]. The mixture was put on the hydrothermal reactor then sealed with 3M amalgamating tape and closed tightly. The reactor was placed inside electric oven, heated for 16 h in 180°C for generated 30 bar hydrothermal pressure inside reactor chamber. The process obtained brownish black constituent that called torrefaction material. The material was transferred from the hydrothermal reactor to microwave furnace and pyrolyzed in domestic microwave oven at 1000 watts of power. 2 g of torrefaction material was mixed with 0.4 g graphite or activated carbon, and contacted microwave for 25 min. In this condition, the temperature inside furnace can reached 800 °C. Graphite was added as microwave adsorbent. The furnace was tapped from the oven after being cold in room temperature. The result of this process was carbon with honeycomb structures.

2.3. Preparation of Electrochemical Double Layer Capacitor (EDLC)

A thin layer electrodes were made by casting carbon:graphite (w/w) on stainless steel sized 206 mm x 36 mm and thick 1 mm. The composite of honeycomb carbon, graphite and polyurethane (1:10:1w/w) was embedded on the stainless steel net sheet. The TLE acted as anode. Above TLE placed dacrone sheet as separator. Above the separator, gel polymer place the same electrode as cathode. Thin metal sheet stick was placed on the corner of TLE as current colector. Strip of metal strand was soldered on current collector as applicator. The electrodes, separator, current and applicator were packaged into cell with polypropylene sheet as cover.

2.4. Measurement and Characterization Method

The surface area of carbon was measured by calculating Metilen Blue and Iodine number, The formula were given below:

$$S_{(m^2g^{-1})} = 2.28 \times 10^{2} - 1{,}01 \times 10^{-1}\ MBN + 3.00 \times 10^{-1}\ IN + 1.05 \times 10^{-4}\ MBN^2 + 2.00 \times 10^{-4}\ IN^2 + 9.38 \times 10^{-4}\ MBN\ IN \quad (1)$$

$$Vm_{(cm^3g^{-1})} = 5.60 \times 10^{-2} - 1.00 \times 10^{-3}\ MBN + 1.55 \times 10^{-4}\ IN + 7.00 \times 10^{-6}\ MBN^2 + 1.00 \times 10^{-7}\ IN^2 - 1.18 \times 10^{-7} MBN\ IN \quad (2)$$

$$Vt_{(cm^3g^{-1})} = 1.37 \times 10^{-1} + 1.90 \times 10^{-3}\ MBN + 1.00 \times 10^{-4} \quad (3)$$

The formula were used to calculate surface area, micro-pores volum and pores volum, respectively. Conductivity with twoprobe method. SEM micrographs were obtained on JEOL JED-2300. X-ray diffraction (XRD) patterns of the carbons were obtained on a Shimadzu X-ray diffractometer XRD 7000 operating at 40 kV and 30 mA, using Cu-K α radiation.. Working electrode is TLE and counter electrode is aluminum foil. EDLC performance was measured by Galvanostatic Charge and Discharge method and direct measurement of capacitance, Charging and discharging was done by function generator with potential difference given by two poles from function generator.

3. Results and Discussion

Electrochemical Double Layer Capacitor (EDLC) required a high adsorption capacity in order to store the electricity inside. Surface area was the most influenced properties related to adsorption capacity. The Metilen Blue and Iodine number calculation resulted a 202 $m^{-1}\ g^{-1}$ surface area, 0.181 $cm^3\ g^{-1}$ of micro pores volum and 0.358 $cm^3\ g^{-1}$ of pores volum. Surface area was the most influenced factor to determine the electrode capacitance. The SWC categorized as a large number of carbon surface area. Thus, the capacity of adsorption in carbon of sengon wood was high enough for EDLC application [5]. High temperature of microwave heating (up to 800°C) has eliminated the acid functional group and resulted on forming a hexagonal structure [6]. The structure with a free electron may increase the ability of carbon to conduct the electricity in surface [7]. In sengon carbon, the conductivity measured was 14.4 S cm^{-1} and suitable for EDLC application as the number was above the average activated carbon monolith as 13.87 S cm^{-1} [8] .

Characterization method by using XRD and SEM proofing the honeycomb structure of Sengon carbon. XRD result (Figure 1a) indicates that the carbon has several crystalline structure The peak of intensity was performed in ranged 29^0 until 48.50° [9]. The highest intensity was at 29.55° and

IOP Conf. Series: Materials Science and Engineering **223** (2017) 012047 doi:10.1088/1757-899X/223/1/012047

indicated a carbon honeycomb structure. The SEM image (Figure 1b) indicates a hydrothermal treatment carbon of sengon with 2000 zooming has a plenty of holes inside the carbon. The structure is closed to honeycomb structure.

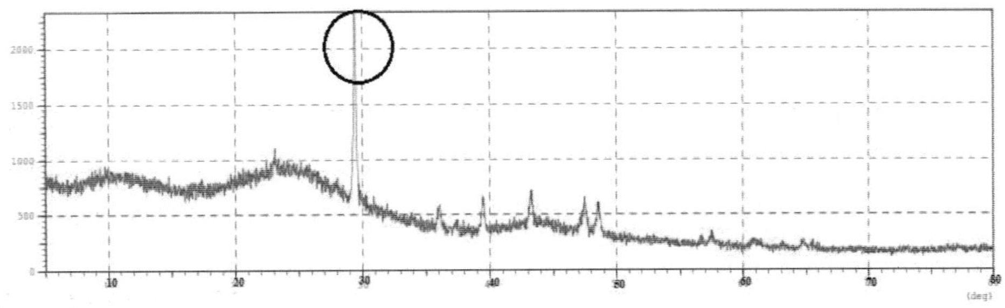

(a)

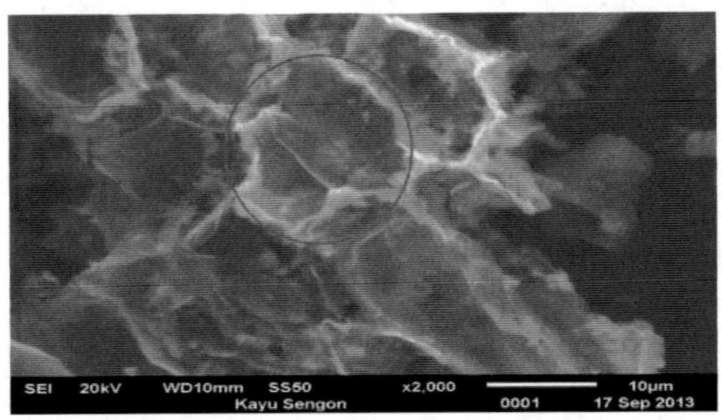

(b)

Figure 1. Characterization result by using (a) XRD (b) SEM

Measurement due to SWC capacitance in two electrolyte percentage, K_2CO_3 and Na_2CO_3, and solid media as shown in Table 1 presented a various number of capacitance. The highest value for two electrolyte were 3.547 μF in Na_2CO_3 15%-PVA and 2,984 μF for K_2CO_3 15%-PVA. Those fulfilled the requirement for electrochemical double layer forming process [9]. However, those electrolyte has low stability in higher temperature so it easily to produces CO_2 then caused a leak of device.

Table 1. Capacitance value of SWC in two electrolyte and solid media

Capacitance (μF/g)	K_2CO_3 10%	K_2CO_3 15%	Na_2CO_3 10%	Na_2CO_3 15%
PVA	1.8	2.98	3.097	3.547
PVAc	2.95	2.09	2.857	3.452

An EDLC performance were measured with variation of weight to weight composites carbon i.e; graphite, SWC, and SWC-graphite (1:10w/w). Each micro structure influenced its voltage and capacitance related to ionic adsorption in carbon molecules. Electrical conductivity also take a role in electron transfer inside the electrodes layer then increased the potential difference among the edges .

Graphite has a 21.2 S/cm of electrical conductivity therefore pure graphite variation has the highest voltage, 0.67 V. Meanwhile, the rest two variation, SWC-SWC and SWC-graphite exhibits a middle low voltage due to decreasing of SWC conductivity. Surface area of carbon affect the capacitance of electrodes. The highest capacitance was 60μF in variation of SWC-graphite composites in $C_8H_{10}NNaS_8$. $3H_2O$ electrolytes (Table 2). SWC was acted as filler between the layer of graphite that might also increased hollow area of composite SWC-graphite.

Surface area of composite was increased capacity of ionic adsoption inside the electrode. These number showed that SWC-graphite was applicable for supercapacitor material. SWC-graphite electrodes structure stored the energy longer and prevent self discharge, respectively. Organic electrolytes, $Na_2B_4O_7.10H_2O$ and $C_8H_{10}NNaS_8$. $3H_2O$. organic electrolyte was chosen related its stability in various temperature and pressure. Both electrolyte has complex functional groups and slowed to be decomposed. Therefore, organic electrolyte might prevent leak of EDLC. The capacitance of each variation was higher in sodium diethythiocarbamate ($C_8H_{10}NNaS_8$. $3H_2O$) electrolyte because the electrolyte was able to have a binding force with electrode interface. The interaction increased a mass transfer inside EDLC then affect the capacitance itself.

Table 2. Variation of Electrode Composite in two organic electrolyte

Electrode	Voltage		Capacitance	
Composite	$Na_2B_4O_7.10H_2O$ (V)	$C_8H_{10}NNaS_8$. $3H_2O$ (V)	$Na_2B_4O_7.10H_2O$ (μF)	$C_8H_{10}NNaS_8$. $3H_2O$(μF)
Graphite-graphite	0.67 V	0.06 V	1.1	59.2
SWC-graphite	0.42 V	0.03 V	4.6	60
SWC-SWC	0.43 V	0.02 V	0.02	30.6

Figure 2 showed the variation of electrode due to the slope of each variant that might indicate the charge-discharge performance.The charge and discharge measurement between variation of electrode composition was shown by the slope of sinusoidal waves.

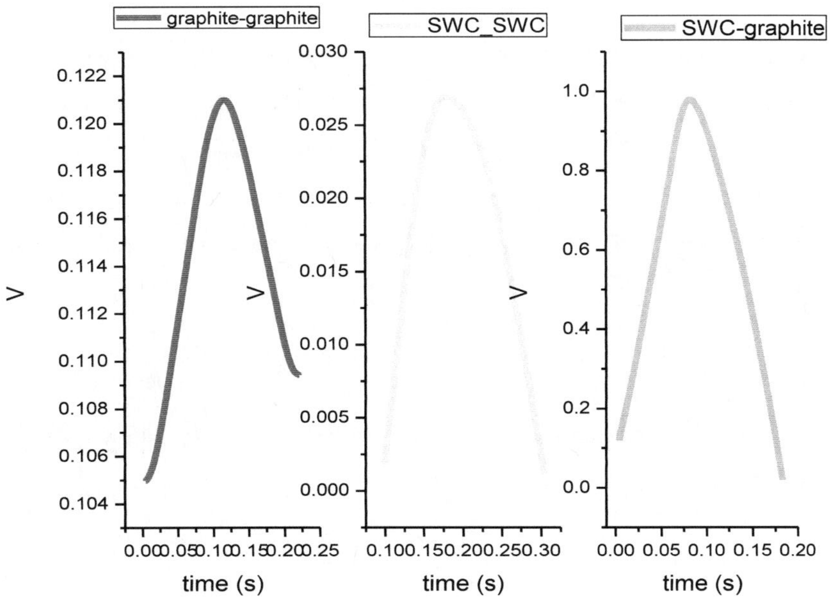

Figure 2. Galvanostatic charge-discharge (GCD) curve

Galvanostatic spectra shown a period for supercapacitor to charge and discharge. The slope was one of parameter to determine the performance of EDLC. The greater slope implied a faster rate whether charging or discharging.

Table 3. Slope of Charge and Discharge in each variation

	Slope	
	Charge	Discharge
Graphite-graphite	0.18513	-0.14622
SWC-SWC	0.39156	-0.30684
SWC-graphite	12.04935	-10.94757

The highest slope, 12.04935 was showed by SWC-graphite composites due to its honeycomb lattice that has a large surface area. The large micro pores surface of SWC caused a double layer effect between electrode and electrolyte interface that increase rates of electron adsorption. Graphite has facilitate the electron transfer so affect to higher charging speed. The surface area of SWC-graphite also influenced discharging curve by having lowest slope that indicated a lower speed. A slow discharge time for SWC-graphite electrode (Table 3) due to honeycomb structure of SWC that able to adsorbed the electron inside the hollow cells. The longer EDLC discharge implied a higher performance of storage device.

4. Conclusion

The results show that honeycomb carbon was able to be an active material for fabricated the EDLC (Supercapacitor). The material was conductive and performed a good electrochemical result when combined with graphite. EDLC from honeycomb, carbon SWC, has granuled structure. EDLC of carbon graphite can store the energy longer as its high capacitance and also has a highest slope that indicate the lower self discharge on EDLC. Overall, SWC was suitable to be built into EDLC.

Acknowledgement

This project was a part of OSN Pertamina Science Project and fully funded by Collaboration International Research Project of Sriwijaya University.

References

[1] Syarif N *et al* 2013 *Electrochem. Sci. Eng.* **2**(3) 37 - 45.
[2] Cui Kehang , *et al*. 2013 *J. Phys. Chem. Letters* **15**(4) 2571-2576.
[3] Liu L *et al. 2007 Carbon* **14**(45) 2836-2842.
[4] Titirici Maria M *et al 2007 Material Chem.* **19** 4205-4212.
[5] Kanawade S M and Gaikwad R W 2011 *Int. J. Chem. Eng. and Application,* **5**(2) 317-319
[6] Syarif N and Pardede M C 2014 *Int. Transaction J. Eng. Management & Appl.* **2**(5) 95-101.
[7] Popov VN 2004 *Matter. Sci. Eng.* **1**(43) 61-102.
[8] Farma R *et al*. 2013 *Int. J. Electrochem. Sci.* **8** 257-273.
[9] Waruwu N I 2015 Sriwijaya: Indralaya, Indonesia.

IOP Conf. Series: Materials Science and Engineering **223** (2017) 012048 doi:10.1088/1757-899X/223/1/012048

Effect of epoxidised soybean oil loading as plasticiser on physical, mechanical and thermal properties of polyvinylchloride

M Rahmah, N Mohd Nurazzi, A R Farah Nordyana, S M Syed Anas
GREEN Polymer Research Group, Faculty of Applied Sciences, Universiti Teknologi MARA (UiTM), 40450 Shah Alam, Selangor, Malaysia

E-mail : rahmahmd@salam.uitm.edu.my

Abstract. The aim of this paper is to study the effect of epoxidised soybean oil (ESO) as an alternative plasticizer on physical, mechanical and thermal properties of plasticised polyvinyl chloride (PPVC). Samples were prepared using 10, 20, 30 and 40% by weight percent of ESO. The samples were characterized for density, water absorption, tensile, hardness and thermal properties. The addition of ESO as plasticizer in PVC had caused significant effect on the physical and mechanical properties of PPVC. Increasing of ESO loading had resulted in decreased density, tensile strength, tensile modulus but increased in elongation at break and shore hardness. From water absorption study, it was observed that the all the samples reached the plateau absorption at days 8 to 10 with absorption percentages of between 1.8 to 2%. In general the crystallinity of PPVC maintained between 10 to 13% with increase in ESO loading while the melting point (T_m) is slightly decreased about 3 to 6°C. In this study, ESO which acts as plasticiser were found to result in lower glass transition temperature (T_g). The enhancements of super cooling with higher ESO loading were found to increase the crystallization temperature, promoting crystallisation and act as nucleating agent.

1. Introduction

Polyvinyl chloride, (IUPAC Poly(chloroethanediyl)) commonly known as PVC had been played an important roles in thermoplastic industries over the many years. However, due to its inherent limitations such as low thermal stability, high melt viscosity and brittleness, it cannot be processed on its own but blended with other additives to achieve desired material quality and application. PVC uses the highest proportion of additives compared to other plastics where it may comprise up to 60% of a PVC products weight [1, 2]. PVC which is compounded without plasticizer is known as unplasticised polyvinyl chloride (UPVC) which is naturally strong and rigid while PVC which is compounded with plasticizer is known as plasticized polyvinyl chloride (PPVC) which flexible and soft.

Currently, the most widely used plasticizer are phthalates due to its high plasticising efficiency, wide availability and low cost [3]. However, the toxicity issue of the phthalates plasticisers had mimic human hormones and also affect various life forms including animals. Medical devices, children toys and items, bottles etc employed plasticisers to increase flexibility of PVC inherent brittleness. Use of phtalates, phosphates sebacates had been established in PPVC compound but they face ban by various countries due to its toxicity. In 2005, the European Commission banned Bis(2-ethylhexyl) phthalate (DEHP), dibutyl phthalate (DBP), and butyl benzyl phthalate (BBP) Di-isononyl phthalate (DINP), diisodecyl phthalate (DIDP), and di-n-octyl phthalate (DNOP) from use in toys and childcare articles [3]. PPVC are also manufactured for medical devices,vinyl flooring and food packaging, which involve sizeable market applications.

Hence, it is necessary to find a new solution for immediate and future requirements. Conventional plasticizers are a rich family opening the door to many innovative opportunities for safe and sustainable plasticisers such as such as adipates, citrates, epoxies, terephthalates, trimellitates, diisononylcyclohexane-1,2-dicarboxylate (DINCH) and phosphate esters polymerics.

There are various possible alternatives for phthalates in plasticized PVC. Citrate-based plasticizers, like ATBC, and have been available for years as plasticizers in medical devices. DINCH is an alternative for applications such as medical devices. Another chemical, trioctyltrimellitate (TOTM) performs as a plasticizer in PVC and it has potential to be use commercially. Promising non-phthalate substitutes for DEHP in vinyl flooring material include DGD (dipropylene glycol dibenzoate) and DEHA (di(2-ethylhexyl) adipate). Other possible phthalate substitutes include: phosphate esters (e.g., tris(2ethylhexyl)phosphate); sebacate and azelate esters (e.g., diisodecylsebacate (DIDS), di-butyl sebacate, and di(2-ethylhexyl) azelate (DOZ)); isosorbide esters (made from renewable resources), and other organic esters such as isobutyratetrimethylpentanediol and diethyl succinate [4].

Polymeric esters or sustainable additives such as isosorbide esters, soya derivatives or hydrogenated phthalate of low molecular weight polymers and reactive plasticizers had been used. Isosorbide diesters, obtained from vegetable oils which are renewable resources, have excellent plasticizing properties for PVC and represent a non-toxic alternative to the phthalates [4]. Table 1 shows the list of vegetables oil and their chemical composition for example palm oil, soybean oil and other ester based oils.

Table 1. The major composition of fatty acids from some of the plant sources [5-12].

Fatty acid	Palmitic (C16/0)	Palmitoleic (C16/1)	Stearic (C18/0)	Oleic (C18/1)	Linoleic (18/2)	Linolenic (18/3)	Iodine value
Palm oil	40.13	-	3.1	43.4	13.2	-	50.4-53.7
Canola oil	4.1	0.3	1.8	60.9	21.0	8.8	110-126
Jatropha oil	14.2	1.4	6.9	43.1	34.4	-	10.362
Soybean oil	11.0	0.1	4.0	23.4	53.2	7.8	125-140
Linseed oil	5.5	-	3.5	19.1	15.3	56.6	175-204
Sunflower oil	4.8	0.8	5.7	20.6	66.2	0.8	123
Flax seed oil	5.0	-	4.0	19.0	14.0	58.0	153-189

Polymeric plasticizers derived from polymeric multifunctional alcohols give long chain molecules whose properties can be determined by the choice of the used alcohols. Molecular weights can be, for example, in ranges from 800 up to more than 8000 versus 390 for the dioctyl phthalate

IOP Conf. Series: Materials Science and Engineering **223** (2017) 012048 doi:10.1088/1757-899X/223/1/012048

(DOP). The precise solution that leads to balance between processing properties, functional characteristics, durability, cost and impact to environment is important. ESO is chosen as it has functional group epoxidised ring triglyceride chain, modified through epoxidation of soybean oil. Studies using ESO for various application in polymers include use of ESO in rubbers in coatings and as plasticisers etc [10, 13].

Hence, in this paper, the studies on physical, mechanical and thermal properties were performed to determine the final properties of PVC, which is applicable as plasticised PVC. Its aim is to replace phthalates and other toxic long ester aromatic chain groups which has high toxicity to human and animals [4]. Using sustainable epoxidised oils for PPVC, it can impart more environmentally safe and sustainable products.

2. Experimental

2.1 Materials

The polyvinyl chloride (PVC), (grade HP65, a product of Malaysian Electrochemicals Industry) Malaysia is used. Epoxidized soybean oil (ESO) EO-132, stearic acid as lubricant, tin as stabilizer and china clay as filler was supplied by Na Ya Plastics Corporation, China. The ESO has chemical composition of acid value below 0.7, Oxirane content 6.4 to 6.8% and iodine value of 5.0 or below. Compound formulation with different loadings is as tabulated in Table 2.

Table 2. Compound formulation used with different loadings of ESO (parts by weight).

PVC (%)	ESO (%)	Stearic acid (%)	Tin stabilizer (%)	China clay (%)
100	0	1	7	20
100	10	1	7	20
100	20	1	7	20
100	30	1	7	20
100	40	1	7	20

2.2 Fabrication of Plasticised PVC

Twin-screw machine model TSE 16TC with machine number 2093-8/R was used for mixing the compound formulation. Counter rotating screw used to suit with heat sensitive of PVC because they allow greater temperature control. A neat PVC compound (0% ESO plasticiser loading) was also made as control sample. The extruded compound was palletized using PRISM cutter machine. All compounded PPVC were microinjection moulding machine (HAAKE MiniJet) for moulding process. The pressure (bar) of the microinjection was maintained at 800 bar. For microinjection process, cylindrical temperature were set to temperature ranges between 195°C and 190°C and mould temperature set at 110°C and 100°C respectively.

IOP Conf. Series: Materials Science and Engineering **223** (2017) 012048 doi:10.1088/1757-899X/223/1/012048

2.3 Physical Characterization

The densities of neat PVC and plasticized PVC were determined using electronic densitometer according to ASTM D1895. Water absorption test were carried out according to ASTM D570. All the specimens were immersed in distilled water at room temperature and were weighed until constant weight was obtained. Water absorption percentage were calculated using equation 1 [14, 15].

Water absorption, WA (%) = $(W_2 - W_1)/ W_2 \times 100$ (1)

Where W_2 represents the weight of the composite after immersion and W_1 is the weight of the composite before immersion. Five replicates were tested for each formulation and test.

2.4 Mechanical Characterization

Tensile test were performed using Universal Tensile machine (Testometric MICR0500) according to ASTM D638. Ten replicates were tested for each formulation. Test was conducted with crosshead speed 2mm/min. Hardness test were performed using shore A hardness tester (HW Wallace) with serial number C30022/19 in accordance with ASTM D2287. Ten replicates were test tested for each formulation.

2.5 Thermal Characterization

The glass transition temperatures (Tg), melting point (Tm), ESO vaporisation temperature, percentage of crystallinity, crystallization temperature (Tc) and supercooling temperature were measured using Perkin Elmer Thermal Analysis Controller (TAC7/DX) DSC-model. The procedure involved rapidly heating the sample from 30 to 220°C at heating rate of 20ºC/min and the flow rate of the nitrogen purge was 40 mL/min then holding under isothermal condition for 1 min at 220°C. The system was then cooled from 220 to 30°C at a heating rate of 20°C/min using liquid nitrogen as a cooling medium.

3. Results and Discussion

3.1 Effect of ESO Loading on Density Properties of PVC

The density of the ESO plasticised PVC with varying ESO loading is given in Table 3. The density of the plasticised PVC decreased with increasing of ESO loading from 1.45 to 1.34g/cm^3. This perhaps because of the formation of links with polymer molecules and ESO act as spacer between molecules of the polymer. Free volume are increase with the addition of ESO to the PVC, increasing of free volume had reduced the density of the PVC compound as ESO loading are increased [16]. Thus, ESO chain can move freely between the interspacing of PVC compounds.

Table 3. Variation of density of plasticised PVC with ESO loading.

Sample	0	10	20	30	40
Density (g/cm3)	1.453	1.414	1.406	1.367	1.336

3.2 Effect of ESO Loading on Water Absorption Properties of PVC

Figure 1 shows the effect of ESO on water absorption of PVC. It can be noted that, the water uptake or absorption of the plasticised PVC increases with the increasing of ESO loading. Initially highest rate of water absorption within 4 days are exhibited by 10% ESO and the lowest rate of water absorption were found for 40% ESO. Upon water absorption after 10 days immersion in water, all reached its maximum water absorption value which range from 1.5 to 2.5%.

PVC itself absorbs water and achieves about 1.8% absorption. Polarity of the PVC and water cause hydrogen bonding linkages to exist form and PVC experienced some water absorption ability. The increases of water absorption for the plasticised PVC could be attributed to ESO that form links with polymer molecules and act as spacer between molecules of the polymer. The bond forces of the polymer weaken due to other linkage of ether to water molecules being established. Thus free volume increases with the addition of ESO to the polymer, increasing of free volume hence increase the ability of compound to absorb water, thus lead to decreasing of water absorption. The water absorption of PVC and plasticised PVC reached a plateau of water absorption at days 8 to 10.

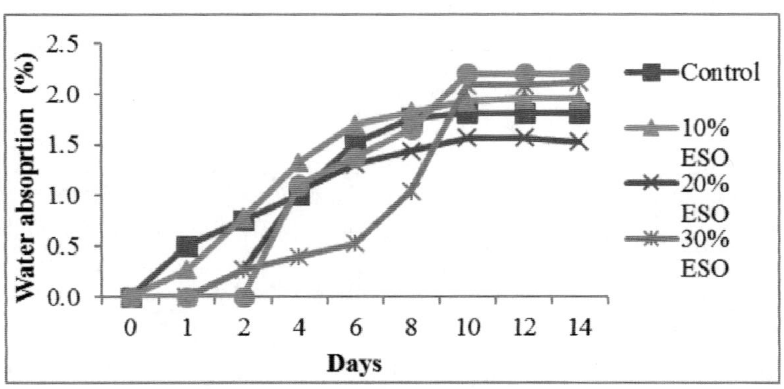

Figure 1. Effect of ESO on water absorption of PVC.

Higher water absorption compared to PVC control were exhibited for sample with 40, 30 and 10%. These ESO with 40% and 30% has slow rate of water absorption initially but upon saturation with water, the higher ESO tend to increase the water absorption after prolong time as ESO may have reduced the interaction and bonding of ESO and PVC itself. The weakening bonding of more hydrophobic ESO with the PVC exceed the interaction of hydrogen bonding capability of water molecules and PVC. Bonding of plasticised PVC with higher amount of ESO surpassed the interaction with PVC water system; hence, a threshold amount of oil content could have acting as spacer between the PVC chains.

3.3 Effect of ESO Loading on Tensile Properties of Plasticised PVC

Figure 2 demonstrates the effect of ESO on the tensile properties of PVC. The total ESO loading of PVC is varied from 10 to 40%. The tensile strength and modulus of the plasticised PVC exhibited similar decreasing trends with increasing ESO loading from 0 to 40%. The tensile strength and tensile modulus of the plasticised PVC ranged from 48 to 17.42 MPa and 2603 to 153MPa, while the addition of ESO caused an increase in elongation from 3 to 63%.

A considerable decrease of tensile strength indicates that the ESO has interdiffuse into the chain system, reduce the interaction and bonding between PVC chain. This was due to action of the plasticizer itself on the molecules of the polymers, which is believed to act as spacer between molecules of the polymers by forming links with polymer molecules. It was mentioned that dipole interaction occurs between polar groups in the polymer and the polar groups in the plasticizer. In such case, the chlorine atom in the PVC resin plays the role of the polar group, while ester group in the epoxidised oil stands for polar group in the plasticizer. More random and long chain ESO oil affects the interaction and bonding. Hence, the bonding forces between polymers weaken due to less interaction within themselves.

The formation of these links weaken the bond forces of the polymer atoms and in addition, incorporating plasticizer to the polymer increases the free volume, which leads to lowering of tensile strength [16, 17].

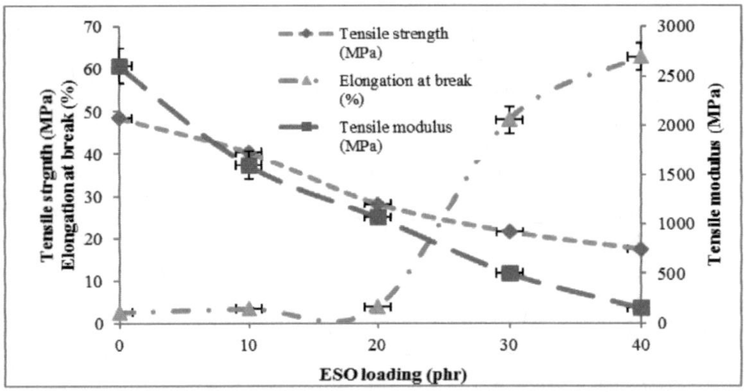

Figure 2. Effect of ESO loading on tensile properties of PVC.

Increasing the ESO loading resulted in decrease of tensile modulus. Unplasticised PVC is rather brittle. Thus by the increasing the weight percentage of ESO has improved flexibility of PVC significantly. It is noticed that the modulus of elasticity decreases with increasing plasticizer. In this work, with variation of ESO from 10 to 40%, the corresponding elasticity varies significantly from 153 to 1599MPa, while for 20 to 30% ESO, the modulus ranges between 502 to 1069MPa. This value showed better modulus as compared to work by Elgozali and Hassan [18], which used di-octyl phthalate (DOP) as plasticiser, where the researcher found that, 21 to 32wt.% of addition of di-octyl phthalate (DOP) as plasticiser has corresponding modulus of elasticity ranged between 8 to 15MPa. On a molecular level, plasticization is the weakening or selective breaking of bonds between molecules, while leaving others intact, to increase intermolecular space, known as free volume. Thus increased space allows for changes in shape, flexing or molding of the final material. The PVC polymer is amorphous rigid material and may be thought of as comprising small areas of order and crystallites among amorphous areas of molecules [19,20]. According to Saeki K *et al.,* (2003), analysis under infrared spectroscopy showed that, the incorporation of plasticizer is into the PVCresin had solvated the amorphous chain segments, but not the crystalline ones and type of plasticiser content in PVC can be identified [21].

Meanwhile, elongation showed increasing trend upon increasing ESO loading. This is also a normal consequence of the increase of ESO loading wt.% which is having a high strain compared to PVC compound without ESO. This is due to presence of ESO acting on PVC as spacer and increase the free volume, which make PVC become softer and weaker and allows it to be elongated longer at low loads [16, 22].

3.4 Effect of ESO Loading on Hardness Properties of Plasticised PVC

Figure 3 demonstrates the effect of ESO loading on the shore hardness of PVC. The hardness showed decreasing trend with increasing of ESO loading with the shore hardness ranged between 99 to 93. ESO is a flexible chain unit which act as spacer between molecules of the polymer and assimilate between the rigid PVC particles.

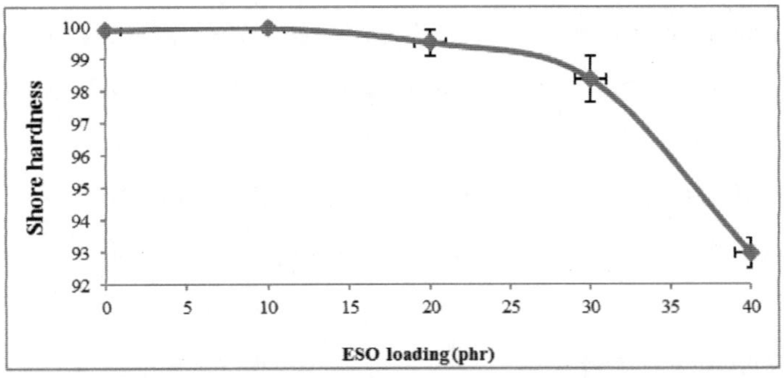

Figure 3. Effect of ESO loading on hardness properties of PVC.

The bonds forces of the polymer weaken due to the link establish and flexibility imparted by ESO thus reduced the brittleness and hardness of the compound. According to the Free Volume Theory, plasticiser molecules will separate the polymer chains and increase the free volume in the polymer matrix [23].

3.5 Effect of ESO Loading on Thermal Properties of Plasticised PVC

The DSC analysis of plasticised PVC with different percentage of ESO loading is showed in Table 4. ESO act as plasticiser had decreased the glass transition temperature and the melting temperature, however it had increased the crystallization temperature. For pure PVC, Tg was 81°C, which is in the temperature range found in literature [24, 25]. The presence of ESO as plasticisers decreased the Tg value of PVC from 81 to 65°C respectively.

Table 4. Effect of ESO loading on DSC analysis of PVC.

ESO (%)	Glass transition temperature, Tg (°C)	ESO Vaporisation (°C)	Melting temperature, Tm (°C)	Crystallization temperature, Tc (°C)	Percentage of crystallinity (%)	Supercooling Tm – Tc (°C)
0	81	-	163	60.93	10	129
10	76	122	166	64.60	12	122
20	71	123	165	64.93	13	121
30	71	123	161	65.27	10	115
40	65	124	160	65.60	10	114

Melting temperature varies to only few degrees, between 3 to 6°C. The melt temperature is the point upon which all crystallite melt. The crystallinity is low for PVC which is only 10% and it varies with amount of plasticiser ESO added from 10 to 13%. Higher amount of ESO had maintained the crystallinity value to 10%. The difference in crystallization temperature between melting temperature can determine how the PVC reacts to cooling phenomenon and this behaviour can determine the supercooling effect brought about by plasticiser.

As proven by many researchers, the function of plasticiser is to make the polymer more pliable and soft and thereby enhancing the flexibility and plasticity to the materials by enhancing the mobility of the polymeric chain and it intensely related to its chemical structure and rate of interaction with the polymeric material present in the formulation [26]. In the PVC formulation, ESO which acts as plasticiser caused Tg to decrease, this is because ESO has high number of non-polar groups caused by reduction in polar forces between polymer chains, thus lowering Tg, and improving low temperature of flexibility.

The crystallization temperature (Tc) increased as ESO loading increased. In the plasticized PVC, the plasticiser molecules were found in the amorphous area. Formation of crystalline structures were found to be faster with high plasticiser content as crystal growth are enhanced , however crystallinity or amount of enthalpy is lower as orientation of the aligned PVC chain is impeded.

Cooling causes the transition of polymers from flow-viscosity or high-elasticity states into glassy states. The transition temperature of an elastic polymer to a glassy state is a parameter of its freezing resistance which meant lower temperature limit of retention of elastic proper [27]. Super cooling exists due to presence of foreign material namely nucleating agents or small crystalline particle dispersed in the crystallizing melt and remain solid at crystallization temperature which can grow and form large number of small crystals formed around matrix. In this work, increase of ESO loading will promote supercooling as lower difference of (Tm – Tc) exhibited, hence the ESO can be the site for nucleation centre. Finally, the presence of ESO can increase supercooling and the rate of crystallization.

4. Conclusions

For plasticised PPVC compounded, the density of plasticised PVC decreased from 1.45 to 1.34 g/cm^3 as compared to neat PVC. Constant water absorption was reached at days 8 to 10 and maximum water absorption value obtained were in ranges from 1.5 to 2.5%. As the ESO loading increased, the tensile strength and tensile modulus decreased, while elongation at break increased. The tensile strength and tensile modulus of the plasticised PVC ranged from 48 to 17MPa and 2603 to 153 MPa, while the elongation at break increased from 3 to 63%. High elongation of PPVC with increase in ESO caused an increase in flexibility, presence of ESO act as spacer which increase the free volume allowing PVC chain to be elongated longer and cause softer and weakening of strength. Shore hardness decreases with increasing ESO loading and lowest reduction of hardness is about 7 for 40% ESO loading. The plasticised PVC decreased the glass transition temperature and the melting temperature, however it had increased the crystallization temperature. Faster formation of crystalline structures with higher plasticizers due to enhancement of crystalline growth and orientation of the aligned PVC chain. Supercooling or difference between melting and crystallization temperature were reduced and this assist in faster crystallization that can be advantages as ESO can function as nucleating agent.

Acknowledgements

The authors thank Research Management Institute (RMI) of Universiti Teknologi MARA (UiTM), Malaysia for financial support and Faculty of Applied Sciences (FSG), Universiti Teknologi MARA (UiTM) Shah Alam, Malaysia for providing good facilities. Our high appreciation is conveyed also to those involved in completing this work, which include the technicians and friends who had assisted in this work.

References

[1] Bishay I K *et al* 2011 *Materials & Design* **32** 62-68.
[2] Norizan M N *et al* 2013 *Adv. Mater. Research* **664** 590-595.
[3] Allard P and Colaiácovo M P 2011 Bisphenol A: metabolism, mechanisms and toxicity. In Reproductive and Developmental Toxicology. Amsterdam: Academic Press.
[4] Markarian J 2007 *Plastics, Additives and Compounding* **9** 22-25.
[5] Haryati T *et al* 1997 *J. the American Oil Chemists' Society* **74** 939-942.
[6] Lang X 2001 *Bioresource Technology* **80** 53-62.
[7] Derksen J *et al* 1995 *Industrial Crops and Products* **3** 225-236.
[8] Wanasundara U N and Shahidi F 1994 *J. the American Oil Chemists' Society* **71** 817-822.
[9] Warwel S 1999 *Industrial Crops and Products* **9** 125-132.
[10] Rahmah M *et al* 2013 *Adv. Mater. Research* **812** 138-144.
[11] Sharma V and Kundu P P 2006 *Progress in Polym. Sci.* **31** 983-1008.
[12] Pim P M N *et al* 2008 *Int .J. Sc. Tech* **13** 5 .
[13] M. Rahmah M *et al* 2013 *Adv. Mater. Research* **812** 204-209.
[14] Ireana Yusra A F *et al* 2015 *BioResources* **10.**
[15] Sahari J *et al* 2012 *Procedia Chemistry* **4** 254-259.
[16] Saunders K J 2013 Organic polymer chemistry: an introduction to the organic chemistry of adhesives, fibres, paints, plastics, and rubbers: Springer Science & Business Media.
[17] Markovitz H 1974 *Polymer science, A Materials Science Handbook* vol 2, AD Jenkins, Ed., North Holland Publishing Company, Amsterdam and London; (American Elsevier Publishing Company, New York) 1822.
[18] Elgozali A and Hassan M 2008 *J.Sci. Tech* **9** .
[19] Shahshoua Y 2001 Inhibiting the deterioration of plasticized poly (vinyl chloride) a museum perspective," Ph.D., Dept. of Chemical Engineering, Danish Polymer Centre, Dept. of Chemical Engineering, the Technical University of Denmark : Dept. of Conservation, the National Museum of Denmark, Denmark.
[20] Stark N M and Rowlands R E 2007 *Wood and fiber science* **35** 167-174.
[21] Tabb D L and Koenig J L 1975 *Macromolecules* **8** 929-934.
[22] Wilkes C E *et al* 2005 *PVC Handbook* (Hanser).
[23] Krauskopf L G and Godwin A 2005 Plasticizers. , PVC Handbook, Hanser, Munich, 173-193., "Plasticizers," in *PVC Handbook*, C. E. Wilkes, J. W. Summers, and C. A. Daniels, Eds., ed: (Hanser Garderer Publisher) 173-193.
[24] Aouachria K and Belhaneche-Bensemra N 2006 *Polymer Testing* **25** 1101-1108.
[25] Arkış E and Balköse D 2005 *Polymer Degradation and Stability* **88** 46-51.
[26] Wypych G 2012 "Effect of plasticizers on properties of plasticized material," in *Handbook of Plasticizers*, 2nd ed Toronto (ChemTec Publishing).
[28] Senichev V Y and Tereshatov V V 2004 Influence of plasticizers on the glass transition temperature of polymers. *Handbook of plasticizers, G. Wypych (ed.).*

IOP Conf. Series: Materials Science and Engineering **223** (2017) 012049 doi:10.1088/1757-899X/223/1/012049

Physicochemical properties of sugar palm starch film: Effect of concentration and plasticizer type

D J Prasetyo[1], W Apriyana[1], T H Jatmiko[1], Hernawan[1], S N Hayati[1], V T Rosyida[1], Y Pranoto[2], C D Poeloengasih[1]

[1]Research Unit for Natural Product Technology, Indonesian Institute of Sciences, Yogyakarta, Indonesia 55861

[2]Faculty of Agricultural Technology, Universitas Gadjah Mada, Yogyakarta, Indonesia 55281

e-mail: crescentianadewi@gmail.com

Abstract. In order to find the best formula for capsule shell production, this present work dealt with exploring physicochemical properties of sugar palm (*Arenga pinnata*) starch film as a function of different kinds and various concentrations of plasticizers. The films were prepared by casting method at different formula: starch 9-11%, glycerol or sorbitol 35-45% and polyethylene-glycol 400 (PEG 400) 5-9%. Appearance, thickness, retraction ratio, moisture content, swelling behavior and solubility of the film in water were analyzed. Both glycerol and sorbitol are compatible with starch matrix. On the contrary, PEG 400 did not form a film with suitable characteristics. The result reveals that glycerol- and sorbitol-plasticized films appeared translucent, homogenous, smooth and slightly brown in all formulas. Different type and concentration of plasticizers altered the physicochemical of film in different ways. The sorbitol-plasticized film had lower moisture content ($\leq$ 10%) than that of glycerol-plasticized film ($\geq$ 18%). In contrast, film plasticized with sorbitol showed higher solubility in water (28-35%) than glycerol-plasticized film (22-28%). As the concentration of both plasticizers increased, there was an increasing tendency on thickness and solubility in water. Conversely, retraction ratio and swelling degree decreased when both plasticizers concentration increased. In conclusion, the sorbitol-plasticized film showed a potency to be developed as hard capsule material.

1. Introduction

For more than a century, gelatin was adopted as the main material of capsule manufacture due to its excellent characteristic including gelling, film forming and surface active properties. Although gelatin has such a wide range of useful applications, big concerns persist among consumers about its usage. Mad cow disease episode in the 1990s, religious and cultural issues, and nature of gelatin capsule are the main reasons to consider the use of gelatin in food and pharmaceutical field. Consequently, development of gelatin alternative is highly desirable because the global halal market is growing rapidly.

Many researchers have been exploring and developing materials to replace gelatin. Numerous studies have been conducted by using starch from plant sources, such as yam [1], sweet potato [2-3], mungbean [3], cassava [4-5], banana [6], corn [7] and sago [8-11]. Moreover, sugar palm starch (SPS) film also have been successfully developed in the last several years [12-18]. Compared to sago starch, SPS has more suitable properties for producing starch gels or dough because of its abilities to form a firmer and more resistant gels at concentrations above the gel point [19].

IOP Conf. Series: Materials Science and Engineering **223** (2017) 012049 doi:10.1088/1757-899X/223/1/012049

Many studies have been conducted to explore film formation from SPS. While earlier studies were focused on developing SPS film with starch concentration of 8-9 % w/w [12-18], this study focuses on developing SPS film with higher starch concentration. Meanwhile, appropriate SPS film for capsule shell formation could be obtained by combining it with plasticizer. The addition of plasticizer would affect physicochemical properties of film, such as appearance, moisture content, and solubility in water. The effects of type and concentration of plasticizer on physicochemical properties of SPS film in high starch concentration were not yet studied. This work focus on studying the effects of plasticizer type and concentration on physicochemical properties of high starch concentration SPS film to provide an appropriate film formula for hard capsule production.

2. Materials and methods

2.1. Materials

SPS with 47.70 % amylose content and 9.54 % water content was obtained from PT Aren Mulya, an SPS industry in Klaten, Central Java, Indonesia in January 2015. Glycerol, sorbitol and polyethylene-glycol 400 (PEG 400) and magnesium nitrate ($Mg(NO_3)_2$) were purchased from Merck.

2.2. Film Formation

The film forming solution was prepared by dissolving an amount of plasticizers (35-45% for glycerol or sorbitol and 5-9% for PEG 400) in distilled water. The desired amount of SPS (9-11%) was added to the aqueous plasticizer dispersion. The mixture was heated above the gelatinization temperature and maintained under continuous stirring for 30 min. The film solution was degassed to remove the bubbles, spread onto the acrylic plate fitted with rims around the edge and dried overnight at 50°C. Film was allowed to cool to room temperature before removing from the plate. All films were stored in the desiccator under 52% RH at 25°C at least for 48 h before analysis.

2.3. Characterization of Film

Visual test and tactile test were performed to observe the effect of various plasticizers on the appearance of SPS film. Film thickness was measured using a digital micrometer (Mitutoyo, Japan) at nine random positions around the film. The measurement was performed on ten replications. The retraction ratio of films was determined according to a method proposed by The *et al.,* (2009) [5] and calculated as the percentage of film shrinkage from the initial film thickness. The initial film thickness (t_0) was obtained after casting (2 mm), whereas the dry film thickness (t) was obtained after drying the film overnight at 50°C in an air circulated oven. The film retraction ratio was calculated by Equation (1).

$$\textit{Retraction ratio}(\%) = \frac{t_0 - t}{t_0} x\ 100\% \quad (1)$$

Moisture content of films was determined gravimetrically. The film was cut into 20 mm x 20 mm and the initial weight (w_0) was measured. The film was conditioned and dried at 105°C until constant weight (w) was obtained. Ten replicates were obtained for each sample. The percentage of moisture content was calculated by following equation:

$$\textit{Moisture content}(\%) = \frac{w_0 - w}{w_0} x\ 100\% \quad (2)$$

The swelling behavior study of film was performed by immersing film sample in distilled water at room temperature following by monitoring the weight change and thickness of film upon soaking in distilled water periodically. Excess water was removed by filtering paper before weighing. Measurements were done in ten replicates. The initial weight (w_0), final weight (w), initial thickness (t_0), and final thickness (t) were measured. The swelling degree and swelling thickness, expressed in percentage, were calculated according to the following equation:

$$\textit{Swelling degree}(\%) = \frac{w - w_0}{w_0} x\ 100\% \quad (3)$$

$$\textit{Swelling thickness}(\%) = \frac{t - t_0}{t_0} x\ 100\% \quad (4)$$

Solubility of film in water was studied and a method proposed by Shih [20] was adopted with slight modification. A film sample was weighed and immersed in 50 mL distilled water. The flask was stored in an orbital incubator at 37°C for one hour with mild agitation at 175 rpm. The specimen was then collected by filtration and dried again in an air-circulating oven at 105°C for 24 h. Ten replicate measurements were taken for each type of film. The initial dry weight (w_0) and final dry weight (w) of film were measured, then solubility of film was calculated by following formula:

$$\textit{Solubility}(\%) = \frac{w_0 - w}{w_0} x\ 100\% \quad (5)$$

IOP Conf. Series: Materials Science and Engineering **223** (2017) 012049 doi:10.1088/1757-899X/223/1/012049

3. Results and Discussion

3.1. Visual and Tactile Test

SPS with PEG 400 as plasticizer did not form film with suitable characteristics. There were many white opaque spots on the film that might be due to the incompatibility of both polymers. It has been referred as blooming or blushing phenomenon. Hence, this film did not characterize during this study.

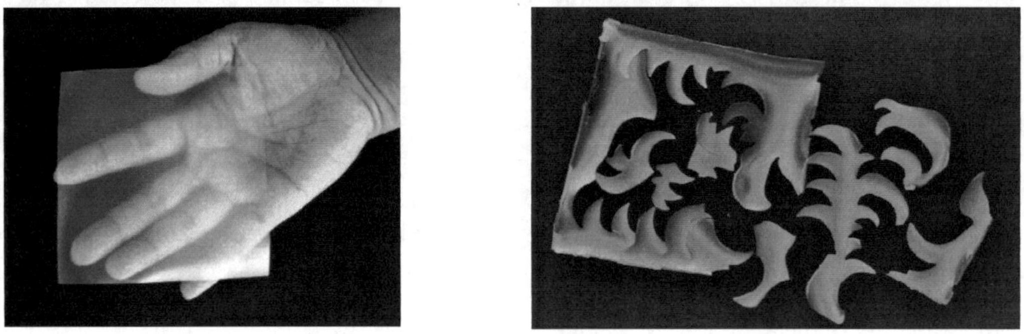

(a) (b)

Figure 1. Visual appearances of homogeneous (a) and brittle-cracked (b) film

Table 1 summarizes the visual observation of films. It was found that both glycerol and sorbitol are compatible with SPS matrix. For glycerol-plasticized film, it was visible that all developed formulas were able to form films with suitable characteristics. However, sorbitol-plasticized films gave different results. It was showed that to produce intact film at 10% starch or above, higher concentration of sorbitol was needed. Since some films were brittle and cracked, it was not possible to analyze other properties. Overall, sorbitol-plasticized film had dry appearance. On the other hand, at higher concentration of glycerol, film would become sticky and hard to handle. Therefore each type of plasticizer has optimum concentration to produce film with expected properties.

Table 1. Appearance of films

Starch Concentration (%w/v)	Plasticizer (%w/w) Glycerol 35	Glycerol 40	Glycerol 45	Sorbitol 35	Sorbitol 40	Sorbitol 45
9*	a	a	a	b	b	b
10	a	a	a	c	b	b
11	a	a	a	c	c	b

[a]Homogeneous, translucent with smooth surface, no tear after peeled, visually looked wet
[b]Homogeneous, translucent with smooth surface, no tear after peeled, visually looked dry
[c]Brittle and cracked
*Data of film formula with 9% (w/v) starch were already presented in International Symposium on Frontier of Applied Physics (ISFAP 2015) held on October 5-7th, 2015 in Indonesia.

3.2. Thickness and Retraction Ratio

The thickness and retraction ratio of film are presented in Table 2. The thickness of glycerol and sorbitol-plasticized films varied from 0.129 to 0.170 mm and from 0.113 to 0.160 mm, respectively. There was an increasing tendency of film thickness as plasticizer concentration increased for all starch concentrations because higher starch and plasticizer concentration would increase the total soluble matter of solution so that it increase film thickness. Furthermore, the thickness of film also affected by retraction phenomenon that occurs in film casting technique. This phenomenon was related to the concentration of dry matter of film and shrinkage during drying period [5]. The retraction ratio of glycerol- and sorbitol-plasticized film varied from 91.45 to 93.55% and from 92.00 to 94.35%, respectively. At the same starch and plasticizer concentration, retraction ratio of sorbitol-plasticized films was slightly higher than that of glycerol-plasticized film. Meanwhile, it was found that the retraction ratio decreased as the plasticizer content increased.

Table 2. Comparison of film thickness, retraction ratio and moisture content of SPS film

Starch concentration (% w/v)	Plasticizer type	Plasticizer concentration (% w/w starch)	Thickness of film (mm)	Retraction ratio (%)	Moisture content (%)
9*	Glycerol	35	0.129	93.55	21.32
		40	0.136	93.20	26.59
		45	0.141	92.95	33.34
	Sorbitol	35	0.113	94.35	8.74
		40	0.125	93.75	8.30
		45	0.134	93.30	8.13
10	Glycerol	35	0.150	92.50	18.46
		40	0.156	92.20	23.61
		45	0.171	91.45	29.33
	Sorbitol	35	NA	NA	NA
		40	0.140	93.00	9.56
		45	0.152	92.40	10.10
11	Glycerol	35	0.152	92.40	21.35
		40	0.166	91.70	24.82
		45	0.170	91.50	30.52
	Sorbitol	35	NA	NA	NA
		40	NA	NA	NA
		45	0.160	92.00	10.20

NA : not available
*Data of film formula with 9% (w/v) starch were already presented in International Symposium on Frontier of Applied Physics (ISFAP 2015) held on October 5-7th, 2015 in Indonesia.

3.3. Moisture Content

Table 2 summarizes the average value of moisture content of the films. Moisture content of films was affected by type of plasticizer. Glycerol-plasticized films had higher moisture content than sorbitol-plasticized films because glycerol is hygroscopic material and can behave as water holding agent. Therefore, moisture content of glycerol-plasticized films would increase as glycerol content increase. On the contrary, sorbitol affected plasticized film differently. Sorbitol-plasticized films had low moisture content, even in film with 9 % of starch concentration there is slightly decreasing tendency as sorbitol concentration increase. Both glycerol (molecular weight 92.09 g/mol) and sorbitol (182.17 g/mol) are polyols which have a similar straight-chain molecule and own molecular structure of glucose. Sorbitol contains more hydroxyl group than glycerol, so its chance to react with starch chain is higher than glycerol. As a result, sorbitol offers higher intermolecular forces with starch molecule thus only fewer hydroxyl groups which are available to interact with water molecules. Hence, sorbitol-plasticized films have lower moisture content than glycerol-plasticized films. This result is in line with the previous study by Godbillot *et al.*, (2006) [21] that film composition affects its moisture content. Meanwhile, by visual observations, there were no differences in opacity among the films. Since in this work the films were only produced of only SPS, this results is in line with the previous study by Fakhouri *et al.*, (2007)[22] that film opacity was affected by amylose content of the starches.

3.4. Swelling Degree and Swelling Thickness

The visual appearances of original and swelled film were shown in Figure 2. The results showed that swelled film had opaque appearances and the original film had transparent appearances.

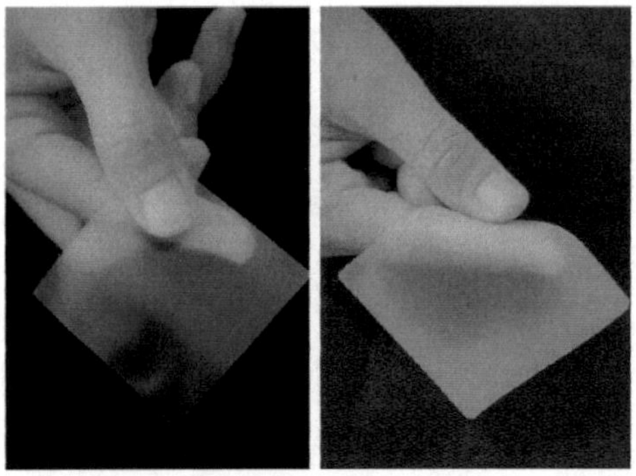

(a) (b)

Figure 2. Visual appearances of original (a) and swelled (b) film

The average value of swelling degree and swelling thickness of the films is presented in Table 3. The results showed that absorption of water was fast at an early stage of immersion and slowly decreased as time increased. Swelling degree and swelling thickness were affected by starch and its interaction in the film matrix. Starch with high amylopectin and low amylose content will produce films with a high swelling degree [23]. Various molecule interaction occurred in film matrix, *i.e.* starch-starch, starch-plasticizer, and plasticizer-plasticizer. As swelling degree affected by the interaction of starch molecule with water molecule, film with less interaction involving starch molecule with others will have high swelling degree. The previous study by Poeloengasih *et al.*, (2016) [16] showed that unplasticized SPS film possessed highest swelling degree compare to other plasticized SPS films. As higher plasticizer content in starch film cause lesser free starch molecules interact with water so it would produce film with lower swelling degree. Since film with higher swelling degree absorbed more water, there is a higher change in film dimension which was shown as a change of film thickness. As swelling degree of film was increased, the swelling thickness also increased. Meanwhile, glycerol-plasticized film had higher swelling degree and swelling thickness than sorbitol ones because in glycerol-plasticized film there is less starch molecule that interacts with plasticizer than in sorbitol-plasticized film.

Table 3. Comparison of swelling degree and swelling thickness of SPS film

Starch concentration (% w/v)	Plasticizer type	Plasticizer concentration (% w/w starch)	Swelling degree (%)		Swelling thickness (%)
			30 min	60 min	
9*	Glycerol	35	130.03	135.87	97.04
		40	104.97	109.31	81.33
		45	80.70	81.81	69.23
	Sorbitol	35	101.53	105.20	93.61
		40	93.95	98.47	86.17
		45	88.74	90.97	82.42
10	Glycerol	35	112.40	119.25	80.03
		40	92.58	95.92	63.27
		45	70.91	73.28	54.71
	Sorbitol	35	NA	NA	NA
		40	93.67	96.77	93.34
		45	86.00	89.41	83.46
11	Glycerol	35	114.67	118.85	97.91
		40	94.37	96.73	82.82
		45	76.22	78.35	62.35
	Sorbitol	35	NA	NA	NA
		40	NA	NA	NA
		45	89.26	91.06	97.24

NA : not available
*Data of film formula with 9% (w/v) starch were already presented in International Symposium on Frontier of Applied Physics (ISFAP 2015) held on October 5-7th, 2015 in Indonesia.

3.5. Solubility in Water

Figure 3a and 3b illustrate the variation in solubility of films for glycerol- and sorbitol-plasticized film respectively (data of film formula with 9% (w/v) starch were already presented in International Symposium on Frontier of Applied Physics held on October 5-7th, 2015 in Indonesia). As the concentration of plasticizer increased from 35 to 45%, solubility of film increased. Both plasticizers reached the highest solubility at 9% of starch and 45% of plasticizer, i.e. 28.22% for glycerol and 34.72% for sorbitol. Since film with formula of 10% starch-35% sorbitol, 11% starch-35% sorbitol, and 11% starch-40% sorbitol were brittle and cracked, the solubility data of those film were not available. This result was in line with the report by Ghasemlou *et al.*, (2011) [24] that higher concentration of plasticizer can increase solubility of the film. Meanwhile, sorbitol-plasticized film showed higher solubility than glycerol-plasticized film.

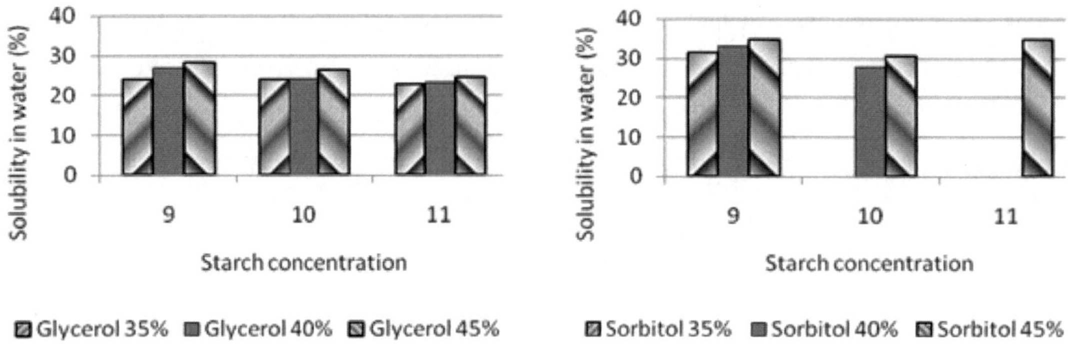

(a) (b)

Figure 3. Solubility of glycerol- (a) and sorbitol-plasticized (b) films at different starch concentration

4. Conclusion

Sugar palm starch is successfully developed as a material for the preparation of a film by combining it with plasticizer. Regardless of type, increasing of plasticizer concentration leads to the increasing of thickness and solubility in water. On the contrary, retraction ratio and swelling degree tend to decrease as plasticizer concentration increase. Meanwhile, glycerol and sorbitol affected physicochemical properties in different ways. It is found that sorbitol-plasticized film had lower moisture content and higher solubility in water than glycerol-plasticized film. In conclusion, sorbitol-plasticized film has more potency than glycerol-plasticized to be hard capsule material.

Acknowledgments

This work was funded by Indonesian Institute of Sciences through Riset Unggulan 2016 program. The authors are grateful to Deputy of Engineering Sciences–Indonesian Institute of Sciences for the support.

References

[1] Mali S *et al* 2005 *Food Hydrocolloids* **19** 157-64.
[2] Zhu F and Wang S 2014 *Trends Food Sci.Tech.* **36** 68-78.
[3] Bae H J *et al* 2008 *Food Chem.* **106** 96-105.
[4] Fama L *et al* 2007 *Carbohyd. Polym.* **70** 265-73.
[5] The D P *et al* 2009 *J. Food Eng.* **90** 548-58.
[6] Pelissari F M *et al* 2013 *Food Hydrocolloids* **30** 681-90.
[7] Bertuzzi M A *et al* 2007 *J. Food Eng.* **82** 17-25.
[8] Abdorreza M N *et al* 2011 *Food Hydrocolloids* **25** 55-60.

[9] Singhal R S *et al* 2008 *Carbohyd. Polym.* **72** 1-20.
[10] Fakharian M H *et al* 2015 *Carbohyd. Polym.* **132** 156-63..
[11] Poeloengasih C D and Anggraeni F D 2014 *Starch/Starke* **66** 1103-8.
[12] Sanyang M L *et al* 2015 *Polymers* **7** 1106-24.
[13] Sahari J *et al* 2013 *Carbohyd. Polym.* **92** 1711-6.
[14] Sahari J *et al* 2014 *Asian J. Chem.* **26** 955-9.
[15] Sahari J *et al* 2012 *Procedia Chem.* **4** 254-9.
[16] Poeloengasih C D *et al* 2016 *AIP Conf. Proc.* **1711** 080003.
[17] Jatmiko T H *et al* 2016 *AIP Conf. Proc.* **1711** 080004.
[18] Apriyana W *et al* 2016 *AIP Conf. Proc.* **1755** 150003.
[19] Adawiyah D R *et al* 2013 *Carbohyd. Polym.* **92** 2306-13.
[20] Shih F F 1996 *Cereal Chem.* **73** 406-9.
[21] Godbillot L *et al* 2006 *Food Chem.* **96** 380-6.
[22] Fakhouri FM *et al* 2007 *Cienc. Tecnol. Aliment.* **27** 369-75.
[23] Chen C H, Kuo W S and Lai L S 2009 *Food Hydrocolloids* **23** 2132-40.
[24] Ghasemlou M *et al* 2011 *Carbohyd. Polym.* **84** 477-83.

IOP Conf. Series: Materials Science and Engineering **223** (2017) 012050 doi:10.1088/1757-899X/223/1/012050

Preparation of immobilized glucose oxidase wafer enzyme on calcium-bentonite modified by surfactant

R K Widi[1], D C Trisulo[2], A Budhyantoro[1], R Chrisnasari[2]
[1] Department of Chemical Engineering, Surabaya University, TG Building 5th floor Jln. Raya Kalirungkut, Surabaya 60293, Indonesia
[2] Department of Biology, Surabaya University, FG Building 2nd floor Jln. Raya Kalirungkut, Surabaya 60293, Indonesia

E-mail: restu@staff.ubaya.ac.id

Abstract. Wafer glucose oxidase (GOx) enzymes was produced by addition of PAH (Poly-Allyamine Hydrochloride) polymer into immobilized GOx enzyme on modified-Tetramethylammonium Hydroxide (TMAH) 5%-calsium-bentonite. The use of surfactant molecul (TMAH) is to modify the surface properties and pore size distribution of the Ca-bentonite. These properties are very important to ensure GOx molecules can be bound on the Ca-bentonit surface to be immobilized. The addition of the polymer (PAH) is expected to lead the substrates to be adsorbed onto the enzyme. In this study, wafer enzymes were made in various concentration ratio (Ca-bentonite : PAH) which are 1:0, 1:1, 1:2 and 1:3. The effect of PAH (Poly-Allyamine Hydrochloride) polymer added with various ratios of concentrations can be shown from the capacitance value on LCR meter and enzyme activity using DNS method. The addition of the polymer (PAH) showed effect on the activity of GOx, it can be shown from the decreasing of capacitance value by increasing of PAH concentration.

1. Introduction

Biocatalysis of redox enzymes has fascinated escalating attention for development of electronic biomaterials and devices for industrial, clinical and environmental applications [1]. One of the enzymes which are being extensively used is glucose oxidase (GOx). The GOx is widely used as a reagent in medical diagnostics [2,3], especially to determination of glucose in blood. However a major problem with the enzymatic methods is the high cost of the enzyme itself. Immobilized enzyme technology has been used to solve the problem [4]. Tests involving immobilized GOx have been used quite frequently in recent year, and one of the tests is immobilized GOx using surfactant-modified bentonite [5].

The method of sol-gel entrapment of biomolecules is a very promising technique of immobilization for biosensors construction, because of its simplicity, low temperature of the process, large amount and low leakage of entrapped material. When the sol-gel technique is used to construct the optical biosensors, the conditions of the process must allow obtaining transparent gels. A simple enzyme electrode can be obtained by immobilization of oxidase on the surface of oxygen electrode. The oxygen concentration depletion is proportional to the concentration of oxidase substrate. Local changes of oxygen concentration caused by enzymatic reaction are measured by oxygen electrode.

Enzyme immobilization matrix using polymer for the development of biosensors has been investigated by many researchers [6-8]. Conducting polymers are capable of incorporating different functionalities in their matrix during or after polymerization. The properties of synthesized polymer films are affected by electropolymerization condition. Many results of the synthesis of polymer films which can be used as a polymer matrix for immobilization of biocomponents have been reported [9,10]. Various conducting polymers have been considered for immobilization of enzymes [11,12]. Amperometric enzymatic electrodes based on GOx, which generates hydrogen peroxide in the presence of oxygen and glucose, are the most widely used for the measurement of blood glucose concentration. Several biosensors with the enzymatic method have been reported [13-17].

However, still it is essential to continue the research in this field with new material and approach, so that the sensitivity and stability of the sensor can be improved. In the present study, we used Ca-bentonite modified by Tetramethylammonium Hydroxide (TMAOH) as an immobilizer. We also used Poly-Allyamine Hydrochloride (PAH) as a polymer GOx matrix. The use of surfactant molecul (TMAH) is to modify the surface properties and pore size distribution of the Ca-bentonite. These properties are very important to ensure GOx molecules can be bound on the Ca-bentonit surface to be immobilized. The addition of the polymer (PAH) is expected to lead the substrates to be adsorbed onto the enzyme.

2. Experimental

Reagents: The enzyme glucose oxidase (GOD) from Aspergillusniger (273 U/mg) was obtained from NacalaiTesque, Ca-bentonit from Pacitan, Indonesia, TMAOH 25%wt (E. Merck), Dinitrosalicylic Acid (Sigma-Aldrich), Sodium sulfite (E. Merck), D-Glucose p.a. (Nacalai Tesque), cellulose acetate (Sigma-Aldrich), PAH (Poly-Allyamine Hydrochloride MW 15.000) (Sigma-Aldrich), reagent Folin-Ciocalteau (E. Merck).

Apparatus: Amperometric (capacitance) measurement was carried out with LCR meter (DEKKO). Preparation of Immobilized GOx Wafer Enzyme: The preparation was divided into two steps; *first step* was preparation of immobilized GOx wafer enzyme in vary of Ca-bentonite:PAH (1:0, 1:1, 1:2, 1:3). The immobilization of GOx was followed the procedure which was described more detailed before [5]. To obtain Tetramethylammonium Hydroxide (TMAH) modified bentonite, 250 mL of several concentrations of TMAH solution (in range 0-5%) were heated until 75 °C and 5 g of bentonite was gradually added to it. Mixtures of heated TMAH solution and bentonite suspension were refluxed for 5 h. The solid phase was separated by filtration and washed with distilled water until the pH become 7.0 to remove the unabsorbed TMAH. The TMAH modified bentonite was dried overnight at a temperature of 100 °C. The modified bentonite powder was sieved with a 140 mesh sieve and the resulting filtrate was used in further experiments. Immobilization of GOx was carried out by dissolving 0.2 g of TMAH-modified bentonite powder and 1 mL of GOx solution (100 IU) in 4 mL 0.1 M phosphate buffer pH 7.0. After being incubated at 20 °C and shaken with rotary shaker overnight, the bentonite enzyme dispersion was centrifuged at 4,000 rpm at 4 °C for 10 min. The supernatant was tested for Hartree Lowry protein assay to determine the amount of un-immobilized enzyme. The pellet was washed several times with phosphate buffer pH 7.0 until no protein was detected in the supernatant. The proteins detected in the supernatants from washing steps werealso considered as un-immobilized enzyme.

Immobilization percentage was calculated usingthe equation below:

% immobilization =(*total amount of immobilized enzyme*)/(*total amount of initial enzyme*) *x* 100%

The total amount of immobilized enzyme was defined as total amount of protein in supernatant beforeimmobilization minus the total amount of protein after immobilization. The total amount of initial enzyme defined as total amount of protein in supernatant before immobilization. The amounts of protein in the supernatant before and after immobilization were determined using Hartree Lowry's method using GOx as the standard. The quantity of protein immobilized on the support wascalculated by subtracting the protein recovered from the combined washings of the modified and unmodified bentonite-enzyme complexes from the total amount of added protein.

The preparation of enzyme wafer was carried out by mixing the suspension of immobilized GOx with PAH solution. The mixture was then centrifuged for 7 h at 4°C. The pellet obtained was swollen in 0.1M phosphate buffer solution (pH 7). The wafer enzyme activity test was measured by DNS method and capacitance values. 500 μL of 1000 ppm of glucose substrate was mixed into 100 uL wafer enzyme, and then incubated for 30 min. After it was centrifuged, the supernatant was measured by DNS method. The second step was pre-design of wafer enzyme device and its performance test by capacitance measurement. The capacitance was measured by adding 100 μL of 1000 ppm of glucose substrate into 100 μL wafer enzyme in the chamber which was connected with LCR meter in room temperature.

3. Results and Discussion

3.1. Activity test of immobilized and free GOx

The use of Ca-bentonite as a support material for adsorption, catalytic and immobilization of enzyme has attracted interest in recent year. Modification of Ca-bentonite is required to increase its capability as a support material. In this study, modification of Ca-bentonite was carried out by intercalation of Ca-bentonite using 5% TMAH. Phenomenon of pores structure formation, very

IOP Conf. Series: Materials Science and Engineering **223** (2017) 012050 doi:10.1088/1757-899X/223/1/012050

important characteristic of Ca-bentonite as a support material, was characterized using FTIR and XRD which has been reported more detail before [18]. Nanostructuring is verified by microscopic inspection allowing studying locally the relevance of the intercalating agents for improvement of Ca-bentonite structure using SEM techniques. The SEM images of TMAH-modified Ca-bentonite in Figure 1 show that the porous structure of bentonite is improved by intercalation process using TMAH. SEM images of Ca-natural bentonite as shown in this figure reveal non-porous structure. On the other hand SEM images of TMAH-modified-Ca-bentonite reveal porous structure. It indicates that TMAH molecules may introduce inside the bentonite layer to improve the pore size distribution of the Ca-bentonite.

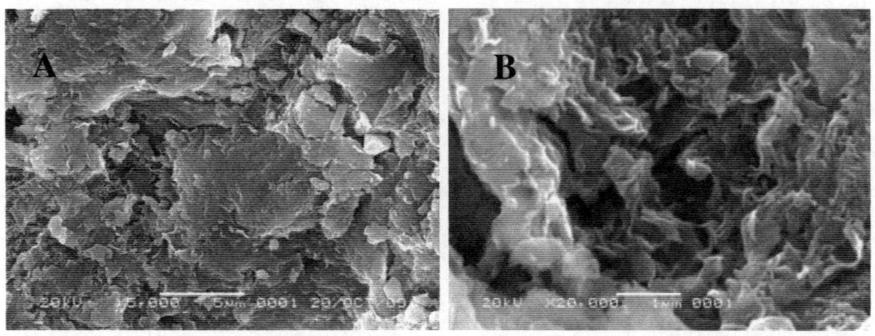

Figure 1. SEM images of Ca-bentonite (A), and intercalated bentonite using TMAH (B)

The modified Ca-benonite was then used as immobilization material for GOx. The result of immobilized GOx is described in Table 1.

Table 1. Percentage of Immobilized GOx on 5% TMAH-Modified Bentonite

Replication	Initial GOx concentration (IU/ml)	Immobilized GOx (IU/ml)	% Immobilization
I	54.3	41.9	77.2
II	56.9	42.8	75.2
III	54.9	42.2	76.9

This result indicates that the immobilization GOx using 5% TMAH give stable and consistent result. The addition of TMAH into bentonite leads increasing the distance of bentonite layer. This is due to ammonium cation from TMAH exchange with potassium and magnesium cation in the bentonite. As a consequence, modified bentonite has higher affinity to organic molecules. It causes the GOx molecules can recline easily into the bentonite pores, and then they form interaction to shape immobilized GOx structure [5].

The free and immobilized GOx enzyme activities can be seen in Figure 2. The data shows that the activity of immobilized GOx enzyme is lower than that of free GOx enzyme. The decreasing of the activity is about 32.5%. The immobilization process may leads to the conformational changes of the GOx, and moreover it also causes some GOx molecules covered by surfactant molecules in the layer of bentonite, which may decrease its affinity and accessibility of substrate molecules thus cause decreasing of enzyme activity [5].

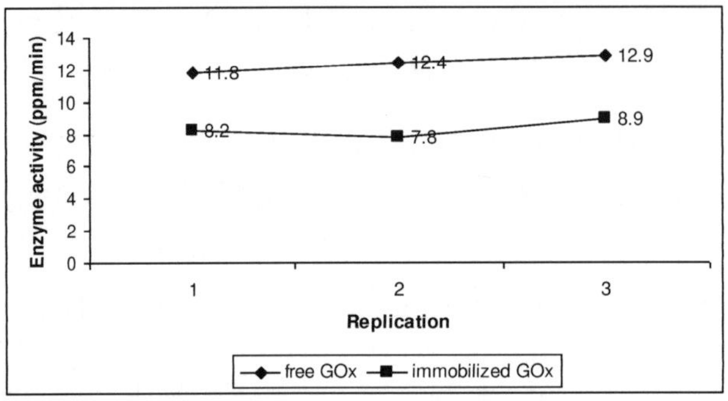

Figure 2. The Activity of Free and Immobilized GOx Enzyme

3.2. Preparation of Immobilized GOx Wafer Enzyme

The main composition of pre-designed device is a wafer. This wafer consists of membranes, immobilized GOx and PAH. This section describes the effect of molar ratio of immobilized GOx:PAH toward the enzyme activity (Figure 3) and its capacitance values (Figure 4).

The addition of the polymer (PAH) plays a role in increasing of the stability of immobilized GOx enzyme. The data shows that by increasing of molar ratio of PAH, the GOx enzyme activity is decreased (Figure 4). This probably increasing of the number of PAH molecules lead to increase the steric hindrance around the sites active of the enzyme which may hinder the substrate molecule interact with the enzyme. As a consequence, the glucose molecule which may oxidize is decreased.

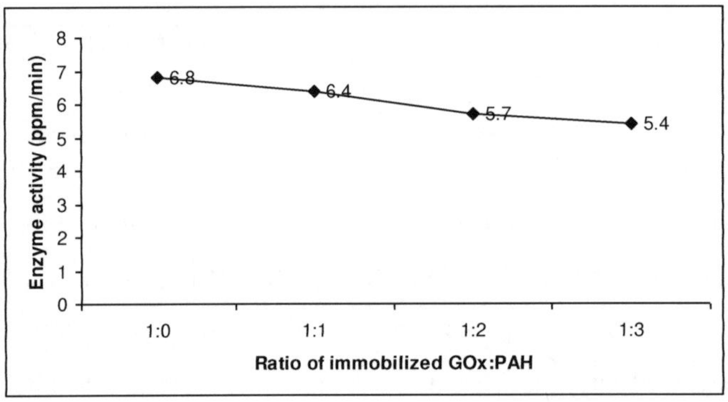

Figure 3. Effect of Ratio of Immobilized GOx and PAH Toward Enzyme Activity

The addition of the polymer (PAH) also plays a role in decreasing of the capacitance of GOx enzyme during the process the substrates oxidize to form gluconic acid and peroxide, and moreover there are some electrons involve in this reaction. The transfer of the electrons can be detected by LCR meter to give its capacitance. The electrons from the reaction is cached by the PAH and then pass through into

IOP Conf. Series: Materials Science and Engineering **223** (2017) 012050 doi:10.1088/1757-899X/223/1/012050

dielectric solution which leads to increase the conductivity. The conductivity value is contrary with capacitance value.

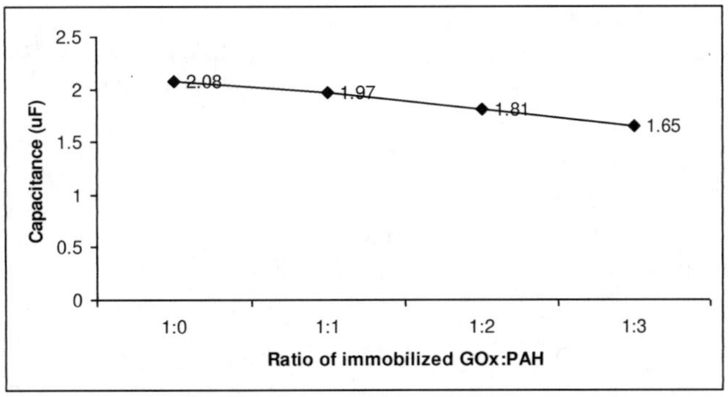

Figure 4. Effect of Ratio of Immobilized Gox and PAH Toward Wafer Capacitance

Thus, the more substrate oxidizes, the more electrons involved the lower capacitance is. It is very clearly described in Figure 4. It can be concluded that the addition of the polymer lead the substrates to be adsorbed onto the enzyme and to be oxidized more easily.

4. Conclusion

In this paper, immobilized glucose oxidase wafer enzyme on Ca-bentonite modified by TMAH has been presented. GOx can be immobilized successfully and still showed very good activity. The wafers consist of immobilized GOx and PAH have also been successfully prepared and show its activity. Increasing of molar ratio of PAH, the GOx enzyme activity is decreased. The present work demonstrates the existence promising applications of immobilized glucose oxidase wafer enzyme on Ca-bentonite modified by TMAH as a health device.

Acknowledgements

We gratefully acknowledge financial support from Ministry of Research, Technology and Higher Education of the Republic of Indonesia (Hibah Penelitian Unggulan Perguruan Tinggi, contract # 015/SP-Lit/LPPM-01/Dikti/FT/V/2016).

References

[1] Katz E *et al* 2004 in *Nanoparticles: From Theory to Applications* (Ed.: G. Schmid), Wiley-VCH, Weinheim, Germany.
[2] Iwuoha E I and Smyth M R 1994 *Anal. Proc.* **31** 19.
[3] Atia K S and Al El-Batal 2005 *J. Chem. Technol. and Biotechnol.* **80** 805.
[4] Ma J *et al* 2007 *J. Sep. Sci* .**30** 3050 .
[5] Chrisnasari R *et al* 2015 *Indones. J. Chem.* **15** 22.
[6] Shirale D J *et al* 2005 *Transactions of the SAEST* **40** 128.

[7] Vidal J C *et al* 2004 *Talanta* **64** 655.
[8] Gaikwad P D *et al* 2006 *Bull.Mater. Sci.* **29** 169.
[9] Shirale D J *et al* 2006 *Mater.Lett.* **60** 1407.
[10] Shirale D J *et al* 2006 *Int. J. Electrochem. Sci.* **1** 62 .
[11] Shirsat M D 2005 *Microwaves and Optoelectronics*, AnshanTunbridge Wells UK 455.
[12] Newman J D and Turner A P F 2005 *Biosens. Bioelectron.* **20** 2435.
[13] Arslan F *et al* 2011 *Sensors* **11** 8152.
[14] Yoo E H, Lee S Y 2010 *Sensors* **10** 4558.
[15] Norouzi P *et al* 2010 *Int. J. Electrochem. Sci.* **5** 1213.
[16] Shan D *et al* 2008. *Mater Sci Eng C* **28** 213.
[17] Budhyantoro A, Restu K W 2013 *Int. J. Mater. Sci.* **8** 91.

IOP Conf. Series: Materials Science and Engineering **223** (2017) 012051 doi:10.1088/1757-899X/223/1/012051

Synthesis and micowave characterization of conductive polyaniline prepared by continous polymerization process

A Manaf[1], A. Bimantoro[1], M AE Hafizah[1], Andreas[2]
[1]Postgraduate Program of Materials Science, Faculty of Mathematics and Natural Science, Universitas Indonesia, Jl. Salemba Raya No.4, Jakarta 10430, Indonesia
[2]Indonesian Institute of Sciences, LIPI, Jl. Sangkuriang, Komplek LIPI, Bandung 40135, Indonesia

E-mail: mas.ayu@alumni.ui.ac.id

Abstract. Results of synthesized polyaniline (PANi) prepared by a continuous polymerization process is reported. The purpose of this research is to obtain material absorber which maximum absorption. FTIR spectrum for synthesized PANi showed no double bond absorption peak at 3442 cm^{-1} which indicates the polymer chain was formed through repetition of single bonds. The results of the electrical conductivity measurement showed the value 17.5 μS/cm for synthesized PANI. The value increased dramatically to 3600, 1520 and 920 μS/cm after the addition of photonic acids of $HClO_4$, H_2SO_4 and HCl respectively. Those values were obtained through sequences process. Results of particle size distribution confirmed that all synthesized PANi have nanoparticles. In addition, all synthesized PANi have shown electromagnetic wave absorbing properties in the frequency range 12-18 GHz. The best peak Reflection Loss (RL) values -10 dB (~ 68 % absorption) and -25 dB (~ 84 %) at 13 and 16.5 GHz respectively obtained on the PANi Emeraldine base (PANi-EB).

1. Introduction

One of technical requirement for Radar Absorbing Materials (RAM) [1,2] in applications is that beside their high ability to absorb the electromagnetic wave, the RAM should also be made of materials with lighter in weight, chemically stable, possesses suitable mechanical and physical properties. It is not surprising to note that the need to develop lighter and thinner microwave absorbing materials with a high ability for absorbing the electromagnetic wave increased nowadays. Several reports have shown that the best materials for microwave absorbers are made of ions substituted M-type hexaferrites[3-4]. M-type hexaferrite is particularly suitable because it has a high magnetization (Ms = 0.48 T), it is an oxide based compound and hence chemically stable, but its magneto crystalline anisotropy constant is a relatively large and hence it possesses a high coercivity (~250 $kA.m^{-1}$). This is the reason why the M-type hexaferrite is widely used as permanent magnets. However, partially ions substituted for Fe for instance, in $BaO.6Fe_2O_3$ was able to reduce the coercivity significantly, but with a little sacrificing in the total magnetization making it suitable as a potential candidate for microwave absorbers. It was reported that doped Barium Hexaferrite (BHF) has a larger value of reflection loss in which more than 51 dB was observed in the frequency ~ 16.7 GHz [1]. The report shown that the material has a wider absorbing width in the frequency range 12 - 20 GHz. However, hexaferrite materials have a relatively

IOP Conf. Series: Materials Science and Engineering **223** (2017) 012051 doi:10.1088/1757-899X/223/1/012051

larger mass density (~ 5.8 gr/mL) which is about the third quarter of the mass density of pure iron (~7.8 gr/mL). Therefore, the use of this kind of materials as the microwave absorber shouldn't be in a large quantity in order to avoid the excessive weight in the applications. One way to achieve this objective is that employing hexaferrites materials as fillers for a composite system. Thus, material selection for matrix is essential to meet therequirement as the lighter microwave absorber. It is considered important to select conducting polymers like conductive polyaniline (PANi) as the alternative for suitable matrix in the composite system. Some reports have also shown the use of PANi as electromagnetic shielding at microwave frequencies [5,6]. PANi which was modified by inorganic filler has shown to improve magnetic and dielectric properties depend on the type of filler. Both properties are essential in the microwave absorber material. Improve electrical permittivity will increase the dielectric constant and hence the capacitance of materials. Whereas, improve magnetic permeability will increase the inductance. The capacitance and inductance together with an electrical conductance are essential for a total impedance of the material. Focus of this research is to explore the properties and characteristics of conductive PANi by additional some strong acids as doping agent to increase the conductivity value then the correlation between electricity property with microwave absorber property will be obtained.

2. Materials and Methods

Aniline monomers was mixed with 1.5 M hydrochloride acid (HCl) at room temperature as monomer solution. Initiator solution was prepared by mixing 10 % of ammonium persulphate (APS) in hydrochloride acid (HCl) 1.5 M. Both solution were introduced into the reactor step by step by controlling flow rate of the feeding. Total duration of reaction time to form Polyaniline emeraldine salt (PANi-ES) was 5 h. The liquid was further treated by subsequence steps starting from filtering, precipitation, washing using ammonia free water and finally drying in the vacuum for 60 h after purging with N_2 gasses. The result of the treatment is a dried PANi-ES. In order to get Polyaniline emeraldine base (PANi-EB), the cleaning of dried PANi-ES were continued using water and methyl alcohol to dissolve residual anilines which formed during the polymerization process. The dry solid, then was dissolved using 0.1 M ammonium hydroxide solution with pH $\geq$ 8. Mixing for 15 h at room temperature to form a complete (PANi-EB) was required. The subsequent treatment as described above was then repeated to PANi-EB. The pure and dry powder of final Polyaniline as stable PANi-ES, then was dissolved by different acids. There are three types of acid are used with different ionic strength those are HCl 37%, $HClO_4$ 98% and H_2SO_4 98%. Each mixture was stirred for 10 h in a separate chamber at room temperature. The precipitant was formed and ready for further characterization. Powders of all synthesized PANi were evaluated by Malvern Particle Size Analyzer

(PSA) and FTIR for identification studies. Finally, the microwave characteristics of all synthesized PANi were evaluated by vector network Analyzer (VNA) through measurements of transmission and reflection of electromagnetic wave in the 12 - 18 GHz frequency range.

3. Result and Discussion

Figure 1 is a FTIR spectrum for aniline monomer. Aniline monomer has specific bands to Infra-Red transmittance at wave numbers 3442 cm^{-1} and 3354 cm^{-1} respectively [7] which are associated with vibration stretching of N-H from primary amine between vibration stretching N-H asymmetric and vibration stretching N-H symmetric. Those bands appear lower than the band of vibration stretching of O-H which appears almost at similar wave number. A shoulder band always appears at 3213 cm^{-1} especially for aniline. The bending vibration of N-H from primary amine appears at 1600 cm^{-1} and 1273 cm^{-1} as stretching vibration of C-N. Another specific band of aniline also are confirmed at 881 cm^{-1} and 748 cm^{-1} as bending vibration of C-H and wag vibration of N-H from primary amine.

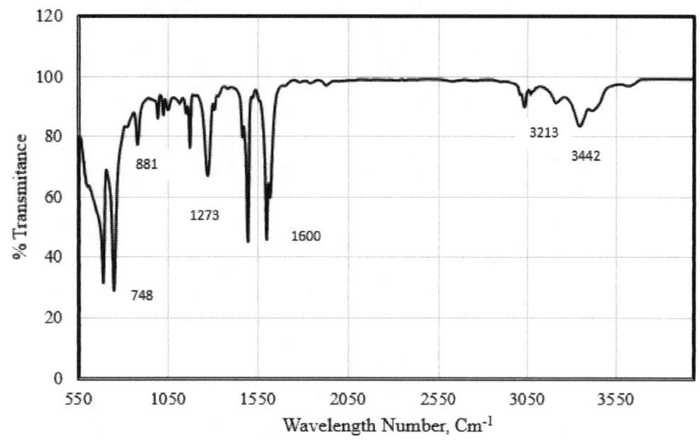

Figure 1. FTIR Spectrum of Aniline Monomer

The FTIR spectrum for synthesized PANi are shown in Figure 2 and Figure 3. The PANi was formed which confirmed by the following results of identification. First spectrum interpretation was a band of Infra-Red transmittance at the wave numbers 1564 cm^{-1} and 1477 cm^{-1} respectively. These are specific band for stretching vibration of C=C from quinoid ring and benzenoid ring as a backbone of PANi [8]. In addition, a transmittance peak at 1435 cm^{-1} was identified and it showed stretching vibration of the aromatic group from ammonium. Other specific bands of PANi can also be confirmed in which stretching vibration of C-N from secondary amine were observed at 1305 cm^{-1} and 1246 cm^{-1} respectively. The strongest peak appears at 1141 cm^{-1} which arise from the stretching vibration of C=N from quinoid ring protonated. Finally, the bands at 989 cm^{-1} and 819 cm^{-1} respectively ensure that the synthesized PANi has really formed which totally different to that of the monomer.

When NH_4OH is added during the polymerization process, proton (H^+) was released and Emeraldine Base (PANi-EB) was formed as a neutral based of PANi. When doped (protonated) by strong acid, Emeraldine Salt (PANi-ES) will be formed as highly electrically conducting structure. Thus, there is a difference in structure and properties between PANi-EB and PANi-ES. Pure PANi will be obtained after filtration, cleaning, and drying as silver brown powder. The propagation process will be formed during reversible reaction formation of PANi-ES and PANi-EB. According to the particle size measurement, it didn't make any effect into particle size of PANi by doping with some different acids. Almost all PANi has particle size below 20 nm, as mentioned by M.A Elita *et al.,* (2016)[9].

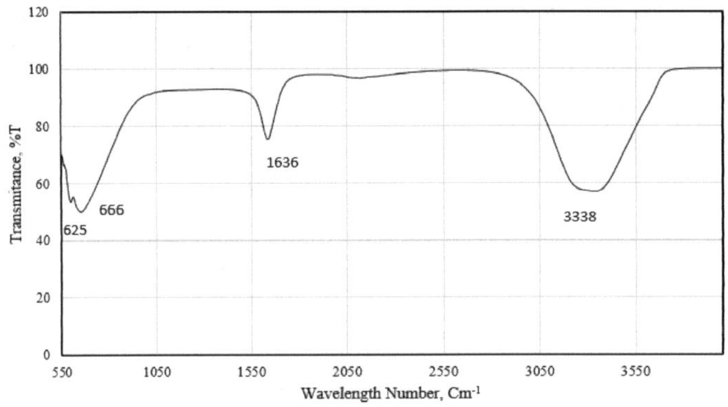

Figure 2. FTIR Spectrum of PANi-ES

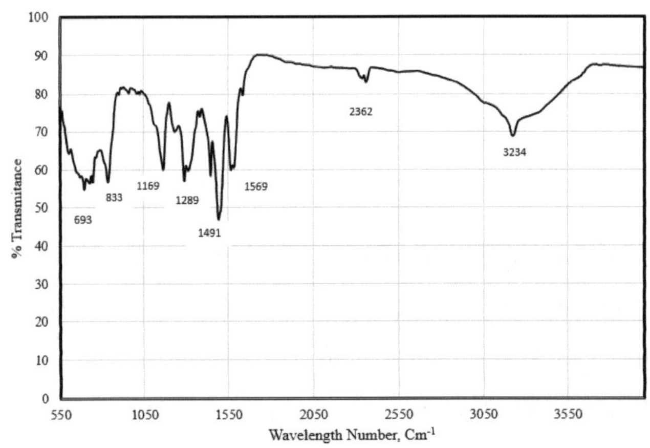

Figure 3. FTIR Spectrum of PANi-EB

It is well known that in general aniline has a zero conductivity value. However, when the aniline is polymerized to form PANi, the conductivity value is raised. In Table 1, the electrical conductivity values for all synthesized PANi are summarized. Results as listed in Table 1 suggested that photonic acid has increased the conductivity value of PANi. PANi which doped by $HClO_4$ has the highest conductivity ~ 3600 μS/cm. The high value for conductivity in PANi-$HClO_4$ could be associated with strongest Brønsted-Lowry acids in which the pKa value is -10 [7]. The conductivity value of PANi is required to obtain high absorption material when polyaniline is used as a matrix. Conductivity (σ), permitivity (ε) and permeability (μ) are important parameters could contribute increasing electromagnetic shielding efficiency [10].

Table 1. Conductivity value of PANi

Material Name	Conductivity Value (μS/cm)
Aniline	0
PANi-EB	17.5
PANi-ES	100
PANi-HCl	920
PANi-H_2SO4	1520
PANi-$HClO_4$	3600

According to above result shown that additional perchloric acid ($HClO_4$) as doping agent with ratio 1 : 10 to PANi produced the highest conductivity value of PANi is ~ 3600 μS/cm. It indicates that perchloric acid is able to influence transition phase from amorphous to crystalline phase of polyaniline molecule. It is also confirmed from the observation result was conducted by Zhang *et al.*, (2012) [11], where is due to of this molecule property caused electron easily moves from the back bond chains inducing enhancement conductivity. The additional of dopant plays role as an attractive potential for the charge carriers and as bridge conduction facilitating electron transport inter-chains. Mechanism electron transport inter-chains of conductive polyaniline is used as basic principle for the mechanism of conducted polyaniline property.

Figure 4 describes the Reflection Loss (RL) values were plotted in the frequency range 12-18 GHz from which all conductive PANi have microwave absorber characteristics at least within the range of applied frequency. The absorbing pattern for all PANi looks similar, in which the highest attenuations occurred at about 13 GHz and 16.5 GHz respectively. However, only PANi-EB has the largest attenuation with RL value ~ -10 dB (~ 68 %) and -25 dB (~ 84 %) at respective frequencies.

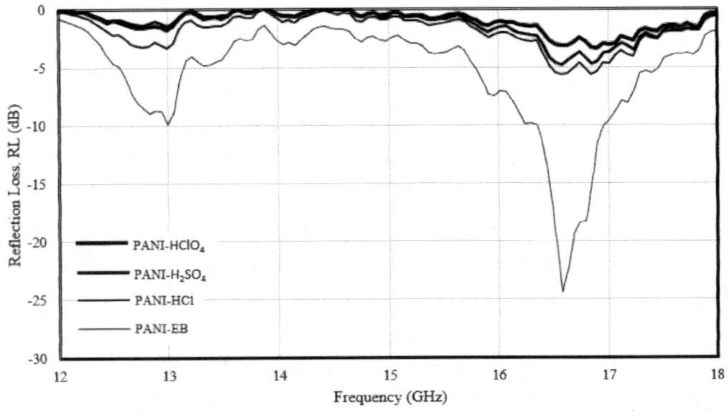

Figure 4. Microwave Absorption Characteristic of Various Synthesized PANi

According to the alternating current basic principle, a part of electrical resistivity and hence the conductivity, a proper matching between dielectric loss and magnetic loss which is the imaginary part of an impedance value determining the losses. When referring to results of Table 1 and Figure 4, the highest conductivity value was obtained in PANi-$HClO_4$ is 3600 μS/cm but its RL value is the lowest. This is in contrary to that of PANi-EB, which has the lowest conductivity value approximately 17.5 μS/cm, hence the largest resistivity. Overall, PANi-EB has the best microwave absorbing characteristic because a high in its resistivity value and hence the largest impedance among the samples. The largest RL values were due to a high dielectric loss only.

4. Conclusion

The electrical conductivity of pure PANi can be increased from zero to 3600, 1520 and 920 μS/cm after the addition of photonic acids of Perchloric acid ($HClO_4$), sulfuric acid (H_2SO_4) and hydrochloride acid (HCl) respectively. All conductive PANi have shown electromagnetic wave absorbing properties in the frequency range 12-18 GHz. The best peak Reflection Loss (RL) values -10 dB (~ 68 % absorption) and -25 dB (~ 84 %) at 13 and 16.5 GHz respectively obtained on PANi Emeraldine base (PANI-EB) which correspond to the largest resistivity value. It indicates that additional doping agent by strong acid can increase the electrical conductivity value of PANi also improve the material absorption property until -25 dB.

Acknowledgements

The authors gratefully acknowledge the support of Postgraduate Program of Materials Science Universitas Indonesia for the research facilities. We are also thankful for the financial support provided by the National Higher Education of the republic of Indonesia under program Penelitian Unggulan Perguruan Tinggi Riset Utama contract no. 3447/H2.R12/HKP 05.00/2014.

References

[1] Ghasemi A *et al* 2006 *J. Magnetism and Magnetic Mater* **302** 429–435.
[2] Ghasemi A *et al* 2007 *J. Magnetism and Magnetic Mater* **316** e105–e108.
[3] Muller K H *et al* 2001 *J. Magnetism and Magnetic Mater* **226** 1370-1376.
[4] Sivakumar M *et al* 2004 *J. Magnetism and Magnetic Mater* **268** 95–104.
[5] Belaabed B *et al* *2012 J. Alloys and Compounds* **527** 137– 144.
[6] Wai Phang S *et al* 2008 *Current Appl. Phys.* **8** 391-394.
[7] Elsayed A H *et al* A 2011 *Int. J. Electrochem. Sci.* **6** 206-221.
[8] Gospodinova N, Terlemezyan 1998 L Prog. Polym. Sci. **23** 1443-1484.
[9] Elita Hafizah M A *et al* 2016 *Procedia Chemistry* **19** 162 – 165.
[10] Ciric G M 2013 Elsevier Synthetic Metals **177** (3) 1–47.

[11] Zhang H *et al* 2011. *Elsevier Polymer* **52** 3059–3064.

IOP Conf. Series: Materials Science and Engineering **223** (2017) 012052 doi:10.1088/1757-899X/223/1/012052

Mechanical strength and ionic conductivity of polymer electrolyte membranes prepared from cellulose acetate-lithium perchlorate

T Sudiarti[1,2], D Wahyuningrum[2], B Bundjali[2], and I Made Arcana[2]
[1]Chemistry Department, State Islamic University Sunan Gunung Djati Bandung, Bandung, Indonesia
[2]Department of Chemistry, Faculty of Mathematics and Natural Sciences, Bandung Institute of Technology, Bandung, Indonesia

E-mail: arcana@chem.itb.ac.id ; s_tety70@yahoo.com

Abstract. The need for secondary batteries is increasing every year. The secondary battery using a liquid electrolyte has some weaknesses. A solid polymer electrolyte is the alternative electrolytes developed to replace the liquid electrolyte type. This study was conducted to determine the effect of lithium perchlorate content on the polymer electrolyte membranes of cellulose acetate-$LiClO_4$. The cellulose acetate-$LiClO_4$ membranes were prepared by mixing cellulose acetate and $LiClO_4$ in various compositions using tetrahydrofurane (THF) as solvent. The effect of $LiClO_4$ ratios on the polymer electrolyte membranes was studied by analysis of the functional groups using FTIR (*Fourier Transform Infrared*) spectroscopy measurement, the ionic conductivity by EIS (*Electrochemical Impedance Spectroscopy*) method, and mechanical properties by tensile tester measurements. The ionic conductivity of the membranes increased with the increasing in the ratios of lithium perchlorate content in the membranes and reached the optimum value at 1.79×10^{-4} S cm^{-1} corresponded to the cellulose acetate doped with 25% (w/w) $LiClO_4$ membrane. The presence of 10% (w/w) $LiClO_4$ content within cellulose acetate membranes can increase the mechanical properties of the membranes from 19.89 to 43.29 MPa for tensile strength, and from 2.55 to 4.53% for elongation at break. However, when the cellulose acetate membranes containing ratio of $LiClO_4$ more than 10% (w/w), consequently the tensile strength tended to decrease and the elongation at break was increased.

1. Introduction

Lithium battery is considered as the most probable candidates for large scale electrical power storage systems due to its many advantages, such as the high specific energy density, highworking voltage, lowself-discharge rate, fast charging, long life usage time, and has no memory effect [1–3]. The main components of the battery consist of three parts, namely anode, cathode, and electrolyte. Electrolyte is a medium for transferring the charge and in the cell is located between the anode and cathode. Electrolytes are divided into two types, namely the solid electrolyte and liquid electrolyte. The liquid electrolyte has some weaknesses, which are prone to leakage and the difficulty in packaging as well as the storage of the liquid electrolyte [4,5]. On the other hand, the solid electrolyte is more secure, easier to use, and can be made with smaller dimensions such as thin layers [4]. The comprehensive requirements for solid polymer electrolyte used in lithium battery include high mechanical strength, excellent flame retardancy, superior thermal stability, high ionic conductivity at ambient temperature, and wide electrochemical window [6-9]. The minimum requirement of all solid polymer electrolytes for ionic conductivity is more than 10^{-5} Scm^{-1} at 25 °C and its tensile strength is more than ≥ 30 MPa [4].

Polyethylene Oxide (PEO) is a synthetic polymer usually used as a solid electrolyte, especially in the lithium-ion battery [4]. The selection of PEO as electrolyte materials because PEO has good ionic conductivity as well as their compatibility with electrode used [10]. However, the PEO has low ionic conductivity at room temperature [10] and difficult to decompose in nature. To overcome these drawbacks, therefore in this research, a new host polymer electrolyte with sufficiently ionic conductivity using biodegradable polymers such as cellulose acetate has been developed to fit the requirement for lithium polymer batteries, taking into account the need for safer practical usage and a more environmentally friendly system.

Recently, the use of biodegradable polymers that have good ionic conductivity and mechanical property has been developed. Among biodegradable synthetic polymers, cellulose acetate-based solid polymer electrolyte is one of the most attractive polymers due to its availability in nature as renewable resources, non-toxic nature, low cost, environmentally friendly, and having good mechanical properties [5, 11]. Furthermore, the research regarding the cellulose acetate-based polymer electrolyte membrane for lithium battery separator was still very few reported. For example, one study reported the plasticizing effect of 1-allyl-3-methylimidazolium chloride for cellulose acetate-based polymer electrolyte membranes , and this polymer electrolyte has the ionic conductivity approximately 1.75 x 10^{-3} Scm^{-1} at room temperature [11]. The cellulose acetate doped with $LiClO_4$ as a biodegradable polymer electrolyte for supercapacitors has the ionic conductivity approximately 4.9 x 10^{-3} Scm^{-1} with lithium salt content of 16% (w/w) at room temperature [5], however, the mechanical properties of the membrane has not been reported yet, even though these properties are very important in polymer electrolyte membranes. Therefore, in order to study the more comprehensive physical properties of cellulose acetate-based polymer electrolyte membranes doped with $LiClO_4$, this research reported the results of mechanical properties and ionic conductivity measurements in various ratios of lithium perchlorate content in cellulose acetate-based polymer elektrolyte membranes. The differences of this study compared to the other reported studies are the more detailed study concerning the mechanical properties and ionic conductivities of cellulose acetate-lithium perchlorate based polymer electrolyte membrane, therefore based on the resulting data then it could be concluded whether this polymer electrolyte membrane has the potent to be applied as electrolyte separator in lithium battery.

IOP Conf. Series: Materials Science and Engineering **223** (2017) 012052 doi:10.1088/1757-899X/223/1/012052

2. Materials and Methods

2.1. Materials

The materials used in preparation of polymer electrolyte membrane were cellulose acetate (Sigma-Aldrich;average molecular weight: 30,000; 39.8 wt% acetyl content) and $LiClO_4$ (Sigma-Aldrich). Tetrahydrofuran (THF) for analysis with purity (GC) ≥ 99.8 % were obtained commercially from E-Merck Chem.Co. All materials were used without further purification.

2.2. Preparation of Polymer Electrolyte Membranes

Polymer electrolyte membranes were prepared by mixing cellulose acetate and lithium perchlorate salt ($LiClO_4$) in various concentrations ranging from 1-25% (w/w) using tetrahydrofuran (THF) as solvent. An amount of cellulose acetate and lithium perchlorate was dissolved in tetrahydrofuran within a ratio of 1% (w/v) and stirred on the hotplate magnetic stirrer for approximately 3 hours to form homogenous solution. The solution was then poured into a petri dish (diameter 10 cm), and the solvent was evaporated to form thin film and dried for approximately 20 h.

2.3. Characterization of Polymer Electrolyte Membrane

2.3.1. Functional Group Analysis

The polymer electrolyte membranes were analyzed by using FTIR (Prestige 21 Shimadzu) spectroscopy method for determining its functional groups with direct firing method. The membranes were scanned within the wavenumber range of 500 to 4000 cm^{-1}.

2.3.2. Ionic Conductivity Measurements

The ionic conductivity of the polymer electrolyte membranes were measured by EIS (Electrochemical Impedance Spectroscopy) method. The membranes were cut rectangular approximately with the dimension of 2 cm × 4 cm, and then the membrane was placed between two platinum wire electrodes within the distance of 1.5 cm and 0.4 cm long. The data of impedance was performed within the frequency range between 20 Hz and 2 MHz at 1 V using the equipment of Agilent™ LCR E4980. The impedance data were analyzed using the *Zview®* software to obtain the resistance value of membranes. The membrane resistance value was taken from the high frequency intercept on the real impedance axis of the Nyquist plot [12]. The ionic conductivity of the membrane was calculated using the following formula:

$$\sigma = \frac{1}{R} . \frac{\sqrt{(d)^2 + (t)^2}}{t.s}$$

IOP Conf. Series: Materials Science and Engineering **223** (2017) 012052 doi:10.1088/1757-899X/223/1/012052

where *σ, R, d, t, s,* respectively, were ionic conductivity (Scm^{-1}), membrane resistance (ohm), thickness of membrane (cm), distance between two electrodes (1.5 cm), and length of electrode that contacts between membrane and electrode (0.4 cm).

2.3.3. Mechanical Strength Analysis

The mechanical strength of the polymer electrolyte membranes were measured using tensile tester (Singel Fiber Tensile Tester (Favigraph-Textechno)). Membrane was cut to rectangular with dimension of 4 cm × 3 mm and the membrane thickness was measured using a micrometer, then the membrane was tested with the gauge length of 2 cm, load cell of 10.000 cN, and test speed of 6 mm/min. The mechanical properties of sample were performed by taking the average value of three repeated measurements.

3. Results and Discussion

3.1. Functional Group Analysis

The analysis of specific functional groups or chemical bonds that exist within a material such as coordination or complexation of Li^+-ion with the cellulose acetate can be determined using FTIR spectroscopy measurement. The changes in the absorption peaks in terms of the wavenumber shifting, relative intensity, disappearance of the existing peak and even appearance of new peaks were evidences from the complexation of these constituents [5,11,13]. FTIR spectra of cellulose acetate and cellulose acetate doped with 25% (w/w) $LiClO_4$ membranes are shown in Figures 1. The widening of absorption peaks was generally observed in the region of 3400–3600 cm^{-1} corresponding to the stretching of intermolecular hydrogen bonds of hydroxyl groups (-OH) [14]. The presence of the wide absorption peaks of pure cellulose acetate FTIR spectrum at 3485.37 cm^{-1} is due to the presence of moisture in the sample, that was not completely removed [5]. The similar wide absortion peaks shifted to 3520.09 cm^{-1} was also observed in cellulose acetate doped with lithium perchlorate (cellulose acetate-$LiClO_4$) membranes FTIR spectrum. The absorption peak at wavenumber of 2949.16 cm^{-1} in pure cellulose acetate FTIR spectrum corresponded to the stretching of –CH- of methyl groups (-CH_3) [5] , which has been shifted to 2945.30 cm^{-1} in FTIR spectrum of cellulose acetate-$LiClO_4$ membrane.. The carbonyl group (C=O) stretching vibration at 1756.22 cm^{-1} for the pure cellulose acetate has been shifted to 1730.15 cm^{-1} for cellulose acetate-$LiClO_4$ membrane. The shifting of these characteristic peaks proves the occurrence of complexation between Li^+ ions and oxygen atoms in polymer backbone. The -CH_2- deformation vibration occurred at 1431.18 cm^{-1} for pure cellulose acetate has been shifted to 1429.25 cm^{-1} for cellulose acetate-$LiClO_4$ membrane.

The occurrence of complexation between cellulose acetate and Li^+ ions was further verified by the broadening peaks at the wavenumber range of 1049–1269 cm^{-1} as result of the simple peaks overlapping. The characteristic peaks of 'C-O-C' anti symmetric stretching vibrations of ester group of pure cellulose acetate at 1240.23 cm^{-1} has been shifted to 1269.16 cm^{-1} in cellulose acetate-$LiClO_4$ membrane. The -C-OH stretching vibration of pure cellulose acetate at wavenumber of 1049.28 cm^{-1} has also been shifted to 1049.28 cm^{-1} for the cellulose acetate-$LiClO_4$ membrane. The presence of the absorption peaks at 904.61 cm^{-1} in pure cellulose acetate could be due to the combination of –C-O- stretching and -CH_2- rocking vibrations that have been shifted to 912.0 cm^{-1} in cellulose acetate-$LiClO_4$ membrane. The symmetric vibration of -ClO^- ion (perchlorate ion) was generally observed at 940 cm^{-1}, which in this study, this peak overlaps with C-O stretching vibration of cellulose acetate, therefore the wavenumber of the same peak at 912.0 cm^{-1} shifted to the lower value in cellulose acetate-$LiClO_4$ membrane [5]. The shifted to the lower value of this wavenumber can be caused by the complexation effect [5,11]. The intensity of all the peaks in the FTIR spectrum of the cellulose acetate-$LiClO_4$ membrane is lower and broader compared to the pure cellulose acetate FTIR spectrum. These effects are generally caused by the intermolecular interactions, as well as due to the presence of the complexation between cellulose acetate and $LiClO_4$ [15,16].

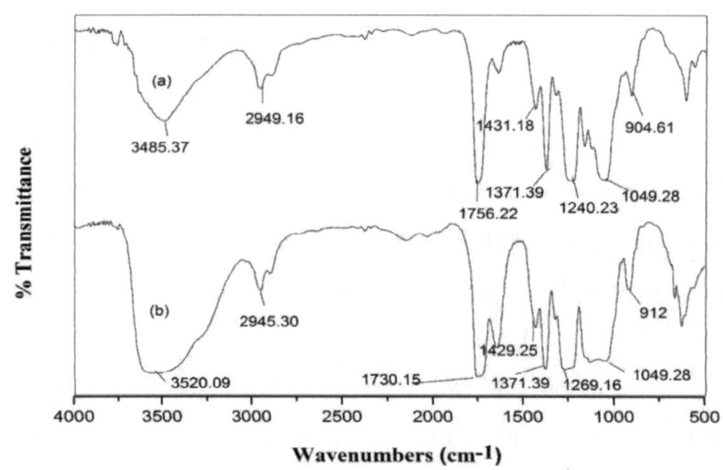

Figure 1. The FTIR spectra of: (a) cellulose acetate; (b) cellulose acetate doped with 25% (w/w) $LiClO_4$ (cellulose acetate-$LiClO_4$) membrane

3.2. Ionic Conductivity Analysis

Figure 2 shows the ionic conductivity for the polymer electrolyte membranes. The ionic conductivity of the membranes increased with the increasing in the ratios of lithium perchlorate content in the membranes and reached the optimum value at 1.79×10^{-4} S cm^{-1} corresponded to the cellulose acetate

IOP Conf. Series: Materials Science and Engineering **223** (2017) 012052 doi:10.1088/1757-899X/223/1/012052

doped with 25% (w/w) $LiClO_4$ membrane. However, the optimum ionic conductivity value in this study was still lower than the result of previous research (4.9×10^{-3} Scm^{-1} for cellulose acetate doped with 16% (w/w) $LiClO_4$ membrane) [5]. This result is likely due to the differences in measurement conditions and the sample preparation method used. However, the optimum ionic conductivity value in this study is still appropriated with the minimum requirement of ionic conductivity for lithium battery, that should be $\geq 10^{-5}$ Scm^{-1} [4]. The effect of amorphicity on the ionic conductivity of the polymer electrolyte membranes doped with metal salts has been extensively studied in the literatures [17–19]. The T_g (glass transition temperature) of the membranes decreased with the presence of lithium salt in the cellulose acetate [5]. The addition of lithium salt in the cellulose acetate can disrupt the intermolecular interactions within the host polymers to make new interactions between molecules, therefore, it can increase the amorphicity of membranes [5]. The decreasing in T_g accompanied by the increase in amorphicity has been reported and has led to the increase of the segmental motion in polymer electrolytes that in turn enhances the ionic conductivity [17–19].

The mechanism for ionic conductivity in polymer electrolyte membranes was developed by Druger, Nitzan, and Ratner with the dynamic percolation model [4,17]. In this case, it is envisaged ionic motion in terms of jumps between neighboring positions, a cation (M^+) such as Li^+ acts as a Lewis acid which is linked to Lewis bases (B^-), such as oxygen atoms associated within a polymer chain. The cation jumps from one site to neighboring site, which takes place by a slow change in the local environment as a lingkage of single M^+-B^- evolves with time. As the polymer gradually changes its conformation due to segmental motion, an accompanying translation of the cation to the neighboring site follows. The increased polymer segmental motion in the membranes of cellulose acetate-lithium salt complex, resulting from the increase of amorphicity or disorder. The increase in the ratios of lithium salt content in the polymer electrolyte membranes led to an increase in the entropy of the system which causes the increase in the ionic conductivity [4, 17–19].

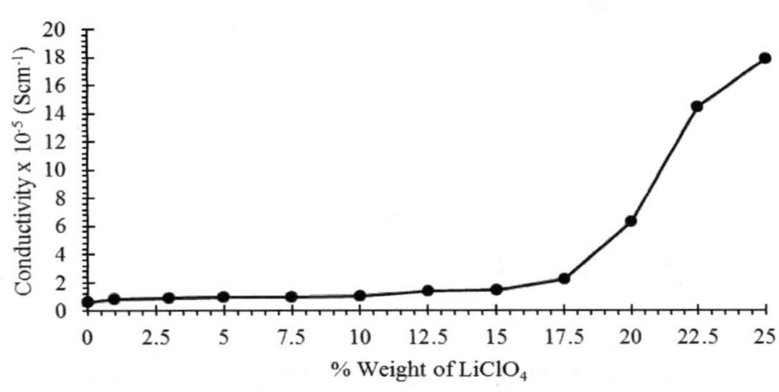

Figure 2. Ionic conductivity of the cellulose acetate membranes doped with various ratio of $LiClO_4$ (in % w/w)

IOP Conf. Series: Materials Science and Engineering **223** (2017) 012052 doi:10.1088/1757-899X/223/1/012052

3.3. Mechanical Strength Analysis

Figure 3 presents the mechanical properties of the cellulose acetate-lithium perchlorate polymer electrolyte membranes. The tensile strength increased with the increasing in the ratios of lithium perchlorate content in cellulose acetate membranes, and reached the optimum value at 43.29 MPa corresponded to the cellulose acetate membrane doped with 10% (w/w) $LiClO_4$ membrane. However, when the polymer electrolyte membranes doped with ratio of lithium perchlorate more than 10%, consequently the tensile strength of the membranes tended to decrease and the elongation at break was increased. The optimum tensile strength value in this study appropriated with the minimum requirement of a solid state polymer electrolyte membranes for lithium battery, that should be ≥ 30 MPa [4]. The tensile strength of the membran increased with the increasing in lithium perchlorate content because it can fill the cavity of the membranes that led to the more dense membranes. However, the addition of higher lithium perchlorate content tends to make the membrane more brittle due to the potent of the salt to disrupt the intermolecular interactions within the host molecule in forming new interactions, which in turn can increase the entropy of mixing between the membranes and lithium perchlorate. This increase in the entropy of mixing increases the segmental movements leading to reduction in the crystallinity [5].

The elongation at break tend increased with the increase in the ratios of lithium perchlorate content in cellulose acetate membranes, and reached the optimum value at 5.51% corresponded to the cellulose acetate doped with 20% (w/w) $LiClO_4$ membrane. The addition of lithium perchlorate caused the reduction in the crystallinity and the enhancement in the flexibility of the molecular system [5]. The deviation in elongation at break corresponded to sample of cellulose acetate doped with 15% (w/w) $LiClO_4$ membrane is due to the absence in the homogeneity of the membrane.

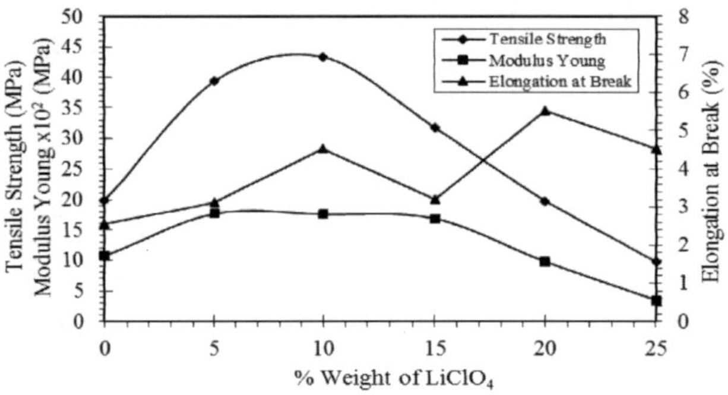

Figure 3. Mechanical properties of the cellulose acetate membranes doped with various ratio of $LiClO_4$

The modulus Young tends to decrease with the increase in the ratios of lithium perchlorate content in cellulose acetate membranes and reached the optimum value at 347.63 MPa corresponded to the cellulose acetate doped with 25% (w/w) $LiClO_4$ membrane. This result is due to the addition of lithium salt can be caused by the increase in the amorphicity and enhancement in the flexibility of the molecular system [5].

4. Conclusion

The ionic conductivity of cellulose acetate membranes increased with the increase in lithium perchlorate content within the membranes and the maximum ionic conductivity at approximately 1.79×10^{-4} Scm^{-1} was observed in the membrane containing 25% (w/w) $LiClO_4$. The addition of lithium perchlorate in the cellulose acetate membrane can disrupt the intermolecular interactions within the host polymers to make new interactions between molecules which led to the increase in amorphicity of membranes. The presence of 10% (w/w) $LiClO_4$ content within cellulose acetate membranes can increase the mechanical properties of the membranes from 19.89 to 43.29 MPa for tensile strength, and from 2.55 to 4.53% for elongation at break. However, when the cellulose acetate membranes containing ratio of $LiClO_4$ more than 10% (w/w), consequently the tensile strength tended to decrease and the elongation at break was increased. This result is due to the increase in the amorphicity and enhancement in the flexibility of the molecular system caused by the addition of lithium perchlorate.

References

[1] Li S *et al* 2012 *Electrochim. Acta.* **65** 221–227.
[2] Xiang H *et al* 2011 *J. Power Sources.* **196** 8651–8655.
[3] Huang X and Hitt J 2013 *J.Membr.Sci.* **425–426** 163–168.
[4] Yue L *et al* 2016 *Energy Storage Materials.* **5** 139–164.
[5] Selvakumar M and Krishna Bhat D 2008 *J. Appl. Polym. Sci.* **110** 594–602.
[6] Fergus J W 2010 *J.Power Sources.* **195** 4554–4569.
[7] Song J Y *et al* 1999 *J.Power Sources.* **77** 183–197.
[8] Ramesh S and Ng H M 2011 *Solid State Ion.***192** 2–5.
[9] Zhang Z and Fang S 2000 *Electrochim.Acta.* **45** 2131–2138.
[10] Ghosh A and Kofinas P 2008 *ECS Trans.* **29** 131-137.
[11] Ramesha S *et al* 2012 *Carbohydr.Polym.* **87** 2624– 2629.
[12] Choudhury N A *et al* 2006 *J. Electrochem. Soc.* **153** A614.
[13] Kamal H *et al* 2014 *J. Radiation Research and Appl. Sci.* **7**146–153.
[14] Lambert J B *et al* 1992 *Introduction to Organic Spectroscopy* (New York:Macmillan).
[15] Rajendran S *et al* 2004 *Solid State Ionics* **167** 335.
[16] Stephen A M, Premkumar T, Renganathan N G, Pitchumani S, Thirunakaran R, Muniyandi N 2000 *J. Power Sources* **89** 80.
[17] Ratner M A and Shriver D F 1988 *Chem. Rev.* **88** 109.
[18] Zheng J P *et al* 1995 *J. Electrochem. Soc.* **142** 2699.
[19] Zheng J P and Jow T R 1995 *J. Electrochem.* Soc. **142** L6.

IOP Conf. Series: Materials Science and Engineering **223** (2017) 012053 doi:10.1088/1757-899X/223/1/012053

Urea-formaldehyde resins: production, application, and testing

A Nuryawan[1], I Risnasari[1], T Sucipto[1], A Heri Iswanto[1], R Rosmala Dewi[2,3]
[1]Department of Forest Products Technology, Faculty of Forestry, University of Sumatera Utara, Medan, Indonesia
[2]Veterinarian, Faculty of Agriculture, University of Sumatera Utara, Medan, Indonesia
[3]Veterinarian, Faculty of Agriculture, University of Tjut Nyak Dhien, Medan, Indonesia

E-mail: arif5@usu.ac.id

Abstract. Urea-formaldehyde (UF) resin, one of the most important formaldehyde resin adhesives, is a polymeric condensation product of formaldehyde with urea, and being widely used for the manufacture of wood-based composite panels, such as plywood, particleboard, and fiberboard. In spite of its benefits such as fast curing, good performance in the panels (colorless), and lower cost; formaldehyde emission (FE) originated from either UF resin itself or composite products bonded by UF resins is considered a critical drawback as it affects human health particularly in indoor environment. In order to reduce the FE, lowering formaldehyde/urea (F/U) mole ratio in the synthesis of the UF resin was done. In this study, synthesis of UF resins was carried out following the conventional alkaline-acid two-step reaction with a second addition of urea, resulting in F/U mole ratio around 1.0, namely 0.95; 1.05, and 1.15. The UF resins produced were used as binder for particleboard making. The board was manufactured in the laboratory using shaving type particle of Gmelina wood, 8% UF resin based on oven dry particle, and 1% NH_4Cl (20%wt) as hardener for the resin. The target of the thickness was 10 mm and the dimension was 25 cm x 25 cm. The resulted particleboard then was evaluated the physical and the mechanical properties by Japanese Industrial Standard (JIS) A 5908 (2003). Further, the resulted particleboard also was used for the mice cage's wall in order to mimic the real living environment. After four weeks exposure in the cages, the mice then were evaluated their mucous organs as well as their blood. The experiment results were as follows: 1) It was possible to synthesis UF resins with low F/U mole ratio; 2) However, the particleboard bonded UF resins with low F/U mole ratio showed poor properties, particularly on the thickness swelling and modulus of elasticity; 3) There was no significant differences among the mucous organs of the mice after a month exposure FE originated from cages wall using histopathology assessment.

1. Introduction

Urea-formaldehyde (UF) resin, one of the most important formaldehyde resin adhesives, is a polymeric condensation product of formaldehyde with urea. Traditionally, UF resins were used as a wood adhesive. Initially, wood working industry used this resin, and in the development, wood based panels industry has been mainly used this resin up to date [1] for manufacturing wood-based composite panels, such as plywood, particleboard, and fiberboard. In other words, the wood panel industry is a major consumer of UF resin adhesive [2].

In spite of its benefits such as good appearance (due to colorless when it is applied in wood), good performance (the strength properties fulfill the required standards), fast curing (as thermosetting polymer, we can apply heat for making faster curing), and economical (the raw material is quite inexpensive and distribute elsewhere, thus large scale production can be applied); formaldehyde emission (FE) originated from either UF resin itself or panel products bonded by UF resins is considered a critical drawback as it affects human health particularly in indoor environment [3-6]. Many countries applied a stringent policy on FE regulation. For instance, United Kingdom, China, and Japan followed a recommendation of World Health Organization (WHO) which adopted limit of 0.1 mg/m^3 or about 0.08 ppm, while Canada, Germany, and Singapore had a slightly higher limit of 0.123 mg/m^3 or about 0.1 ppm [7]. Either manufacturers of UF adhesive or researchers responded to produce ideal resin to satisfy both the regulations as well as public awareness in health regarding the FE using various methods, for instance modifying UF resins itself [8-14] or modified UF resin bonded wood products [5, 15-23].

It is believed that producing resins which fulfilled these requirements could be done by decreasing the formaldehyde/urea (F/U) mole ratio [3]. Extensive works have been reported that the properties of UF resins are significantly influenced by the mole ratio of F/U, and it is often done by the incorporation of excess urea during the condensation step [24]. This procedure can remove most of free formaldehyde species and hemiformals when hot pressing stage. Further, Pizzi *et al.,* (1994) [25] strengthen that theoretically, low mole ratio of UF resin can deal with the physical and mechanical properties on the wood panels bonded by UF resins.

Yet, despite aforementioned studies in low mole UF resins; there is still an information gap on the comprehensive report on UF resins. Therefore, this study prepared synthesis of low F/U mole ratio liquid UF resins, their application on particleboard, and then rendered them to expose into mice in order to imitate the real condition of FE in building. The objective of this study is to evaluate the synthesis of low F/U mole of UF resins, the application to panel products particularly particleboard, and the effect of formaldehyde release originated from particleboard bonded low F/U mole of UF resin into mice.

2. Materials and Methods

In this study, synthesis of UF resins was carried out following the conventional alkaline-acid two-step reaction (the pH of addition and condensation reaction was 7.8 and 4.6, respectively) with an addition of second urea, resulting in F/U mole ratio around 1.0, namely 0.95; 1.05, and 1.15. For higher mole, UF resin with F/U mole of 2.0 was also prepared.

Then, UF resins produced were used as binder for particleboard making. The board was manufactured in the laboratory using shaving type particle of Gmelina wood, 8% UF resin based on oven dry particle, and 1% NH_4Cl (20%wt) as hardener for the resin. The target of the thickness was 10 mm and the dimension was 25 cm x 25 cm. The resulted particleboard then was evaluated the physical and the mechanical properties by Japanese Industrial Standard (JIS) A 5908 (2003) [26], including their class emissions.

Further, the resulted particleboard was used for the mice cage's wall and floor in order to imitate the real living environment. The mice were acclimatized prior to enter the cages for a week. During four weeks exposure in the cages, the mice were examined their behavior (movement and feeding manners) including the body weight. Observation on tissue organs changes was evaluated by histopathology anatomy for evaluation the effect of FE releasing from the particleboard. Additional analysis has been studied on the hematology of the mice.

3. Results and Discussions

3.1. Synthesis of Low F/U mole of UF Resins

UF resins are synthesized by the reaction of formaldehyde and urea. Usually, technical grade urea granules (99%) and formalin (37%) are used for the raw materials and aqueous solutions of both formic acid (20%wt) and sodium hydroxide (20%wt) are used to adjust the pH level during and the final of reaction.

The reaction is influenced by several parameters. It requires precise control of purity, amount and sequence of addition of the raw materials, alkaline and acid catalysts. The preparation conditions are adjusted and monitored with respect to temperatures, pH, and concentration of the reactants. Initially, the resin was reacted with an alkaline catalyst to initiate the addition reaction, and at the appropriate reaction time the resin solution was converted to the acid side to promote the condensation reaction [27]. Even though there were three possible procedures according to reaction pH conditions for synthesis, namely alkaline, weak acid, and strong acid [28-29], however, synthesis of UF resin in alkaline condition resulted in the best internal bond strength on particleboard [29].

Synthesis of UF resins in the laboratory was described elsewhere [29-37]. It is usually based on conventional two step reactions, consisted of methylolation or addition reaction and condensation reaction. Briefly, the formalin is placed in the glass cooking reactor with the mantle heater and the condition is adjusted into pH 7.8~8.0. Then, the 1st urea is placed into the reactor, yielding a molar ratio F/U of 2.0. The mixture is heated to 90°C for one hour to allow the methylolation reaction. The temperature is then adjusted to 80°C and the condensation reaction is carried out in the pH 4.6 up to target viscosity. When the condition is reached, the 2nd urea is added in order to consume excess of formaldehyde and determine the F/U mole ratio of final UF resins. After all of the urea dissolve, the UF resins are cooled to room temperature, and the pH is subsequently adjusted to 8.0 for terminating the reaction. Figure 1 showed the synthesis of UF resin and its result.

IOP Conf. Series: Materials Science and Engineering **223** (2017) 012053 doi:10.1088/1757-899X/223/1/012053

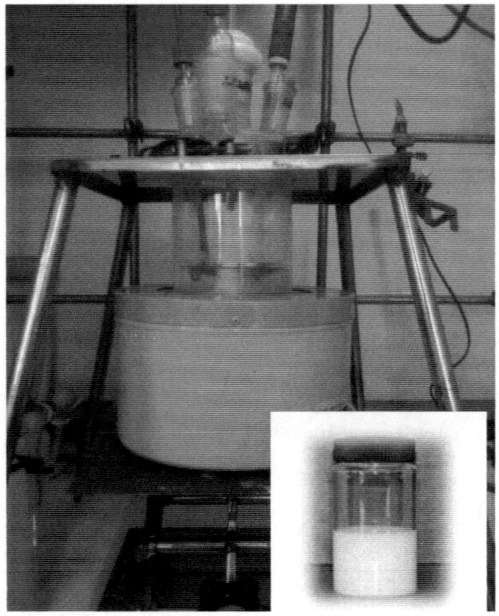

Figure 1. Synthesis of UF Resin in the Laboratory and its Results (inset)

In this work, it was not possible to synthesis UF resins with low F/U mole ratio, *i.e.* 0.90. Even though Pizzi *et al.,* (1994) [25] stated that theoretically F/U mole ratio of 0.70 can give adequate balance between strength and emission, but practically only F/U mole ratio in between 0.90 and 1.0 was effective for industrial application. Therefore, in this study, synthesis on UF resins was carried out to the final F/U mole ratio of 0.95; 1.05; and 1.15. For comparison, high F/U mole of 2.0 was also made.

Table 1 showed the characteristics of UF resins synthesized in this work. Reducing the F/U mole ratio resulted in increasing of solid content of the resin due to the addition of the second urea used in raw materials. Moreover there was a tendency of the density; the lower F/U mole ratio of UF resin the higher of the density. UF resin with the lowest F/U mole ratio gave the highest density probably because of the sol fraction containing in the resin. Previous work on film UF resin exhibited abundant sol fraction in low F/U mole UF resin compare to in high F/U mole UF resin [38].

In this work, the viscosity performed fluctuate but the trend showed the highest F/U mole ratio of UF resin has had the most viscous, indicating that the reactivity of high F/U mole ratio UF resin adhesive was greater compared to that of the low F/U mole ratio. A lower viscosity indicated a low molecular weight of UF resin adhesives because the viscosity of liquid UF resin adhesives is proportional to their molecular weight [39]. Related to gel time measurements, indeed UF resin with high F/U mole ratio was more reactive contrasted by low F/U mole ratio. The milky white color of the resin is due to ageing or further advancement of the resin by condensation reaction [40-42].

Table 1. Characteristics of UF Resins Synthesized in this Study

No	F/U mole ratio	Solid Content (%)	Density (g/cm^3)	Viscosity (MPa.s)	Gel time (minutes)	Color
1	0.95	52.65 ± 0.38	1.54	215	13.13	Dispersed, milky white
2	1.05	48.64 ± 0.08	1.48	144	11.41	Opaque, milky white
3	1.15	46.99 ± 0.07	1.47	131	9.45	Milky white
4	2.00	37.17 ± 0.38	1.46	444	4.41	Dispersed, milky white

3.2 Properties of Particleboard Bonded by low F/U mole of UF Resin

Comparison among the properties was made not only using particleboard bonded by low F/U mole ratio UF resins data but also using particleboard bonded by higher mole UF resin, commercial and JIS standard as well. Figure 2 showed the performance of particleboard bonded by either UF resin synthesized in this work or commercial one.

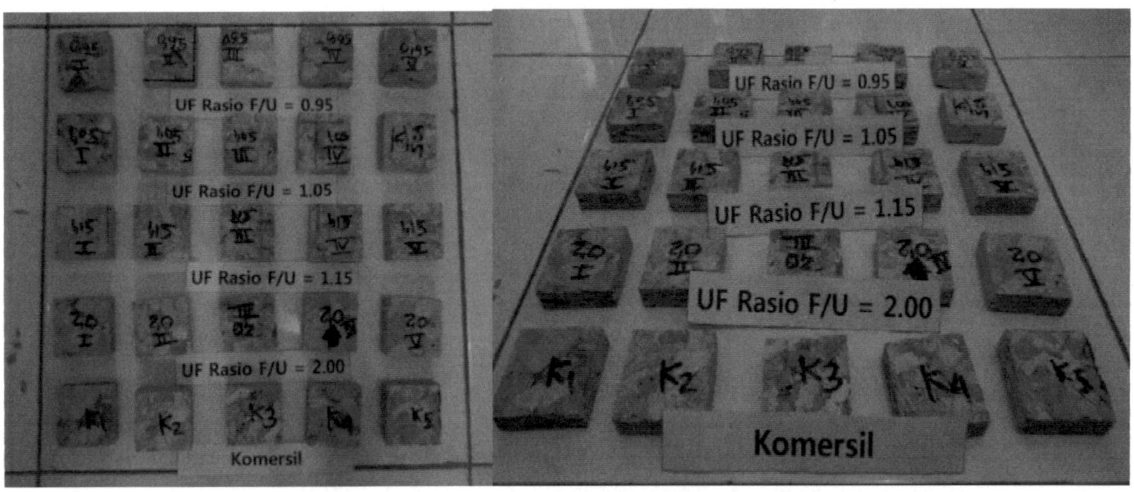

Figure 2. Performance of Particleboard Bonded by UF Resins in this Study

Generally, the performance among the particleboard was similar. Because of the excess of thickness (more than 10 mm) in consequence of spring back effect [43], the density of the particleboard was obviously below the target of 0.75 kg/cm^3 as shown in Table 2. Fortunately, the average values of the density have been fulfilled the JIS standard. This means in this case the UF resin adhesives with low F/U mole were able to compete with higher F/U mole (2.0) and commercial one to reach the density target. Thus the particleboard can be used as raw material of mice cage.

Table 2. The Properties of Particleboard Bonded by UF Resin with Various F/U mole Ratio

Property	Unit	UF resin with F/U mole ratio					JIS A 5908 (2003)
		0.95	1.05	1.15	2.0	Commercial	
Density	g/cm^3	0.54 (0.05)	0.50 (0.02)	0.51 (0.02)	0.57 (0.03)	0.54 (0.04)	0.40-0.90
Internal Bonding	kgf/cm^2	1.06 (0.84)	0.31 (0.09)	0.45 (0.40)	2.76 (0.53)	3.07 (1.23)	> 1.5
Modulus of Elasticity (MOE)	kgf/cm^2	296 (77)	359 (176)	246 (211)	99 (25)	154 (36)	> 20000
Modulus of Rupture (MOR)	kgf/cm^2	45 (26)	36 (13)	49 (25)	145 (35)	134 (19)	80

Remarks: The data were mean values based on three replications from three different particleboard and data in the parentheses were the standard deviation.

The mechanical properties the particleboard bonded UF resins with low F/U mole ratio comprising of internal bonding, MOE, and MOR, were not satisfactory (Table 2). Even though the physical property *i.e.* density, has met the JIS standard, combination these two resulted in dimensional instability, *viz.* greater thickness swelling and water absorption. The similar phenomenon has been occurred in many previous works [43-46].

By contrast, both higher mole F/U of 2.0 UF resin and commercial one exhibited reasonable mechanical properties particularly for internal bonding and MOR. However, values of MOE were far away from the target. This condition was prevalent in particleboard products particularly which was made in laboratory scale. Therefore, there are many ways to improve these circumstances such as making oriented particleboard [47] or covering the particleboard using veneer [48]. In this study, the particleboards were used as cage mice's wall which did not need structural requirements. Therefore, the lower values of MOE were not a barrier for this necessity.

According to the JIS standard, the formaldehyde release originated from all the observed particleboard showed range values between 0.1032 and 0.3741 mg/L except commercial one. These values were categorized as class emission of F**** or very low concentration. We did not include the commercial one because we cannot determine the initial formalin used in this type of adhesive. From this point, we can assume that the particleboard used in this experiment have very low emission of formaldehyde even though many factors affected this situation, namely relative humidity, temperature, ventilation, periods of storage, type of adhesive, type of the wood used, and condition of production [10].

3.3 Effect of FE Release from Particleboard bonded by low F/U mole of UF Resin on Mice

Figure 3 shows mice in the cage with particleboard wall except the control. During observation for a month, there were no mortality rate and no differences on the behavior between treated mice and the control. The behavior including normal movement, normal feeding behavior, and no weight body loss which indicated the mice were not stress.

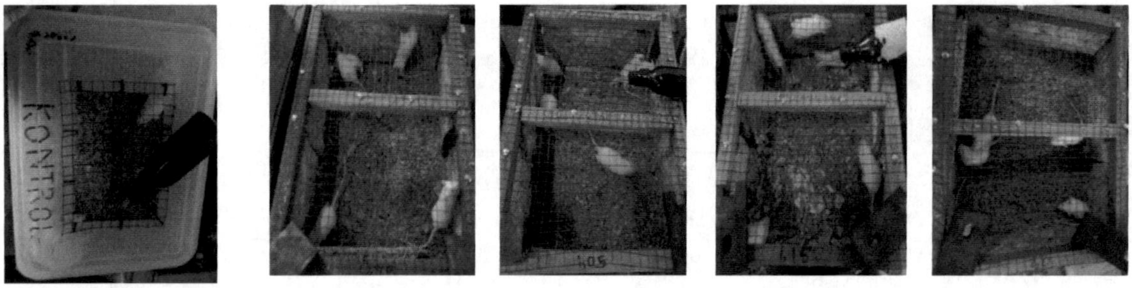

(a) (b) (c) (d) (e)

Figure 3. Typical mice in their cage with particleboard wall when four weeks exposed (a) control; (b) using particleboard wall bonded by UF resin of 0.95 (c) using particleboard wall bonded by UF resin of 1.05 (d) using particleboard wall bonded by UF resin of 1.15, and (e) using particleboard wall bonded by UF resin of 2.0

Table 3 showed the body weight of mice before, after and the changes during observation. There was addition of mice's body weight around 18-36% which indicated the mice did not suffer and their nutrition was appropriate..

Table 3. Mice's body Weight during Observation on Exposing Formaldehyde release from Particleboard Bonded UF Resins

UF resin adhesive used on particleboard wall	Body weight			%
	Initial (g)	After exposed (g)	Changes (g)	
Control (without)	28.6	38.8	10.2	35.66
Mole F/U of 0.95	27.6	35.4	7.8	28.26
Mole F/U of 1.05	27.2	32.2	5.0	18.38
Mole F/U of 1.15	25.6	34.4	8.8	34.38
Mole F/U of 2.0	24.2	31.4	7.2	29.75

Related to the influence of formaldehyde release originated from particleboard bonded of UF resins, investigation on organ's tissue alteration such as lung and liver by histopathology anatomy examination have to be carried out.

According to the aforementioned mice's behavior, presumably there were no differences on histopathology mucous membrane (trachea) among the examined mice because it needs longer time periods [49]. Therefore, in this experiment the first investigation on histopathology mucous membrane was carried out on the trachea. Trachea is the first area that contact with the air. Figure 4 showed the example of the photomicrograph of mice's trachea tissue which exposed to formaldehyde emission, with the magnification of 40x and 400x, respectively.

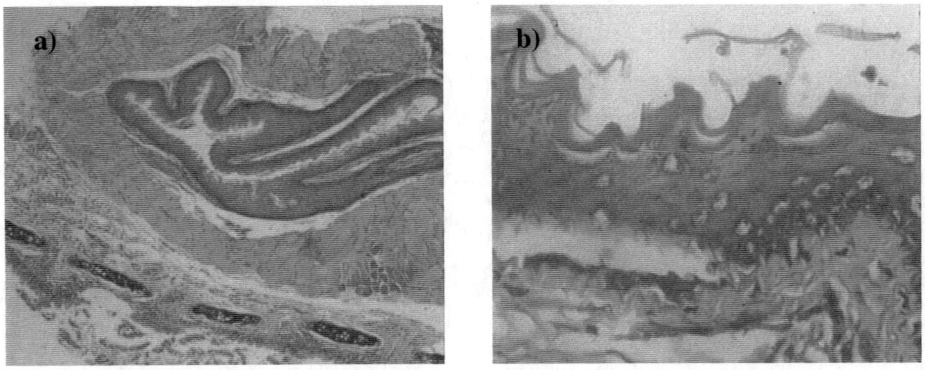

Figure 4. Photomicrograph of observed mice's trachea tissue which exposed to formaldehyde emission from particleboard a) on mol of formalin 0.95 with magnification of 40x b) on mol of formalin 2.0 with magnification of 400x.

As demonstrated in Figure 4, there was no specific lesion between control (without exposure on formaldehyde emission) and treatment (on mol of formalin 2.0; the highest exposure). In other words, the histological examination of the trachea revealed normal condition. Therefore, additional testing, such as blood analysis [50] has been also recommended in order to investigate further.

The blood was examined and then was analyzed on the differences of hemoglobin, white blood cell (WBC), red blood cell (RBC), and platelets as shown in Table 4.

Table 4. Results of hematology parameters of control mice and those exposed to particleboard bonding of UF resin on various mol of formalin

No	Sample code	Hemoglobin (g/dl)	WBC (10^9/L)	RBC (10^{12}/L)	Platelets (10^9/L)
1	0.95 I	15.7	2.6	7.23	212
2	0.95 II	12.4	3.9	6.29	378
3	1.05	16.5	2.9	7.00	132
4	1.15 I	14.1	4.5	6.39	174
5	1.15 II	14.1	3.7	7.17	322
6	2.0 I	14.1	5.2	4.28	196
7	2.0 II	16.9	8.1	6.23	392
8	control	15.5	3.5	6.93	158

The observed increase in WBC as the increasing of mol of formalin in exposed mice in this study may be attributed to the primarily function of the WBC in body defense against foreign bodies including emissions of formaldehyde [50].

4. Conclusion

In this work, it was not possible to synthesis UF resins with low F/U mole ratio *i.e.* 0.90. Therefore, in this study, synthesis on UF resins was successfully carried out to the final F/U mole ratio of 0.95; 1.05; 1.15 (for low mole) and 2.0 (for high mole)

All the UF resins produced can be applied as adhesive for making particleboard. Even though the quality of particleboard was poor, particularly in MOE, in this study the particleboards were only used as cage mice's wall which did not need structural requirements. Therefore, at least the particleboard can be utilized for this purpose.

According to the mice behavior, presumably there were no significant alterations among the mucous membrane of the mice because it is not adequate time in exposing mice into particleboard bonded by UF resins.

Acknowledgments

This work was supported by the Competitive Grant Program through Directorate Research and Community Service funded by Ministry of Research, Technology and Higher Education, Republic of Indonesia (Number: 017/SP2H/LT/DPRM/II/2016, date of February 17, 2016). Thanks also for Kartini Dewi Sartika Tambun, Safrizal Saragih, and Rosli Delina Batubara for laboratory working helps.

References

[1] Dunky M 1998 *Int. J. Adhes Adhes* **18** 95.
[2] Meyer B, Nunlist R 1981 *ACS Nat Meeting, Div Polym Chem, Symp on Aqueous polym syst* (Atlanta GA).
[3] Myers GE 1984 *For Prod J.* **34** 35.
[4] Myers GE, Koutsky JA 1990 *Holzforschung* **44**(2) 117.
[5] Park BD *et al* 2006 *J.App.Polym Sci* **101** 1787.
[6] Jin S *et al* 2013 *Infor Tech J.* **12**(24) 8527.
[7] Tang X *et al* 2009 *Environ Int* **35** 1210.
[8] Dutkiewicz J 1983 *J.App.Polym Sci* **28** 3313.
[9] Dutkiewicz J 1984 *J.App.Polym Sci* **29** 45.
[10] Tohmura S *et al* 2000 *J. Wood Sci* **46** 303.
[11] Ringena O *et al* 2006 *Holz Roh-und Werkstoff* **64** 321.
[12] Abdullah ZA, Park BD 2009 *J.App.Polym Sci* **114** 1011.
[13] Park BD, Jeong HW 2011 *Int. J. Adhes Adhes* **31** 524.
[14] Park BD, Jeong HW 2011 *J.Korean Wood Sci Tech* **39**(2) 179
[15] Ebewele RO *et al* 1991 *J.App.Polym Sci* **43** 1483
[16] Ebewele RO *et al* 1993 *J.App.Polym Sci* **49** 229
[17] Ebewele RO *et al* 1994 *J.App.Polym Sci* **52** 689.
[18] Tohmura S *et al* 2001 *J. Wood Sci* **47** 451.
[19] Rachtanapun P, Heiden P 2003 *J.App.Polym Sci* **87** 898.
[20] Abdullah ZA, Park BD 2010 *J.App.Polym Sci* **117** 3181.
[21] Roffael E, Hüster HG 2012 *Eur J Wood Wood Prod* **70** 401.

[22] Zhang H *et al* 2013 *J. Adhes Sci Tech* **27(9)** 1023.
[23] Li B *et al* 2014 *Bioresources* **9(3)** 5125.
[24] Kim M G 1999 *J. Polym Sci Part A: Polym Chem* **37** 995.
[25] Pizzi A *et al* 1994 *Holzforchung* **48** 254.
[26] Japanese Industrial Standard JIS A 5908 (2003) for Particleboard (Japan).
[27] Hse CY *et al* 1994 *Holzforchung* **48(6)** 527.
[28] Park B D *et al* 2003 *J. App. Polym Sci* **88** 2677.
[29] Park B D *et al* 2006 *J. App. Polym Sci* **100** 422.
[30] Park B D *et al* 2008 *J. App. Polym. Sci* **110** 1573.
[31] Park B D *et al* 2009 *Eur J Wood Wood Prod* **67** 121.
[32] Park B D *et al* 2011 *J. App. Polym. Sci.* **120** 1475.
[33] Park B D, Kim JW 2008 *J.App.Polym Sci* **108** 2045.
[34] Park B D, Jeong HW 2011 *Int. J. Adhes Adhes* **31** 524.
[35] Park B D, Jeong HW 2011 *J.Korean Wood Sci Tech* **39(2)** 179.
[36] Park B D, Jeong HW 2011 *J.App.Polym Sci* **122** 3255.
[37] Park B D, Causin V 2013 *Eur. Polym. J.* **49** 532.
[38] Nuryawan A *et al* 2014 *J Adhes* **91(9)** 677.
[39] Kim MG 2001 *J.App.Polym Sci* **80(14)** 2800.
[40] Zanetti M, Pizzi A 2004 *J. App. Polym. Sci.* **91** 2690.
[41] Despres A, Pizzi A 2006 *J.App.Polym Sci* **100** 1406.
[42] Wieland S *et al* 2007 *J.App.Polym Sci* **104** 2633.
[43] Nuryawan A *et al* 2009 *J Ilmu Teknologi Hasil Hutan* **2(1)** 57.
[44] Haligan AF 1970 *Wood Sci Tech* **4(4)** 301.
[45] No Y B, Kim M G 2007 *J. App. Polym. Sci.* **106(6)** 4148.
[46] Papadapolous, AN (2006) *BioResources* **1(2)** 201.
[47] Nuryawan A *et al* 2008 *J Ilmu Teknologi Hasil Hutan* **1(2)** 60.
[48] Nuryawan A, Massijaya M 2004 *Kultura* **39(1)** 32.
[49]Pitten F A *et al* Pathol Res Pract **196** 193.
[50] Akpan K V *et al* 2014 Cur Adv Env Sci **2(2)** 46.

IOP Conf. Series: Materials Science and Engineering **223** (2017) 012054 doi:10.1088/1757-899X/223/1/012054

Incorporation of titanate nanosheets to enhance mechanical properties of water-soluble polyamic acid

Christian Harito, Dmitry V Bavykin, Frank C Walsh
Faculty of Engineering & the Environment, University of Southampton, Southampton, United Kingdom, SO17 1BJ

E-mail : ch8g13@soton.ac.uk

Abstract. Pyromeliticdianhydride (PMDA) and 4',4'-oxydianiline (ODA) were used as monomers of polyimide. To synthesise a water soluble polyimide precursor (polyamic acid salt), triethylamine (TEA) was added to polyamic acid with a TEA/COOH mole ratio of 1:1. Titanate nanosheets were synthesised by solid-state reaction, ion-exchanged with acid, and exfoliated by TEA. Exfoliated titanate nanosheets were mixed with water soluble polyamic acid salt as reinforcing filler. Drop casting was deployed to synthesise polyamic acid/titanate nanosheet nanocomposite films. Scanning electron microscopy (SEM) and transmission electron microscopy (TEM) were employed to study the morphology and dispersion of nanosheets in the colloidal dispersion and the solid film composite. Modulus and hardness of nanocomposites was provided by nanoindentation. Hardness increased by 90% with addition of 2% TiNS while modulus increased by 103% compared to pure polymer. This behaviour agrees well with Halpin-Tsai theoretical predictions up to 2 wt% filler loading; agglomeration occurs at higher concentrations.

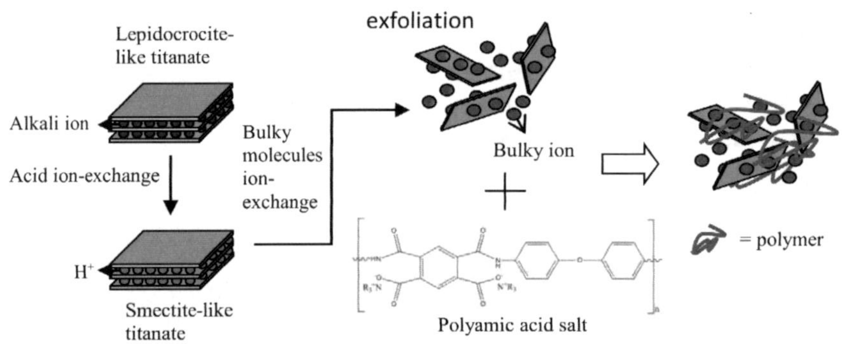

Figure. Illustration of titanate nanosheets exfoliation and incorporation in polyamic acid salt

1. Introduction

Polymer nanocomposites consist of a hybrid organic matrix containing a dispersed nanostructured filler. The filler can provide a significant improvement in polymer properties at low filler loading, due to the high degree of contact between filler and polymer. For example, small addition of exfoliated clay (4.7 wt%) can increase flexural modulus of nylon 6 by four times at 120°C [1]. Successful application of a clay filler to improve the mechanical properties of polymer requires further research on other nanosheet based polymer nanocomposites containing, e.g., graphene [2], boron nitride nanosheet [3]. Despite their increasing importance, there are few recent studies of recently titanate nanosheets for the reinforcement of polymer nanocomposites [4].

IOP Conf. Series: Materials Science and Engineering **223** (2017) 012054 doi:10.1088/1757-899X/223/1/012054

Titanate nanosheets (TiNS) are single sheet inorganic materials created by exfoliation of lepidocrocite-like titanium oxide. Due to their unique properties, this so-called graphene analogue attracts international research interest. It has been applied in some polymer nanocomposites for functional purposes. Titanate nanosheets act as a photo-initiator for water-soluble vinyl monomers owing to the photocatalytic properties of titanate nanosheets [5]. Proton-donating monomers such as N-isopropylacrylamide (IPAAm), acrylamide (AAm), and acrylic acid (AAc) can strongly bond to the negatively charged titanate nanosheets and polymerize with TiNS as a physical photo-crosslinker. The polymerization occurs near the TiNS and is trapped by adjacent TiNS, producing a highly sensitive optical response to thermal-stimuli. The proton conductivity of sulfonated poly(ether ether ketone) (SPEEK) membrane can be doubled by a low loading of titanate nanosheets (1.67 wt%) [6] and the composite is stable at high temperature (140 °C) and in a wet environment (100% relative humidity). Adding TiNS to polyethylene naphthalate (PEN) film significantly decrease helium gas permeability [7]. However, systematic study about mechanical reinforcement effects of titanate nanosheets on the polymer nanocomposite is still limited [8].

In this work, titanate nanosheets (TiNS) have been used to enhance the mechanical properties of polyamic acid salt (PAAS). The morphology of titanate nanosheets is imaged by scanning electron microscopy (SEM) and transmission electron microscopy (TEM). Nanoindentation has been used to examine the reduced modulus (E_r) and hardness (H) of the composites. Experimental results are compared with theoretical models due to Halpin-Tsai [9] and Halpin-Kardos [10]. The dispersion of titanate nanosheets in polyamic acid is studied by TEM to determine its effect on the mechanical properties of resultant titanate nanosheets-polyamic acid salt composites.

2. Experimental Methods

Synthesis method of titanate nanosheets, polyamic acid salt, and composites are considered.

2.1. Synthesis of titanate nanosheets

A synthesis method for single-layer titanate nanosheets adapted from Sasaki *et.al.* [11] Sasaki's single-layer titanate nanosheets are based on two starting materials as alkali precursors (K, Li, Rb, Cs) and TiO_2 powder. For synthesis of titanate nanosheets by the caesium route, TiO_2 (Degussa P25) from Aeroxide and Cs_2CO_3 from Sigma-Aldrich were used without further purification. There are three steps to synthesis single-layer titanate nanosheets. Firstly, solid-state reaction of TiO_2 and Cs_2CO_3 was conducted based on the work of Grey *et.al.* [12] 2.6 g of Cs_2CO_3 and 2 g of TiO_2 (molar ratio 1: 5.3) was mixed, ground and put in a platinum crucible, an inert crucible being crucial to prevent reaction of alkali precursor and crucible, for one hour heating at 800 °C. The mixture was allowed to cool spontaneously. It was ground then heated at 800 °C for 20 h. The material was ground then the heat treatment cycle was repeated. The resulting product was a white powder of lepidocrocite-like caesium titanate ($Cs_{0.7}Ti_{1.825}\Upsilon_{0.175}O_4$) with ϒ representing a vacancy [13,14].

Secondly, acid ion-exchange was used to create smectite-like acid titanate ($H_{0.7}Ti_{1.825}\Upsilon_{0.175}O_4 \cdot H_2O$) [13,14]. 2.4 g of $Cs_{0.7}Ti_{1.825}\Upsilon_{0.175}O_4$ was stirred with 96 mL of 1 M HCl for 4 days. The acid solution was replaced with a fresh acid solution every day to maintain the amount of H^+ ions for acid leaching of Cs^+ ions. Then, the solution was decanted with deionized (DI) water three times followed by vacuum filtration using 0.2 μm nylon membrane and washed with distilled water until the conductivity around 10 μS cm^{-1}.

Thirdly, bulky molecules ion-exchange was demonstrated by exfoliating 0.08 g of $H_{0.7}Ti_{1.825}\square_{0.175}O_4 \cdot H_2O$ with 20 mL of distilled water and 400 μL triethylamine (TEA) with bath ultrasonicator for 9 h. The solution was kept for 4 days to separate non-exfoliated sheets at the bottom. Finally, the upper, exfoliated nanosheets were directly used to make composites.

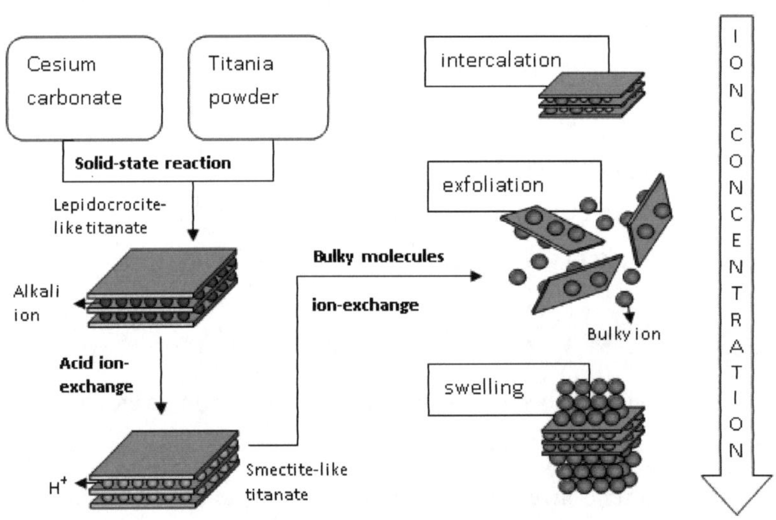

Figure 1. Schematic representation of titanate nanosheet synthesis

2.2. Synthesis of polyamic acid salt

The work of Lee, *et.al.* [15] has been used as a guide to synthesise polyamic acid salt. An aromatic polyamic acid salt were prepared from pyromelliticdianhydride (PMDA) and 4,4'-oxydianiline (ODA) as monomers and N,N-dimethylacetamide (DMAc) as a solvent. All the reactions were shielded by nitrogen gas. As a typical example, a mixture of 1.201 g of ODA

(1 mol) and 22.59 g of DMAc was stirred in a 250 mL round bottom flask at room temperature (22 °C). After ODA had dissolved, 1.309 g of PMDA (1 mol) was added to the solution followed by stirring for 1 day, yielding a 10 wt% DMAc solution of polyamic acid. 1.656 mL of triethylamine (TEA; 2 mol of monomer) was added. Local white precipitation was visible upon addition of TEA to the solution as stated by Kreuz, *et.al.* [16] and disappeared after approximately 30 min. The resulting solution was poured into acetone and filtered with a 0.2 μm pore size nylon membrane, followed by washing twice with acetone. After drying at 22 °C for 24 h, light yellow PAAS powder was collected and stored at 22 °C and 40±5% relative humidity.

2.3. Synthesis of polyamic acid salt-titanate nanosheet composites

All composites in this work were made by solution mixing. A known concentration (determined by UV-vis spectrometry) of titanate nanosheet colloidal suspension was mixed with PAAS powder to produce 5 wt% of PAAS-TiNS aqueous solution with various concentration of TiNS (0.5, 1, 2, 3, and 5 wt% of solid). For pure polymer, 0.215 g of PAAS was stirred in 4 mL distilled water and 80 μL TEA to assist dissolution of PAAS, yielding 5 wt% of PAAS aqueous solution. The solutions were stirred for 3 days and drop cast on glass. The drop cast solutions were kept at 22±2 °C and 40±5% relative humidity for 4 days to dry and achieve an equilibrium moisture content in the films.

2.4. Characterisation

The morphology of titanate nanosheets was examined by transmission electron microscopy (TEM) (JEOL-3010) and field emission scanning electron microscopy (FE-SEM) (JEOL JSM-6500). To prepare TEM samples, a diluted colloidal suspension was dropped on a copper grid containing perforated carbon film and dried for 1 day in 22 °C. A diluted colloidal suspension of titanate nanosheets was dropped on top of a silicon wafer for SEM characterisation. The concentration of titanate nanosheet suspension was determined with UV-vis spectroscopy. TEM (JEOL-3010) was employed to study dispersion of titanate nanosheets in polyamic acid salt. Samples were dropped on a copper grid with perforated carbon film and spin coated at 4300 rpm for 1 min to achieve a very thin film for TEM. The optical properties of polymer composites were evaluated with a 20.1 megapixel camera.

The reduced modulus and hardness were obtained with a nanoindenter (Nanotest Platform 3, Micro Materials Ltd, UK). Reduced moduli were converted to Young's moduli using the Oliver-Pharr equation [17]. To prepare the samples, a sliver of the film (around 0.5 cm x 0.5 cm) was glued by epoxy glue to the soda-lime glass with the smoother surface (cast side) on top to be measured. Then, the glass was glued with acrylic glue to cylinder holder. The cylinder holder with glass and samples on top was put inside the nanoindenter to be measured. A Berkovich (3-side pyramidal) diamond tip was employed for nanoindentation. A constant loading and unloading rate of 20 mN s^{-1} were applied with 30 seconds holding time after reaching the maximum depth. 20 indentations were performed on each sample at a 0.5 mN applied load.

3. Results and Discussion

3.1. Morphology of titanate nanosheets

The structure of the nanosheets is shown in Figure 2.

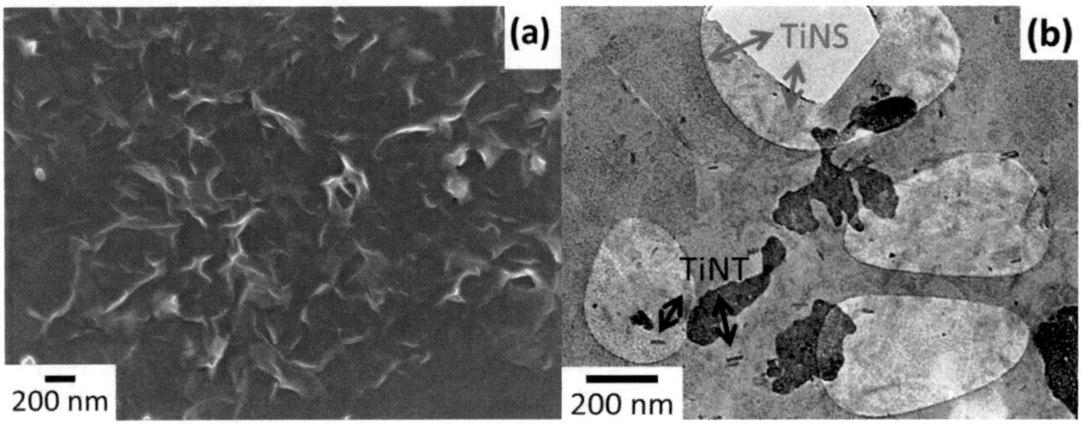

Figure 2. Morphology of titanate nanosheets: (a) SEM Images; (b) TEM Images

Here we report the use of triethylamine to exfoliate titanate nanosheets. The morphology of titanate nanosheets is shown in Figure 2a. The nanosheets tend to restack into a large platelet when dried, creating a tactoid phase or 'skewed' agglomerates [18]. TEM is needed to identify the size of the nanosheets. TEM images of TiNS can be seen in Figure 2b. The extremely thin nanosheets result in a very feint contrast, making it hard to distinguish nanostructured material from the background substrate[19]. Instead of single layers, some nanosheets stack horizontally on the substrate and become more prominent than single layer nanosheets. The colloidal suspension consists mainly of small, irregularly shaped nanosheet particles with a diameter of *ca.* 100 nm. However, there are some large sheets with a diameter of *ca.* 300 nm together with very small nanotubes (*ca.* 50 nm in length). According to our previous work [20] titanate nanosheets tend to scroll into nanotubes in an alkaline environment. Such scrolling may also occur with very small nanosheets in triethylamine solution.

3.2. Experimental and theoretical analysis of the mechanical properties of polyamic acid salt-titanate nanosheet composites

IOP Conf. Series: Materials Science and Engineering **223** (2017) 012054 doi:10.1088/1757-899X/223/1/012054

Nano indentation and transmission electron microscopy were used to study the behaviour of titanate nanosheets embedded in polymer at controlled concentrations (0.5, 1, 2, 3, 5 wt%). TEM images of PAAS/TiNS are provided in Figure 3. At low concentration (1 wt%, Figure 3a), titanate nanosheets tend to retain their exfoliated form inside the polyamic acid salt. Very thin and small nanosheets (*ca.* 100 nm long) were mostly well dispersed, with some traces of few layer nanosheets. This shows that the polymer may interact well with the titanate nanosheets preventing agglomeration. However, some nanosheets lay down on the substrate in a very thin layer which were difficult to observe. At 3 wt% of titanate nanosheets, the distance between adjacent nanosheets became closer and van der Waal's forces between nanosheets became predominant, resulting in agglomerates (Figure 3b). Such agglomerates are likely to form elongated, 'skewed' assemblies of densely packed nanosheets.

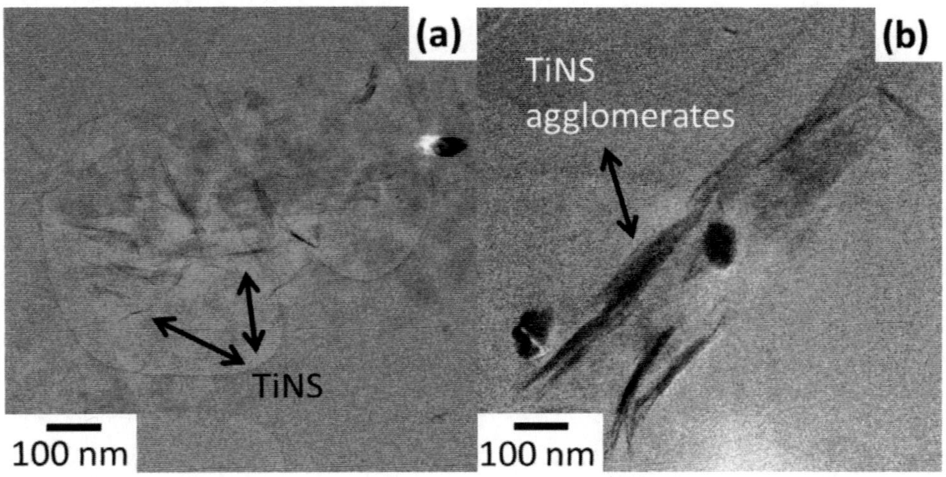

Figure 3. TEM images of titanate nanosheets incorporated into the polyamic acid salt matrix at (A) 1 wt% and (B) 3 wt% of TiNS.

To study and analyse the reinforcement effect of titanate nanosheets, nanoindentation data of the polyamic acid salt-titanate nanosheet (PAAS/TiNS) were compared with theoretical predictions due to Halpin-Tsai [9] and Halpin-Kardos [10]. Halpin-Tsai is a well-known model to estimate reinforcement of unidirectional oriented short fibres and has been successfully used to predict the reinforcement of carbon nanotubes at low filler content (>1 wt%) [21]. It can also be adapted to predict the modulus of polymer nanocomposite with nanosheets (e.g., clay and graphene) as filler [22,23]. The modified Halpin-Tsai equation is:

$$\frac{E_c}{E_m} = \frac{1+2A_f\mu\phi_f}{1-\mu\phi_f} \quad (1)$$

Where E_c and E_m are the Young's modulus of composite (PAAS/TiNS) and matrix (PAAS), respectively. The Young's modulus of matrix (PAAS) is adapted from nanoindentation measurements of pure PAAS. A_f is the filler aspect ratio (l/h), which in this case (l) is the nanosheet length (100 nm) and (h) is the nanosheet thickness (0.75 nm, assuming all the nanosheets are completely exfoliated to single layers). Φ_f is the volume fraction of filler which obtained by converting the weight filler fraction, knowing the density of TiO_2 (4.23 g mL^{-1}) and polyamic acid (1.04 g mL^{-1}). μ is a geometric factor, which is

$$\mu = \frac{E_f/E_m - 1}{E_f/E_m + 2A_f} \quad (2)$$

Where E_f is modulus of TiO_2 (282.76 GPa, by assuming TiNS as TiO_2) as taken from the CRC [24]. The Halpin-Tsai equation is also able to predict hardness of micro and nanocomposite by simply changing the modulus with hardness [25]. The hardness of the matrix (H_m) is determined by nanoindentation of pure PAAS while the hardness of TiO_2 (10.99 GPa) is taken from literature [24].

The Halpin-Kardos equation can also be used for comparison. Originally, this equation was applied to randomly oriented short fibres with a quasi-isotropic laminate assumption involving the $[0/+45/90/-45]_n$ configuration [10]. However, the expression can be modified and has been successfully used to predict the reinforcement of clay up to 2 wt% in a polymer blend [23]. The authors argued that the tactoid phase of clay may act in a similar fashion to short fibres. The modified Halpin-Kardos equation is:

$$\frac{E_c}{E_m} = \frac{3}{8}\left[\frac{1+2A_f\eta_L\phi_f}{1-\eta_L\phi_f}\right] + \frac{5}{8}\left[\frac{1+2\eta_T\phi_f}{1-\eta_T\phi_f}\right] \quad (3)$$

Where η_L and η_T can be determined with the following equations

$$\eta_L = \frac{E_f/E_m - 1}{E_f/E_m + 2A_f} \quad (4)$$

$$\eta_T = \frac{E_f/E_m - 1}{E_f/E_m + 2} \quad (5)$$

IOP Conf. Series: Materials Science and Engineering **223** (2017) 012054 doi:10.1088/1757-899X/223/1/012054

To obtain the hardness, this equation was modified by substituting modulus for hardness. Hardness and modulus of polyamic acid salt-titanate nanosheet (PAAS/TiNS) nanocomposites is shown in Figure 4.

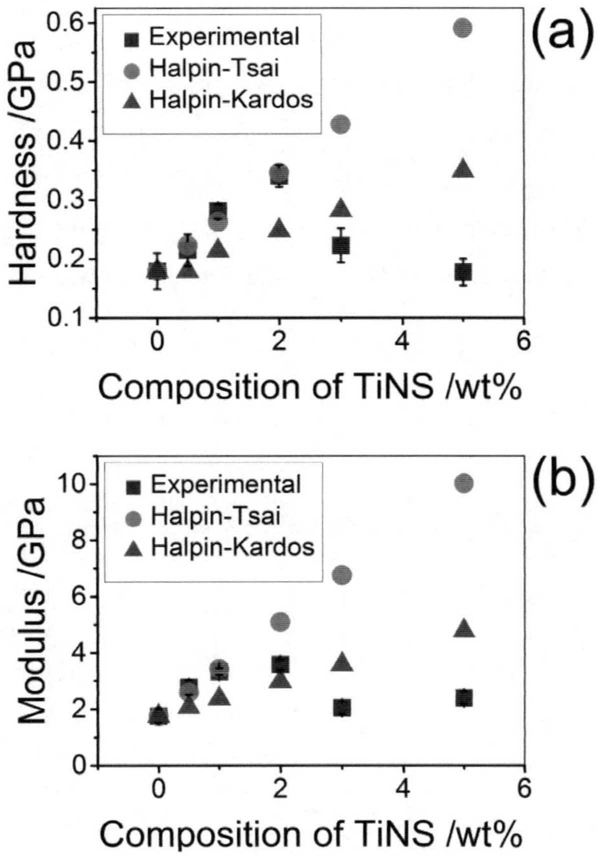

Figure 4. The effect of titanate nanosheets on the mechanical properties of polyamic acid salt compared to the predictions of theoretical models: (A) hardness; (B) modulus.

The addition of titanate nanosheets significantly enhanced the hardness and modulus of water-soluble polyamic acid salt. Hardness increased 90% with addition of 2% TiNS, while the modulus increased 103% compared to the pure polymer. The improvement in hardness follows the Halpin-Tsai theory up to 2 wt% while the modulus follows up to 1 wt%. This might happen due to the nanosheets at low filler concentration being fully exfoliated in Figure 3a) maintaining the same aspect ratio and polymer-filler interaction. Moreover, titanate nanosheets have an anisotropic reinforcement which is highest when the loading is perpendicular to the nanosheet [5]. Figure 3a) shows that most of the nanosheets lay flat on the substrate, rendering them perpendicular to the direction of nanoindentation loading. This is consistent with the Halpin-Tsai theory which assumes that the reinforcement direction is unidirectional to the loading. At higher loading of filler, 'skewed' platelets or tactoid phases were created (Figure 3b). Such elongated agglomerates may act as short fibres with random orientation, in agreement with the Halpin-Kardos prediction. Similar effects occur with clay fillers [23]. However, experimental results do not follow set theories at higher filler concentration (5 wt%), possibly due to severe agglomerations acting as stress concentration sites [23].

Addition of nanostructures may induce brittleness in polymers [26]. Nanoindentation is also able to give an indication of brittleness via the plasticity index. This index is the ratio of plastic work done on the material to total work (plastic work and elastic work) applied [27]. The index ranged from 0 to 1; brittle materials, such as glass, tend to have low value (0.38) while ductile steel has a higher value (0.72). Table 1 shows that the addition of titanate nanosheets up to 5 wt% did not significantly change the brittleness of polyamic acid salt.

Table 1. Plasticity index of the pure polymer (PAAS) and the composites (PAAS/TiNS) for various loading of TiNS

Sample name	wt% of TiNS	plasticity index
Pure PAAS	0	0.63208
PAAS/TiNS/0.5	0.5	0.691766
PAAS/TiNS/1	1	0.660274
PAAS/TiNS/2	2	0.636735
PAAS/TiNS/3	3	0.622159
PAAS/TiNS/5	5	0.69291

Figure 5 shows that the transparency of polyamic acid salt is reduced when the concentration of titanate nanosheets exceeds 1 wt%. The agglomeration may start to form at a higher titanate loading in Figure 3b) causing scattering of light inside the polyamic acid salt.

Figure 5. A photograph showing the effect of titanate nanosheets on the optical properties of polyamic acid salt (titanate content from left to right: 0, 0.5, 1, 2, 3, 5 wt%)

4. Conclusions

Polyamic acid salt-titanate nanosheet composites have been successfully prepared by solution mixing. Triethylamine (TEA) can be used as a bulky molecule, instead of the more common tetrabutyl ammonium hydroxide, to exfoliate titanate nanosheets. Ultrasonication can assist intercalation of TEA into the nanosheets and their delamination into thin nanosheets. Addition of titanate nanosheet to water soluble polyamic acid significantly enhanced hardness and modulus of the polymer. The hardness increased by 90% on addition of 2% TiNS while the modulus increased by 103% compared to pure polymer. A high aspect ratio (l/d = 133.33) and uniform distribution of titanate nanosheets provides significant reinforcement of polymer. This improved behaviour is consistent with Halpin-Tsai theory at low filler content, assuming that the hardness and modulus of titanate nanosheet are the same as TiO_2. At high filler loading (>2 wt%), the mechanical properties of composites do not fit established theories due to agglomeration of titanate nanosheets. These sheets were stacked on each other, creating several layers of 'skewed' agglomerates or tactoid phases. This causes poor distribution of nanosheets and might disturb stress distribution between nanosheet and polymer. The incorporation of titanate nanosheets has no significant impact on the plasticity of polymer, as indicated by the plasticity index of composite; this assures flexibility of the composites. However, the inclusion of the nanosheets slightly reduces the transparency of polymer, particularly when the titanate nanosheet weight percentage exceeds 1 wt%, due to agglomeration.

Acknowledgements

The authors are grateful to the Indonesian Endowment Fund for Education (LPDP) for funding.

IOP Conf. Series: Materials Science and Engineering **223** (2017) 012054 doi:10.1088/1757-899X/223/1/012054

References

[1] Yoshitsugu K *et al* 1993 *J. Mater. Res.* **8** 1185-9.
[2] Xin Z *et al* 2010 *Macromolecules* **43** 2357–63.
[3] Kirana M S R N *et al* 2011 *Scripta Materialia* **64** 592–5.
[4] Lianzhou W and Takayoshi S 2014 *Chem. Rev.* **114** 9455−86.
[5] Liu M *et al* 2013 *Nat Commun.* **4** 2029.
[6] Marani D *et al* 2010 *Chem. Mater.* **22** 1126.
[7] Ratanatawanate C *et al* 2012 *Jr. Mater. Lett.* **66** 242.
[8] Harito C *et al* 2017 *Composite Part B: Engineering* **124** 54-63.
[9] Halpin J C 1969 *Journal of Composite Materials* **3** 732.
[10] Halpin J C and Kardos J L 1976 *Polym. Eng. Sci.* **16** 344-52.
[11] Sasaki T and Watanabe M 1998 *J. Am. Chem. Soc.***120** 4682–9.
[12] Grey I E *et al* A 1987 *J. Solid State Chem.* **66** 7–19.
[13] Ma R *et al* 2003 *Chemical Physics Letters* **380** 577-582.
[14] Ma R *et al* 2005 *J. Phys. Chem. B* **109** 6210-6214.
[15] Lee T *et al* 2009 *European Polymer Journal* **45** 19-29.
[16] Kreuz, J A *et al* 1996 *J Polym Sci Part A: Polym Chem* **4** 2607–16.
[17] Oliver W C and Pharr G M 1992 *J. Mater. Res.***7** 1565.
[18] Paula D R and Robeson L M 2008 *Polymer* **49** 3187–204.
[19] Sasaki T *et al* 2001 *J. Phys. Chem. B* **105** 6116–21.
[20] Bavykin D V *et al* 2006 *Advanced Materials* **18** 2807–24.
[21] Bokobza L 2007 *Polymer***48** 4907–20.
[22] Wan C and Chen B 2012 *J. Mater. Chem.***22** 3637-46.
[23] Ebadi-Dehagani H *et al* 2015 *Composite Part B: Engineering* **69** 133-44.
[24] Shackelford J F and Alexander W 2001 *CRC Materials Science and Engineering Handbook* (Boca Raton: CRC Press LLC) chapter 6 474.
[25] Goyal R K *et al* 2008 *Materials Science and Engineering A* **491** 230-6.
[26] Schadler L S2004 *Nanocomposite Science and Technology,* ed P M Ajayan P M, Schadler L S and Braun P (Weinheim: Wiley-VCH) 77.
[27] Beake B D *et al* 2010 *Nanostructured Thin Films and Coatings: Mechanical Properties,* ed Zhang S (Boca Raton: CRC Press Taylor & Francis Group) chapter 6 206-11.

IOP Conf. Series: Materials Science and Engineering **223** (2017) 012055 doi:10.1088/1757-899X/223/1/012055

Transesterification of sago starch and waste palm cooking oil in densified CO_2

H Muljana[1], A K Sugih[1], N Christina[1], K Fangdinata[1], J Renaldo[1], Rudy [1], H J Heeres[2], F Picchioni[2]
[1] Department of Chemical Engineering, Parahyangan Catholic University, Ciumbuleuit 94,Bandung 40141, West Java Indonesia
[2] Department of Chemical Engineering, ENTEG, University of Groningen, Nijenborgh 4, 9747 AG Groningen, the Netherlands

E-mail : hmuljana@gmail.com

Abstract. In this work, the synthesis of biodegradable and yet renewable thermoplastics materials through a transesterification reaction of sago starch and waste palm cooking oil (WPO) in densified CO_2 as the solvent is reported. The aim of this research is to investigate the potential used of sago starch and WPO as raw materials in the thermoplastics starch synthesis. The starch esters was successfully synthesized using sago starch and WPO as reagent in densified CO_2 as shown from the presence of carbonyl group (C=O, 1743 cm^{-1}) in the FT-IR spectra coupled with the presence of the proton (^{1}H-NMR) of the fatty acid in the starch backbone (0.8 – 2 ppm). The product crystallinity decreases as shown in XRD results and resulting with a change in the thermal properties (melting and crystalline temperature) of the products. In addition, the products show a different granular morphology and a higher hydrophobicity compared with native sago starch. This research shows the potential used of sago starch and WPO in the thermoplastics starch synthesis and opens a new perspective on the product application.

1. Introduction

Plastics have become indispensable material for daily human activities. This is supported by the fact that number of plastics production increases significantly in the past 20 years, from only about 100 million tons in 1990 and already reach 311 million tons in 2014 [1]. The huge amount of plastics production and their utilization have caused two major problems. The first problem relates with the availability of the raw material. As known, most of plastics are synthesized using precursor derived from the fossil fuel. Approximately 100 million barrels of oil are used to synthesize 8 million tons of plastics [1]. The high demand of fossil fuel for plastics production may cause a problem due to the limited amount of fossil fuel available in the world.

The second problem relates with the high amount of plastics waste in which approximately 50% of the plastic waste ends up in the landfill [1]. This causes serious environmental problem because of high volume to weight ratio and their resistance to degradation [2]. Therefore, the incentive to synthesis a 'green' plastics has significantly arisen in order to overcome the problems described before. The term 'green' in the plastics synthesis is strongly related with the use of biodegradable materials, renewable feedstocks and safer solvent for the environment [3].

Starch fatty acid ester (FASE) is known as a potential biodegradable thermoplastics material [2]. FASE shows a higher mechanical and thermal properties and has a higher hydrophobicity compared to those of their native counterparts [3,4]. Furthermore, the presence of fatty acid chain in the starch backbone may act as internal plasticizer which beneficial for further processing of the products [4]. FASE is synthesized via esterification of starch with various fatty acid esters such as fatty acid vinyl esters (FAVE) i.e: vinyl laurate,vinyl stearate [3-5], fatty acid methyl esters (FAME) i.e methyl laurate, methyl oleate [6,7] and vegetable oils i.e.: oleic sunflower oil, linseed oil, palm oil [8,9]. In addition, various solvent can be used in the reaction such as DMSO [3,5-6], ionic liquid [7] and densified/supercritical CO_2 [4].

The aim of this research is to investigate the potential used of sago starch and waste cooking oil as raw materials in the thermoplastics starch synthesis in densified CO_2. To the best of our knowledge, there is no publication available on the utilization of waste cooking oil in the transesterification of sago starch using densified CO_2 as a solvent. Sago starch is used due to its high production in Indonesia (approximately 200.000 tons/year) [10] while its utilization is still limited. In the other hands, waste cooking oil is inexpensive and also abundantly available as waste from food industries, restaurants and many similar facilities [11]. Indonesia was estimated to produce 6.5 million ton of waste cooking oil on 2013 and this is only less than 10% of the worldwide waste cooking oil production [12]. Furthermore, densified CO_2 is selected as the solvent not only due to its green properties, but since it can act as a plasticizer for most polymer/biopolymer [13].

This paper deals with a systematic study to evaluate the influence of several process variables such as pressure (15-21 MPa), and catalyst intake (K_2CO_3, 0.1 – 0.4 mol/mol anhydroglucose unit (AGU)) on the reactivity as indicated by the Ester Content in the starch ester product (EC). Relevant product properties were determined with Scanning Electron Microscope (SEM), X-Ray Diffraction (XRD), Contact Angle measurement, Thermogravimetric Analysis (TGA) and Differential Scanning Calorimetry (DSC).

2. Materials and Method

2.1 Materials

Sago starch (water content of 16.9 % wt/wt) was purchased from Bina Sago Lestari (Indonesia). Palm Cooking Oil was chosen as the material and was purchased from the local market. The waste palm cooking oil (WPO) was prepared in the laboratory by frying 1 kg of tempeh with 1 kg commercial PO. Prior the experiments, both PO and WPO were analyzed for their fatty acid

IOP Conf. Series: Materials Science and Engineering **223** (2017) 012055 doi:10.1088/1757-899X/223/1/012055

composition. The fatty acid composition is comparable with the values reported in literature [14,15]. Analytical grade potassium carbonate (K_2CO_3) was purchased from Merck. Technical grade methanol and ethanol was purchased from Bratachem (Indonesia). All chemicals were used as received without further purification. High purity CO_2($\geq$ 98 % volume) and N_2 ($\geq$ 98% volume) were purchased from Sangkuriang (Indonesia).

2.2 *Method*

2.2.1 Experimental Procedures

A series of experiments were done by reacting sago starch with waste PO (WPO) at various pressures (15 – 21 MPa), and various K_2CO_3 intake (0.1 – 0.4 mol/mol AGU) at a fixed temperature (120 °C) in densified CO_2. Prior the experiment, the free fatty acid (FFA) content of WPO were analyzed according to the procedure published in literature [16]. The FFA contents of WPO was 0.19% wt/wt, suggesting that WPO could directly be used in the experiment without any necessary pretreatment. All the materials (sago starch (2.5 g, dry basis), WPO (3 mol/mol AGU), and K_2CO_3 (0.1 – 0.4 mol/mol AGU)) were charged to the autoclave. The starch and WPO intakes were chosen based on the values used in previous experiments [17]. CO_2 was introduced inside the reactor using a positive displacement pump equipped with integrated cooling system (Chrom Tech). The reactor was heated to the desired temperature (120 °C). Additional CO_2 was added to increase the pressure to the desired level (depending on the variation, 15 – 21 MPa). After reaction (1.5 h), the reactor was cooled down to room temperature and depressurized. The solid products were filtered to separate from unreacted WPO. The solid products were washed with ethanol (200 mL) and dried in a vacuum oven at temperature of 70 °C until constant weight.

2.2.2 Determination of the Ester Content (EC) of the Products

A hydrolysis method is used to determine the EC values according to a published procedure [8,9]. The following equation was used to calculate the EC values [8,9]:

$$EC = \frac{[(V\ HCl_{blanko} - VHCl_{sampel}) \times M\ HCl \times 1000]}{m} \quad (1)$$

where V_{NaOH} is the volume of NaOH (mL), V_{HCl}blanko and V_{HCl} sample are the the titration volume of acid for the blank and fatty acid starch sample (mL) respectively, M_{HCl} molarity of used HCl and m the mass of the fatty acid starch sample (g). V_{HCl} blanko is obtained by the same procedure for native sago starch.

2.2.3 Analytical Equipment

Fourier Transform Infrared (FTIR) spectra were performed on a FTIR (IRTracer100 Shimadzu) ATR (attenuated total reflectance) golden gate apparatus (Graseby-Specac Ltd, Orpington, UK). A total 64 scans with a resolution of 4 cm^{-1} were applied. ^{1}H-NMR spectra were acquired (64 scans, relaxation times of 1.0 s) on an Oxford NMR AS 400 spectrometer operated at 400 MHz using DMSO-d_6 as solvent. TGA measurements were performed on a Perkin Elmer TGA 7 Thermogravimetric Analyzer. The samples were heated to 900°C in an inert atmosphere with a heating rate of 10°C min^{-1}. DSC analyses were performed using a TA Instrument DSC 2920. The sample (about 10 mg) was heated from room temperature to 200°C (heating rate of 10°C min^{-1}) and was cooled to -50 °C (cooling rate of 10 °C min^{-1}) and repeated in two cycles [4].

Water contact-angle measurements were performed using an optical contact angle instruments (OCA 15EC, Dataphysics) equipped with SCA software (Dataphysiscs). The morphology were determined using Scanning Electron Microscope on JEOL T330A. X-ray Diffraction (XRD) analyses were done in a diffractometer type PW1710 BASED with Anode Cu. The degree of crystallinity of samples (X_c) was quantified as the ratio of the area of the crystalline (A_c) reflections and the overall area (A_t) following the method of Nara and Komiya (1983) [13,18].

3. Results and Discussion

3.1 Starch Modification with WPO

A systematic study were done in order to investigate the effect of pressure and catalyst ratio on the reactivity as shown by the changes in the ester content (EC). Within the experimental range (*vide supra*), a broad Ester Content value is obtained (127.6 – 820.1 meq/kg). The detailed EC obtained in the experiments is shown in Figure 4.

3.1.1 Proof of Principle

The FT-IR spectra of native sago starch and starch ester products (EC = 212.6 meq/kg and 820.1 meq/kg)) were acquired and the results are shown in Figure 1. All spectra show several strong absorption bands at 3300 – 3600 cm^{-1} (OH stretching), 2925 – 2950 cm^{-1} (CH stretching), 1609 - 1612 cm^{-1} (OH bending), 1420 cm^{-1} (CH bending), and 900- 1300 cm^{-1} (highly coupled C-O and C-C vibrational modes) [4,7]. These absorption bands are typical bands appeared in FT-IR spectra of starch.

Additional peak at 1743 cm^{-1} is occurred in the spectra of the products corresponds to the presence of carbonyl (C=O) group in the product [4]. This implies the presence of the fatty acid chain in the starch backbone. The presence of the fatty acid chain also confirmed by the stronger absorption band at 2929 cm^{-1} (CH stretching) compared with the one of native starch as observed in the spectra [4].

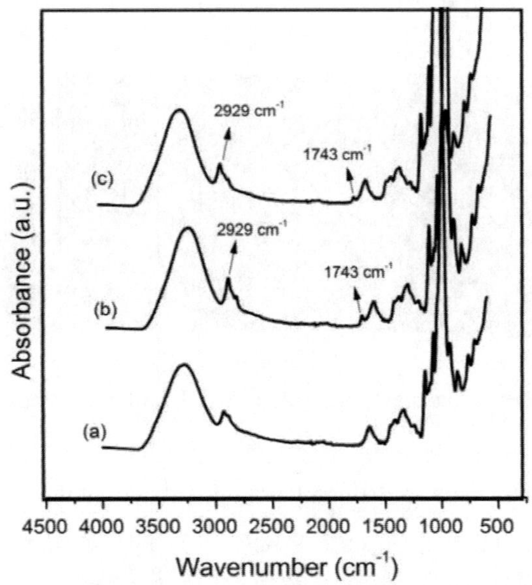

Figure 1. FT-IR spectra of native sago starch (a), FASE with EC of 212.6 meq/kg (b) and 820.1 meq/kg (c)

Figure 2 shows the typical ^{1}H-NMR spectra of the native starch and the products (EC = 212.6 meq/kg and 820.1 meq/kg). In comparison with the ^{1}H-NMR spectra of the native starch, it is clear that all spectra show a broad and overlap peaks in the δ 3.2 to 5.5 ppm range. These peaks are assigned to starch protons [4]. Moreover, additional peaks in the δ 0.8 to 2.5 ppm range are presence in the ^{1}H-NMR spectra of the products. These peaks are corresponded to the aliphatic hydrogen atoms of the fatty acid chain. In agreement with FTIR spectra, these observations confirm that the transesterification reaction of sago starch and WPO in densified CO_2 was successful.

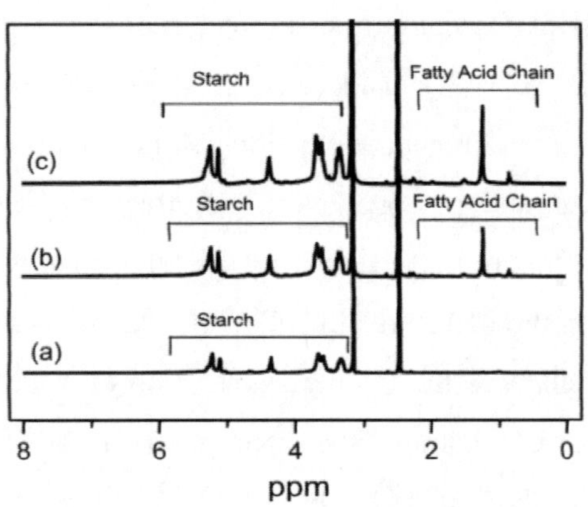

Figure 2. ^{1}H-NMR spectra of native sago starch (a), FASE with EC of 212.6 meq/kg (b) and 820.1 meq/kg (c)

IOP Conf. Series: Materials Science and Engineering **223** (2017) 012055 doi:10.1088/1757-899X/223/1/012055

The change in the morphology of the products is observed using the SEM. The results are shown in Figure 3. The pictures show the change in the granular structure of the products (Figure 3). The native sago starch has an oval granular shape as in agreement with the one reported in literature [19].

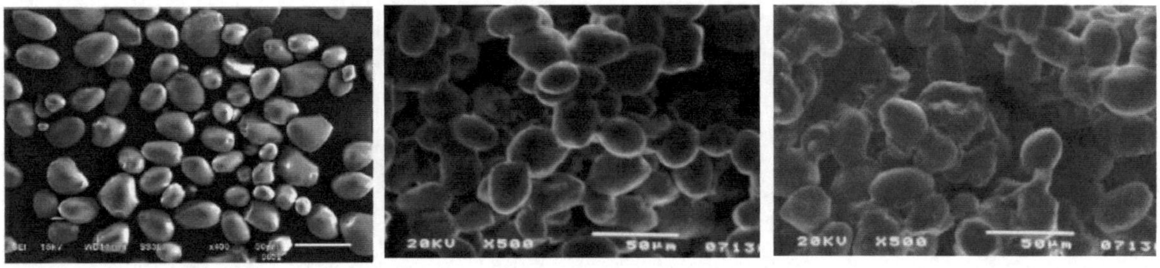

(a) (b) (c)

Figure 3. SEM pictures of native sago starch (a), FASE with EC of 212.6 meq/kg (b) and 820.1 meq/kg (c)

The products, in the other hands, show a different granular shape as the native. The granules tend to agglomerate and fuse with each other (Figure 3b and 3c). The similar observation is reported in literature, the agglomeration of the esterified products is possible due to an increase in the hydrogen bond because of the presence of the carbonyl group in the starch backbone [20].

3.1.2 The Effect of Pressure and K_2CO_3 Intake on The Reactivity

Figure 4 shows the EC values at various pressure and K_2CO_3 intake. In all cases, EC reach a maximum values at pressure of 15 MPa and decreases with pressure. The decrease in reactivity with pressure may be related with two competitive effect of pressure on the reaction. The first is plasticizing effect of CO_2 on the starch matrices [13]. At higher pressure, more CO_2 can be dissolved inside the starch granules, follow by the swelling of the starch matrices and increases the free volume in the starch structure. This leads to a higher diffusion rate of WPO inside the starch granules and increase the reactivity. In contrary, the second effect, the compressive effect, reduces the free volume of the starch matrices and inhibits the diffusion of the WPO inside the starch granules [4]. The two effects plays important role and influenced the diffusion rate in the reaction and the final EC values.

Similar trend as the pressure effect can also be observed in the effect of K_2CO_3 (Figure 4). It is clear that changes in EC values is a strong function of catalyst intake. The trend is somewhat in agreement with our previous result as reported in the literature [21]. This may be related with the possibility that alkaline salt base catalyst (K_2CO_3) enhances the hydrolysis reaction of WPO as well and reduce the transesterification reaction rate (lower EC value). The assumption can be supported by the fact that indeed alkaline base plays role in hydrolysis of triglycerides which is reported as side reaction in the biodiesel production [22].

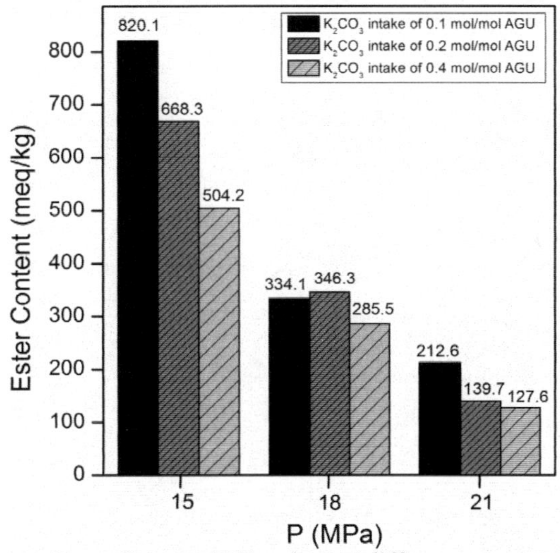

Figure 4. Ester Content (EC) values at different pressure and K_2CO_3 ratio

3.2 Product Characterization

3.2.1 X-Ray Diffraction Analysis

The change in the crystallinity of native sago and the products were measured with the XRD analysis. (see Figure 5). It is apparent that all spectra show the characteristics for C-type crystalline patterns of starch with strong diffraction peaks at diffraction angle (2θ) of 15.08° – 15.28°, 17.24 ° – 17.3°, 18.04°-18.09 °, 23.12° – 23.18° and 26.7 ° – 26.8° [19].

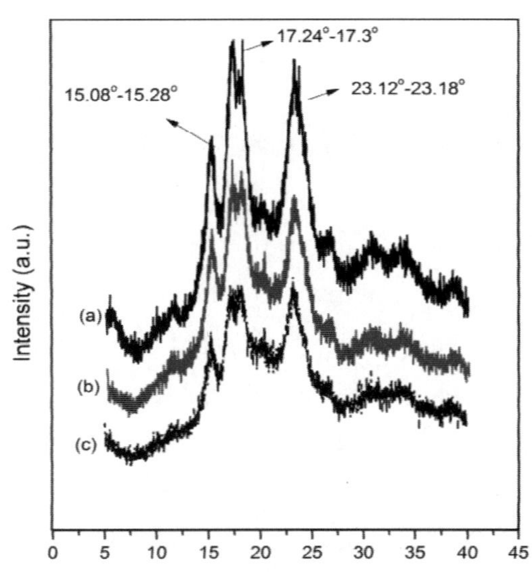

Figure 5. X-Ray Diffractogram of native sago starch (a), FASE with EC of 212.6 meq/kg (b) 820.1 meq/kg (c)

IOP Conf. Series: Materials Science and Engineering **223** (2017) 012055 doi:10.1088/1757-899X/223/1/012055

The degree of crystallinity of the native sago and the products is given in Table 1. It is clear that the degree of crystallinity (Xc) decreases at higher Ester Content. This is also confirmed by the result from the DSC measurement (see below). The decrease in crystallinity of the esterified starch products is also reported in our previous work with palm cooking oil (PO) and various fatty acid ester in literature [4], [9]. Within the same range of EC (around 210 meg/kg), FASE synthesized from PO gives the comparable degree of crystallinity (Xc = 0.27) as the one obtained from WPO [9].

The decrease in the crystallinity is happening due to the presence of CO_2 in the starch matrices leading to an enhanced plasticizing effect, induces the gelatinization of the starch and eventually reduces the crystallinity of starch [13]. In addition, when starch is gelatinized, the OH group is more susceptible to react with fatty acid chain in the triglycerides. Thus, it is plausible to assume that at higher degree of gelatinization, the degree of crystallinity decreases, starch granules become more reactive and eventually lead to higher EC in the products.

Table 1. Degree of crystallinity of native sago and the products

Sample	Ester Content (meq/kg)	Ac	At	Degree of Crystallinity (Xc)
Native Sago		2718.64	9624.26	0.28
FASE	212.6	1479.95	5621.89	0.26
FASE	820.1	1767.23	7423.85	0.24

3.2.2 Thermal Properties of The Products

Another important feature from the product characterization is their thermal properties. The DSC and TGA results of the native sago and the products are shown in Table 2. Changes in the transition properties of the products compared with the native sago are shown in Table 2.Clearly, the melting enthalpy (ΔH_m) and crystallization enthalpy (ΔH_c) decreases with the EC value. This suggests that the product has a lower crystallinity compared with the one of native starch and the change in crystallinity is a strong function of EC. This is also confirmed by the lower melting temperature (T_m) and crystallization temperature (T_c) of the products compared with the ones of native starch. Unfortunately, native sago starch displays the transition temperatures (T_m and T_c) near or higher than the degradation temperature and therefore cannot be determined accurately. The thermal stability of native sago and the products as obtained from TGA are given in Table 2. It is apparent that both products display a slightly higher thermal stability compared with the native starch [2]. In addition, thermal stability increases at higher EC suggesting that the thermal stability is caused by the presence of fatty acid chain in the starch backbone.

Table 2. Thermal Properties and Water Contact Angle results of the native sago and the products

No	Samples	Ester Content (meq/kg)	Contact Angle	Degradation Temperature Range (°C)	T_m (°C)	ΔH_m (J/g)	T_c (°C)	ΔH_c (J/g)
1	Native Sago	-	59.2°	270 – 275		nd*		
2	FASE	212.6	88.4°	278 – 280	10.64	32.22	5.8	12.39
3	FASE	400	91.1°	nm**	5.9	26.94	1.5	13.08
4	FASE	820.1	89.2°	282 - 285	4.04	15.3	1.33	7.14

*nd : not detectable
** nm : not measured

3.2.3 Water Contact Angle Measurements

Water contact angle measurements were done on native sago and the products with a range of EC (212.6 – 820.1 meq/kg). For each sample, the measurements were repeated six times to ensure the reproducibility. The contact angle of native sago and the products are shown in Table 2.

In comparison with the native sago starch (59.2°), it is clear that all of the products show a higher contact angle (88.4° – 91.1°). This means that all of the starch products have a higher hydrophobicity compared with native sago starch. The hydrophobicity increases with EC except at EC value of 820.1 meq/kg where small reduction of contact angle is occurred. This can be related to an internal plasticizing effect of fatty acid chain on the starch matrices which expected to increase the starch chain mobility and eventually to a lower surface hydrophobicity [4].

4. Conclusions

An exploratory study on the synthesis of fatty acid starch ester with sago starch and waste palm cooking oil (WPO) in densified CO^2 is described. The proof of principle is verified by the presence of carbonyl group (C=O, 1743 cm^{-1}) in the FT-IR spectra and with the presence of the proton from the fatty acid in the ^{1}H-NMR spectra (0.8 – 2 ppm). The EC of the products is a strong function of pressure and K_2CO_3 intake. The morphology, crystallinity, hydrophobicity, and thermal properties of the product are also influenced by the presence of fatty acid chain in the starch backbone and depending on the product EC. The products show a lower crystallinity, higher thermal stability and a higher hydrophobicity compared with native sago starch. This research demonstrates the potential used of sago starch and WPO in the thermoplastics starch synthesis and opens a new perspective on the product application.

Acknowledgments

The authors thank G.O.R Alberda van Ekenstein (Polymer Department, University of Groningen) for his help with the DSC and TGA measurements. Henky Muljana acknowledges Indonesia Ministry of Research, Technology, and Higher Education (RISTEKDIKTI) for providing a grant through Joint Research and International Publication Scheme (1012/K4/KM/2022) to carry out this research.

References

[1] Association of Plastic Manufacturer 2015 *An analysis of European Plastics, Production Demand and Waste Data Report* (Plastic Europe) http://www.plasticseurope.org (accessed Oct 6, 2016).
[2] Junistia L *et al* HJ 2009 *Starch- Starke* **61**(2) 69-80.
[3] Kerton FM 2008 *Introduction to Chemicals from Biomass*, *ed* J Clark and F Deswarte (Chichester : John Wiley & Sons) 47.
[4] Muljana H *et al* 2010 *Carbohydr. Polym.* **82**(2) 346-54.
[5] Winkler H *et al* 2013 *Carbohydr. Polym.* **98**(1) 208-16.
[6] Barrios SE *et al* 2013 *Int. J. Biol. Macromol.* **59** 384-90.
[7] Xie WL, Wang YB 2011 *Starch-Starke* **63**(4) 190-7.
[8] Aburto J *et al* 2005 *Starch-Starke* **57**(3-4) 145-52.
[9] Muljana H 2015 *Proc. Int. Conf. on Starch Technology* (Bangkok, Thailand) 71.
[10] The Sago Palm, The Food and Environmental Challenges of the 21st Century, ed The Society of Sago Palm Studies (Kyoto University Press) 2015.
[11] Knothe G *et al* 2009 *Bioresour. Technol.* **100**(23) 5796-801.
[12] Toop G *et al* 2014 Report Trends in the UCO Market, Ecofys.
[13] Muljana H *et al* 2009 *Carbohydr. Polym.* **78**(3) 511-19.
[14] Verma P, Sharma MP 2016 *Renew. Sustainable Energy Rev.* **62** 1063-71.
[15] Maneerung T *et al* CH 2016 *Energy Convers. Manage.* **123** 487-97.
[16] Hammond EW 2003 *Encyclopedia of Food Science and Nutrition*, *ed* B Caballero, P Finglass, and F Toldra (United States : Academic Press (Elsevier)) 2311 .
[17] Hermawan E *et al* 2015 *Int.J Chem.Eng.Appl.* **6** 152–55.
[18] Wang SJ *et al* 2006 *Carbohydr. Res.* **341** 289-93.
[19] Pukkahuta C, Varavinit S 2007 *Starch-Starke* **59**(12) 624-31.
[20] Muljana H Manuscript LPBM 2010 *Starch-Starke* **62**(11) 566-76.
[21] Muljana *et al* 2010 *Carbohydr. Polym.* **82**(3) 653-62.
[22] Pecha J 2016 *Chem. Eng. J.* **288** 680-8.

IOP Conf. Series: Materials Science and Engineering **223** (2017) 012056 doi:10.1088/1757-899X/223/1/012056

Physicochemical properties and characteristics of microcrystalline cellulose derived from the cellulose of oil palm empty fruit bunch

H Nasution[1], Yurnaliza[2], Veronicha[1], Irmadani[1] and S Sitompul[1]
[1]Department of Chemical Engineering, Faculty of Engineering, Universitas Sumatera Utara, Padang Bulan, Medan 20155, Indonesia.
[2]Department of Biology, Faculty of Mathematics and Natural Sciences, Universitas Sumatera Utara, Padang Bulan, Medan 20155, Indonesia.

E-mail : h_dahliana@yahoo.com

Abstract. Cellulose from oil palm empty fruit bunch was hydrolyzed with different concentration of hydrochloric acid (HCl) viz. 2; 2.5; 3; 3.5 N to prepare microcrystalline cellulose (MCC). The temperature of hydrolysis process was at 75 °C. The physicochemical properties such as organoleptic characteristic, water soluble substance, and loss on drying were conducted during this study. Infrared, x-ray diffraction, and scanning electron microscope (SEM) image were also performed to investigate the effect of hydrolysis on the molecular structure, crystal structure and morphology of the MCC, respectively. The results have showed, in term of the physicochemical properties, for all concentrations of HCl the MCC was obtained. The FTIR results showed the –OH functional group tended to reduced as alpha cellulose has changed to MCC. The x-ray diffraction pattern revealed that the crystal structure was traced on MCC where the highest was observed at hydrolysis process of 2.5N HCl. The image from SEM displayed an individualized and fibrous MCC on the treatment of 2.5N HCl.

1. Introduction

The main content which has a potential to be used as raw material of microcrystalline cellulose is fibrous material having cellulose content. Oil palm empty fruit bunches (EFB) containing cellulose w/w 35% which is high enough to be used as raw material for the manufacture of microcrystalline cellulose [1]. The selection of EFB rather than palm fronds, due to the age factor of oil palm. Oil palm plants have a lifespan that is long enough to remain productive of fruit. If the palm fronds used as the main raw material, the possibility of getting palm fronds of plants that are not productive is small. As for the EFB, it is still be generated during oil palm industry in operation [2].

Microcrystalline cellulose has been made from several natural sources, such as from kenaf fibers [3], saw dust [4], rice hull and bin hull [5], corn cobs [6], groundnut shell [7], cotton linters [8], rice straw [9], sisal fibers [10], cotton rags [11] and oil palm empty fruit bunch [12]. Some methods have been related to the manufacture of microcrystalline cellulose such as reactive extrusion [13], enzymatic hydrolysis [14], partial hydrolysis with oxygen and/or carbon dioxide [15], microorganism [3] and acid hydrolysis [16-19]. From all of methods for manufacturing of microcrystalline cellulose, acid hydrolysis

IOP Conf. Series: Materials Science and Engineering **223** (2017) 012056 doi:10.1088/1757-899X/223/1/012056

has been chosen because it is shorter time reaction and economic. Therefore, in this study, the effect of variations of hydrochloric acid (HCl) concentration on the physicochemical properties and characteristics of microcrystalline cellulose were investigated.

2. Method

2.1 Materials

Raw material used in this study was cellulose derived from oil palm empty fruit bunches, supplied by Oil Palm Research Center in Medan, Indonesia. Other chemicals such as HCl (37%), NaOH, NaOCl were supplied Sigma-Aldrich and used as received.

2.2 Isolation of α-cellulose

A total of 75 g cellulose derived from oil palm empty fruit bunches added with 500 mL of 17.5% NaOH and heated at 80°C for 30 min. The results are filtered using a cloth napkin, washed with water and squeezed. The solid mixture is added to 500 mL of 3.5% NaOCl and water (1:1) and heated at 100°C for 5 min, then filtered using a cloth napkin, washed with water until the filtrate is clear. The solid were then squeezed and dried at 60°C for 4-5 h in the oven and then α-cellulose was obtained. A method of Ohwoavworhua *et. al.,* (2009) was adopted for this isolation [20].

2.3 Procedure of Making Microcrystalline Cellulose

A total of 1 g of α-cellulose put in glass Beaker and hydrolyzed with HCl with concentration varied viz. 2 N; 2.5 N; 3 N and 3.5 N of 20 mL by boiling with a temperature 75 °C for 15 min. After cold, the hydrolized α-cellulose was added with 50 mL aquadest while stirring strongly with a spatula and let stand for overnight. The solution was filtered through Whatman filter no.1 and then dried at a temperature of 57-60 °C for 3 h. After the microcrystalline cellulose crushed dried and stored at room temperature in a desiccators. A method of Yuvraj *et al.,* (2009) was adopted for this preparation [21].

2.4 Characteristics of Microcrystalline Cellulose (MCC).

The physicochemical properties such as organoleptic characteristic, pH, water soluble, and loss on drying were conducted during analysis. Other characterization such as x-ray diffraction (XRD), infrared diffraction (FTIR) and scanning electron microscope (SEM) image were also investigated. XRD was recorded in 6100 Shimadzu. FTIR were carried out using Shimadzu IR-Prestige 21. SEM images were analyzed using EVO MA 10 ZEISS.

3. Results and Discussion

3.1 Physicochemical

The results of the effect HCl concentration on the mycrocrystalline cellulose product show that, for all HCl concentrations, the physicochemical properties are in accordance with the standards of British Pharmacopeia 2009 specification [22]. Table 1 shows some physicochemical properties of microcrystalline cellulose derived from oil palm EFB.

Table 1. Physicochemical properties of microcrystalline cellulose

TEST	MCC
Organoleptic	Odorless, tasteless and white fine powder
pH	6-7
Water soluble substances	≤ 0.25%
Loss on drying	0 – 1.64%

From the table, the organoleptic observation shows odorless, tasteless and white fine powder, whereas the value of pH on microcrystalline cellulose is in range 6-7. The results of water soluble substances for each varied HCl concentration on microcrystalline cellulose product in have been meet the requirements of British Pharmacopeia 2009 i.e ≤ 0.25% [22]. This value giving an indication of crystal phase was appeared in microcrystalline cellulose. This could be attributed to the process where during hydrolysis process the glucose content in α-cellulose was diluted. Consequently, in microcrystal cellulose the soluble substance was low. On the other hand, the value obtained for loss on drying is 0-1.64% which is in accordance with British Pharmacopeia 2009 i.e ≤ 6% [22]. Here, crystalline phase was difficult to absorb store water as compared to amorphous phase; so that the value of loss on drying would be low because of store water was not too high due to its crystalline structure. Since water contains OH functional group, the evidence of the appearance of crystal in MCC can be explaining using FTIR spectra. The molecular structure of MCC as compared to two other celluloses (cellulose and α-cellulose) is show in Figure 1.
It shows the shift of maximum absorbance of functional group of OH decreased (3340 cm^{-1}) as compared with other two samples of cellulose (3363 cm^{-1} and 3446 cm^{-1} for α- cellulose and cellulose, respectively). Moreover, the functional OH group of MCC is less intense than those of two other celluloses. This shift and intensity have proved that MCC is more crystalline than other samples.

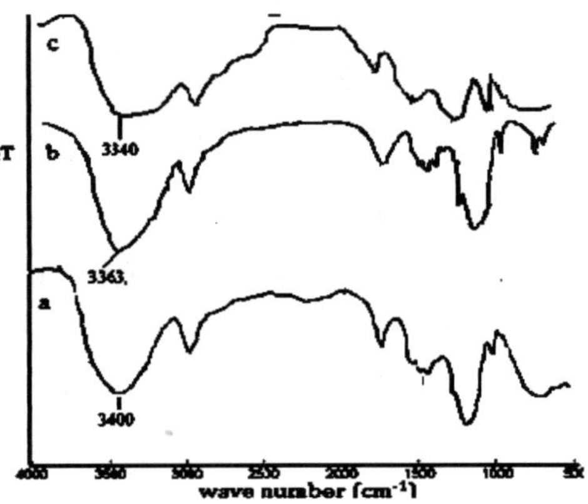

Figure 1. Infra-red spectra of a. cellulose; b. α- cellulose; c. MCC (2.5N HCl)

All observations of physicochemical properties viz. organoleptic, pH, water soluble subtances and loss of drying have shown that for all variations of acid concentrations (2; 2.5; 3; and 3.5N) the results meet the requirements of British Pharmacoea (2009) as microcrystalline cellulose. However, due to the efficiency concern of using acid (HCl) during hydrolysis process, the use of concentrations 2 and 2.5N have been selected.

3.2 X-ray Diffraction (XRD)

The x-ray diffraction spectra of cellulose, α- cellulose, MCC (2.5N HCl) and MCC (2N HCl) is shown in Figure 2 below.

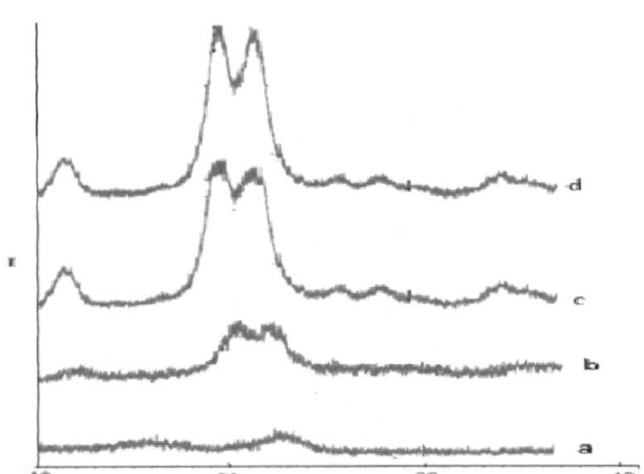

Figure 2. X-ray diffractograms of a. cellulose; b. α- cellulose; c. MCC (2N HCl); d. MCC (2.5 N HCl)

IOP Conf. Series: Materials Science and Engineering **223** (2017) 012056 doi:10.1088/1757-899X/223/1/012056

The figure above exhibits different diffraction pattern amongst samples where cellulose derived from oil palm empty fruit bunch was highly amorphous as indicated by the least peak in the diffractogram (Figure 2a). The peaks that appeared at diffraction angels (2theta) ranging from 20° to 24°. This behaviour was attributed to the presence an amorphous aromatic compounds such as lignin and hemicellulose. The crystallinity has been already traced on the α-cellulose (Figure 2b) due to removal of hemicellulose and lignin which existed in amorphous region during delignification process. In this case, the diffraction peaks of the 2 theta angels at 18° to 24°. However, these two peaks are smeared and appeared as one broad peak. Figure 2c and 2d show MCC product diffractograms using 2N HCl and 2.5N HCl, respectively. These diffractions revealed similar patterns, where the appearance of two peaks are an evidence of the coexistences of cellulose I and cellulose II allomorphs. The appearance of two significant peaks indicating the highly crystalline structure of MCC. The higher crystal structure was obtained for MCC due to removal of the amorphous phase in α-cellulose by acid hydrolysis. During hydrolysis acid process, an amorphous cellulose near surface is hydrolized and followed by slower rate of hydrolysis of amorphous in deeply microfibrils. Here, the diffraction peaks of the 2theta angles for MCC are shown at 12°; 20° and 22°. However, the peaks of MCC (2.5N HCl) (Figure 2d) is slightly higher intensity than MCC (2N HCl) (Figure 2c).

3.3 Scanning Electron Microscopy (SEM)

To determine the characteristics at each stage of cellulose into α-cellulose and microcrystalline cellulose product, the SEM images can be viewed in Figure 3 below. Figure 3a shows macrofibril cellulose molecules bound to one another to form long polymer chains and interconnected. There is an insulation on the surface macrofibrils due to the presence of amorphous phase such as lignin and hemicellulose. Here, during first alkali treatment, the removal of an amorphous structure such as cellulose and lignin was not perfect. Therefore, it is clearly shows the shape and distribution of the macrofibrils in cellulose was not uniform. However, for the α-cellulose (Figure 3b), the macrofibrils structure shows amount of naked macofibril bundles. The insulation contains lignin and hemicellulose which have appeared in macrofibrils was removed due to the subsequent alkali treatment. On the other hand, the MCC image shows an individualized and fibrous structure (Figure 3c). In this case, after macrofibrils were hydrolyzed and rinsed, the volume of amorphous cellulose was occupied by water molecule and then removed. However, the image of MCC (2N HCl) (Figure 3d) shows the structure of microfibrils is slightly different where there is an insulation on the crystalline surface; this is due to the amorphous phase could not well dissolved during hydrolyzing process.

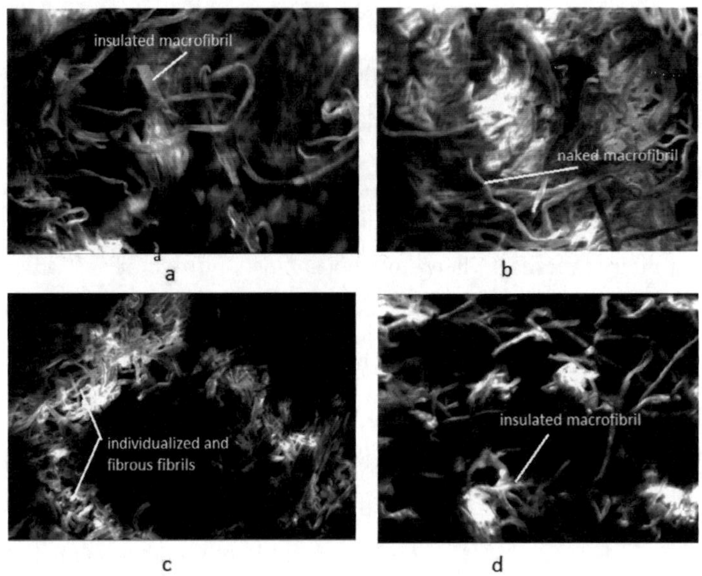

Figure 3. SEM with mag. of 500x a.cellulose; b. α- cellulose; c. MCC (2.5N HCl); d. MCC (2N HCl)

From FTIR, XRD and SEM characterizations, the results have indicated the crystal part was traced on MCC, where the –OH groups tended to reduced as alpha cellulose has changed to MCC and the appearance of two peaks in diffractogram. Moreover, SEM image have shown the individualized and fibrous structure which indicated the appearance of crystal stucture.

4. Conclusion

The results showed for all concentration of HCl in term of the physicochemical properties the MCC have been obtained. The organoleptic characteristics such as odorless, tasteless, white and fine powder; the pH value of 6-7; the value of water soluble substances $\leq 0.25\%$ and the value of loss on drying 0-1.64% have meet in *British pharmacopeia* 2009. From XRD spectra and SEM image, the results have indicated the crystal structures were occurred in MCC especially in 2.5N HCl of hydrolyzed process.

Acknowledgements

The authors gratefully acknowledge that the present research is supported by Ministry of Research and Technology and Higher Education Republic of Indonesia. The support is under the research grant BP-PTN

USU of year 2016 Contract Number 6049/UN5.1.R/PPM/2016.

References

[1] Rosnah M S *et al* 2002 *Proc of the Research and Consultancy Seminar*. Faculty of Mechanical Engineering, UiTM Shah Alam, Malaysia. 140–145.
[2] Li X, Tabil L G and Panigrahi S C 2007 *J. Polym. Environ.* **15** 25–33.
[3] Keshk S M A S and Haija M A 2011 *Carbohydr. Polym.* **84** 1301–1305.
[4] Oyeniyi Y J and Itiola O A 2012 *Int J Pharm Pharm Sci* **4** 197–200.
[5] Adel A M *et al* 2011 *Carbohydr. Polym.* **83** pp 676–687.
[6] Azubuike C P and Okhamafe A O 2012 *Int. J. Recycling of Organic Waste in Agriculture* 1–9.
[7] Azubuike C P *et al* 2012 *J. Excipients and Food Chem.* **3** (3).
[8] Nada A M A *et al* 2009 *Bio Resources* **4** (4) 1359–1371.
[9] Nawar G A M *et al* 2010 *J. American Science* **6** (10) 226–231.
[10] Bhimte N A and Tayade P T 2007 *AAPS Pharm Sci Tech* **8** 1–7.
[11] Chauhan Y P R *et al* 2009 *Int. J. Chem. Sci.* **7**(2) 681–688.
[12] Mat Soo R *et al* 2009 *J. Oil Palm Research* **21** 613–620.
[13] Hanna M *et al* 2001 *Industrial Agricultural Products Center Publications & Information.* **9**.
[14] Maha M I *et al* 2013 *Cellulose* 10.1007/s10570-013-9992-5.
[15] Asko K, Veli V, Leif R, Kari P, Olli D, Kari V 2013 *United States Patent Application Publication* 0112193.
[16] Hafiz M K M *et al* 2013 *Carbohydr. Polym.* **93** 628– 634.
[17] Ejikeme P M 2008 *Cellulose* **15** 141–147.
[18] Laka M and Svetlana C 2007 *Bio Resources* **2** (3) 583–589.
[19] El-Sakhawy M and Hassan M L 2007 *Carbohydr. Polym.* **67** 1–10.
[20] Ohwoavworhua F O *et al* 2009 *Processing pharmaceutical grade Microcrystalline Cellulose from groundnut husk : Extraction method and characterization:* 10.4103/0973-8258.54895.
[21] Yuvraj P C *et al* 2009 *Int. J. Chem. Sci.*: **7** (2) 681–688.
[22] British Pharmacopoeia. The Commission office London 2009. **111** 6578–6585.

IOP Conf. Series: Materials Science and Engineering **223** (2017) 012057 doi:10.1088/1757-899X/223/1/012057

Preparation of micro-fibrillated cellulose from sorghum fibre through alkalization and acetylation treatments

Ismojo[1,2], P H Simanulang[1], A Zulfia[1], M Chalid[1]
[1]Department of Metallurgical and Materials Engineering, Faculty of Engineering, Universitas Indonesia, West Java, Depok 16424, Indonesia;
[2]Department of Automotive Mechanical Engineering diploma III, Institute Technology Indonesia, South Tangerang, Serpong, Indonesia.

Email: chalid@metal.ui.ac.id

Abstract. Recently, the pollution due to non-degradable materials including plastics, has led to needs on the development of environmental-friendly material. Owing to its biodegradability nature, sorghum fibres are interesting to be modified with petro-polymer as a composite. These materials are also expected to reduce the impact of environmental pollution. Surface modification of sorghum through chemical treatment was aimed to enhanced crystalline part of micro-fibrillated cellulose, thus increased compatibility to petro-polymer, as mean to improve composite properties. The experiments were conducted by alkalization process (10% NaOH) followed by acetylation with acetic acid glacial and acetic anhydride $(CH_3CO_2)_2$ with additions of 1 and 2 drops of 25% H_2SO_4. Fourier transform infra-red (FTIR) spectroscopy, field-emission scanning electron microscope (FE-SEM) and x-ray diffraction (XRD) were used to characterize the treated and untreated fibres. The results of investigation showed that the chemical treatments have effectively produced MFC with the smallest fibre size around 5.5 – 6.5 microns and reduced lignin and hemicellulose where the highest crystalline part up to 80.64% was obtained through acetate acid treatment of 17.4 M, followed acetic anhydride with 1 drop of H_2SO_4 addition. Based on the current results, it is promising that the synthesized composites can be improved for their compatibilities.

1. Introduction

Cellulose, available on earth, is the most abundant natural polymer that renewable, biodegradable in nature, has a lower density than synthetic filler and inexpensive so hence it is considered an alternate to non-degradable fossil-based polymers. In the most recent decades, several examinations have spent their considerations on how to obtain scale micro fibrillated cellulose (MFC) from natural fibres. This objective has been driven by finding suitable uses of MFC in material designing. The inherent stiffness and high level of crystallinity make it preferably suited for strengthening and load bearing application in composite [1-3].

Sorghum is an ancient, old world cereal that was domesticated in Africa, became an important food crop and ranked after wheat, rice, corn, and barley, and it is cultivated in many countries [4]. One of the distinctive properties of sorghum is its resistance to drought and tolerance to water logging. Sorghum also has a great potential to be developed in Indonesia because of its natural adaptation. It can be used as industrial raw materials such as sugar ants, amino acids, starch, etc., or fuel as bioethanol, and animal feed [5].

Previous studies have proposed that natural fibres like flax, hemp, jute and sisal can provide great potential as reinforcements in composite materials [6]. In 2015, E Yuanita *et al.*, (2015) [7] investigated the preparation of MFC obtained from Arenga Pinata or "ijuk". Another research, in the same years, reported that sorghum has higher cellulose and lowest lignin contents compared to the other agro-waste materials such as bagasse and rice straw [8]. Based on the result investigations be related to sorghum , which might a chance to be candidate as source to obtain MFC as reinforcing fibres in polymer composite.

Among all techniques preparation of MFC, chemical treatments such as alkalization and acetylation are known as simple and cheap methods. The alkalization using sodium hydroxide solution is aimed at reducing the impurities on the surface of the fibre including lignin, wax oils, and hydrophilic functional group (-OH), in order to increase fibre resistance to water. On the other hand, the acetylation or esterification using a solution containing acetyl functional group (CH_3CO^-) aims to reduce the hygroscopic properties of natural fibres and to increase the dimensional stability of the fibre with the result that expectable increasing compatibility in the composite [9].

The solid waste from sorghum extraction process is abundant; however its usage for reinforcing material is still limited. Therefore, the current research is focused on obtaining micro-fibrillation cellulose fibres from sorghum to be used further as bio-composites. More specifically, the investigation aims to determine the effect of alkalization followed by acetylation on several properties of the fibres, such as sorghum bicolor fibres, compound composition, morphology, size and the crystallinity of the fibre.

2. Materials and Methods

2.1. Materials

Sorghum waste was obtained provided from a traditional market. Sorghum fibre was dried in room temperature (25°C) for 3 days. The dried sweet sorghum were cut into ± 5 cm and crushed to pass a 40 mesh. Sodium hydroxide in pellets form, acetic acid, acetic anhydride and sulfuric acid solution was purchased from Merck.

2.2. MFC preparation

The alkalization samples were prepared by soaking 7 g of size reduced sorghum fibre in 50 mL of 10 % sodium hydroxide agent solution with continuous agitation. The mixture was kept at 80°C for 1 h. Acetylation treatment was carried out after alkalization by soaking in 50 mL of 17.4 M acetic acid in agitated condition at 25°C for 1 h. Sample from acetylation treatment was filtered and soaking in 20 mL acetic anhydride solution for 10 min with adding one and two drops of 25% sulfuric acid as a catalyst, and sample AGA2 was followed by soaking in 25% sulfuric acid at room temperature during 2 h. After each cycle of the treatment, sample were washed in three times with tap water and then dried in air at 25°C for 3 days to ensure the wash medium already reached neutral and filtered. Aside from hydrolysis, the sample was soaking in water distillation while 1 h for left out acid concentration which still adhere. The experiment procedure in this work is summarized in Table 1.

Table 1. Summarized of step experiment and sample code sweet sorghum fibres preparation.

Sample code	Alkalization	Acetylation	
SV	Un-treated		
AGA1			$(CH_3CO_2)_2O$ + 1 drop H_2SO_4 25 %;10 min;25°C
AGA2	10 % NaOH; 1 hour;80°C	CH_3COOH 17,4 M;1 hour;25°C	$(CH_3CO_2)_2O$ + 2 drop H_2SO_4 25 %;10 min;25°C
AGA2H			$(CH_3CO_2)_2O$ + 2 drop H_2SO_4 25 %;10 min;25°C + hydrolysis 2 h;25°C

Note: SV: Sorghum Virgin; AGA: Acetic Glacial Anhydride; H: Hydrolysis

2.3. *MFC characterization*

The characterizations were carried out in the Universitas Indonesia, Department of Metallurgy and Materials Engineering's, using Spectrum Two™ Infrared Spectrometer Perkin Elmer (ASTM E 1252) was used to determine composition of untreated and treated sorghum with chemical treatment. The morphologies of used untreated and treated sorghum fibres were analyzed using Field-Emission Scanning Electron Microscope (FE-SEM) FEI Inspect F50 with gold using a vacuum sputter coater. An X-Ray Diffraction (Shimadzu XRD-700 X-Ray Diffractometer) was used to study crystallinity of untreated and treated sorghum fibres. Crystalline parts of the fibres were determined based on reflected intensity data from method of Segal *et al* [10]:

$$Cr.I = \frac{(I_{002} - I_{amp})}{I_{002}} \; x \; 100 \qquad (1)$$

Where I_{002} is the maximum intensity which located at angle around 2θ = 22° and I_{amp} is the intensity scattered by the amorphous part of the sample at the diffraction angle around 2θ =18°.

3. Result and Discussion

3.1. Compound in Un-treated and Treated Sweet Sorghum Fibres

The FTIR spectra of untreated sweet sorghum are presented in Figure 1. As shown in Figure 1, FTIR spectrum of untreated sweet sorghum fibre (SV) shows some peaks. The peaks are 1245, 1510, and 1607 cm^{-1} wavenumbers that indicate the occurrence of phenolic C-O stretch, aromatic rings, and C=C chemical bond of lignin. Peak at 1735 cm^{-1} wavenumber, indicates the occurrence of chemical bonding C=O stretch of hemicellulose and at 3344 cm^{-1} wavenumber indicates the occurrence of chemical bond O-H stretch part of lignin, hemicellulose and cellulose fibres that indicate sorghum fibres hydrophilic properties [11-14].

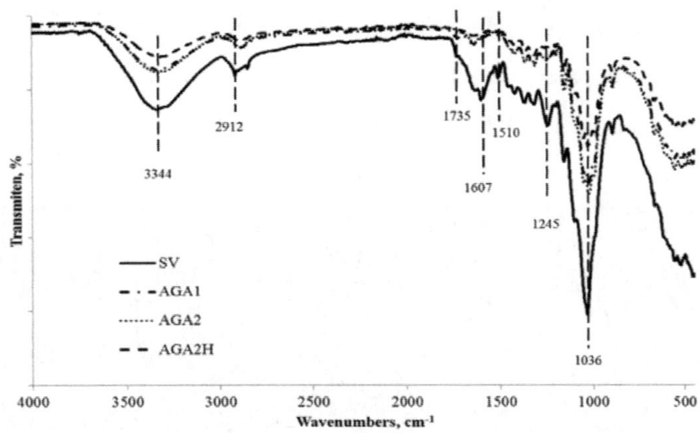

Figure 1. FTIR spectra of untreated sorghum and alkalization-acetylation glacial-anhydride-catalyst –with and without hydrolysis treated sorghum

Figure 1 also shows the treatment of sorghum with alkalization-acetylation and alkalization-acetylation-hydrolysis. The purpose of treatments are removing impurities such as a wax and oil that adhere on the surface of the fibre, as well as number of amorphous fractions (lignin and hemicellulose) in the cell wall of sorghum. Significant change in peak IR-spectra occurred at 1735 cm^{-1}, which is indicates the occurrence of chemical bonding C=O stretch of hemicellulose, disappear after alkalization-acetylation and alkalization-acetylation-hydrolysis treatment. The same occurrence at peak around 1245, 1510, 1610 cm^{-1} which is associated with phenolic C-O stretch, aromatic rings, and C=C chemical bond of lignin, also disappear after alkalization-acetylation-hydrolysis. Another important peak are at 1036, showing a decrease in intensity which is indicate reducing of lignin and hemicellulose after alkalization-acetylation and alkalization-acetylation-hydrolysis.

3.2. Morphology in Untreated and Treated Sweet Sorghum Fibres

Structure of sorghum fibre are constitutes of cellulose microfibrils are binding together with lignin, pectin and hemicellulose, as well as joining adjacent fibres together to form fibre bundles. SEM evaluation were applied to untreated sorghum fibre and MFC that was obtained from the modification treatments of sorghum fibres are displayed in Figure 2. The micrograph in Figure 2(a) implies the untreated sorghum fibres (SV) are covered with a layers, it is likely that waxy composition substances, evenly that the layers not distributed along the fibres surface but its thickness varies from point to point. The diameter of untreated sorghum fibre around 100μm. on the other hand, alteration in the morphology structure sorghum virgin occurred after preparation with 10 % NaOH solution and 17.4 M acetic acid followed soaking in acetic anhydride with or without hydrolysis solution. It was observed that the treatments removal of the layers on surface fibres as seen in Figure 2 (b) and Figure 2 (c), with the result that, the surface of treated fibres became smoother as compared to that untreated fibres. A loss of the waxy substance on the surface of lignocelluloses materials and substitution of surface hydroxyl groups by acetyl groups could make clear something of the fibre surface after acetylation. It is according with Tserkia *et al.,* (2005) and Zafeiropoulos *et al.,* (2002) that study of the effect of acetylation surface treatments on natural fibres [15,16].

(a) (b) (c)

Figure 2. SEM image of sorghum fibres; (3a) untreated sorghum fibre surface (SV), (3b) alkalization-acetylation glacial anhydride-2 drop catalyst sorghum fibre surface (AGA2) and alkalization-acetylation glacial anhydride-2 drop-hydrolysis (AGA2H). White bar scale represents 200μm.

Figure 2 (b) and 2(c) affirmatively display the fibre became fibrillated after treatments. It is peer that sorghum fibre bundle separate into individual fibres. It was also observed from Figure 2 (c) that hydrolysis with 25 % sulfuric acid during 2 h after acetylation make fibre damage and crack. The diameter of treated sorghum fibres decreased to an average of around 5.5-6.5 μm resulted from acetylation without hydrolysis treatment. It indicated that the waxy which bind the fibril structure were removed after treatment. It was verified also that with removed of shallow layer can improve increased contact area which make more interesting for application as reinforcement in composite materials.

3.3. Crystallinity in Untreated and Treated Sweet Sorghum Fibres

Most of the natural fibre are polymer that having amorphous and crystalline parts, where the lignin and hemicellulose as amorphous and cellulose as semi-crystalline. Only the crystal polymorph I of cellulose was appear in both untreated and treated fibres, as usual for native cellulose [15]. The

IOP Conf. Series: Materials Science and Engineering **223** (2017) 012057 doi:10.1088/1757-899X/223/1/012057

XRD pattern of untreated and alkalization-acetylation-hydrolysis sorghum fibres are showed in figure 3. As seen in Figure 3 the pattern treated fibres implies four peaks at 15.1, 22.1, 34.0 and 47.5°. The 15.1° reflections correspond to the $(1\bar{1}0)$ crystallographic planes. The other peaks at 22.1 and 34.0° correspond to the (200) and (023) or (004) planes, respectively [15]. Referred to Segal's empirical equation (1), I_{002} is the maximum of the lattice diffraction peak for plane (002) which is lie at a diffraction angle around $2\theta = 22°$ (for this case the peak is at $2\theta = 22.1°$) and the lowest of the intensity scattered by the amorphous part is measured at the diffraction angle around $2\theta = 18°$ (for this case the peak is at $2\theta = 15.1°$) [10, 17]. If the fibre within high amounts of amorphous materials, as in this case of sorghum fibres, such as lignin, hemicellulose, and amorphous cellulose, the peak at around 15°-16° appearing as one broad peak [15].

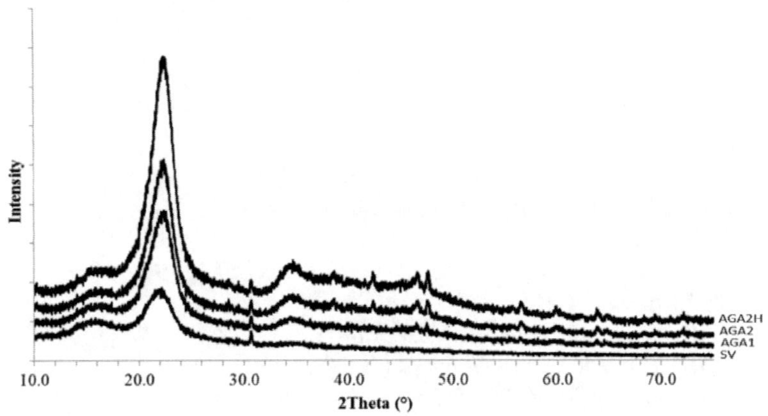

Figure 3. X-ray diffraction patterns of untreated and alkalization-acetylation-hydrolysis sorghum fibres

Table 2 was presenting the calculated results of crystallinity index which extracted from XRD curve for the both untreated and alkalization-acetylation-hydrolysis treatment, based on the equation (1). As seen in Table 2, that the alkalization-acetylation-hydrolysis treatment point out significantly remove the amorphous parts in sorghum fibres and the alkalization-acetylation with acid hydrolysis treatment has the best effectiveness in removing the amorphous part. This result is according with the data obtained from FTIR spectroscopy.

Table 2. Crystallinity Index of untreated and chemical treated sorghum

	SV	AGA1	AGA2	AGA2H
Crystallinity Index (%)	41.12	80.64	76.18	82.61

IOP Conf. Series: Materials Science and Engineering **223** (2017) 012057 doi:10.1088/1757-899X/223/1/012057

4. Conclussion

This study has effectively acquired micro-fibrillated cellulose from sorghum as a matter of fact potential source which is large in number in nature. The MFC was set up by alkalization and acetylation. The both modification treatments demonstrated that the chemical treatments have adequately diminished lignin and hemicellulose where the most astounding crystallinity up to 82.6% was acquired through alkalinization and acetate acid treatment of 17.4 M acetic anhydride with 2 drop of H_2SO_4 addition followed by 25 % sulfuric acid hydrolysis and the treatments result MFC with the lowest fibre sized around 5.5-6.5 μm. In view of the present results, it is promising that the integrated composites can be enhanced for their compatibilities.

References

[1] Lavoine N *et al* 2012 *Carbohydr. Polym.* **90** 735-764.
[2] Bhattacharya D *et al* 2008 *Carbohydr. Polym.* **73**, 371-377.
[3] George J, Sabapathi SN 2015 *Nanotechnol. Sci. and Appl.* **8** 45-54.
[4] Dahlberg J *et al* *Maydica* **56** 1750.
[5] Bambang I, Nana S 2011 *Forum Penelitian Argo Ekonomi* **29** (2) 99-113.
[6] Bakker B *et al* 2013 *Eng. J.* **4** 265.
[7] Yuanita E *et al* 2015 *Procedia Chemistry* **16** 608-615.
[8] Kumar A *et al* 2014 *Mat. Phys. and Chem.* **2** (1) 1-8.
[9] Xue L *et al* 2007 *J. Polym. Environ.* **15** 25-33.
[10] Segal L *et al* 1962 *Text Res. J* **29** 786-794.
[11] Yan Z *et al* *Fuel* **158** 152-158.
[12] Ridzuan M J *et al* 2016 *Mater. and Design* **89** 839-847.
[13] Kabir M M *et al* Cordana F 2013 *Appl. Surface Sci.* **276** 13-23.
[14] Bledzki A K *et al* 2008 *Express Polym. Let.* **2**(6) 413-422.
[15] Tserki V *et al* 2005 *Composites Part A* **36** 1110–1118
[16] Zafeiropoulos N E *et al* 2002 *Composites Part A* **33** 1083.
[17] Uma C *et al* 2012 *Biomat. Bioener.* **46** 555-563

IOP Conf. Series: Materials Science and Engineering **223** (2017) 012058 doi:10.1088/1757-899X/223/1/012058

Performance comparison of plastic shopping bags in modern and traditional retail

F A Radini, R Wulandari, S J A Nasiri, D A Winarto
Agency for Assessment and Application for Technology, Banten, Indonesia

E-mail : fitri.ayu@bppt.go.id

Abstract. Followed by implementation of paid plastic bag policy in Indonesia's modern and traditional retail, community question related to plastic shopping bag performance arise. But, there is limited information about it. Therefore, the assessment of the performance to compare between plastic shopping bags in modern retail and traditional retail should be interesting. The observation performance of plastic shopping bag were weight holding capacity, tear resistant and elongation. This performance were tested using Universal Testing Machine. Physical and physico-chemical properties also identified to determine factor affecting the performance of plastic shopping bag. The physical properties were analysed using visual and thickness gauge to see the colour and measure the thickness. The analysis of physico-chemical properties were carried out using DSC (Differential Scanning Calorimetry), TGA (Thermal Gravimetry Analysis), Furnace and FTIR (Fourier Transform Infra Red Spectroscopy) to identify the materials, also its melting and decomposition temperature. The result showed that the performance difference between modern retail plastic bag with traditional retail plastic bag appears only in the performance of elongation. The elongation of modern retail plastic bag is 121 - 413%, while traditional has 170 - 609%. According to physico-chemical test result, modern retail and traditional retail plastic bag contain polyethylene as main material and has melting temperature in the range of High Density Polyethylene (HDPE) melting temperature. However, modern retail plastic bag has 18.31 - 33.87% of inorganic filler percentage, whereas the traditional retail plastic bag has 0.35 - 9.85%. This inorganic filler percentage probably a contributing factor in the elongation performance difference between modern retail plastic bag with traditional retail plastic bag.

1. Introduction

Plastic bag are functional, lightweight and cheap [1]. Therefore, it used broadly over the retail sector, both in modern and traditional retail as single use plastic shopping bag [2,3]. Based on research done by Yustia (2011), 75% out of 100 responden often receive plastic bag from retailer when shopping [4]. In order to reduce the usage of plastic shopping bag, paid plastic shopping bag policy implemented in selected area of Indonesia's modern and traditional retail since February 2016. Then the question arise in community related to the performance of plastic shopping bag in modern and traditional retail.

IOP Conf. Series: Materials Science and Engineering **223** (2017) 012058 doi:10.1088/1757-899X/223/1/012058

Performance of plastic shopping bag is determined by its function, which is carrying and transporting goods such as food, merchandise or groceries out of the store. Some of the expected performance of a plastic shopping bag are [5] :

1) How much the maximum weight of goods which can be held by a plastic bag (Weight Holding Capacity)
2) How much the maximum load imposed on plastic bags that do not undergo a tear (Tear Resistant)
3) How much plastic shopping bag can be stretch before it breaks (Elongation).

Certain properties can affect the mechanical performance. Plastic bag contain biodegradable polymer , organic filler or inorganic filler will give change in mechanical performance compare to polyolefin bag [6-9]. At present, there is limited information regarding plastic shopping bag performance in Indonesia. The M. Cornelia, et al compare only tensile strength and elongation of several types of plastic shopping bags from domestic market [6]. Considering there is no availability of weight holding capacity and tear resistant data of plastic shopping bags in Indonesia's modern and traditional retail also the need to respond the community question about plastic bag performance, it is interesting to assess overall performance of modern retail and traditional retail plastic shopping bag.

In this research, mechanical performance of modern retail plastic shopping bag and traditional retail plastic shopping bag was assessed and compared. The physical and physico-chemcal properties of plastic bag was also identified to determine factor affecting the performance difference between modern retail plastic bag with traditional retail plastic bag.

2. Methodology

2.1 Sampling Method

This study uses 10 samples of plastic shopping bags from Jakarta and Tangerang. Five samples of plastic shopping bags were taken from modern retail and the others five are from traditional retail. Modern retails that selected for sampling are members of Indonesia Retailers Association (APRINDO) which support the paid plastic shopping bag policy by charges Rp 200,- for a plastic bag. Plastic shopping bag samples from traditional retail are given free of charge to the costumer, also obtained from street vendors and traditional retail traders in Jakarta and Tangerang. Ten samples of plastic shopping bags are shown in Figure 1.

IOP Conf. Series: Materials Science and Engineering **223** (2017) 012058 doi:10.1088/1757-899X/223/1/012058

Figure 1. Ten Samples of Plastic Shopping Bags

2.2 Physical Properties Testing

The physical properties were analyzed using visual and thickness measurement. Visual analysis was used to see the colour. Thickness gauge Preisser Digi-Met was used to measure the thickness of the plastic bags.

2.3 Physico-chemical Properties Testing

2.3.1 FTIR (Fourier Transform Infra Red).

FTIR analysis was done to identify type of polymer on plastic shopping bags. FTIR spectra were collected on Bruker Tensor 27 FTIR using transmission FSF (Free Standing Film) method.

2.3.2 DSC (Differential Scanning Calorimetry).

DSC analysis was performed on Perkin Elmer DSC 8000 to identify melting point of plastic shopping bag. DSC result also used to support FTIR test result in polymer identification. Nitrogen gas was purged at a rate of 20 mL/min during testing. Scanning was carried out in a heated – cooled – heated program at temperature ranging from 30 - 150 °C at 10 °C/min.

2.3.3 TGA (Thermogravimetric Analysis).

Equipment used for measuring decomposition temperature of plastic shopping bag was Netzsch Tarsus TG 209. Temperature rate was 10 °C/min from range of temperature 30 up to 900 °C. Nitrogen rate flown during experiment was 20 mL/min while oxygen flown during temperature 600 - 900 °C with rate 20 mL/min.

2.3.4 Filler Content.

Determination of total filler content was done using ASTM D5630-13 method. This method involves placing 5 g plastic shopping bag samples in a crucible and heating it in an air atmosphere furnace to a temperature 800 °C for 30 min. Residue remaining in the crucible was considered filler unless the residue less than 1 % were typically the result of inorganic compound from additives burnt off.

2.4 Mechanical Performance Testing

The analysis of mechanical performance were done using Shimadzu AGS – 10kN Universal Testing Machine to measure the tensile strength (ASTM D882), elongation and tear resistant (ASTM D1004) at machine direction. The resulted force from tensile strength was used to calculate weight holding capacity of plastic shopping bags. Tensile strength force (Newton) was divided by specimen width (millimeter) then multiplied by total width of four handle in plastic shopping bags (millimeter). After that, the force was convert into mass (kilogram) to obtain weight holding capacity. Speed during tensile strength and elongation testing were 500 mm/min, while 50 mm/min for tear resistant testing. All experiment were conducted at room temperature.

3. Results and Discussion

3.1 Physical Properties

Colour, thickness and dimension of plastic shopping bags from modern retail are shown in Table 1, while Table 2 present colour, thickness and dimension of plastic shopping bag from traditional retail. Plastic shopping bag sample used in this study was consists of small to medium sizes. Based on dimension data from Table 1 and 2, medium sizes of traditional retail plastic bag (plastic C1-E1) has slightly bigger dimension than all of modern retail plastic bag. A probability plot of the thickness of modern and traditional retail plastic bag in Figure 2 a - b showed that both of modern and traditional retail plastic bag have population being sampled are normally distributed. However, the bigger p-value of modern retail plastic bag thickness showed that modern retail plastic bag data is normally distributed better than traditional retail plastic bag [10].

Table 1. Physical Characteristic of Modern Retail Plastic Shopping Bags

Characteristic	A	B	C	D	E
Colour	Milky White	Milky White	Milky White	Milky White	Milky White
Thickness (mm)	0.015	0.011	0.012	0.019	0.009
Width (mm)	270	275	210	300	280
Length (mm)	386	355	265	405	400
Holding Length (mm)	115	130	70	125	130

Based on colour characteristic, plastic A1-B1 of traditional retail plastic bag more clearly or transparent than plastic C1-E1. This probably due to inorganic filler contained in plastic C1-E1, since inorganic filler tend to decrease transparency of polymer [11-12]. For modern retail plastic bag, all of samples tend to be more opaque than traditional retail plastic bag. The better opacity can also be contributed by less branches in chemical structure leading to more cristallinity [11,13]. This suggest that modern retail plastic bag probably have more inorganic content or have less branches in chemical strucutre than traditional retail plastic bag.

Table 2. Physical Characteristic of Traditional Retail Plastic Shopping Bags

Characteristic	A1	B1	C1	D1	E1
Colour	Clearly Blue	Clearly White	Black	Red	Milky White
Thickness (mm)	0.006	0.009	0.033	0.013	0.012
Width (mm)	150	150	400	300	317
Length (mm)	213	234	430	405	425
Holding Length (mm)	80	90	145	125	130

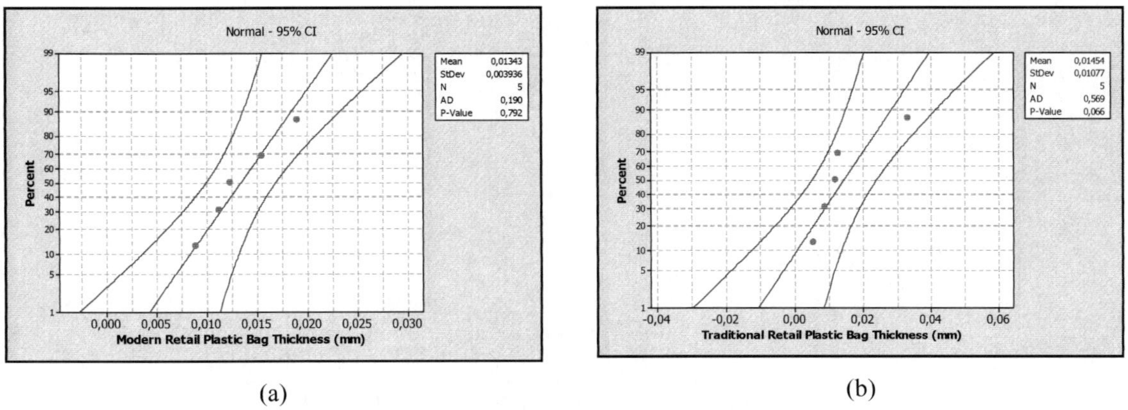

Figure 2. Normal Probabillity Plot of The Thickness of (a) Modern Retail Plastic Bag ; (b) Traditional Retail Plastic Bag

3.2 Physico-chemical Properties

As seen in Figure 3 and 4, all plastic shopping bag samples have polyethylene characteristic peak. It shows antisymmetric and symmetric CH stretching vibration of CH_2 at 2919 – 2934 cm^{-1} and 2850 – 2851 cm^{-1}, CH_2 scissoring vibration at 1466 – 1467 cm^{-1} and 1472 cm^{-1}, CH_2 rocking vibration at 720 – 725 cm^{-1}. FTIR spectrum of traditional retail plastic shopping bags also show peak at around 1370 cm^{-1} which is CH_3 symmetric deformation, and its indicates the presence of methyl end groups and chain branches. Side chain will cause less crystalline, more amorphous structures to occur and lower density [11]. This can be seen that some traditional retail plastic shopping bags less opaque than others. FTIR spectrum of all modern and some traditional retail plastic shopping bag samples show broad peak at around 1466 – 1428 cm^{-1}. The broad peak indicated an inorganic filler [11]. The presence of inorganic filler in FTIR sepctrum of all modern and some traditional retail plastic shopping bag samples were seen by the additional FTIR peak at 876, 1430, 1796 and 2515 cm^{-1} . Also, peak around 720 cm^{-1} in modern retail plastic shopping bags (E, D and B) had shifted to around 710 cm^{-1}. The additional peak is estimated belonging to calcium carbonate peak characteristic, which is :

1) Carbonate peak in 1430 cm^{-1} for C=O stretch and 875 cm^{-1} for C=O bend [11].

2) Calcite (crystalline phase of calcium carbonate) peak in 2512, 1796, 1434 and 712 cm^{-1} [14].

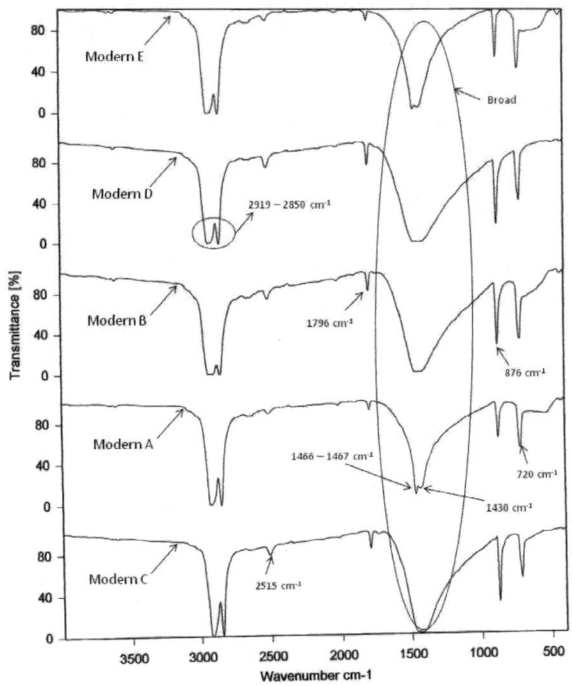

Figure 3. Overlay FTIR spectrum of modern retail plastic shopping bags

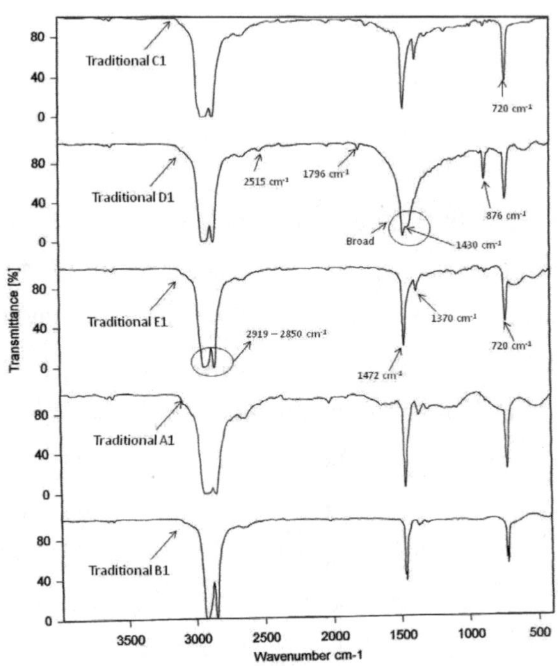

Figure 4. Overlay FTIR Spectrum of Traditional Retail Plastic Shopping Bags

IOP Conf. Series: Materials Science and Engineering **223** (2017) 012058 doi:10.1088/1757-899X/223/1/012058

However, this estimation of inorganic filler need to be confirmed by testing the filler using FTIR and SEM (Scanning Electron Microscope).

Table 3. DSC, TGA and Filler Content Analysis Results of Modern Retail Plastic Shopping Bag

Characteristic	A	B	C	D	E
Melting Temperature (°C)	126.06	126.37	127.91	126.38	126.43
Filler Content (%)	33.87	21.24	18.31	26.93	30.37
Decomposition Temperature (°C)	460.6	*	442.9	468.2	*

* Test is not done

Table 4. DSC, TGA and Filler Content Analysis Results of Traditional Retail Plastic Shopping Bag

Characteristic	A1	B1	C1	D1	E1
Melting Temperature (°C)	127.8	128.53	126.54	127.16	129.23
Filler Content (%)	0.35	0.07	3.28	9.85	2.87
Decomposition Temperature (°C)	*	*	474.4	494.4	448

* Test is not done

The presence of inorganic filler in all modern and some traditional retail plastic shopping bag samples were also indicated by the results of filler content analysis shown in Table 3 and 4. It showed that modern retail plastic shopping bags have more inorganic filler content than traditional retail plastic shopping bags. This is suitable with physical properties and FTIR result, where the modern retail plastic bag has more opacity and less side chain than traditional retail plastic bag.

Table 4 showed that the traditional retail C1, D1 and E1 plastic shopping bag containing inorganic filler. However, only FTIR spectrum of D1 plastic shopping bags that indicates the presence of inorganic filler because it has broad peak at around 1466 – 1428 cm^{-1} and calcium carbonate peak characacteristic (876, 1430, 1796 and 2515 cm^{-1}). This is probably caused by the fewer content of inorganic fillers in C1 and E1 plastic shopping bags.

Although FTIR spectrum of traditional retail plastic bag indicates some chain branching leading to less crystalline, DSC results analysis in Table 3 and 4 showed that melting temperature of all plastic shopping bag samples are in High Density Polyethylene (HDPE) melting temperature range, which is 125 – 137 °C [15]. However, melting temperature range of modern retail plastic bag is slightly lower than melting temperature range of traditional retail plastic bag. This probably due to inorganic filler content of modern retail plastic bag higher than traditional retail plastic bag. The increasing of inorganic filler loading might restrict the crystallization process so the heat required to melting is reduced [16].

Decomposition temperature analysis by TGA were carried out by sampling 6 samples from 10 plastic shopping bag samples. Table 3 showed the modern retail plastic shopping bags had decomposition temperature range 442.9 – 468.2 °C, while Table 4 showed that traditional plastic shopping bags had 448 – 494.4 °C.

3.3 Mechanical Properties

Figure 5 -7 showed the mechanical performace of modern retail and traditional retail plastic bag. According to Figure 5a - 6b, weight holding capacity and tear resistant of modern and traditional retail plastic bag have slightly different range. Weight holding capacity of modern plastic bag is 13 – 23 kg, while traditional retail plastic bag is 13 – 28 kg. Tear resistant of modern retail plastic bag in the range of 3 – 5 N, while traditional retail plastic bag is 2.5 – 7 N . The big difference of performance appears only in elongation. Figure 7a-b showed that elongation of modern retail plastic bag is 121 – 413%, while traditional retail plastic bag is 170 - 608%.

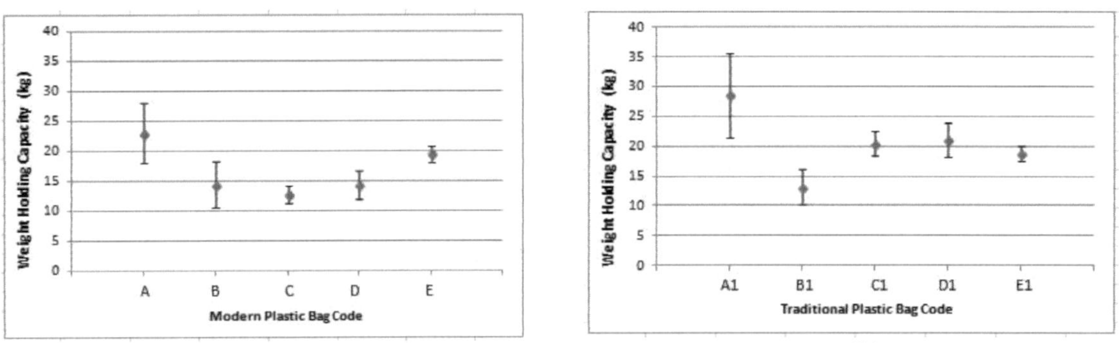

(a) (b)

Figure 5. Weight Holding Capacity of Modern (a) and Traditional (b) Plastic Shopping Bags

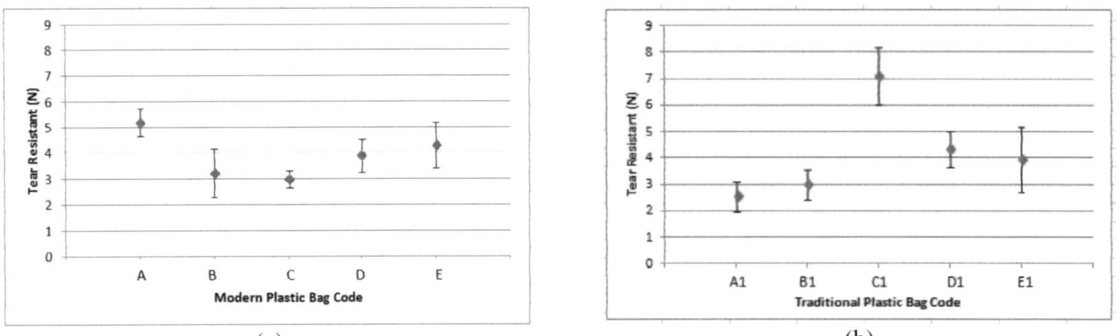

(a) (b)

Figure 6. Tear Resistant of Modern (a) and Traditional (b) Plastic Shopping Bags

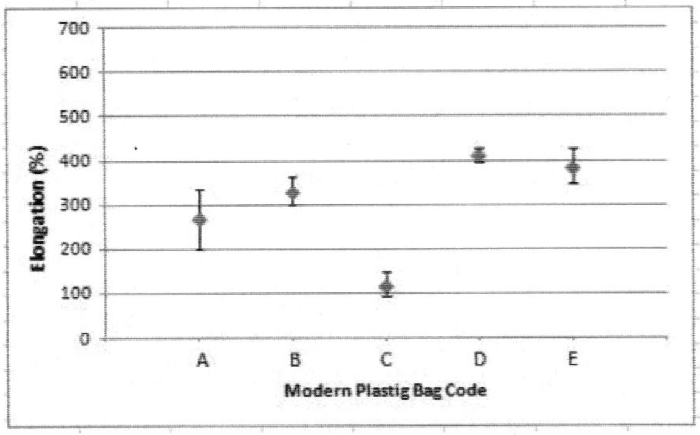

(a)

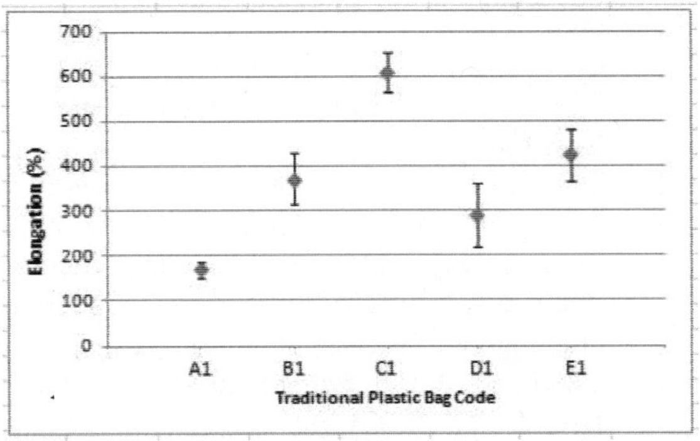

(b)

Figure 7. Elongation of Modern (a) and Traditional (b) Plastic Shopping Bags

Based on the physical and physico-chemical explanation above, the important difference betwen modern retail plastic bag and traditional retail plastic bag relies on inorganic filler content. The inorganic filler content also affect the mechanical performance. Increasing inorganic filler load will improves stiffness of plastic bag hence elongation percentage of modern retail plastic bag lower than traditional retail plastic bag. The slightly difference in tensile strength and tear resistant is probably due to agglomeration of inorganic filler at higher concentration so that lowering the reinforcing ability of filler [7].

According to mechanical performance test result of plastic bag, increasing inorganic filler load in plastic bag will result decreasing mechanical performance. Increasing inorganic filler load more leads to lowering the cost of material by replacing expensive polymer [17].

4. Conclusion

The assessment and comparison performance between ten plastic shopping bags from modern and traditional retail stores in Jakarta and Tangerang were already done. The performance difference clearly visible in elongation performance. Modern retail plastic bag has lower elongation than traditional retail plastic bag. This difference probably caused by higher inorganic filler load in modern plastic shopping bag.

Acknowledgement

This research was funding by Indonesian Center of Excellence (PUI) in 2016. We are thankful to our colleague Arief Rachman Hakim who provided expertise in mechanical testing. We are also grateful to Syuhada as Testing Manager in Sentra Teknologi Polimer that greatly assisted the testing.

IOP Conf. Series: Materials Science and Engineering **223** (2017) 012058 doi:10.1088/1757-899X/223/1/012058

References

[1] Gibson C *et al* 2013 *Household Sustainability: Challenges and Dilemmas in Everyday Life*: Edward Elgar Head, L., & Waitt, G. Household Sustainability: Challenges and Dilemmas in Everyday Life.
[2] Tolinski M 2011 *Plastics and Sustainability: Towards a Peaceful Coexistence between Bio-based and Fossil Fuel-based Plastics* Wiley.
[3] Environment Protection Authorithy (February 2016). Plastic Shopping Bags Option Paper : Practical Action for Plastic Shopping Bags. Retrieved from http://www.epa.nsw.gov.au/publications/waste/160143-plastic-shopping-bags-options.htm
[4] Yustia N 2011 *Perancangan Media Kampanye Mengurangi Penggunaan Kantong Plastik.* JBPTUNIKOMPP Bandung.
[5] Muthu S S and Li Y 2013 *Assessment of Environmental Impact by Grocery Shopping Bags: An Eco-Functional Approach* Springer Singapore.
[6] Cornelia M *et al* 2013 *Adv. Mater. Res.* **789** 300-304.
[7] Khalaf and Moayad N 2015 *J. Saudi Chem. Soc.* **19** 88-91.
[8] Zaman Haydar U and Beg M D H 2014 *Fibers and Polymers* 15 (4) 839-46.
[9] Oragwu Ifeoma P and Igwe Isaac O 2013 *American J. Eng. Res.* **02** (07) 22-27.
[10] Mathews P G 2005 *Design of Experiments with MINITAB* ASQ Quality Press.
[11] Mayo D W *et al* 2004 *Course Notes on the Interpretation of Infrared and Raman Spectra* Wiley.
[12] Elleithy 2010 J. Appl. Polym. Sci. **117** 2413-2421.
[13] De Santis *et al* 2013 *The Scientific World J.* **7.**
[14] Akyuz S 2009 *J Molecular Structure* **924-926** 400-403.
[15] SNI 2012 *Bijih Plastik Polietilena* (7808:2012). Indonesia : National Standardization Agency of Indonesia.
[16] Chafidz 2012 *J. Polym. Res.* **19** 9860.
[17] Xanthos M 2010 *Functional Fillers for Plastics* Wiley.

IOP Conf. Series: Materials Science and Engineering **223** (2017) 012059 doi:10.1088/1757-899X/223/1/012059

Swelling and tensile properties of starch glycerol system with various crosslinking agents

R Mohamed, N Mohd N Nurazzi, M I Siti Aisyah, F Mohd Fauzi
GREEN Polymer Research Group, Faculty of Applied Sciences, Universiti Teknologi MARA (UiTM), 40450 Shah Alam, Selangor, Malaysia

E-mail : rahmahmd@salam.uitm.edu.my

Abstract. Brittle properties of starch had been overcome by the modification process. In this work, sago starch is being modified with variable amount of plasticiser, namely glycerol at 20 and 40% and crosslinking agent had been added to the system. The film of the modification and characterizations of the starch glycerol system with various crosslinking systems were produced by casting method. The film properties of the starch glycerol system were then characterized by tensile strength (mechanical properties) and swelling (physical properties). The modification of the starch glycerol had improved that system by increasing the tensile strength, modulus however lowering its elongation. The increasing in percentage of the water absorption and also swelling are due to the intrinsic hydroxyl groups presence from the starch and glycerol itself that can attract more water to the system. Upon crosslinking, films casted with chemicals namely, glyoxal, malonic acid, borax, PEG were characterised. It was found that, all the film of sago starch crosslinked and undergoing easy film formation. From this modification, borax and malonic acid crosslinking agent had been determined as the best crosslinking agent to the starch glycerol system.

1. Introduction

Sago is a starch extracted from the pith inside stems of the sago palm metroxylon sagu. It is composed primarily of amylose and amylopectin. Several minor components such as lipids, protein and phosphates are also present within the granules but at low concentrations. Amylose is an amorphous polymer linkage with chain lengths of 6,000 D-glucopyranose units and molecular weight between 150,000 and 600,000 Da. Amylopectin is a very large and highly branched polymer linked. There are many types of starch especially for commercial use, they are derived from a variety of cereals like rice, wheat, sorghum, corn and tubers like potato, tapioca, sweet potato and etc. Internationally popular forms of starch are mostly derived from corn and tapioca due to their easy availability. However, tapioca starch content of less amylose than sago starch.

Since sago starch is abundant in Malaysia, it can be as an alternative source of material with lower cost [1]. Starch glycerol and system can create variety of products in various field including packaging, medicinal, food and pharmaceutical industries. However, the drawback of starch is having the brittle properties. In order to overcome this problem, the modification process should be used by adding with glycerol as a plasticizer. It can be defined by two purposes, which are to improve the flexibility, extensibility and overcomes the film brittleness. Film forming capacity of different chemical modified corn starches was evaluated by Lopez *et al.,* (2010). The effect of glycerol as plasticizer on film properties depend on its concentration, low film solubility in water and a more compact structure than those of unplasticized films. In their findings, mechanical behaviour of plasticized acetylated starch films depends on glycerol concentration, being rigid and brittle than the unplasticized ones, more ductile and became very flexible for higher plasticizer content) [2].

Starch had been modified for emulsifiers, as food thickeners while as hydrogels, super absorbents, edible films, adhesive, film forming and packaging in biodegradable packaging industries also exploited modification of starch technology to develop products of high performance. In packaging, interest in environmental friendly products has created a significant market opportunity for starch. The goal has been to increase the amount of starch in thermoplastic composites designed for various packaging materials [3, 4].Modification of starch are well established using different types of starches and types of crosslinkers and agents [5]. Chemical modifications of starch generally involve esterification, etherification or oxidation of the available hydroxyl groups on the B-D-glucopyranosyl units that make up the starch polymers. Reactions used to produce most commercially modified starches have been reviewed by [6, 7].

Starch that are unplasticized are rigid and brittle [8]. The strength of starch with glycerol is quite low especially in aqueous solution. In order to improve the strength of the film, crosslinker can be added in the system. Crosslinking is common approach to improve the performance of starch. Crosslinking agent concentration, starch content, pH, treatment temperature and treatment time are factors that affect crosslinking. In order to crosslink polymers containing hydroxyl groups, such as starch, the crosslinking agent must be able to react with at least two hydroxyl groups in a single polymer molecule. Crosslinking density affects the gel content and water absorption or swelling of starch molecules. Not only types of crosslinkers and amount of crosslinkers used, the synthesis techniques and pH media with incorporation of other additives and temperature affect final product performances

According to Singh and Nath (2012) had synthesised and evaluated the physicochemical properties of crosslinked sago starch. The properties were investigated using scanning electron micrograph (SEM), x-ray powder diffractometer (XRD) and thermogravimetric analysis (TGA). It was revealed that crystalline nature of native sago starch was transformed after cross-linking. TGA report exhibited higher thermal stability, which makes it suitable for various industrial applications [9].

There are many types of crosslinking agents that are well established. However in this work the use of selected crosslinkers and chemicals namely malonic acid, borax, glyoxal, dicyandiamide (DCN) and poly(ethylene glycol)(PEG) have been investigated. The different types of crosslinking agent can be analysed according to its ability to crosslink or enhanced flexibility of the polysaccharide chain.

IOP Conf. Series: Materials Science and Engineering **223** (2017) 012059 doi:10.1088/1757-899X/223/1/012059

2. Experimental

2.1 *Materials*

Sago starch (SS) used in this research was supplied by Bor Giap Sdn Bhd of Batu Pahat, Johor. Sago starch supplied by has specification composition states its moisture content in the sago starch varied between 10.6 and 16.0%, ash about 0.1% maximum, fiber about 0.2% maximum while the amylose content varied between 24 and 30%. Its pH is within 5.5 to 6.5. Glycerol, crosslinking agent; malonic acid, glyoxal, was supplied by MERCK (M) Sdn Bhd, while dicyandiamide (DCN), borax, Poly (ethylene glycol) (PEG) was supplied by SYSTERM (M) Sdn Bhd.

2.2 *Preparation of Samples*

There are two steps of the process:

Step I: Gelatinization of starch

10g of SS was weighed with analytical balance and placed in a 150mL beaker. SS then added with 20% of glycerol that act as the plasticizer based on the weight of SS. The mixture was then added with 100mL of distilled water at 90-100°C to gel the sago/glycerol system. Mixture is kept agitated in water bath at 80°C until it turns sticky. Glycerol SS is then added with crosslinking agent, at 2% based on the weight of SS/glycerol; similar procedure was followed with other crosslinking agents. The procedure is repeated for 40% glycerol and crosslinking done with malonic acid, glyoxal, dicyandiamide, borax and PEG.

Step II: Casting Process

After the gelatinization of the starch, films were fabricated by solution casting method in the aluminium moulds that have been provided. Then, the samples prepared put in the oven about 80°C in order to form film samples.

2.3 *Characterizations*

Swelling test was performed by immersed the samples in deionized water at room temperature. The initial thickness of all the samples was taken and recorded. The swelling properties of the starch films observation was done within ten days.

$$\text{Swelling } \% = 100 \times (\text{Final thickness} - \text{Initial thickness})/(\text{Initial thickness}) \qquad (1)$$

Tensile properties were performed by using Instron Universal Testing Machine (UTM). The samples were cut with 50mm length and 4.025 mm width of dumbbells shape by using the dumbbells cutter.

3. Result and Discussions

3.1 Swelling Properties

Figure 1 shows the swelling percentage of samples with different crosslinker and percentage of glycerol. From the Figure 4, it found that the malonic acid with 40 % glycerol has the highest swelling percentage at 34.09 % and followed by borax at 30.56 %.

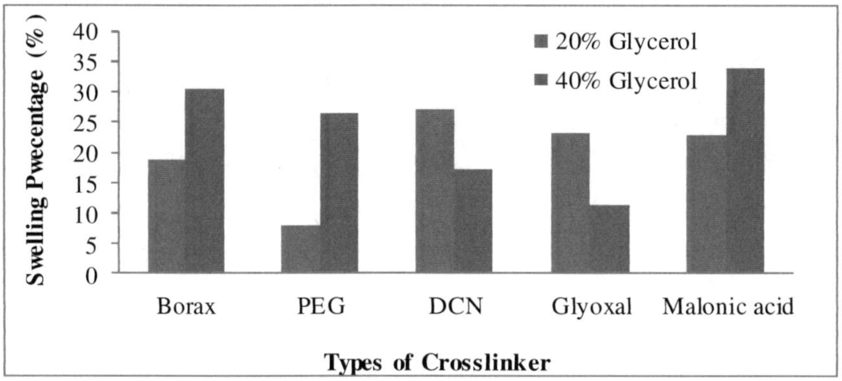

Figure 1. Graph of swelling percentage (%) of starch films with different crosslinker and percentage of glycerol.

According to Mali *et al.*, (2005), the addition of plasticizer provides more active sites by exposing its hydrophilic hydroxyl groups in which the water molecules could be adsorbed. Besides that, higher levels of plasticizer increased the films moisture affinity and these results could be attributed to the hydrophilicity of the plasticizers, which presented hydroxyl groups capable to interact with water by hydrogen bonds. Eventhough the glycerol molecules are small, they present high capability to interact with starch chains hence enhancing the molecular mobility. This increases the free volume in the film matrix with consequence of increases the water molecules absorbed [8].

3.2 Tensile Properties

Figure 2 shows the tensile strength with various crosslinker and different percentage of glycerol. From both percentage of glycerol, the starch film with 40% of glycerol crosslinked by malonic acid shows the highest tensile strength. High in tensile values indicates the films are stronger with high strength by greater interaction and bonding. According to Mali *et al.*, (2005), the unplasticized starch film represents low value of tensile. According to Sangeeta (2007), the tensile strength will increased due to the crosslinking and/or modification by glycerol [8, 10].

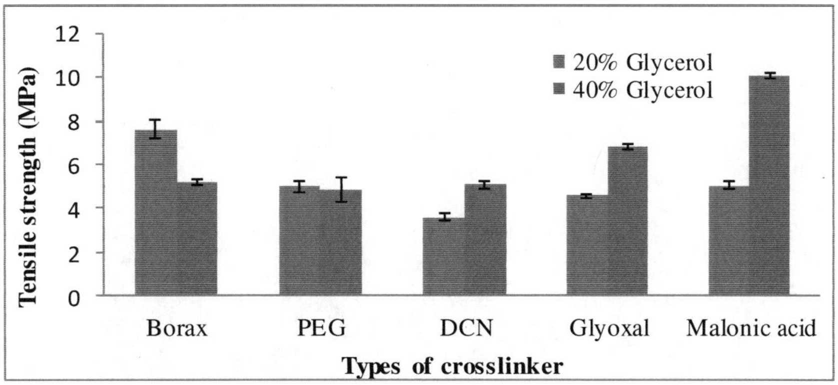

Figure 2. Tensile strength of starch film with various crosslinker.

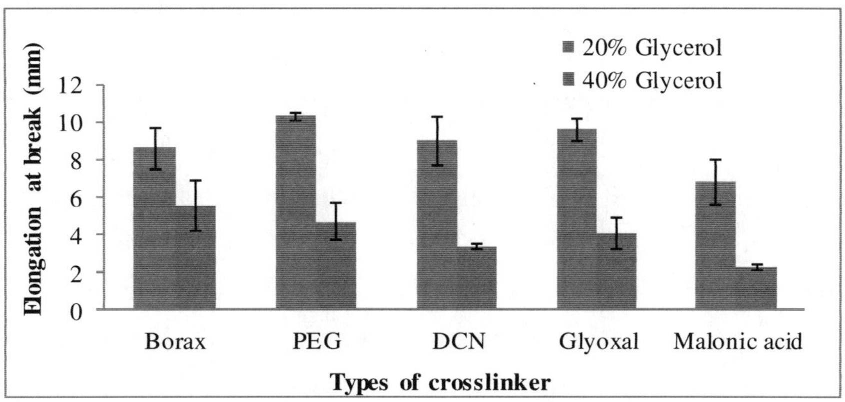

Figure 3. Elongation at break of starch film with various crosslinker.

From Figure 3, about similar trend were observed for 20 and 40 % addition of glycerol. Film at 20% of glycerol crosslinked with PEG crosslinker has the highest elongation with 10.29 mm and followed by the glyoxal crosslinker with 9.59 mm of elongation. However, film at 40 % of glycerol crosslinked with borax crosslinker has the highest elongation with 5.51mm of elongation and followed by PEG crosslinker with 4.64 mm of elongation, while film that crosslinked by malonic acid still the lowest elongation at break as both 20 and 40% of glycerol addition with 6.77 and 2.22 mm of elongation respectively.

According to Lopez *et.al.*, (2008), unplasticized starch is brittle and by the addition of the plasticizer (glycerol), tensile strength was decreased but increase in the elongation at break. This is due to the plasticizer interfering with the polymeric chain associating their slipping and thus enhances the film flexibility [2]. However Malonic acid decreases in flexibility of starch/glycerol system while PEG showed opposite trend. Effect of glycerol on physical and mechanical properties of wheat starch edible films had been studied by Farahnaky *et al.*, (2013), they found glycerol did not change X-ray patterns of starch films; however, the degree of crystallinity were reduced [11].

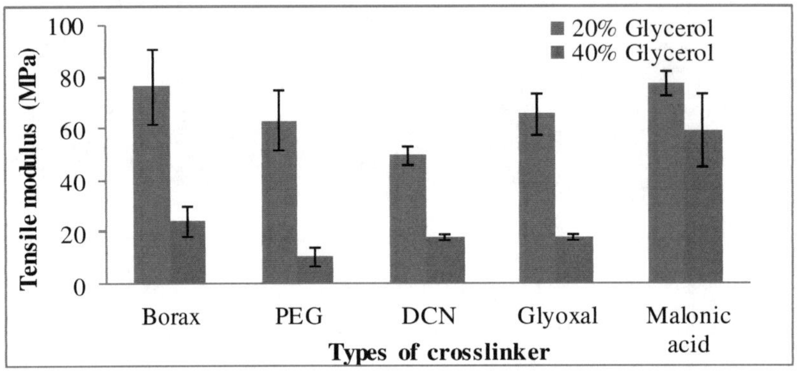

Figure 4. Tensile modulus of starch film against varies of crosslinker.

Figure 4 shows the film crosslinked by malonic acid give the highest rigidity through tensile modulus. When compared 20 % of glycerol with the 40% of glycerol, film crosslinked by borax has reduced its tensile modulus by 300 % or three times reduction upon increasing glycerol content from 20 to 40 %. Film that crosslinked by malonic acid does not affect significantly on starch modulus compared to film with borax crosslinker. It only decreased by 23 % from 20 to 40 of glycerol.

With glycerol content, the strain at break increased considerably, however the breaking stress and elastic modulus decreased. However, with an increase in the glycerol content, was shown to results in decrease of the tensile modulus as happens with plasticizers used with synthetic polymers in general (Briassoulis (2004)) [12]. When considering on the ability of the plasticizer, according to Mali *et al*., (2005), the plasticizer can increase the film flexibility due to their ability to reduce internal hydrogen bonding between polymer chains while increasing molecular space. The most effective plasticizers will generally resemble most closely the structure the polymer they plasticized, thus, the most commonly plasticizers used in starch-based films are polyols, such as sorbitol and glycerol. They avoid cracking of the film during handling and storage, affect of gas, water vapor and solute permeability and sorption characteristic [8].

4. Conclusions

In conclusion, SS/glycerol system with various crosslink agents had increased tensile, modulus but decrease in elongation; meaning SS system is less flexible. Considering the characterization from swelling test, the highest percentage of swelling for 20% glycerol with a crosslinker DCN and 40% glycerol with a crosslinker malonic acid. Higher levels of plasticizer increased the films moisture affinity and these results could be attributed to the hydrophilicity of the plasticizers, which presented hydroxyl groups capable to interact with water by hydrogen bonds. While the best crosslinking agent which highest performance in terms of tensile and modulus are borax and malonic acid. Lastly, it can be conclude that, crosslink agent can be easily modified to obtain desired strength, rigidity for specific drug delivery application based on the variable of glycerol percentages and also variable types of crosslinker of =C–H out of plane bend is appeared.

Acknowledgements

The authors thank Faculty of Applied Sciences (FSG), Universiti Teknologi MARA (UiTM) Shah Alam, Malaysia for providing good facilities. Highly appreciation is conveyed to those involved in completing this work.

References

[1] Ahmad F B and Williams P A 1998 *Journal of Agricultural and Food Chemistry* **46** 4060-4065.
[2] López O V *et al* 2008 *Carbohydr. Polym.* **73** 573-581.
[3] Doane W M 1994 *Cereal foods world (USA)* **39** 556-563.
[4] Poutanen K and Forssell P 1996 *Trends in Polymers Science* **4** 128-132.
[5] Chiu C and Solarek D 2009 *Starch: Chemistry and Technology* **3** 629-655.
[6] Shah U *et al* 2016 *Comprehensive Reviews in Food Sci. and Food Safety.*
[7] Ashwar B A *et al* 2016 *Food Hydrocolloids* **56** 108-117.
[8] Mali S *et al* 2005 *Carbohydr. Polym.* **60** 283-289.
[9] Singh A V and Nath L K 2012 *Int. J. Biological Macromol.* **50** 14-18.
[10] Sangeeta G and Asim Kumar J 2007 *Eur. Polym. J.* **43** 3976-3987.
[11] Farahnaky A *et al* 2013 *J.Texture Studies* **44** 176-186.
[12] Briassoulis D 2004 *J. Polym. Environment* **12** 65-81.

IOP Conf. Series: Materials Science and Engineering **223** (2017) 012060 doi:10.1088/1757-899X/223/1/012060

Analysis physical properties of composites polymer from cocofiber and polypropylene plastic waste with maleic anhydrate as crosslinking agent

E Pelita[1], T R Hidayani[1], A Akbar[2]
[1]Analysis Chemistry, Polytechnic of ATI Padang,Bungo Pasang Tabing, Padang, 25147, Indonesia
[2] Chemical Engineering, Polytechnic of ATI Padang,Bungo Pasang Tabing, Padang, 25147, Indonesia

E-mail : rachmihidayani@yahoo.com

Abstract. This research was conducted with the aim to produce composites polymer with polypropylene plastic waste materials and cocofiber which aims to produce wood replacement material in the home furnishings industry. This research was conducted with several stages. The first stage is the process of soaking coco fiber with detergent to remove oil and 2% NaOH. The second stage is to combine the polypropylene plastic waste with cocofiber is a chemical bond, modification by adding maleic anhydride as a crosslinking agent and benzoyl peroxide as an initiator each as much as 1%. Mixing materials done by reflux method using xylene solvent. In this study, carried out a wide range of weight variation of coco fiber are added to the 10, 20, 30, 40 and 50%. The third stage is a polymer composite molding process using hot press at a temperature of 158°C. The results of polymer composites Showed optimum condition on the addition of 40% cocofiber with supple tensile strength value of 90.800 kgf /cm^2 and value of elongation break at 3.6726 x 104 (kgf/cm^2), melting point at 160.02°C, burning point 463.43°C, residue of TGA is 19%, the density of 0.84 g/mL. From these data, conclude that the resulting polymer composites meet the SNI 03-2105-2006 about ordinary composite polymer and polymer composite structural type 8 regular types from 17.5 to 10.5.

1. Introduction

Wood needs increase rapidly while availability decreases, especially for furniture, so it is necessary to find a solution scarcity of wood [1]. Today the wood plastic composite (Wood-Plastic Composite) is one of the most dynamic sectors of the plastics industry. This material consists of a mixture of wood fibers or the like with a polymer that is a thermoplastic such as polyethylene (PE), polypropylene (PP) and so on [2]. Thermoplastic polymer would be soft when heated and hardens when cool. These properties allow other materials such as wood particles or the like can be mixed with this type of plastic to form a composite material [3]. Composite is a material system consist of mixture or combination from two or more different macro constituents that are not mutually dissolve [4]. Composite consist of matrix and reinforcement. Matrix is the major constituent of composite while the minor constituent called reinforcement [5]. PP is a plastic that is solid, hard, and water-resistant, which is difficult to be degraded naturally, and has a strong potential of environmental pollution [6]. The utilization of plastic waste is an effort to surpress the plastic waste in extent to save resources and reduce dependence on imported raw materials. The utilization of plastic waste devided into reuse and recycle. Almost all types of plastics waste (80%) can be recycle and reuse back into the original item, although it must be mixing with new raw materials and additives to improve its quality [7]. The utilization of polypropylene waste for biodegradable plastic has been done with the addition of starch from durian seed [8]. Utilization of plastic to be particle board was done [9] which utilizes sawdust from the forest with a mixture of plastic, particle board produced has less than maximum adhesion between the plastic with wood shavings. Coco fiber is a substantial part of the coconut fruit, namely 35% of the total weight of the fruit. Coconut fiber consist of fiber and cork conecting one fiber with other fibers. Each coconut contains 525 g of fiber (75% of the coco fiber), and cork 175 g (25% of coco fiber) [10]. Today coco fiber is only used as a medium of plants (cocopeat) and compost. The use of composite boards for home furniture which are widely used by community such as *Olympic* and *Ligna*, generally using a water-soluble adhesive, so the product can easily be damaged if exposed to water, especially for the people who live in coastal areas [11]. Manufacture of composite particle board by using matrix of recycled plastic, in addition to improving the efficiency of wood utilization that can be substituted for other fibers, can also reduce the environmental burden from plastic waste and delivering innovative products as a substitute for wood furniture. The advantages of this product include: cheaper production costs, abundant raw material, flexible manufacturing process, low density, biodegradable (rather than plastic), better properties than the raw material, as well as can be recycled [12]. The advantage of wood plastic composites compared with natural wood is consistency and uniform shape, not decayed and eaten by insects, not absorb water and not require periodic painting.

From all the above issues then sought an alternative to particle board of wood plastic composite that has a good mechanical strength, does not absorb water, anti-microbial, abundant raw materials and readily available, inexpensive and environmentally friendly [13]. Therefore, the researchers want to make particle board from plastic waste PP and coconut fiber with a cross linking agent maleic anhydride and benzoyl peroxide as initiator. The use of maleic anhydride as a cross linking agent aims to produce chemical interaction between coco fiber with plastic so that it can improve physical strength composite board [14]. The use of maleic anhydride as a cross linking agent with the aid of benzoyl peroxide initiator has been done in the research manufacture of biodegradable plastics from starch durian seeds and waste polypropylene plastic with optimum composition of maleic anhydride and benzoyl peroxide as much as 1% [15, 17].

From all the above issues then sought an alternative method to make particle board of wood plastic composite that has a good mechanical strength, does not absorb water, anti-microbial, raw materials are plentiful and readily available, inexpensive and environmentally friendly. Therefore, the researchers want to make particle board manufacture from plastic waste from the packaging of mineral water with coconut coir fiber which is the result of a major agricultural region of Pariaman, West Sumatra. PP plastic waste powder preparation was conducted by reflux with the solvent xylene, coco fiber preparation by washing with water to remove the oil, and soaking with 2% NaOH to remove lignin and carbohydrates [13]. Mixing materials done by the reflux method and crosslinking agents maleic anhydride and benzoyl peroxide initiator and solvent xylene. The use of maleic anhydride as a crosslinking agent aims to produce chemical interaction between coco fiber with plastic so that it can improve physical strength composite board. The use of maleic anhydride as a crosslinking agent with the aid of benzoyl peroxide initiator has been done in the research manufacture of biodegradable plastics from starch durian seeds and waste polypropylene plastic with optimum composition of maleic anhydride and benzoyl peroxide as much as 1% [11]. The particle board printing method using a hot press. To know the characteristics of particle board conducted tests on physical properties include density, mechanical properties include tensile strength, modulus of elasticity, DTA and SEM. The resulting composite board is expected to have strong physical properties, it can be a composite board as the basic material of home furnishings at a cheap price compared to particle board of wood powder.

2. Materials and Method

2.1 Equipment and Materials

The materials used in this study are as follows: Waste plastic polypropylene from mineral water packing, Coconut Fiber (agricultural product waste from Pariaman, Sumatera Barat), NaOH, Maleic Anhydride PA (Peekay Agencies Pvt. Ltd) , Benzoyl Peroxide (Merck), Methanol (Merck), Xylene (Merck). The equipment used in this study are as follows : DTA and TGA DT-30 Shimatzu Japan, Universal testing Machine tool sets, tool sets hot-press Type: IL.70.110 / 220V.

2.2 Procedure

2.2.1 Treatment on Coconut to Extract Cocofiber

Cocofiber is separated from the hard, cut or chopped to form short fibers and weighed as much as 2.5 kg, then soaked with detergent for ± 2 h to remove the oil content. Then the fiber is washed and dried under sun burned. Cocofiber is separated from the foam (cork) with a metal brush. Furthermore, weighed 80 g NaOH and dissolved into 5 L of distilled water and stir until dissolved NaOH. 2% NaOH solution was poured into coco fiber and stir until completely submerged and left for 24 h to eliminate the levels of carbohydrates in fiber, after the immersion of coco fiber - NaOH washed with clean water several times until the washing water is clear. Then coco fiber is dried in an oven at a temperature of 50-60°C blower to dry, then cut into small coco fiber size of 1-1.5 cm and is ready for use [14].

2.2.2 Grafted Polypropylene Plastics Waste and mixing with Coconut Fiber

Polypropylene plastic waste to be cleaned using water mixed with detergent, rinsed with water and then dried in the sun. Then PP plastic waste that has a clean cut to size ± 0.5cm x 0.5cm. Waste plastics PP, maleic anhydride 1% and 1% benzoyl peroxide is refluxed using xylene solvent of 500 mL for 2 h at 165°C, to produce PP-g-MA in a solvent mixture of xylene. Coco fiber is inserted into the flask reflux and then refluxed for 15 min and then precipitated with 100 mL of methanol. A mixture of PP-g-MA and coconut coir fiber (polyblend) is dried in the oven at a temperature of 80°C.

2.2.3 Pressing Composites Polymer With Molding Method

Compression molding tool temperature is set at a temperature of 158°C, further polyblend placed into a mold in a frame made of steel with the size of the 25.0 cm x 15.0 cm x 1.0 cm where the steel plate was coated with aluminum foil. Polyblend sample is then placed between two heating to a temperature of 163°C for ± 10 min without pressure from the tool. Then proceed with the pressure bit by bit for ± 20 min. After it was given a maximum pressure of 40 bar (4 MPa) for ± 20 min. Then the hot-press switched off and allowed to cool to a temperature of ± 90°C [3]. Furthermore, the samples are removed from the hot-press and put into the water and released from the mold, then allowed to reach room temperature. Composites polymer ready to be cut in accordance with the needs of characterization.

2.3 Characterization

2.3.1 Mechanical Test Analysis

Specimens particle board with a thickness of 1 cm cut into specimens for tensile test. Tensile strength testing by ASTDM D-1822 type L with a load of 100 kgf and a rate of 50 mm/min with a specimen thickness of 1 cm. Initially turned *Torsee*'s Electronic System and allowed to stand for 1 h. Specimens clipped to tensile test equipment then the strain, stress and the unit determined by using griff. Then the recorder is turned on, pressed the start button and the value of the stroke and the load is made under conditions of zero. Record the value of load and stroke when the sample broke up. Do the same treatment for each sample. Rated load and stroke are acquired, used to calculate the value of tensile strength and elongation of the specimen [7].

2.3.2 Thermogravimetric Analysis

Thermogravimetic analysis (TGA) studies on the samples were carried out on Q-500 series TGA Analyzer (TA Instruments, USA). The weight of each sample was 5–10 mg and temperature range was 35–750 °C with a heating rate of 20 °C/min. An inert atmo- sphere was maintained by the continuous flow of nitrogen at 50 mL/min [6].

2.3.3 Differential Thermal Analysis (DTA)

Differential thermal analysis (DTA), which measures the difference in temperature (T), between the referent sample with inert material as a function of temperature. DTA can be used as a characterization tool or material analysis. In a sample whose identity is not known, the use of DTA alone will not be much help in identification. In the field of polymer blends (polyblends) observation glass transition temperature (Tg) is very important to predict the interaction between the chain and the mechanisms of blending polymers. Homogeneous polymer blend will show a single sharp Tg peak (exothermic) as a function of the composition. Tg mixture usually located between Tg of both components, because the homogeneous mixing is used to lower the Tg, as well as a plasticizer with a liquid plasticizer [18].

3. Result and Discussion

3.1 Analysis of Tensile Strength and Modulus of Elasticity Bending

Analysis of tensile strength and modulus of elasticity with different variations of filler composition is an important factor for determining the mechanical properties of the desired material. From the data Table 1 can be seen that the variations of the composition of the particle board that meets the standards of particle boards SNI 03-2105-2006 [16] are PP waste-g-MA : Coco fiber (90:10), (80:20), and (60:40). The rest of variation not meet the standards. This is probably caused by the value of the tensile strength of the resulting particle board.

Table 1.The result of Analysis of Bending Modulus and Modulus of Elasticity Bending.

No	Ratio (Composition and mass) (%)	Load (kgF)	Deflection (cm)	Tensile Strength(kgf/cm^2)	Modulus of Elasticity(kgf/cm^2) (10^4)
1	PP Waste (100)	7.397	0.058	118.352	3.136
2	PP Waste-g-MA:Coco fiber (90 : 10)	7.165	0.062	114.640	2.476
3	PP Waste-g-MA : Coco fiber (80 : 20)	6.300	0.058	100.800	2.693
4	PP Waste-g-MA : Coco fiber (70 : 30)	6.079	0.033	97.264	0.412
5	**PP Waste-g-MA :** Coco fiber **(60 : 40)**	**5.675**	**0.054**	**90.800**	**3.673**
6	PP Waste-g-MA : Coco fiber (50 : 50)	1.675	0.155	26.800	2.047

PP Waste-g-MA : Coco fiber (60 : 40) has a value of tensile strength 90.8 kgf/cm^{2}and modulus of elasticity $3.6726x10^4$ kgf/cm^2 that meet the standards SNI 03-2105-2006 [16] about general particle board type 8 and general particle board structural type 17.5 -10.5. The other types did not reach the optimum value because less of mixture homogenity. On a particle board with other comparisons, a low

IOP Conf. Series: Materials Science and Engineering **223** (2017) 012060 doi:10.1088/1757-899X/223/1/012060

mixing rate may occur because of the ratio of the composition between the matrix with the unequal filler, or more dominantly the syntethic polymer (polyprophylene).

3.2 Heat Resistance Analysis

Table 2. Result of DTA Analysis Composite Board

No	Ratio (Composition and Mass) (%)	Melting Point (°C)	Flash point (°C)
1	PP Waste (100)	164.62	462.03
2	PP Waste-g-MA : Coco fiber (90 : 10)	163.21	460.55
3	PP Waste-g-MA : Coco fiber (80 : 20)	162.43	461.65
4	PP Waste-g-MA : Coco fiber (70 : 30)	162.09	462.65
5	PP Waste-g-MA : Coco fiber **(60 : 40)**	**160.02**	**463.43**
6	PP Waste-g-MA:Coco fiber (50:50)	161.47	462.19

Analysis of thermal properties provides information about physical changes in particle board, for example, the melting point, the chemical process that includes polymerization, degradation and decomposition. The result show that the addition of coco fiber raise the flash point of particle board due to mixing with other compounds in this case coco fiber. The melting point of board particle after mixing with coco fiber is decrease compare with before mixing. The change of melting point and flash point of board particle from the initial state indicate that the interaction between waste PP and coco fiber. Expected positive result is the addition of coco fiber as filler does not significantly change the initial properties of polypropylene, but produces an increase in the value of the flash point. The optimum condition of board particle can be seen in the variation of waste PP : coco fiber (60 : 40) wherein the melting point is 160.02 °C and the flash point is 160.02°C while the other variations did not reach the optimum value caused by less of homogenity. The level of homogeneity of particle board based on the data that shows the SEM image of particle board electromagnetic plastic mixture with coconut coir (Figure 1-6).

3.3 TGA Result

Table 3. Result of TGA Analysis Composite Board

No	Ratio (Composition and Mass) (%)	Decomposition point (°C)	Residue (%)
1	PP Waste (100)	471.38	0.01
2	PP Waste-g-MA:Coco fiber (90:10)	470.59	2.82
3	PP Waste-g-MA : Coco fiber (80:20)	471.00	5.46
4	PP Waste-g-MA : Coco fiber (70:30)	471.12	4.68
5	PP Waste-g-MA : Coco fiber **(60:40)**	**471.89**	**19.00**
6	PP Waste-g-MA : Coco fiber (50:50)	472.03	14.14

TGA is typically used in research to determine the characteristics of materials such as polymers, temperature decreases, absorbed material content, inorganic and organic components in materials, decomposition of explosive materials, and solvent residues. In the mixing of polymeric materials of the natural polymer type and synthetic polymer, the expected percentage of residual carbon atoms is greater than 10% to show the involvement of natural polymers in the mixing process.

From the TGA result can be seen that the more coco fiber added so the residue also increased. The optimum result can be seen in variation PP waste-g-MA : Coco fiber (60 : 40) with 19 % residue. In variation PP waste : Coco fiber (50 : 50) has the lower residue. This is caused by plastic waste polypropylene which minimum number cut into short chain, so that the fibers bonded to plastic, thus lowering the value of residuals.

3.4 SEM Result

SEM analysis is used in order to homogeneity between a mixture of fibers and plastics can be seen. any variation blend homogeneity can be seen from the surface structure using SEM so that later can be seen best from mixing composition. The result indicate that increasing the number of coco fiber is added to the plastic PP waste, so the structure more evenly.

Figure 1. PP waste 100%

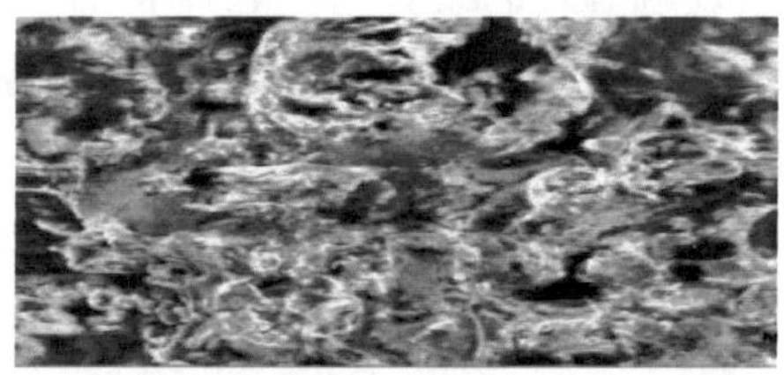

Figure 2. PP waste-g-MA : Coco fiber (90:10)

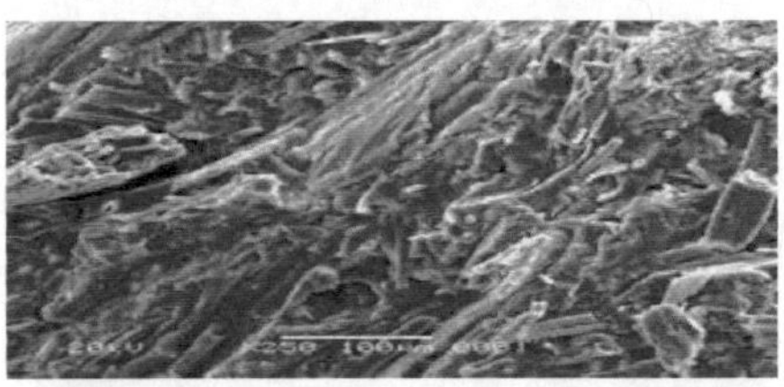

Figure 3. PP waste-g-MA : Coco fiber (80:20)

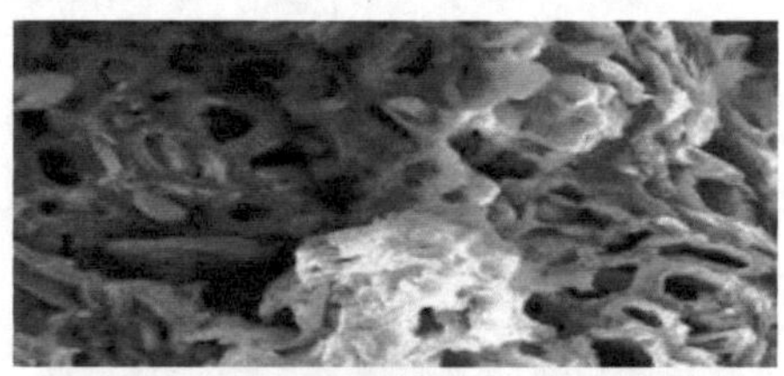

Figure 4. PP waste-g-MA : Coco fiber (70:30)

Figure 5. PP waste-g-MA : Coco fiber (60:40)

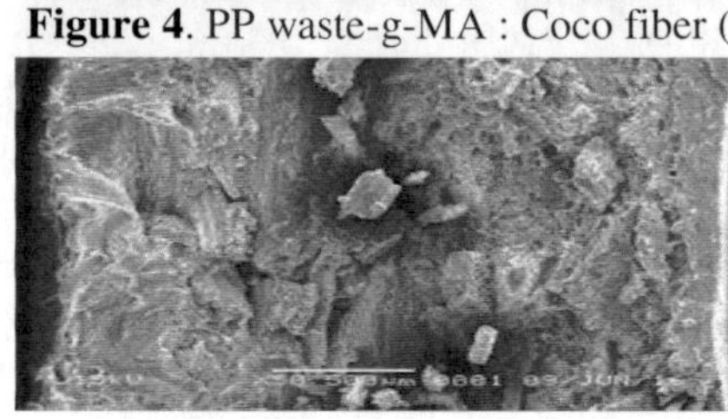

Figure 6. PP waste-g-MA : Coco fiber (50:50)

This happens because of their chemical interactions between synthetic polymers and natural polymers with the addition of maleic anhydride. n Particle Board with the addition of coconut fiber can be

IOP Conf. Series: Materials Science and Engineering **223** (2017) 012060 doi:10.1088/1757-899X/223/1/012060

differences in color and structure of the SEM results due to the addition of coconut fiber. However, the structure of the SEM results Particle Board unnoticeable among coconut coir fiber with plastic waste (yield homogeneous) with the help of maelat anhydride as a crosslinking agent between coconut coir dust and waste polypropylene plastic.

This is in accordance with the explanation [8] that the addition of maleic anhydride as a crosslinking agent in plastics waste polirpopilena with coconut coir fiber, preventing the formation of hydrogen bonds between coco fiber and causes the surface properties of coco fiber and polypropylene waste is becoming more homogeneous. The addition of maleic anhydride as crosslinking agents facilitate direct contact between coco fiber and matrix waste polypropylene, as well as increasing the spread of the matrix phase when the entire fiber has been covered by a layer of polypropylene matrix materials.

3.5 Density Analysis Composite Polymers

Table 4. Density Analysis Result

No	Ratio (Composition and Mass) (%)	Density (g/cm^3)	Water Content (%)
1	PP Waste (100)	0.8713	0.1527
2	PP Waste-g-MA : Coco fiber (90:10)	0.82793	0.6610
3	PP Waste-g-MA : Coco fiber (80:20)	0.80659	0.7342
4	PP Waste-g-MA : Coco fiber (70:30)	0.85819	0.9786
5	**PP Waste-g-MA : Coco fiber (60:40)**	**0.83631**	**1.1387**
6	PP Waste-g-MA : Coco fiber (50 : 50)	0.8499	1.590

The addition of coconut fiber to the matrix of PP waste result in a decrease density, but the value of the decrease is not too large. From the density test showed that the entire value of the density of the resulting polymer composites meet SNI 03-2105-2006 where values ranging from 0.40 - 0.90 g/cm^3, while the density value of the resulting polymer composite is between 0.8 g/cm^3.

Table 4 also shows the effect of fiber on water content. The result indicate that the addition of coconut fiber will increase the percent moisture content of polymer composites. This happens because the coco fiber has the large pores so it can absorb water well. Water content meet the requirements SNI 03-2105-2006 [16] about water content in the polymer composites that no more than 14%, where in the table can seen the water content in the resulting polymer composite is in the range between 0.1 to 1.6%.

4. Conclusion

From this research it has been found a good particle board composition, in order to overcome the problems of increasingly high price of wood for home furnishings materials, while utilizing waste plastics and waste oil. Particle board produced has the advantage of light weight properties, resistance to hot water and good mechanical properties that meet the standards set on particle board

The results of polymer composites showed optimum condition on the addition of 40% cocofiber with supple tensile strength value of 90.800 kgf / cm^2 and value of elangation break at 3.6726 x 104 (kgf/cm^2), melting point at 160.02°C, burning point 463.43°C, residue of TGA is 19%, the density of 0.83631 g/cm^3. Reviews from these data, conclude that the resulting polymer composites meet the SNI 03-2105-2006 about ordinary composite polymer and polymer composite structural type 8 regular types from 17.5 to 10.5.

References

[1] Talens P *et al* 2016 *J Food Hydrocolloids* **56** 9-19.
[2] Parra D F *et al* 2016 *J Radiation Phys. Chem.* **118** 11-15.
[3] Bhunia H *et al* 2016 *J Radiation Phys. Chem.* **123** 37-45.
[4] Vasile C *et al* 2016 *J Radiation Phys. Chem.* **125** 134-144.
[5] Rhee K Y *et al* 2016 *J Industrial Eng. Chem.* **34** 250-257.
[6] Parnas R S *et al* 2012 *J Composites : Part A.* **43** 1160-68.
[7] Parnas R S *et al* 2014 *Composites* : Part A. **7** 5890-97.
[8] Liu J *et al* 2016 *Composites :* Part A. **87** 29-45.
[9] Nakatani H et al 2012 J. Polym. Degradation and Stability **97** 2177-84.
[10] Orefice R L *et al* 2016 *J. Carbohydr. Polym.* **136** 210-215.
[11] Bagheri R *et al* 2016 *J. Carbohydr. Polym.* **141** 75-81.
[12] Ummartyotin S and Pechyen C 2016 *J. Carbohydr. Polym.* **142** 133-140.
[13] Rujiravanit R *et al* *J. Carbohydr. Polym.* **148** 335-344.
[14] Beke S *et al* 2015 *J. Mater. Sci. and Eng* **C.56** 305-310.
[15] Liao J *et al* 2016 *J. Mater. Sci. and Eng.* **C.63** 285-291.
[16] Innocenti F D *et al* 2015 *J Cleaner Productuon* **102** 493-500.
[17] Su H et al 2016 J. Int. Biodeterioration & Biodegradation **11** 154-61.

IOP Conf. Series: Materials Science and Engineering **223** (2017) 012061 doi:10.1088/1757-899X/223/1/012061

The microwave assisted-synthesis of carboxymethyl cellulose from nata de-coco bacterial cellulose

L O A N Ramadhan[1], M Nur Rahmat[1], P E Susilowati[1], L O Ahmad[1], U Edy Rusbandi[2]

[1]Department of Chemistry, Faculty of Mathematics and Natural Sciences, Halu Oleo University, Kendari, Indonesia, 93232

[2]School of Life Sciences and Chemical Technology, Ngee Ann Polytechnic, Singapore, 599489

E-mail : ramadhan305@gmail.com

Abstract. Bacterial cellulose (BC) is one of natural biopolymers which can be derivatized to make functionalized materials. Carboxymethyl cellulose (CMC) is a candidate derivative for such a direction. The aim of the present study is to investigate the usability of microwave energy to transform BC into CMC. The results showed that CMC was produced in a yellowish white powder by a short irradiation for 30 s at 650 W. The best combination of monochloroacetic acid and BC as anhydrogucose unit was found at the molar ratio of 1:5. The obtained CMC is soluble in distilled water, and aqueous NaOH solution. The highest degree of substitution, viscosity, and molecular weight of the CMC are 0.263, 15.61 Pa·s and 197,187, respectively. This study showed the usefulness of the microwave-assisted reaction to transform BC rapidly into water-soluble ionized derivative.

1. Introduction

Cellulose is the most abundant biomass material on Earth. Cellulose forms the basic scheletal structure of plant cell walls. It is mainly used in the textile and paper industries, and it is derived from vegetable products. Bacterial cellulose (BC) is a type of cellulose synthesized by some bacteria. BC has a crystaline polyform Iα unlike plant cellulose which is present mainly in Iβ structure. Traditionally, BC frequently used comes from nata de-coco [1].

Nata de-coco BC is formed as a gellous sheet on the surface of an acidic sugary medium by the action of *Gluconacetobacter xylinus*. This BC is the most promising of biopolymers and used in a number of applications such as hydrogels [2], and medical materials such as wound dressing [3], and vascular prosthetic devices [4]. BC can be chemically modified to produce cellulose derivatives such as carboxymethylcellulose (CMC). It contained carboxymethyl groups attached to the polysaccharide backbone making it a polyelectrolyte. The conventional process to produce CMC is characterized by a slow reaction process [5], thus requiring the development of a technology that allows the reaction to progress at a higher rate. One of the identified technologies is to utilize microwave to enhance the productivity of the process. The microwave technology has been widely used in the chemical industry and is able to speed up the processing time [6].

IOP Conf. Series: Materials Science and Engineering **223** (2017) 012061 doi:10.1088/1757-899X/223/1/012061

The purpose of this study was to investigate the application of microvawe power to obtain the CMC from nata de-coco BC by reaction with monochloroacetic acid. Characterization of the CMC produced was also made by FTIR spectroscopy, solubility test, and measurement of degree of substitution, viscocity, and molecular weight. This study presented a rapid process to transform BC into water-soluble ionic derivative.

2. Materials and Methods

2.1 *Materials*

BC was purchased from a local nata de-coco industry (Alfafood, Poasia, Kendari, Indonesia). Analytical grade reagents used in this paper were NaOH, ethanol, methanol, urea, monochloroacetic acid glacial acetic acid and HCl, purchased from the Emsure Merck, Jakarta.

2.2 *Pretreatment of Bacterial Cellulose*

Pretreatment of BC was adopted according to the method described by Suparjo and Nelson, (2011) [7]. BC obtained from local nata de-coco industry was pulverized in the size appoximately 0.297 mm. Then, it was pasteurized repeatedly until the acidic smell was eliminated and dried by heating 1 h at 105 °C. The dried BC was demineralized by treatment with aqueous 1N HCl solution at the solid to liquid ratio of the mass of 1:15 (w/v). The mixture was allowed to stand for 30 min at room temperature, and filtrated to recover the treated BC. After repetition of this treatment for more than 3 times, BC was washed with distilled water until neutral pH, and dried again as described above. BC for derivatization was finally given as a yellowish white powder.

2.3 *Synthesis of Carboxymethyl Cellulose*

Synthesis of carboxymethyl cellulose (CMC) is based on the modified method described by *Qi et al.,* (2009) [8]. Two point six g of BC was dissolved in 100 mL of a solution containing NaOH (7 g), urea (12 g) and distilled water (81 g). A 200 mL Erlenmeyer glass flask was basically used for solubilization and further reaction. The mixture was stirred for 90 min at 100 °C using a shaker to produce a slurry of alkali-cellulose. This slurry was subsequently frozen at 0 °C and thawed using a shaker at room temperature to obtain a transparent cellulose solution.

The flask containing cellulose solution was set inside a modified microwave oven (1 kW) equipped with condenser as shown in Figure 1. Monochloroacetic acid was added to this solution slowly at room temperature. External addition of NaOH in the reaction medium is not necessary because the reactant BC is already solubilized in the aqueous NaOH/urea solution. Microwave-assisted carboxymethylation reaction took place for 30 s at 5 different power condition of 180, 300,

450, 650, and 800 W. For determination of the adequate combination of BC as anhydroglucose and monochloroacetic acid, microwave treatment was done under 5 different molar ratios of 1:1, 1:2, 1:3, 1:4 and 1:5. The reaction produced sodium glycolate as a byproduct by substitution of Cl with OH when concentration of monochloroacetic acid is low, leading to decrease in degree of substitution. After the reaction, acetic acid (90% v/v) was added to make a pH close to neutral for removal of excess amount of NaOH. The precipitated CMC was then recovered by vacuum filtration and washed 6 times with 70% aqueous ethanol, and dried for 5 h at 70 °C. The remaining sodium glycolate if present can be solubilized away by washing with methanol. Reaction condition and abbreviation of the reaction products are listed in Table 2.

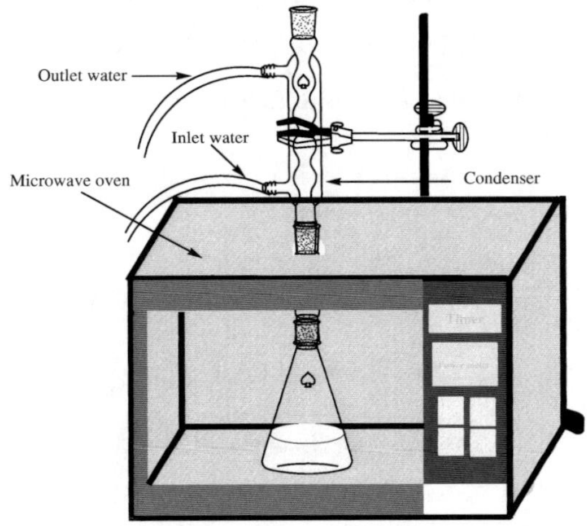

Figure 1. Schematic diagram of the modified microwave oven

2.4 Analytical Methods of Carboxymethyl Cellulose

Carboxymethyl functional group analysis was conducted by FTIR as KBr discs prepared with KBr (100-200 mg) and CMC (5 mg). The absorption spectra were measured under transmittance mode by using FTIR (Shimadzu 8400, UPI, Bandung) in the wavelength range 4000-400 cm^{-1}. The analysis of the functional groups was carried out at room temperature with a resolution of 8.0 cm^{-1}.

Solubility test was carried out using distilled water, 0.1 N HCl, and 0.1 N NaOH as the solvents. About 1 mg of CMC was loaded into an Eppendorf tube containing 1 mL of solution, and the mixture was then stirred using a vortex until the solids completely dissolved in solution.

The degree of substitution of CMC was determined by the method as described by Bono et al. (2009) [9]. A total of 2 g of CMC was dissolved in 40 mL of demineralized water. Hydrochloric acid was added into the solution to adjust the pH < 2. This solution was titrated with 0.1 N NaOH solution and the pH was recorded at fixed intervals using pH meter. The degree of substitution can be calculated using equation 1.

IOP Conf. Series: Materials Science and Engineering **223** (2017) 012061 doi:10.1088/1757-899X/223/1/012061

$$DS = \frac{162 \times A}{m_{CMC} - 58 \times A} \quad (1)$$

Where:

DS = degree of substitution; A = V_{NaOH}(mL) × C_{NaOH} (M); V_{NaOH} = volume of sodium hydroxide solution in mL; C_{NaOH} = concentration of sodium hydroxide solution in M; m_{CMC} = weight CMC from nata de-coco (g); 162 = molecular weight of anhydroglucose; 58 = molecular weight of carboxymethyl group

The viscosity of CMC solution was measured at 0.6, 0.7, 0.8, 0.9 and 1.0% w/v by Ostwald Viscometer at room temperature. The sample solution was supplied to the Viscometer at least 3 times. Time (*t*) (s) recorded were then inserted into the equation (2) to obtain the viscosity of Mark-Houwink equation the solution. The viscosity value is used to determine the molecular weight through the Mark-Houwink equation (3) [10].

$$\eta = \eta_o \frac{\rho t}{\rho_o t_o} \quad (2)$$

$$[\eta] = KM^{\alpha} \quad (3)$$

Where:

η = solution viscosity (Pa·s); α = coefficient of CMC, 1.00; K = coefficient, 7.92 × 10^{-5};
t = flow time of solution(s); M = molecular weight; ρ = solution density (g/cm^3);
η_o = solvent viscosity (Pa·s); ρ_o = solvent density (g/cm^3); t_o = flow time of solvent

3. Results and Discussion

3.1 Synthesis of Carboxymethyl Cellulose

The experimental results given by the microwave-assited carboxymethbylation of BC are listed in Table 1.

Table 1. Selection of the reaction condition

Product	Molar ratio (monochloroacetic acid /anhydroglucose repeating unit)	Power (W)	Mass (g)	Degree of substitution	Yield (%)
CMC1	1 : 1	650	4.14	0.142	188
CMC2	1 : 2	650	5.07	0.165	184
CMC3	1 : 3	650	5.23	0.201	189
CMC4	1 : 4	650	5.33	0.253	188
CMC5a	1 : 5	650	5.73	0.263	201
CMC5b	1 : 5	180	3.72	0.117	136
CMC5c	1 : 5	300	4.63	0.124	168
CMC5d	1 : 5	450	5.13	0.139	182
CMC5a	1 : 5	650	5.73	0.263	201
CMC5e	1 : 5	800	5.24	0.165	184

IOP Conf. Series: Materials Science and Engineering **223** (2017) 012061 doi:10.1088/1757-899X/223/1/012061

The effect of microwave irradiation in organic synthesis is a combination of thermal effects, arising from the heating rate, superheating or ''hot spots'' and the selective absorption of radiation by polar substances [6].

Based on the generated yield (Table 1), the amount of monochloroacetic acid tend to affect the CMC mass produced. The CMCs producted have at least three variations of mono substitution on the atom C-6, C-2 and C-3 (Figure 2).

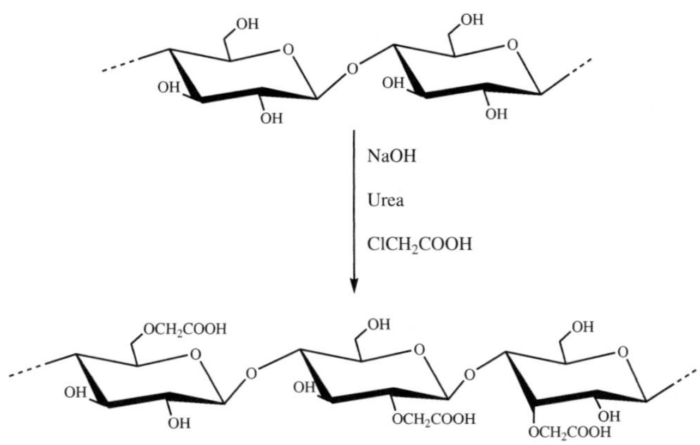

Figure 2. Reaction scheme of formation of CMC [8]

3.2 Analysis of Functional Groups

FTIR spectra of the native BC cellulose, and CMC1-CMC5 are shown in Figure 3.

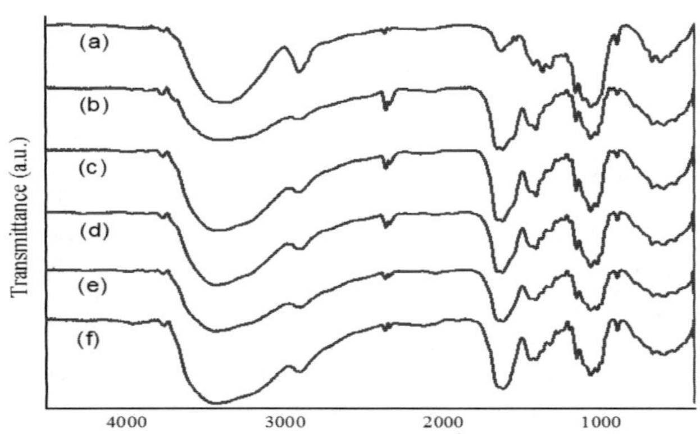

Figure 3. FTIR spectra of (a) Cellulose (b) CMC1 (c) CMC2 (d) CMC3 (e) CMC4 (f) CMC5

The results indicate the overall similarity in the spectra of CMC1-CMC5, but intencity of absorptions at about 1643 cm^{-1} due to carboxylic acid (COO^-) and 1420 cm^{-1} ($-CH_2$) strengthed in these spectra compared with the native cellulose. The differences can also be seen in the absorptions at 1054 cm^{-1} ($>CH-O-CH_2$), 2889 cm^{-1} (C-H), and 3449 cm^{-1} (-OH) which tend to increase. On account of solubility all samples of CMC1-CMC5 were soluble in distilled water-and 0.1 N NaOH solution. Thereby DS can be determined by the potentiometric titration method.

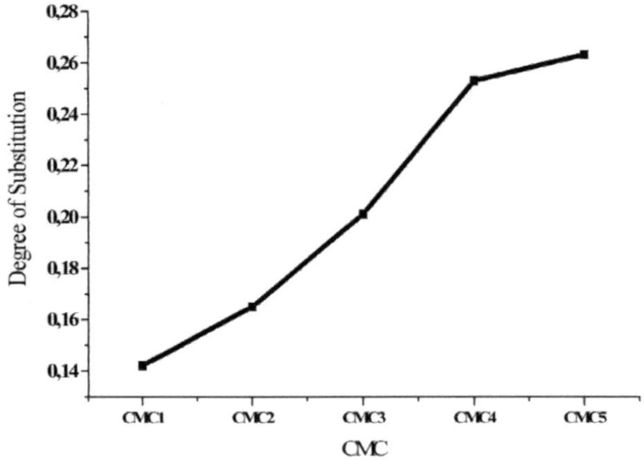

Figure 4. Effects of the amount of monochloroacetic acid on degree of substitution CMC

As shown in Figure 4, the carboxymethylation reaction progressed with increase in concentration of monochloroacetic acid, and DS attained at a value of 0.263. DS values are small, so it does not have a significant influence on the outcome of the yield. In terms of quality, the greater the degree of substitution, the better for the quality of CMC because of greater solubility in water.

The molecular weight of the products was determined by using Ostwald Viscometer. Intrinsic viscosity values of CMC1, CMC2, CMC3, CMC4, and CMC5 were 3.21, 9.57, 11.10, 12.76 and 15.61 Pa·s, respectively. The molecular weight of each CMC was calculated and the results are summarized in Figure 5.

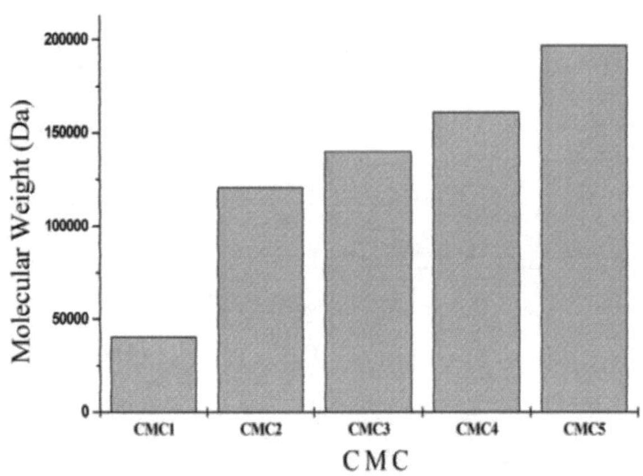

Figure 5. Molecular weight of CMC

The highest molecular weight was obtained as 197,187. The increase of molecular weight revealed that the increase of concentration of the reactants will make the molecular size higher.

IOP Conf. Series: Materials Science and Engineering **223** (2017) 012061 doi:10.1088/1757-899X/223/1/012061

4. Conclusion

CMC was synthesized from nata de-coco BC by using microwave energy. The reaction condition to achieve the DS of 0.263 (molecular weight of 197,187) was found to use 5 moles of monochloroacetic acid for 1 mole of anhydroglucose unit of cellulose. FTIR analysis of the reaction products showed the typical absorptions due to carboxymethyl group at 1620 cm^{-1} and 1423 cm^{-1}. All products can be dissolved in distilled water and 0.1 N NaOH solution. The obtained results showed that the microwave-assisted method was useful for preparation of CMC from nata de-coco.

Acknowledgement

The first author is thankful to Directorate of Higher Education, Ministry Research, Technology, and Higher Education for financial support of competitive research grant under contract number 063-3/PPK/UHO/IV/2014.

References

[1] Budhiono A *et al* 1999 *Carbohyd. Polym.* **40** 137–143.
[2] Halib N *et al* *Appl. Polym. Sci.* **116** 2920-2929.
[3] Czaja W *et al* 2006 *Biomaterials* **27** 145-151.
[4] Bäckdahl H, *et al* 2006 *Biomaterials* **27** 2141-2149.
[5] Joshi G *et al* 2015 *Waste Management* **38** 33-40.
[6] Wang L Y 2004 *Dyes Pigments* **62** 21–25.
[7] Suparjo dan Nelson 2011 *Agrinak* **1**:1-10.
[8] Qi H, Liebert T, Meister F and Heinze T 2009 *React. Funct. Polym.* **69** 779–784.
[9] Bono A *et al* 2009 *Adv. in Natural and Appl. Sci.* **3**: 5–11.
[10] Yang F *et al* 2009 *Carbohyd. Polym.* **78** 95–99.

IOP Conf. Series: Materials Science and Engineering **223** (2017) 012062 doi:10.1088/1757-899X/223/1/012062

Effect of bench time polymerization on depth of cure of dental composite resin

K Harahap, A Yudhit, F Sari
Faculty Dentistry, Universitas Sumatera Utara, Medan, Indonesia

Email: kholidinaimandaharahap@gmail.com

Abstract. The aim of this research was to investigate the effect of bench time before light cured polymerization on the depth of cure of dental composite resin.Nanofiller composite resin (Filtek Z350 XT,3M, ESPE,China) was used in this study. Sixty samples of nanofiller composite resin were made and divided into control and test groups with bench time for 0, 15, 30, 45, and 60 min. For the test group, composite resins were stored in refrigerator with 4°C temperatures. Meanwhile, for the control groups, the composite resin was stored at room temperature. The samples were prepared using metal mould with size diameter of 6 mm and 4 mm in thickness. Samples were cured for 20 s by using visible blue light curing unit. Part of samples that unpolymerized were removed by using a plastic spatula. The remaining parts of samples were measured by digital caliper and noted as depth of cure (mm). Data were analyzed to one-way ANOVA and LSD tests ($p \leq 0.05$). Results showed there was no significance differences between test groups ($p=0.5$). A 60 minutes bench time group showed the highest depth of cure value among test group, and it was almost similar with control group value. It can be concluded that longer bench time can increase the depth of cure of composite resin.

1. Introduction

A common practice done by dentists is storing the composite resin in a refrigerator at 4°C if it is not used for a long time. This is also suggested by the manufacturer of composite resin andit aims to maintain the stability and prolonged the shelf life of the composite resin. But it did not explain any further about when is the right time to use after removing it from the refrigerator. Dentists might usually use it wait for a while or immediately after it is removed from the refrigerator.

There are several parameters that affect the polymerization of the composite resin, which are composition, color and translucency, light devices characteristics, irradiation distance, exposure time and temperature of the composite resin [1]. Providing temperature treatment on the composite resin before curing can improve its properties. Some studies prove that the temperature of composite resin before curing affects its properties. The temperature can affect the degree of polymerization of the composite resin. Awliya (2007) found that higher temperature increases the polymerization of composite resin [2] Muniz *et al.,* (2013) found that heating the composite resin before curing can increase the hardness [3].

IOP Conf. Series: Materials Science and Engineering **223** (2017) 012062 doi:10.1088/1757-899X/223/1/012062

Composite resin polymerization is related to the depth of cure. The depth of cure of composite resin depends on the light penetration into the composite resin. Depth of cure is related to material thickness, wavelength of light, irradiation distance, and spread of light. The intensity of the beam on the surface of the material is very important in improving the polymerization of composite resin [1,4]. Munoz (2008) found that preheating resin composites with a composite warmer increases the monomer conversion rate and increases the depth of cure and hardness of the tested composites [5].

From the explanation above, it can be seen that temperature can affect the polymerization of the composite resin. With the suggestion to store the composite resin in a refrigerator at 4° C, it may affect the depth of cure. Previous study explained that preheating composite resin by using a composite warmer can increase depth of cure and hardness of composite resin. But not yet study before, that explained about allow the dental composite resin at room temperature for several minutes after storage in refrigerator to increase the depth of cure. The aim of this study is to investigate the effect of bench time before light cured polymerization on the depth of cure of dental composite resin.

2. Materials and Methods

2.1 Sample Preparation

Sixty samples were prepared using nano filler composite resin (Filtek Z 350XT, 3MESPE, China) with the measurements of 6.0 mm in diameter and 4.00 in height. Samples were divided into control group and test group. Materials in control group were stored at 24°C and test group were stored at 4°C for 10 days. For the test groups, samples were divided based on bench time of 0, 15, 30, 45 and 60 min at 24°C after removed from refrigerator. Nano filler composite resin was filled into steel molds, which were previously coated with silicon oil. The mould was placed on a strip of a transparent film covering a filter paper and was filled with the test material until the mould is slightly overfilled. Then a second strip of the transparent film was placed on top. The mould and strip of the transparent film were pressed between glass slides to exude the excess material. The microscope slide was removed and gently placed at the exit window of the external energy source against the strip of film. The material was irradiated for 20 s by using light curing unit. The sample was removed from the mould after completion of light exposure.

2.2 Evaluation of Depth of Cure

After the samples were cured, the uncured material was gently removed by using a plastic spatula. The height of the cured samples was measured by using a digital caliper to an accuracy of ± 0.1 mm. This value was recorded as the depth of cure of samples.

IOP Conf. Series: Materials Science and Engineering **223** (2017) 012062 doi:10.1088/1757-899X/223/1/012062

2.3 Statistical Analysis

Data were analyzed using one way ANOVA and Least Significance Data test for finding significant differences between groups ($p \leq 0.05$).

3. Results and Discussions

The results of depth of cure are summarized in Table 1. The highest value of the depth of cure is control group (3.90±0.07 mm). The depth of cure of composite resin with a bench time of 60 min (3.897±0.09 mm) almost similar with control group. Whereas the lower depth of cure was composite resin with a bench time of 0 mm (3.8254±0.17mm) among the test group. One way ANOVA revealed a non significant differences among test group (p=0.504).

Table 1. Depth of cure of dental composite resin

Group	Depth of cure (mm)	p value
Control	3.9 ± 0.07	
Bench time of 0 min	3.8 ± 0.17	
Bench time of 15 min	3.8 ± 0.10	
Bench time of 30 min	3.85 ± 0.10	0.504
Bench time of 45 min	3.9 ± 0.07	
Bench time of 60 min	3.9 ± 0.09	

Composite resins are not stable at room temperature and therefore require refrigerated storage below 8°C [12]. Some manufacturers recommended to storing composite resin under refrigeration to make it more durable. If such materials are used at this temperature, one could suppose that retardation of the reaction speed could occur, resulting in a cure beyond the ideal. This is because the refrigeration can increase the viscosity of the material, lowering movement of monomer and altering polymerized composite resin. So it can be presumed that the materials used directly in the teeth after patching procedures when stored in the refrigerator will have an inadequate polymerization [10]. The results in this study showed that samples with the bench time of 60 min after removal from the refrigerator has the greatest depth of cure. The depth of cure value that obtained from samples with a bench time of 60 minutes was also almost similar to the value of the composite resin that was stored at room temperature. While the samples with a bench time of 0 min. showed the lower value of depth of cure. The data of this study showed the longer the bench time will increase depth of cure of composite resin. Therefore, the lower the temperature during the time of polymerization of the composite resin, the lower the energy

IOP Conf. Series: Materials Science and Engineering **223** (2017) 012062 doi:10.1088/1757-899X/223/1/012062

molecule initiator to initiate the polymerization process. This will certainly decrease the depth of cure of composite resin as obtained in this study. However, there are no significant differences statistically between test groups. Resin composites in the control group had a depth of cured almost similar to composite resin with a bench time of 60 min. This is probably due to the temperature of the composite resin with a bench time of 60 min is almost close to room temperature.

Temperature plays an important role in the process of composite resin polymerization because it is related to the viscosity of composite resin. High temperature will decrease the viscosity and low temperature may increase the viscosity of composite resin. Viscosity is associated with the speed of molecules movement during polymerization. High temperature will increase the movement of monomer and free radicals. Whereas low temperature will lower the movement speed of monomer molecules to react during polymerization [9,10].

Adequate polymerization of composite resin depends on depth of cure. The depth of cure is the distance from the surface to the base of composite resins that has undergone curing. The depth of cure depends on the light penetration that is related to the thickness of the material, the wavelength, irradiation, and the spread of light. The intensity of the beam on the surface of the material is very important in improving the polymerization of composite resin [1,11]. Several factors affect the depth of cure of composite materials with shade and translucency being the most significant ones [12].

Adequate polymerization is an important factor in obtaining optimal physical properties and good clinical performance of composite resin restorative materials [6]. Some problems related to inadequate polymerization are decrease in physical properties, solubility in the oral environment, and increase in occurrence of micro leakage that causes recurrent caries and irritation of the pulp tissue [7]. Three basic components required for adequate polymerization are radiant intensity, correct wavelength of visible blue light and curing time. Other factors include type of composite resin, color and translucency, temperature of composite resin, material thickness, distance of light exposure and exposure time [8].

Temperatures which was related to the viscosity of the composite resin, that determines the speed of movement of the composite resin molecules during polymerization. Monomer require the movement to be able to convert forming polymer. But since storage at lower temperature will increase the viscosity of the resin composite and decrease the mobility of monomers, it would interfere with the polymerization. So it takes time to increase the temperature of the new composite resin removed from the refrigerator before it is polymerized [10].

IOP Conf. Series: Materials Science and Engineering **223** (2017) 012062 doi:10.1088/1757-899X/223/1/012062

The heat effect on the monomer that may contribute to enhance the degree of curing and mixing so thus mechanical strength can be improved. Heating reduces the resin viscosity with the activation energy ranging from 21.9-27.6 kJ/mol [13].

Some factors have been proposed for the increased conversion of preheated composites. An elevated composite temperature leads to an increase in molecular mobility. Therefore, the propagation stage takes longer time without becoming diffusion controlled. Furthermore, temperature rise below the glass transition improves the mobility of polymer chain, postponing the reaction diffusion-controlled termination. By improving the monomer conversion the glass transition temperature will be increased inducing a greater amount of conversion at higher polymerization temperatures [14].

Several previous studies suggest that the use of composite resins at higher temperatures will trigger greater monomer conversion. An increase in temperature causes an increase in the composite resin molecular mobility. Therefore, the propagation phase will last longer without their diffusion controlled. Furthermore, the increase in temperature below the glass transition will increase the mobility of the polymer chain, delaying the termination of diffusion controlled reactions. By increasing the glass transition temperature of the monomer conversion will increase the number of conversions at high polymerization temperature [14]. Additionally, because the composite resin temperature increase may accelerate the conversion of the monomers, both the preheating and the light curing unit may induce a fast increase in the material toughness, leading to greater shrinkage stress [15]. It is known that the curing process is not only break chemical chain reaction, where the double carbon bond to break down into molecules that bind to each other to form a polymer chain. It is also known that the higher temperature will shorten the time of induction, in which the initiator molecules become more energized or activated and initiate the transfer of energy to the monomer molecule [14].

4. Conclusion

It can be concluded that that longer bench time at room temperature after storage the composite resin in refrigerator can increase the depth of cure of composite resin. A bench time is necessary if dental composite resin was stored under refrigerator before use as restorative material to obtain optimum depth of cure.

Acknowledgement

The authors would like to express our thanks to Universitas Sumatera Utara for the Grant tofund this research.

References

[1] Anusavice K J *et al* 2013 *Philip's Science of Dental Materials*. 12th ed. St.Louis, Elsevier: 286-90.
[2] Awliya W Y 2007 *J. Contemp. Dent. Pract.* **8**(6):9-16.
[3] Muniz G L *et al* 2013 *Indian J Dent Res* **24** 708-12.
[4] Munoz C A *et al* 2008 *American J. Dentistry* **21**(4) 215-222.
[5] EvicKÏÂ *et al* 2001 *J. Oral Rehabilitation* **28** 586-591.
[6] Tarle Z *et al* 1995 *Acta Stomatol Croat* **29(1)** 9-14.
[7] Saade E G *et al* 2009 *Laser Physics* **19(6)** 1–6.
[8] Calheiros F C *et al* 2014 *Dent Mater* http://dx.doi.org/10.1016/j.dental
[9] Osternack F H R *et al* 2009 *Braz. Dent J* **20(1)** 42-47.
[10] Powers J M, Wataha JC 2008 9th ed. Missouri: Mosby Elsevier:24-6.
[11] Mahn E. 2013 *Rev Clin PeriodonciaImplantol Rehabil Oral* **6**(3): 148-153.
[12] Kashi T S J *et al* 2015 *Biomedical Technol.* **2(1)** 15-22.
[13] Ekworapoj P *et al* 2002 *J. Metal, Materials and Minerals* **12(1)** 39-50.
[14] Torres C R G *et al* 2011 IJCD **2(3)** 41-5.
[15] Osternack F H *et al* 2013 *Operative Dentistry* **38** 50-57.

IOP Conf. Series: Materials Science and Engineering **223** (2017) 012063 doi:10.1088/1757-899X/223/1/012063

Effect of chitosan addition to characteristic and antimicrobial activity of zinc doped hydroxyapatite

A Rasyida, S T Wicaksono, N N Pradita, H Ardhyananta, A Purnomo
Department of Materials and Metallurgical Engineering, Faculty of Industrial Technology, Institut Teknologi Sepuluh Nopember (ITS) Surabaya

E-mail : amaliyarasyida@gmail.com, amaliya@mat-eng.its.ac.id

Abstract. Hydroxyapatite (HAp) doping with zinc was prepared using sol gel method; different chitosan content were further added to prepare the composite, namely 10, 15 and 20% wt. The samples were characterized using FTIR, XRD, SEM-EDX, and AAS. In vitro antimicrobial activities of the composite were evaluated against gram positive and negative bacteria. FTIR results revealed that there were no important changes in the structure of composite, while 10% wt of chitosan in composite shows the highest inhibition zone against Escherichia coli after 24 h incubation. In addition, after 7 days of immersion in simulated body fluid, there were apatite formations in the surface of the composite. These might indicate that this composite could be used as a material candidate for bone substitute applications.

1. Introduction

Biocompatibility and osteo-conductivity of hydroxyapatite [$Ca_{10}(PO_4)_6(OH)_2$, HAp] are well known and utilized in dentistry as well as orthopedic application due to its chemical similarity to human bone. The emerging trend of bone tissue engineering with hydroxyapatite has attracted more significant due to its bio-functional properties, such as bioactivity and biocompatibility. Recently, since HAp has high dissolvability in the body and lower mechanical properties as compare to human bone, the development HAp with various additives has been one of the primary aims in the field of biomaterials to be used for bone substitutions [1]. Zinc doped HAp has a stimulatory effect of bone formation tough both in vitro and in vivo, triggers metabolism and bone development promote bone density, and prevent bone losses [2]. In addition, it can improve dissolvability, in small concentrations serve as micronutrient for some processes in the body, but in large concentrations can be toxic to the body [3]. Concerning to the antimicrobial properties, the ability of HAp-Zn in inhibiting the bacteria proved to be limited to gram-positive bacteria such as *Staphylococcus aureus*, and very little observed in gram-negative bacteria one is *Escherichia coli* [3].

On the other hand, chitosan (CS) is one of the most abundant natural polymeric materials. Chitosan is a deacetylated derivative of chitin with amine and hydroxyl groups. Those structures bring chitosan to have tremendous ability to form metal complexes. Moreover, zinc is one of the metal ions that can easily coordinate with chitosan [4]. It is well known that both CS and Zn have the properties

IOP Conf. Series: Materials Science and Engineering **223** (2017) 012063 doi:10.1088/1757-899X/223/1/012063

of disinfection and bactericide [5, 6]. In addition, chitosan is known to have a good anti-bacterial activity against some bacteria, both gram-positive and gram-negative.

In our previous work [7] we reported the morphology of chitosan and each composite was determined by X-ray diffraction (XRD), Fourier transform infrared spectroscopy (FTIR), and scanning electronmicroscopy (SEM). XRD and FTIR results indicate that there were interactions between chitosan and HAp presented by the decreasing of intensity and broadening of the peak, moreover SEM micrographs showed that there wasthe presence of particles that agglomerated HA-Zn presenting the ability of chitosan to bind HA-Zn particles.

Considering this, the aim of this work was to study if the presence of chitosan can affect considerably to biological properties of zinc doped HAp, presenting potentially interesting properties for some medical applications such as bone substitute applications.

2. Materials and Method

2.1. Materials

Chitosan derived from crab with a degree of deacetylation 84% obtained from CV. Biochitosan Indramayu, West Java. Kalsium nitrat tetrahidrat and zinc nitrat tetrahidrat were supplied by PT. Sumber Ilmiah Persada Surabaya.

2.2. Preparations

HAp doped Zn was prepared by sol gel method in accordance with the reference [8]. The synthesis process of composite Chitosan / HA-Zn is performed using a wet mixing method with weight percent ratio of chitosan 10, 15, and 20 %wt. Chitosan was first dissolved in 100 ml of 3% acetic acid solution. Then stir with a constant temperature of 700°C for one hour. HA-Zn powder is then added and stirred for 1 h at 700°C.

2.3. Characterizations

Evaluation of antibacterial activity in vitro was measured by the amount of inhibition zone diameter on antibacterial testing. Escherichia Coli and Staphylococcus Aureus first dissolved in 2 mL of 0.85% NaCl (1.5 x 108 CFU / ml; equated to a 0.5 McFarland standard solution). 1 mL of bacterial suspension was dropped onto Mueller-Hinton Agar (MHA) and spread using sterile spatula. The characterization is performed using paper disc diffusion method by dipping a paper disc diameter of 0.6 cm into the composite mixture and using pH 7.2 phosphate buffer solutions with a concentration of 800 mg/mL. Observations were carried out after 24 and 48 h incubation.

Soaking in Simulated body fluid (SBF) was performed to evaluate in vitro bioactivity. SBF using artificial solution that has ion concentrations nearly resembles human blood plasma. Cylindrical samples soaked for 7 days in 60 mL of SBF. On third, fifth, and seventh day samples were taken from the solution, dried and weighed to determine the mass percent lost during immersion using the formula [9]:

$$W = (ma - mb)/mb \times 100 \quad \text{..............................(1)}$$

ma and mb represent mass after and before immersion respectively.

Atomic Absorption Spectroscopy (AAS) was carried out to determine the concentration zinc and calcium dissolved. Characterization of functional groups of composite were performed using the Thermo Scientific Nicolet iS 10 (Thermo Fisher Scientific.Inc, MA, USA) with a wavelength of 4000-400 cm^{-1} to evaluate the structure after SBF immersion. Scanning electron microscopy with Energy-dispersive X-ray spectroscopy (SEM-EDAX) was performed to analyze the surface morphology and the constituent elements contained in the composite using machine type PhenomProX (Phenom-World, Eindhoven, NL). This characterization was carried out on both sample, before and after immersion of SBF.

3. Results and Discussions

3.1. Characteristics of Product

In vitro evaluation using *Simulated Body Fluid (SBF)* presented in Figure 1, it shows that the composited has a higher solubility as compared to pristine HAp-Zn. It can be due to chitosan is not able to envelop HAp particles. Chitosan serves as the glue which holds the particles of HAp [15]. In addition, the presences of porosity facilitate SBF solution to go into and effects on chitosan to be more easily degraded. This is strengthened by the results of FTIR after 7 days of immersion in which the OH peak at 3000-3400 cm^{-1} and amide peak at around 1600 to 1650 cm^{-1} in Figure 2 decreased intensity [9].

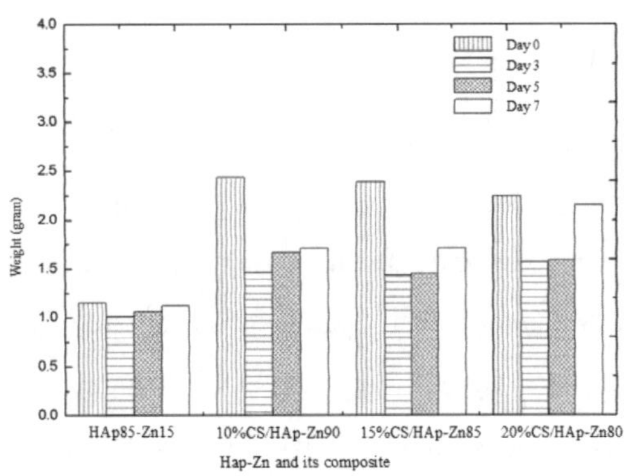

Figure 1. Weight loss curve

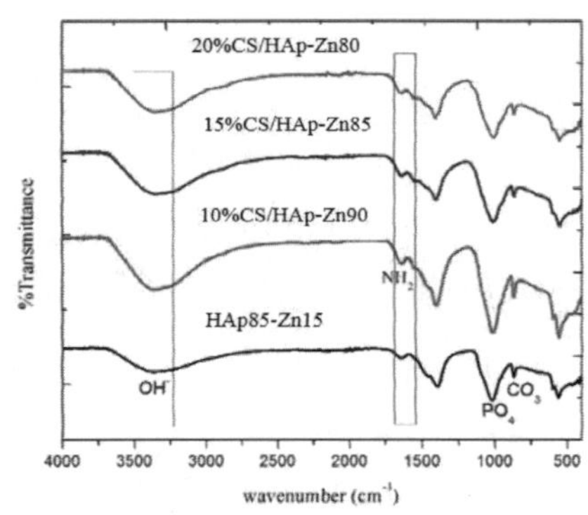

Figure 2. FTIR pattern after SBF immersion for 7 days

Moreover, SEM micrographs, before and after 7 days of immersion, support the previous results. In Figure 3, marked with a red circle, the chitosan was no longer visible. Before soaking structure exists as the glue which holds the grains of HAp-Zn. However, after soaking the adhesive is lost, leaving the porosity between the grains HAp-Zn one another, so that a similar structure as the adhesive can be assumed as chitosan.

Figure 3. SEM micrographs of composite containing 15% before (a) and after (b) soaking in SBF for 7 days

Although chitosan is expected to be first degraded, it does not rule out the possibility of calcium and zinc in the composite subsequently degraded after chitosan which serves as an adhesive merged in SBF. From the test results AAS, in Table 1, Zinc dissolved in SBF increase in line with the addition of chitosan, but still within safe levels of Zinc contain in blood. The concentration of zinc in normal human red blood cells is in the range of 3.6 - 25.4 ppm, while the blood plasma ranging from 0.49-7.70 ppm. We have to keep in mind that the range Zinc concentration levels in the blood are different for each population [10]. This difference can be due to geographical factors, environment and diet.

Table 1. The results from AAS for Zinc dissolved after 7 days of immersion SBF

Sample	Zinc dissolved (ppm)
HAp85-Zn15	3.25±2
10%CS/HAp-Zn90	3.45±2
15%CS/HAp-Zn85	3.44±3
20%CS/HAp-Zn80	3.25±2

Bone substitutes material must have good bioactivity and the ability to trigger the growth of bone cells, as in vitro it can be showed through the formation of apatite layer on the surface of the sample. Table 2 present the concentration (ppm) of calcium dissolved in the SBF solution after soaking the samples for 7 days. It can be observed that the addition of chitosan lead to enhancement of Ca ions. Calcium ions are used in the formation of apatite layer, resulting in the decrease in Ca concentration in solution, thus the possibility of the apatite layer found on the surface is greater [9].

Table 2. The results from AAS for Calcium dissolved after 7 days of immersion SBF

Sample	Calcium dissolved	
	0 day	After 7 days
HAp85-Zn15	0.69	0.28
10%CS/HAp-Zn90	8.55	12.41
15%CS/HAp-Zn85	47.72	12.83
20%CS/HAp-Zn80	42.01	5.46

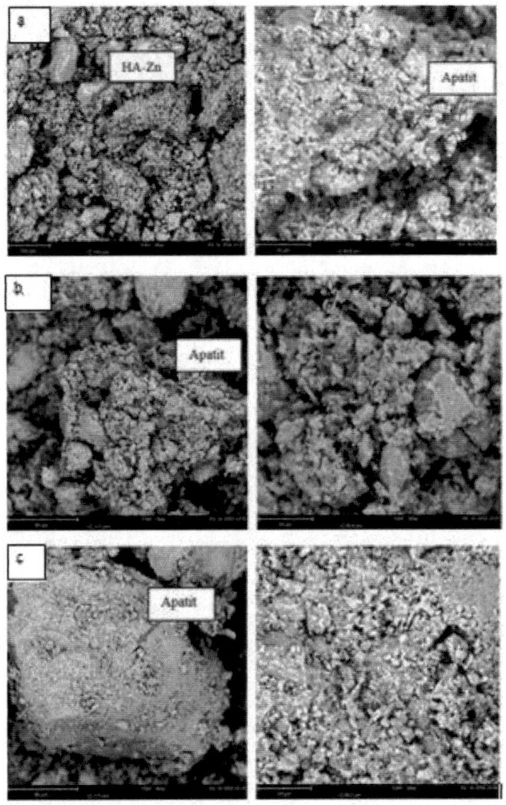

Figure 4. SEM micrographs of composite with 10%wt (a), 15%wt (b) and 20% (c) of chitosan after 7 days immersion in SBF

Those results are in accordance with SEM micrographs on the surface of the composite shown in Figure 4. The white layer on the surface of the composite prior to soaking was not detected. However, after immersion, many similar layers appear. Thus, from the weight gain started on the fifth day and reducing the concentration of calcium in SBF solution, this white coating is assumed to be the apatite layer. Similar results were found in agreement with previous research by Machou *et al.,* (2008) [9].

3.2 Antimicrobial Activities

The bacteria used were *Escherichia coli* (gram negative) and *Staphylococcus aureus* (gram positive) which is a common cause of bacterial pathogens Orthopedic Implant Infections [11, 12].

IOP Conf. Series: Materials Science and Engineering **223** (2017) 012063 doi:10.1088/1757-899X/223/1/012063

Observations were carried out on *E. coli* during the first 24 h showed that the addition of chitosan on HAp-Zn affect the antibacterial activity of composite. However, the presences of chitosan narrow the inhibition zone of bacteria. HAp-Zn have better inhibition zone of 0.25 mm, while for the composition 10% K, 15% K, and 20% of K has 3 mm, 1.16 mm and 0.5 mm respectively. Meanwhile, antimicrobial activity against S. aureus present reducing the ability of HAp-Zn. It was measured 6.65 mm, while the addition of 10% K, 15% K, and 20% K is 0.5; 0.25; and 0.4 mm respectively (see Table 3). From the picture it can be seen that the inhibition zone of HA-Zn found to be greater against *E. coli* then *S. aureus*. This result is in agreement with previous research [3,13] showed that S. aureus are more susceptible to HA-Zn as compared to *E.Coli* supported by the greater percentage reduction in bacterial colonies on S.aureus. This difference occurs due to the gram-positive bacteria have membrane cells structure composed of peptidoglycan and sour Teihoic (TA) which is easier to be penetrated by zinc causing damage to DNA and RNA and inhibit the reproduction of bacteria that can lead to death [13].

Table 3. Antibacterial activity measured by inhibition zone of Hap-Zn and its composite

Composition	Inhibition zone (mm)			
	S. Aureus		E. Coli	
	24 h	48 h	24 h	48 h
HAp85-Zn15	1.43	1.23	0.65	0.00
10%CS/HAp-Zn90	0.63	0.61	0.70	0.77
15%CS/HAp-Zn85	0.64	0.61	0.70	0.73
20%CS/HAp-Zn80	0.61	0.00	0.67	0.00

Meanwhile, the addition of chitosan gives significant effect on the inhibitory activity of *E. coli* but not so with *S. Aureus* [14]. However, the decrease can be due to lower amount of HAp-Zn with the addition of chitosan. The presences of Zn strengthen the ability of chitosan in inhibiting bacteria. Zn binds to the active site of chitosan, the amide group at C2 position resulting from protonation;thus can facilitate interaction with the anionic components on the surface of cells such as LPS and proteins. This interaction alter the structure of the outer membrane protein causes the release of a number of cells [15].

Observations for 48 h present the antibacterial activity of composite decrease. Those results confirmed that the chitosan has a bacteriostatic effect which is inhibiting for a certain span of time in this case the first 24 h post-incubation. It can be due to subpopulation emergence of resistant bacteria as a result of physiological adaptation to exposure chitosan cells [9].

4. Conclusions

Zinc doped HAp and its composite based on different weight of chitosan namely 10, 15, and 20 %wt were prepared in order to evaluate if the presence of chitosan could affect the antimicrobial activity as well as in vitro bioactivity properties of HAp-Zn. The additions of chitosan in the composite affect in antibacterial properties. Greater inhibitory activity of bacteria found against E. coli than S. aureus with inhibition zone diameter of 3 mm. In addition, after 7 days of immersion in SBF it was observed a white apatite layer on the surface of the composite, pointed by the reducing of calcium ions in SBF solution. From the test results, the optimal composition is obtained when the chitosan addition of 10% by weight in terms of antibacterial activity and apatite layer formed.

Acknowledgements

The author would like to acknowledge PENELITIAN PEMULA PNBP ITS 2016 for funding.

References

[1] Deepa C *et al* 2013 *Nanosystems: Physics, Chemistry, Mathematics.* **4** (3) 370–377.
[2] Fuzeng R *et al* 2009 *Acta Biomaterialia* **5** 3141–3149.
[3] Radovanovic S *et al* 2012 *J The Serbian Chemical Society* 77 (2012) 1787–1798.
[4] Wang X *et al* 2004 *Carbohydr. Polym.* **56** 21–26.
[5] Jeon Y J, Kim S K 2000 *Carbohydr. Polym.* **41** 133–141.
[6] Takai K *et al* 2002 *Microbiol. and Immunol.* **46** 75–81.
[7] S T Wicaksono *et al* 2017 IOP Conf. Ser.: Mater. Sci. Eng. 202 012080 doi:10.1088/1757-899X/202/1/01208.
[8] Norhidayu D 2008 Development of Zinc Doped Hydroxyapatite for Bone Implant Application ICCBT - F 24 257-270.
[9] Maachou H *et al* 2008 *Trends Biomater.* Artif. Organs **22** (1) 16-27.
[10] Lisa C *et al* 2012 *J. Infections* **64** 169-175.
[11] Byrne F M dan Wilcox M H 2011 *Int. J. Care Injured* S3-S6.
[12] Kolmas *et al* 2014 *Bio. Med Research Int.* 1-15.
[13] Xiaohui W *et al* 2004 *Carbohydr. Polym.* **56** (1) 21-26.
[14] Raafat, Dina 2008 *Chitosan as an antimicrobial compound: Modes of action and resistance mechanisms.* Dr.rer.nat Dissertation, Bonn: Universität Bonn.
[15] Zhang L *et al* 2005 *J. Mater Sci. : Mater. in Medicine* **16** 213-219.

IOP Conf. Series: Materials Science and Engineering **223** (2017) 012064 doi:10.1088/1757-899X/223/1/012064

Effect of oil palm empty fruit bunches fibers reinforced polymer recycled

B Hermawan[1], S Nikmatin[1], Sudaryanto[2], H Alatas[1] and S G Sukaryo[2]
[1]Applied Physic Division, Department of Physics, Bogor Agricultural University, Jl. Meranti, Kampus IPB Darmaga, Bogor,16680, Jawa Barat, Indonesia.
[2]Centre for Science and Advanced Materials Technology, National Nuclear Energy Agency of Indonesia, South Tangerang, Indonesia

E-mail : sitinikmatin@yahoo.co.id

Abstract. The aim of this research is to process the OPEFB to become fiber with various sizes which will be used as a filler of polymer matrix recycled acrylonitrile butadiene styrene (ABS). Molecular analysis and mechanical test have been done to understand the influence of fiber size toward material capability to receive outer deformation. Single screw extruder formed a biocomposites granular continued with injection moulding to shaped test pieces. Maleic anhydride was added as coupling agent between filler and matrix. Filler concentration were 10 and 20% in fiber size respectively with constant additif. Two kind of fiber glass (10%) were used as comparator. In order to analyze the results of the mechanical test Fisher least significant difference (LSD) in ANOVA method was performed (-with α=0,05-).

1. Introduction

Nowadays a lot of special interest in environmental friendly material developing especially wasted natural products processing with a touch of technology. Oil palm empty fruit bunches (OPEFB) were one of the solid waste produced from crude palm oil (CPO) manufacturing and its biomass were totally plentiful [1]. Concern increase in polymer waste is a starting point to reduce and reuse it. The Acrylonitrile Butadiene Styrene (ABS) plastics are found in numerous items, such as household items (appliances, piping, luggage, etc), automobile parts, and aircraft interior structure [2].

In term of reinforcement material two fiber categories has been defined, first are fibers from natural on the other hand is sintetic material [3]. Several study has explored the using of natural fiber as a reinforcement agent in biocomposites, the material of wich include rattan [4], banana [5], nut-shells [6], almond shells [7].

In this research a mechanical milling without chemicals assistance has been processed OPEFB to become a various fiber [8]. It is a great challenge in order to screening the composition that have a good mechanical properties, and a statistical model was performed to make a clear difference among all of the treatments. Because of the abundant of the OPEFB and its sustainability, it would be efficient to utilize them as a biocomposites application to reinforced matrix recycled polymer ABS. Other than that, the use of of this natural fiber can replace the need of synthetic fiber as a

IOP Conf. Series: Materials Science and Engineering **223** (2017) 012064 doi:10.1088/1757-899X/223/1/012064

reinforcement agent. Thus, the aim of this research is to process the OPEFB to become fiber with various sizes which will be used as a filler of polymer matrix acrylonitrile butadiene styrene (ABS) and also can replace the used of fiberglass as a synthetic filler.

2. Materials and Method

OPEFB was taken from PKS-PTPN VIII Cikasungka, Bogor, West Java. The research was held at laboratory of physics department Bogor Agricultural University (IPB), National Nuclear Energy Agency of Indonesia (BATAN) Serpong and PT MUB Jaya Cibinong.

In this research the OPEFB immediately taken from the conveyor before it entered the dump truck then brought it to the laboratory to be cleaned in a container full of tap water. All cleaned OPEFB has been chopped to form a chip afterward dried under the sun about 2 days and continued by oven (YENACO tipe YNC-OV) at 100°C about 6 h, m= 2 kg. Into a mechanical milling (FOMAC type FCT-Z2000) a dried chips milled about 10 min, v= 25000 rpm, m= 0.5 kg. A score of times sieve-shaker was done to obtained a various fiber size using a 20 mesh and 100 mesh sieve.

For the matrix recycled ABS supplied by PT MUB Jaya, Bogor was then used. The additives used in this study were maleic anhydride from Germany. A prepared mixture stirred up by a mixer machine afterward get it into the single screw extruder (SSE) machine hopper, m= 25 kg to form a biocomposites granular. In the barrel samples blended with a gradient temperature of 195–215–220–220–220–225–225–225°C. Biocomposites granular entered the moulding machine (Hwa Chin, HC-250, China), m= 5 kg with a gradient temperature of 170-185- 200°C to shaped a test pieces.

2.1. Fiber Size Measurement

In order to make a measurement of the real fiber size samples were put under a light microscope (OLYMPUS BX-51) connected to a computer installed with DP2-BSW software to help carry out the measurements. The t-student distribution was performed to make the 95% confidence interval of the fiber size using Minitab 17 software.

2.2. Molecular Analysis

OPEFB fibers, ABS polymers, and biocomposites granular then analyzed further related to functional groups of constituent materials using Spectrophotometer Fourier Transform Infra Red (ABB, model MB300, Canada). OPEFB short fiber powder made into pellets with KBr method, while granular ABS and biocomposites is made into a thin film (± 1 μm).

2.3. Mechanical Test

Rockwell test method has been done according to ASTM D-785 using Jinan hensgrand instrument, model XHRD-150 from China to test the sample hardness. Whereas izod impact test also has been done according to ASTM D-256A using Jinan hensgrand instrument, model XJU-22 from China to determined the samples ability to received outer impact. Both of the test performed in five repetition. All of the results were analyzed by Fisher least significant difference (LSD) test in ANOVA one way method (with α=0.05) using Minitab 17 software.

3. Results and Discussions

3.1. Fiber size

Prior to analyzed both of outlier test and normality test was performed. Outlier test use to identify a single outlier in a sample, Dixon ratio help to detect outlier in the presence of other suspected outlier. Dixon noted that larger samples from normal population are more likely to include extreme values. Whereas Normality test used to determine wether the data follow a normal distribution, because tests that require normally distributed data could produce misleading results if used with non normal data.

Kolmogorov-Smirnov (K-S) test can be used for detecting differences of any kind between the population from wich samples have been drawn [9]. The distribution size of fiber obtained from the microscope observation (Figure 1) were shown in Table 1. In terms of size the mention of medium and short fiber based on the fiber separation using a sieve.

Table 1. t-student distribution of fiber size from microscope observation.

Sieve	Fiber Size		Class
	Length (µm)*	Diameter (µm)*	
Passed through20 mesh	339 - 1746	104.6 - 191.0	Medium
Passed through100 mesh	161.9 - 298.3	42.05-75.02	Short

* 95% confidence interval

a

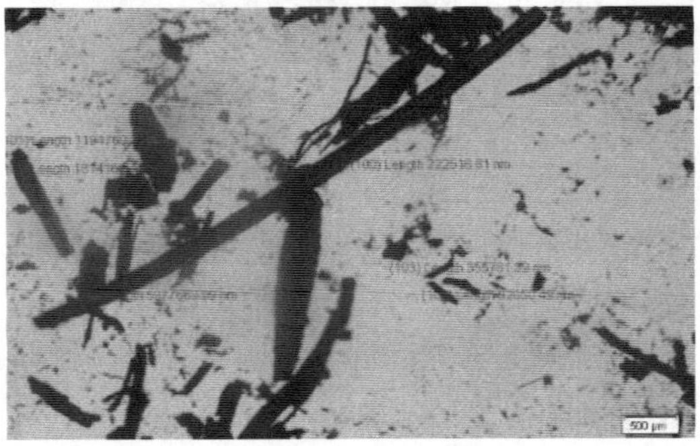

b

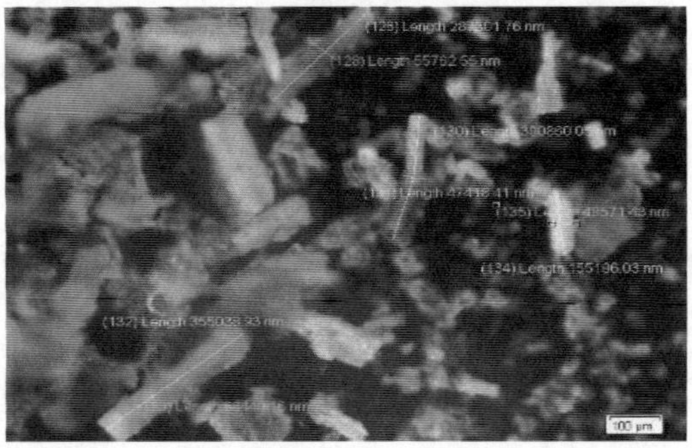

Figure 1. OPEFB fiber image (a) medium (b) short

Table 2 shows the wide variety of treatments based on the size of the fibers to produce biocomposites. Fiberglass composites used as comparison because it is used by companies to supply the demand for raw materials for making other products. Concentrations of 10 and 20% used to be able to see the extent to which the composition OPEFB as natural fibers can replace the use of fiberglass as synthetic fibers. Coupling agent were added in a fixed amount in order to see the effects of fiber size and fiber concentration.

Table 2 Various biocomposites treatment.

Treatments	Fiber conc (%)	Coupling agent (%)	ABS recycled (%)
Fiber glass 1 (FG1)*	10	-	-
Fiber glass 2 (FG2)**	10	-	-
Medium Fiber 10% (MF10)	10	2	88
Medium Fiber 20% (MF20)	20	2	78
Short Fiber 10% (SF10)	10	2	88
Short Fiber 20% (SF20)	20	2	78

* Fiberglass-ABS composites from factory 1
**Fiberglass-ABS composites from factory 2

3.2. Molecular Analysis

A common way of polymer blends is adding a coupling agent to reduce the interfacial tension, increase the adhesion, and obtain better properties. We shall be concerned mainly with organic molecules, both of butadiene and styrene in ABS matrix and maleic anhydride. This three molecules, because of the C=C (Carbon double bond) were extremely reactive (**Figure 2**). Methacrylic acid, acrylic acid, maleic anhydride (MAH), and long chain unsaturated monomers are among the monomers most commonly grafted onto polymers containing butadiene [10-11].

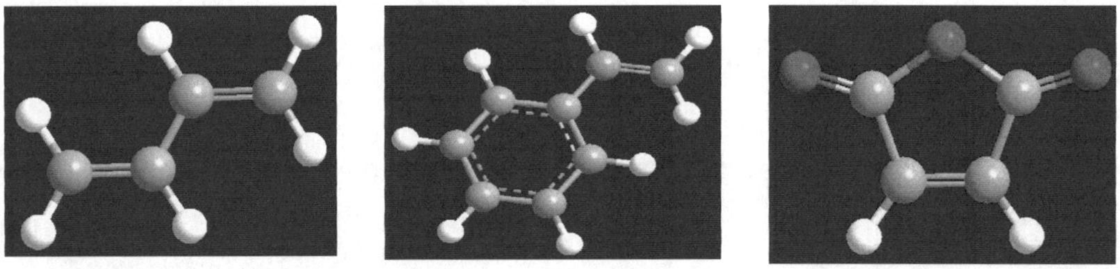

(a) (b) (c)

Figure 2. Double bond conjugated system (a) butadiene (b) styrene (c) Maleic anhydride ; grey = Carbon; red = Oxygen; white = Hydrogen

In this case, the two double bonds are separated by one σ bond. Due to the close proximity of the two π bonds, they can conjugate with one another, for which we shall develop the Hückel molecular-orbital theory. A molecular orbital (MO) may be considered to be built up from a combination of atomic orbitals (AOs), and to extend over the whole molecule. The σ frame- work (using the 2s, $2p_x$ and $2p_y$ orbitals on carbon with the 1s orbitals on hydrogen) separated completely from the π system made up from the $2p_z$ orbitals.

Since this conjugated double bond consists of 4 2p orbitals of carbon to house 4 electron delocalized, by appropriately constructing LCAO-MO (linear combination atomic orbital-molecular orbital) wave functions from 2pz AOs and subjecting these to the secular determinant [12]. Through the Hückel theory, two p orbitals are described as being conjugated with each other in making the π bond. Electrons are delocalized in conjugated systems. This generally lowers the overall energy of the molecule and increase the stability and are subject to the symmetry of the molecule, its point group. Conjugated bonds are stabilizing because they increase the number of resonance stabilized structure that can form. Highest occupied molecular orbital (HOMO) and the lowest unoccupied molecular orbital (LUMO) are the most important orbitals with respect to reactivity [13-16].

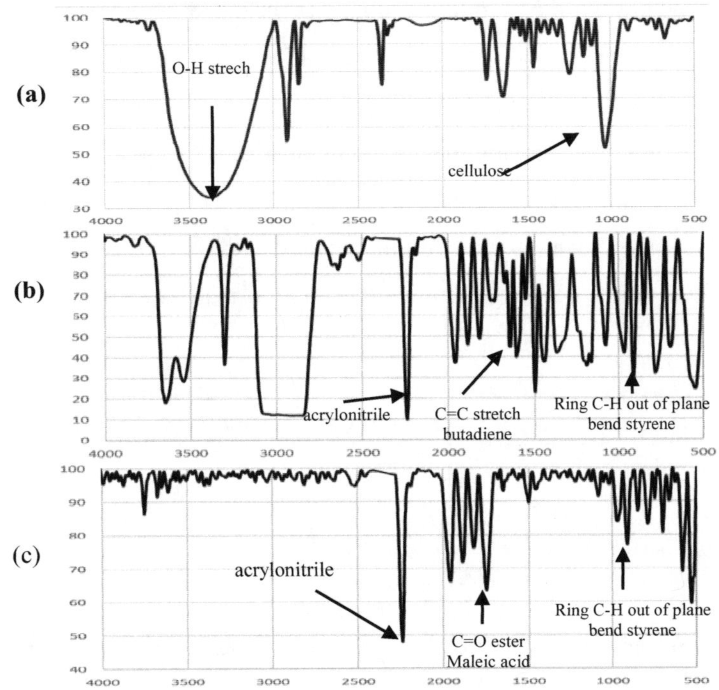

Figure 3. FTIR peaks of (a) OPEFB fiber (b) ABS recycled (c) SF20

Upon analysis of the FTIR spectrum in Figure 3a we see that there is a peak associated with the functional groups of cellulose at peak 1034 cm^{-1} [17-18]. The spectrum related to ABS (Figure 3b) showing the peak marker associated to acrylonitrile appear at 2237 cm^{-1}. The peak at 1639 cm^{-1} associated to C=C for butadiene and styrene, whereas the C=C at 1590 cm^{-1} is a marker for butadiene.

Determination of the peaks associated with styrene is a bit more difficult, therefore the focused only on the aromatic ring of benzene in styrene. Ring C-H in plane bending appear at the 1157 cm^{-1}, ring C-C stretch at 1083 cm^{-1}, and ring C-H out of plane bend peaks appear at 910 cm^{-1} and 849 cm^{-1}. While a peak 910 cm^{-1} related to butadiene, the aromatic band at 1605 cm^{-1} is free from interference, and is utilized as a marker of styrene [19-21].

There is a sharp difference between ABS recycled before and after the addition of filler as we can see in Figure 3b, 3c. The peaks of styrene and butadiene either change its shape and size, on the contrary the peak of acrylonitrile unchanged. This is could be linked to the reactivity of the two monomer molecules due to π bond the C=C, and single torsional bond [22]. Likewise if figure 3a and 3c compared, will be seen a dramatic difference, such as the loss or alteration of the peaks of the filler. Since intensity of the peak at 3379 cm^{-1} in the figure 1c was reduced obviously, indicating that more intermolecular hydrogen bond is destructed compared to the intramolecular hydrogen bond [23]. The loss in butadiene peaks and increasing of ester peak in the FTIR spectra indicating there is a bond between matrix and filler through the coupling agent. Hence, the short fiber 20% (SF20) has a large surface area and high concentration, we concluded that all of the fiber sizes in all concentration we used in this study could be reacting with the coupling agent.

Figure 4 shows the possible reaction mechanisms for reaction that occurs by the addition of the MAH to the ABS, however we still not convince the bond between MAH and cyclic of the styrene. The hydroxyl group of cellulose can interact with this maleic acid to form a bond through the carboxyl group, therefore hydroxyl groups at C-2, C-3 and C-6 of cellulose are reactive and can react with it.

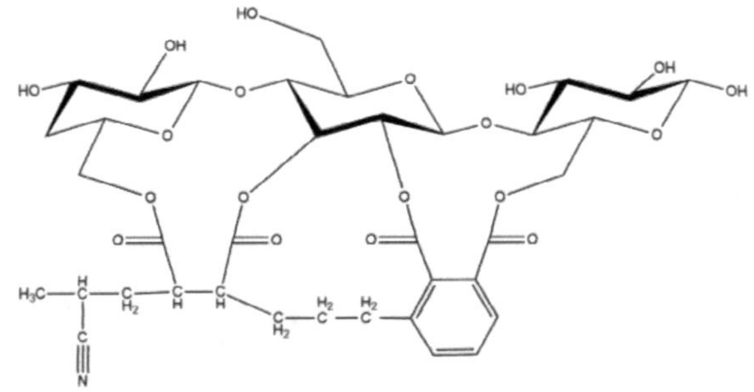

Figure 4. The possible mechanism reaction for cellulose with maleic anhydride and ABS

3.3. Mechanical Test

When several means must be compared we used one-way ANOVA (analysis of Variance), because it is allows us to analyze and interpret observation from several populations. The ANOVA calculation is based on taking the total variation among all observation. This process combines degree to which the confidence intervals overlap into a single measure which, under the null hypothesis of equal population means, has an F-distribution. The level of uncertainty is often expresses as p-value, equivalent to the level of significance at which the null hypothesis would be rejected [9].

Test of the null hypothesis that the means are equal is performed by comparing the observed F-value with the F-tabulated value. The F test will determine if significant differences exist between the sample means, if there is really no difference in the mean obtained from the test, the F-value will be close to zero or close to F-tabulated value, and the *p-value* is higher than uncertainty level (in this study we used $\alpha=0.05$).

Table 3 and 4 shows test of the null hypothesis that the means are equal. Because both of the mechanical test have the observed F-value (18.85 and 23.52) are exceeded from the F-tabulated value (2.64), and the p-value is less than 0.05 (0.000), so we can conclude that there is a significant difference among the population means [24].

Table 3. ANOVA table for impact izod test (KJ/m).

Source	Degree of freedom (df)	Sum of Squares	Mean Square	F-Value	P-Value
Factor	5	1644.1	328.81	18.85	0.000
Error	23	401.1	17.44		
Total	28	2045.2			

Table 4. ANOVA table for hardness test (HRR).

Source	Degree of freedom (df)	Sum of Squares	Mean Square	F-Value	P-Value
Factor	5	3560.7	712.13	23.52	0.000
Error	23	696.4	30.28		
Total	28	4257.1			

ANOVA F-test for equality means is only the initial step in our analysis, our goal should be more than the merely conclude that the treatment differences are indicated by the data. Rather, we must detect likenesses and differences among the treatments. Fisher least significant difference (Fisher LSD) method used in ANOVA to create confidence intervals for all pairwise differences between factor level means while controlling the individual error rate to a level specify [25]. All pairwise were grouped using a letters, if there were a differences or claimed as significantly different it would be shown by a different letters, otherwise if there is a likeness or claimed as significantly not different it would be shown by sharing same letters.

Table 5 shows the impact izod output test that has been performed statistically using Fisher LSD method. MF20 (73.98 kJ/m) was significantly different as a toughest material from all treatments. Both of fiber glass (FG1 and FG2) treatments and MF10 claimed as significantly not different because of sharing same letters. This is due to medium fiber that able to give a good impact because more of the fiber strands inside the test pieces has an effective stress transfer between filler and matrix. An impact strength is defined as how well the material can resist fracturing when force is applied [26-27].

A good bonding between fiber and ABS at high concentration of medium fiber (MF20) has effected the impact strength. Toughness requires strength as well as ductility, which allows a material to deform before fracturing. The impact strength of the biocomposites are strongly influenced by the nature of the fiber, polymer, fiber–matrix interfacial bonding, and testing condition [28].

Table 5. Fisher LSD (n=15, df=5, α=0.05) for the izod impact test.

Factor	Mean	Grouping*
MF20	73.98	A
FG1	59.63	B
MF10	58.35	BC
FG2	57.74	BC
SF20	53.02	CD
SF10	50.85	D

*Means that do not share a letter are significantly different.

A materials ability to withstand local deformation, essentially abrasion resistance, by an indenter is known as hardness [29]. Fisher LSD for hardness test in Table 6 shows that FG1 and FG2 significantly not different with SF20 as a hardest treatments material. This is due to the surface of short fiber at 20% were bonded well to the matrix through coupling agent, the higher value in the hardness in composites indicating good bonding between filler and matrix [30]. Furthermore, the hardness of composites increases along with the increased of particle loading, since the fillers have much larger modulus than the matrix composites stiffness depends significantly in particle loading not particle/matrix adhesion [31-32].

Table 6. Fisher LSD (N=15, df=14, α=0.05) for the hardness test.

Factor	Mean	Grouping*
FG1	113.76	A
FG2	113.20	A
SF20	107.20	AB
SF10	101.80	BC
MF20	96.50	C
MF10	81.60	D

* Means that do not share a letter are significantly different.

4. Conclusion

OPEFB medium fiber and short fiber can be obtained by the mechanical milling and a score of times sieve shaker using 20 and 100 mesh sieves. ABS and OPEFB fiber can be mixed using MAH as a compatible agent. Possible Grafting reactions of Maleic anhydride molecules occur at the double bond of butadiene and styrene. The hydroxyl group of cellulose molecules can interact with maleic acid molecules through the carboxyl group to form a bond. A better properties of OPEFB fiber could be used to replace fiberglass as a reinforcement agent. Thus the OPEFB waste can be reduced to save the environment.

References

[1] BPS-Statistics Indonesia 2014 Indonesian Oil Palm Statistics (www.bps.go.id).
[2] Rutkowsky J V, Levin B C 1986 *Fire and Mater.* **10** 93-105.
[3] Fowler P A *et al* 2006 *J. Sci. Food Agric.* **86** 1781-1786.
[4] Nikmatin S *et al* 2015 *Jurnal Teknologi* **77**(16) 181-187..
[5] Paul S A *et al* 2008 *Compos. Part. A-Appl. S.***39** 1582- 1588
[6] Bahraoui I *et al* 2016 *J. Mechanical and Civil Eng. (IOSR-JMCE)* **13** 43-50.
[7] El Mechtali F Z *et al* 2015 *J. Bionic Eng.* **12** 483-494.
[8] Iwamoto S *et al* 2005 *Appl.Phys.* **A81** 1109-1112.
[9] Wolstenholme L C 2000 *Mechanical Testing of Advanced Fibre Composites* ed Hodgkinson J M (Woodhead Publishing Limited, Abington Hall, Abington.Cambridge CB1 6AH, England) 314-339.
[10] Lohbeck K *et al* 2000 *Ullmann's Encyclopedia of Industrial Chem.* **22** 145-155.
[11] Qi R, Qian J Zhou C 2002 *J. Appl. Polym.* Sci. **90** 1249-1254.
[12] Yates K 1978 *Huckel Molecular Orbital Theory* (New York: Academic press) chapter 2 27-87.
[13] Beez M *et al* 1973 *Helvetica Chimica Acta* **56** 1028-1046.
[14] Skaarup S, Boggs J E *Tetrahedron* **32** 1179-1181.
[15] Ladd M 1998 *Symmetry and group theory in chemistry* (Chicester: Horwood publishing).
[16] Fleming I 2010 *Molecular Orbitals and Organic Chemical Reactions*, Refference edition (Chicester: Wiley).
[17] Stuart B H 2004 *Infrared Spectroscopy: Fundamentals and Applications* (Chicester: John Wiley & Sons).
[18] Kumar A *et al* 2014 *J. Mater. Phys. and Chem.* **2** 1-8.
[19] Hergeth W D 1997 Optical Spectroscopy on Polymer Dispersions. *Polymeric Dispersions: Principles and Application* ed Asua J M (Kluwer Academic Publishers, Dordrecht) 243-256.
[20] Nishikida K, Coates J 2003 *Infrared and Raman Analysis of Polymers Handbook of Plastics Analysis* ed(s) Lobo H, Bonilla J V (Marcel Dekker, Inc., 270 Madison Avenue, New York. USA).
[21] Craig N C *et al* 2005 *J. Molecular Structure* **742** 21-29.
[22] Choi C H, Kertesz M 1997 *J. Phys. Chem.* A **101** 3823-3831.
[23] Zhang Y *et al* 2014 *Carbohydr. Polym.* **99** 126-131.
[24] Johnson R A, Bhattacharyya G K 2010 *Statistics Principles & Methods.*6th ed (John Wiley & Sons,inc., 222 Rosewood drive, Danvers, MA 01923, United States of America).
[25] Williams J, Abdi H 2010 *Encyclopedia of research design*Leevers P S 1999 *Impact Strength Mechanical Properties and Testing of Polymers an A-Z reference* ed. Swallowe G M (Springer science+business media Dordrecht, Kluwer Academic Publishers) 127-150
[26] Leevers P S 1995 *Int. J. Fracture* **73** 109-127.
[27] Joseph P V *et al* 2003 *Compos. Part. A-Appl.S.* **34**(3) 275-290.
[28] Briscoe B J, Shinha S K 1999 Hardness and Normal Indentation of Polymers *Mechanical Properties and Testing of Polymers an A-Z reference* ed. Swallowe G M (Springer science+business media Dordrecht, Kluwer Academic Publishers) 113-122.
[29] Jagannatha T D, Harish G 2015 *Int. J. Mech. Eng & Rob.Res.* **4**(2) **.**
[30] Aruniit A *et al* 2011 *Agronomy Res. Biosystem Eng. Special* **1** 23-29.
[31] Fu S Y *et al* 2008 *Composite Part* B **39** 933-961.

IOP Conf. Series: Materials Science and Engineering **223** (2017) 012065 doi:10.1088/1757-899X/223/1/012065

Effects of additional nanosilica of compressive strength on mortar

N Retno Setiati
Institute Road of Engineering, Research and Development Agency, The Ministry of Public Works and Public Housing, Indonesia

E-mail: retno.setiati@pusjatan.pu.go.id

Abstract. The use of nanosilica as one of the innovations in concrete technology has developed very rapidly. Some research mentioned that nanosilica obtained from the synthesis process silica sand is a type of material that is as pozolan when added to the concrete mix, so as to accelerate the hydration process in concrete. With the addition of nanosilica into the concrete mix, the compressive strength of the concrete can be increased and it has a high durability. This study aims to determine the effect from the addition of nanosilica on mechanical properties of concrete. Laboratory testing is conducted by making the mortar test specimen size of 50 mm x 50 mm x 50 mm. The material used is composed of silica sand, nanosilica, gravel, superplasticizer, cement, and water. Nanosilica percentage amount is added as much as 5, 10, and 15% by weight of cement. Testing of mechanical properties such as compressive strength mortar done at age 3, 7, 14, and 28 days. Based on the analysis and discussion obtained that at 28 days, mortar with the addition of 5% and 15% nanosilica has the compressive strength of 23 MPa. Addition nanosilika into the mortar to improve the mechanical properties by increasing the compressive strength of mortar. The compressive strength of mortar with the addition of 10% nanosilica is 19 MPa. The increase in compressive strength of mortar with the addition of 5% and 15% nano silica is 21% larger than the mortar with the addition of 10% nanosilica and without nanosilica. Nanosilica addition of more than 10% can cause agglomeration when mixed into the mortar so that the impact on the compressive strength of mortar.

1. Introduction

The use of concrete technology that is applied to the bridge structure has developed very rapidly. Several studies have related to high-performance concrete has been done in some countries, one of them is with the nano technology. In Indonesia, the research on nanotechnologies for concrete construction began in 2009. Some research states that nanotechnologies applied to the concrete structure can improve the compressive strength and durability so that it can extend the service life of the bridge. The concept of nano technology aims to make concrete becomes denser. Very high density of the concrete is obtained because the empty spaces that exist between particles of relatively large such as cement particles can be filled by a nanometer-sized grains of fine dust from reactive. Accordingly forming the concrete with a very dense arrangement of the structure, where the pores are formed are in the nanometer size, smaller than the size of the capillary or practically no longer containing capillary-sized pores. With the being filled pores of the concrete by nanomaterial cause the mechanical properties and durability of concrete to be better than conventional concrete [4]. In general, concrete is a mixture of water, cement, fine aggregate, and coarse aggregate. The addition of

IOP Conf. Series: Materials Science and Engineering **223** (2017) 012065 doi:10.1088/1757-899X/223/1/012065

nanomaterial as a cement substitute material causes the strength and durability of concrete is higher. In a study conducted a set of test specimen making and testing of compressive strength of mortar. Nanosilica as the material is smaller than 100 nm have the chemical and physical properties that are superior to large-size material, so that the resulting concrete is expected to have better characteristics. The use nanosilica as a partial cement substitute material capable of increasing the density and improves the mechanical properties of concrete, especially in the area of interface between coarse aggregate and mortar to form a compact structure. The development of nano technology allows nanosilica used as concrete forming materials. However, before being used in the concrete mix, the nanosilica important to know that the performance characteristics of concrete formed according to strength of plan and has no negative impact on the establishment of concrete materials [11]. Experimental test of concrete with nano-scale material substitutions have been carried out both in terms of physical properties and mechanical properties. Some of the studies that have been done related to the addition of nanomaterials to a mortar mix which are:

Byung *et.al.,* (2007) evaluated the compressive strength of mortar with the addition nanosilica 0; 3, 6, 9, and 12%. Variations of water cement ratio used is 0.23; 0.25; 0.32; 0.35 and 0.48. The compressive strength of mortar evaluated with the addition of 5%, 10%, and 15% silica fume. The results of the study mortar at 28 days showed that the strength of mortar with nanosilica higher than the compressive strength of mortar with silica fume [1] . Khanzadi *et al.,* (2010) to test the SEM on the concrete with and without the addition of nanosilica. SEM test results indicate that ettringite crystals that has been infiltrated by the crystal C-S-H to form a slab that is more dense, so concrete with nanosilica more homogeneous than plain concrete [2]. Nittaya and Apinon (2008) carried out tests SEM nanosilica mortar with the addition of 4% and 10%. The addition in the cement paste nanosilica cause microstructural differences pasta [7]. Nili *et al.,* (2010) examines that the use nanosilica 6.0% as a partial replacement of cement in the concrete mix with f.a.s = 0.45 able to increase the compressive strength and reduces absorption [8]. Sobolev *et al.,* (2006) study the mechanical properties of mortar with nanosilica were synthesized using sol-gel method. The results showed an increase in the compressive strength of the mortar in the early stages of hardening followed by a decrease in strength in the elderly. The addition of superplasticizer done to overcome these obstacles. With the addition of superplasticizer in mortar, an increase in the compressive strength of (15-20)% [10]. Luciano *et al.,* (2009) conducted a study on the effect of nanosilica (nano SiO_2) against setting time in the cement paste. Samples made with nanosilica composition of 0%; 1%, 2% and 5% of the weight of the cement. Consistency pasta adjusted by adding superplasticizer in different amounts. In general it was observed that the addition in the cement paste nanosilica affect the process of setting time[3].

IOP Conf. Series: Materials Science and Engineering **223** (2017) 012065 doi:10.1088/1757-899X/223/1/012065

Sadrmomtazi *et al.,* (2009) conducted a study with nanosilica and silica fume added to the mortar mix with a composition of 0; 3; 5%; Nanosilica 7% and 10% silica fume. Based on the results of the study concluded that absorption value was lowest for the mixture with nanosilica as much as 7% [9].
Configuring the microstructure of concrete is influenced by the composition of the mixture, the water-cement ratio (f.a.s), diameter, and the gradation of its constituent materials [5]. The way to determine the microstructure of concrete is cut transversely and elongated so that the configuration between fine aggregate, coarse aggregate, mortar and air pores clearly visible. If the cross section was enlarged with the test SEM (Scanning Electron Microscope), then the chemical elements C-S-H, C-A-S-H and C-H contained in the concrete microstructure can be seen clearly. Repair concrete microstructure can improve the performance of concrete. Concrete ways to improve performance is to increase or decrease the density of pores, increasing the strength of the bond between the mortar and the mortar and coarse aggregate [11]. Interface Transition Zone (ITZ) is the weakest area on the concrete material and often decline and extinction of first. This is caused by the bond between coarse aggregate and mortar are not perfect. Therefore, efforts to strengthen Interface Transition Zone is one of the solutions to obtain high-strength concrete. The concrete given uniaxial load will experience the crack perpendicular to the direction of loading. When cracks occur, the strength will be reduced very rapidly which is characterized by the instability of the concrete after the peak load. Behavior of concrete begins to deviate from linear conditions along with the start cracking in concrete, which initially raised in the transition area aggregates and pasta. The crack propagation resistance weakens the concrete to the work load, thus forming a stress strain diagram of the curve. Mortar is the material used for the construction which consist of a mixture of cement and fine aggregate. The mix of cement and aggregates using specific comparison so that the mortar resistance to stress and pull will be higher or maximum. Based on data specifications, the largest content of silica sand as fine aggregate used is a compound with a composition of 99.09% SiO_2. In addition as a mixture of mortar, sand is used as nanosilica material processed through the synthesis process and nanosiasi. In normal concrete, mortar or bonding strength between the mortar and aggregate are relatively weaker than the strength of aggregate. This study aims to determine the increase in compressive strength when added to nanosilica as a partial cement substitute material

2. Methodology

The methodology used in this research begins with the synthesis and characterization of nanosilica conducted at the Center for Ceramics. Making test specimen carried in Bridge Structures Laboratory. Test specimen was made in the form of mortar size of 50 mm x 50 mm x 50 mm. Making mortar is done with addition of nanosilica variation of 5, 10, and 15% by weight of cement OPC (Ordinary

IOP Conf. Series: Materials Science and Engineering **223** (2017) 012065 doi:10.1088/1757-899X/223/1/012065

Portland Cement). Mortar with addition of 5, 10, and 15% respectively nanosilica identified as NS5, NS10, and NS15. Preparation of test specimens for each mortar is 3 specimen. Fine aggregate used was silica sand from Bangka island with a mesh size 325. Each variation uses a mixture of silica fume (SF) with a percentage of 5% of the weight of cement. The addition of superplasticizer conducted to ease the workability of mortar mix with a maximum use of 2% to weight of cement. The testing of mortar compressive strength is done at the age of 3, 7, 14, and 28 days with the amount each of the 3 test specimen. Testing equipment used in making mortar consists of ovens, scales, mixers, vicat, measuring cup, cone mold, leveling instrument, vibrators, dryers fine aggregate, mortar mold, and other equipment. Mortar mix made with the following composition:

Cement OPC	= 475	g (21%)
Silica sand from Bangka	= 1392	g (60%)
Silica Fume (5%)	= 25	g (1%)
Water	= 399.8	g (18%)

SEM testing aims to determine the configuration of the mortar structure microscopically and element content in mortar. SEM testing for mortar without nanosilika done with a magnification of 1000 times, 5000 times, 10000 times, 15000 times, 20000 times, and 50000 times. Making the mortar mix made with reference to the SNI-03-6825-2002 (method of testing the compressive strength of Portland cement mortar for civil works).

3. Results and Discussion

Penambahan nanosilika ke dalam campuran mortar dimaksudkan untuk mengisi rongga atau pori yang terkandung dalam mortar. By being filled voids or pores are expected structural configuration within the mortar could be more solid so that the compressive strength of the mortar is higher than conventional mortar. Nanosilica are put into a mortar mix should be reactive, because in addition to functioning as a filler, it acts as a binder nanosilica. To determine the level of reactivity nanosilica, it must be tested XRD. If the test result shows the XRD amorphous chart pattern, then the nanosilica are reactive and can be added to the mortar to improve the compressive strength than mortar. Nanosilica used in this study are shown in Figure 1.

Figure 1. Nanosilica were added into the mortar

Based on the test results of XRD, the diffraction pattern of the resulting nanosilica is amorphous as shown in Figure 2. Nanosilica obtained from the synthesis when it is separated from impurities (ie Al_2O_3) will produce the compound SiO_2 greater than 99%. If it does not done the continuation of the synthesis process, the test results will show the XRD diffraction pattern of crystalline. The sharp diffraction patterns indicate that the process of synthesis which do not produce the amorphous nanosilica.

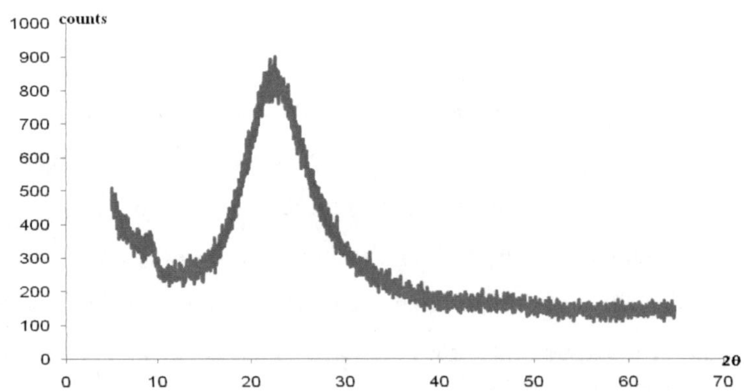

Figure 2. Nanosilica diffraction pattern (XRD test results) (Source: ITB 2015).

With the diffraction pattern as shown in Figure 1 is proven that nanosilica used as a cement substitute material is reactive, so that when added to the mortar will cause an increase in the compressive strength. The amorphous properties of that shown in Figure 2 indicates that the levels of SiO_2 compound contained in such nanosilica in greater than 99%. At the time of the hydration process, a mortar mix consisting of silica sand, cement, and water will react to primary products calcium silicate hydrate (3CaO.2SiO2.3H2O) gel and calcium hydroxide. The main elements of this paste directly influence the strength of the mortar and to function as a space filler between the cement grains, filling the pores of the capillaries thus diminishing the porosity and increasing strength.

The test results of mortar to the age of 3 days, 7 days, 14 days and 28 days are shown in Table 1.

Table 1. The compressive strength of mortar with addition of variations nanosilica.

No	Sample	Compressive strength (MPa)			
		3 days	7 days	14 days	28 days
1	NS0 (1)	13	16	19	21
2	NS0 (2)	14	15	17	20
3	NS0 (3)	14	15	19	19
	Average	14	15	18	20
	Standard deviation	0.64	1.01	1.50	0.63
1	NS5 (1)	15	17	20	23
2	NS5 (2)	15	18	21	24
3	NS5 (3)	14	17	20	24
	Average	15	17	20	24
	Standard deviation	0.79	0.73	0.59	0.61
1	NS10 (1)	12	18	16	21
2	NS10 (2)	11	17	14	20
3	NS10 (3)	11	15	16	18
	Average	12	17	15	20
	Standard deviation	0.20	1.57	0.93	1.19
1	NS15 (1)	15	16	23	27
2	NS15 (2)	13	17	25	26
3	NS15 (3)	11	17	21	18
	Average	10	11	16	17
	Standard deviation	4.97	6.89	8.90	8.70

Remarks:
NS0 is a mortar without nanosilica
NS5 is a mortar with the addition of 5% nanosilica
NS10 is a mortar with the addition of 10% nanosilica
NS15 is a mortar with the addition of 15% nanosilica

Based on Table 1 the value of the average compressive strength of mortar N0, N5, N10, and N15 at 28 days respectively was 20 ; 24 ; 20 and 17. Barchart against the concrete compressive strength test results in Table 1 can be seen in Figure 3.

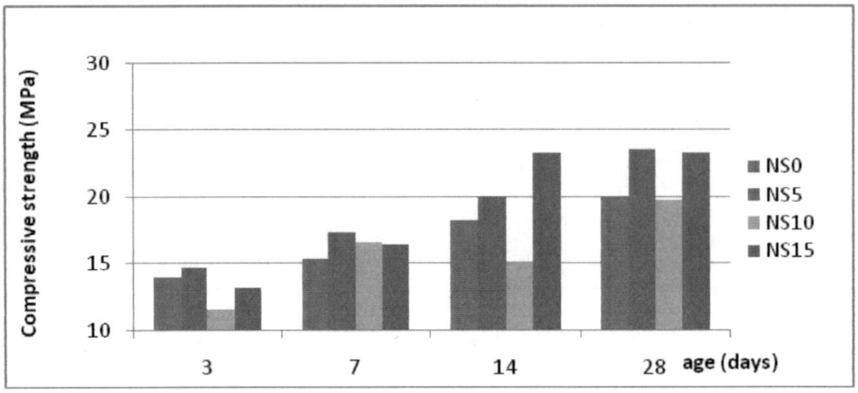

Figure 3. Barchart of compressive strength for mortar.

Figure 2 shows the development of the compressive strength of mortar by age. All mortar composition has the behavior of different compressive strength at age 3, 7, 14, and 28 days. NS10 mortar compressive strength at age 3 days is the most low of 12 MPa and the highest is NS5 at 15 MPa. At the age of 3 days, mortar with the addition of 5% nanosilica (NS5) 15 MPa compressive strength is almost the same as the mortar without nanosilica amounted to 14 MPa. Furthermore, at the age of 7 to 28 days of mortar that contains nanosilica progressing compressive strength faster than the compressive strength of mortar without nanosilica. Nanosilica addition of more than 10% of the compressive strength will not increase because during the process of mixing the mortar will occur agglomeration. This is caused by nanosilica is fine and reactive material so that when mixed with calcium oxide (CaOH) free produce secondary reactions that form new cement paste. The development compressive strength of mortar with 10% nanosilica at the age of 3 days up to 28 days has a different strength than other mortar. At the age of 3 day compressive strength of 12 MPa was the lowest compared to other mortar. After reaching the age of 7 days, the compressive strength of mortar (NS10) an increase of 17 MPa. But at the age of 14 days, compressive strength being dropped 15 MPa and at 28 days to 20 MPa. The increase in compressive strength of mortar with the addition of 15% nanosilica (NS15) occurred in the age of 7 to 28 days. At the age of 3 days, the compressive strength of mortar NS15 is smaller (i.e 13 MPa) than without nanosilica (14) MPa. It is caused the hydration reactions at beginning of age is still dominated by the cement hydration reaction with water. At the age of 28 days, the compressive strength of mortar with the addition of 15% nanosilica (NS15) at 23 MPa less than the compressive strength of mortar with 5% nanosilica (NS5) is 24 MPa. But at the age of 14 days, the compressive strength of mortar NS15 larger (ie 23 MPa) compared with 5% nanosilica mortar (NS5) with a compressive strength of 20 MPa. The increase in compressive strength of mortar with the addition of 5% nanosilica is the most optimum for the increase is continuous from the age of 3, 7, 14 up to 28 days. Mortar with the addition of a 10% increase in compressive strength nanosilica experience a discontinuity at the age of 14 days. This could be caused during the process of mixing mortar occur agglomeration nanosilica so it does not react in the optimum. From some of the tests conducted, if nanosilica mixed with water and water will immediately be absorbed. While the percentage increase of 15% nanosilica to mix mortar for compressive strength increase is not significant at 28 days. To determine microscopic structure of the mortar test SEM (Scanning Electron Microscope)/EDS (Electron Dispersion Spectrometry). Testing SEM/EDS was conducted to determine the composition and structure of chemical compounds in micro mortar. Figure 4 shows the test results of SEM/EDS mortar without addition of nanosilica. Magnification of mortar structure made up to 1000, 5000, 10000, 15000, 20000 and 50000 times.

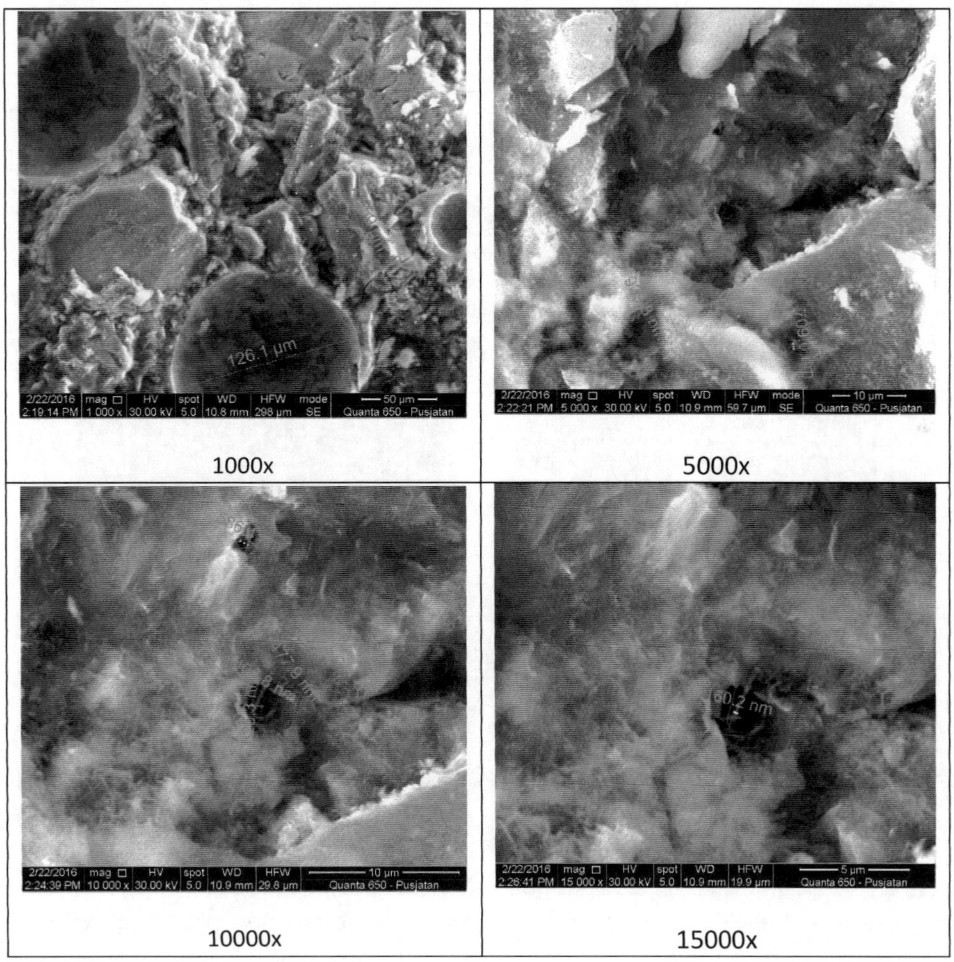

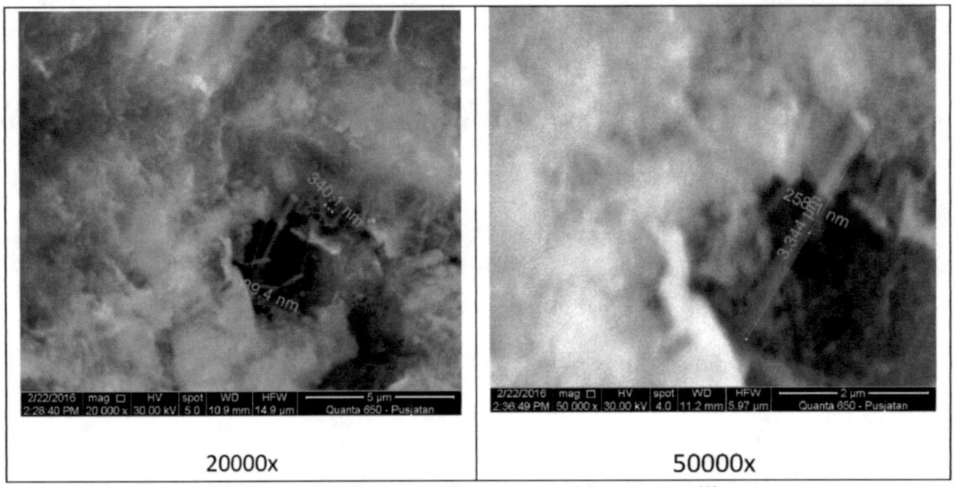

Figure 4. SEM test results mortar without nanosilica.

Based on Figure 4 can be seen that the mortar is still a small pipes at a magnification of 20000 times. Small pipes have a hollow (or pore) that causes mortar not be impermeable. The composition of elements and chemical compounds contained in mortar shown in Figure 5.

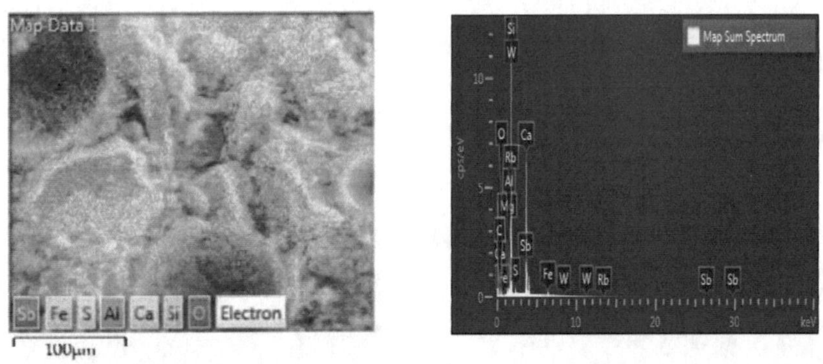

Figure 5. The composition of the configuration (a) and the number of element content in mortar without nanosilica (b).

The number of elements and chemical compounds in the mortar of Figure 4 can be seen in Table 2.

Table 2. Total composition of elements and compounds in a mortar without addition nanosilica.

Element	Line Type	Apparent Concentration	k Ratio	Wt%	Wt% Sigma	Standard Label	Factory Standard
1	K series	4.75	0.04746	10.54	0.43	C Vit	Yes
O	K series	76.77	0.25833	55.07	0.34	SiO2	Yes
Mg	K series	0.96	0.00636	0.67	0.03	MgO	Yes
Al	K series	1.65	0.01186	0.91	0.03	Al2O3	Yes
Si	K series	38.85	0.30783	18.04	0.15	SiO2	Yes

S	K series	0.96	0.00829	0.47	0.02	FeS2	Yes
Ca	K series	34.74	0.31037	12.34	0.09	Wollastonite	Yes
Fe	K series	1.59	0.01591	0.66	0.02	Fe	Yes
Rb	K series	0.03	0.00025	0.01	0.07	Rb (v)	Yes
Sb	L series	3.26	0.03259	1.30	0.07	Sb	Yes
W	L series	0.00	0.00000	0.00	0.00	W	Yes
Total:				100.00			

Based on Table 2, the element oxygen (O) contained in the compound in a mortar SiO_2 is the highest of 55.07% and element content of silica (Si) in the compound was 18:04%. Total SiO_2 composition of each of the elements oxygen and silica amounted to 73.11%. The percentage of the element silica comes from silica sand used as fine aggregate. While the composition of the structure in a mortar with the addition of nanosilica (for the number of nanosilica 5% by weight of the binder) can be seen from the test results of SEM as shown in Figure 6.

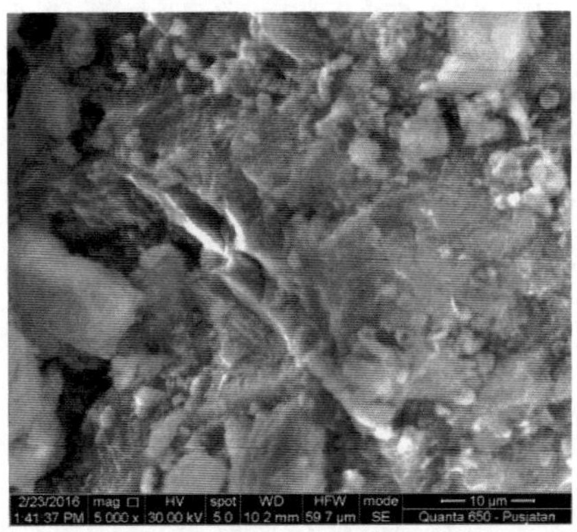

Figure 6. SEM test results mortar with the addition of 5% nanosilica.

Based on Figure 6, the structural arrangement mortar with the addition of 5% nanosilica there are the pipes as the capillary cavity. At a magnification of 20000 times, look more resistant mortar. The composition of elements and compounds contained in the mortar can be seen from the test results of EDS as shown in Figure 7.

IOP Conf. Series: Materials Science and Engineering **223** (2017) 012065 doi:10.1088/1757-899X/223/1/012065

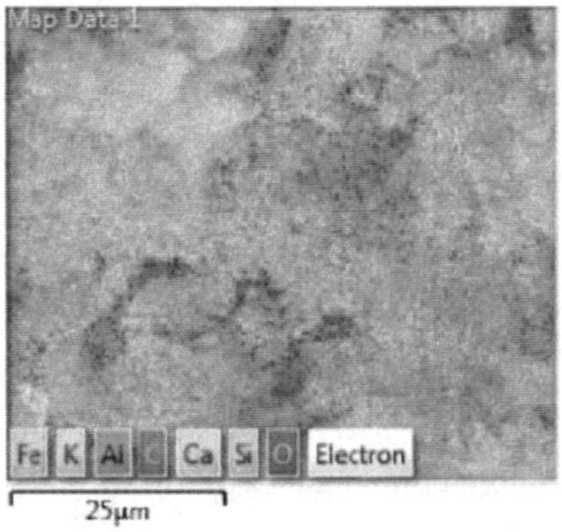

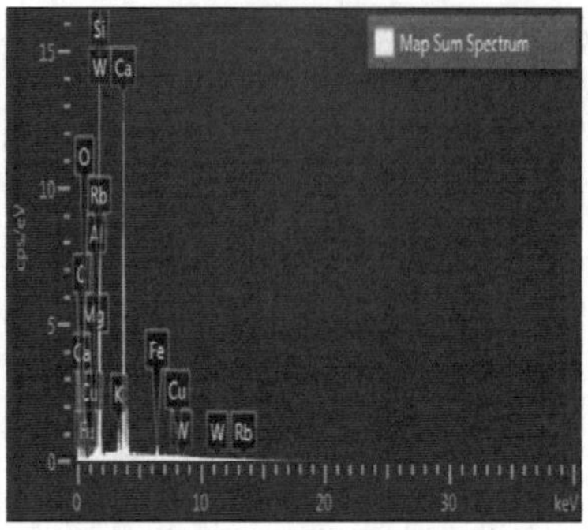

Figure 7. The composition of the configuration (a) and the number of element content in mortar with 5% nanosilica (b).

Based on Figure 7, the element oxygen (O) contained in the mortar is the most dominant element. Total the composition each of the elements and compounds contained in a mortar with the addition of 5% nanosilica shown in Table 3. From Table 2, the element oxygen (O) contained in the compound in a mortar SiO_2 is the highest of 48.56% and element content of silica (Si) in the compound was 14:03%. Total SiO_2 composition of each of the elements oxygen and silica amounted to 62.59%. The percentage of the element silica comes from silica sand used as a fine aggregate and nanosilica. This is less than the amount of SiO_2 mortar composition without nanosilica. By the SiO_2 content as the dominant chemical compounds, nanosilica potentially reducing calcium hydroxide in mortar.

Table 3. Total composition of elements and compounds in a mortar with the addition of 5% nanosilica.

Element	Line Type	Apparent Concentration	k Ratio	Wt%	Wt% Sigma	Standard Label	Factory Standard
C	K series	9.18	0.09181	18.13	1.09	C Vit	Yes
O	K series	52.36	0.17621	48.56	0.70	SiO2	Yes
Mg	K series	0.76	0.00507	0.54	0.03	MgO	Yes
Al	K series	2.95	0.02122	1.66	0.04	Al2O3	Yes
Si	K series	29.10	0.23057	14.03	0.22	SiO2	Yes
K	K series	3.26	0.02761	1.16	0.03	KBr	Yes
Ca	K series	40.37	0.36070	14.75	0.22	Wollastonite	Yes
Fe	K series	2.28	0.02283	0.99	0.04	Fe	Yes
Cu	K series	0.41	0.00408	0.18	0.04	Cu	Yes
Rb	K series	0.00	0.00000	0.00	0.00	Rb (v)	Yes
W	L series	0.00	0.00000	0.00	0.00	W	Yes
Total:				100.00			

This reduction continued because the chemical reaction between compounds of silica dioxide (SiO_2) with calcium hydroxide to form a calcium silicate hydrate (CSH) new. It can be concluded that nanosilica freely mixed with calcium hydroxide will happen further reaction that forms a new cement paste. This gives effect to the relationship between the surface of the mortar with the zone to be stronger and provide increased compressive strength in mortar.

4. Conclusion

The highest mortar compressive strength at 28 days resulting in a mixture with nanosilica 5% ie 23.5649 MPa.

Nanosilica addition of more than 10% can cause agglomeration when mixed into the mortar so that tje impact on then compressive strength of mortar. It was proven that the compressive strength of the mortar with the addition of more than 10% nanosilica be lower than conventional mortar. Mortar compressive strength with the addition of 15% nanosilica at 17 MPa lower than conventional mortar which is 20 MPa. The decrease in the compressive strength can be caused when the distribution mixing of nanosilica were added to the mortar uneven, in part, on the agglomeration and therefore can not fill the pores in the mortar.

Acknowledgments

The author would like to thank Mr. DR. Ir. Herry Vaza, M.Eng.Sc. as the Head of Institute Road of Engineering and Mr. Setyo Hardono, ST., MT which has provided support for this research activity. Furthermore, the authors also wish to thank Mr. M. Ikhlasul Amal, PhD., Mr. DR. Jonbi, and Miss. DR. Saloma as a resource person who has given the science is very helpful to the author.

References

[1] Byung *et al* 2007 *KSCE Journal of Civil Engineering* **11** 37-42.
[2] Khanzadi M *et al* 2010 *Second International Conference on Sustainable Construction Materials and Technologies.*
[3] Luciano S et al 2009 3rd *International Conference on Concrete and Development,* Teheran, Iran.
[4] M. Schmidt E Fehling 2007 *10 Years of Research and Development at the University of Kassel, Kassel* University Press, Schriften Reihe Baustoffe und Massivbau Heft 7.
[5] Mehta P K and Monteiro 2006 *Concrete Microstructure, Properties, and Materials The Mc Graw Hill Companies, Inc.,* New York, NY.
[6] Nazari A *et al* 2010 *J. American Sci.* 8 550-554.
[7] Nittaya T and Apinon N 2008 *J, Nat, Sci, Special Issue on Nanotechnology*,7.
[8] Nili *et al* 2010 *Second International Conference on Sustainable Construction Material and Technology*, Italy.
[9] Sadrmomtazi *et al* 2009 *3rd International Conference on Concrete and Develpoment,* Tehran, Iran.
[10] Sobolev K *et al* 2006 *Proceedings of ACI Session on "Nanotechnology of Concrete: Recent Developments and Future Perspective,* Denver, USA.
[11] Saloma 2014 *Sifat Fisik dan Mekanik beton Nanomaterial Disertasi program Doktor* , Institut Teknologi Bandung.

IOP Conf. Series: Materials Science and Engineering **223** (2017) 012066 doi:10.1088/1757-899X/223/1/012066

Water absorption and its effect on the tensile properties of tapioca starch/polyvinyl alcohol bioplastics

H Judawisastra[1], R D R Sitohang[2], L Marta[2] and Mardiyati[1]
[1] Materials Science and Engineering Research Groups, Faculty of Mechanical and Aerospace Engineering, Institut Teknologi Bandung, Ganesha 10, Bandung 40132.
[2]Alumni of Materials Science and Engineering Study Program, Institut Teknologi Bandung, Bandung, Indonesia

E-mail : hermawan.judawisastra@material.itb.ac.id

Abstract. Tapioca is one of the largest sources of starch and makes it suitable to be used for bioplastic material. Addition of polyvinyl alcohol (PVA) has been shown to successfully reduce the brittleness of starch bioplastic. This study aims to investigate the influence of PVA addition to water absorption behavior and its effect on the tensile properties of tapioca starch/PVA bioplastics, which are still not yet fully understood until now. The bioplastics were prepared by solution casting method at gelatinization temperature, with PVA addition from 0 to 29 wt%. Examinations were carried out by means of water absorption test, tensile test and Fourier Transform Infrared (FTIR) Spectroscopy. Increasing content of PVA, up to 29 wt%, was found to decrease the water absorption of the bioplastics, with the lowest water saturation point of 251%. This is due to the interaction between starch and PVA which reduces the free OH groups in the resulting bioplastics. Consequently, this led to a decrease in water absorption-related deterioration, i.e. tensile properties degradation of the bioplastics. The addition of 29 wt% resulted into the lowest degradation in tensile strength (6%) and stiffness (30%), while accompanied with the highest elongation increase (39%) after water immersion.

1. Introduction

There has been a growing interest in the application of bioplastics, which are plastics material that are either biobased, derived from biomass; biodegradable; or features both properties [1]. This is mainly driven by the major issues involved in the use of conventional petrochemical plastic that are non-biodegradable and ultimately result into several environmental problems. Starch-based plastics are one type of bioplastics that are both biobased and biodegradable. The raw material for this type of plastics is sustainable and abundant, and thus make them a promising candidate as an alternative to conventional plastics. Starch can be found in the seeds, leaves, fruits, stems, roots, and tubers of most plants, such as maize, rice (paddy), wheat, potatoes, cassava, bananas, yams, and sorghum [2].

Despite the growing interest in bioplastics, the use of starch-based plastic is still limited due to its brittleness and moisture sensitivity when compared to conventional synthetic plastics [3]. The hydrophylic nature of starch makes the plastic based on this material have a poor long-term stability and its mechanical properties are sensitive to moisture [1, 3]. One strategy that has been demonstrated to improve starch-based plastic performance is by adding polyvinyl alcohol (PVA). PVA is a water-soluble synthetic polymer that is made from petrochemicals, and hence is biodegradable [4]. Therefore, the blends composed of PVA and starch are biodegradable. The addition of PVA is of particular interest, since it has been shown to have an excellent compatibility with starch [5], and in fact has successfully reduced the brittleness and the water uptake of starch-based plastics [6].

Tapioca (cassava) is one of the largest source of starch, particularly in tropical environments [2]. This type of plant is abundantly available in Indonesia as it is one of the main food crops in the country. Hence, it is believed that tapioca has a great potential to be used for bioplastic material. Several studies on this topic have been carried out for the past years, i.e. to develop plastic foams [7], edible films [8], and thermoplastic starch [9]. To present, however, the effect of water absorption in tapioca starch/PVA bioplastic, particularly to its tensile properties degradation, is still not yet fully understood. Therefore, this study aims to investigate the influence of PVA addition to water absorption behavior and its effect on the tensile properties of tapioca starch/polyvinyl alcohol bioplastics.

2. Research Methodology

2.1. Materials

A commercially available tapioca starch, trade name Gunung Agung supplied by PT. Budi Acid Jaya Tbk., was used in this study. The polyvinil alcohol (PVA) polymers used were purchased from Central Kimia, Bandung. High water content was used to produce starch-based bioplastic film.

2.2. Bioplastic Film Preparation

The bioplastic films were prepared by solution casting method, with PVA addition at different concentrations of 0 (starch reference material), 17 and 29 wt%. Pure PVA was also investigated as PVA reference material. The weight percentages of PVA were calculated with respect to the total weight of dry starch and PVA. The sample nomenclature is referenced according to the tapioca starch (TS) and PVA contents, i.e. TS/PVA x/y for a film with x wt% TS and y wt% PVA. All films were made with constant amount of water and dry tapioca starch, i.e. 140 g and 10 g, respectively. PVA was first added into a beaker containing water. The mixture was then stirred with a magnetic stirring bar and heated to 85°C. Once the desired temperature of 85°C was reached, the mixture was immediately cooled down to 40°C using a cooling fan. Tapioca starch was then added to the mixture, stirred and heated in a water bath to the gelatinization temperature of 68°C. Once the gelatinization temperature was reached, the blend was cast onto the aluminium mould and let dried at room temperature for about 1 to 2 days.

2.3. Water Absorption Tests

Water absorption tests were performed on the samples following ASTM D570-98 with modification, i.e. immersion method in distilled water at 21.5°C for the duration up to 60 min. The samples were taken out periodically at different intervals, i.e. at 1, 3, 5, 10, 15, 20, 25, 30, and 60 min. After being taken out from the water, all of the samples were weighed immediately after wiping out the water on the surface of the sample. The amount of water absorbed was determined by calculating the weight gain after immersion to the nearest 0.0001 g.

2.4. Bioplastic Film Characterizations

Chemical structure of the bioplastics was characterized by Fourier Transform Infrared (FTIR) Spectroscopy in order to examine the mixing of the tapioca starch and PVA, and the effect of water absorption. The FTIR characterizations both before and after water immersion, with the samples in dry condition, were carried out in order to investigate the effect of water absorption.

Tensile tests were carried out according to ASTM D-882-02 for thin film with a thickness of 0.1 mm. Tensile properties were measured using a Tensilon RTF-1310 equipped with a 1 kN load cell, with a 10 cm gage length at a crosshead speed of 5 mm/min. Two replicates of the samples before and after water immersion were tested in dry condition for each type of material and the average value was reported. The wet samples after immersion were first dried using cooling fan prior to the tensile test. The samples were dried until the initial weight before water immersion was reached.

3. Results and Discussion

3.1. Tapioca Starch/Polyvinyl Alcohol Bioplastics Chemical Structure Analysis

FTIR characterization results of the pure tapioca starch (TS) and pure polyvinyl alcohol (PVA) bioplastics are shown in Figure 1a and b, respectively. Functional groups assignment of TS and PVA has been reported in many other studies [10-13].

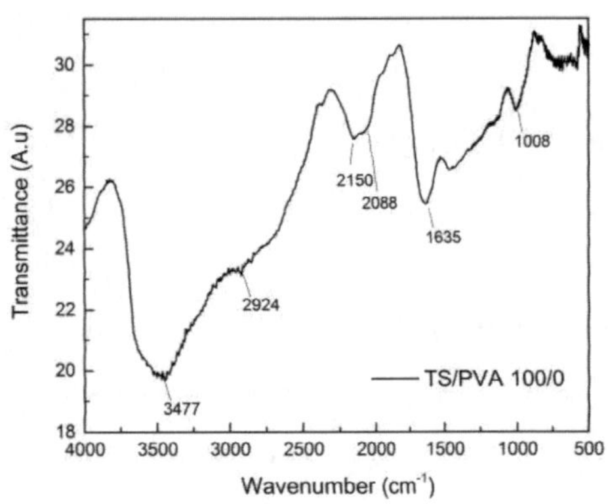

(a)

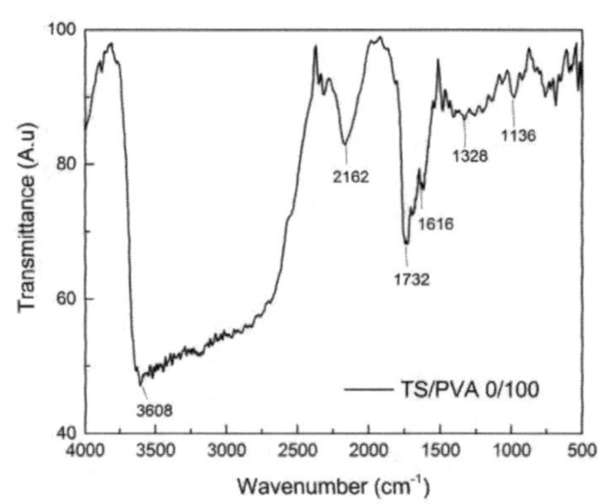

(b)

Figure 1. FTIR spectra of: (a) pure tapioca starch (TS) and (b) pure polyvinyl alcohol (PVA) bioplastic reference materials.

The characteristic peaks assignment for the pure TS and pure PVA are summarized in Table 1. Compared with those already reported by other research groups [10-13], some differences are evident from the FTIR result in the present work. In the pure TS, peaks at 2150 and 2088 cm^{-1}, were not evident in the other reported studies [10, 11]. Those peaks are indicators for combination of O-H stretching vibration that represent the degree of hydrogen bonding in TS. The adjacent peaks to the peaks at 2150 and 2088 cm^{-1} indicate that there are hydrogen bonds among amylose chains, and between amylose and amylopectin.

In the pure PVA, peak at 2162 cm^{-1} were not evident in the other reported study on PVA [12, 13]. This is again the peak which is indicator for combination of O-H stretching vibration. The single peak in PVA indicates that the hydrogen bonds are only formed among PVA chains. Further, the peak at 3000 - 2800 cm^{-1}, which is assigned for C-H stretching of CH_2 [12, 13], was not obviously seen in the FTIR spectra of pure PVA (see Figure 1b). This may occur due to the presence of water in the pure PVA sample which then makes the broad O-H stretching peaks to overlap with the C-H stretching peak.

Table 1. FTIR assignment of pure tapioca starch and pure polyvinyl alcohol bioplastic reference materials

Assignment	Tapioca starch	Polyvinyl alcohol
O-H stretching	3477 cm^{-1} (broad)	3608 cm^{-1} (broad)
C–H stretching of CH_2	2924 cm^{-1}	-
Combination of O-H stretching	2150 and 2088 cm^{-1}	2162 cm^{-1}
C=O stretching of PVA residue	-	1732 and 1616 cm^{-1}
O-H bending	1635 cm^{-1}	-
C-H wagging of CH_2	-	1328 cm^{-1}
C-O stretching	1008 cm^{-1}	1136 cm^{-1}

Figure 2 presents the Fourier transform infrared spectroscopy (FTIR) spectroscopy characterization result for all tapioca starch/polyvinyl alcohol (TS/PVA) bioplastics. Pure TS and pure PVA reference materials are added for comparison. There are two regions of interest shown in Figure 2 i.e. the first region (I), broad peaks in the range 3500 to 3400 cm^{-1}, which are assigned for O-H stretching vibration, and the second region (II), from 2200 to 2000 cm^{-1}, where the O-H stretching combination peak can be found.

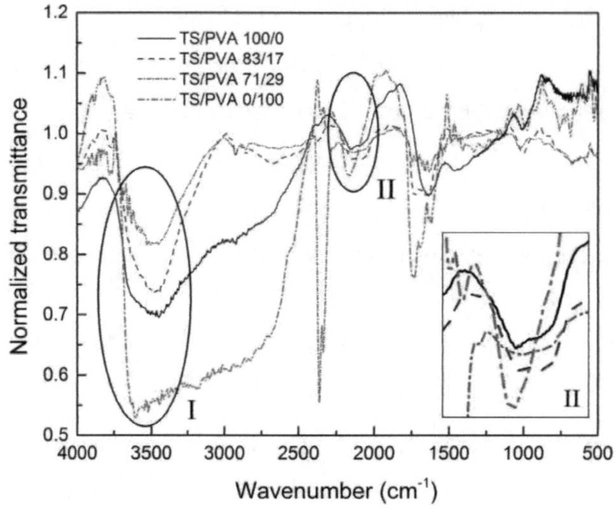

Figure 2. Combined FTIR spectra of pure tapioca starch (TS/PVA 100/0), pure polyvinyl alcohol (TS/PVA 0/100) and tapioca starch/polyvinyl alcohol bioplastics

It can be seen from region I that the highest O-H stretching peak can be found in the pure PVA and immediately followed by pure TS. The addition of PVA appears to reduce the amount of O-H bonds in the pure TS, which indicates that free OH groups in the pure TS are forming hydrogen bonds with the OH groups in the pure PVA. The increasing amount of hydrogen bonding between TS and PVA leads to the decrease in the detected amount of O-H stretching.

The O-H stretching combination found in region II represents the degree of hydrogen bonding in the samples. There are two adjacent peaks at 2150 and 2088 cm^{-1} in the pure TS, while in the pure PVA, there is one peak at 2162 cm^{-1}(see Figure 2). In the bioplastics with the addition of 17 and 29 wt% PVA, there exists two adjacent peaks, i.e. at 2146 and 2073 cm^{-1}, and at 2135 and 2071cm^{-1}, respectively. Note that the peak at 2017 cm^{-1}in the higher amount of PVA, at 29 wt%, is not as evident as the peak at 2073 cm^{-1}seen in the 17 wt% PVA sample. This shows that the amount of peaks in the resulting bioplastics gets more closely resemble that of the pure PVA with the addition of PVA. Therefore, the observations in region I and II (as shown in Figure 2) confirms the formation of bonds between PVA and tapioca starch.

3.2. Water Absorption Behaviour

Figure 3 shows the amount of water absorbed by the bioplastic film samples at three different PVA contents. All of the films absorbed water very rapidly during the initial stage and later gradually level off reaching the saturation level after around 30 min of immersion. The pure TS film showed higher water absorption compared to the TS/PVA films. Water absorption increased with increasing immersion time, but decreased with increasing PVA content, with the lowest water saturation point of 251% from the addition of 29 wt% PVA. This is similar to the behavior seen in another starch blend films, i.e. polyvinyl alcohol/corn starch blends [14]. The starch is mainly responsible for the water absorption of the blends due to its inherent hydrophilic nature.

A general rule of mixtures can be used to explain the decrease of water absorption with increasing PVA content. The rule is that the final properties of a blend consisted of two or more polymers are a weighted mean of the material constituents' properties. The tapioca starch reference material has been shown to have a maximum water absorption of 495% (see Figure 3, TS/PVA 100/0), while PVA has been reported to have a maximum of 43% [14]. The addition of 17 and 29 wt% PVA resulted into water saturation point of 271% and 251%, respectively. These values are in fact still in the range between the saturation points of each pure material constituent.

As previously shown in the FTIR result in section 3.1. the interaction between PVA and TS leads to the decrease of free OH groups in the TS/PVA bioplastics. This will ultimately make the resulting TS/PVA bioplastics become less hydrophilic, hence less sensitive to the presence of water.

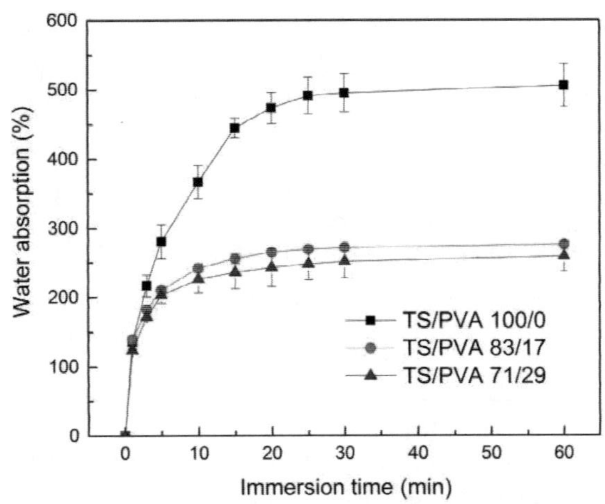

Figure 3. Water absorption of tapioca starch/ polyvinyl alcohol (TS/PVA) bioplastics.

3.3. Fourier Transform Infrared Spectroscopy (FTIR) Spectroscopy Analysis after Water Absorption

Figure 4 presents the FTIR characterization results for pure TS (see Figure 4a) and TS/PVA (see Figure 4b) bioplastics in two conditions: before and after water immersion. There are four regions of interest which show the influence of water absorption on the chemical structure of the bioplastics. A similar tendency of the influence of water absorption was observed for both the pure TS and TS/PVA bioplastics.

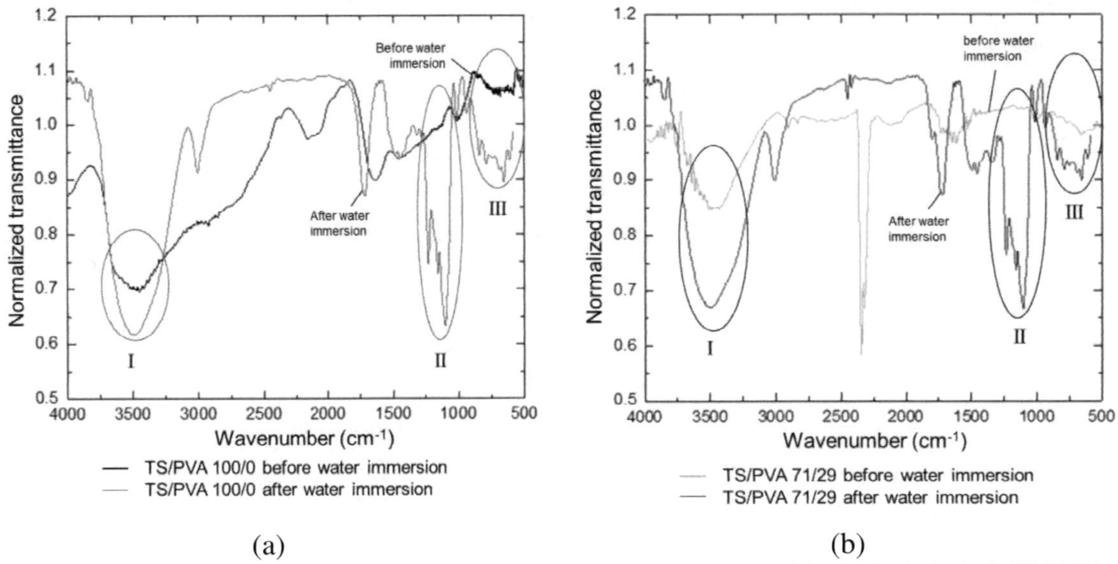

Figure 4. Effect of water absorption on the chemical structure of tapioca starch/ polyvinyl alcohol (TS/PVA) bioplastics samples (tested in dry condition).

The observation in the three regions indicates that hydrolysis of starch occurs due to the interaction with water. After water immersion, the O-H stretching peak, broad peak in the range from 3500 to 3000 cm^{-1} (see region I), became stronger. Further, strong peaks from 1300 to 1000 cm^{-1} (see region II) and around 840 and 915 cm^{-1} (see region III) became evident after water immersion. The strong peaks in region II are assigned for C-O stretching, while the strong peaks in region III are characteristic for α-D glucose. After water absorption, hydrolysis may occur in the bioplastics when water attacks C-O-C bonds in starch and forms O-H bonds, leading to the breaking down of starch into smaller chains and hence degrades the tapioca starch [15]. As indicated in the FTIR result, the breaking down of C-O-C bonds in starch leads to the formation of C-O bonds and α-D glucose monomers. This degradation in starch chemical structure may ultimately affect the tensile properties of the bioplastics, as will be shown later in the following section.

3.4. Effect of Water Absorption on Tensile Properties of Tapioca Starch/Polyvinyl Alcohol Bioplastics

Figure 5 shows how the tensile properties of the tapioca starch/polyvinyl alcohol bioplastics are affected by water absorption. When looking at each tensile property individually, the tensile strength after water absorption appears to be similar for all types of TS/PVA bioplastics. On the other hand, the tendency in the change of both the tensile stiffness and elongation after water absorption were still similar with that of before water absorption. After water absorption, the bioplastics with 29 wt% PVA still have a lower stiffness but with a higher elongation compared to the ones with lower PVA content.

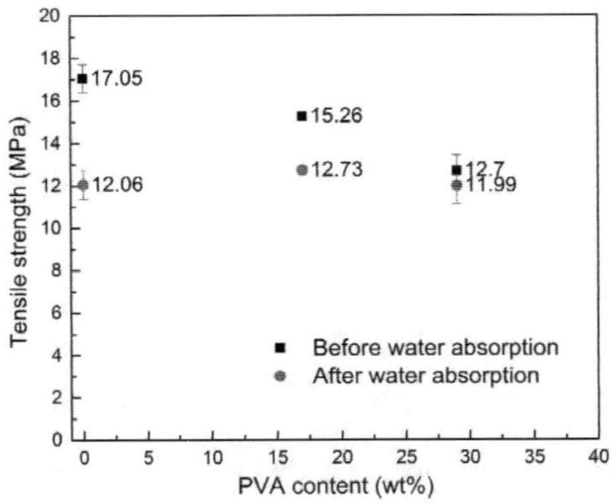

(a)

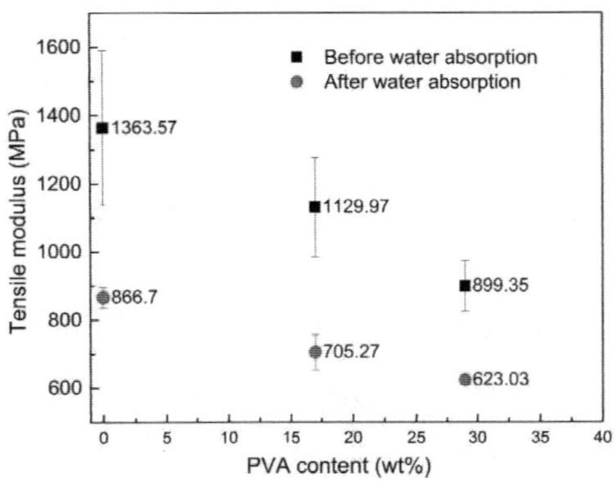

(b)

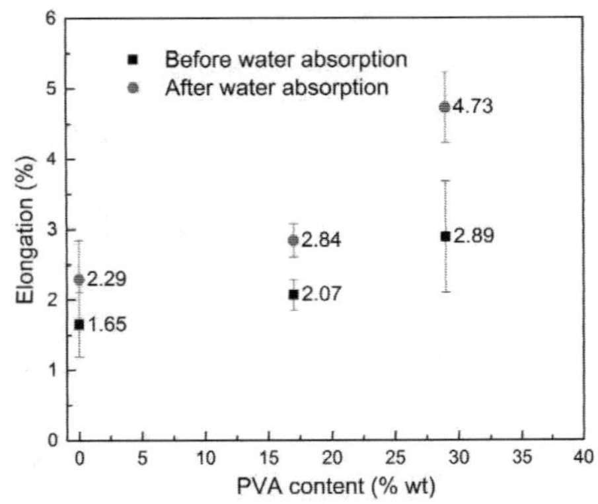

(c)

Figure 5. Effect of water absorption on the tensile properties of tapioca starch/ polyvinyl alcohol bioplastics: (a) tensile strength, (b) stiffness and (c) elongation.

It is immediately clear from Figure 5a and b that the absorbed water degrades the tensile strength and stiffness of the bioplastics. Increasing PVA content up to 29 wt% appears to decrease the tensile strength and stiffness degradation which occurs after water immersion. The lowest degradation in tensile strength and stiffness of 6% and 30%, respectively, were observed at the highest PVA addition (29 wt%). On the other hand, Figure 5c shows that the absorbed water increases the elongation. Increasing PVA content leads to a higher elongation increase which occurs after water immersion. The highest increase in elongation of 39% (compared to the dry condition) was also observed at 29 wt%. This reduction in brittleness after water absorption is particularly advantageous for bioplastics application, for example as packaging material, in aqueous environment.

From the observations in Figure 5, it can be concluded that the addition of PVA to TS bioplastics leads to a better tensile strength and stiffness retention, and a higher elongation increase after water immersion. This may be attributed to the reduction of water absorption by the bioplastics with PVA addition, which was caused by the less hydrophilicity of these bioplastics, as confirmed in the previous data (see section 3.2.). The reduction in absorbed water then leads to a decrease in the amount of hydrolyzed bioplastic polymer chains.

4. Conclusion

Tapioca starch/polyvinyl alcohol bioplastics were made with the addition of PVA from 0 to 29 wt%. Increasing content of PVA was found to decrease the water absorption of the bioplastics, with the lowest water saturation point of 251% observed at the highest PVA addition (29 wt%). This is due to the interaction between the starch and PVA which reduces the free OH groups in the resulting bioplastics. Consequently, this also led to a decrease in water absorption-related deterioration, where a better retention of tensile strength and stiffness, and a higher elongation increase after water immersion was found. The addition of 29 wt% resulted into the lowest degradation in tensile strength (6%) and stiffness (30%), while accompanied with the highest elongation increase (39%) after water immersion.

References

[1] Laycock B G and Halley PJ 2014 *Starch applications: State of market and new trends*. Elsevier: Burlington, MA.
[2] Carvalho A J 2008 Polymers and Composites from Renewable Resources 321.
[3] Peelman N *et al* 2013 Trends in Food Sci. Technol **32** 128-41.
[4] Goodship V and Jacobs D 2009 Smithers Rapra Technology: Shrewsbury, Shropshire.
[5] Leja K and Lewandowicz G 2010 Polish J. Environmental Studies **19** 255-66.
[6] Follain N *et al* 2005 *Carbohydr. Polym.* **60** 185-92.
[7] Bhatnagar S and Hanna M 1995 Transactions of the ASAE **38** 567-71.
[8] Parra D *et al* 2004 *Carbohydr. Polym.* **58** 475-81.
[9] Versino F *et al* 2015 Industrial Crops and Products **65** 79-89.
[10] Arrieta A A *et al* 2011 J. the Brazilian Chem. Society **22** 1170-6.
[11] Kaewtatip K and Tanrattanakul V 2008 *Carbohydr. Polym.* **73** 647-55.
[12] Shehap A 2008 *Egypt J Solids* **31** 75-91.
[13] Pawde S 2008 *J. Appl. Polym. Sci.* **109** 1328-37.
[14] Azahari N *et al* 2011 *J. Phys. Sci.* 22 15-31.
[15] Dumitriu S 2004 Marcel Dekker: New York.

Institute of Physics
Dirac House, Temple Back
Bristol BS1 6BE UK

ISSN: 1757-8981
ISBN 978-1-5108-4645-6
9 781510 846456

CURRAN ASSOCIATES INC.
proceedings.com